# Lecture Notes in Electrical Engineering

## Volume 597

The book series *Lecture Notes in Electrical Engineering* (LNEE) publishes the latest developments in Electrical Engineering - quickly, informally and in high quality. While original research reported in proceedings and monographs has traditionally formed the core of LNEE, we also encourage authors to submit books devoted to supporting student education and professional training in the various fields and applications areas of electrical engineering. The series cover classical and emerging topics concerning:

- Communication Engineering, Information Theory and Networks
- Electronics Engineering and Microelectronics
- Signal, Image and Speech Processing
- Wireless and Mobile Communication
- Circuits and Systems
- Energy Systems, Power Electronics and Electrical Machines
- Electro-optical Engineering
- Instrumentation Engineering
- Avionics Engineering
- Control Systems
- Internet-of-Things and Cybersecurity
- Biomedical Devices, MEMS and NEMS

For general information about this book series, comments or suggestions, please contact leontina.dicecco@springer.com.

To submit a proposal or request further information, please contact the Publishing Editor in your country:

**China**

Jasmine Dou, Associate Editor (jasmine.dou@springer.com)

**India**

Aninda Bose, Senior Editor (aninda.bose@springer.com)

**Japan**

Takeyuki Yonezawa, Editorial Director (takeyuki.yonezawa@springer.com)

**South Korea**

Smith (Ahram) Chae, Editor (smith.chae@springer.com)

**Southeast Asia**

Ramesh Nath Premnath, Editor (ramesh.premnath@springer.com)

**USA, Canada:**

Michael Luby, Senior Editor (michael.luby@springer.com)

**All other Countries:**

Leontina Di Cecco, Senior Editor (leontina.dicecco@springer.com)

**** Indexing: The books of this series are submitted to ISI Proceedings, EI-Compendex, SCOPUS, MetaPress, Web of Science and Springerlink ****

More information about this series at http://www.springer.com/series/7818

Pradeep Kumar Singh · Arpan Kumar Kar ·
Yashwant Singh · Maheshkumar H. Kolekar ·
Sudeep Tanwar
Editors

# Proceedings of ICRIC 2019

## Recent Innovations in Computing

Springer

*Editors*
Pradeep Kumar Singh
Department of Computer Science
and Engineering
Jaypee University of Information
Technology
Waknaghat, Himachal Pradesh, India

Yashwant Singh
Central University of Jammu
Jammu, Jammu and Kashmir, India

Sudeep Tanwar
Institute of Technology
Nirma University
Ahmedabad, Gujarat, India

Arpan Kumar Kar
Indian Institute of Technology Delhi
New Delhi, Delhi, India

Maheshkumar H. Kolekar
Indian Institute of Technology Patna
Patna, Bihar, India

ISSN 1876-1100 ISSN 1876-1119 (electronic)
Lecture Notes in Electrical Engineering
ISBN 978-3-030-29406-9 ISBN 978-3-030-29407-6 (eBook)
https://doi.org/10.1007/978-3-030-29407-6



This Springer imprint is published by the registered company Springer Nature Switzerland AG
The registered company address is: Gewerbestrasse 11, 6330 Cham, Switzerland

# Preface

The Second International Conference on Recent Innovations in Computing (ICRIC 2019) targeted researchers from different domains of advanced computing, intelligent networking, image processing and computer vision, e-learning, cloud and big data, security and privacy, and Digital India on a single platform to showcase their research ideas. The conference aims to be an annual ongoing event inviting researchers to exchange their ideas and thoughts. We hope that it will continue evolving and contributing in the field of computing technologies. The Second International Conference on Recent Innovations in Computing (ICRIC 2019) was hosted by Central University of Jammu, J&K, India, during March 8–9, 2019. We are thankful to our valuable authors for their contribution and our Technical Program Committee for their immense support and motivation toward making the 2nd ICRIC 2019 a grand success. We are also grateful to our eminent speakers: Prof. Dr. Devesh C Jinwala, Professor, IIT Jammu, J&K; Mr. Amit Sharma, Additional Director, DRDO, India, for his Cybersecurity Talk; and Dr. Zdzislaw Polkowski, Jan Wyzykowski University, Polkowice, Poland, for delivering his talk on importance of cloud technologies in SMEs. We are also thankful to the vice chancellor of the university, Prof. Ashok Aima, for extending his support to make this event happen.

We are also thankful to our various session chairs for sharing their technical sessions and enlightening the delegates of the conference. We want to express our thanks to Dr. Harsh K. Verma, Prof. Suchita Upadhaya, Dr. Arvind Selwal, Dr. Manoj Kr. Gupta, Dr. Deepti Malhotra, Dr. Jitender Manhas, Dr. Bhavna Arora,

Dr. Sudhanshu Tyagi, Prof. Devanand, and many more professors for spending their valuable time during the paper presentations. Selected papers were presented in various parallel tracks in six sessions during two days of conference.

Waknaghat, India Pradeep Kumar Singh
New Delhi, India Arpan Kumar Kar
Jammu, India Yashwant Singh
Patna, India Maheshkumar H. Kolekar
Ahmedabad, India Sudeep Tanwar
March 2019

# Contents

# Advanced Computing

# Predictive Analysis of Absenteeism in MNCS Using Machine Learning Algorithm

**Krittika Tewari, Shriya Vandita and Shruti Jain**

**Abstract** Absenteeism has become a severe problem for many organizations. The problem posed in this paper was to build a predictive model to predict the absenteeism for MNCs by previously recorded data sets. This exercise not only leads to prevent or lower absenteeism but forecast future workforce requirements and suggests ways to meet those demands. For faster processing of massive data set, the data was analyzed efficiently so that we get the minimum response time and turn-around time, which is only possible when we use the right set of algorithms and by hard wiring of the program. Different machine learning algorithms are used in the paper that includes linear regression and support vector regression. By analyzing the results of each technique, we come across that the age parameter mainly affects the absenteeism that is linearly related to absenteeism.

**Keywords** Absenteeism · Machine learning · Linear regression · Support vector regression

## 1 Introduction

Absenteeism is a vital issue that requires immediate attention by both, the employee and the employer. If the employee enjoys the work they do, then they will not take leave. The employers are expected to keep their employees in good spirit and motivated so that the employees deliver their best to the organization for the benefit of both. It is a habitual pattern of the absence from duty. Absenteeism not only affects the cost but is also an indicator of the poor morale of the employees. Absenteeism can be a result of depression, personal stress, and anxiety which can lead to an employee being detached and unable to cope up with the work and interaction at workplace, burnout, heavy workloads, stressful meetings/presentations, bullying at workplace,

K. Tewari · S. Vandita · S. Jain (✉)
Department of Electronics and Communication Engineering, Jaypee University of Information Technology, Solan, Himachal Pradesh, India
e-mail: jain.shruti15@gmail.com

S. Vandita
e-mail: simranshriya@gmail.com



P. K. Singh et al. (eds.), *Proceedings of ICRIC 2019*, Lecture Notes in Electrical Engineering 597, https://doi.org/10.1007/978-3-030-29407-6_1

childcare, illness, and unequal treatment that leads to heart burning and feelings of being unappreciated (non-recognition of his/her contribution). Absenteeism has become a severe problem for many organizations. Obviously, it has been an undeniable issue faced by companies which can result in serious financial and non-financial losses [1]. Because of the negative consequences of employee absenteeism, it is important that the organization keep the absenteeism low [2]. The following may be adopted by the organization to check or reduce absenteeism:

1. Adopting a clear attendance policy.
2. Providing healthy and safe working environment to the employees.
3. Encouraging the employees by visiting the workplace during working hours by the higher management.
4. Celebrating the success of any project as a team.
5. Providing some reward to the employee for excess extra paid time off.
6. Maintaining a flexible work schedule if an employee comes late, he/she may be allowed but should be asked to put extra working hours in the week to compensate.

Employees can cope with stress much better when they are given some autonomy for self-governance, social support, and opportunities for personal growth. It has been seen that the rate of absenteeism does not follow a normal distribution, but it is skewed and truncated distribution that denies the beliefs of conventional statistical methods such as correlation analysis and ordinary least square (OLS) regression [3]. It has been seen that correlation and multiple regression dominate absence research. Managers and organizational practitioners need a detailed method for measuring absenteeism loss as well as other measures needed for managerial evaluation to decrease absenteeism rate and compare the effectiveness of absence/attendance policy from period to period. Since the 1980 s, the companies have collected abundant amount of customer data stored in databases [1]. The data is collected by the companies and the thought process on how to provide additional benefits or to improve the operations. This type of thought process formed a natural progression toward the use of improving estimates, forecasts, decisions, and efficiency [2]. These databases grew to such a large extent that it becomes difficult for humans to analyze on their own. Predictive analytics is an answer on how to handle such large databases. It is a procedure that incorporates the use of computational methods to determine important and useful patterns in large data. It was conceived from the study of related areas like artificial intelligence, statistics, machine learning, pattern recognition, and data mining [2].

We can use either supervised learning based models or unsupervised learning models for prediction. The basic idea behind supervised learning based models is to predict a target variable. Supervised learning is also referred to as predictive modeling. Classification is a popular predictive modeling algorithm while dealing with categorical variable [4–8]. Another type of supervised learning is regression where we predict continuous outcomes [9, 10]. This procedure determines the computational methods and incorporates the patterns in large data. In descriptive modeling or unsupervised learning based approaches, a model is always constructed through clusters of the data [11].

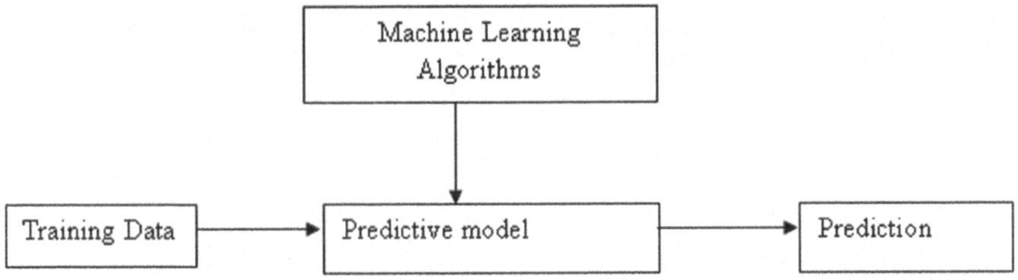


**Fig. 1** Steps using machine learning algorithm

Authors in [12] explain machine learning is a scientific discipline which focuses on automatically recognizing complex patterns and making intelligent decisions based on available data. This branch of study evolves behavior that helps in developing an algorithm for the computer. Figure 1 signifies the typical machine learning algorithm [11, 12]. Machine learning focuses on the development of computer programs that can change when exposed to new data. It is the process of converting experience into expertise or knowledge [12]. There are different machine learning algorithm which includes linear regression, decision tree, logistic regression, Naïve Bayes, support vector machine (SVM), k-nearest neighbor (kNN), and random forest.

Predictive model is made to train data that helps in analyzing the parameters which affect the absenteeism in MNCs and how to reduce that absenteeism. Data is collected from online source which is preprocessed by removing outliers. Feature engineering is applied to the data. There are various parameters which affect the absenteeism. Using feature selection, best features were selected and different machine learning algorithms like linear regression and support vector regression are applied.

In this paper, Sect. 2 defines the predictive analysis using machine learning algorithm; Sect. 3 explains the different implementation steps which were concluded at the end of the paper.

## 2 Predictive Analysis Using Machine Learning Algorithm

Predictive analysis is used to predict unknown events or unobserved events by analyzing the existing data set with the help of machine learning techniques, statistical modeling, and data mining. For predictive analysis, objectives are defined, and then, the data set is prepared. Based upon the prepared data, a model is laid down for deployment and monitoring. Predictive analysis identifies the cause–effect relationship across the variables from the given data set and discovers hidden patterns with the help of data mining techniques. It may apply to the observed patterns to unknowns in the past, present, or the future. The predictive models have a clear focus on what they want to learn and how they want to learn. The models which are used for the prediction of target features of categorical values are known as classification models [4–7]. Predictive models may also be used to predict numerical values of the target feature based on the predictor features. The models which are used for the prediction of the numerical values of the target feature of a data instance are known as regression models. There are different types of regression models (linear, logistic, support vector regression). Mainly, they are categorized as linear and nonlinear. In linear regression model, the dependence of the response on the regressors is defined by a linear function, which makes their statistical analysis mathematically tractable. On the other hand, in nonlinear regression model, this dependence is defined by a nonlinear function, hence, the mathematical difficulty in their analysis.

In this paper, we are working on linear regression and support vector regression (SVR).

## 2.1 Linear Regression

For linear regression, dependent variable ($y$) is continuous and independent variable ($x$) may be continuous or discrete. Figure 2 shows the algorithm illustrating the linear regression.

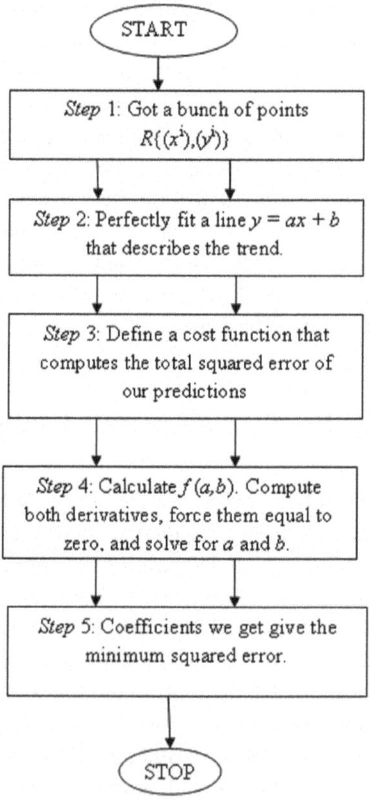


**Fig. 2** Algorithm illustrating linear regression model

For linear regression, the shape of regression line is linear whose slope of line is $b$ and intercept is $a$. Linear regression is expressed as:

$$\underbrace{y}_{\text{actual (observed)}} = \underbrace{ax + b}_{\text{explained (prdeicted)}} + \underbrace{\varepsilon}_{\text{error}} \tag{1}$$

In Eq. (1), $e$ is the error term. Linear regression can also be expressed by Eq. (2)

$$\underbrace{y}_{\text{observed}} = \underbrace{\hat{y}}_{\text{predicted}} + \underbrace{\varepsilon}_{\text{error}} \tag{2}$$

For predicted values, Eq. (1) can be written as:

$$\hat{y} = \hat{a}\,x + \hat{b} \tag{3}$$

where slope is represented as:

$$\hat{a} = \frac{SS_{xy}}{SS_{xx}} = \frac{\sum(x - \bar{x})(y - \bar{y})}{\sum(x - \bar{x})^2} \tag{4}$$

where $\bar{x}$ and $\bar{y}$ are the sample means and intercept is represented by Eq. (5),

$$\hat{b} = \bar{y} - \hat{a}\,\bar{x} \tag{5}$$

If the independent variable ($x$) is more than one (i.e. $x_1, x_2, x_3 \ldots$) than regression is known as multiple regression. For multiple regressions, Eq. (1) can be expressed as

$$y = a + b_1x_1 + b_2x_2 + b_3x_3 + \cdots b_nx_n + e \tag{6}$$

Methods like partial least square analysis (PLS) and OLS are used for calculation of linear regression, while methods like stepwise approximation (SWA), forward selection (FS), and backward elimination (BE) are used in multiple regression analysis.

## *2.2 Support Vector Regression (SVR)*

A nonlinear function is leaned by linear learning machine mapping into high-dimensional kernel-induced feature space containing all the main features that characterize maximum margin algorithm. The system capacity is controlled by parameters that are not affected by the dimensionality of feature space. The main idea is to optimize the generalization bounds and rely on defining the loss function that

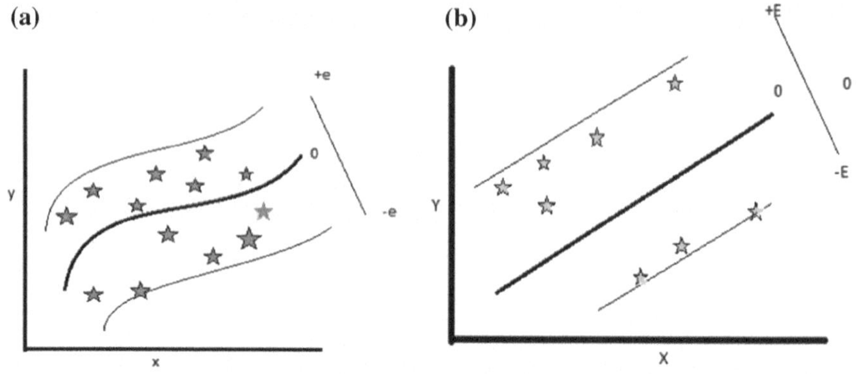


**Fig. 3** SVM **a** nonlinear and **b** linear hyperplane

ignores errors that are located within the certain distance/range of the true value for regression. SVR is also used for making prediction models. Another widely used and powerful learning algorithm is a support vector machine (SVM). In SVM, the objective is to maximize the margin which is defined as the distance between the separating hyperplane and training samples that are closest to hyperplane. SVM can be applied as regression and not limited to classification only. SVR works on same principle as SVM works with only a few differences. Figure 3a shows the nonlinear plane of SVM, and Fig. 3b shows linear plane of SVM.

The main idea is to maximize the margin by individualizing the hyperplane. The results of linear regression and linear SVR are mostly similar. The main focus of linear regression is to maximize the conditional likelihoods of the training data from the source, which makes it more susceptible to outliers as compared to SVM. Furthermore, linear regression models can be easily updated, and the SVM focuses on the points that are closest to the decision boundary (support vectors).

## 3 Implementation of Predictive Model for Absenteeism

Our aim is to predict the absenteeism for MNCs by the previous recorded data sets. We analyzed the data from online source www.kaggle.com, which is a platform for predictive modeling and analytics competitions. Kaggle is an online community of machine learners and data scientists. It allows users to find, explore, and publish data sets. The availability of advanced machines and special tools has led to the analysis of big data. It also broadened our horizons of looking at an unknown data and trying to find useful features and patterns. We programmed our analysis on Python language. Figure 4 shows the proposed methodology employed in the paper to make a predictive model that helps in predicting the absenteeism. A significant amount of time and effort was spent in organizing, cleaning, and redefining variables in the data.

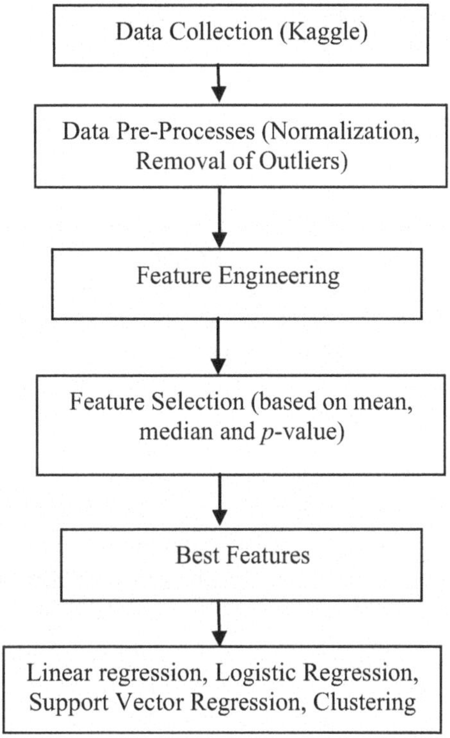


**Fig. 4** Proposed methodology

Different steps were followed for implementation of our algorithm:

1. *Data Exploration*: In this section, data set is explored. This is an important step in the machine learning process as firstly we need to know more information about the data we are using and secondly we need to make a few alterations to the data itself. In this paper, we are using 741 observations and 21 parameters before data extraction and selection out of which we are considering the following different parameters: absent (1 = YES, 0 = NO), employee class (1 = 1st, 2 = 2nd, 3 = 3rd), name, age, sex, distance from home, seasons transport, service time, day shifts, workload, hit target, and education.
2. *Data Preprocessing*: Raw data can be transformed into understandable format is called data preprocessing. Data preprocessing is required because real-world data is often inconsistent, noisy, incomplete, or lacking in certain behaviors or trends. The missing values can be identified through data preprocessing. The identification of missing values is important in successful management of data. Missing values, if not handled properly, lead to inaccurate inference about the data. The result obtained will differ from ones where the missing values are present due to improper handling of data.
3. *Feature Engineering*: Since the data can have missing fields, incomplete fields, or fields containing unknown information, a fundamental step in building any

**Table 1** Title from the field Name

| Index | Title | No of occurrence |
|---|---|---|
| 1. | MR | 757 |
| 2. | MRS | 198 |
| 3. | MS | 2 |

prediction system is feature engineering. Many times the data set contains highly varying features in magnitudes, units, and range. The field "Name" contained employee's title: Mr., Mrs., Ms. Since name is unique for each employee, it is not useful for our prediction system. However, an employee's title can be extracted from his or her name. In our data set, we found three titles which are shown in Table 1. Title indicates employee's sex (Mr. and Mrs.) and age (Ms. and Mrs.).

4. *Machine Learning Algorithm*: It explains how we have applied different machine learning algorithms (linear and SVR) to our data set in order to build a model. Figure 5 elaborates the machine learning algorithm that we have incorporated in our paper. We have split the data set into training data set and testing data set in the ratio of 70:30, and using training data set, we came up with a model to be deployed. To this model, we applied test data set that yields predictive model.

In this paper, we have used linear regression and SVR to get predictive model. Linear regression requires minimal adjustments to the data provided by Kaggle. We have performed linear regression of all the parameters (shown in Fig. 6) and find that absenteeism is linearly varying with age parameter only.

Figure 7 shows that the *age of the employee* being our predictor variable and *number of days our employee is absent* as our response variable using linear regression.

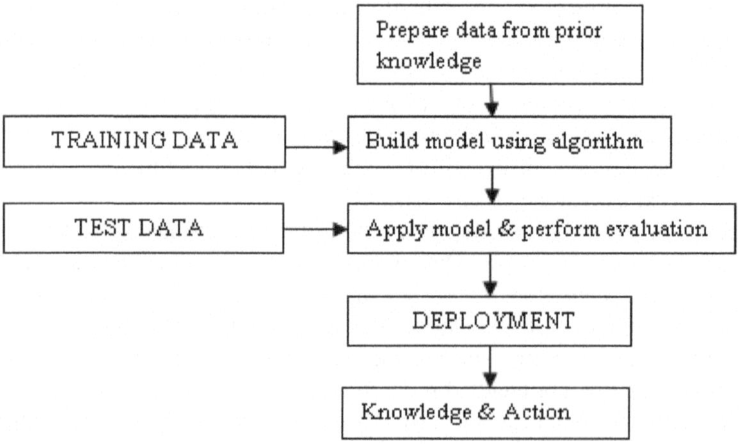


**Fig. 5** Machine learning algorithm

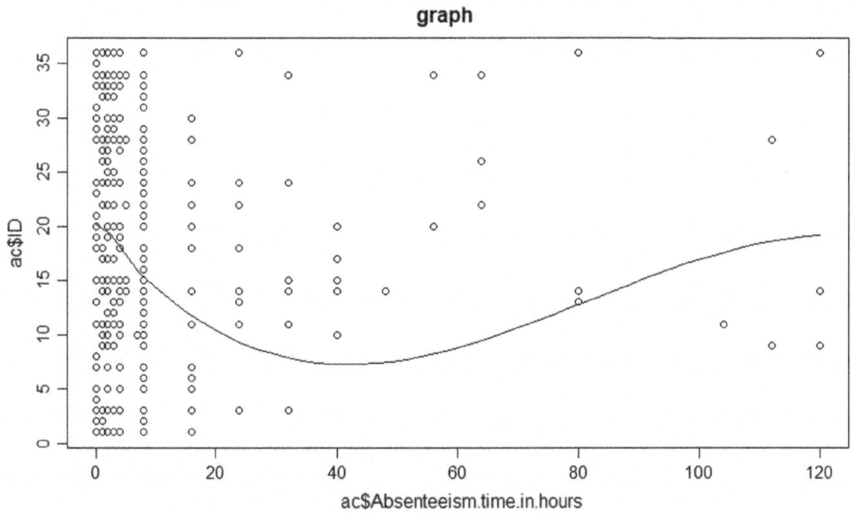


**Fig. 6** Linear regression on different variables our data set

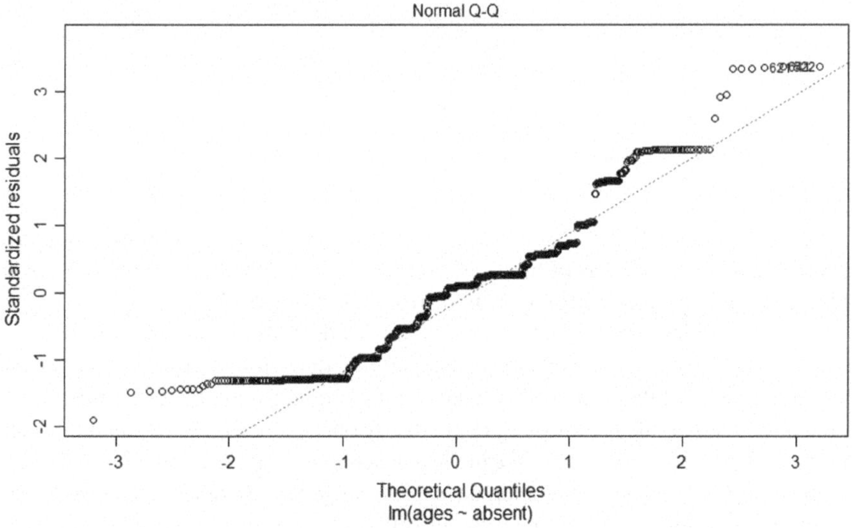


**Fig. 7** Residuals versus age using linear regression

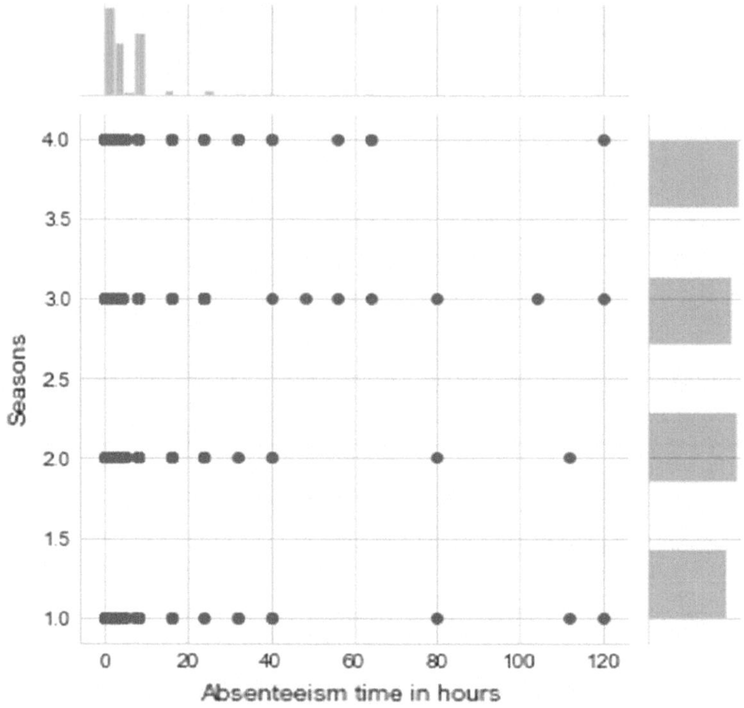


**Fig. 8** Graph between seasons of the year versus absenteeism using SVR

We have also applied SVR on our data sets. For SVR, we have considered two parameters age and seasons of the year. Figures 8 and 9 depict the graph for season of the year and age versus absenteeism respectively using SVR.

We have divided the year into four major seasons with an interval of 0.5 but as shown in Fig. 8, we can infer that absenteeism rate does not get much affected by a particular season of the year. The absenteeism rate remains unaffected by this factor as the absenteeism rate is almost equal in all seasons. From Fig. 9, we can infer that the age group of 35–40 years has the highest absenteeism rate as compared to the age group of 50–60. With age parameter, days of the week with hours is also considered.

Figure 10 depicts the absenteeism on different days of the week. Here, each day starting from Monday to Saturday is assigned a different color and the gradual decrease in absenteeism on 2nd day of the week is observed as the age increases, and the absenteeism on the 6th day (Saturday) is fairly high than any other day, reason being the end of the week.

From the experiments, we come to know that age parameter of the age group of 35–40 is linearly related to absenteeism and the maximum absenteeism is on the 6th day of the week, and minimum absenteeism is on the 2nd day of the week. There is no much affect of season on absenteeism.

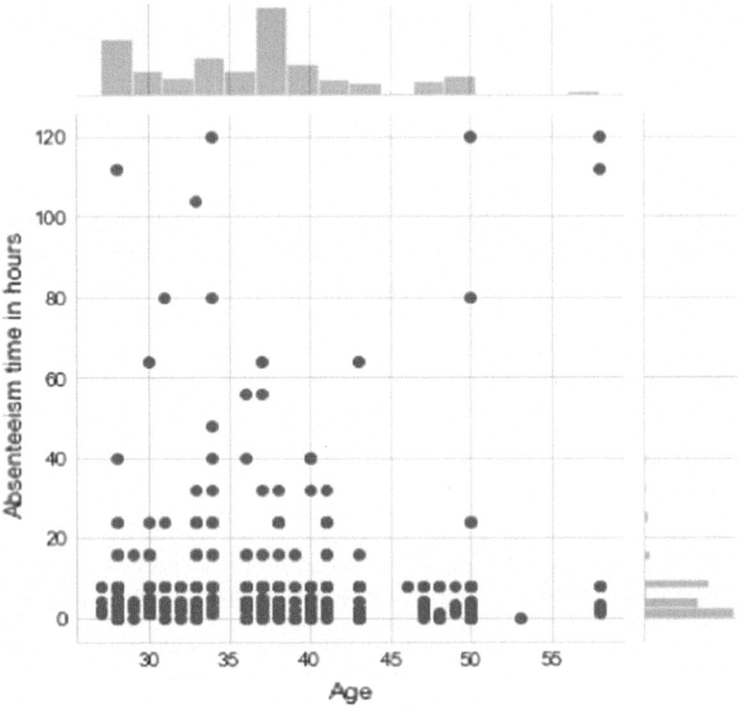


**Fig. 9** Graph between age versus absenteeism using SVR

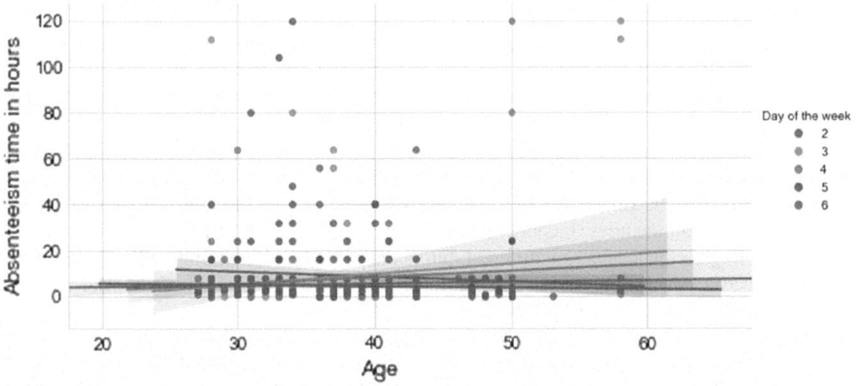


**Fig. 10** Age and absenteeism in hours considering days of the week. *Note* 2 defines 2nd day, 3 defines the 3rd day of the week

## 4 Conclusion and Future Work

Predictive analysis of absenteeism in MNCs served as a framework for introductory predictive analytic methods. Managers and organizational practitioners need a detailed method for measuring absenteeism loss as well as other measures needed for managerial evaluation to decrease absenteeism rate and compare the effectiveness of absence/attendance policy from period to period. The motivation of this study developed from a desire to learn, understand, and apply linear, logistic, and SVM regression. It is apparent that the effort put forth when working on the absenteeism in MNCs problem has achieved our aims and goal of this study by linear regression and SVR. In the future, we will encode categorical values of parameters such as month of the absence.

## References

1. Delen, D., Zaim, H., Kuzey, C., Zaim, S.: A comparative analysis of machine learning systems for measuring the impact of knowledge management practices. Decis Support Syst **54**(2), 1150–1160 (2013)
2. Witten, I.H., Frank, E., Hall, M.A., Pal, C.J.: Data Mining: Practical Machine Learning Tools and Techniques, p. 560. Morgan Kaufmann Publishers, San Francisco (2005)
3. Faber, F.A., Hutchison, L., Huang, B., Gilmer, J., Schoenholz, S.S., Dahl, G.E., Vinyals, O., Kearnes, S., Riley, P.F., von Lilienfeld, O.A.: Prediction errors of molecular machine learning models lower than hybrid DFT error. J. Chem. Theory Comput. **13**(11), 5255–5264 (2017)
4. Jain, S.: Classification of protein kinase B using discrete wavelet transform. Int. J. Inf. Technol. **10**(2), 211–216 (2018)
5. Jain, S., Chauhan, D.S.: Mathematical analysis of receptors for survival proteins. Int. J. Pharma Bio Sci. **6**(3), 164–176 (2015)
6. Bhusri, S., Jain, S., Virmani, J.: Classification of breast lesions using the difference of statistical features. Res. J. Pharm., Biol. Chem. Sci. (RJPBCS), 1366 (2016)
7. Rana, S., Jain, S., Virmani, J.: SVM-based characterization of focal kidney lesions from B-mode ultrasound images. Res. J. Pharm., Biol. Chem. Sci. (RJPBCS) **7**(4), 83 (2016)
8. Sharma, S., Jain, S., Bhusri, S.: Two class classification of breast lesions using statistical and transform domain features. J. Glob. Pharma Technol. **9**(7), 18–24 (2017)
9. Jain, S.: Regression analysis on different mitogenic pathways. Netw. Biol. **6**(2), 40–46 (2016)
10. Jain, S.: System modeling of AkT using linear and robust regression analysis. Curr. Trends Biotechnol. Pharm. **12**(2), 177–186 (2018)
11. Zhang, L., Tan, J., Han, D., Zhu, H.: From machine learning to deep learning: progress in machine intelligence for rational drug discovery. Drug Discov. Today **22**(11), 1680–1685 (2017)
12. Borchers, M.R., Chang, Y.M., Proudfoot, K.L., Wadsworth, B.A., Stone, A.E., Bewley, J.M.: Machine-learning-based calving prediction from activity, lying, and ruminating behaviors in dairy cattle. J. Dairy Sci. **100**(7), 5664–5674 (2017)

# IoT Based Healthcare Kit for Diabetic Foot Ulcer

**Punit Gupta, Anushka Pandey, Poonia Akshita and Anshul Sharma**

**Abstract** IoT helps the society to solve various problems in medicine. There are many problems, which can be solved using IOT healthcare such as detection of diabetes, detection of diabetic foot ulcer, abnormality in heart rate and many more. This paper presents the design and implementation of IoT-based system to be used in healthcare for detection of diabetic foot ulcer. The model will monitor the health of diabetic foot ulcer patient and will send alerts if found any abnormality. The development of this model will be done on Node MCU development board. This model enables users to record, analyze and send large data to the users in real time and efficiently. This will help in reducing visits to doctors and will help in live alerts and the abnormality in the patient.

**Keywords** Internet of things (IoT) · Foot pressure · Diabetic foot ulcer · Health monitoring and alerts

## 1 Introduction

IoT is a system of interconnected devices with the Internet, each device having a unique identifier and have the ability to transfer the data over a network without human interaction. The connection of devices with Internet makes the device smarter as it can send the data as well as receive the data from the Internet and respond accordingly. IoT in healthcare has brought many changes to the world. IoT helped

P. Gupta (✉) · A. Pandey (✉) · P. Akshita (✉) · A. Sharma (✉)
Department of Computer and Communication Engineering, Manipal University Jaipur, Jaipur, Rajasthan, India
e-mail: punitg07@gmail.com

A. Pandey
e-mail: anushkapandey51777@gmail.com

P. Akshita
e-mail: akshitaiodine@gmail.com

A. Sharma
e-mail: anshulsharma325@yahoo.com



P. K. Singh et al. (eds.), *Proceedings of ICRIC 2019*, Lecture Notes in Electrical Engineering 597, https://doi.org/10.1007/978-3-030-29407-6_2

to save a lot of time of patients as well as doctors and helped patients to spend less on healthcare. IoT also helps doctors to analyze the patient whenever he feels so.

IOT in healthcare has brought a lot of advancement and changes. Management and analyzing of monitored results have lowered the risk rates and cost of treatment. The patient need not visit the doctor on regular basis for checkups and reports. It helps the doctor to analyze the patient whenever his readings vary abnormally. To make healthcare sector more efficient and accurate for the patients and doctors, IoT is a solution to provide efficient techniques for diagnosis and prediction system.

Diabetes mellitus is one of the major causes of high blood sugar, which includes insulin that moves sugar from the blood into the cells, which can be used to store energy. Diabetic foot ulcer is one of the major sources of toxicity and leading cause of hospitalization. Every year, almost 20% of hospitalizations are due to diabetic foot ulcers since it can lead to infection, gangrene, amputation and possibly death. It is found that limb amputation in diabetic patients is ten times higher than the non-diabetics. In majority of cases, the cause for the foot ulcer is altered architecture of the foot due to neuropathy resulting ion abnormal pressure points.

In this paper, we have proposed an IOT-based smart sole shoe to detect the health of foot ulcer. The work is divided into four sections related work, proposed work, results and conclusion.

This paper proposes an IoT-based smart sole shoe to detect the health of foot ulcer. The work is divided into four sections related work, proposed work, results and conclusion.

## 2 Related Work

IOT health care has been emerging a very beneficial in the world of medicines and healthcare. Therefore, many IOT-based healthcare systems are becoming functional such as, blood pressure monitor, heartbeat detectors and pulse audio meters. Hence, the research is getting advance with emergence of various healthcare machinery. In this section, few works related to health care are presented.

Milka D Madhale has proposed "foot pressure scanner" [1] for study of dynamic foot pressure measurement in a diabetic patient. In the present study, a low-cost foot pressure scanner has been developed and dynamic plantar pressures in a group of 110 Indian patients with or without diabetes are measured.

Practical implementation: If these pressure points can be detected, ulcer can be prevented by providing offloading footwears. Originality/value: differences are found in dynamic foot pressures in different study groups namely diabetic patients with foot ulcer and non-diabetics the differences are significant which showed the validity of the tool. The consistency of the tool was proved by retest method.

Luay Fraiwan, Mohanad AlkKhodari, Jolu Ninan have proposed "Diabetic foot ulcer mobile detection system using smartphone thermal camera" [2]. The proposed system embraced a thermal camera which was connected to a smartphone, which was used to acquire thermal images using MATLAB platform. The thermal imaging system possesses a stimulated temperature gradient greater than 2.2 °C which can indicate probability of development of ulcers using basic image processing techniques.

Implementation: The analysis and interpretation were based on two techniques namely point-to-point main difference and Otsu thresholding technique; with these two techniques, the system could successfully detect temperature changes.

Killeen AL, Walters JL has proposed Remote Temperature Monitoring in Diabetic Foot Ulcer Detection [3]. This paper proposed the use of RTM foot mat for the early detection and prevention of DFUs. Three patients with a history of diabetes and DFUs were given RTM foot mat and instructed its working. The persistence increase in temperature greater than 1.75° between the left and the right feet indicated to call the patient for further diagnosis.

Kulesza [4] proposes the Internet of things (IoT)-based information system for indoor and outdoor use [4]. This paper was based on a conducted survey of related works indicating failure of methodological approaches compared to the design process. Hence, a designed methodology fulfills the required target according to the outlook of the stakeholders and potential users. Many other existing health care solution also exist for disease prediction [5–7].

The solution makes use of the three axial accelerometer and magnetometer including the decision trees algorithm. This architecture enabled positioning of a monitored person in a four room zones accurately. It also identifies the movement activity based on these activities the system differentiates them as normal suspicious or dangerous which is used to prompt the healthcare staff for various possible problems. This proposed solution is validated by day-to-day life scenario.

## 3 Proposed System

We have proposed a robust health monitoring system for diabetics that is intelligent enough to monitor the patient automatically IOT that collects in information through these systems that is patient's.

In this section, model and system architecture is proposed for diagnosis of diabetic ulcer using IoT and collect the information for further analysis by logging parameters like heart rate, body temperature, dynamic foot pressure and blood pressure. This would help the patient and the doctor to predict the symptoms of diabetic foot ulcers and send an emergency alert to the patients-trusted doctor with his current status and full medical information. Diabetic foot ulcer is a major source of morbidity and a leading cause of hospitalization. Milka D. Madhale, in his research paper "The study of dynamic foot pressure measurement in diabetic patients," stated the difference in foot pressures of diabetic foot ulcer and a normal person according to their age.

The main objective is to develop a complete healthcare kit which could detect diabetic foot ulcer using 6-foot pressure point measured by flexi-force sensor of capacity 100 lbs. The pressure sensor measures the foot pressures six times a day whenever the patient is standing along with heart rate, blood pressure and body temperature which could result into diabetic complications. These readings are sent using Bluetooth connection to the patient's smart phone and then stored in cloud using Internet.

## 3.1 Hardware

The IoT module of our model is the node MCU board is an open source IoT platform. It includes firmware which runs on the ESP8266 Wi-Fi SOC from Espressif Systems and comes with Wi-Fi and inbuilt Bluetooth. The term "Node MCU" by default refers to the firmware rather than the development kits. The firmware uses the Lua scripting language. The sensor attached to the patient's foot is the flexi-force sensor of capacity 100 lbs. The device also has a heartbeat sensor (XD58C pulse sensor) and a temperature sensor (DHT11) to detect the patient temperature. Blood pressure sensor is added to detect the blood pressure of the patient over the period of time which defines the health of the patient. All the data collected from the sensor is recorded over the cloud of think spark for analysis over the time to find any abnormality in the patient health and lifestyle.

The cloud would analyze the data received from the hardware and if found any abnormality would alert the doctor and the patient through a Web application.

The system is divided into three phases:

1. Sensing
2. Harvesting and analysis
3. Alerts and reports.

## 3.2 Sensing

In this phase, data is collected from various sensing devices and collected over the IoT board for filtration and then to be transmitted to cloud through the Internet connection of smart phone.

## 3.3 Harvesting and Analysis

In this phase, the data from many such devices is collected over the cloud. Data is collected three times a day to check the health of the patient. Then starts the analysis process to analyze the data and finds any abnormality in health by comparing the trained healthy record of the patient with current and previous record. Figure 1a, b shows the relation of age and the diabetes which may result in foot ulcer. The cloud is trained with the basic parameters like body temperature, blood pressure and foot

**(a)**

| Gender | Age groups | Normal (%) | DM (%) | DM with DPN (%) | Foot ulcer (%) | Total (%) |
|---|---|---|---|---|---|---|
| Male | <50 | 4 (26) | 3 (30) | 3 (23) | 5 (22) | 15 (24.5) |
| | 50-60 | 11 (73) | 6 (60) | 5 (38) | 9 (39) | 31 (50.8) |
| | >60 | 0 | 1 (10) | 5 (38) | 9 (39) | 15 (24.5) |
| Female | <50 | 7 (46) | 2 (10) | 1 (17) | 3 (37) | 13 (26.5) |
| | 50-60 | 6 (40) | 15 (75) | 2 (33) | 2 (25) | 25 (51) |
| | >60 | 2 (13) | 3 (15) | 3 (50) | 3 (37.5) | 11 (22.4) |
| Total | <50 | 11 (36.6) | 5 (16.6) | 4 (21) | 8 (25.8) | 28 (25.4) |
| | 50-59 | 17 (56.6) | 21 (70) | 7 (36.8) | 11 (35.4) | 56 (50.9) |
| | >60 | 2 (6.6) | 4 (13.3) | 8 (42.1) | 12 (38.7) | 26 (23.6) |
| | Total | 30 | 30 | 19 | 31 | 110 (100) |

DM = Diabetes mellitus, DPN = Diabetic peripheral neuropathy

**(b)**

| Study groups | Dynamic foot pressure points | Number of participants | Dynamic foot pressures (K pascal) Mean±SD |
|---|---|---|---|
| Normal | P0 | 30 | 87.82±26.29 |
| | P1 | 30 | 76.63±25.83 |
| | P2 | 29 | 66.55±26.51 |
| DM | P0 | 30 | 105.32±23.56 |
| | P1 | 30 | 91.52±25.13 |
| | P2 | 30 | 76.98±25.73 |
| DM with neuropathy | P0 | 19 | 105.32±23.56 |
| | P1 | 19 | 91.52±25.13 |
| | P2 | 19 | 76.98±25.73 |
| Foot ulcer | P0 | 31 | 92.7±33.253 |
| | P1 | 31 | 76.31±35.363 |
| | P2 | 31 | 80.25±34.124 |
| Total | P0 | 110 | 95.42±28.919 |
| | P1 | 110 | 81.07±29.823 |
| | P2 | 109 | 75.45±29.57 |

DM = Diabetes mellitus, P0 = 1st metatarsal, P1 = 2nd metatarsal, P2 = 5th metatarsal

**Fig. 1** **a** Age-wise study of ulcer and **b** foot pressure study of ulcer

pressure for a healthy person and also the rages when the system should generate the alerts.

## 3.4 *Alert and Reporting*

In this phase, abnormality, i.e., gradual increase in temperature blood pressure or foot pressure records is reported in form of alert which means active response is required and n form of report which shows an record of abnormality in any parameter which may result in future health degradation. These report helps to take preliminary actions rather than waiting for the big problem (Figs. 2 and 3).

The patient and the doctor would get weekly, monthly and annual reports regarding their medical conditions, and that would help to know their progress and distortion. This would help to change their medication and living conditions (Fig. 4).

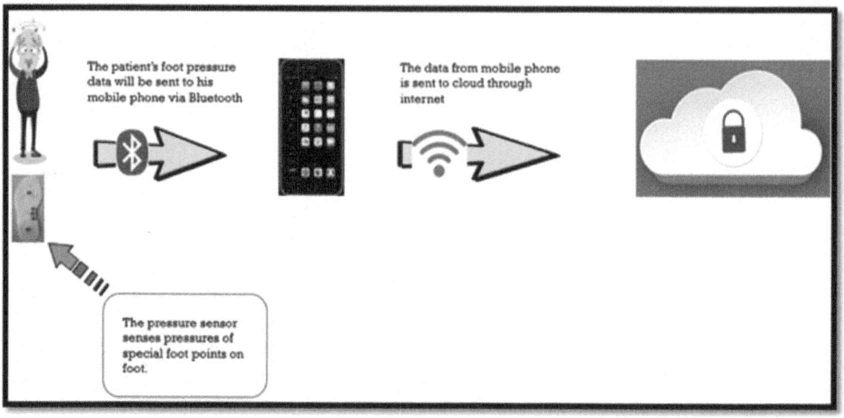


Fig. 2 Input collection model

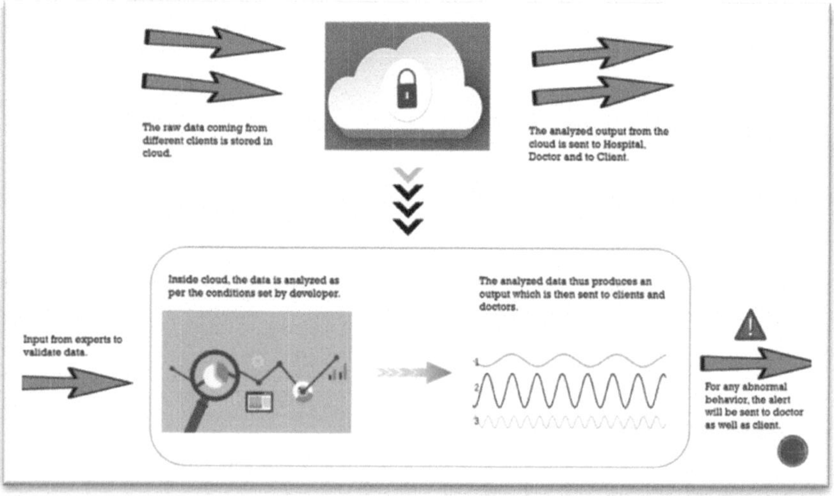


Fig. 3 Analysis module

## 4 Results

The aim of this solution to predict any abnormal behavior in patient health before it becomes a problem. Cloud help in computing the large data, comparing them with original one to find an abnormality in health and take action by changing the diet or few medicines.

- Our solution is better than the above-related work as it helps to predict the symptoms of diseases and alerts the patient as well as the doctor.

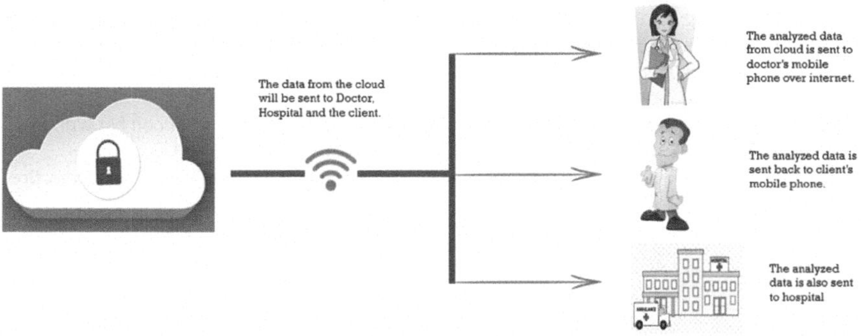


**Fig. 4** Report delivery and alert module

- It helps in analyzing the patient's medical conditions and keeps a track of them by providing time to time medical reports.
- It is a complete healthcare kit for a diabetic patient which would help him know his other diabetic complications.

## 5 Conclusion

Solution for foot ulcer will help the patient and doctor to track the health of the patients and find any abnormality in the health. The system is developed and is been tested over the dataset provided in research work. This work has almost covered all the work done in the field of foot ulcer and its symptoms and has reviewed all the work. The work has done on open-source hardware which may help it to improve further and study the work. The work will help the patient to track abnormality in heart or blood pressure.

## References

1. Madhale, M.D., Tyagi, N.K., Godhi, A.S.: A study of dynamic foot pressure measurement in diabetic patients. J. Sci. Soc. **44**(2), 76 (2017)
2. AlKhodari, M., Fraiwan, L., Saleh, A., Ghazal, M., Mustafa, B., Ninan, J.: Diabetic foot ulcer mobile detection system using smart phone thermal camera: a feasibility study. Biomed. Eng. Online **16**(1), 117 (2017)
3. Liu, X., Yang, T., Yan, B.: Research on the Architecture of Wildlife Observation and Communication System. Computer Network Information Center (CNIC) Chinese Academy of Sciences IEEE, Beijing, China (2015)
4. Dziak, D., Jachimczyk, B., Kulesza, W.J.: IoT-based information system for healthcare application: design methodology approach. Appl. Sci. 8.**7**(6) (2017)

5. Chen, Q., Zheng, L., Tian, J., Dubrova, E., Pang, Z.: Ecosystem analysis in the design of open platform-based in-home healthcare terminals towards the internet-of-things. In: Advanced Communication Technology, 2013 15th International Conference on 2013 Jan 27, pp. 529–534. IEEE (2013)
6. Ellwood, S.A., Wilson, R.P., Addison, A.C.: Technology in conservation: a boon but with small print. In: Key Topics in Conservation Biology, pp. 105–119. Oxford (2007)
7. Walters, J.L., Killeen, A.L.: Remote Temperature Monitoring in Diabetic Foot Ulcer Detection. Wounds: Compend. Clin. Res. Pract. **30**(4), 44–48 (2018)

# Context—Aware Smart Reliable Service Model for Intelligent Transportation System Based on Ontology

**M. Swarnamugi and R. Chinnaiyan**

**Abstract** IoT-based transportation system is getting smarter and smarter to provide quick, safe and reliable services to the user. This smarter transportation system is called Intelligent Transportation System (ITS). ITS incorporates wired and wireless communication, electronic technologies, computational technologies, cloud platforms, GPS and sensor to assist user to be informed on road safety and make safer, coordinated, comfort and 'smarter' use of transportation medium. ITS is an advanced IOT application that connects huge number of objects to communicate with each other. As number of objects connected to ITS application increases, we face with a challenge of adding value to raw sensor data. The focus of this paper is to address this challenge with a context—aware model. Also, the effectiveness of context—aware in ITS is illustrated by discussing different real time scenarios.

**Keywords** Intelligent transportation system (ITS) · Context—aware · Ontology · Sensors

## 1 Introduction

The major issue urbanization face today is the rapid vehicular growth. As the population in the urban area of our country increases, it naturally increases the traffic congestion. This has put massive focus on transforming the traditional transportation network to Intelligent Transportation System. The main objective of this transformation is to minimize and provide optimal solution to traffic-related problems. ITS is an advanced IoT application aims to afford services relating to different modes of transportation and assist user to be better informed on road safety and make safer,

---

M. Swarnamugi (✉)
Department of MCA, Jyoti Nivas College, Bengaluru, India
e-mail: swathidevan@gmail.com

R. Chinnaiyan
Department of Information Science and Engineering, CMR Institute of Technology, Bengaluru, India
e-mail: vijayachinns@gmail.com



P. K. Singh et al. (eds.), *Proceedings of ICRIC 2019*, Lecture Notes in Electrical Engineering 597, https://doi.org/10.1007/978-3-030-29407-6_3

coordinated, comfort and 'smarter' use of transportation medium. ITS provides solutions to minimize traffic congestion and achieve traffic efficiency, allows vehicles to communicate, provides smart parking, enables dynamic traffic signal management, monitors and controls variable speed limits, surveillance of roads and highways, utilizing energy conservation, enabling collision avoidance and enhancing comfort and road safety to the users with the help of sensor, information and communication technologies [1]. As number of objects connected to ITS increases, it generates enormous amount of raw sensor data. The increase in raw sensor data creates a new challenge in delivering effective services to the user. Therefore, the data have to be analyzed and reasoned to fix the challenge of adding value to sensor data. Fixing this challenge ensures effective delivery of services to the user of ITS.

For effective service delivery, this paper focuses on using context—awareness to enrich semantics of the raw information collected. In general, context means the ability of a system to describe the state of an entity [2]. An entity is a person, a device, place, an environment relevant to user and application. Context—aware—a key component of ubiquitous computing focuses on giving a transformation to raw information by enriching its semantics and inferring its contextual description. Many different methods are employed in the literature [3] to infer context description. For effectiveness and reliability of the enormous amount of context information collected from different distributed sensors and sources, an ontology-based model is considered to be more suitable for deployment [4].

The objective of this paper is twofold: (1) For an IoT paradigm, which service model is best applicable? (2) How an Intelligent Transportation System provides effective services to the users using context—aware model.

This paper is organized as follows: The background study is described in Sect. 2. The context abstraction in ITS model is described in Sect. 3. Section 4 describes the effectiveness of using context—aware in ITS through various use case scenarios and Sect. 5 with conclusion.

## 2 Background Study

The urban cities in India face serious traffic-related problems due to increase in vehicular growth. The growth rate of these vehicles will grow double or even more as the day passes. There are many emerging technologies such as Google map and Microsoft autoroute to assist the user of the vehicle to alleviate traffic and transportation problems. This technology widely uses satellite network for GPS devices, cellular services such as GSM/3G/4G for smart phones and other navigation devices for data collection. To provide effective services to the users and to ensure the accuracy of the services provided, many researchers have proposed different methods and algorithms in the literature. In Web paradigm, the two widely used methods are similarity-based and content-based methods.

The traditional similarity-based method delivers services to users based on the similarity that exists among the consumers, who have used or initiated the same query.

Bruke [5] addressed collaborative and content-based recommendation system in this paper. The cold start problem in similarity based (or collaborative) is identified, and he proposed a new hybrid recommendation system to overcome cold start issue. The content based method uses information about the user history to provide new and relevant services. The main problem with this method is it suggests same services that are similar and used by the user in the past. Zhao et al. [6] proposed a content-based recommendation for traffic signal control. In this paper, the author used k-nearest neighbor method to find the similar traffic condition in the past history.

For IoT Intelligent Transportation System, where numerous objects are connected on the fly, the traditional methods or recommended systems can only address explicit services relevant to user without giving preference to user context and environment context. For IoT paradigm, the recent development is on context aware methods to capture high-level descriptions of traffic conditions and user centric observation. Deepti et al. [4] proposed a context—aware service framework for traffic management using multimedia ontology to extract descriptions of traffic details and vehicles across the transportation network. To handle real-time contextual observation, the authors have used MOWL with dynamic Bayesian networks. The main objective of their paper is, with the analyzed and extracted contextual information on how to predict the congestion and show an alternative congestion free route to the user.

Davy et al. [7] proposed an architecture on ontology-based policy relations of traffic engineering to deliver user context—aware and QoS based services. Nisha et al. [8] proposed on ontology-based context—aware framework for traffic surveillance. Phlippe and Fawzi used a semantic aware ontology to prepare a high-level representation on vehicles and environment to assist users in taking relaxation decision. Abayomi and Gyu proposed a framework for exploiting IoT for context—aware trust-based services. The author focused much on context classification and reasoning to improve context aware trust based services. Aditya et al. [9] proposed a multi-level smart city architecture based on semantic Web technologies and Dempster–Shafer theory of evidence. This paper focused on giving a transformation to raw information by enriching its semantics and inferring it by customized applications.

## 3 Ontology Based Context—Awareness in ITS

Context in general means the ability of a system to gather information about its environment at any given time and adapt behaviors according to generate dynamic decision [10]. Of the many possible techniques available to model ITS, this paper focuses on using ontology or cognitive reasoning technique to enable high level, semantic knowledge representation to perform context—awareness on ITS.

Any context—aware system model built for ITS focuses on collecting, analyzing and inferring high-level context information from environment. Ontologies for ITS addresses the following key features.

(i) The ITS should be adaptive to automate the changes in the system with respect to contextual situations. For example, in a four-cross traffic signal management, the system automates the signal lights based on sensing the vehicle presence. The traditional way of regulating signal light is based on timer (equal time distribution). Let us assume, there is no vehicle presence on one cross when signal shows green. This blocks the other crossers having more number of vehicles. A context—aware system at this scenario should be adaptive enough to take quick decision on automating changes in the system.
(ii) Anticipating preferences—ready to serve services. ITS is itself dynamic in nature at any time. There is no guarantee the same level of traffic density exists all the time. For example, if raining during peek hours, the context—aware ITS has to dynamically decide on regulating the traffic signal based on density of the vehicles at each signal cross.
(iii) Context—aware to recognize situations—enabling the system to take immediate decision in an emergency situations like accidents.
(iv) Giving preference to user context—this enables personalized service. For example, during peak hours, the traffic is always congested. If any path is blocked, having context—aware system will assist user a personalised service to travel to his/her destination by suggesting different path.

Figure 1 illustrates the context abstraction in IoT ITS model proposed in our previous work [11]. At end points, exists the object embedded with hardware, software and various types of sensors like accelerometer sensor, proximity sensor, motion detector sensor etc.

At fog and cloud layer, the main focus is on analyzing the data collected from end points [12]. The context—aware system using ontologies plays an important role

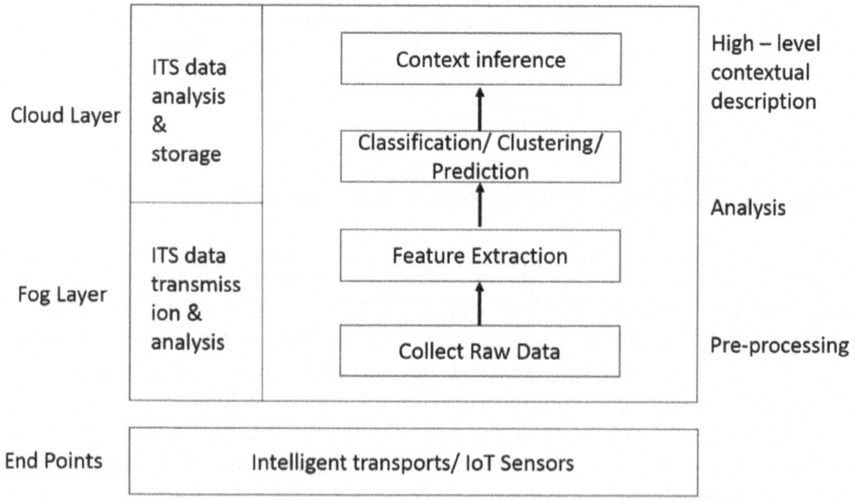


**Fig. 1** Context abstraction in ITS model

to infer high-level contextual situation (or) description from the raw data collected from the end points.

(i) Pre-processing—low-level data emitted by sensors are transformed into information to be used for further analysis.
(ii) Feature extraction—focus on filtering relevant data depending on the activities generated. The extracted feature vectors are then used in the next process.
(iii) Classification/clustering/prediction—depending on the task or activities, any of the mentioned analysis is applied to derive the context knowledge. Machine learning algorithms such as ANN, SVM and Bayesian classifier are used to derive context knowledge from feature vector extracted.
(iv) Context inference—to derive cognitive meaning of context information, context—aware system requires a cognitive model. Many models have been proposed in the literature to infer high-level description. Among all, ontology modeling for infer of high-level cognitive meaning has yielded better accuracy than other methods.

Figure 2 depicts context inference model for ITS. The first layer in the inference model is the domain ontology. With ITS, the important contexts are user context, device context, environment context, weather context, parking context, time, etc.

The second layer in the model is specific to ontologies of ITS. With ITS, it is ontology for smart parking, ontology for route prediction, ontology for anticipating preference, etc. The main focus is to integrate reasoning and rules for inferring contexts specific to ITS services. The context rules are defined in the third level.

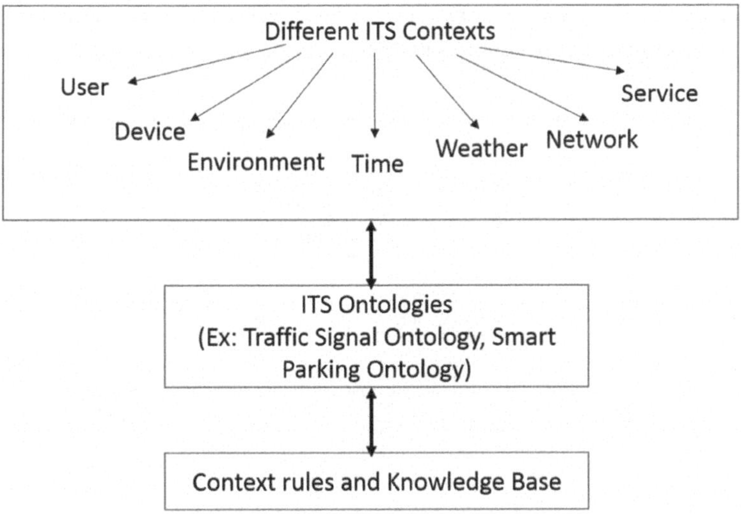


**Fig. 2** Context inference model

## 4 Context—Aware ITS: Use Case Scenarios

This section discusses the context—aware ITS through two use case scenarios.

### 4.1 *Use Case 1: Dynamic Traffic Signal Management (DYSM)*

The traditional traffic signal operates based on a timer. For a four-cross traffic signal management, the red, green and yellow signals will be open and closed for a minimum of 30 s to maximum of 90 s. Let us assume no presence of vehicle is found in one cross and as per the timer the signal is opened to green. In this case, vehicles waiting on the other crosses have to wait for no vehicles to cross the lane.

To improve the effectiveness of this scenario, we can infer the context that if no vehicles are on a cross, then the system should dynamically decide on extending the particular cross to be closed and open green signal to other crosses depending on the vehicle density. This reduces the time delay in dynamic traffic signal management. Figure 3 depicts the ontology for the scenario dynamic traffic signal management.

### 4.2 *Use Case 2: Smart Parking System*

The primary focus of context—aware ITS is the user. For context—aware ITS, preference is the crucial component to satisfy the user need.

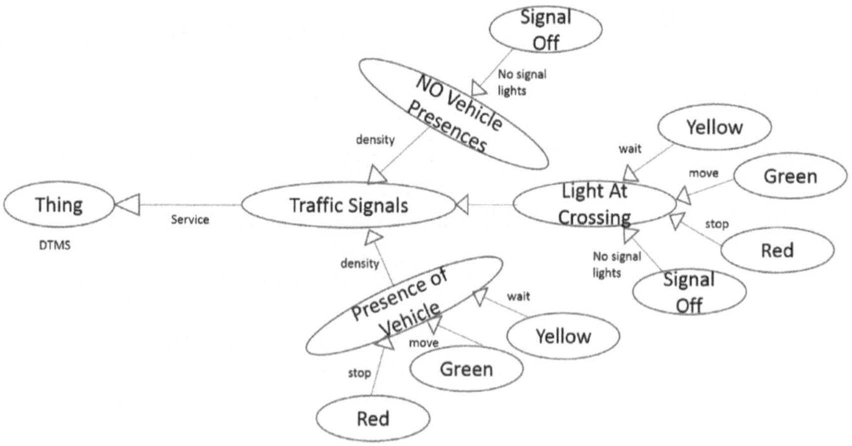


**Fig. 3** Ontology for dynamic traffic signal management

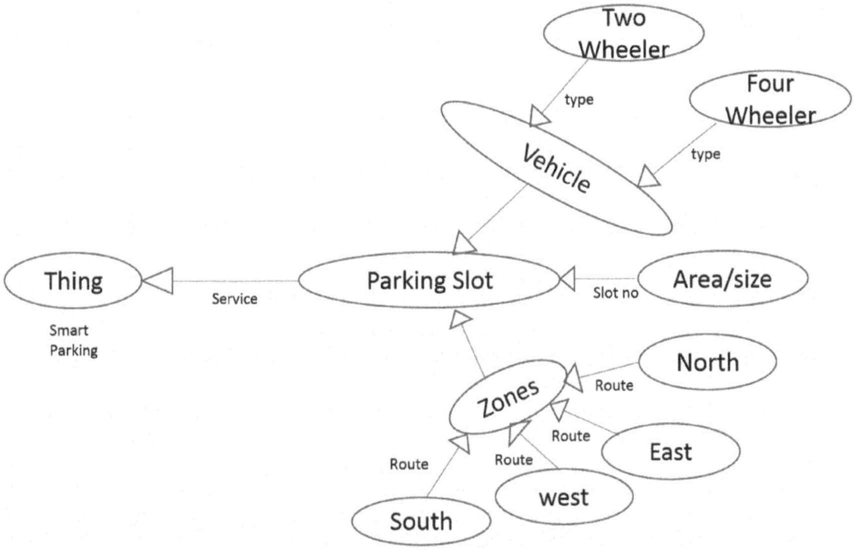


**Fig. 4** Ontology for smart parking system

Let us consider a smart parking system in a public forum that allocates parking slot to the user to park his/her vehicle. The traditional parking allocates a random slot or free slot at time instant, and this message is given to the user to park the vehicle. If the user is intended to go to a particular shop located on the north zone, the system has to allocate a slot on north zone or at least nearer to north zone. The context of the user can be inferred by getting his preference though a mobile app. Figure 4 depicts the ontology smart parking system.

## 5 Conclusion

Intelligent transport systems are widely transforming the current transportation system to improve road safety, traffic efficiency and driving experience. IoT handshakes with ITS to provide smart and quick services to users. To enable effective service delivery to the users of ITS, this paper addressed the use of context—awareness using ontology. To understand the high-level description of collected raw sensor data, context abstraction proposed in ITS highlights the importance of context inference model with the identified context of ITS. For the identified ITS context, we have also presented two use case scenarios to understand the inference model used.

## References


1. Zhu, Y., Zhu, X., Zhu, S., Guo, S.: Intelligent transportation system based on internet of things. In: World Automation Congress (WAC), IEEE Explore, Oct 2012
2. Perera, C., Zaslavsky, A., Christen, P., Georgakopoulos, D.: Context aware computing for the internet of things: a survey. IEEE Commun. Surv. Tutor. (2013)
3. Olabiyi, O., Martinson, E., Chintalapudi, V., Guo, R.: Driver Action Prediction Using Deep (Bidirectional) Recurrent Neural Network. Intelligent Computing Division, Toyota InfoTechnology Center USA
4. Goel, D., Chaudhury, S., Ghosh, H.: IoT Approach for Context—Aware Smart Traffic Management Using Ontology, WI'17. ACM, Leipzig, Germany (2017)
5. Burke, R.: Hybrid Web recommender systems. In: Brusilovsky, P., Kobsa, A., Nejdl, W. (eds.) The Adaptive Web: Methods and Strategies of Web Personalization, LNCS, vol. 4321, pp. 377–408. Springer, Heidelberg, Germany (2007)
6. Zhao, Y.F., Wang, F.Y., Gao, H., Zhu, F.H., Lv, Y.S., Ye, P.J.: Content-based recommendation for traffic signal control. In: The Classic Webster Method, IEEE 18th International Conference on Intelligent Transportation Systems, 15–18, 2015
7. Davy, S., Barrett, K., Jennings, B., van der Meer, S., Serrano, M., Strassner, J.: Policy Interactions and Management of Traffic Engineering Services Based on Ontologies, pp. 95–105 (2007). https://doi.org/10.1109/LANOMS.2007.4362464
8. Pahal, N., Mallik, A., Chaudhury, S.: An ontology-based context—aware IoT framework for smart surveillance. In: Proceedings of the 3rd International Conference on Smart City Applications, ACM, 10–11 Oct 2018
9. Gaura, A., Scotneya, B., Parra, G., McCleana, S.: Smart city architecture and its applications based on IoT. Procedia Comput. Sci. **52**(2015), 1089–1094 (2015)
10. http://www.whatis.com
11. Swarnamugi, M., Chinnaiyan, R.: IoT hybrid computing model for intelligent transportation system (ITS). In: IEEE Second International Conference on Computing Methodologies and Communication (ICCMC), 15–16 Feb 2018
12. Swarnamugi, M., Chinnaiyan, R.: Cloud and fog computing models for internet of things. Int. J. Res. Appl. Sci. Eng. Technol. December (2017)

# Static, Dynamic and Intrinsic Features Based Android Malware Detection Using Machine Learning

**Bilal Ahmad Mantoo and Surinder Singh Khurana**

**Abstract** Android is one of the smartest and advanced operating systems in the mobile phone market in the current era. The number of smartphone users based on the Android platform is rising swiftly which increases its popularity all over the world. The rising fame of this technology attracts everyone toward it and invites more number of hackers in Android platform. These hackers spread malicious application in the market and lead to the high chance of data leakage, financial loss and other damages. Therefore, malware detection techniques should be implemented to detect the malware smartly. Different techniques have been proposed using permission-based or system call-based approaches. In this paper, a hybrid approach of static, dynamic and intrinsic features based malware detection using k-nearest neighbors (k-NN) and logistic regression machine learning algorithms. The intrinsic feature contribution has also been evaluated. Furthermore, linear discriminant analysis technique has been implemented to evaluate the impact on the detection rate. The calculation uses a publicly available dataset of Androtrack. Based on the estimation results, both the k-nearest neighbors (k-NN) and logistic regression classifiers produced accuracy of 97.5%.

**Keywords** Dynamic analysis · Static analysis · Intrinsic features · Logistic regression · k-NN

## 1 Introduction

Malware is defined as a program which contains malicious code for performing vulnerable activities in any platform. These coded programs are used to perform operations that are usually used for personal interest like accessing to secret information, stealing password or credit card details. Security to computer assets and

B. A. Mantoo (✉) · S. S. Khurana
Department of Computer Science & Technology, Central University of Punjab, Bathinda, Punjab, India
e-mail: bilalbashir136@gmail.com

S. S. Khurana
e-mail: surinder.seeker@gmail.com



P. K. Singh et al. (eds.), *Proceedings of ICRIC 2019*, Lecture Notes in Electrical Engineering 597, https://doi.org/10.1007/978-3-030-29407-6_4

other Internet gadgets has always been a question of serious interest for security experts. Since the current generation is doing most of its task via mobile phones using Internet, its security has become a prime importance. Classic mobile phone has been superseded by smartphone with Android operating system. Different types of applications are present in this platform which are used for myriad number of day-to-day activities. These applications can be used for personal activities like photography, saving contact details and connecting with different types of people all over the world via e-mail. Same device can be used in combination with different persons in any organization via the applications Whatsapp, Facebook, LinkedIn, etc. Due to its high growing market and fame, hackers always try to spoof information from these devices by inserting different types of malwares. And these malwares are increasing day by day; as reported by AV-TEST, the number of malwares increases from 29.4 million to 847.34 million since last ten years.

Android market is considered as one of the famous platforms for mobile phone users in status quo. The Android market is increasing with steady speed, and it is now at 84.7% expansion of all mobile technology. The main reason is its wide functionality and best human interface. Further, it lets its users to download the application from many sources like torrents, Google play store, third-party markets, etc. [1]. But this openness of downloading from different sources leads to high chances of downloading malwares [2]. Programs that contain malicious payloads are used in getting access to other applications [3] which leads to confidential breaks, earning money through premium calls and SMS and other high-risk malicious attacks. Therefore, its detection should be handled safely and securely.

### *1.1 Android Malware Detection*

Android malicious applications installed in the smartphone can be recognized together with genuine applications by their permissions present in the Android application shown in Fig. 1 that the application requests during installation.

There are two approaches to detecting an Android malware application. First, the permissions are extracted from the Android manifest file of the application. Dataset of all the permissions present in the benign and malware applications is created, and finally, the machine learning algorithms are used to classify and identify the malware in Android platform. This comes in the approach of static malware detection, where we statically take out the features from an Android app. Second approach is dynamic behaviors of an application are observed by the system call series at the end.

### *1.2 Approaches to Detecting Malware*

Malware applications of Android have developed rapidly in a very short interval of time and would keep increasing as mobile phone users are increasing with high rate.

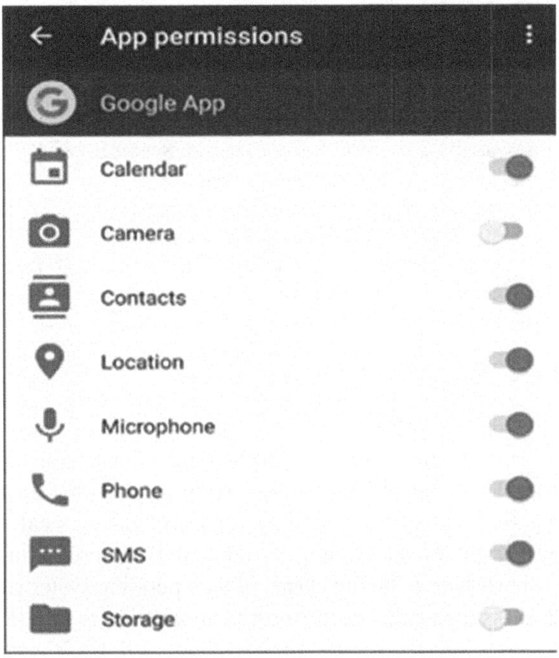


**Fig. 1** Application permissions

It becomes a very serious issue to make a distinction between Android malware from benign applications. There are two approaches to detecting a malware on Android operating systems.

I. Static Analysis Approach.
II. Dynamic Analysis Approach.

I. **Static Analysis Approach**

In this approach, a predefined Android application that is present in the smartphone comes in an Android package (.apk) archive. This .apk folder consists a zip package of AndroidManifest.xml and other important files and folders. In order to dig out the features, we have to perform reverse engineering of all .apk files. This is done using the apktool [4]. The Manifest.xml file of all Android applications consists of large number of features that can be helpful for performing static analysis. One such iconic and important feature used is the permissions granted by the application during its installation stage. The AndroidManifest.xml encompasses the number of permissions prerequisite by the Android application. To extract these permissions, the Android package has to be de-packaged using any available Android emulator. Figure 2 shows the snapshot of Android manifest file.xml of YouTube app.

II. **Dynamic Analysis Approach**

Dynamic analysis means observing the behavior of applications in a simulated context to build a model that characterizes the application's behavior. In this approach,

```
</application>
<uses-permission android:name="com.android.browser.permission.WRITE_HISTORY_BOOKMARKS"/>
<uses-permission android:name="android.permission.INTERNET"/>
<uses-permission android:name="android.permission.ACCESS_WIFI_STATE"/>
<uses-permission android:name="android.permission.ACCESS_NETWORK_STATE"/>
<uses-permission android:name="android.permission.READ_PHONE_STATE"/>
<uses-permission android:name="android.permission.SYSTEM_ALERT_WINDOW"/>
<uses-permission android:name="android.permission.GET_TASKS"/>
<uses-permission android:name="android.permission.INTERNET"/>
<uses-permission android:name="android.permission.ACCESS_WIFI_STATE"/>
<uses-permission android:name="android.permission.ACCESS_NETWORK_STATE"/>
<uses-permission android:name="android.permission.READ_PHONE_STATE"/>
```

**Fig. 2** Android manifest file of youtube.apk

run-time behavior of an Android application is constantly observed by installing the app in an Android emulator. Therefore, dynamic analysis methods cannot be simply sidestepped because they consist of features called system call which give the best result on detection. A system call is the programmatic way in which a program requests help from the kernel of the operating system to perform any task. To extract these system calls, each Android application is installed in Android emulator which is prepackaged with the Android Studio [5]. Each Android application is installed in separate emulator called Genymotion and its system call gets logged.

## 2 Machine Learning Classifiers

Machine learning is a branch of artificial intelligence (AI) that lets the system to learn and improve automatically the experience without being programmed. This concept can be simply explained with a simple example. We all have come up with the experience of putting on a new show in our life; at the initial stages, the show does not get fit to our foot, but when the time passes, the show comes in a right manner to our feet. Same is the case with the machine learning, it gets fit with the data as it experiences the data couple of times. Thus, machine learning algorithms learn from experiences without being programmed. The most widely used machine learning classifiers or algorithms for analysis training include the following classifiers.

### *2.1 Support Vector Machine*

It is also called the support vector networks. It is a non-probabilistic binary linear classifier but can be also used as a non-linear classifier using implicit mapping. It demonstrates linear separability in high dimensions by using hyperplanes. When used for classification, the SVM classifier builds an ideal and a unique hyperplane in high-dimensional feature space. In this way, it works like a maximal marginal classifier. When used for regression, SVM classifier performs linear regression in the high-dimensional vector space, and further, it does not penalize small.

SVM works to maximize the margin between different classes. Hence, it plays the role of a discriminative classifier [6]. SVM is a supervised learning method that creates projected plotting functions from a dataset which is labeled training data. To compute the SVM classifier, we need to minimize the SVM formula:

$$\left[1/n \sum_{i=1}^{n} \max(0, 1 - y_i(w, x_i - b))\right] + \gamma \|w\|^2 \tag{1}$$

## 2.2 *k-Nearest Neighbor*

k-NN [7] is a relatively easy vigorous and multipurpose supervised learning algorithm that works in a non-parametric and instance-based way. An advantage of using k-NN classifier is that it takes into account the local nature of the data. In k-NN, the training dataset is used to perform the classification of each of the members of the target dataset, i.e., the set to be classified. For each of the entities in the target dataset, the k-closest neighbors are selected, and then the distance of the entity from each of these is measured mainly by using the Euclidean distance to ascertain how these neighbors will be affecting the classification of the entity in question. k-NN algorithm can be used for both classification and regression problems, but it is most widely used for classification. In our case, the algorithm is used for the classification problem.

## 2.3 *Logistic Regression*

When the variables are dual or one is depending on another, in such scenarios logistic regression is the best way to classify them. Logistic regression is used to find the dependency of binary variables designate data and to elucidate the association between one dependent binary variable and one or extra minimal variables. The dependent variable should be dichotomous (e.g., present vs. absent). There should be no outliers present in the dataset.

$$P(C_k|w, x_i) = 1/(1 + \exp(-w, x_i)) \tag{2}$$

We know that logistic regression always searches for to minimize the likelihood $\mathcal{L}$ as loss function.

$$\min\mathcal{L}(w) = \prod_{i}^{n} p(Ck|w, xi) \quad (3)$$

In practical approaches, loss function $\mathcal{L}$ is log-likelihood:

$$\min\ \mathcal{L}(\mathrm{w}) = \log \mathrm{L}(\mathrm{w}) = \prod_{i}^{n} \log P(C_k|w, X_i) \quad (4)$$

In case of dataset which consists of only two classes like in our case, the algorithm changes from $C_0$ and $C_1$ to 0 and 1.

## 2.4 *Random Forest*

Random forest is flexible easy to implement supervised machine learning algorithm, which gives better result all the time without using parameter tuning. As the name indicates, it creates a forest and makes it a bit random. Therefore, the approach should be such that more number of trees should be present in the forest, so that the accuracy increases. Random forest pseudocode is given below:

1. From the dataset, select the "*k*" features from the total features say "*m,*" where $k \ll m$.
2. From the selected features, compute the value of node "*d*" using split point.
3. Divide the node into child node with best split.
4. Repeat the steps from 1 to 3, until "*I*" number of nodes will be created.
5. Then create a forest by reiterating the steps 1–4 for *n* times to create forest with "*n*" number of trees.

The initiation of the algorithm commences with choosing the value of "*k*" features from the total dataset. In the next step, we use these randomly selected "*k*" features to create a root node with the help of best split technique. The next step is creating the child nodes of the root node using the same spilt approach. Repeat from step 1 to step 3 until we came up with "n" number of trees. This comes in the training phase of the algorithm after this testing or we can say prediction phase starts. Random forest prediction pseudocode is given below.

1. Take the test features and use them in the random forest algorithm to create a decision tree to predict the output and store it in the prediction matrix.
2. Calculate the votes of each output value.
3. Predicted value which gets the high rates will be considered as final prediction from the random forest.

# 3 Proposed Methodology

The technique used for performing the static and dynamic analysis of different Android applications is based on hybrid approach. The representation of the proposed approach is given below. It consists of the following steps.

## *3.1 Data Collection*

The dataset comprises of different application of benign and malware applications of Android. We collect the data from free data resources like Androtrack, Google play store, etc.

## *3.2 Feature Collection*

The features are collected using different tools that are presently free of cost, like Gennymotion used for extracting the dynamic features and APK tool for extracting static features. The dataset contains 20 features consists of static, dynamic and intrinsic features of 600 Android malware and benign applications containing equal proportions of both the classes. The dynamic features consists of system calls that an Android application uses during the entire process are shown in Table 1.

Static features that are used are extracted from manifest file of an Android application and are shown in Table 2.

Intrinsic features used in the dataset are shown in Table 3.

**Table 1** Dynamic features used in our work

| S. No. | Feature | Type |
|---|---|---|
| 01 | Close | Int64 |
| 02 | Read | Int64 |
| 03 | Get time of day | Int64 |
| 04 | Futex | Int64 |
| 05 | Clock get time | Int64 |
| 06 | Mprotect | Int64 |
| 07 | Epoll_pwait | Int64 |
| 08 | Receive from | Int64 |
| 09 | Send to | Int64 |
| 10 | Ioctl | Int64 |
| 11 | Write | Int64 |
| 12 | Getuid | Int64 |

**Table 2** Static features used in our work

| S. No. | Feature | Type |
|---|---|---|
| 01 | SMS | Int64 |
| 02 | Phone | Int64 |
| 03 | Storage | Int64 |
| 04 | Contacts | Int64 |
| 05 | Location | Int64 |
| 06 | Camera | Int64 |
| 07 | Microphone | Int64 |

**Table 3** Intrinsic feature used in our work

| S. No. | Feature | Type |
|---|---|---|
| 01 | Size | Int64 |

The dataset has been created composing of both malware and benign applications. A total of 600 applications have been downloaded from the Androtrack malware project and the Google play store.[1] The downloaded application from the platforms is first scanned via virus total antivirus[2] search engine.

### 3.3 *Feature Extraction*

To implement any machine learning algorithm, it is mandatory that relevant data should be extracted in a systematic manner. The data here, in our approach, are static, dynamic and intrinsic features of all the applications. These features are extracted from the applications by simply using apktool for extracting the static features. Dynamic features are extracted by installing each application in restricted environment of Genny motion studio using *strace* command [8].

The system calls are extracted using *strace –p id –c* Linux command in the Genny motion shell after its process id is obtained. Process id is obtained by running the Linux command *ps* in the root shell of emulator. These system calls are recorded according to the frequency of each system call in an excel file for later processing. The permissions are stored as binary values "0" for absence and "1" for presence. The analysis shows that the size of malware is very less as compared to the benign applications shown in Fig. 3.

[1] www.malgenomeproject.org.

[2] www.virustotal.com.

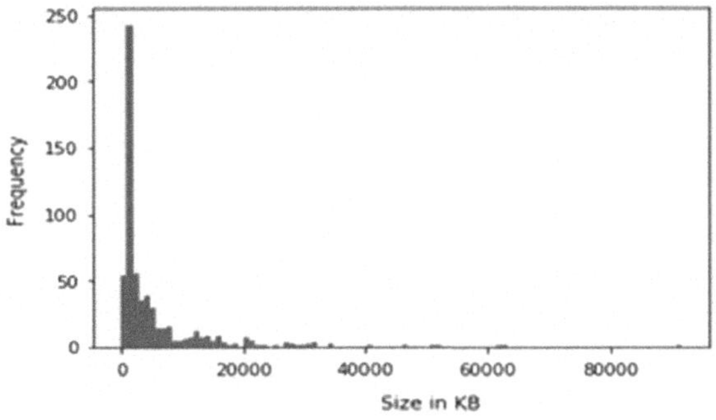


**Fig. 3** Statistics of feature "size"

## *3.4 Data Preprocessing*

This phase involves initial preprocessing of the raw data. Preprocessing refers to the transformation applied to our data, so that our machine learning algorithm performs efficiently. For achieving better outcomes in machine learning, format of the data should be well defined, and the null values should not be present in the dataset as most of the machine learning classifiers does not work with the null values, so it is essential to have null value-free dataset.

Further, the data should be formatted in such a way that different algorithms should be implemented on a single dataset, and best out of them should be retrieved. Figure 4 shows the detailed steps in malware detection. In the first step, Android apk is decompiled for static features, and then it is installed in the Android emulator for dynamic features and dataset of features is created. The dataset is then given to the machine learning algorithm. The machine learning phase divides the dataset into two sections—one for training and other for testing. The testing data are then used in detection for predicting each application

### 3.4.1 Linear Discriminant Analysis

In our approach, the data preprocessing technique used is linear discriminant analysis. This technique is the most commonly used technique in data preprocessing for the classification problems in machine learning. The aim of this technique is to project the high-dimensional data onto a lower-dimensional space with high-class separation. The general approach of linear discriminant analysis is perfectly matching with the dimensionality reduction algorithm PCA (principle component analysis). So in simple words, the aim is to project the large space of features in a small subspace. General steps used in linear discriminant analysis are summarized below.

1. Figure out the d-dimensional mean vectors from different classes from the dataset, that is $\boldsymbol{m}_i$ $(i = 1, 2)$, for two classes. $S_i$
2. Calculate the scatter matrices from the given data.

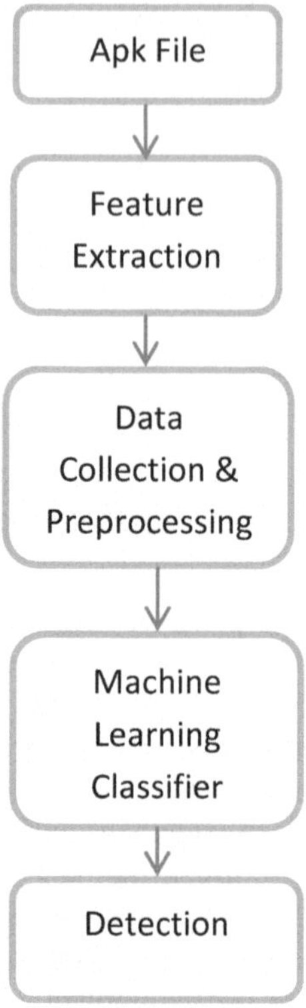


**Fig. 4** Diagramatic view of malware detection

$$S_w = \sum_{i=1}^{c} S_i (2)$$

where $S_w$ is within class scatter matrix.

3. Find the eigenvalues $(\boldsymbol{e}_1, \boldsymbol{e}_2, \boldsymbol{e}_3 \ldots, \boldsymbol{e}_d)$ and corresponding eigenvectors (λ1, λ2,..., λd) for the scatter matrix.
4. The eigenvectors obtained are sorted according to their decreasing order and choose the largest eigenvector k with the value of eigenvalue to be high to form a new matrix $d * k$ dimensional matrix W.
5. Now the matrix $d * k$ eigenvector is transformed into new subspace.

## 4 Experiment and Results

After the dataset is created, it is fed to the two machine learning algorithms. The dataset is imported to the Python-based Jupyter notebook installed on 4 GB RAM laptop. The dataset is first spitted into training and testing data using Python inbuilt module. The tested data are fed to the machine learning algorithms one after another. The first algorithm is logistic regression, and the second one is K-nearest neighbor (discussed above). In k-NN algorithm bases on the number of neighbors taken in the algorithm. To find the exact value of "*k*," simple Python script has been written which gives the better plot of choosing the value of neighbors. The plot is shown in Fig. 5.

As shown in the plot, the perfect value for the dataset lies in the range of 2–5. Therefore, $k = 5$ is chosen in our experiments as shown in Fig. 6.

We have also checked at different values of $k$ by either increasing or decreasing the value of $k$, but the results came was not up to the mark. Also the experiment shows the frequency of system call of the dataset, in which the malware application frequency rises up to the 75 K lac Fig. 8, which is far more than the benign frequency Fig. 7.

The experiment emanates that out of 300 instances of malware applications, 293 applications have been correctly classified as malware and 307 as benign applications using k-NN algorithm, which gives the accuracy of 97.5%. The error percentage was

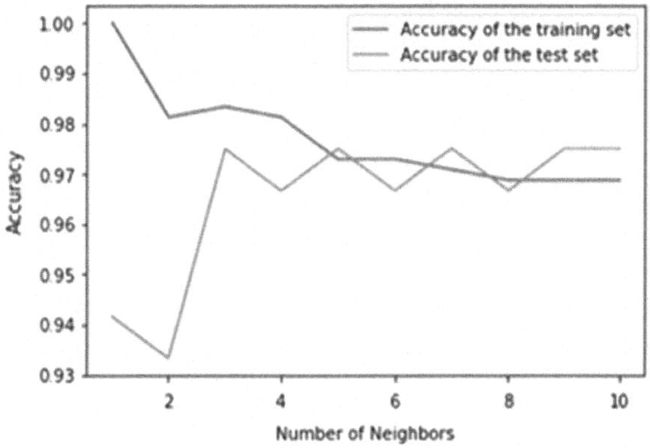


**Fig. 5** Checking for the best value for k

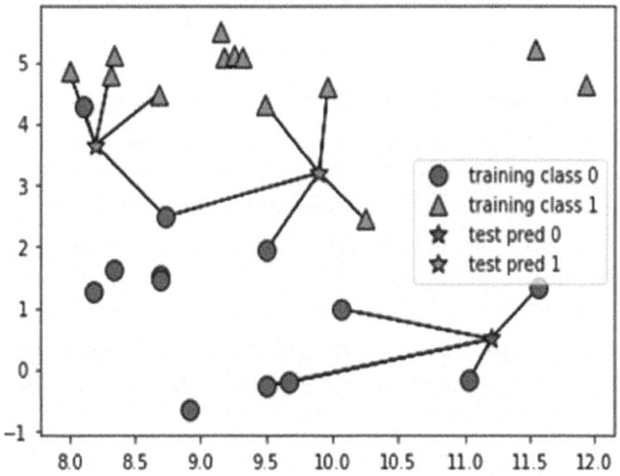


**Fig. 6** Plot of neighbors in k-NN

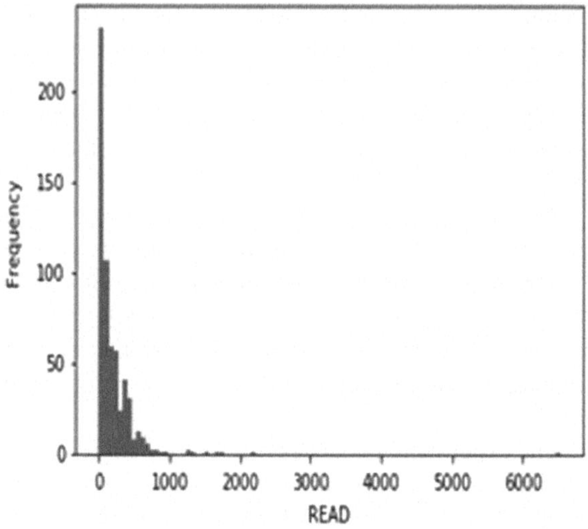


**Fig. 7** Plot for frequency of feature read

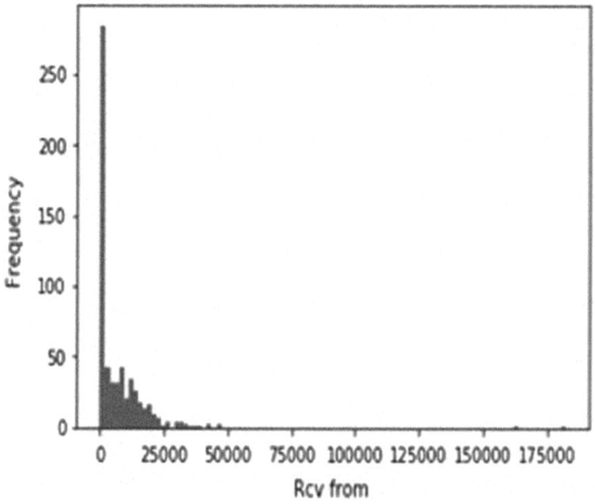


**Fig. 8** Frequency of feature Rcv from

7%. The same results have been achieved using logistic regression. The results show that both the algorithms perform better and can be used for malware detection.

This work was further evaluated by changing the preprocessing approach, i.e., with LDA, but the results obtained were slightly below. Also the impact of newly introduced intrinsic feature size helps in increasing the accuracy of the algorithms. By removing the feature size, the accuracy shows the dip of 1%.

As shown in Table 4, the accuracy of the algorithms is increasing when we are including the feature size and decreases slightly on its removal. Furthermore, the use of LDA shows negative impact on the detection rate. The confusion matrices

**Table 4** Accuracy summary of algorithms

| Algorithms | Accuracy | LDA used | Size used |
|---|---|---|---|
| Logistic regression | 97.0 | Yes | Yes |
| Logistic regression | 96.7 | Yes | No |
| Logistic regression | 97.5 | No | Yes |
| Logistic regression | 95.3 | No | No |
| k-NN | 97.0 | Yes | Yes |
| k-NN | 96.8 | Yes | No |
| k-NN | 97.5 | No | Yes |
| k-NN | 95.3 | No | No |

for both the algorithms with different combinations of intrinsic feature and linear discriminant algorithm are shown in Tables 5, 6, 7, 8, 9, 10, 11 and 12.

**Table 5** Confusion matrix for k-NN algorithm with intrinsic feature but without LDA

| Class | Precision | Recall | F1-score |
|---|---|---|---|
| 0 (malware) | 0.90 | 1.00 | 0.95 |
| 1 (benign) | 1.00 | 0.91 | 0.95 |
| Avg/total | 0.95 | 0.95 | 0.95 |

**Table 6** Confusion matrix for k-NN algorithm without intrinsic feature and LDA

| Class | Precision | Recall | F1-score |
|---|---|---|---|
| 0 (malware) | 0.99 | 0.94 | 0.96 |
| 1 (benign) | 0.94 | 0.99 | 0.97 |
| Avg/total | 0.97 | 0.96 | 0.96 |

**Table 7** Confusion matrix for k-NN algorithm without intrinsic feature but with LDA

| Class | Precision | Recall | F1-score |
|---|---|---|---|
| 0 (malware) | 0.93 | 0.94 | 0.93 |
| 1 (benign) | 0.94 | 0.95 | 0.94 |
| Avg/total | 0.96 | 0.96 | 0.96 |

**Table 8** Confusion matrix for k-NN algorithm with intrinsic feature and LDA

| Class | Precision | Recall | F1-score |
|---|---|---|---|
| 0 (malware) | 0.94 | 0.94 | 0.95 |
| 1 (benign) | 0.94 | 0.93 | 0.93 |
| Avg/total | 0.96 | 0.96 | 0.96 |

**Table 9** Confusion matrix for logistic regression with intrinsic feature and LDA

| Class | Precision | Recall | F1-score |
|---|---|---|---|
| 0 (malware) | 0.92 | 0.95 | 0.93 |
| 1 (benign) | 0.95 | 0.96 | 0.94 |
| Avg/total | 0.96 | 0.96 | 0.96 |

**Table 10** Confusion matrix for logistic regression with intrinsic feature but without LDA

| Class | Precision | Recall | F1-score |
|---|---|---|---|
| 0 (malware) | 0.99 | 0.94 | 0.96 |
| 1 (benign) | 0.94 | 0.99 | 0.97 |
| Avg/total | 0.97 | 0.96 | 0.96 |

**Table 11** Confusion matrix for logistic regression without intrinsic feature and LDA

| Class | Precision | Recall | F1-Score |
|---|---|---|---|
| 0 (malware) | 0.96 | 0.95 | 0.96 |
| 1 (benign) | 0.96 | 0.96 | 0.96 |
| Avg/total | 0.96 | 0.96 | 0.96 |

**Table 12** Confusion matrix for logistic regression without intrinsic feature but with LDA

| Class | Precision | Recall | F1-score |
|---|---|---|---|
| 0 (malware) | 0.93 | 0.95 | 0.92 |
| 1 (benign) | 0.94 | 0.94 | 0.95 |
| Avg/total | 0.96 | 0.96 | 0.96 |

## 5 Conclusion and Future Scope

In this paper, two machine learning algorithms k-NN and logistic regression have been implemented over a dataset of 600 Android applications to identify malicious Android application and assess the performance of each algorithm. Here, we implemented a simple approach for classifying Android applications. The dataset consists of 20 features of static, dynamic and intrinsic features. The dataset is then divided into training and testing sets. The training data are used to train the model. Both the algorithms show the accuracy of 97.5%, but the logistic regression shows high accuracy over testing data. The addition of intrinsic feature also helps in increasing the accuracy. The improvement in this area is increasing the dataset and also using different machine learning classifiers. The same work could be evaluated using another dimensionality reduction algorithm (principle component analysis) in future to check its impact on accuracy.

## References

1. Handa, A., et al.: Malware detection using data mining techniques. Int. J. Adv. Res. Comput. Commun. Eng. **5**, 2015 (2015)
2. Xialoeiwang, Y.Z.: Accurate malware detection in cloud. SpringerPlus, 123 (2015)
3. Hiranwal, et al.: A survey on techniques in detection and analysing malware executables. Int. J. Adv. Res. Comput. Commun. Eng. **3**(4), 422–428 (2013)
4. Virus Total. Retrieved from virustotal.: https://www.virustotal.com/#/home/upload (2015)
5. Install Android Studio. Retrieved from Developer Android: https://developer.Android.com/studio/install (2018)
6. Cortes, C.: Support-vector networks. Mach. Learn. **20**, 273–297 (1995) (Kluwer Academic Publishers, Boston. Manufactured in The Netherlands)
7. Blog, K.Z.:. A complete guide to K-nearest-neighbors with applications in Python and R. Retrieved from kevinzakka.github: https://kevinzakka.github.io/2016/07/13/k-nearest-neighbor (2018)
8. How to use strace and itrace commands in Linux. Retrieved from the geek dairy: https://www.thegeekdiary.com/how-to-use-strace-and-ltrace-commands-in-linux (2018)

# Machine Learning: A Review of the Algorithms and Its Applications

**Devanshi Dhall, Ravinder Kaur and Mamta Juneja**

**Abstract** In today's world, machine learning has gained much popularity, and its algorithms are employed in every field such as pattern recognition, object detection, text interpretation and different research areas. Machine learning, a part of AI (artificial intelligence), is used in the designing of algorithms based on the recent trends of data. This paper aims at introducing the algorithms of machine learning, its principles and highlighting the advantages and disadvantages in this field. It also focuses on the advancements that have been carried out so that the current researchers can be benefitted out of it. Based on artificial intelligence, many techniques have been developed such as perceptron-based techniques and logic-based techniques and also in statistics, instance-based techniques and Bayesian networks. So, overall this paper produces the work done by the authors in the area of machine learning and its applications and to draw attention towards the scholars who are working in this field.

**Keywords** Algorithm · Machine learning · Data · Artificial intelligence

## 1 Introduction

Machine learning is a study of computer science in which the machine is trained to perform some specific task. It has originated from computational learning and pattern recognition in the area of AI (artificial intelligence). The machine is trained on some data sets, and then, the algorithms are applied, so that the machine can make predictions and learn, respectively, on the given data sets. A model is programmed, procedures are operated upon it, and inputs are taken to obtain data-driven predictions instead of firm static program instructions. Each instance in the data set that is used

D. Dhall · R. Kaur (✉) · M. Juneja (✉)
University Institute of Engineering and Technology, Punjab University, Chandigarh, India
e-mail: ravinder.kaur7@yahoo.com

M. Juneja
e-mail: mamtajuneja@pu.ac.in

D. Dhall
e-mail: devanshidhall12@gmail.com



P. K. Singh et al. (eds.), *Proceedings of ICRIC 2019*, Lecture Notes in Electrical Engineering 597, https://doi.org/10.1007/978-3-030-29407-6_5

**Table 1** Comparison between supervised and unsupervised learning

| Parameters | Supervised learning | Unsupervised learning |
|---|---|---|
| Definition | Supervised learning is defined as that type of machine learning in which there is input and output variables and an algorithm helps to understand the mapping function from input to output | Unsupervised learning is a type of machine learning that includes only input and no output variables |
| Type of data | The data in supervised learning is labelled, and it predicts the output from the given input data | The data in unsupervised learning is unlabelled, and prediction is made by inheriting structure from the given input data |
| Accuracy | Supervised learning provides more reliable and accurate results | Unsupervised learning produces average results as compared to the supervised learning |
| Complexity | The computational complexity in the case of supervised learning is too complex | The computational complexity in the case of unsupervised learning is less |

by the machine learning algorithms is depicted by features belonging to the same set [1]. The features can be categorical, binary or continuous. If data is labelled, then it is supervised learning; otherwise, it is unsupervised learning in which there is unlabelled data as shown in Table 1. Various applications are based on supervised learning. Let us take the example of recognising spoken speech in which conversion of the acoustic speech signal to ASCII text is taken place. There may be differences in pronunciation of words because of variations in gender, age or pronunciation. Consequently, the idea in machine learning is to collect an extensive collection of data from people and plot them to words. The algorithms of machine learning are non-interactive because the model is built from the results gathered by the data set [2]. The past observations are studied to produce predictions. Examples include differentiating which emails are spammed and non-spammed with the help of machine learning algorithms. This could be achieved by gathering examples of non-spammed and spammed emails. Further, algorithms are applied to these samples to differentiate between spammed and non-spammed email[2].

The formal definition of machine learning states that "A computer program learns from experience E for some performance measure P and some task T, if its performance on T, as measured by P, improves with experience E". Machine learning is required in the following occurrences:

1. Navigation on Mars where there is no human expertise.
2. When humans are not able to demonstrate their expertise as in speech recognition.
3. Also, when the solution differs concerning time. The example is temperature control.
4. In biometrics where the solution needs to be changed for particular cases.

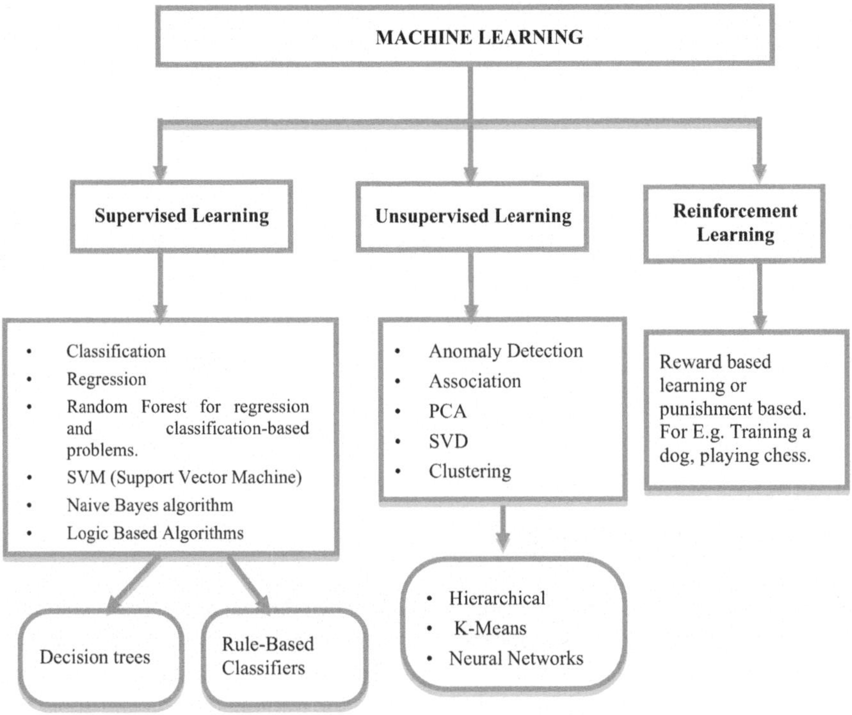


**Fig. 1** Flow chart of machine learning and the algorithms used for training the machine [3]

Hence, machine learning is applied to those problems where their theoretical knowledge is not complete, but there exists an ideal number of results and observations. The primary objective of machine learning is that once the algorithm learns what to do with the given data, it performs its task automatically [3] (Figs. 1 and 2).

## 2 Machine Learning

Machine learning (ML) is an area of computer science and a subset of artificial intelligence which provides the computers with the capability to think and learn without being programmed explicitly. Machine learning is used in various computational tasks and its root objective is to train the machine with the help of data provided, the data can be labelled in case of supervised learning and unlabelled in case of unsupervised learning in order to produce better outcomes for the specified problem. The main focus is to make computers learn from past experience [1].

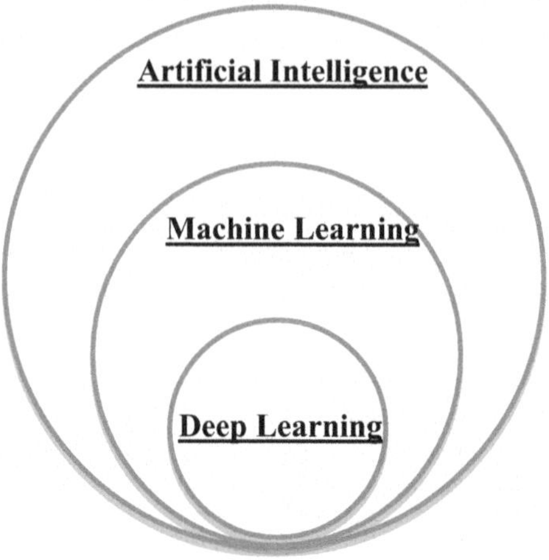


**Fig. 2** Representation of artificial intelligence, machine learning and deep learning

## 2.1 *Supervised Learning*

Supervised machine learning algorithms need labelled data, and the data is split into two parts, one is testing data set, and the other is training data set [4]. The trained data set has some output that needs to be predicted. The task is to make the machine learn from some similar kind of patterns obtained from the training data set and apply the same on the data set to be tested to predict the real-valued output [5]. The algorithms are listed below.

### 2.1.1 Support Vector Machine (SVM)

Support vector machine is a supervised learning formed on the classification method. The underlying idea of SVM is based on margin calculation. It is a non-probabilistic linear classifier. This algorithm is called as the state-of-the-art machine learning algorithm. In this type of algorithm, every data item is marked as a point in n-dimensional space, where n is the number of features, each feature being the value of the particular coordinate. This method is used to analyse the vectorised data and find a hyperplane which classifies between the two inputs [3] (Figs. 3 and 4).

Different margins are drawn between various classes, and a hyperplane drawn is such in which the mean-squared error is minimized, and the distance between the margin and the classes is maximised (Fig. 5).

### 2.1.2 Naive Bayes

Naive Bayes (NB) is a probabilistic classifier technique which mainly focuses on text classification industry. It is also known as the independent feature model. The

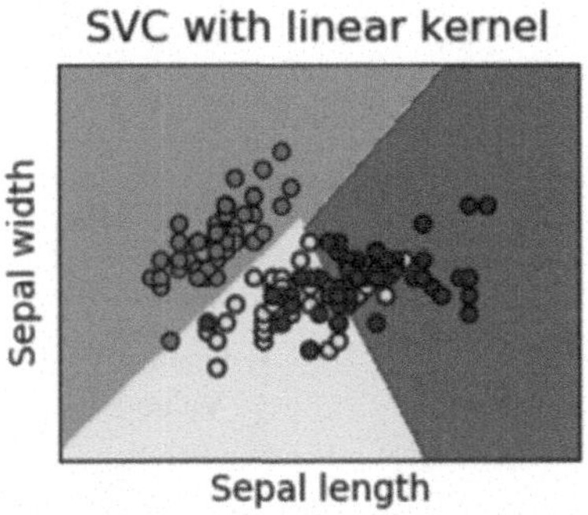


**Fig. 3** Support vector machine algorithm in a linear fashion [6]

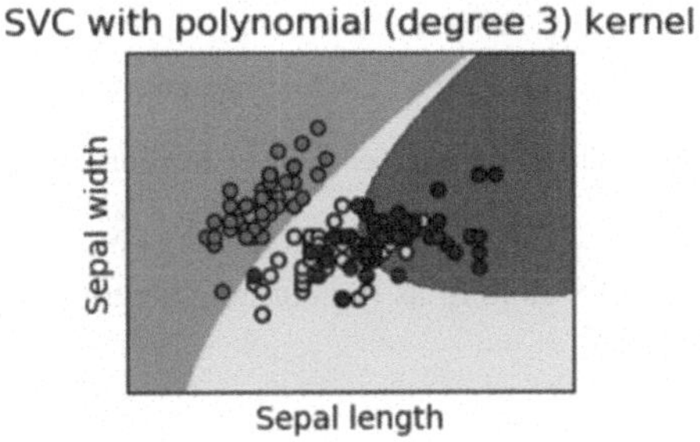


**Fig. 4** Support vector machine algorithm in a nonlinear fashion [6]

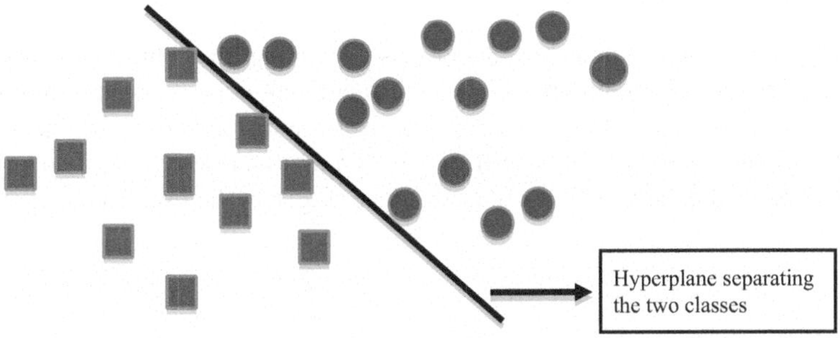


**Fig. 5** Representation of the hyperplane separating the two classes

main idea behind Naive Bayes revolves around the conditional probability. Trees are created based on their probability of occurrence and are known as Bayesian network. The algorithm is based on Bayes' theorem and depending upon the nature of the probability model; the classifiers can be trained systematically by needing a reduced amount of training data to decide the parameters required for classification. In Naive Bayes classification method, the only variance of each class needs to be found out, not the entire covariance matrix because of the assumption of independent variables. The main advantage of the algorithm is that only a small amount of data (that is to be trained) is required to decide the parameters for classification. This algorithm works

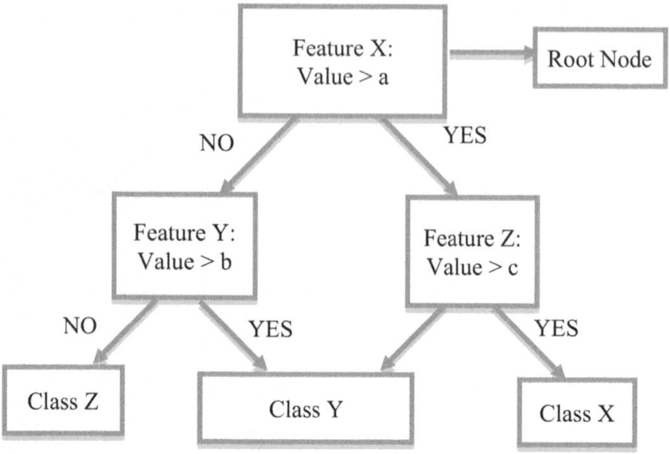


**Fig. 6** Graphical representation of decision tree

correctly as long as the correct category is more probable as compared to the other ones [3].

### 2.1.3 Decision Tree

The decision tree is a technique of a supervised learning algorithm that is used for classification. The algorithm groups attribute depending upon the values in order of their ascending or descending order [3]. The decision tree consists of branches and nodes where the node represents attributes of a group that is to be classified, and the branch displays the value which a node can take (Fig. 6). The pseudocode is shown in Fig. 7.

### 2.1.4 Regression

Linear regression, a supervised learning algorithm aims to predict the real-valued output. It is used to predict a response B from the predictor variable A. The assumption is that there exists a linear relationship between A and B. This algorithm is a parametric method and gives numerical values as the output. The cost function is an example of regression where we train the machine with the labelled data, and it predicts the next cost of the house as the real-valued output (Fig. 8).

The mathematical form of the equation is as follows:

$y = \beta_0 + \beta_1 x_1 + \beta_2 x_2 + \cdots + \beta_n x_n$, where $y$ represents the predicted value.

```
Tree_Learn(Training_example, Label, Attributes)
{
    A root node is created for the tree.
    If Training_example is identical to the label attribute value T_a,
        Then return the root node with the label= T_a,
    If Label = empty ( No attributes present)
        Then return the root, along with the most common value of Label in Training_example.
    Else
    {
        Choose attribute R from attributes that best classifies Training_example based on measure of entropy
        Set R as the attribute for Root.
        For every value of R, X_i, do
        {
            Add a root node above the branch, corresponding to R=X_i
            Let Training_example_ti be the subset of Training_example such that R=X_i
            If Training_example_ti is empty,
                Then a leaf node is added under the branch with label value = most frequent value of Label in Training_example.
            Else under the branch, join the subtree learned
                Tree_Learn(Training_example, Label, Attributes)
        }
    }
    Return (Root Node)
}
```

**Fig. 7** Pseudocode of decision tree

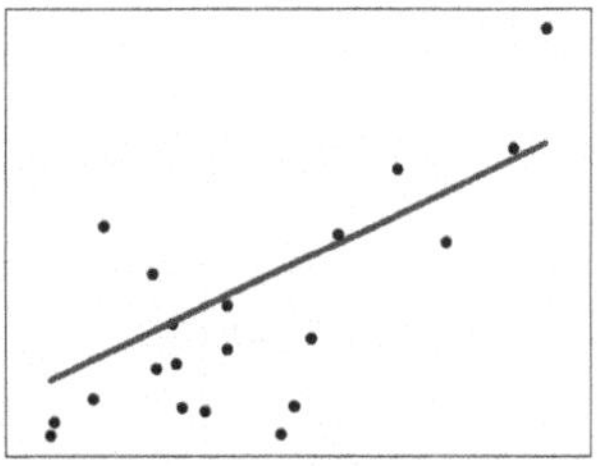

**Fig. 8** Graph of linear regression [7]

### 2.1.5 Classification

Logistic regression is known as classification, which is a kind of supervised learning technique used for classifying between two or more classes. Classification of email whether it is spam or not and classification between edible and poisonous fruits, etc., are all examples of classification. The algorithm makes use of a linear equation with independent predictors to predict the value, and the result is either a yes (1) or a no (0) [3] (Fig. 9).

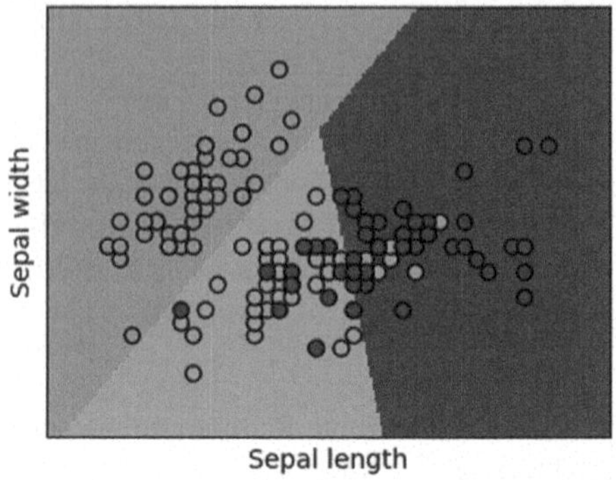


Fig. 9 Graph of classification [8]

## 2.2 Unsupervised Learning

Unsupervised learning is a form of machine learning that requires no labelled data to the machine. An algorithm is made based upon the input data, and then, the algorithm is analysed on a set of data. The training data set is used in creating and training of the model, whereas the testing data set helps in predicting the correct values [3]. The machine predicts the outcome based on past experiences and learns from the previously introduced features to predict the real-valued outcome (Fig. 10).

### 2.2.1 K-Means Clustering

K-means clustering or grouping is an unsupervised learning algorithm which makes groups or clusters automatically when initiated. The data which possesses similar features is grouped in the same cluster. The algorithm is named K-means because it creates K-distinct groups. The objective is to present K-centroids, one centroid for

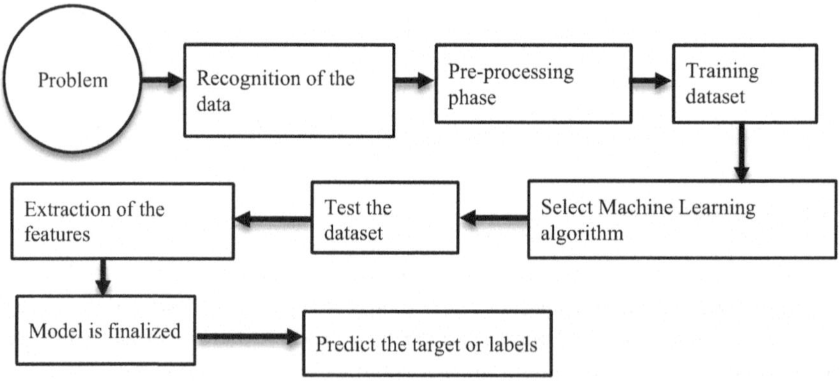


Fig. 10 Workflow of unsupervised learning

1. First, arrange k points which are represented by the objects that have to be grouped.
2. Then assign every object to the corresponding group which has the closest centroid.
3. After assigning all the objects, again calculate the positions of k centroids.
4. Then repeat step 2 and step 3 till the centroids don't move.

**Fig. 11** Pseudocode for K-means clustering

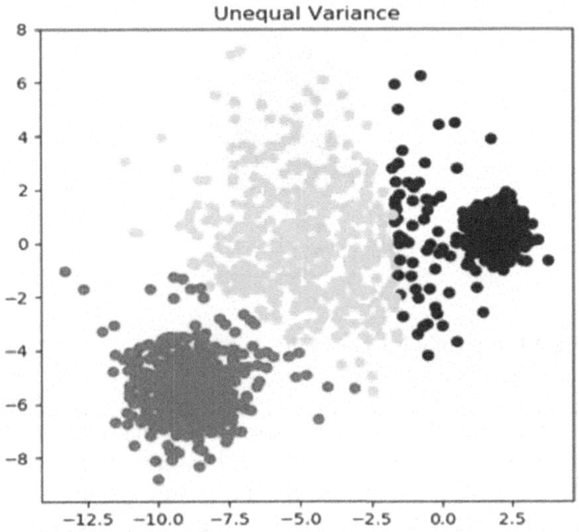


**Fig. 12** Unequal variance [9]

every cluster and to minimise the square error function. The mean value lies in the centre of the particular cluster [3] (Figs. 11, 12 and 13).

### 2.2.2 Principal Component Analysis (PCA)

The principal component analysis is an unsupervised machine learning algorithm in which the data is reduced, so that computations are more accessible and faster. PCA converts the two-dimensional data into one-dimensional data. It is done by transforming the set of variables to new ones known as principal components (PC), and PCs are orthogonal. The data set on which the PCA algorithm is applied must be scaled as the results are sensitive to the relative scaling. This algorithm is used in summarisation of data [3] (Fig. 14).

### 2.2.3 Neural Networks

Artificial neural network (ANN), an unsupervised learning, works on the concept of neurons. Neural networks comprise of three things, one is the input layer, second

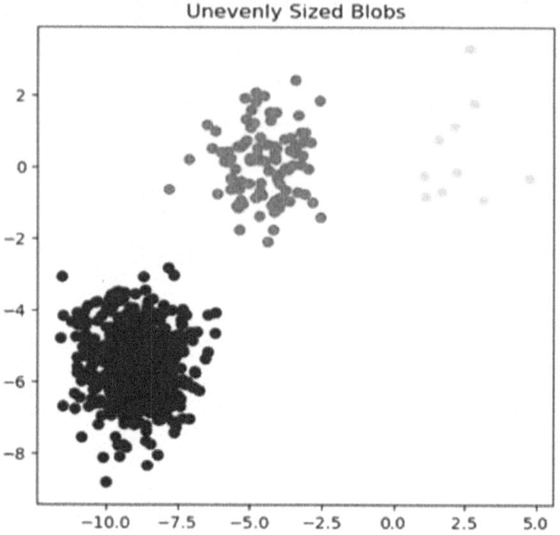


**Fig. 13** Unevenly sized blobs [9]

1. First, take the complete dataset of S samples. Ignore the targets.
2. Then compute the S dimensional mean vector.
3. Calculate the scatter matrix of the entire dataset.
4. Then calculate the eigenvalues and the eigenvectors ($e_1, e_2, \ldots, e_S$).
5. Decrease the eigenvalues and sort the eigenvectors .
6. Choose V eigenvectors within the largest eigenvalues to shape a SxT matrix U.
7. Then SxT eigenvector matrix should be used to transform the sample on a new subspace.

**Fig. 14** Pseudocode for principal component analysis [3]

is the hidden layer and third is the output layer [10]. The input is provided by the input layer, and then, it is processed by the hidden layer. The output is calculated and sent to the output layer. In the unsupervised neural network, there is no hint for the output corresponding to the input provided. The data is put together based on similar characteristics. The task of the neural network is to establish the relation between inputs and cluster them (Fig. 15).

## 2.3 *Deep Learning*

Deep learning, also known as hierarchical learning, is a part of ML (machine learning) algorithms and architectures. It includes all machine learning methods which are based on learning data representations [11]. The learning can be categorised as

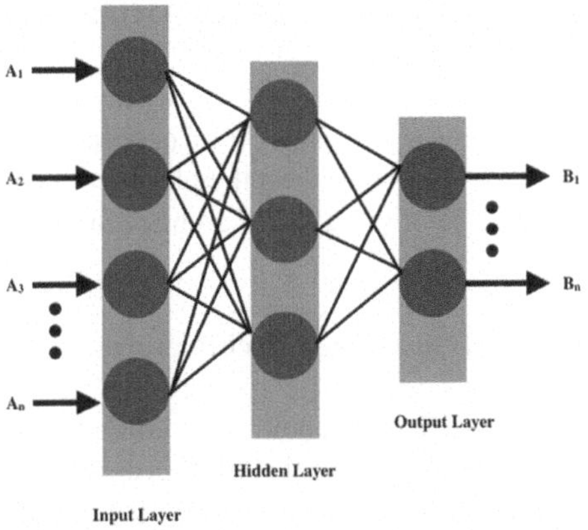


**Fig. 15** Graphical representation of neural networks [3]

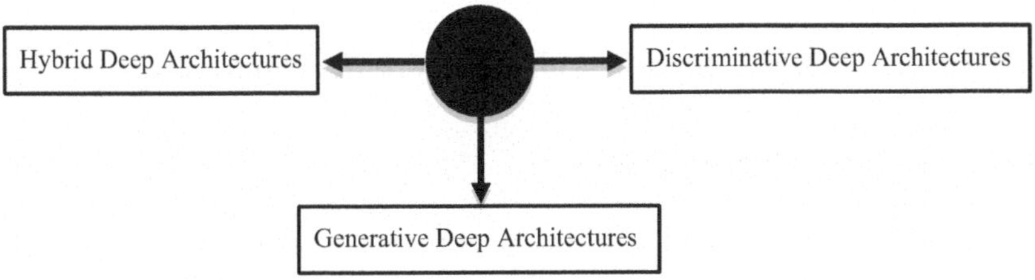


**Fig. 16** Classes of deep architecture

supervised, semi-supervised or unsupervised. In supervised learning, classification is done, and in unsupervised learning, similar features or characteristics are grouped [12]. Deep learning algorithms extract features implicitly, and the significance of the word deep means the number of layers throughout from which the data is to be transformed. Algorithms of deep learning are applied to both supervised and unsupervised learning. In unsupervised learning, there is more amount of unlabelled data as compared to supervised learning; hence, this is more beneficial. Deep learning extracts the best features, and the solution is an end-to-end method [10, 12, 13] (Fig. 16).

### 2.3.1 Generative Deep Architectures

Generative deep architectures are deep architectures that distinguish the high-order properties of the visible or observed data for analysis of patterns and characterise the statistical distributions of the linked classes and the observed data. This architecture can be converted into discriminative one by Bayes rule [10].

### 2.3.2 Discriminative Deep Architectures

As the name suggests, discriminative deep architectures directly provide discriminative power for analysis of patterns by distinguishing the posterior divisions of various classes based on the observed data. CNN (convolutional neural network) is an example of this type of architecture [10].

### 2.3.3 Hybrid Deep Architectures

Hybrid deep architecture is a combination of generative and discriminative deep architectures. The goal is to discriminate but effectively produce better outcomes of generative architectures through better optimisation techniques [10].

## *2.4 Various Deep Learning Algorithms or Technologies*

### 2.4.1 Deep Neural Network (DNN)

The deep neural network is an artificial neural network (ANN) consisting of many input and output layers. The neural network is deep because of multiple layers inside it. In this, the machine extracts the features implicitly from the data. This technique finds the most optimal way to convert the input into the output. It can be in any form whether linear or nonlinear relationship. In DNNs, the data moves from the input layer to the output layer without looping and is known as feedforward networks. The deep neural network works on a huge amount of data and makes use of training parameters like learning rate, size and the initial weights [10, 13–15].

### 2.4.2 Convolutional Neural Network (CNN)

Convolutional deep neural networks are the neural networks which work on the same principle of neural networks and are used in speech recognition and computer vision. It is a deep and feedforward neural network [10].

### 2.4.3 Recurrent Neural Networks (RNN)

Recurrent neural networks are the ones in which data can flow in any direction, and its application includes language modelling [10].

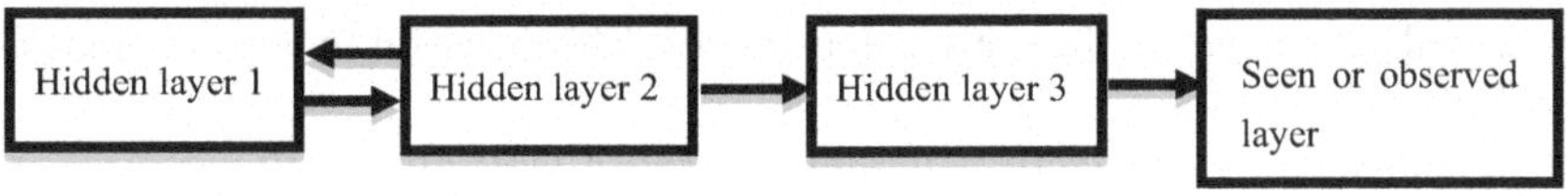


**Fig. 17** Graphical representation of deep belief network

**Table 2** Difference between machine learning and deep learning

| Machine learning | Deep learning |
|---|---|
| Machine learning algorithms enable the machine to make decisions | Deep neural networks work on the principal of artificial neural networks (ANNs) |
| ML algorithms work on less amount of data | Deep learning techniques require a large amount of data to learn |
| In ML, features need to be identified | In DL, the machine learns the features automatically from the given data |
| The problem solved by machine learning algorithms is divided into small parts, and then, each part is solved individually | Problems in deep learning are solved by neural networks |

### 2.4.4 Deep Belief Network (DBN)

Deep belief network in machine learning (ML) is a deep neural network that comprises of various layers with hidden units. There exists a connection between the layers but not between the units within every layer. It is also known as the generative graphical model. The deep belief network can be seen as a composition of unsupervised, simple networks like auto-encoders or restricted Boltzmann machines (RBMs) wherein every sub network's hidden layer acts as the visible layer for the next one [10, 16] (Fig. 17 and Table 2).

## *2.5 Applications*

This section demonstrates the application areas of machine learning under supervised, unsupervised and reinforcement learning.

### 2.5.1 Supervised Learning

Data of an Email

(a) *Answering of incoming messages that are automatic*—Rather than typing the similar answer each time when someone sends the emails with the similar problems and queries, machine learning techniques and algorithms study those emails and generate the reply automatically. This is useful in large companies [1].

(b) *Organisation of automatic mail into folders*—A large number of messages are coming daily, so it is difficult for users to separate the messages manually. In this case, machine learning is most efficient by classifying email into many user-defined tabs such as social, primary, forums. If the specific message from the particular sender is transferred from the update to the primary tab, then the remaining future messages from that particular user will be shown in the primary tab [1].

(c) *Thread and email sum-up*—The incoming messages are taken into consideration, and then, the most relevant sentences are taken from the email thread and translated into a summary. The summary is constructed on the unique features of an email [1].

(d) *Spam Filtering*—Spam filtering is done to separate or filter-unsolicited bulk email (UBE), unsolicited commercial email (UCE) or junk mail from a large number of emails. It can also learn from the emails that you mark as spam and the ones you mark as non-spam [1].

Recognition of Handwriting

These days it is not so expensive to send an email across various countries because when the address is mentioned on the envelope, it means that the learning algorithm had learned how to recognise and read the handwriting so that the envelope can be sent at the appropriate route and the cost is also less [1].

Recognition of the Face

The human face is not distinct; various factors make variations in the appearance. There are large application fields where this technique is utilised such as in areas of surveillance, image database investigation, security measure and the criminal justice system [1, 17, 18].

Recognition of Speech

The software used for speech recognition makes use of machine learning. It uses two different learning stages: first is before the software is transported and the second stage is after the user has bought the software [1].

### 2.5.2 Unsupervised Learning

Classification of DNA

The aim behind DNA classification is to make a group consisting of different individuals in such a way that each has a gene. The clustering algorithm is applied to form a group of individuals into a variety of people. This application is under unsupervised learning because no information is provided in advance [14, 19, 20].

Analysis of Social Network

This is one of the most useful applications under unsupervised learning. It forms links for the people to connect across the globe and can automatically determine the mutual friends within the friend circle in Google or Twitter. It also tells about the mutual relationships among various people [1].

To Organise Large Clusters

Data centres that consist of large clusters of computers unsupervised learning aims at identifying which machines can perform together so that if the machines are kept together, then the centres of data can work effectively [1].

Segmentation of Market

Almost every company has a vast database about customer's information. Unsupervised learning algorithms can analyse the customer data and identify the segments of the market, therefore grouping customers into different market segments so that it is easier for the company to sell its products [1].

### 2.5.3 Reinforcement Learning

Games in Computer

The gaming industry has achieved much success in recent years. Artificial intelligent agents are employed to develop interactive and exciting gaming experience. There can be a variety of roles for the agents such as teammates, the opponent of the player or any non-player person [1].

Applications of the Machine

Machinery applications mean which cannot be performed by hand. The computer has to learn by itself to solve the problem. The example is autonomous helicopters [1, 21].

Services Provided in Traffic

Controlling traffic is a major concern these days. This can be resolved by training the machine to reach a particular conclusion. The example includes those systems which overlay predictions for future traffic conditions for a traffic flow map [1].

## 3 Conclusion

This paper explains machine learning algorithms and lists all the recent applications in the research area. In today's world, every person is making use of machine learning techniques whether unknowingly or knowingly. The paper illustrates the supervised and unsupervised techniques in detail and explains where machine learning is applied. Based on the analysis and observations, a new model is developed. It also talks about deep learning, which is a part of machine learning, its architecture and a brief introduction of deep learning algorithms used to solve the given problem as shown in Table 2. Further advancements should be made to achieve the goal and new trends display that machine learning is growing with time and has shown continuous growth in countries like Singapore and India. Hence, it has come out to be the most certain field of technology in today's world.

## References

1. Das, S., Dey, A., Pal, A., Roy, N.: Applications of artificial intelligence in machine learning: review and prospect. Int. J. Comput. Appl. **115**(9) (2015)
2. Angra, S., Ahuja, S.: Machine learning and its applications: a review. In: 2017 International Conference on Big Data Analytics and Computational Intelligence (ICBDAC), pp. 57–60. IEEE (2017)
3. Dey, A.: Machine learning algorithms: a review. Int. J. Comput. Sci. Inf. Technol. **7**(3), 1174–1179 (2016)
4. Kotsiantis, S.B., Zaharakis, I.D., Pintelas, P.E.: Machine learning: a review of classification and combining techniques. Artif. Intell. Rev. **26**(3), 159–190 (2006)
5. Simon, A., Singh, M.: An overview of M learning and its Ap. Int. J. Electr. Sci. Electr. Sci. Eng. (IJESE) **22** (2015)
6. Support Vector Machine, https://scikit-learn.org/stable/modules/svm.html. Last accessed 27 Feb 2019
7. Linear Regression, http://scikitlearn.org/stable/auto_examples/linear_model/plot_ols.html#sphx-glr-auto-examples-linear-model-plot-ols-py. Last accessed 11 May 2018

8. Logistic regression 3 class-classifier, http://scikitlearn.org/stable/auto_examples/linear_model/plot_iris_logistic.html#sphx-glr-auto-examples-linear-model-plot-iris-logistic-py. Last accessed 11 May 2018
9. Demonstration of K-means assumption, http://scikitlearn.org/stable/auto_examples/cluster/plot_kmeans_assumptions.html#sphx-glr-auto-examples-cluster-plot-kmeans-assumptions-py. Last accessed 11 May 2018
10. Deng, L.: Three classes of deep learning architectures and their applications: a tutorial survey. APSIPA Trans. Signal Inf. Process. (2012)
11. LeCun, Y., Bengio, Y., Hinton, G.: Deep learning. Nature **521**(7553), 436 (2015)
12. Deng, L., Yu, D.: Deep learning: methods and applications. Found. Trends® Signal Process. **7**(3–4):197–387 (2014)
13. Schmidhuber, J.: Deep learning in neural networks: an overview. Neural Netw. **61**, 85–117 (2015)
14. Kaur, R., Juneja, M.A.: Survey of different imaging modalities for renal cancer. Indian J. Sci. Technol. **9**, 44 (2016)
15. Bhatia, N., Rana, M.C.: Deep learning techniques and its various algorithms and techniques. Int. J. Eng. Innov. Res. **4**(5) (2015)
16. Kaur, R., Juneja, M., Mandal, A.K.: A comprehensive review of denoising techniques for abdominal CT images. Multimedia Tools Appl. **77**(17), 22735–22770 (2018)
17. Valenti, R., Sebe, N., Gevers, T., Cohen, I.: Machine learning techniques for face analysis. In: Machine Learning Techniques for Multimedia, pp. 159–187. Springer, Berlin, Heidelberg (2008)
18. Wang, J., Yuille, A.L.: Semantic part segmentation using compositional model combining shape and appearance. In: Proceedings of the IEEE Conference on Computer Vision and Pattern Recognition, pp. 1788–1797 (2015)
19. Kaur, R., Juneja, M.: Comparison of different renal imaging modalities: an overview. In: Progress in Intelligent Computing Techniques: Theory, Practice, and Applications, pp. 47–57. Springer, Singapore (2018)
20. Cho, S.B., Won, H.H.: Machine learning in DNA microarray analysis for cancer classification. In: Proceedings of the First Asia-Pacific bioinformatics conference on Bioinformatics 2003, vol. 19, pp. 189–198. Australian Computer Society, Inc. (2003)
21. Kaur, R., Juneja, M.: A survey of kidney segmentation techniques in CT images. Curr. Med. Imaging Rev. **14**(2), 238–250 (2018)

# Deep Neural Networks for Diagnosis of Osteoporosis: A Review

**Insha Majeed Wani and Sakshi Arora**

**Abstract** Osteoporosis, a pathological disorder of bones affects millions of individuals worldwide and is the most common disease of bones after arthritis. It is caused due to a decrease in mineral density of bones leading to pain, morbidity, fractures and even mortality in some cases. It is diagnosed with DXA, but its high-cost, low-availability and inconsistent BMD measurements do not make it a promising tool for diagnosis of osteoporosis. The computer-aided diagnosis has improved the diagnostics to a large extent. Deep learning-based artificial neural networks have shown state-of-the-art results in the diagnostic field leading to an accurate diagnosis of the disease. This paper reviews the major neural network architectures used for diagnosis of osteoporosis. We reviewed the neural network architectures based on the questionnaires and the deep neural architectures based on image data implemented for diagnosis of osteoporosis and have summarized the future directions which could help in better diagnosis and prognosis of osteoporosis.

**Keywords** Osteoporosis · DXA · Neural networks · Deep learning

## 1 Introduction

Accurate and decisive diagnosis and prognosis in the medical field remains a challenge. Diagnostic problems are unique in the sense that they require highly accurate results. Mostly, the interpretations of the medical data are done by the medical experts. Human diagnostic leads to many errors and social expectations are not fulfilled. The advent of the computer vision techniques has improved the interpretation of the medical data, thus helping in the better diagnosis of the diseases. The medical data has changed with time from just mere reports to the data in the form of images as images can now be stored and transferred easily [1]. Medical image analysis using

I. M. Wani · S. Arora (✉)
School of Computer Science Engineering, Shri Mata Vaishno Devi University, Katra, Jammu and Kashmir, India
e-mail: sakshi@smvdu.ac.in

I. M. Wani
e-mail: insha333@gmail.com



P. K. Singh et al. (eds.), *Proceedings of ICRIC 2019*, Lecture Notes in Electrical Engineering 597, https://doi.org/10.1007/978-3-030-29407-6_6

computer-aided techniques can be treated as the combination of the computer and medical experts and are used in medical applications like for detection of breast cancer mammography [2], detection of thin lung nodules [3], CAD based on contour models [4], diagnosis of Alzheimer's disease [5], etc. The output of these systems is treated as the second opinion by radiologists [6]. The need for the automatic systems which have an intelligence of their own and can learn the features by themselves has led to the concept of self-learning which is termed as machine learning in computer terminology.

Machine learning is based on the theory of biological learning of brain neurons. As in the human brain, different neurons learn from the different types of senses by themselves and work according to the situation. The same terminology is being mimicked by the computers in machine learning by forming the neural networks which are given some sort of input at input layer and the neurons present in the different layers of neural network learn the features by themselves and give us the required output at the output layer. These networks have given success in many fields: speech recognition [7, 8], character recognition [9, 10], self-driving systems [11–13] as well in the medical field [14–19]. These networks take the data in the numeric as well as the image data. Medical images are currently been used to train the more complex neural networks with many added intermediate layers to get the required information from the medical images, such type of learning is termed as deep learning. Deep learning has enabled the computers to perform its tasks without being explicitly programmed by humans and can make the decisions or prediction based on the data by constructing the self-learning algorithms [20]. It has given state-of-the-art results in many medical applications like detection of cellular and anatomical structures [14, 17], tissue segmentation [15, 19], computer-aided diagnosis and prognosis [16, 18].

In this review paper, the role of deep learning for the diagnosis of osteoporosis is discussed. Osteoporosis [21] is the pathological disorder of bones. It is the result of the decreased bone mineral density [22] leading to the fracture, pain, musculoskeletal morbidity [2] and even to mortality in elderly, especially in postmenopausal women [23]. Osteoporosis leads to the fractures in hip, vertebrae, pelvis, humerus, distal forearm, etc. Using population demographic data, it was projected that the number of hip fractures worldwide will increase from 1.66 million in 1990 to 6.26 million by 2050 [24]. So, it is desired to diagnose the disease in advance so that it can be treated properly. WHO [25, 26] has recommended the T-score of bone mineral density to distinguish between the normal and osteoporotic bone: T-score $\geq -1$ = normal bone, T-score is between $-1$ and $-2.5$ = osteopenia and T-score $\leq -2.5$ = osteoporosis. The gold standard in medical terminology for detection of osteoporosis is the dual-energy X-ray absorptiometry (DXA) [27] by assessing the mineral density of bones. DXA is the two-dimensional imaging technology which can scan the whole body and produces scans in the form of X-rays. Though DXA is a medically acceptable technique, it suffers from various limitations like it gives a real measurement and gives different values for variable bone sizes. It has a very costly setup which low economies may not afford and also it is not widely available. These limitations pave a need to find some alternate diagnostic system for osteoporosis. CAD techniques like deformable models [28], texture-based models [29], thresholding models, etc., have been used by many researchers to solve the problem. According to [1], deep learning has beaten other CAD techniques and given the state-of-the-art results in diagnostic fields, so we will see its role in the diagnosis of osteoporosis.

The objective of this review was as follows:

To analyze the role of deep neural networks for the diagnosis of osteoporosis.
To identify the different challenges to analyze better future solutions for early diagnosis of osteoporosis.
To investigate the neural networks based on simple clinical data, to the deep neural networks based on image data.

This review is structured as followed. In Sect. 2, the overview of deep learning methodology is discussed. Section 3 describes the role of deep learning for diagnosis of osteoporosis which is divided into two sections, Sect. 3.1 describes the role of neural networks based on clinical data, and Sect. 3.2 describes the deep neural networks based on image data for osteoporosis diagnosis. Section 4 discusses the different challenges in building the diagnostic system and future directives. Finally, in Sect. 5, the review is concluded.

## 2 Overview of Deep Learning

Deep learning is the form of machine learning that uses models of the deep neural network which is the variation of artificial neural network [30] but with a large approximation to human brain using advance mechanism as compared to simple neural networks. The neuron [31] which is a basic computational unit in a neural network is a concept inspired by the study of the human brain. Its input is in the form of multiple signals that are combined linearly using weights and finally through nonlinear operations passes the combined signals to generate the output signals.

Conventional neural networks have got great progress in prediction and diagnosis from medical data like questionnaires, clinical findings, laboratory results, etc. Neural networks' general architecture is depicted in Fig. 1. Hinton's deep belief network [32], a deeply structured learning architecture, was a breakthrough in deep learning. The considerable gain in computational power [33] is also one factor for increased use of deep learning as it provides multiple levels of abstraction for learning the different representations of data because of the presence of the multiple processing layers for computational models [34]. Deep learning uses general-purpose learning procedures to learn from the data instead of the human engineered features used in conventional neural networks. Deep learning employs supervised as well as unsupervised learning in its architectures.

A deep neural network forms a hierarchical feature representation by hierarchically stacking the multiple layers of neurons. It forms a gigantic model with the number of layers extending up to thousand [35], classifying millions of images. With

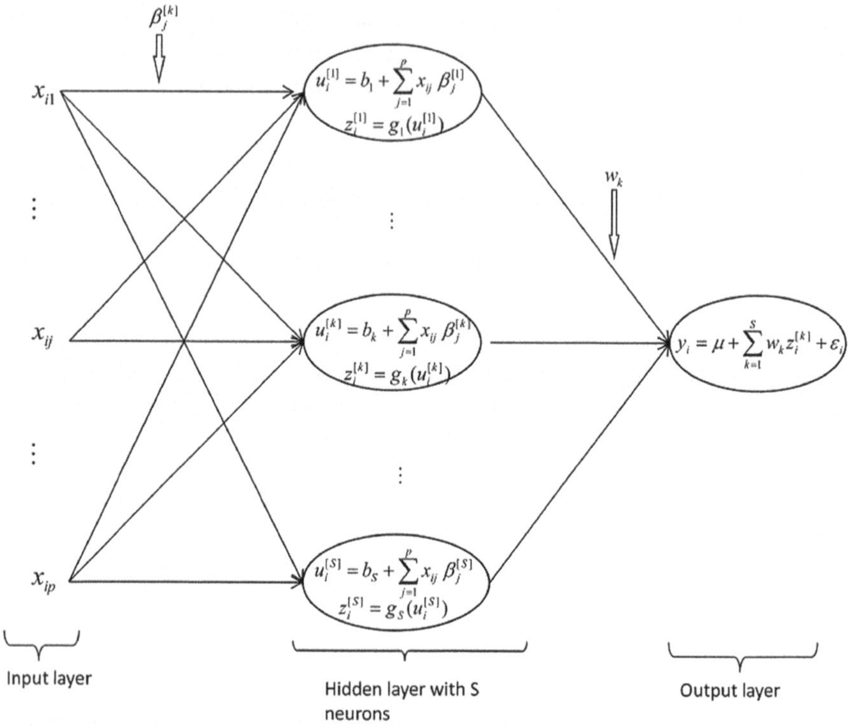


**Fig. 1** General architecture of a neural network

such a gargantuan modeling capacity, the deep network models have got the power to memorize all feasible mappings when trained with an adequately large knowledge database and thus making the intelligent predictions (interpolations and/or extrapolations) for unseen cases. These intelligent features have made deep learning to generate a large impact on computer vision modalities as well as in medical imaging. Deep learning architectures have many variations as described in Table 1. Convolutional neural networks have shown the significant performance for images and videos as

**Table 1** Deep learning architectures used for medical image analysis

| Architectures | Description |
|---|---|
| DXS | Deep neural network artificial neural network with deep layers |
| CNN | Convolutional neural network, many convolutional layers for better feature extraction |
| AE/SAE | Auto-encoder and stacked auto-encoder, feature learning by encoding decoding |
| RBM | Restricted Boltzmann machine is a special type of Markov random field |
| DBN | Deep belief networks. SAEs with AE layer replaced by RBMs |
| RXS | Recurrent neural networks, used as autoregressive generative models |

they can process two-dimensional data present in the form of grids. Deep learning architectures require very less or no preprocessing of images and can work on raw image data [36].

# 3 Deep Learning for Diagnosis of Osteoporosis

This section is divided into two subsections: In the first section, the conventional neural networks based on general clinical data or questionnaires used for diagnosis of osteoporosis are discussed, and in the second section, the deep neural networks based on medical images are discussed.

## *3.1 Neural Networks Based on Clinical Observations*

Bone mineral density is considered to be the predictor for the onset of osteoporosis. As osteoporosis is caused when there is a decrease in the level of mineral density in bones. The level of BMD in bones is calculated by taking its T-score values. T-score predicts whether bones have undergone or not in the stage of osteoporosis. For this purpose, DXA and sometimes QCT are used but these techniques cannot be relied upon fully. BMD measurements of patients having a fracture and those not having fracture may overlap. Statistical techniques cannot perform well in such noisy environments. With the rising use of artificial neural networks in the medical field, the researchers tried to use neural networks for predicting the fracture risk and osteoporosis. In 1996, Jensen et al. [37] and Redei et al. [38] used ANN for predicting the osteoporotic fractures by feeding the ANN with the BMD measurements taken at distal forearm, lumbar spine L II-L IV, the trochanter, ward triangle, and the femoral neck and vertebral height measurements and XY-coordinates from T4-L4 derived from lateral spine radiographs, respectively. The input variables for the ANN were also taken from the different questionnaires filled by either the patients or some expert medical practitioners to predict the osteoporosis. Questionnaires were based on the risk factors [39, 40], anthropometric measurements (i.e., age, waist to hip ratio, sex, body mass index, etc.) [41, 42], demographic characteristics [43, 44], clinical data [43], nutritional habits [45–47], lifestyle [45–47], general factors [44], medical conditions [46, 47], medications [49], diagnostic factors [48]. The neural networks in which the inputs were taken from the questionnaires showed good results, but they cannot be relied upon in medical terminology due to various limitations like data filled by the patients in the questionnaires may not always be right. Due to human negligence, some data may be written wrong, or while copying the data from the questionnaire to the NN, some data may be missed or can be some typing errors. So, there is a need for some other form of medical data which can be fully relied upon. We have medical images which can be used for this purpose. In the next section, deep neural networks with medical images are being discussed.

Deep neural networks are the neural networks with more added layers and every layer containing the more neurons to make the complex decisions and handle the situations more intelligently by directly learning from the data. In osteoporosis, the micro-architecture of bones is affected and image analysis can be used to study the changes in micro-architecture of bone with the succession of osteoporosis. Trabecular bone comprises of rod-shaped and plate-shaped elements and can be described by 2D surfaces. The deterioration of the micro-architecture leads to the fracture of bones. Trabecular bone is shown in Fig. 2. We have divided this section into two sections. The first section covers the conventional artificial neural network architectures used on image data, and the second section covers the deep convolutional neural network architectures used on image data.

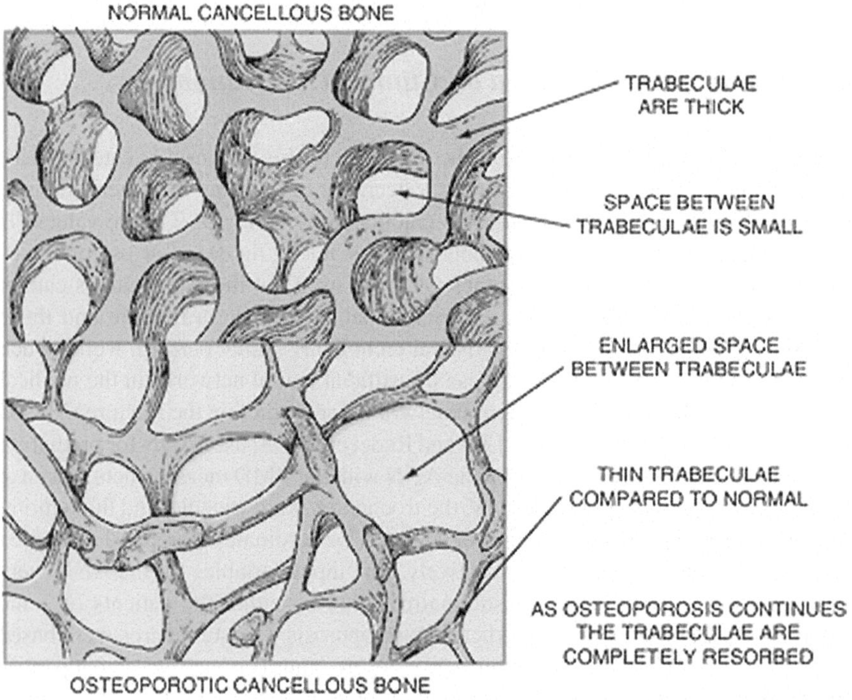


**Fig. 2** Structure of trabecular bone before and after osteoporosis [50]

## 3.2 *Deep Neural Networks Based on Image Analysis*

### 3.2.1 Conventional Neural Network Architectures

The neural networks like feedforward neural networks (FFNN), regression-based artificial neural networks (RANN) and neural network classifiers are discussed in this section. Akgundogdu et al. [49] in their work implemented ANFIS (adaptive neuro-fuzzy inference system), a combination of FIS (fuzzy inference system) and ANN to identify the osteoporotic samples (in form of images) but due to fewer samples (only nine in this work) available its success was poor.

They also implemented SVM and GA and got a great success. They used HSGA [51], which generates structural models of an object, to measure the morphological parameters, topological parameters and mechanical parameters from the micro-architecture of the bones.

Then in 2014, Lee et al. [52] studied the trabecular micro-architecture from hip radiographs for the diagnosis of osteoporosis using neural networks. A two-layered feedforward artificial neural network was used for the classification purpose with six input neurons, ten neurons in the hidden layer and two neurons in the output layer. This work showed good performance and provided the best classification results. Regression artificial neural network classifier-based computer-aided detection system is proposed in [53] for the detection of osteoporosis from digital calcaneus radiographic images. The radiographic images were stored in DICOM format and after selecting of the region of interest the radiographs were preprocessed like denoising, etc., and used Otsu's method for binarizing the images. Then, the features like entropy, contrast, energy, convex hull, eccentricity, etc., were extracted to be fed to the neural network for classification purpose. The approach showed good performance for the diagnosis of osteoporosis. In Sect. 3.1, we have seen that the clinical examinations also play an important role in the diagnosis of osteoporosis. So the combination of clinical examinations with the image analysis and also the gold standard for osteoporosis diagnosis, DXA may result in a better diagnosis. The combination of clinical data and DXA findings with features from X-ray radiographs was applied by Yu et al. [54] in their work and showed very effective results for the diagnosis of osteoporosis. Singh et al. [55] also used neural networks for automatic processing of X-ray images to diagnose the osteoporotic patients from trabecular bone structure. The DXA-DICOM images were analyzed in [56], and features based on the histogram and binary algorithms were extracted and fed to the multilayer feedforward neural network for total BMD estimation and was observed that it could be used for standardizing all DXA machines for bone measurements irrespective of the manufacturer. The hand and wrist radiographs are used by [57] to get the cortical bone indices and cancellous features for detecting the low bone mass by training the neural network classifier with these features. This could be used as the automatic diagnostic technique for early diagnosis of the onset of osteoporosis.

### 3.2.2 Deep Convolutional Neural Network Architectures (DCNN)

The deep convolutional architectures used for diagnosis of osteoporosis from image data are discussed. The first work using DCNN for osteoporosis detection was done in 2016 in which a classifier based on deep convolutional neural network (CNN) was implemented by Hatano et al. [58] on computed radiography (CR) images for the diagnosis of osteoporosis. The phalangeal CR images were used as input images, and the performance was evaluated by the threefold cross-validation method and a good ratio of diagnosis was achieved. Osteoporosis causes the fractures of bones; one of the main fractures which cannot be examined by computed tomography examinations are the vertebral fractures. Tomita et al. [59] implemented the deep convolutional neural networks for detecting the osteoporotic vertebral fractures from CT scans by extracting logical features from each slice of CT scan. The results matched the performance of practicing radiologist for their test set which improves its chances of using it in actual clinical set up for detecting osteoporotic vertebral fractures. The recent study [60] on detection of osteoporosis shows that the osteoporosis can be detected from the dental panoramic radiographs with the help of the deep convolutional neural networks. The system showed a high agreement with experienced maxillofacial and oral radiologists for detecting the osteoporosis. Deep convolutional network-based segmentation also has shown that MRI images can be used to measure the bone quality and for assessment of fracture risk. Deniz et al. [61] shows that the MR images of the proximal femur can be automatically segmented using the DCNN for assessment of fracture risk and bone quality. From this section, we can conclude that osteoporosis like other diseases can also be detected by different image analyses systems based on neural networks. And the deep neural networks show very similar results to that of practicing radiologists. Deep neural networks could prove to provide the promising diagnostic systems which could be used in clinical practicing and are easily available and not incurring the high cost as that of the current diagnostic technique of DXA.

## 4 Challenges and Future Perspectives

We have divided the future directives based on the challenges in three parts.

1. **Generalized osteoporosis database**

Challenge: As we have seen in Tables 2 and 3 that the datasets used by researchers are local datasets either of a particular community or a particular country.

Future direction: There is a need to make a generalized dataset which could be used universally by the researchers and studied to get more generalized features improving the accuracy of diagnostics.

2. **Cost-effective and reliable system**

Challenge: The main drawback of DXA [27] used clinically for osteoporosis detection is its high cost and sometimes gives confusing results.

**Table 2** Accuracy of different neural network architectures based on different types of questionnaires

| S. No. | Ref. No. | Type of questionnaire | No. of input variables | Age group (in years) | ANN architecture | Accuracy | Dataset | |
|---|---|---|---|---|---|---|---|---|
| | | | | | | | Size | Type |
| 1 | [39] | Risk factor | 20 | Mean age = 61.2 | Multivariate systems | 73.10% | 274 | Local |
| 2 | [40] | Risk factors | to | >60 | BP ANN | AUC = 0.55 ± 0.029 | 725 | Taiwan |
| 3 | [41] | Anthropometric measurement | 4 | 13–79 | Multilayer perceptron | – | 1154 | Rome |
| 4 | [42] | Anthropometric measurement | 10 | 41–86 | ANN | Spine-0.774, Femoral Neck-0.867 | 2158 | Tehran |
| 5 | [43] | Demographic characteristics, anthropometric measurement and clinical data | 7 | 50–91 | Multilayer perceptron | AUC = 0.52 ± 0.03 | 1403 | Taiwan |
| 6 | [44] | Demographic characteristics | 7 | 20–79 | ANN | 86.60% | 2000 | Egyptian |
| 7 | [45] | Nutrition and lifestyle | 10 | 500–69 | Feedforward ANN | – | 200 | Local |
| 8 | [46] | Diet, lifestyle, medical conditions and medication nutritional habits | 15 | – | Feedforward ANN | – | – | Local |
| 9 | [47] | Lifestyle and previous fracture | 27 | >45 | SVM & ANN | 80% | 60 | Italy |
| 10 | [48] | Diagnostic factors | 33 | – | Multilayer perceptron | – | 589 | Local |

**Table 3** Accuracy of neural networks and deep neural network architectures based on image data

| S. No. | Ref. No. | Image type | No. of inputs | Age group (years) | ANN architecture | Accuracy | Dataset | | Bone Type |
|---|---|---|---|---|---|---|---|---|---|
| | | | | | | | Size | Type | |
| 1 | [49] | Micro-architecture images | 27 | – | ANFIS | – | 9 | Francs | – |
| 2 | [52] | Radiographs | 7 | >50 | Two-layered FFNN | 98.1 | 53 | Local | Hip |
| 3 | [53] | Radiographic images | 12 | – | Regression ANN | 90% | – | Local | |
| 4 | [54] | X-ray | | Avg. age-65 | ANN | 95% | 119 | Local | Vertebra |
| 5 | [55] | X-ray | | Avg. age-60 | SVM, ANN | 98% | 17 | 4 Paris | Total body scan |
| 6 | [56] | DXA–DICOM | Input neuron: 256 (for histogram) 77321 (binary algorithm) | 22–49 years | MLFF-ANN | 100% | 3000 | Local | Hand and wrist |
| 7 | [57] | Radiographs | 83 | 30–35 years | NN Classifier | 88.50% | 13 5 | Indian | |
| 8 | [58] | Computed radiography | 32 ROIS | – | DCNN | TPR: 64.7% FPR = 6.51% | 101 | Local | |
| 9 | [59 | CT scans | – | – | Deep CNN | 89.20% | 1432 | Local | |
| 10 | [60] | Panoramic radiographs | – | (52.5 ± 22.3) years | Deep CNN | 0.9991 | 1263 | Local | Too th |
| 11 | [61] | MRI | 32 initial maps | >50 years | DCNN-Unet | AUC: 0.998 ± 0.001 | 36 | Local | Proximal |

Future direction: Deep learning architectures which have shown state-of-the-art results in many medical applications can be used for diagnosis of osteoporosis. With the advent in technology, these systems have become economical to use in clinical practice making it available for every sect of society.

3. **Fusion of clinical and image data**

Challenge: The review shows that the osteoporosis can be detected from images, but at the same time, some clinical and other factors can contribute to the diagnosis of osteoporosis.

Future directions: The dataset formed for deep learning architectures should combine both clinical and image data. This will help to build a more sophisticated system for automatic diagnosis of osteoporosis.

## 5 Conclusion

Osteoporosis is a serious disease of bones affecting millions of people worldwide, especially the aged one. It is caused by the decrease of bone mineral density and currently has an only single acceptable way for its detection which is dual-energy X-ray absorptiometry (DXA). It makes the bones weak, thus reducing its strength and can cause a fracture in bones like hip fracture, the fracture in the pelvis, fracture in foot joints, etc. It can make a person bedridden forever and even can cause its morbidity. It poses a great socioeconomic burden for the low economies. This disease has to be diagnosed in its early stages, so that proper treatment can be given. So, there is a need to devise the cost-effective, automatic and reliable system for the diagnosis of the disease which can be used in clinical practice. Many computer-aided diagnostic systems have been devised but are not clinically accepted. Deep learning methods deep learning architectures have shown the state-of-the-art results and thus accepted as a promising, cheap and reliable tool for diagnosis in medical field, henceforth it can also verify for osteoporosis. Although in recent years deep learning methods like convolutional neural networks have been used by many researchers but not extensively. So, there is a good scope of using the automatic computer-aided diagnostic system based on deep learning methodologies over generalized datasets for osteoporosis.

## References

1. Litjens, G., Kooi, T., Bejnordi, B.E., Setio, A.A.A., Ciompi, F., Ghafoorian, M., Van Der Laak, J.A., Van Ginneken, B. and Sánchez, C.I.: A survey on deep learning in medical image analysis. Med. Image Anal. **42**, 60–88 (2017)
2. Birdwell, R.L., Bandodkar, P., Ikeda, D.M.: Computer-aided detection with screening mammography in a university hospital setting. Radiology **236**(2), 451–457 (2005)

3. Li, Q.: Recent progress in computer-aided diagnosis of lung nodules on thin-section CT. Comput. Med. Imaging Graph. **31**(4–5), 248–257 (2007)
4. Ryu, J.H., Kim, H.S., Lee, K.H.: Contour-based algorithms for generating 3D CAD models from medical images. Int. J. Adv. Manuf. Technol. **24**(1–2), 112–119 (2004)
5. Padilla, P., López, M., Górriz, J.M., Ramirez, J., Salas-Gonzalez, D., Álvarez, I.: NMF-SVM based CAD tool applied to functional brain images for the diagnosis of Alzheimer's disease. IEEE Trans. Med. Imaging **31**(2), 207–216 (2012)
6. Doi, K.: Computer-aided diagnosis in medical imaging: historical review, current status, and future potential. Comput. Med. Imaging Graph. **31**(4–5), 198–211 (2007)
7. Hinton, G., Deng, L., Yu, D., Dahl, G.E., Mohamed, A.R., Jaitly, N., Senior, A., Vanhoucke, V., Nguyen, P., Sainath, T.N., Kingsbury, B.: Deep neural networks for acoustic modeling in speech recognition: the shared views of four research groups. IEEE Signal Process. Mag. **29**(6), 82–97 (2012)
8. Deng, L., Hinton, G., Kingsbury, B.: New types of deep neural network learning for speech recognition and related applications: an overview. In: IEEE International Conference on Acoustics, Speech, and Signal Processing 2013, pp. 8599–8603. IEEE (2013)
9. Chen, L., Wang, S., Fan, W., Sun, J. and Naoi, S.: Beyond human recognition: A CNN-based framework for handwritten character recognition. In: 3rd IAPRAsian Conference on Pattern Recognition 2015, ACPR, pp. 695–699. IEEE (2015)
10. Netzer, Y., Wang, T., Coates, A., Bissacco, A., Wu, B., Ng, A.Y.: Reading digits in natural images with unsupervised feature learning. In: NIPS Workshop on Deep-Learning and Unsupervised Feature Learning, vol. 2011, no. 2, p. 5 (2011)
11. Teichmann, M., Weber, M., Zoellner, M., Cipolla, R., Urtasun, R.: Multinet: real-time joint semantic reasoning for autonomous driving. In: IEEE Intelligent Vehicles Symposium 2018, vol. IV, pp. 1013–1020. IEEE (2018)
12. Chen, C., Seff, A., Kornhauser, A., Xiao, J.: Deep driving: learning affordance for direct perception in autonomous driving. In: Proceedings of the IEEE International Conference on Computer Vision, pp. 2722–2730 (2015)
13. Huval, B., Wang, T., Tandon, S., Kiske, J., Song, W., Pazhayampallil, J., AndrilukaM., Rajpurkar, P., Migimatsu, T., Cheng-Yue, R., Mujica, F.: An empirical evaluation of deep learning on highway driving. arXiv preprint arXiv:1504.01716 (2015)
14. Pereira, S., Pinto, A., Alves, V., Silva, C.A.: Brain tumor segmentation using convolutional neural networks in MRI images. IEEE Trans. Med. Imaging **35**(5), 1240–1251 (2016)
15. Esteva, A., Kuprel, B., Novoa, R.A., Ko, J., Swetter, S.M., Blau, H.M., Thrun, S.: Dermatologist-level classification of skin cancer with deep neural networks. Nature **542**(7639), 115 (2017)
16. De Fauw, J., Ledsam, J.R., Romera-Paredes, B., Nikolov, S., Tomasev, N., Blackwell, S., Askham, H., Glorot, X., O'Donoghue, B., Visentin, D., van den Driessche, G.: Clinically applicable deep learning for diagnosis and referral in retinal disease. Nature Med. **24**(9), 1342 (2018)
17. Rajpurkar, P., Irvin, J., Zhu, K., Yang, B., Mehta, H., Duan, T., Ding, D., Bagul, A., Langlotz, C., Shpanskaya, K., Lungren, M.P.: Chexnet: Radiologist-level pneumonia detection on chest x-rays with deep learning. In: arXiv preprint arXiv:1711.05225 (2017)
18. Bien, N., Rajpurkar, P., Ball, R.L., Irvin, J., Park, A., Jones, E., Bereket, M., Patel, B.N., Yeom, K.W., Shpanskaya, K., Halabi, S.: Deep-learning-assisted diagnosis for knee magnetic resonance imaging: Development and retrospective validation of MRNet. PLoS Med. **15**(11), e1002699 (2018)
19. Zhang, Y., Yang, L., Chen, J., Fredericksen, M., Hughes, D.P., Chen, D.Z.: Deep adversarial networks for biomedical image segmentation utilizing unannotated images. In: International Conference on Medical Image Computing and Computer-Assisted Intervention, pp. 408–416. Springer, Cham (2017)
20. Liu, W., Wang, Z., Liu, X., Zeng, N., Liu, Y., Alsaadi, F.E.: A survey of deep-neural network architectures and their applications. Neurocomputing **234**, 11–26 (2017)
21. Osteoporosis. Park Ridge, IL, American Academy of Orthopaedic Surgeons, 1986

22. NIH Consensus Development Panel on Osteoporosis Prevention, Diagnosis, and Therapy. Osteoporosis prevention, diagnosis, and therapy. JAMA **285**(6), 785–795 (2001)
23. Kaplan, F.S.: Osteoporosis. Women's Health **10**(2/3), 95–114 (1985)
24. Cooper, C., Campion, G. and Melton, L. J.: III. Hip fractures in the elderly: a world wide projection. Osteoporos. Int. **2**, 285–289 (1992)
25. World Health Organization (ed.): Assessment of Fracture Risk and Its Application to Screening for Postmenopausal Osteoporosis. World Health Organization, Geneva (1994)
26. Kanis J.A.: Assessment of fracture risk and its application to screening for postmenopausal osteoporosis: synopsis of a WHO report. WHO study group, Osteoporos. Int., 368–814 (1994)
27. Kanis, J.A., McCloskey, E.V., Johansson, H., Oden, A., Melton III, L.J., Khaltaev, N.: A reference standard for the description of osteoporosis. Bone **42**(3), 467–475 (2008)
28. Mclnerney, T., Terzopoulos, D.: Deformable models in medical image analysis: a survey. Med. Image Anal. **1**(2), 91–108 (1996)
29. Southard, T.E. and Southard, K.A.: Detection of simulated osteoporosis in maxillae using radiographic texture analysis. IEEE Trans. Biomed. Eng. **43**(2), 123–132 (1996)
30. Hinton, G.E., Salakhutdinov, R.R.: Reducing the dimensionality of data with neural networks. Science **313**(5786), 504–507 (2006)
31. Schmidhuber, J.: Deep learning in neural networks: an overview. Neural Netw **61**, 85–117 (2015)
32. Hinton, G.E., Osindero, S., Teh, Y.W.: A fast learning algorithm for deep beliefnets. Neural Comput. **18**(7), 1527–1554 (2006)
33. Badea, M.S., Felea, I.I., Florea, L.M., Vertan, C.: The use of deep learning in image segmentation, classification, and detection. In: arXiv preprint arXiv:1605.09612 (2016)
34. LeCun, Y., Bengio, Y., Hinton, G.: Deep learning. Nature **521**(7553), 436(2015)
35. Krizhevsky, A., Sutskever, I., Hinton, G.E.: Imagenet classification with deep convolutional neural networks. In: Advances in Neural Information Processing Systems, pp. 1097–1105 (2012)
36. Han, J., Zhang, D., Wen, S., Guo, L., Liu, T., Li, X.: Two-stage learning to predict human eye fixations via SDAEs. IEEE Trans. Cybern. **46**(2), 487–498 (2016)
37. Jensen, J.E., Sharpe, P.K., Caleb, P., Sørensen, H.A.: Fracture prediction using artificial neural networks. Osteoporos. Int. **6**, 132 (1996)
38. Redei, J., Ouyang, X., Countryman, P.J., Wang, X., Genant, H.K.: Classification of vertebral fractures: landmark-based shape recognition by neural networks. Osteoporos. Int. **6**, 126 (1996)
39. Rae, S.A., Wang, W.J., Partridge, D.: Artificial neural networks: a potential role in osteoporosis. J. R. Soc. Med. **92**(3), 119–122 (1999)
40. Liu, Q., Cui, X., Chou, Y.C., Abbod, M.F., Lin, J., Shieh, J.S.: Ensemble artificial neural networks applied to predict the key risk factors of hip bone fracture for elders. Biomed. Signal Process. Control **21**, 146–156 (2015)
41. Mohamed, E.I., Maiolo, C., Linder, R., Pöppl, S.J., De Lorenzo, A.: Artificial neural network analysis: a novel application for predicting site-specific bone mineral density. Acta Diabetol. **40**(1), s19–s22 (2003)
42. Sadatsafavi, M., Moayyeri, A., Soltani, A., Larijani, B., Nouraie, M., Akhondzadeh, S.: Artificial neural networks in prediction of bone density among post-menopausal women. J. Endocrinol. Invest. **28**(7), 425–431 (2005)
43. Chiu, J.S., Li, Y.C., Yu, F.C. and Wang, Y.F.: Applying an artificial neural network to predict osteoporosis in the elderly. In: Studies in Health Technology and Informatics, vol. 124, p. 609 (2006)
44. Abdel-Mageed, S.M., Bayoumi, A.M., Mohamed, E.I.: Artificial neural networks analysis for estimating bone mineral density in an Egyptian population: towards standardization of DXA measurements. Am. J. Neural Netw. Appl. **1**(3), 52–56 (2015)
45. de Cos Juez, F.J., Suárez-Suárez, M.A., Lasheras, F.S., Murcia-Mazón, A.: Application of neural networks to the study of the influence of diet and lifestyle on the value of bone mineral density in post-menopausal women. Math. Comput. Model. **54**(7–8), 1665–1670 (2011)

46. Shaikh, A.B., Sarim, M., Raffat, S.K., Ahsan, K., Nadeem, A., Siddiq, M.: Artificial neural network: a tool for diagnosing osteoporosis. Res. J. Recent Sci. ISSN **2277**, 2502 (2014)
47. Bortone, I., Trotta, G.F., Cascarano, G.D., Regina, P., Brunetti, A., De Feudis, I., Buongiorno, D., Loconsole, C., Bevilacqua, V.: A Supervised approach to classify the status of bone mineral density in post-menopausal women through static and dynamic baropodometry. In: International Joint Conference on Neural Networks (IJCNN) 2018, pp. 1–7. IEEE (2018)
48. Iliou, T., Anagnostopoulos, C.N., Stephanakis, I.M., Anastassopoulos, G.: A novel data preprocessing method for boosting neural network performance: a case study in osteoporosis prediction. Inf. Sci. **380**, 92–100 (2017)
49. Akgundogdu, A., Jennane, R., Aufort, G., Benhamou, C.L., Ucan, O.N.: 3D image analysis and artificial intelligence for bone disease classification. J. Med. Syst. **34**(5), 815–828 (2010)
50. Taber's online homepage, https://www.tabers.com/tabersonline/repview?type=539-126&name=o120p0
51. Aufort, G., Jennane, R., Harba, R. and Benhamou, C.L.: Hybrid skeleton graph analysis of disordered porous media. Application to trabecular bone. In: 2006 IEEE International Conference on Acoustics, Speech and Signal Processing, 2006. ICASSP 2006 Proceedings, vol. 2, pp. II–II. IEEE (2006)
52. Lee, J.H., Hwang, Y.N., Park, S.Y. and Kim, S.M.: Diagnosis of osteoporosis by quantification of trabecular microarchitectures from hip radiographs using artificial neural networks. In: Bio-Inspired Computing-Theories and Applications, pp. 247–250. Springer, Berlin, Heidelberg (2014)
53. Vishnu, T., Saranya, K., Arunkumar, R., Gayathri Devi, M.: Efficient and early detection of osteoporosis using trabecular region. In 2015 Online International Conference on Green Engineering and Technologies (IC-GET), pp. 1–5. IEEE (2015)
54. Yu, X., Ye, C., Xiang, L.: Application of artificial neural network in the diagnostic system of osteoporosis. Neurocomputing **214**, 376–381 (2016)
55. Singh, A., Dutta, M.K., Jennane, R., Lespessailles, E.: Classification of the trabecular bone structure of osteoporotic patients using machine vision. Comput. Biol. Med. **91**, 148–158 (2017)
56. Mohamed, E.I., Meshref, R.A., Abdel Mageed, S.M., Moustafa, M.H., Badawi, M.I., Darwish, S.H.: A novel morphological analysis of DXA-DICOM images by artificial neural networks for estimating bone mineral density in health and disease. J. Clin. Densitometry (2018)
57. Areeckal, A.S., Jayasheelan, N., Kamath, J., Zawadynski, S., Kocher, M.: Early diagnosis of osteoporosis using radiogrammetry and texture analysis from hand and wrist radiographs in Indian population. Osteoporos. Int. **29**(3), 665–673 (2018)
58. Hatano, K., Murakami, S., Lu, H., Tan, J.K., Kim, H., Aoki, T.: Classification of osteoporosis from phalanges CR images based on DCNN. In: 2017 17th International Conference on Control, Automation, and Systems (ICCAS), pp. 1593–1596. IEEE (2017)
59. Tomita, N., Cheung, Y.Y., Hassanpour, S.: Deep neural networks for automatic detection of osteoporotic vertebral fractures on CT scans. Comput. Biol. Med. **98**, 8–15 (2018)
60. Lee, J.S., Adhikari, S., Liu, L., Jeong, H.G., Kim, H. and Yoon, S.J.: Osteoporosis detection in panoramic radiographs using a deep convolutional neural network-based computer-assisted diagnosis system: a preliminary study. Dentomaxillofacial Radiol. **48**(1), 20170344 (2019)
61. Deniz, C.M., Xiang, S., Hallyburton, R.S., Welbeck, A., Babb, J.S., Honig, S., Cho, K., Chang, G.: Segmentation of the proximal femur from MR images using deep convolutional neural networks. Sci. Rep. **8**(1), 16485 (2018)

# Predicting Drug Target Interactions Using Dimensionality Reduction with Ensemble Learning

**Kanica Sachdev and Manoj K. Gupta**

**Abstract** Drug target interaction is one of the most significant fields of research for drug discovery. The laboratory experiments conducted to identify the drug target interactions are tedious, delayed, and costly. Hence, there is an urgent need to develop highly efficient computational methods for identifying potential drug target interactions that can limit the search space of these laboratory experiments. The existing computational techniques for drug target interaction have been broadly classified into similarity-based methods and feature-based methods. In this paper, a novel feature-based technique to predict drug target interactions has been proposed. The technique uses ensemble learning to determine drug target interactions. Ensemble learning offers greater accuracy in comparison with the traditional classifiers. Thus, the proposed technique aims to improve accuracy using ensemble learning. Also, dimensionality reduction of drug target features is performed using principal component analysis so that the computational time of the method can be reduced. The results indicate an improved performance in comparison with the state-of-the-art methods in the field.

**Keywords** Drugs · Drug target interaction · Proteins · Targets

## 1 Introduction

Prediction of the drug target interactions is an emerging research field in the area of drug invention [1]. Drugs are the chemical compounds that bind with proteins in our body to create a change [2]. The various proteins in our body, to which the drugs attach themselves, are known as targets. Drug target interaction refers to the process of the drug attaching itself to a particular target to effectuate a change. Drug target interaction prediction is the mechanism to identify whether a particular drug will interact with a particular target or not [3].

K. Sachdev (✉) · M. K. Gupta
Shri Mata Vaishno Devi University, Katra, J&K, India
e-mail: kanica.sachdev@gmail.com

M. K. Gupta
e-mail: manoj.gupta@smvdu.ac.in



P. K. Singh et al. (eds.), *Proceedings of ICRIC 2019*, Lecture Notes in Electrical Engineering 597, https://doi.org/10.1007/978-3-030-29407-6_7

Drug target interaction prediction has various applications. It helps in the process of drug discovery. Identifying the interaction aids in discovering novel compounds that can be used for a particular disease. It also helps in drug repositioning. Drug repositioning is a technique of utilizing already existing drugs for treating some other disease [4–6]. The study of drug target interaction is also useful in identifying the drug side effects. The negative interaction between the drugs and the targets can be investigated to recognize the potential side effects of a drug compound [7–9].

In order to effectively predict drug target interactions with high accuracy, it is important to gather the maximum possible data regarding drugs, protein targets as well as their interactions. During the past decade, there has been a massive growth in the amount of chemical and biological data that has been accumulated in different databases. This has been made possible due to the growing technology and various open data initiatives that have been developed. The PubChem dataset [10], for instance, has assembled more than two million compounds that have been tested in biochemical or cell-based assays. More than fifty thousand compounds have been examined that contains 4000 drugs. These drugs have been analyzed with eleven thousand targets, and the various identified drug target interactions have been determined. The extensive chemical and biological data has created ample opportunities for discovering new frameworks and tools for identifying drug target interactions.

Various techniques have been proposed to identify the potential drug target interactions. These techniques have been broadly classified into ligand-based methods, docking-based methods and chemogenomic methods. The ligand-based methods predict the interactions using the similarity between the various protein ligands. The performance of these methods, however, decreases considerably when the number of known ligands is less [11]. The docking-based methods utilize the 3D structure of the proteins to predict whether they will interact with the drug compounds or not [12, 13]. However, the 3D structures of all proteins are not known presently, and hence, it can be used for only protein targets with known structure [14]. To overcome the shortcomings of the ligand-based and the docking-based approaches, chemogenomic approaches were proposed. The chemogenomic methods form predictions based on the chemical and the genomic information [15–21]. The chemical information of the drug compounds is used in conjunction with the genomic information of the various protein targets to identify the drug target interactions.

The chemogenomic approaches have further been divided into similarity-based approaches and feature-based approaches. The similarity-based approaches use the drug similarity matrix as well as the target similarity matrix for the drug target interaction prediction [15, 18]. The prediction can be made using various techniques like network diffusion-based techniques, neighborhood model-based techniques, matrix factorization techniques, etc. The feature-based approaches first encode the drugs and target proteins as features. These feature sets are the used to identify interactions using various classifiers [22–24].

Various similarity-based drug target interaction prediction methods have been proposed that include nearest neighbor (NN) [25], weighted nearest neighbor (WNN) [25] and network-based inference (NBI) [4] techniques. The nearest neighbor technique uses the information of the drug or the target nearest to the drug target pair.

For instance, the interaction matrix for a new drug $d_n$ is computed as follows:

$$Y(d_n) = S_d(d_n, d_{\text{nearest}}) \times Y(d_{\text{nearest}}) \tag{1}$$

where $Y(d_n)$ is the interaction matrix of a new drug, $d_{\text{nearest}}$ is the drug compound that is most similar to the new drug $d_i$ and $S_d$ represents the similarity matrix of the drugs. Similarly, the nearest targets can be used to infer the interaction profile of a new target.

The weighted nearest neighbor also uses the most similar drugs and compounds to make interaction predictions. The key difference is that it uses the average of the interactions of the most similar drugs and targets to infer interaction profiles. The weighted interaction profile for a new drug $d_n$ can be calculated as follows:

$$Y(d_n) = \frac{\sum_{i=1}^{m} S_d(d_n, d_i) \times Y(d_i)}{\sum_{i=1}^{m} S_d(d_n, d_i)} \tag{2}$$

In both of the above-mentioned methods, the average of the interaction score from the drug side as well as the target side is used to calculate the final prediction output of the drug protein pair.

In the network-based inference model, a bipartite graph is constructed for the drug target interaction using the interaction matrix $I$. The network diffusion is implemented as follows:

$$I' = WI \tag{3}$$

where $W$ is the weight matrix that can be computed as

$$W_{mn} = \frac{1}{R_{mn}} \sum_{i=1}^{x} \frac{Y_{mi} Y_{ni}}{k(t_i)} \tag{4}$$

In the above equation, $R$ represents the diffusion rule and $k(y)$ represents the degree of the node $x$ in the drug target bipartite network graph. The rule $R$ is given as $R = k(d_n)$.

The state-of-the-art feature-based methods include prediction using decision trees (DT) [26], support vector machine (SVM) [27] and rotation forest (RF) [28]. Decision tree is a kind of a flowchart where each node represents a test, and the successive children represent the outcomes of the test. Each of the leaf nodes depicts the labels of the resulting class.

SVM has been used in the field of drug target interaction in various works. It had already been used previously to predict protein–protein interactions and the classification of chemicals into drugs and non-drugs. SVM classifier can be used to classify the drug target pair into interacting and non-interacting class. The drug feature vector and the target feature vector are concatenated to form a drug target pair vector. These vectors are then fed to the SVM classifier to make predictions.

Rotation forest has also been used to identify drug target interactions. This ensemble classifier uses the entire training data. The training data is split into n equal subsets. Principal component analysis (PCA) [29] is then applied to each of these n subsets. These subsets are then used to train k decision trees. If T is the total number of features, each of these decision trees contains T/n features. The columns from the resulting data are used to form the new interaction matrix. Rotation forest has been shown to have superior accuracy in comparison with SVM-based techniques.

The rest of the paper is organized as follows: Sect. 2 explains the steps of the proposed methodology in detail. Section 3 describes the drug target interaction data and the experimental results. Section 4 concludes the study and proposes future work in the field.

## 2 Proposed Methodology

The proposed methodology involves the use of ensemble learning to identify drug target interactions. The ensemble framework has been depicted in Fig. 1. Ensemble learning uses multiple models or classifiers collectively to make predictions [30]. The ensemble of classifiers aims to improve the accuracy and performance of prediction in comparison with a single classifier. The various base models or classifiers are combined in order to produce an optimal final model. Various ensemble models like Bayes optimal classifier, Bootstrap aggregating, stacking model, etc., have been proposed to construct optimal ensemble models.

Ensemble learning is mostly applied in order to increase the efficiency of a model that may be used for prediction, function estimation, etc. The efficiency is increased as it lessens the chance of selection of a poor classifier. There are various other uses of ensemble classifiers that range from data fusion to error correction, etc.

The proposed model has been shown in Fig. 2. It uses the drug and target features as the input. The encoding of drugs and targets as features has been explained in Sect. 2.1. The features are then used to construct subsets of data that can be fed to various decision trees. The dimensionality of the input data is reduced using PCA.

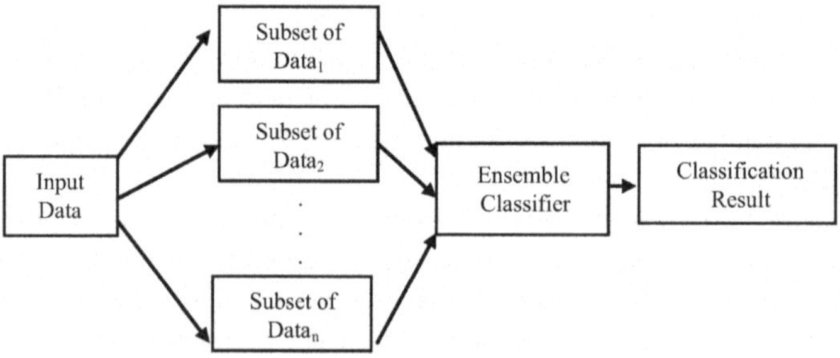


**Fig. 1** Framework of ensemble classifier

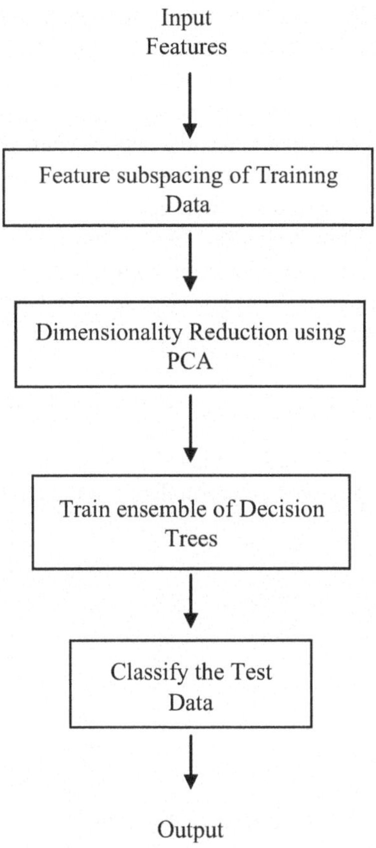


**Fig. 2** Proposed methodology

The subsets of training data are then fed to the ensemble of decision trees. The output of the ensemble of decision trees is then combined to form the final output result. The proposed method has been explained in detail in the following subsections.

## *2.1 Feature Encoding*

The drugs are encoded as features using the Rcpi package [31]. These drug features include structural properties, physicochemical properties, molecular properties, constitutional descriptors, etc. For the targets, PROFEAT package [32] was used to compute the protein properties. Although it has been suggested that the entire protein information is contained in the protein sequences [23], they cannot be used to encode target vectors. This is because the protein sequences are of varying length, and hence, the resulting features will also have different lengths. Thus, PROFEAT package encodes the target proteins as fixed-length vectors with various properties.

The target features that have been generated include amino acid composition, amino acid properties, autocorrelation, etc.

After encoding the drugs and targets as features, certain features were removed. There exist certain features which were absent in certain drugs or targets. Also, certain features had a similar value over the entire range of drugs and targets. These features were also omitted as they were not responsible for affecting the drug target interaction prediction.

The drug target pair is represented by concatenating the drug vector and the target vector. Let the drug vector be represented as $[d_1, d_2, \ldots, d_n]$ and target vector be represented as $[t_1, t_2, \ldots, t_m]$. The drug target pair vector is formed by concatenating the two vectors as

$$d \oplus t = [d_1, d_2, \ldots, d_n,\ t_1, t_2, \ldots t_m] \tag{5}$$

To avoid the bias, the vectors are normalized in the range [0, 1] using the min–max equation. The normalization is performed as

$$\forall i = 1, 2, \ldots, n, d_i = \frac{d_i - \min(d_i)}{\max(d_i) - \min(d_i)} \tag{6}$$

$$\forall j = 1, 2, \cdots \mathrm{m}, \mathrm{t}_j = \frac{\mathrm{t}_j - \min(\mathrm{t}_j)}{\max(\mathrm{t}_j) - \min(\mathrm{t}_j)} \tag{7}$$

## 2.2 *Feature Subspacing*

In order to form a subset of data for each of the base learners of the ensemble, feature subspacing is performed. A random subset of features ($F_x$) is extracted from the feature set. From this random subset, a positive interaction data is obtained. Let this be denoted as $P_x$. This positive subset $P_x$ is oversampled. The positive data is oversampled in order to reduce any kind of bias. There exist particular types of interactions in the positive dataset that are underrepresented in comparison with the other interactions. This causes the prediction techniques to be biased toward the better represented interaction types. Thus, oversampling the positive dataset helps to attain better generalization of the prediction performance.

Since the number of unknown or negative interactions is greater than the positive interactions, the technique is expected to be biased toward the negative class. Hence, the number of positive and negative samples is equalized to reduce any biased results. Thus, for the negative subset, a random sample $N_x$ is selected such that the number of samples in $P_x$ is equal to the number of samples in $N_x$, i.e., $|P_x| = |N_x|$. These samples of $N_x$ are removed from the original data. Finally, feature subspacing is performed on $N_x$ using the random subset $F_x$.

### 2.3 Dimensionality Reduction

The dimensions of the features are further reduced using PCA. Dimensionality reduction helps to decrease the complexity as well as the computational time of the proposed method.

PCA transforms a set of possibly correlated observation values into linearly uncorrelated variables using orthogonal transformation [29]. Let there be a total of $x$ observations with $y$ variables. The total number of principal components formed will be equal to min $(x - 1, y)$. The conversion using PCA is performed in a manner such that the first principal component has the maximum variance. All the following principal components have the largest variance under the condition that they are orthogonal to the previous components. Thus, the resulting principal components are an orthogonal set of uncorrelated values.

Hence, PCA is used to form a smaller number of uncorrelated valued from a larger amount of data. Thus, it can be used as a dimensionality reduction technique to reduce the drug target feature dimension.

### 2.4 Ensemble Learning

An ensemble of decision trees is trained in order to form the predictive model. This machine learning technique combines several base decision trees in order to form an optimal decision model [33].

A base decision tree concludes the output class on the basis of certain conditions or questions. It can also be used in a similar manner to find solutions for quantitative problems. However, various factors like the type of features, threshold at each node, etc., define the performance of the decision trees. Ensemble learning, on the other hand, helps to resolve such issues. Instead of relying on the performance of just one decision tree and trying to optimize it, ensemble learning helps to improve performance by considering a collection of decision trees. The final output is based on the aggregation of results of all the decision trees.

The proposed method uses the subset of positive and negative datasets obtained by feature subspacing as explained in Sect. 2.2. The dimensionality of the obtained subsets of data is reduced using PCA as described in Sect. 2.3. Using these subsets of data, an ensemble of decision trees is obtained. The ensemble is then used to make drug target interaction prediction on the test data.

## 3 Experiments

The proposed method was evaluated by comparing the performance of the method with six state-of-the-art methods mentioned in Sect. 1. These methods are nearest neighbor (NN), weighted nearest neighbor (WNN), network-based inference (NBI), decision trees (DT), support vector machine (SVM) and rotation forest (RF). The parameters of these methods have been set to the default values in MATLAB. The details of the dataset and the results have been described in the following subsections.

### 3.1 Data

The drug target interaction data used for the analysis of the proposed method has been taken from the DrugBank database [34]. The data consists of 12,674 drug target interactions. A total of 5877 drug compounds and 3348 protein targets are considered. As mentioned in Sect. 2.1, Rcpi package is used to construct drug features and PROFEAT server is used to encode protein features. The total features for drugs and targets are 193 and 1290, respectively.

### 3.2 Experimental Results

The sensitivity analysis for the parameter $N$ (number of decision trees to be trained) has been shown in Fig. 3. It can be seen that the performance of the proposed technique increases as the number of decision trees in the ensemble increase. The improvement of performance of the model is at a steady rate beyond 80. The performance may increase further on increasing the number of decision trees beyond 100. However, the computational time will also increase.

To compare the performance of the proposed method with the state-of-the-art methods, fivefold cross-validation experiments were carried out. For each of the cross-validation experiment, area under ROC curve (AUC) value was calculated.

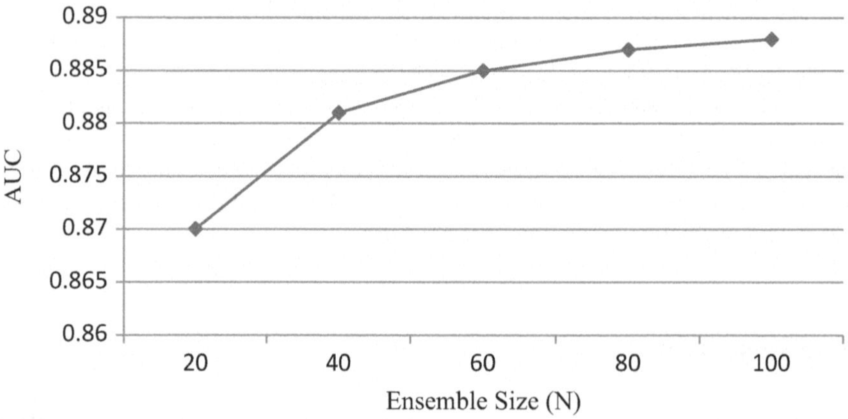


**Fig. 3** Sensitivity analysis for the parameter $N$

**Table 1** AUC results of state-of-the-art methods

| Methods | AUC |
|---|---|
| Nearest neighbor (NN) | 0.679 |
| Weighted nearest neighbor (WNN) | 0.793 |
| Network-based inference (NBI) | 0.606 |
| Decision trees (DT) | 0.760 |
| Support vector machine (SVM) | 0.804 |
| Random forest (RF) | 0.855 |
| **Proposed method** | **0.888** |

AUC is a standard metric for measuring performance that depicts the aggregate of the classifier performance over all possible threshold values. AUC score is not sensitive to skewed class distributions [35]. Hence, it is an appropriate metric for measuring the drug target interaction prediction performance as the dataset in this study is highly skewed (the number of negative interactions is much more than the positive interactions). For each of the methods, five AUC values were computed corresponding to the fivefold. These values are averaged to obtain the final AUC score.

The AUC scores of the state-of-the-art methods and the proposed technique have been shown in Table 1. As shown in the table, the proposed method outperforms the other methods. It achieves an AUC of 0.888. The second best method is random forest with an AUC of 0.855. Random forest is also an ensemble-based method where an ensemble of decision trees is constructed. The output of this ensemble is the mode of the classes or the mean value of the base trees. Based on the experimental results, it is evident that ensemble methods help to achieve better performance than simple classifiers like SVM or DT. Also, the performance of the ensemble-based methods is significantly better than the similarity-based methods, i.e., nearest neighbor (NN), weighted nearest neighbor (WNN) and network-based inference (NBI) techniques.

## 4 Conclusion

This paper proposes a novel technique for predicting drug target interactions using ensemble learning. The technique constructs subsets of data for the various base decision trees. Also, PCA has been applied to the data to decrease the dimensionality of the data and hence the computational time. A comparison of the AUC scores shows that the proposed model outperforms the state-of-the-art techniques.

The future work in this field can focus on exploring various other dimensionality reduction techniques to further improve the performance. Also, various hybrid ensembles can also be constructed to increase the prediction accuracy.

## References

1. Chen, X., et al.: Drug–target interaction prediction: databases, web servers and computational models. Briefings Bioinf. **17**(4), 696–712 (2016)
2. Paul, S.M., et al.: How to improve R&D productivity: the pharmaceutical industry's grand challenge. Nat. Rev. Drug. Discov. **9**(3), 203 (2010)
3. Ezzat, A., et al.: Computational prediction of drug-target interactions using chemogenomic approaches: an empirical survey. Briefings Bioinf. bby002–bby002 (2018)
4. Cheng, F., et al.: Prediction of drug-target interactions and drug repositioning via network-based inference. PLoS. Comput. Biol. **8**(5), e1002503 (2012)
5. Emig, D., et al.: Drug target prediction and repositioning using an integrated network-based approach. PLoS. One. **8**(4), e60618 (2013)
6. Jin, G., Wong, S.T.: Toward better drug repositioning: prioritizing and integrating existing methods into efficient pipelines. Drug Discov. Today **19**(5), 637–644 (2014)
7. Atias, N., Sharan, R.: An algorithmic framework for predicting side-effects of drugs. In: Annual International Conference on Research in Computational Molecular Biology. Springer (2010)
8. Pauwels, E., Stoven, V., Yamanishi, Y.: Predicting drug side-effect profiles: a chemical fragment-based approach. BMC Bioinf. **12**(1), 169 (2011)
9. Yamanishi, Y., Pauwels, E., Kotera, M.: Drug side-effect prediction based on the integration of chemical and biological spaces. J. Chem. Inf. Model. **52**(12), 3284–3292 (2012)
10. Bolton, E.E., et al.: PubChem: integrated platform of small molecules and biological activities. In: Annual Reports in Computational Chemistry, pp. 217–241. Elsevier (2008)
11. Jacob, L., Vert, J.-P.: Protein-ligand interaction prediction: an improved chemogenomics approach. Bioinformatics **24**(19), 2149–2156 (2008)
12. Li, H., et al.: TarFisDock: a web server for identifying drug targets with docking approach. Nucleic. Acids. Res. **34**(suppl_2), W219-W224 (2006)
13. Xie, L., et al.: Drug discovery using chemical systems biology: weak inhibition of multiple kinases may contribute to the anti-cancer effect of nelfinavir. PLoS. Comput. Biol. **7**(4), e1002037 (2011)
14. Mousavian, Z., Masoudi-Nejad, A.: Drug–target interaction prediction via chemogenomic space: learning-based methods. Expert. Opin. Drug. Metabol. Toxicol. **10**(9), 1273–1287 (2014)
15. van Laarhoven, T., Nabuurs, S.B., Marchiori, E.: Gaussian interaction profile kernels for predicting drug–target interaction. Bioinformatics **27**(21), 3036–3043 (2011)
16. Bleakley, K., Yamanishi, Y.: Supervised prediction of drug–target interactions using bipartite local models. Bioinformatics. **25**(18), 2397–2403 (2009)
17. Zheng, X., et al.: Collaborative matrix factorization with multiple similarities for predicting drug-target interactions. In: Proceedings of the 19th ACM SIGKDD International Conference on Knowledge Discovery and Data Mining, ACM (2013)
18. Gönen, M.: Predicting drug–target interactions from chemical and genomic kernels using Bayesian matrix factorization. Bioinformatics **28**(18), 2304–2310 (2012)
19. Wang, W., Yang, S., Li, J.: Drug target predictions based on heterogeneous graph inference. In: Biocomputing, pp. 53–64. World Scientific (2013)
20. Tabei, Y., Yamanishi, Y.: Scalable prediction of compound-protein interactions using minwise hashing. BMC. Syst. Biol. **7**(6), S3 (2013)
21. Mei, J.-P., et al.: Drug–target interaction prediction by learning from local information and neighbors. Bioinformatics **29**(2), 238–245 (2012)
22. Nagamine, N., Sakakibara, Y.: Statistical prediction of protein–chemical interactions based on chemical structure and mass spectrometry data. Bioinformatics **23**(15), 2004–2012 (2007)
23. He, Z., et al.: Predicting drug-target interaction networks based on functional groups and biological features. PloS. one. **5**(3), e9603 (2010)
24. Perlman, L., et al.: Combining drug and gene similarity measures for drug-target elucidation. J. Comput. Biol. **18**(2), 133–145 (2011)

25. Yamanishi, Y., et al.: Prediction of drug–target interaction networks from the integration of chemical and genomic spaces. Bioinformatics **24**(13), i232–i240 (2008)
26. Safavian, S.R., Landgrebe, D.: A survey of decision tree classifier methodology. IEEE. Trans. Syst. Man. Cybern. **21**(3), 660–674 (1991)
27. Cristianini, N., Taylor, J.: Support vector machines and other Kernel-based learning algorithms. Technical Report (2000)
28. Breiman, L.: Random forests. Mach. Learn. **45**(1), 5–32 (2001)
29. Wold, S., Esbensen, K., Geladi, P.: Principal component analysis. Chemometr. Intell. Lab. Syst. **2**(1–3), 37–52 (1987)
30. Dietterich, T.G.: Ensemble methods in machine learning. In: International Workshop on Multiple Classifier Systems. Springer (2000)
31. Cao, D.-S., et al.: Rcpi: R/Bioconductor package to generate various descriptors of proteins, compounds and their interactions. Bioinformatics **31**(2), 279–281 (2014)
32. Li, Z.-R., et al.: PROFEAT: a web server for computing structural and physicochemical features of proteins and peptides from amino acid sequence. Nucleic. Acids. Res. **34**(suppl_2), W32–W37 (2006)
33. Dietterichl, T.G.: Ensemble learning (2002)
34. Law, V., et al.: Drugbank 4.0: shedding new light on drug metabolism. Nucleic. Acids. Res. **42**(D1), D1091–D1097 (2013)
35. Fawcett, T.: An introduction to ROC analysis. Pattern. Recogn. Lett. **27**(8), 861–874 (2006)

# Integration of Fog Computing and Internet of Things: An Useful Overview

**G. Rekha, Amit Kumar Tyagi and Nandula Anuradha**

**Abstract** In the past decade, the evolution of computing has moved from distributed, parallel, grid, cloud, and now to fog computing. The massive amount of data generated by Internet of Things (IoT) devices is growing up exponentially. The flood of information (generated by those IoT/Internet-connected devices) becomes troublesome for data processing and analytical prediction functionality using cloud computation. Several problems have been investigated with cloud computing with respect to latency, limited bandwidth, low Internet connectivity, etc. Here, solution to such problems can be solved by introducing fog computing with powerful functionality of cloud framework, i.e., based on the deployment of fog nodes called microclouds at nearest edge of data sources. Fog computing for big data/IoT data analytics is in evolving phase and requires extensive research to produce more knowledge and smart decisions. This article discusses several basic facts related to fog commuting, challenges in fog computing and opportunities in the near future, in the context of fog big IoT data analytics. In addition, this work also emphasizes the key characteristics in some proposed research works, those make the fog computing as a suitable (useful) platform for new proliferating IoT devices, services, and applications. Most significant fog applications (e.g. healthcare monitoring, smart cities, connected vehicles, and smart grid, etc.) will be discussed here to create a well-organized green and quantum computing paradigm to support the next generation of IoT applications.

**Keywords** Internet of Things · Cloud computing · Fog computing · Big data · Data analytics

G. Rekha · N. Anuradha
Department of Computer Science and Engineering, Koneru Lakshmaiah Education Foundation, Guntur Andhra Pradesh, 522502, India
e-mail: gillala.rekha@klh.edu.in

N. Anuradha
e-mail: nandulaanuradha@gmail.com

A. K. Tyagi (✉)
Department of Computer Science and Engineering, Lingaya's Vidyapeeth, Faridabad, Haryana 121002, India
e-mail: amitkrtyagi025@gmail.com



P. K. Singh et al. (eds.), *Proceedings of ICRIC 2019*, Lecture Notes in Electrical Engineering 597, https://doi.org/10.1007/978-3-030-29407-6_8

## 1 Introduction

Fog computing is a pushing edge for processing computer applications, data, and services in close proximity of end-user. Fog Computing or Fog Networking or Edge Computing, and it is also known as Fogging, provides computing facilities away from centralized cloud to the logical stream of the network edge. Fog computing is started with the word 'fog', means 'closer to someone'. Fog computing is closed to cloud and nearer to end-users. It extends the traditional cloud computing architecture to the edge of the network architecture. The fog extends the basic building block of the cloud like compute, storage, and networking services to the edge of the network so-called fog nodes, near the end devices (like IoTs, smartphones, GPS, wearable devices, etc.). Fog offers distributed points for assembling data generated by the end devices via proxies, access points, and routers placed at the network edge, near the source devices. Though now days cloud computing is widely used a backbone to Internet of Things, and provide resource utilizations (accessing) from anywhere, anytime. But still cloud computing has some limitations. The fundamental limitation of cloud computing is the connectivity between the cloud and the end devices. Such connectivity is set over the Internet, not suitable for a large set of cloud-based applications [1], such as connected vehicles, smart grid [2], re-detection [3], and content delivery [4]. Further, the distribution-based cloud application made up of multiple components, sometimes deployed separately over multiple clouds, may deteriorate the latency due to the overhead induced by inter-cloud communications [5]. Yet, as another limitation of cloud computing is, the regulations may suggest processing at locations where the cloud provider may have no data centre [6]. In literature [7], it is widely recognized that cloud computing is not feasible for most of the Internet of Things (IoT) applications and fog could be used as an alternative solution to solve this problem (feasibility with IoTs). Nowadays, IoT becomes an important and essential device that promises a smart, easier, longer life to human beings. IoT-enabled devices allow communication between the machines (or devices), objects, and everything connecting together with people. The need to build that IoT is to dig up, access, and analyse the information that provides valuable insight. The information generates digital data by connecting the IoT devices over the Internet (or World Wide Web), for example, online transaction, downloading of applications, retail, etc.

The IoT systems consist of things or devices in the real world and sensors are embedded with devices which are communicated to the Internet (via wired or wireless networks as a medium). The variety of powerful devices like servers, smart access devices such as smartphone, tablets, smart sensors, smart appliances in home, and many more. Therefore, all connecting things together formulate an IoT ecosystem, used to solve real-world applications. The various types of connections used by IoT sensors are RFID, Wi-Fi, Bluetooth, and ZigBee, and for wide area connectivity, GSM, GPRS, 3G, and LTE are used. As reported by Ovum and CISCO [1] in March and June 2017, respectively, at present the main areas of investment are the Industrial Internet of Things (IIoT) [8], by deployment of IoT devices in industries include manufacturing operations, transportation, smart grid technologies, smart buildings and increasingly, consumer IoT, and smart home automation. Moreover, in general, some applications of fog computing are connected cars, smart cities and smart grids, real-time analytics (e.g. mobile big data analytics, etc.), water pressure at dams, health data, and any other smart utility services. Hence, an architecture of fog computing or fog network with consist cloud environment is depicted as Fig. 1. Irrespective of cloud computing, fog computing provides a better-analysed experience to the end-user by generating some real-time network information (because it is nearer much to end-users than cloud with a high Internet connectivity, i.e., with a low latency, refer Fig. 2). Hence, big IoT data analytics with fog computing provides better, efficient insights by enabling devices to make smart and intelligence-based decisions without the interaction of human. These data or information analyses by big data can be used in solving or predicting several real-world problems in the near future.

Hence, the rest of this paper is organized as follows: Sect. 2 deals with the overview of fog computing and its related work. Further, Sect. 3 discusses the interfacing fog, with fog, cloud, IoT devices/end-user devices and Sect. 4 presents the salient features of fog computing. Real-time use cases are discussed in Sect. 5. Further, some challenges with respect to fog computing have been discussed in Sect. 7, and in last, this work is concluded with some future remarks/enhancements in Sect. 7.

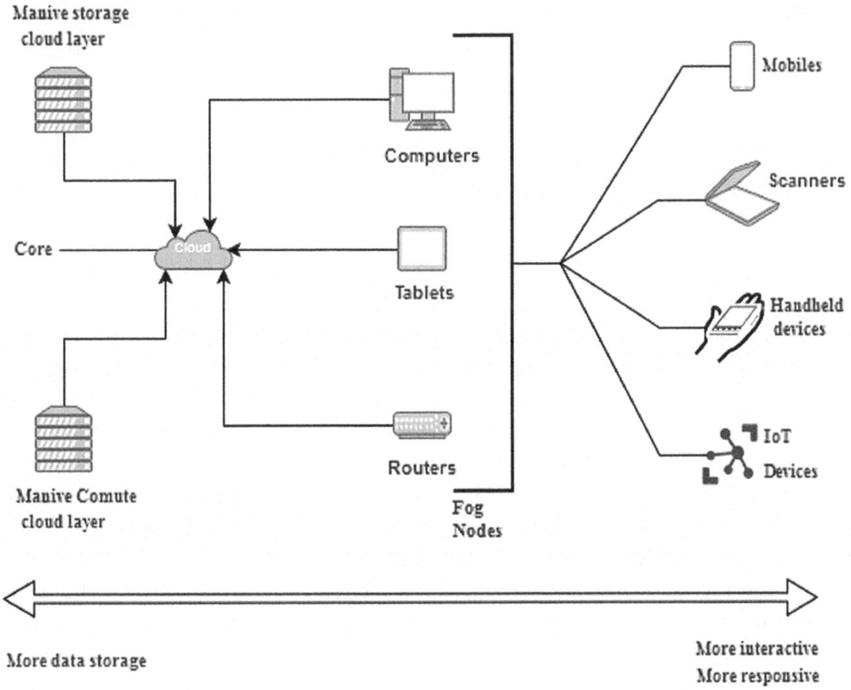


**Fig. 1** Architecture of fog computing

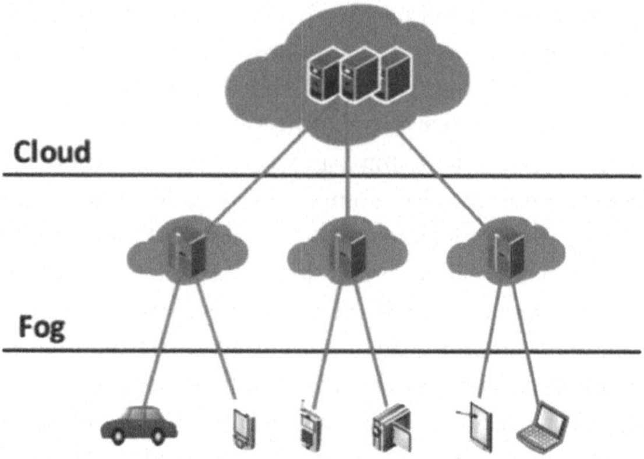


**Fig. 2** A generic convergence of IoT devices and fog devices [2]

## 2 Related Work

As data is increasingly generated by the IoT devices at the core network end, to efficiently process all data at the close proximity of IoT, CISCO in 2012 introduced a concept called fog computing. It is an extension of cloud computing architecture from the cloud to the edge of the network. This section provides an overview of fog definition and discusses the similarities and differences between the related concepts and its interface with cloud, IoT, and other fog nodes in brief.

### *2.1 Fog Computing—Definition*

Fog computing is an extension of cloud computing paradigm by enabling computation at the edge of the network, closer to IoTs and/or the end-user devices (i.e. to provide reliable and efficient services to users). It (i.e. fog computing) presents a distributed hierarchical architecture, offering more flexibility to the system. It supports integration with various technological components and services like smart cities, smart homes, and smart grid system [7]. Fog computing provides pool of resources with one or more decentralized nodes which cooperate and communicate with each other or in a group at the edge of the network. Figure 2 shows a fog system with three-tier architecture consisting of IoT/end-user devices, edge-level nodes or fog nodes and the cloud-level nodes. Fog computing provides services at the edge of IoT devices and the communication happens via local area network (LAN). The main focus of the nodes at the edge is to sense the collection of raw data, process, and control of end-user or IoT devices as shown in Fig. 2.

**Table 1** A brief description of cloud computing, cloudlets, mobile edge computing

| | Cloud computing | Cloudlets | MEC |
|---|---|---|---|
| Definition | It includes software and applications running on a central server having built-in local networks with big data centres | It provides a small cloud computing architecture by inheriting the features from centralized cloud computing | MEC enables technologies to be available for cloud computing to provide QoS at the edge of the network |
| Uses | SaaS, PaaS, IaaS | Time-sensitive and limited bandwidth applications | Mobile applications |
| Mode of operation | Connected | Stand-alone or connected to cloud | Stand-alone |
| Applications | Any application | Mobile offloading application | Mobile offloading application |
| Difference | Mainly driven by R and D | Mainly driven by R and D | Mainly driven by industry |

### 2.2 *Similarities and Differences with Cloud Computing, Cloudlets, and Mobile Edge Computing*

There are some similar terms that come along with fog computing called cloud computing, cloudlets, and mobile edge computing (MEC). Even though cloud computing, cloudlets, and MEC aim at computing at the edge, few differences exist in each new technology. Table 1 illustrates the definition and the difference among all three.

Hence, this section discusses the related work to fog computing, like definition of fog computing, similarities, and difference in fog computing technology/environment with existing (new) technologies like cloud computing, cloudlets, and mobile edge computing. Now, the next section will discuss some interacting scenarios, i.e., with cloud, and IoT devices, or as cloud based IoT/ IoT based Cloud.

## 3 Interfacing Fog with Fog, Cloud, Internet of Things Devices/End-User Devices

As discussed in Sect. 2, fog computing develops from cloud computing, and this term was coined in 2012 to provide flexible, reliable services to end-users. It provides functionality with more flexibility for computation at the edge of the network and shared same processing strategies, visualization, and multi-tendency features. The more significant features support by fog computing are: (a) supports high-speed moving applications like smart connected rail, smart vehicles, etc., (b) low latency, (c) provide distributed environment in large scale, (d) supports geographically distributed applications using sensor networks to monitor the different environment. The

**Table 2** Interfacing of fog to cloud, fog to fog, fog to IoT

| Interface | | | |
|---|---|---|---|
| Fog to cloud and vice versa | It is considered compulsory to support collaboration between fog to cloud and cloud to fog services | It also supports functionalities like supervision at fog end | It also transfers data for processing, comparing between each other |
| Fog to fog | It must have pool of resources to support processing with each other | Multiple fog nodes act together to support backups for each other | All the fog nodes deployed can share their data, computation, and processing for one or several applications |
| Fog to IoT | Fog computing provides services to widely distributed IoT devices like smart devices and sensors | It provides IoT devices to access fog services in a user-friendly environment | Fog computing provides high QoS to the IoT devices or end-users |

most fundamental components in fog computing architecture are called fog nodes. They are set of hardware and software elements that can be configured together to perform specific functions like providing resources for services such as cloudlet. Generally, cloudlet is a cloud data centre located at the edge of the Internet. As fog computing is an emerging technology and in infant stage (i.e. just started, not developed completely). Therefore, further research is needed in this area/domain. The architecture of fog computing allows processing, networking and storage services to dynamically allocate at the fog nodes, cloud, or IoT systems. However, the interface of fog with other devices is must to enable well-suited end-user assessment and also to provide effective QoS management. Table 2 illustrates the different interfacing between the fog, cloud, and IoT.

Hence, this section discusses the interfacing of fog with fog, cloud, IoT devices, or end-user devices. Now, the next section will deal with the salient features of fog computing in brief.

## 4 Salient Features or Characteristics of Fog Computing

The movement of computational load to fog networking from cloud enables the development of IoT structure for diversity of applications and services. The fog computing characteristics are as follows:

(a) Cognition: Cognition means thinking and perceiving the client's intentions. Fog-based data processing offers a more lucid picture about the client's interests.

For example, getting insight into how to store, control, and transfer data from cloud to IoT and vice versa. The devices at the edge respond more quickly to the client's requests as they are at the ground level than the remote cloud-based services.

(b) Heterogeneity: Fog computing is an integration of various computing platforms offering a variety of computing infrastructure. It bridges the cloud server and the senor enabled devices. Fog computing servers utilize federation of clouds in a distributed environment.

(c) Geographical environment distribution: Fog computing has the capability of providing Quality of Service (QoS) in both mobile and stationery environments by distributing the fog nodes so that they can communicate with the nearest gateways. E-heath, m-health, traffic monitoring are some examples of location-aware services.

(d) Edge location with low latency: Fog nodes (existed in a fog computing environment/fog network) give quick response than the cloud, thus improving the Quality of Service (QoS) in terms of low latency. This is an important requirement of any application that involves streaming of data.

(e) Real-time interaction: Real-time interaction is the most crucial aspect in fog applications where the edge devices should transmit real-time information such as airflow inside mines, radiation levels in nuclear power plants, leakage of poisonous gases inside chemical factories, etc.

(f) Support for mobility: Multiple geographically distributed fog nodes can communicate with the mobile devices in applications such as vehicle networks and efficient logistic management.

(g) Large-scale sensor network: Fog computing has the capability of forming a hierarchical network architecture by combining data from multiple gateways for global data processing.

(h) Widespread wireless access: In fog computing, Wireless Application Protocol (WAP) and mobile gateways enable users to connect to the closest fog nodes.

(i) Interoperable Technology: Fog computing infrastructure can be considered as an interoperable system which can be physical, virtual, or a hybrid consisting of a collection of sensor devices, cloud services, communication protocols that facilitates dissimilar components to generate solutions to specific issues.

Hence, this section discusses several characteristics that make the fog computing essential to end-users and non-trivial extension of the cloud computing, that is, fog computing provides service to users like edge location, location awareness, low latency, providing hierarchical organization and proximity of data to end-users, dense geographical distribution, large-scale sensor networks, large number of nodes, support for mobility, real-time interactions, predominance of wireless access, heterogeneity, federation and interoperability, and integration with the cloud and support for online analytic. Now, the next section will discuss some real-time uses cases in brief.

## 5 Real-Time Use Cases with Fog Computing

In this section, we discuss the role the fog computing plays in IoT environment or IoT ecosystem with explaining four interesting scenarios, i.e. video streaming data, healthcare monitoring systems, gaming, and smart traffic light (STL) system. Hence, each use case/scenario is discussed as follows:

(a) Video streaming data: The characteristic features such as low response time, mobility, position awareness and real-time data processing capabilities of fog nodes benefit the real-time video streaming applications. Aqua computing is a mobile edge network that enables the computing resources to be integrated with communication resources, thereby improving the connected users' experience. It also optimizes the computing resources. Here, the fog nodes act as clones at the edge and as buffers at the user's desktop. Also, systematic deployment of fog nodes enhances real-time on-demand video streaming.
(b) Healthcare monitoring systems: Many health and fitness applications are built to continuously monitor individual's health. The captured data is analysed and diagnostic predictions can be made to take appropriate decisions by healthcare applications. This can be achieved with the integration of massive number of fog nodes that can create a real-time transmitting system than a cloud which has high response latency. Cao et al. [9] have proposed a real-time monitoring application, U-Fall, which can be separated into three major sections, front-end, back-end, and communication module, where front-end and back-end both make independently detections results. Thus, eliminating the false alarm rates and improving the detection accuracy. The other significant factor of U-Fall application is that it discovers the exhaustive fall using data of mild strokes. Hence, the experimental findings showed an increased sensitivity. It can be understood that the application of fog computing gateway improves complex techniques such as data mining and priority-based storing. FAST [10] is a distributed data analytics application with fog sensors that monitor cardiac strokes. It was shown that a fog-based edge network offered enhanced QoS by identifying the location of the patients quickly.
(c) Gaming: Cloud computing has increased the popularity of online multiple player computer games without any kind of hardware infrastructure. Wang and Dey [11] described mobile gaming application framework based on the cloud servers, where centralized server works on all possessing load of the game. But due to inherent latency of the cloud computing, the gaming commands issued by the server take much time, thereby disrupting the game. Hence, the application of fog computing in online gaming enhances the Quality of Service (QoS) and Quality of Experience (QoE). The experiment showed that the collected data encoding improves the playback consistency and end-time-driven buffer forecast approach.
(d) Smart traffic light (STL) System: Traffic lights are equipped with sensors that can sense, process, and communicate. These STLs can monitor the ongoing traffic. This data can be analysed and sent to traffic controller to avoid accidents

and traffic congestions. Also, it helps an ambulance to find a traffic-free route or clears the route for ambulance. Thus, STLS requires high data transmission rates from one fog node to another fog node for making quick decisions for choosing alternative routes.

Hence, this section discusses several real-time use cases of fog computing in IoT. Now, the next section will briefly explain the challenges of fog computing for IoT.

## 6 Challenges in Fog Computing with Internet of Things

Fog nodes (in a fog networking) mostly provide localization, therefore enabling low latency and context awareness, the cloud provides global centralization. Many applications require both fog localization and cloud globalization, particularly for analysing the data generated by IoT devices. Hence, some challenges have been investigated with fog computing, which are included as follows:

(a) Security: As fog nodes are distributed throughout heterogeneous networks and platforms, adhering to a specific set of security protocols is not feasible. So, the fog nodes at the gateways may be compromised. Security is also not the first topic in the current IoT discussions and is still largely treated as a compulsory yet secondary subject. Such disregard can be attributed to the lack of organizational policies and the ambiguities in government laws [12].
(b) Data processing, integrity, and quality: Huge number of interconnected IoT devices transmits continuous streams of data. The data collected is not only huge, but generated at various rates and dynamic in nature. The quality of data collected determines QoS. Error in data collection, noise, and error in measurement will affect the quality of services. SLAs play an important role in achieving the required QoS. Also, in a distributed environment data integrity is an important consequence. Data integration provides a single view of the data arriving from different sources and combines the view of data [13].
(c) Real-time data analytics: As discussed in [14], Internet of Things produces a lot of data (called big data), which need to be analysed smartly with more meaningful and useful information, which requires algorithms that are capable of analysing continuous streams of data generated by various sources in real time. Choosing a specific algorithm for a specific IoT application is crucial to make real-time decisions. Big data implementations must perform analytics with real-time queries to help organizations obtain insights quickly, rapidly make decisions, and interact with people and other devices in real [8].
(d) Privacy: Preserving network privacy, data privacy, location privacy, and identity privacy is a biggest challenge (or concern) in fog computing (due to connected via wireless). The leakage of private data or information from fog node is receiving attention from many malicious users while using networks. As solution to this problem to preserve privacy, encryption methods like HAN (Home-Area Network) can be used to counter these issue/challenges.

(e) Resource management: As discussed above, fog servers and data storage facilities require in a fog computing environment (at the edge of the network, to increase processing). The management of large number of fog nodes and providing storage to each fog node increase additional computing a fog computing environment, i.e. increase management costs. We (i.e. researchers) need to provide properly analysed services, effective management to fog servers in the near future.
(f) Heterogeneity: Several types of IoT devices, sensors, etc., are communicating in fog environment (i.e. with fog servers). To handle all devices, sensors, etc. (they have different protocols, storage capabilities, sensor characteristics, etc.), will create a massive problem, i.e., to provide a reliable coordination between these devices and fog server is a big challenge.
(g) Latency: Latency is an essential parameter in fog computing. It was a reason why it (fog computing) was used over cloud computing, but again latency (in fog computing) is not improved till end user's expectation level. Hence, if latency requirement is not satisfied, the performance would be degraded resulting in user dissatisfaction, and interest of users will be decreased.
(h) Complexity: In fog computing, large number of sensors and IoT devices are connected together to provide flexible service to end-users. But, providing such services with an optimized route/device is really a critical issue/challenging task. Sometimes, used software and hardware in IoTs may also increase complexity in fog environment, i.e. decrease performance.

In fog computing environments, nodes are higher and making communication continuously, and then computation is distributed among all nodes (in parallel) and can be less energy-efficient. Note that energy consumption or battery is a big issue with IoT Devices. Fog computing is closer to end-user with IoT devices, so improving battery power and saving energy in IoTs (during making a communication with devices or human being) is an essential work, it is a bigger challenge to overcome in the near future. The next section will conclude this work with some future enhancements in brief.

## 7 Conclusion with Future Research Directions

Fog computing performs some computing and storage at the edge of the network rather than relying entirely on cloud. This paper presents a few application areas such as health care, smart traffic light system (STLS), real-time gaming, etc., where fog computing and cloud computing go together. The key-enabling technologies underlying the fog computing are computation offloading, latency management, communication interfaces such as network function virtualization (NFV), software-defined networking (SDN), 5G, etc., and pre-cache system storage and storage expansion. The computation offloading can be achieved by provisioning live VM migration. SDN and NFV offer network scalability. Although there are many favourable areas for the application of fog computing, there are still challenges such as safety and security risks, energy-efficiency issues to be resolved. This work discussed cloud and fog computing in detail. Cloud computing offers massive data handling capability, scalability and flexibility, whereas fog computing offers user-centric services with low latency, location awareness, mobility, real-time response. This work provides a general overview to understand more about fog computing, and also some other new technologies like cloudlet, mobile edge computing, etc. (i.e. with similarities and dissimilarities). As future work with fog computing, we can try to build a smart gateway between networks to realize cloud with fog computing. Also, reduction of energy consumption in fog computing (including IoT devices) is always an essential and mandatory issue to overcome.

**Acknowledgements** This research is funded by the Koneru Lakshmaiah Education Foundation, Lingaya's Vidyapeeth, and Anumit Academy's Research and Innovation Network (AARIN), India. The authors would like to thank Koneru Lakshmaiah Education Foundation, Lingaya's Vidyapeeth, and AARIN, India, an education foundation body and a research network for supporting the project through its financial assistance.

## References

1. Jiao, L., Friedman, R., Fu, X., Secci, S., Smoreda, Z., Tschofenig, H.: Cloud-based computation offloading for mobile devices: state of the art, challenges and opportunities. In: Future Network and Mobile Summit (FutureNetworkSummit), pp. 1–11. IEEE (2013)
2. Stojmenovic, I.: Fog computing: a cloud to the ground support for smart things and machine-to-machine networks. In: 2014 Australasian Telecommunication Networks and Applications Conference (ATNAC), pp. 117–122. IEEE (2014)
3. Yangui, S., Ravindran, P., Bibani, O., Glitho, R.H., Hadj-Alouane, N.B., Morrow, M.J., Polakos, P.A.: A platform as-a-service for hybrid cloud/fog environments. In: 2016 IEEE International Symposium on Local and Metropolitan Area Networks (LANMAN), pp. 1–7. IEEE (2016)
4. Zhu, X., Chan, D.S., Hu, H., Prabhu, M.S., Ganesan, E., Bonomi, F.: Improving video performance with edge servers in the fog computing architecture. Intel Technol. J. **19**(1) (2015)
5. Pop, D., Iuhasz, G., Craciun, C., Panica, S.: Support services for applications execution in multi-clouds environments. In: 2016 IEEE International Conference on Autonomic Computing (ICAC), pp. 343–348. IEEE (2016)
6. Massonet, P., Naqvi, S., Ponsard, C., Latanicki, J., Rochwerger, B., Villari, M.: A monitoring and audit logging architecture for data location compliance in federated cloud infrastructures. In: 2011 IEEE International Symposium on Parallel and Distributed Processing Workshops and Phd Forum (IPDPSW), pp. 1510–1517. IEEE (2011)
7. Bonomi, F., Milito, R., Zhu, J., Addepalli, S.: Fog computing and its role in the internet of things. In: Proceedings of the first Edition of the MCC Workshop on Mobile Cloud Computing, pp. 13–16. ACM (2012)
8. Gantz, J., Reinsel, D.: The digital universe in 2020: big data, bigger digital shadows, and biggest growth in the far east. IDC iView: IDC Analyze the future **2007**(2012), 1–16 (2012)
9. Cao, Y., Hou, P., Brown, D., Wang, J., Chen, S.: Distributed analytics and edge intelligence: pervasive health monitoring at the era of fog computing. In: Proceedings of the 2015 Workshop on Mobile Big Data. pp. 43–48. ACM (2015)
10. Zaharia, M., Chowdhury, M., Das, T., Dave, A., Ma, J., Mccauley, M., Franklin, M., Shenker, S., Stoica, I.: Fast and interactive analytics over Hadoop data with spark. Usenix Login **37**(4), 45–51 (2012)

11. Wang, S., Dey, S.: Cloud mobile gaming: modeling and measuring user experience in mobile wireless networks. ACM SIGMOBILE Mobile Comput. Commun. Rev. **16**(1), 10–21 (2012)
12. Wang, Y., Uehara, T., Sasaki, R.: Fog computing: Issues and challenges in security and forensics. In: 2015 IEEE 39th Annual Computer Software and Applications Conference (COMPSAC), vol. 3, pp. 53–59. IEEE (2015)
13. Giang, N.K., Blackstock, M., Lea, R., Leung, V.C.: Developing IoT applications in the fog: a distributed dataflow approach. In: 2015 5th International Conference on the Internet of Things (IoT), pp. 155–162. IEEE (2015)
14. Tyagi, A.K., Anuradha, N., Rekha, G., Sharma, S., Niladhuri, S.: How a user will look at the connection of internet of things devices?: a smarter look of smarter environment. In: ICACSE: 2019: 2nd International Conference on Advanced Computing and Software Engineering, KNIT Sultanpur, 2019, India, 8–9 February 2019

# Review of Machine Learning Techniques in Health Care

**Rohan Pillai, Parita Oza and Priyanka Sharma**

**Abstract** Health care is an emerging industry with all our lives dependent on it. With the advancement of technology, the health care industry is also advancing. With better prevention, diagnosis and treatment options, technology is rapidly integrating itself with medical sciences for the betterment of humankind. However, it is time to delve deeper into this integration. One upcoming technical aspect known as machine learning is a very useful tool when it comes to its application in the health care industry. Machine learning algorithms such as support vector machines and artificial neural networks when combined with the existing medical infrastructure prove not only to be at par with the state-of-the-art technology but also prove to be more efficient and faster than them. This paper looks at the possible applications as well as the current progress of the integration of machine learning algorithms in the health care industry.

**Keywords** Algorithms · Support vector machine · Naïve Bayes · Neural networks · Training · Testing · Data

## 1 Introduction

With the advancement in technology, most of the health care industry has started embracing the fact that technology will be a helping hand in their work. Most medical records are now dealt with electronically. All types of body scans—CT scan, MRI scan, etc. are a result of the integration of technology with medical sciences. But the current technology is now outdated and needs to be upgraded. With so much data flowing in, there needs to be a proper way to manage, analyze, secure and store data

---

R. Pillai · P. Oza (✉) · P. Sharma
Nirma University, Ahmedabad, Gujarat, India
e-mail: parita.prajapati@nirmauni.ac.in

R. Pillai
e-mail: rpillai1997@gmail.com

P. Sharma
e-mail: priyanka.sharma@nirmauni.ac.in

 
P. K. Singh et al. (eds.), *Proceedings of ICRIC 2019*, Lecture Notes in Electrical Engineering 597, https://doi.org/10.1007/978-3-030-29407-6_9

electronically and yet be efficient and fast enough. Thus, the integration of machine learning with medical sciences is only for the better to come. Machine learning algorithms can be and have been applied to solve several problems in the health care industry. Not only have they stood the test of efficiency but also turned out to be better than the current technology.

# 2 Literature Survey

## *2.1 Health Care*

Health care is a very broad term encompassing prevention of a disease, diagnosis of a disease as well as its treatment. In other words, anything related to improvement in a person's health is put under the umbrella of health care. Its definition as well as its access varies from region to region depending on the social and economic policies in place. It is generally a very important aspect of a region's economy. According to a report by OECD among its 34 members, health care was responsible for an average 9.3% of the GDP, with the USA being the highest with 17.7% and Indonesia being the last in the list with a negligible rate [1] (see Fig. 1).

## *2.2 Machine Learning*

Machine learning is the ability of a system to automatically analyze and learn from a set of inputs provided to it as experience without requiring any explicit assistance. The system should then be able to infer from its experiences and be able to predict the output based on the input. In other words, a system should be able to learn on its own. There are two phases in a machine learning system—training and testing.

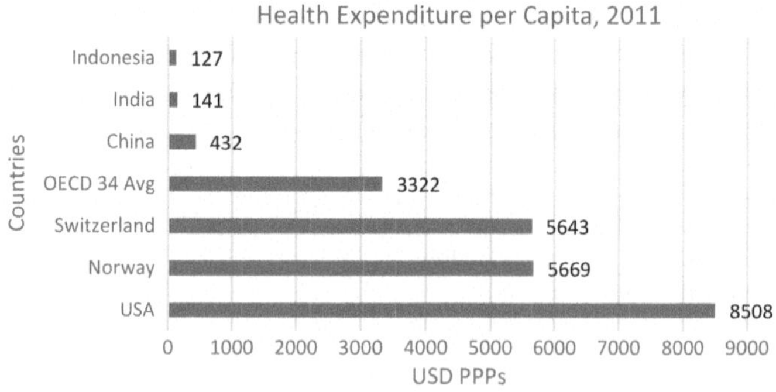


**Fig. 1** Health expenditure per capita, 2011 (adapted from Fig. 7.1.1, p. 157 [1])

Training is the process where a set of either labeled or unlabeled input is provided to the system. The system needs to analyze, which means to extract relevant data and parameters from the dataset and store it as a model for future prediction. Training is the most resource-intensive phase of the process. The next and final phase is the testing phase. In testing, an alien input is provided to the system and the system from its pre-trained model needs to predict the output. A trained system's accuracy depends on the accuracy of its output.

Machine learning algorithms are basically classified into four categories (see Fig. 2):

**Supervised Learning**: In this category, labeled input is provided to the system. The system needs to provide an output based on these labels.
**Unsupervised Learning**: In this category, the input provided is neither classified nor labeled. The system doesn't need to find the correct output; it creates its own inferences in order to successfully produce an output.
**Semi-Supervised Learning**: This category is a mix of supervised and unsupervised learning. This means such a system uses both labeled and unlabeled data for training.
**Reinforcement Learning**: A system implementing a reinforcement learning algorithm produces actions based on its interaction with the environment and tries to achieve the highest reward.

Some machine learning algorithms are:

**K-Means Clustering**: Clustering is a concept where a certain n number of input points are to be classified into k categories or clusters. This involves calculating the root mean squared distance from the centers which are initially chosen randomly but are updated with every iteration based on the average of the cluster members [2].

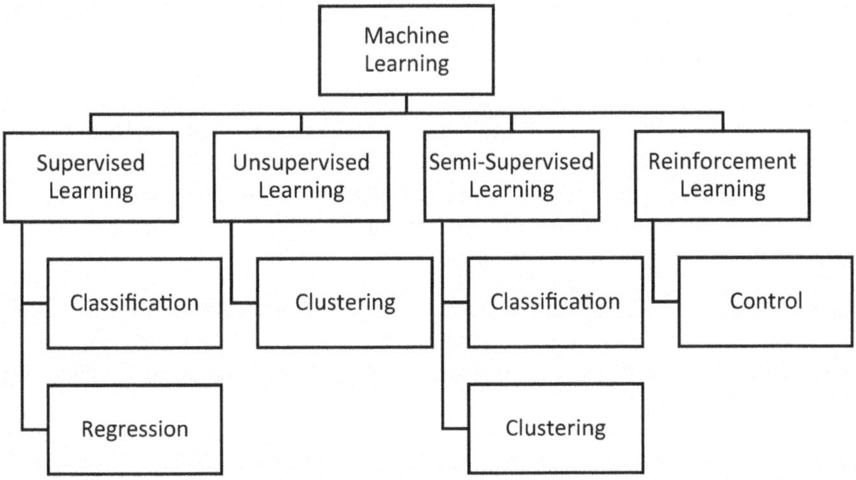


**Fig. 2** Machine learning algorithms classification

**Support Vector Machines**: Support vector machines (SVMs) are another form of classifiers where the equation of a plane is calculated that separates the two categories or clusters [3].
**Naïve Bayes Clustering**: Naïve Bayes classifier is a type of classifier that classifies labeled data. They require high number of parameters in order to determine the probabilistic classification value [4].
**Neural Networks**: Neural networks are a model developed on the ideas of a human brain. It has its own set of neurons that calculate the output based on the input they receive. Neural networks have several applications such as image classification and image recognition [5].
**Decision Trees**: Decision trees are one kind of classifiers that work on categorical data. It lists the possible consequences of certain events by calculating their probabilistic value [6].

## 3 Applications and Current Integration of Machine Learning in Health Care

Machine learning algorithms are quite useful in many areas. Not only are they useful in their particular domain such as image processing and artificial intelligence, but they can most certainly be used in the health care industry for improving the current state of technology [7]. According to a report by McKinsey, the integration of machine learning in the pharma and health care industry can generate up to $100 billion per annum [8]. Some of the applications are:

### *3.1 Medical Imaging*

Medical imaging is the process of generating representations of the insides of human body parts for medical examinations. The main purpose of this is to seek more information regarding the internal parts of the human body such as the tissues or organs. Medical imaging is basically categorized into three major categories: Optical imaging, radiology and nuclear medicine.

**Optical Imaging**: Optical imaging is the process of creating representations of hollow organs at a very cellular level. Example of this is the optical coherence tomography (OCT). These cellular level representations help understand any irregularities present in the organs [9].
**Radiology**: CT scan, MRI scan and X-Ray scans are examples of the radiology technique. In this, in a very spatial resolution, detailed body parts are represented.
**Nuclear Medicine**: Human body's metabolism is represented by nuclear medicine imagine techniques. Examples of this technique include PET scan.

Since the dataset we're dealing with here are images, machine learning could play a major role in helping with diagnosis. Machine learning algorithms like neural networks can be used to analyze medical images provided by the medical imaging techniques. In paper [10], segmentation of brain tumor was conducted with deep neural networks. The system was developed for analyzing high grade as well as low-grade glioblastomas in images. Deep neural networks were found to be more efficient and consistent in analyzing medical images. Not only did such an architecture improve the current technology, it was approximately 30 times faster, when tested on the 2013 BRATS test data. In an improved version [11], deep neural networks combined with discrete wavelet transform and principal component analysis were used to improve the classifier used in medical images.

### 3.2 Diagnosis of Diseases

Disease diagnosis is the process of examinations of a person's symptoms and health signs in order to find the cause of it. It is used to correlate the symptoms to the disease a person may be suffering from. A diagnosis can consist of examinations of one's medical history or a physical examination, or even a diagnostic test. Since, a particular symptom can occur in various circumstances, correlating it with a particular disease is quite challenging. Thus, several possible options need to be evaluated in order to correctly diagnose a disease. This is known as differential diagnosis. In differential diagnosis, many possible diseases are identified based on a set of symptoms and this is then followed by a process of elimination, until a point is reached where only one option seems viable.

However, according to a report by the National Academies of Sciences, Engineering, and Medicine (NASEM) published in 2015, majority of people will be a part of at least one diagnostic error in their life [12]. Incorrect or errors in diagnosis can be harmless in many but lethal in some cases. Incorrect diagnosis of a rare disease may also lead to death, or in other case, prolonged ignorance of a disease can turn a harmless disease to a life-threatening disease as well. Some diagnosis errors can result from human errors. However, machine learning can be very helpful in such cases. It can assist, if not replace human intervention in disease diagnosis.

For instance, Quest Diagnostics' partnership with IBM led to the birth of IBM Watson Genomics which is the first such device to introduce machine learning in diagnosis. It can study the unique genetic differences of tumor biopsies in order to suggest a diagnosis.

### 3.3 Behavior Modification or Treatment

Predictive analysis can be used to treat a patient's individual health by pairing it with his or her medical data. This form of machine learning is known as supervised learning. In supervised learning, a supplied input is mapped to an output by inferring its training data. This labeled training data helps it to categorize the input to one of the categories. IBM Watson for Oncology is one such example that uses a patient's medical history in order to devise a personalized cancer treatment for them. However, one obvious problem in this application is the availability of abundant relevant medical data. A patient's medical data is not readily available, and even if available, there is always a question on its reliability. One probable solution to this can be the increased use of tracking devices and micro biosensors. Increased use of such devices will allow for interrupted collection of data. Many such devices already exist. Some smartphone manufacturers provide these trackers inbuilt in their phones. This integration of tracking devices with mobile phones and applications will pave way for more reliable tracking, measurement and monitoring ability. Some startups are already working on this application:

**SkinVision**: It is a mobile application that allows you to track your moles and analyze them for possible cancer symptoms. They also claim to be the first and only CE certified online assessment.
**Somatix**: Somatix is a New York-based startup that tracks a person's gestures in real time for health enhancement.

### 3.4 Clinical Trial Research

Clinical trial research is defined as the studies performed on individuals in order to test a new medical technology. In other words, it is used to make sure a new drug, treatment or a medical device is effective and does not have any harmful side effects than an already existing treatment. Generally, there are four phases in a clinical trial research experiment. The first phase involves testing on a small group to judge its safety. The second phase deals with judging its effectiveness. The third phase is related to finding more about its safety, and the last phase is testing on a diverse population. Machine learning can be introduced in such trial researches to monitor and track data from the participants remotely for better safety. For example, a person's mental and biological signs can be remotely tracked and analyzed for signals of any harm to the participant [13].

### 3.5 Smart Electronic Health Records

Electronic health records are medical data of patients stored in a digital format. This data spans across various different parameters such as age, gender, weight, test results, medical history, immunization status, and bills. Earlier these data were stored on paper, but with the advancement of technology, they are now stored digitally for better security and management. Machine learning can be used for document as well as query classification. Algorithms such as support vector machines (SVMs) and naïve Bayes classifier can be helpful in better classification of data and improved retrieval systems. Optical character recognition (OCR) can also help in information retrieval.

Google's Cloud Vision provides both the ability to train models and pre-trained models for better analysis of images. Using machine learning, it is able to provide various features such as classification of images, text extraction, powerful web image searches, content modification and moderation.

### *3.6 Epidemic Outbreak Prediction*

A sudden outbreak of communicable diseases among a large group of people is known as an epidemic. Epidemics can occur via airborne transmission or biological transmission, or even fecal-oral transmission. Epidemics have existed since the beginning of humanity. The first recorded epidemic occurred in 429 BC in Greece when approximately 80,000 people died due to typhus. Some of the worst epidemics include the influenza epidemic that broke out during 1918–1920 that had an estimate death toll of 75 million. Another such epidemic is the ongoing HIV/AIDS epidemic that commenced in the Congo basin and spread worldwide. It has an estimated death toll of 30 million. Although the death toll as well as the number of epidemics are gradually reducing, it wouldn't be such a bad thing to be prepared for one than face the consequences later [14].

Machine learning can help play a major role in this. Machine learning algorithms can be applied to analyzing, monitoring and speculating possible epidemic outbreaks around the globe. This will be possible by collecting relevant data from real-time social media posts, data from satellites, historical data, anonymous medical data, etc. Such a concept is not foreign in the health industry. Studies and experiments have been conducted before to test machine learning algorithms like SVM and artificial neural networks (ANNs) to predict outbreaks. One such study was the malaria outbreak prediction model studied on ANN and SVM. The dataset consisted from 2011 to 2014 of 35 districts in Maharashtra, India. The research took under consideration various natural factors such as humidity, average monthly rainfall, air quality, and temperature. The research concluded that the SVM model was able to predict a possible outbreak at least 15–20 days in advance [15].

Another way to use machine learning algorithms is to use sentiment analysis on real-time social media posts. In paper [16], a system was proposed for detecting tweets related to influenza. This was implemented using SVM classifier. Based on a dataset of approximately 5000 tweets from November 2008, there were approximately 40% negative tweets containing the term "influenza" with a correlation of 0.89. Thus, tweets are a good resource to utilize for such purposes.

In another such experiment with tweets [17], a weekly analysis of Twitter data was done and compared with a week before in order to detect dengue epidemic. Since not all tweets containing the terms related to the disease suggest an epidemic, the tweets

**Table 1** Summary of all types of epidemics, algorithms and datasets discussed

| Sr. no. | Types of epidemics | Algorithms | Dataset used |
|---|---|---|---|
| 1 | Malaria outbreak prediction | Artificial neural networks and support vector machines | Data of 35 districts of Maharashtra, India from 2011 to 2014 |
| 2 | Influenza surveillance | Support vector machine | Twitter |
| 3 | Dengue detection | Spatiotemporal analysis | Twitter |
| 4 | Flu detection | Automatic regression | Twitter |
| 5 | H1N1 surveillance | Decision tree, support vector machines, Naïve Bayes, random Forest | Twitter and News agencies |

were divided into five categories: Marketing tweets, personal opinion, resource, personal experience and ironic/sarcastic tweets. The training set was assembled by selective sampling. To detect the epidemic, spatiotemporal analysis was done.

In research paper [18], tweets related to flu were analyzed. The research found a high correlation between the number of cases of flu and the tweets posted between 2009 and 2010. It used an automatic regressive model to predict the cases based on the Twitter data.

In paper [19], feed from Twitter, other social media posts as well as data from broadcasting news agencies were used to create the training set for an H1N1 surveillance system. It compared various machine learning algorithms like decision tree, support vector machines, Naive Bayes and random forest. Although SVM outperformed the other algorithms, it was proven that machine learning algorithms are a good way to predict potential epidemic outbreaks [20] (Table 1).

## 4 Conclusion

Machine learning algorithms when applied in the health care industry help us in several ways. Not only do they help in better prevention, diagnosis and treatment of a disease but are at the same time more efficient and faster than the current state-of-the-art technologies. They have been proven not just theoretically but also practically. Thus, such an integration should be encouraged for the betterment of humanity.

## References

1. Health at a Glance. OECD Indicators, OECD (2013)
2. Kanungo, T., Mount, D.M., Netanyahu, N.S., Piatko, C.D., Silverman, R., Wu, A.Y.: An efficient k-means clustering algorithm: analysis and implementation. IEEE Trans. Pattern Anal. Mach. Intell. **24**(7) (2002)

3. Yang, Y., Li, J., Yang, Y.: The research of the fast SVM classifier method. In: 12th International Computer Conference on Wavelet Active Media Technology and Information Processing (ICCWAMTIP), Chengdu, pp. 121–124. https://doi.org/10.1109/iccwamtip.2015.7493959 (2015)
4. Rish, I.: An empirical study of the naive Bayes classifier (2001)
5. Maind, S.B., Wankar, P.: Research paper on basic of artificial neural network. Int. J. Recent Innov. Trends Comput. Commun. IJRITCC **2**(1), 96–100 (2014). ISSN 2321-8169
6. Gupta, B., Rawat, A., Jain, A., Arora, A., Dhami, N.: Analysis of various decision tree algorithms for classification in data mining. Int. J. Comput. Appl. (0975 – 8887) **163**(8) (2017)
7. Sennaar, K.: Machine Learning for Medical Diagnostics—4 Current Applications, from https://emerj.com/ai-sector-overviews/machine-learning-medical-diagnostics-4-current-applications/ (2018)
8. Cattell, J., Chilukuri. S., Levy, M.: from http://www.mckinsey.com/industries/pharmaceuticals-and-medical-products/our-insights/how-big-data-can-revolutionize-pharmaceutical-r-and-d
9. Dhawan, A.P., D'Alessandro, B., Fu, X.: Optical imaging modalities for biomedical applications. IEEE Rev. Biomed. Eng. **3**, 69–92 (2010). https://doi.org/10.1109/RBME.2010.2081975
10. Havaeia, M., Davy, A., Warde-Farley, D., Biard, A., Courville, A., Bengio, Y., Pal, C., Jodoin, P., Larochelle, H.: Brain tumor segmentation with deep neural networks. Med. Image Anal. **35**, 18–31 (2017)
11. Mohsen, H., El-Dahshan, E.A., El-Horbaty, E.M., Salem, A.M.: Classification using deep learning neural networks for brain tumors. Future Comput. Inform. J. **3**(1), 68–71 (2018)
12. National Academies of Sciences, Engineering, and Medicine: Improving Diagnosis in Health Care. The National Academies Press, Washington, DC (2015). https://doi.org/10.17226/21794
13. National Institute of Aging: Clinical Trials, from https://www.nia.nih.gov/health/clinical-trials (2017)
14. Wikipedia. List of epidemics: from https://en.wikipedia.org/wiki/List_of_epidemics (2019)
15. Sharma, V., Kumar, A., Panat, L., Karajkhede, G., Lele, A.: Malaria outbreak prediction model using machine learning. Int. J. Adv. Res. Comput. Eng. Technol. (IJARCET) **4**(12) (2015). ISSN 2278-1323
16. Eiji, A., Maskawa, S., Morita, M.: Twitter catches the flu: detecting influenza epidemics using Twitter. In: Proceeding of the Conference on Empirical Methods in Natural Langauge Processing, Association for Computational Lingusitics (2011)
17. Gomide, J., et al.: Dengue surveillance based on computational model of spatio-temporal locality of Twitter. In: Proceedings of the 3rd International Web Science Conference (2011)
18. Achrekar, H., et al.: Predicting flu trends using Twitter data. In: 2011 IEEE Conference on Computer Communications Workshop (INFOCOM WORKSHOPS). IEEE (2011)
19. Kumar, J.V., Kumar, S.: An effective approach to track levels of Influenza-A [H1N1] pandemic in India using Twitter. Procedia Comput. Sci. **70**(2015), 801–807 (2015)
20. Singh, R., Singh, R., Bhatia, A.: Sentiment analysis using Machine Learning technique to predict outbreaks and epidemics. Int. J. Adv. Sci. Res. **3**(2), 19–24 (2018). ISSN 2455-4227

# A Review of IoT Techniques and Devices: Smart Agriculture Perspective

**Deep Rani and Nagesh Kumar**

**Abstract** Internet of things (IoT) is the hot point in the Internet field. The concepts help to intercommunicate physical objects furnished with sensing, actuating, computing power and hence connect to Internet. With the help of sensor, actuators and embedded microcontrollers, the verdict of smart object is realized. Wherein these smart objects colligate data from the environment of development, process them, and take reasonable actions. Thus, the IoT may generate unbelievable benefits and helps human beings in living a smart and luxurious life. Due to the potential utilizations of Internet of things (IoT), it has ended up being an unmistakable subject of logical research. The significance and the utility of these advances are in sizzling exchange and research, yet on the field of agribusiness and ranger service, it is very less. In this paper, utilizations of IoT on farming and silviculture has been well perused and broke down; additionally, this paper briefly presented the innovation IoT, agribusiness IoT, rundown of some potential applications areas where IoT is exercisable in the horticulture part, advantages of IoT in farming, and displays a survey of some literature survey.

**Keywords** IoT · Smart agriculture · Sensors · Actuators · Intercommunication

## 1 Introduction

Agriculture is the procedure of germinating food or tillage of plants and other organisms for genesis of food, fiber. It comprehends irrigation, fertilizers, organic farming, and harvesting. Defiance is measuring the capability of producing crops in a fixed land from year to year, understanding the change in the silt structure, optimize water use, and improve plant health. By the use of IoT, the humanity has been benefiting from the automation. IoT is a beyond liberal of promising dynasty of technologies

D. Rani (✉) · N. Kumar
School of Electrical and Computer Science, Shoolini University of Biotechnology and Management Sciences, Solan, Himachal Pradesh, India
e-mail: akaurdp@gmail.com

N. Kumar
e-mail: nageshkumar@shooliniuniversity.com

 
P. K. Singh et al. (eds.), *Proceedings of ICRIC 2019*, Lecture Notes in Electrical Engineering 597, https://doi.org/10.1007/978-3-030-29407-6_10

which is competent of offering solutions near to the modernization of agriculture. All the research institutions and scientific groups, as well as the industries, are in a race trying to exuberant IoT products to the agricultural profession stakeholders, and, in conclusion, lay the subtraction to have a apparent preface when IoT takes place a mainstream technology. In India, the people are servicing the analogical with agriculture industry [1]. Agriculture is the main constituent of the Indian economy. The advancement of agricultural industry is good for the entire nation, and this can help to improve economy progressively. Indian farmers use conventional methods for farming. The use of IOT in farming will help enhancement the yield as well as fruitful for the farmers [2].

## *1.1 Internet of Things (IoT)*

It is a propensity to utilize information technology to headship the evolution of recent agriculture. According to food and agriculture organization. The population of India increase day by day so the world will exigency to prolific 70% more food in 2050 to forage the growing inhabitant of the earth. To interfuse this demand, farmers and agriculture companies are diverting them from simple farming's to the IoT for analytics and greater production capabilities. The IoT enlightens to the expeditiously growing network of confined objects that are capacitated to collect and exchange data using embedded sensor. At the same moment, cloud computing, which is so temporal, and fog computing endow adequate resources and solutions to strut, store, and decompose the enormous amounts of data germinated by IoT devices. The way of managing and analysis of IoT data ("Big Data") might be used in auto-mode processes, forecast circumstances, and renovate many activities, even in real-time. Additionally, the belief of interoperability amongst heterogeneous gadgets motivates the conformation of the appropriate tools, by which new utilities and services can incite and endue an added value to the data stream which is produced at the strand of the network.

The agricultural sector was beyond measure affected by Wireless Sensor Network (WSN) technologies and is anticipatory to be uniformly benefited by the IoT. The IoT is situated to flicking the forthcoming generation of farming to the next level. IoT technology constituted a variety of technologies such as sensors, automation, telecommunications, computers, agriculture machinery, agriculture products, and plant science. The IoT is the torrid point in the Internet field. The conviction help to appertain physical objects furnished with sensing, actuating, computing power and hereby lends them the potential to conjunction on an act work in unison remnant related to the Internet termed as the "IoT (Internet of things)." By the use of sensor, actuators and embedded microcontrollers, the conviction of smart object is realized. In this paper, applications of IoT on agriculture has been studied and explored, also this paper succinct introduced the technology IoT, agriculture IoT, list of some feasible applications domains where IoT is exercisable in the agriculture sector, profits of Iot agriculture, and expresses the reviews of some literature [3].

## 2 Applications of IoT in Agriculture

### 2.1 *Sensor Technology*

A variety of electronic sensors are used in agricultural production in farms to detect various activities such as soil moisture sensors, water-level sensors, instrument used for sampling the state of the troposphere at a given time meteorological sensors (monitors the current state of atmosphere), heavy metal detection sensors for lead mercury biosensors (detection of an Analyte), and gas sensors (detects presence of gas) [4].

### 2.2 *Soil Moister and Water-Level Sensor*

This sensor is used to measure the water content in soil. Soil moisture sensor measures the moisture in different environmental factors like as soil type and temperature, basically used in agriculture purpose [5]. This sensor has a sensing probe element, which senses the surface level of nearly any fluid, which includes water, saltwater, and oils. The main advantage of the sensor is it does not get corrode easily [6] (Figs. 1 and 2).

### 2.3 *RFID Technology*

European Union was first who introduced RFID technology in agriculture sector in the late 1990s and shortly thereafter many other countries introduced it, the Australian government was the most passionate in implementing RFID technology can be a dominant tool for obtaining food safety and managing agriculture and livestock

**Fig. 1** Soil moisture sensor [4]

**Fig. 2** Water-level sensor [4]

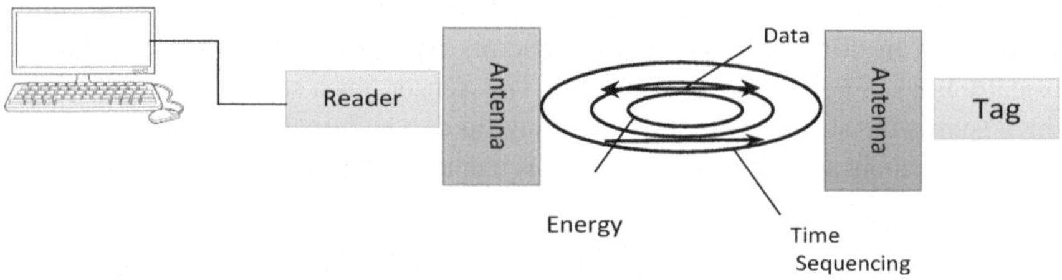


**Fig. 3** Frequency identification system a basic overview [7]

[7]. RFID is extensively used in animal tracking and identification. RFID is extensively used in animal tracking and identification in the farms and other agro-based industries like milking and meat production. It helps to achieve intelligent monitoring, recognizing, traceability of animals, and their management [8, 9] (Fig. 3).

RFID works a follows: tags enter the magnetic field; get signals discharged by the reader; send the item data put away in the chip through the actuated flows gotten by the vitality (Passive Tag), or send a frequency signal actively (Active Tag); reader read the data and decoded; send the data to the focal and send the information to be handled on central system. The fundamental model of radio frequency distinguishing proof framework is shown as in the figure.

## *2.4 Radio Transmission Technology in Agriculture*

Zigbee is superlative prevailing industry wireless mesh technology yardstick for blending sensors, instrumentation and control system. Zigbee, a specification for communication in a wireless personal area network, has been known as the "IoT."

**Fig. 4** Zigbee [4]

Zigbee is an open global. Packed-based protocol contemplated to provide an easy-to-use architecture for secure, reliable. Self-organizing wireless data transmission can be achieved with Zigbee wireless sensor networks. In large-scale farming, it has been widely used for data transmission [6] (Fig. 4).

## 2.5 *Intelligent Irrigation Technology*

Based on satellite positioning network and "shallow wells underground cables and field or automatic irrigation system pipe" technology, it can accumulate irrigation water, irrigation, electricity, and time data to accomplish automation of farmland irrigation and through an absolute dissection of information technology software to monitor irrigation.

## 2.6 *Technical Quality Safety of Agricultural Products*

Supply chain visibility and efficiency is a giant part of running a advantageous agricultural business. In the agricultural industrial chain (production–circulation–sales), recording and monitoring of the chain can understand the entire procedure of regulation (Fig. 5).

Linking diverse facet of the supply chain using IoT-based solutions can have a huge influence on how things are finished in agriculture. Following are some of the ways through which the agriculture supply chain will change.

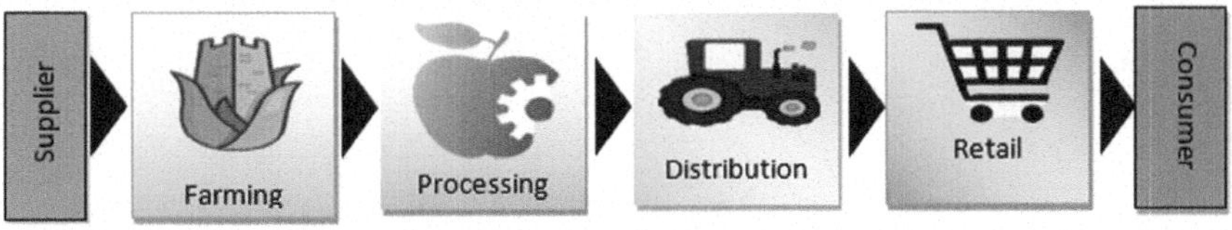


**Fig. 5** Steps for agriculture supply chain [10]

**Incident Management**. Unpredictable supply shift like procrastinations or accidents will be soon recognized by the use of real-time sensors or stave off altogether depending on predictive models. Regulation: Strict food quality, protection, and sustainability needs will be met in a more easily and documented with granular, end-to-end data.

**Transportation**. End customers will perceive where their food proceeds from fastigial information if required via a more transparent supply chain.

**Consumer Transparency**. Optimal pathway can be determined using traffic, weather, and other environmental factors data to agglomeration fuel and detract redundancies.

**Packaging**. Dynamic best prior dates and soon caveat of food incidents based on present and historical conditions.

**Inventory.** Fewer non-availability occurrences in warehouses with dynamic inventory monitoring.

**Process**. Ultimately, most facet of the supply chain will be in auto-mode, so goods can be administered in real-time rooted on supply and demand [7].

## *2.7 Precision Seeding and Spraying Techniques*

The evolution and implementation of accuracy agriculture or site-specific farming have been made eventual by modulating the Global Positioning System (GPS) and geographic information systems (GIS). Such type of technologies authorizes the coupling of real-time data accumulation with induration position information, salient to the accomplished tempering and abstraction of spacious amounts of geospatial data. GPS-based utilities in exactitude farming are being employed for farm preparative, mapping of field, sampling of soil, guidance of tractor, scouting of crop, variable-rate consumptions, and mapping of yield. Such type of modern technologies like GPS allows farmers to work throughout low visibility field conditions such as heavy rain, heavy dust, light or heavy fog, and darkness. Depending on the technology combined with Global Positioning System (GPS) navigation technology, seeding technology, and fertilization at a variable rate, it can cognizable identical implementation of the spraying, planting, and refining the consumption of pesticides, seeds, and so on [10].

# 3 Benefits of IoT in Agriculture

## *3.1 Data Collected by Smart Agriculture Sensors*

IoT sensors can collect data about weather circumstance, quality of soil, crop's growth headway, or health of cattle. These types of data can be practiced to examine the state of your business in general, as well as staff execution, equipment competence, etc.

## *3.2 Superior Control Over the Internal Processes and, as a Result, Scarce Production Risks*

The caliber to foresee the yield of our production allows us to scheme for superior product distribution. If farmers know precisely how much crops they are going to harvest, they can make sure that the end product will not lie around unsold.

## *3.3 Cost Management and Waste Shortage Felicitation to the Enhanced Control Over Production*

Being capacitate to see any discrepancy in the crop evolution or livestock health, we will be capacitating to slacken the danger of detriment your yield.

## *3.4 Increased Business Competence Through Process Automation*

Using these types of smart devices, we can automate several processes across the production cycle.

## *3.5 Irrigation, Fertilizing, or Pest Control. Enhanced Product Quality and Volumes*

It achieves better control upon the production process and sustains higher standards of crop quality and growth capacity through automation [11].

## 4 Discussions

Fortunately, advancing of technologies, especially the combination of the Internet of things and precise agriculture, is paving the road for reaching the goal. Human now is on the extremity of the second green revolution, which is largely built on the IoT and related technologies. IoT-based productivity assistants promise to structure the farm of the future more productive and deficient with less labor work needed. It is established in the utilization of data to form more efficient and effective farming practices and drive-associated environmental and social benefits. With the help of internet of things and associated potentially farmers to treat crop and animals more precisely. The future implications of data collected through these technologies also allow farmers to make much more strategic and effective decisions to increase creativeness with fewer environmental impacts [12].

Issues in regards to horticulture, agriculture, rural zone, and farmers have been continually deflecting India's development. Horticultural modernization is the main answer for these three issues. All things considered, India's horticulture is far from modernization. The utilization of IoT in rural modernization will conceivably take care of the issues. In light of the highlights of IoT and distributed computing, cloud services, Service Oriented Architecture (SOA) and perception innovations can create gigantic information engaged with horticultural generation [5].

RFID with IoT innovations can manufacture a plant processing plant that can control agrarian creation consequently. Ideal utilization of present-day innovation and IoT and mix of them can animate the fast improvement in the modernization of agriculture framework. Utilization of savvy IoT in agribusiness could viably comprehend the issues concerning farmers, horticulture, and rustic territory. As indicated by the above investigation, data innovation faculty and agriculture researcher ought to be exchange of ideas to trade thoughts. Particularly, those personals comprehend planting and comprehend IT can improve and advance the modernization of cultivation and farming. Modernization of cultivating can improve farming creation and the board, the objective of ecological assurance, and energy consumption; environmental protection could be accomplished. By utilizing IoT in horticultural, farmers would almost certainly comprehend the present decision of rural soil, they would most likely realize which crops are proper for cultivating in the present stage, other ecological data of farmland, through intelligent analysis on the agriculture data and better administration. Meanwhile, the accompanying situation could be seen: Instead of drudging the yield in hot water, farmers would almost certainly control on PCs like a cell phone or on some monitoring devices apparatuses, to comprehend watering, developing, seeding, and harvesting. The proceeded and quickly improvement of microelectronic innovation, organize innovation is an open door for experts to effectively investigate the innovation in agriculture [3].

In this research paper, authors have analyzed that to change the agriculture field as automation by the use of Wireless Sensor Networks. A system is evolved, which is outgo to monitor the quality of soil, selection of crop, and evolution of the land

and then intimates the farmers about their field through IOT by the help of different sensors interfaced with microcontrollers. By using this channel, authors have enhanced the creativeness of food and wealth of the farmers. Also, this paper enucleated landslides through clinometers and rainfall sensor in the mountain region and caveats the people to spare their lives from calamity by sending messages to concern area's higher authorities. Through this paper have made an effort to minimize the manual work of the farmers and make them a smart farmer. Authors have executed a framework that will accumulate information from different electronic sensors, and utilizing this sensor information for farmers will almost certainly control the tasks in the agriculture fields. The control will be finished by remotely from anyplace and whenever. This type of mode will give management given to the ranchers to digitalizing agribusiness. The future conceivable work on this framework may incorporate bringing together the information and giving administrations to each harvest individually [13].

In this article, aspects related to WSN were approached, as well as the needed technologies for their implementation and simulation. A simulation type of environment which enables its users to look over and illustrate the data gathered by sensors was developed. These data are produced by events procreated by the interaction between the users and the application. Hence, upon, it becomes perceptible that TinyOS, TOSSIM, Python, and Pygmy technologies procure the robust and veteran development of many WSN applications. Distinct to the soccer field irrigation case, the objective is to cognize areas with low humidity levels through data captured by sensors that are geographically spread and irrigate only those areas, by saving resources and avoiding water wastes, when applied to a real situation. As future works, authors have the intention to renovate the environment with enhancements for WSN application development as, for example, enable the instantiation of new network sensors through the interface, as well as permit the structuring of such sensors with the use of drag and drop tools and assigning their configurations by using these components properties [13].

In this paper, author develops a functional prototype of an agriculture monitoring system. The final product is a device that can be placed in an agriculture land in different places reading the water level, soil moisture, pH of the soil, and the temperature of the surrounding; the read data is sent to the main control; this data can be read by the user through a Web page, and the water pump can be controlled using the switch given on the same Web page [6].

The valuables monitoring system based on RFID technology, using UHF frequency band in the range between 300 MHz (megahertz) and 3 GHz (gigahertz), which travels almost entirely by line-of-sight propagation (LOS) and ground reflection, supports accuracy identification and multi-passive tags to meet the monitoring requirements completely, and hence accomplish the real time and accuracy. The purpose of applying this system is to monitor the warehouse agricultural material products in the real life to know the actual and accurate details so that a corrective and well-planned steps will be taken by considering all parameters also which gives us the correct recognition rate up to 99.5%, which is our primarily demand. The system follows the most enhanced international standards and can be widely applied in the future [14].

## 5 Conclusion

Issues in regards to horticulture, country region, and ranchers have been continually dissuading India's development. Agriculture modernization is the novel correction to these three issues. In any case, India's horticulture and agriculture are far from modernization. The utilization of IoT in farming modernization will conceivably take care of the issues. In view of the highlights of IoT and distributed and scaled cloud computing, cloud administration services, Service Oriented Architecture (SOA), and representation advancements can create immense amount of information engaged with agriculture and horticulture. RFID with IoT advances can help in establishment of a plant manufacturing unit that can control the development of plant naturally. Ideal utilization of current innovation and IoT and mix of them can animate the improvement in the modernization of rural framework and the given area. Utilization of keen IoT in horticulture could efficaciously explain the issues concerning ranchers, farming, and country territory. As per the above examination, data innovation workforce and farming researcher ought to be urged to exchange of ideas. Particularly, those personals comprehend planting and comprehend IT can develop and advance the modernization of cultivating and farming also. Modernization of cultivating can improve farming creation and the executives; the objective of ecological security and vitality sparing could be accomplished. By utilizing IoT in agriculture, farmers would most likely comprehend the present decision of horticultural soil; they would almost certainly realize which crops are suitable for cultivating in the present stage, other natural data of farmland, through keen investigation and better management of cultivated crops. Meanwhile, the accompanying situation could be seen: Instead of working the field in hot water, farmers would almost certainly control on personal computer, a cell phone, or on some other electronic monitoring devices, to comprehend watering, developing, seeding, and procuring, at that point they can, without much of a stretch completion overwhelming, homestead work. The advancement of microelectronic innovation, organize innovation is an open door for experts to effectively investigate the mechanical elaboration of present-day agribusiness. Utilization of the IoT is assuming critical job in the movement of the world's cutting edge and horticulture, which sets an establishment for agriculture industrial improvement.

## References

1. Swapnal, B., Manivannan S.: Analysis: smart agriculture and landslides monitoring system using internet of things (IoT). Int. J. Pure Appl. Math. 118(24), 1–7 (2018)

2. Aher, A., Kasar, J., Ahuj, P., Jadhav, V.: Smart agriculture using clustering and IOT. Int Res J Eng Technol **5**(3), 4065–4068 (2018)
3. Lakhwani, K., Gianey, H., Agarwal N., Gupta S.: Development of IoT for smart agriculture a review. In: Emerging Trends in Expert Applications and Security, pp. 425–432. Springer, Singapore (2019)
4. Lakshmisudha, K., Hegde, S., Kale, N., Iyer, S.: Smart precision based agriculture using sensors. Int. J. Comput. Appl. 146(11) (2011)
5. VimalJerald, A., Rabara, S.A., Bai, T.D.P.: Internet of things (IoT) based smart environment integrating various business applications. Int. J. Comput. Appl. **128**, 32–37 (2015)
6. Jagadesh, M., Rajamanickam, S., Saran, S.P., Sai, S.S., Suresh, M.: Wireless sensor network based agricultural monitoring system. Int. J. Creat. Res. Thoughts **6**(1), 502–509 (2018)
7. Abbasi, A.Z., Islam, N., Shaikh, Z.A.: A review of wireless sensors and networks' applications in agriculture. Comput. Stand. Interface **36**(2), 263–270 (2014)
8. Ruiz-Garcia, L., Lunadei, L.: The role of RFID in agriculture: applications, limitations and challenges. Comput. Electr. Agric. **79**(1), 42–50 (2011)
9. Du, J., Li, P., Gao, W., Wang, D., Wang, Q., Zhu, Y.: An application of rfid in monitoring agricultural material products. In: International Conference on Computer and Computing Technologies in Agriculture, pp. 254–259. Springer, Berlin, Heidelberg (2009, October)
10. Balafoutis, A., Beck, B., Fountas, S., Vangeyte, J., Wal, T., Soto, I., Eory, V.: Precision agriculture technologies positively contributing to GHG emissions mitigation, farm productivity and economics. Sustainability **9**(8), 1339 (2017)
11. Shao, C., Roh, H., Lee, W.: Next-generation RF-powered networks for Internet of Things: architecture and research perspectives. J. Netw. Comput. Appl. **123**, 23–31 (2018)
12. Hassan, Q.F.: Internet of Things A to Z: technologies and applications. Wiley (2018)
13. de Lima, G.H.E.L., e Silva, L.C., Neto, P.F.R.: WSN as a tool for supporting agriculture in the precision irrigation. In: Sixth International Conference on Networking and Services, pp. 137–142. IEEE (2010)
14. Ping, H., Wang, J., Ma, Z., Du, Y.: Mini-review of application of IoT technology in monitoring agricultural products quality and safety. Int. J. Agric. Biol. Eng. **11**(5), 35–45 (2018)

# A Review of Scheduling Algorithms in Hadoop

**Anil Sharma and Gurwinder Singh**

**Abstract** In this epoch of data surge, big data is one of the significant areas of research being widely pondered over by computer science research community, and Hadoop is the broadly used tool to store and process it. Hadoop is fabricated to work effectively for the clusters having homogeneous environment but when the cluster environment is heterogeneous then its performance decreases which result in various challenges surfacing in the areas like query execution time, data movement cost, selection of best Cluster and Racks for data placement, preserving privacy, load distribution: imbalance in input splits, computations, partition sizes and heterogeneous hardware, and scheduling. The epicenter of Hadoop is scheduling and all incoming jobs are multiplexed on existing resources by the schedulers. Enhancing the performance of schedulers in Hadoop is very vigorous. Keeping this idea in mind as inspiration, this paper introduces the concept of big data, market share of popular vendors for big data, various tools in Hadoop ecosystem and emphasizing to study various scheduling algorithms for MapReduce model in Hadoop and make a comparison based on varied parameters.

**Keywords** Big data · Hadoop · TaskTracker · JobTracker · Scheduling · MapReduce

## 1 Introduction

Nowadays large amount of data is generating at a rapid-fire speed and in vast and diverse formats which are non-manageable by traditional databases. Hence, the term 'Big Data' is coined which made it promising to store, process, access and analyze this enormous measure of data. 'Big Data' does not only mean only large quantity of data, it also includes the pace at which data is being generated, transferred, organized and re-organized. It also backs diversified formats of data like text, audio, video, images,

A. Sharma · G. Singh (✉)
School of Computer Applications, Lovely Professional University, Punjab, India
e-mail: gurwinder.11@gmail.com

A. Sharma
e-mail: anil.1956@lpu.co.in



P. K. Singh et al. (eds.), *Proceedings of ICRIC 2019*, Lecture Notes in Electrical Engineering 597, https://doi.org/10.1007/978-3-030-29407-6_11

machine logs, etc. Factors like volume, velocity and variety make it radically different and divergent from conventional data. Furthermore, the word 'Big Data' is somewhat fresh and novel in field of IT and business industry. In the recent literature, various experts and researchers have used the term and highlighted its use in a huge amount of scientific data for visualization [1]. A plethora of definitions of 'Big Data' is extant. Thus, big data is "the amount of data just beyond technology's capability to store, manage, and process efficiently" [2]. 'Big Data' is primarily characterized by three Vs: volume, variety and velocity [3, 4]. These terms were initially used by Gartner to define big data features. But some researchers do believe that big data should not be categorized by only three Vs given by Gartner but one more V, i.e., value should also be added in its definition [5, 6].

Study done by McKinsey [2] revealed that big data can enhance productivity, upsurge competitive advantage and upturn economic surplus for consumers. As per a white paper of International Data Corporation [7], data generation by year 2025 is all set to reach the level of 163 zettabytes as depicted in Fig. 1.

Various IT giants like YouTube, Facebook, Twitter, Google, Instagram, Flickr, LinkedIn and Amazon are playing a hyperactive role in evolution of unstructured data. The categories of computing resources accountable for creation of data are Core, Edge and Endpoints [7]. 'Core' consists of datacenters of computing for cloud and enterprises. 'Edge' is name comprises to those devices which are not typical datacenters, i.e., local datacenters created for fast responding time, server rooms and field servers. 'Endpoint' refers devices like cameras, sensors, computers, etc.

Among the vendors of big data, according to Wikibon [8], as depicted in Fig. 2 are: IBM, HP, Teradata, CSC, Dell, EMC, Hitachi, Atos S. A., Huawei, Siemens, Xerox,

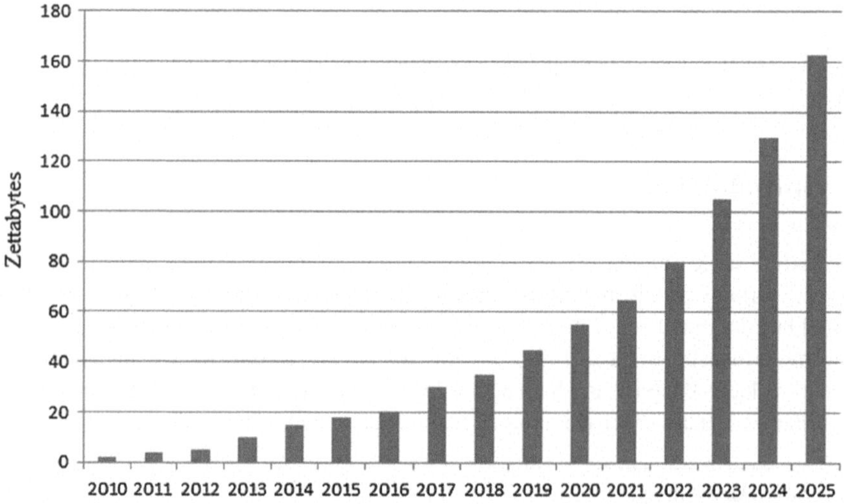


**Fig. 1** Projection big data 2010–2025, IDC's data age 2025 study [7]

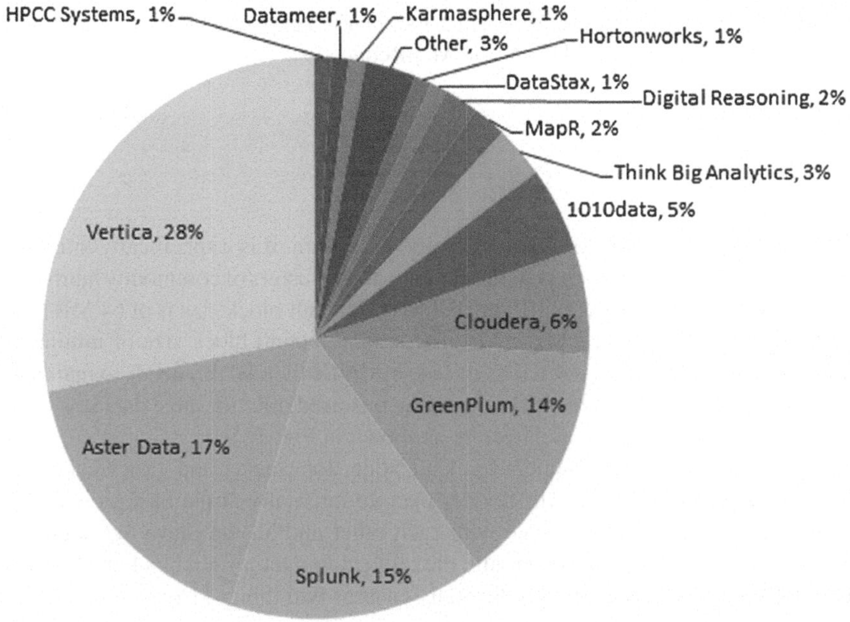


**Fig. 2** Market shares of big data vendors [8]

Tata Consultancy Services, Seagate, SGI, Logica, Intel, Microsoft, Splunk, MarkLogic, Cloudera, Red Hat, Accenture, Informatica, SAS Institute, Amazon Web Services, ClickFox, Super Micro, SAP, Think Big Analytics, MapR, Digital Reasoning, Oracle, Pervasive Software, Hortonworks, DataStax, Capgemini, Attivio, QlikTech, HPCC Systems, Datameer, Fujitsu, Karmasphere, Tableau Software and NetApp.

The major challenges in the sphere of big data are: data representation, life cycle management of data, data compression, analytical mechanism, privacy and confidentiality of data, reduce redundancy and scalability [6]. There is variety of tools available in market for big data like: Apache Hadoop, Pentaho Business Analytics, Karmasphere Studio and Analyst, Jaspersoft BI Suite, Tableau Desktop and Server, Skytree Server, and Splunk. Among these tools, the most prevalent tool for storing and processing of big data is Hadoop [8].

## 2 Hadoop

It is an open-source framework of Apache Software foundation used for storage and processing of large and huge data sets (however for small datasets it is not recommended) with clusters of commodity hardware. Hadoop is written in Java and was originally conceived and designed by Dough Cutting in year 2005 [9]. Hadoop Ecosystem contains various tools such as HDFS, HBase, MapReduce, HCatalog,

Pig, Mahout, Flume, Chukwa, Avro, Hive, Oozie, Zookeeper and Kafka. Two central components of Hadoop are: HDFS and MapReduce [9].

## 2.1 HDFS

The objective of HDFS is to store and preserve the data. It is a specifically intended file system utilized for storage gigantic data sets with clusters of commodity hardware with streaming access pattern [10]. In HDFS, the default block size is of 64 MB (can assign also 128 MB or 512 MB) [11]. The reason behind block size of minimum 64 MB is that if block size is 4 KB as in case of UNIX then HDFS needs to maintain more metadata for small size blocks, i.e., why it is used only for huge data sets [12]. HDFS follows master-slave architecture as shown in Fig. 3.

NameNode, Secondary NameNode, DataNode, JobTracker and TaskTracker are five services of HDFS [13]. All services operate internally at the background. All Master Services can communicate with each other and all the Slave Services can converse to each other [14]. By default, the replication factor for a block in HDFS is three. HDFS default replication policy only ensures two things [15]:

i. A DataNode cannot have more than one replica of any block.
ii. A rack cannot have more than two copies of same block.

All DataNodes send 'heartbeat message' and block report to NameNode for every short period of time to say that certain clients have stored some blocks in local DataNode and are still alive, processing and working properly. If any of the DataNode is not giving proper heartbeat in-time, then NameNode may think that DataNode has become dead and is removed from metadata and then some other DataNode is chosen to store the copy of file to maintain the default replication property of HDFS. DataNode which was declared dead may start storing the data freshly [14].

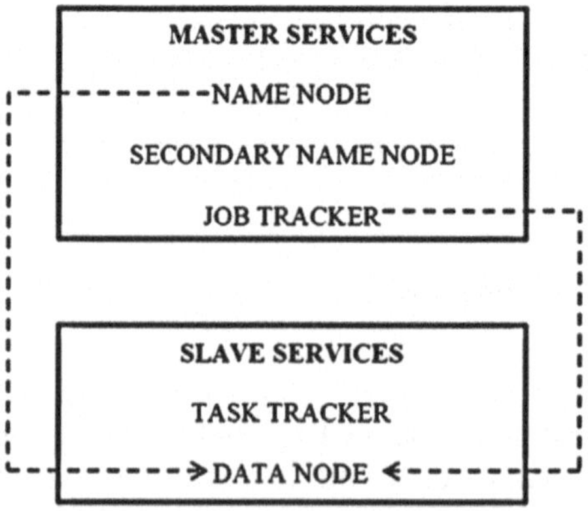


**Fig. 3** HDFS architecture

## 2.2 *MapReduce*

After storing data, the next thing is processing the data which is performed using MapReduce. It consists of two major components: JobTracker and TaskTracker. JobTracker act as master and TaskTracker acts as workers [16]. JobTracker does not know which data is placed on which DataNode because there is no one-to-one communication between JobTracker and DataNode. JobTracker requests NameNode which in turn goes through its metadata to get the locations [16, 17]. Client requests JobTracker to accept jobs which are in the form of MapReuce programs comprising mapper and reducer methods, and set of input/output files. Location of these input files is retrieved from the metadata of NameNode by JobTracker. TaskTrackers are assigned tasks by JobTracker, applying suitable scheduling algorithm. The TaskTrackers apply the designated operations encoded in reducer, on the data elements spilled from mapper and generates the outcomes in the form of key-value pairs to write it to HDFS [16].

# 3 Literature Review

Several regions of attempt have concerns with enormous information and more or less traditional business applications have confronted enormous information for quite a while, such as aircraft reservation frameworks, and more modern business applications to exploit massive information are under development (e.g., information-sharing hubs and groups of databases). Big data problem can be split into two individual issues: Big data objects and big data collections [1]. Scientists in computer field as well as biologists are struggling with more and more huge data sets these days [18].

Various problems that surface and required to be surmounted while processing big data includes timeliness, scalability, privacy, error handling, visualization and heterogeneous data and the same must be taken into consideration for highly effective and accomplished processing of big data. This paper also seeks to compare some of the Hadoop's components like Hive, HBase, Cassandra, MongoDB and Redis of Hadoop [19]. Public key-based extensions: Public Key cryptography for Initial Authentication in Kerberos (PKINIT), Public Key utilizing Tickets for Application servers (PKTAPP) and Public Key Cryptography for Cross-Realm Authentication in Kerberos (PKCROSS) in [20] provides extra support to public-key cryptography in Kerberos framework at various stages which results in enhanced and improved security and as well as scalability.

Comparison of performance of Hadoop clusters in homogeneous and heterogeneous environments was made and a new load balancing framework for MapReduce was for chalked out much more efficacious performance better in both environments. Furthermore, instead of bifurcating the tasks for participating nodes equally, as in conventional framework of MapReduce, MARLA ("MApReduce with adaptive Load balancing for heterogeneous and Load imbalAnced clusters") forms a large number of tasks conceived from input divisions, more noteworthy in number than the summation of its nodes. Thus, this load balancing technique permits the contributing nodes to themselves appeal the tasks as and when previous task is completed [21].

Load balancing scheme in [22] has taken advantage of Cloud Storage's grey prediction theory to predict the load rate for various replicas at some particular instance for a node to dynamically perform different operations. Yet another forsighted and pioneering dynamic strategy envisaged to balance the load in HDFS is by placing blocks on DataNodes considering metrics bandwidth of network and fault tolerance [23].

Information dispersal algorithm in [24] along with data placement strategy is used for encoding the files to the blocks and for storing the blocks in cluster to improve storage efficiency to enhance availability of data and for better network load both the storage load of DataNode and usage of network were combined. Data placement solution in [25] improves data transfer time in Hadoop by calculating the bandwidth among the DataNode and client periodically, and exhausting the DataNode having maximum bandwidth placing data blocks. By combining disk utilization and service blocking rate Model's new improved technique in [26] has been proposed to balance the load in HDFS as compared to load balancing method used in existing HDFS.

Workload imbalance of unbalanced node can be balanced by minimizing the data transfer time by computing the capacity of each node and balancing the data stored in each node dynamically for heterogeneous Hadoop environment [27]. Load balancing can be improved by heightening the execution of reduce task as referred to [28] by accomplishing all reduce tasks at same time based on performance of nodes by utilizing the historical information and assigning the tasks to the nodes as per performance, i.e., the input to nodes with poor execution is diminished. In [29], the authors introduced a new node called BalanceNode with which DataNodes having Heavy-load and light-load can be compared, to enable the light loaded nodes to share a portion of load from heavy loaded nodes, consequently to minimize the blocks movement cost. Based on system information, a load balancing method envisaged by Kim et al. [30] undertakes total available CPU cores and the memory size at disposal to identify the performance of each DataNode and implements the algorithm with sufficient and insufficient memory environment.

Triple-H hybrid design for heterogeneous storage architecture given by Islam et al. [31] hides the disk access cost for Read/Write operations, cutting down I/O bottlenecks, ensures the trustworthiness by using SSD-based staging for HDFS and efficient use of different types of devices on High Performance Computing Clusters. Multiple partitioning technique [32] resolves the problem of load imbalance at MapReduce stage, instigated by Hadoop's default partition algorithm, by refining the tasks and balancing the reducer input in the map phase to improve job scheduling and resource utilization.

To deal with problem of maintaining privacy while executing the sensitive and critical data on unsecured cluster [33] introduces a dynamic scheduling algorithm to balance the load and transfer data among Hadoop racks on the basis of log files without revealing the private information. Anonymization algorithm for probabilistic inference attack and similarity attack by Nayahi et al. [34] provides better and effective privacy and data utility on the basis of resilient and clustering. Data encryption scheme for HDFS given by Song et al. [35] for ARIA and Advanced Encryption Standard algorithms together on Hadoop provides splitting of blocks and data processing component for variable-length blocks to implement encryption and decryption in efficient way for various applications, such as sorting, word counting, k-Means and hierarchical clustering.

Dynamic Information as a Service architecture for Hadoop cluster by Tao et al. [36] using scheduling, monitoring, virtual machine management and virtual machine migration modules highly useful in providing load balancing which in turn enhances data locality, resource scheduling. For maximum utilization of the resources which are held by idle slots [37] held out BASE (Benefit Aware Speculative Execution) as resource stealing scheduling algorithm which improves execution time of job and reduces the speculative tasks which are not advantageous in MapReduce by stealing the resources from idle slots. Longest Approximate Time to End scheduling in [38] overcomes the weaknesses of both progress-rate-based and threshold-based scheduling by detecting straggler tasks early and improves the response time of Hadoop by factor of 2 on a cluster of 200 machines on Amazon's Elastic Compute Cloud.

Deadline Constraint algorithm envisaged in [39] for scheduling confirms execution of jobs for which deadlines can be encountered through a cost model for executing the job, but it does not pay any head to features like estimating filter ratio and run time for MapReduce task, distribution of data. Dynamic Priority Multiqueue algorithm [40] renders the tasks near to finish the job on a priority to adorn the response time for Hadoop jobs in MapReduce phase.

## 4 Findings

By default Hadoop attained three configurable scheduler policies: First Come First Serve (FCFS), Hadoop Fair Scheduler (HFS), and Capacity Scheduler policy. FCFS scheduler processes the jobs in accordance with their submission. Its major demerit is low resource utilization. It does not pave the way for reasonable sharing among users and furthermore deficit in response time for processing of minor jobs. Capacity Scheduler was envisioned and propounded by Yahoo to render the partaking of cluster possible among organizations. Therefore, to achieve unbiased sharing of cluster a minimum guaranteed capacity of the queues was set. Further, Facebook planned HFS to moderately segment the cluster among different applications and users. Consequently, it ensures an evenhanded segmentation of the capacity of cluster over the time [9, 15, 19].

Drawback in default scheduling algorithms of MapReduce is that these algorithms assume that the nodes and environment are homogeneous. These scheduling algorithms randomly select the DataNodes for processing and storage. The period of time for MapReduce jobs fluctuate from seconds to days [34].

Based on up to examined literature, authors comprehend that there are a glut of challenges which are being encountered by Hadoop. The major challenges in Hadoop are query execution time, data movement cost, selection of best cluster and racks for data placement, preserving privacy, overall load distribution to handle: Imbalance in input splits, computations, partition sizes and heterogeneous hardware, and scheduling. The average interval of map and reduce tasks is different for each task and job depending upon the available number of TaskTrackers. The JobTracker assigns TaskTracker a task which is nearby locality to the TaskTracker. Each application is taken as job in MapReduce framework and a lot of map and reduce tasks constitute a job. A variety of factors like number of running jobs, wait time, response time and run time plays an essential role while inducing the load in MapReduce. Hadoop scheduler makes use of queue data structure to assigns the tasks [16]. Table 1 shows comparison on features of default and some other scheduling algorithms.

**Table 1** Scheduling policies comparison

| S. No. | Scheduling Policy | Key Features | Challenges |
|---|---|---|---|
| 1 | First Come First Serve (FCFS) Scheduler [9, 15, 19, 39] | • Resources are allocated as per arrival<br>• Jobs are executed in the similar manner in which they are submitted<br>• High Throughput | • Data Locality and starvation are reduced<br>• Resources are not fairly distributed which leads to low resource utilization<br>• Non-preemptive<br>• Suitable to clusters having Homogeneous Environment |
| 2 | Capacity Scheduler by Yahoo [9, 15, 19, 39] | • It brings about fare distribution of resources among different users with minimum assured capacity<br>• High resource utilization and more rapidly response time | • Maximum capacity is need to set, to limit the number of users who can access the resources<br>• It does not make sure the fairness and stability for pending jobs in queue<br>• Performance decreases in Heterogeneous Environment |
| 3 | Hadoop Fair Scheduler (HFS) by Facebook [9, 15, 19, 39] | • Individual task is confirmed with a rational portion of the resource<br>• Provide reasonable share of the cluster capacity over time | • Does not consider the job weight for individual which results in unbalanced performance of nodes<br>• Restriction on number of jobs to be placed in pool<br>• Performance decreases in Heterogeneous Environment |

(continued)

**Table 1** (continued)

| S. No. | Scheduling Policy | Key Features | Challenges |
|---|---|---|---|
| 4 | Longest Approximate Time to End (LATE) [38] | • Ponders heterogeneity of cluster<br>• Flourishes in refining the data locality | • Slightly negotiates fairness and reliability<br>• Static scheduler |
| 5 | Delay Scheduler [40] | • Ponders data locality issue<br>• For execution of complex calculations no overhead is required | • If majority of tasks are considerably more than an average job, then it is not effective<br>• Not suitable for Heterogeneous environment<br>• No resource sharing<br>• Static scheduler |
| 6 | Deadline Constraint Scheduler [39, 41] | • Originate least count criteria for map and reduce task<br>• Specifies the deadlines to improve system utilization | • Not considered aspects runtime estimation for map and reduce task, filter ratio approximation, distribution of data and execution of more than one MapReduce tasks<br>• Identical nodes are required which leads to more cost<br>• Constraint about deadline for individual job is to be stated by user |

## 5 Conclusion

Because of tremendous measure of information around us, the current database tools face issues identified with colossal measure of information, speed of data, data sharing, scalability, efficiency, privacy and security of data. This paper demonstrates a review of various scheduling algorithms used in MapReduce phase of Hadoop. The default schedulers of Hadoop: FIFO, Capacity Scheduler and Fair Scheduler accept the cluster environment to be homogeneous and work viably in homogeneous condition. In case the cluster is heterogeneous, the execution of Hadoop is significantly cut down. Various scheduling algorithms have been envisioned but they do bargain on some attributes or metrics to improve one or more attribute. To meet the needs related to diverse objectives like improving data locality, resource utilization, amount of data processed, fairness and so forth, there is need to device scheduling algorithms for complex decision making.

## References


1. Cox, M., Ellsworth, D.: Managing big data for scientific visualization. ACM Siggraph. 97, 5.1–5.17 (1997)
2. Manyika, J., Chui, M., Brown, B., Bughin, J., Dobbs, R., Roxburgh, C., Byers, A.H.: Big Data : The Next Frontier for Innovation, Competition, and Productivity (2011)
3. Zikopoulos, P.C., DeRoos, D., Parasuraman, K., Deutsch, T., Corrigan, D., Giles, J.: Harness the Power of Big Data. The McGraw-Hill Companies (2013)
4. Berman, J.J.: Principles of Big Data : Preparing, Sharing, and Analyzing Complex Information. Morgan Kaufmann Elsevier (2013)
5. Gantz, J., Reinsel, D.: Extracting Value from Chaos (2011)
6. Chen, M., Mao, S., Liu, Y.: Big Data: A Survey. Mob Netw Appl **19**, 171–209 (2014)
7. Reinsel, D., Gantz, J., Rydning, J.: The Digitization of the World- From Edge to Core (2018)
8. Kelly, J., Vellante, D., Floyer, D.: Big Data Market Size and Vendor Revenues (2012)
9. White, T.: Hadoop: The Definitive Guide. O'Reilly Media (2015)
10. Saraladevi, B., Pazhaniraja, N., Paul, P.V., Basha, M.S.S., Dhavachelvan, P.: Big Data and Hadoop-A Study in Security Perspective. Procedia Comput. Sci. **50**, 596–601 (2015)
11. Ji, C., Li, Y., Qiu, W., Awada, U., Li, K.: Big data processing in cloud computing environments. In: 2012 International Symposium on Pervasive Systems, Algorithms and Networks. pp. 17–23. IEEE (2012)
12. Song, Y.: Storing Big Data—The Rise of the Storage Cloud (2012)
13. Ghazi, M.R., Gangodkar, D.: Hadoop, MapReduce and HDFS: a developers perspective. Procedia Comput. Sci. **48**, 45–50 (2015)
14. Shvachko, K., Kuang, H., Radia, S., Chansler, R.: The Hadoop distributed file system. In: 2010 IEEE 26th Symposium on Mass Storage Systems and Technologies, MSST2010, pp. 1–10 (2010)
15. Martha, V.: Big Data processing algorithms. In: Mohanty, H., Bhuyan, P., Chenthati, D. (eds.) Studies in Big Data, pp. 61–92. Springer (2015)
16. Raj, E.D., Dhinesh Babu, L.D.: A two pass scheduling policy based resource allocation for mapreduce. In: Procedia Computer Science, International Conference on Information and Communication Technologies (ICICT 2014), pp. 627–634. Elsevier B.V. (2015)
17. He, B., Fang, W., Luo, Q., Govindaraju, N.K., Wang, T.: Mars. In: Proceedings of the 17th International Conference on Parallel Architectures and Compilation Techniques—PACT '08, p. 260 (2008)
18. Marx, V.: Technology feature: the big challenges of Big Data. Nature **498**, 255–260 (2013)
19. Bhosale, H.S., Gadekar, D.P.: A review paper on Big Data and Hadoop. Int. J. Sci. Res. Publ. **4**, 1–7 (2014)
20. Al-janabi, S.T.F., Rasheed, M.A.: Public-key cryptography enabled kerberos authentication. In: 2011 Developments in E-systems Engineering Public-Key, pp. 209–214. IEEE (2011)
21. Fadika, Z., Dede, E., Hartog, J., Govindaraju, M.: MARLA : MapReduce for heterogeneous clusters. In: 12th IEEE/ACM International Symposium on Cluster, Cloud and Grid Computing, pp. 49–56. ACM (2012)
22. Mao, Y., Ling, J.: Research on load balance strategy based on grey prediction theory in cloud storage. In: 2nd International Conference on Electronic & Mechanical Engineering and Information Technology (EMEIT-2012), pp. 199–203. Atlantis Press, Paris, France (2012)
23. Ye, X., Huang, M., Zhu, D., Xu, P.: A novel blocks placement strategy for hadoop. In: Proceedings—2012 IEEE/ACIS 11th International Conference on Computer and Information Science, pp. 3–7. IEEE (2012)
24. Ling, J., Jiang, X.: Distributed storage method based on information dispersal algorithm. In: Proceedings—2013 2nd International Symposium on Instrumentation and Measurement, Sensor Network and Automation, IMSNA 2013, pp. 624–626. IEEE (2013)
25. Kumar, S.D.M., Shabeera, T.P.: Bandwidth-aware data placement scheme for Hadoop. In: 2013 IEEE Recent Advances in Intelligent Computational Systems (RAICS), pp. 64–67. IEEE (2013)

26. Fan, K., Zhang, D., Li, H., Yang, Y.: An adaptive feedback load balancing algorithm in HDFS. In: 2013 5th International Conference on Intelligent Networking and Collaborative Systems, pp. 23–29. IEEE (2013)
27. Lee, C.W., Hsieh, K.Y., Hsieh, S.Y., Hsiao, H.C.: A dynamic data placement strategy for Hadoop in heterogeneous environments. Big Data Res. **1**, 14–22 (2014)
28. Gao, Z., Liu, D., Yang, Y., Zheng, J., Hao, Y.: A load balance algorithm based on nodes performance in Hadoop cluster. In: APNOMS 2014—16th Asia-Pacific Network Operations and Management Symposium, pp. 1–4. IEEE (2014)
29. Lin, C.Y., Lin, Y.C.: A load-balancing algorithm for Hadoop distributed file system. In: Proceedings—2015 18th International Conference on Network-Based Information Systems, pp. 173–179. IEEE (2015)
30. Kim, D., Choi, E., Hong, J.: System information-based hadoop load balancing for heterogeneous clusters. In: RACS '15 International Conference on Research in Adaptive and Convergent Systems, pp. 465–467. ACM (2015)
31. Islam, N.S., Lu, X., Shankar, D., Panda, D.K.D.K.: Triple-H : A hybrid approach to accelerate HDFS on HPC clusters with heterogeneous storage architecture. In: 15th IEEE/ACM International Symposium on Cluster, Cloud and Grid Computing Triple-H, pp 101–110. ACM (2015)
32. Wang, S., Zhou, H.: The research of MapReduce load balancing based on multiple partition algorithm. In: IEEE/ACM 9th International Conference on Utility and Cloud Computing, pp. 339–342. IEEE/ACM (2016)
33. Hou, X., Pal, D., Kumar T.K.A., Thomas, J.P., Liu, H.: Privacy preserving rack-based dynamic workload balancing for Hadoop MapReduce. In: IEEE 2nd International Conference on Big Data Security on Cloud, IEEE International Conference on High Performance and Smart Computing, IEEE International Conference on Intelligent Data and Security, pp. 30–35. IEEE (2016)
34. Nayahi, J.J.V., Kavitha, V.: Privacy and utility preserving data clustering for data anonymization and distribution on Hadoop. Futur. Gener. Comput. Syst. **74**, 393–408 (2016)
35. Song, Y., Shin, Y., Jang, M., Chang, J.: Design and implementation of HDFS data encryption scheme using ARIA algorithm on Hadoop. In: 4th International Conference on Big Data and Smart Computing (BigComp 2017), pp. 84–90. IEEE (2017)
36. Tao, D., Lin, Z., Wang, B.: Load feedback-based resource scheduling and dynamic migration-based data locality for virtual Hadoop clusters in OpenStack-based clouds. Tsinghua Sci. Technol. **22**, 149–159 (2017)
37. Guo, Z., Fox, G., Zhou, M., Ruan, Y.: Improving resource utilization in MapReduce. In: IEEE International Conference on Cluster Computing, pp. 402–410. IEEE (2012)
38. Zaharia, M., Konwinski, A., Joseph, A.D., Katz, R., Stoica, I.: Improving MapReduce performance in heterogeneous environments. In: 8th USENIX Symposium on Operating Systems Design and Implementation, pp. 29–42. USENIX Association (2008)
39. Kc, K., Anyanwu, K.: Scheduling Hadoop jobs to meet deadlines. In: 2nd IEEE International Conference on Cloud Computing Technology and Science Scheduling, pp. 388–392. IEEE (2010)
40. Dai, X., Bensaou, B.: Scheduling for response time in Hadoop MapReduce. In: IEEE ICC 2016 SAC Cloud Communications and Networking, pp. 3627–3632. IEEE (2016)
41. Cheng, D., Rao, J., Jiang, C., Zhou, X.: Resource and deadline-aware job scheduling in dynamic Hadoop Clusters. In: Proceedings—2015 IEEE 29th International Parallel and Distributed Processing Symposium, IPDPS 2015, pp. 956–965 (2015)

# Cellular Level Based Deep Learning Framework for Early Detection of Dysplasia in Oral Squamous Epithelium

**Rachit Kumar Gupta, Mandeep Kaur and Jatinder Manhas**

**Abstract** Over the past few decades, the artificial intelligence is being employed in diverse fields like pattern classification, image processing, object identification, recommender systems, speech recognition, etc. Machine learning has made it possible to develop intelligent systems through training that equip machines to handle different tasks, exactly on the analogy similar to humans. In medical field, machine learning algorithms are being used for prediction, early detection and prognosis of various diseases. These algorithms suffer a certain threshold due to their inability to handle large amount of data. Deep learning based techniques are emerging as efficient tools and can easily overcome the above difficulties in processing data related to medical imaging that includes mammographs, CT scans, MRIs and histopathology slide images. Deep learning has already achieved greater accuracy in early detection, diagnosis and prognosis of various diseases especially in cancer. Dysplasia is considered to be a pathway that leads to cancer. So, in order to diagnose oral cancer at its early stage, it is highly recommended to firstly detect dysplastic cells in the oral epithelial squamous layer. In our research work, we have proposed a deep learning based framework (convolutional neural network) to classify images of dysplastic cells from oral squamous epithelium layer. The proposed framework has classified the images of dysplastic cells into four different classes, namely normal cells, mild dysplastic cells, moderate dysplastic cells and severe dysplastic cells. The dataset undertaken for analysis consists of 2557 images of epithelial squamous cells of the oral cavity taken from 52 patients. Results show that on training the proposed framework gave an accuracy of 94.6% whereas, in testing it gave an accuracy of 90.22%. The results produced by our framework has also been tested and validated by comparing the manual results recorded by the medical experts working in this area.

R. K. Gupta (✉)
Department of Computer Science and IT, University of Jammu, J&K, India
e-mail: gupta.rachit1990@gmail.com

M. Kaur
Department of Oral Pathology, Indira Gandhi Government Dental College, J&K, India
e-mail: dr_mandeep_kaur@yahoo.com

J. Manhas
Department of Computer Science and IT, Bhaderwah Campus, University of Jammu, J&K, India
e-mail: manhas.jatinder@gmail.com



P. K. Singh et al. (eds.), *Proceedings of ICRIC 2019*, Lecture Notes in Electrical Engineering 597, https://doi.org/10.1007/978-3-030-29407-6_12

**Keywords** Deep learning · Machine learning · Dysplasia · Oral cancer · Oral squamous epithelium · Dysplastic cells · Convolutional neural network

## 1 Introduction

Oral cancer is one of the most commonly occurring cancers in India. In fact, it is one of the top three cancers among Indians. The mortality of oral cancer has increased seven times in the last 50 years in spite of millions worth research going on globally. India has one of the highest rates in oral cancer in the world partly attributed to the high prevalence of tobacco chewing. Alarmingly 194 million people aged 15 years and older have a tobacco habit. In parallel to the increase in oral cancer, borderline malignant lesions which range from epithelial dysplasia to intraepithelial carcinoma have also increased in numbers [1]. In the past few years, deep learning has proven to be efficient in many classification problems, especially in medical imaging data. Due to technological advancement in medical sciences and computer science, a huge volume of imaging data related to cancer and pre-cancerous symptoms has been collected, hence making deep learning a popular tool for researchers. In the last few years, a large number of machine learning based frameworks especially deep learning are being employed by researchers to detect and classify oral cancer. These deep learning based systems have proved to be very efficient and have achieved accuracy almost equal to the specialist pathologist that does its work manually and has ample of experience in his field [2].

A brief introduction to dysplasia and deep learning is given in this section.

### *1.1 Dysplasia*

The term dysplasia was introduced by Reagon in 1958 in a study where he described the features of dysplasia with relation to exfoliated cells of the epithelium of cervical lesions [3]. In medical terms, dysplasia means an abnormal development in cells while histomorphologically any cellular or structural change in the epithelium is dysplasia. The individual change in the cell is known as atypia, while a general disturbance in the epithelium is designated as dysplasia. Dysplasia is characterized by following major pathological microscopic changes at the cellular level as well as tissue level [4] (Table 1).

Dysplasia is a part of the pathway to malignancy. Dysplastic cells cannot be called malignant until and unless it invades the connective tissue or metastasize. The dysplastic cells can be classified into mild, moderate and severe dysplasia. The progressive change in dysplastic cells from mild to moderate and then from moderate to severe can be seen in the epithelium. The invasive activity of dysplastic cells starts from the lower layer, then progress to the middle layer and finally full thickness of epithelium. Next, it will invade the basement membrane [5] (Fig. 1).

**Table 1** Major changes in dysplasia at the cellular and tissue level

| Cellular level changes | Tissue level changes |
|---|---|
| Abnormal variation in cell size (Anisocytosis)<br>Increased nuclear/cytoplasmic ratio<br>Increased mitotic figures<br>Abnormal mitotic figures (abnormal in shape or location)<br>Nuclear and cellular pleomorphism<br>Enlarged nuclei and cells<br>Abnormal variation in nuclear size (anisonucleosis)<br>Hyperchromatic nuclei<br>Increased number and size of nucleoli | Loss of polarity<br>Disordered maturation from basal to squamous cells<br>Includes top-to-bottom change of carcinoma in situ<br>Increased cellular density<br>Basal cell hyperplasia<br>Dyskeratosis (premature keratinization and keratin pearls deep in epithelium)<br>Bulbous drop-shaped rete pegs<br>Secondary extensions (nodules) on rete tips |

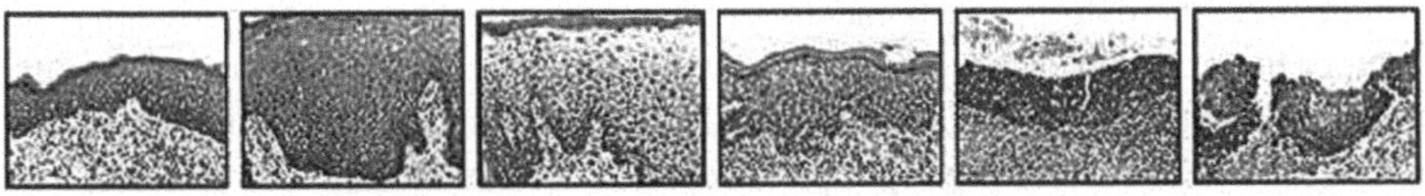

**Fig. 1** Progression of dysplasia from normal epithelium to in-situ carcinoma

As the dysplasia arises in the basal layer of the epithelium and extends, with progression, to the upper epithelial layers, the scheme classifies mild dysplasia as involvement of the lower third of the epithelium only, moderate dysplasia as extension to the middle third and severe dysplasia as extension to the superficial third of the epithelium [6].

Early detection of cancer symptoms plays a key role in cancer diagnosis. This can improve long-term survival rates. With the advent of visualization in technology, medical imaging has become a very important technique for early cancer detection and diagnosis, monitoring and follow-up after the treatments [7].

## *1.2 Deep Learning*

Extensive research has been going on cancer for the past two decades. Many researchers have been doing research in the field of early cancer detection, diagnosis, prognosis using machine learning techniques. Among other techniques, deep learning has proved efficient in different applications. The success of deep learning in natural scene image classification and segmentation stimulates the research of adopting it in image-based cancer detection and diagnosis. Deep learning has been heavily employed in the following fields: colorization of black and white images; adding sounds to silent movies; automatic machine translation; object classification in photographs; automatic handwriting generation; character text generation; image caption generation; and automatic game playing. One major advantage of deep learning is that it automatically extracts features from input data, which is one of the most complicated and time-consuming tasks in machine learning. This task of feature extraction and refining becomes more complicated and troublesome with redundant image input data. Deep learning architectures that are already existent can be used for new applications easily and efficiently with minor changes in architecture or no change at all. Despite the advantages of deep learning, there are also some disadvantages for applying deep learning, deep learning models require a huge amount of data for training data in order to be effective than other machine learning techniques. Due to large training data, the training process is computationally expensive because of additional GPU hardware and it is very time-consuming. Also, there is no methodology that could explain exactly what is happening inside deep layers of deep complex models [8]. Deep learning is mainly based upon the huge artificial neural network architecture, where deep simply means a higher number of layers, deep learning can be supervised, unsupervised or can be even reinforced. The input layer takes the input from the environment or the user and passes to the next layers until it reaches the output layer. This output layer provides the result based upon the calculations in the hidden layers. In deep learning, features are automatically extracted, and it generally needs very large dataset. Deep learning models tend to perform well with an increasing amount of data, whereas old machine learning models stop improving after a saturation point. One of the differences between machine learning and deep learning model is on the feature extraction area. Feature extraction is done by human in machine learning, whereas deep learning model figures out by itself [9]. Medical imaging has always been very helpful for diagnosis of cancer, monitoring and prognosis, survival rates and reoccurrence of cancer. Since medical images interpretation by a human is subjected to bias and erroneous interpretation. Therefore, from the early 1980s, computer-aided diagnosis systems have been deployed to help doctors to efficiently diagnose the medical images [10]. Computer-aided diagnosis (CAD) systems that use medical imaging employ many machine learning techniques for cancer detection and diagnosis. Though machine learning has been used for classification of medical images, it struggled to cross the threshold. Deep learning is finally enabling us to cross that threshold and achieving better performance than before [11]. Traditional machine learning techniques depend upon feature engineering for their working, but this feature extraction has many weaknesses that limit their performance. So to overcome this weakness of traditional machine learning techniques more emphasis is given to representation learning than feature engineering. One advantage of deep learning is that it can generate high-level feature representation directly from the raw images. Recent development in parallel computing architecture, graphics processing units (GPUs), deep learning techniques have performed very well in many fields like image recognition, object detection and speech recognition [12]. This paper aims to classify oral dysplasia by using popular deep learning technique, namely convolutional neural network. Our data is in form of images that's why CNN is preferable over other deep learning techniques. Recent studies show that CNN has achieved better results in cancer detection and diagnosis [9, 12]. The basic architecture of CNN is shown in Fig. 2.

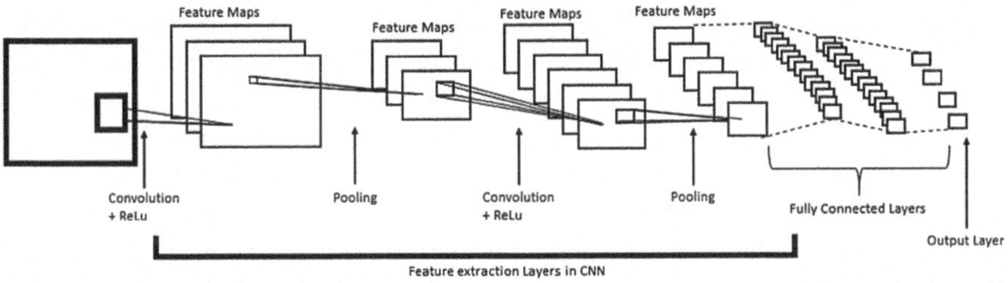


**Fig. 2** Basic structure of a CNN

CNN belongs to feedforward neural networks which can be expressed as [13]

$$f(x) = f_N(f_{N-1}(f_{N-2} \ldots (f_1(x)))) \tag{1}$$

where N denotes the number of hidden layers, and $f(x)$ represents the function in the corresponding layer. In atypical CNN model, the main functional layers include a convolutional layer, an activation layer, pooling layer, fully connected layer and predication layer. In the convolutional layer, is composed of multiple convolution kernels ($K^1$, $K^2$, …, $K^{M-2}$, $K^M$). Each $K^M$ represents a linear function in the Mth kernel, which can be represented as:

$$K^M(x, y) = \sum_{p=-t}^{t} \sum_{q=-u}^{u} \sum_{r=-v}^{v} W_M(p, q, r) I(x - p, y - q, z - r) \tag{2}$$

where (x, y, z) denotes the position of the pixel in input I, $W_M$ denotes the weight for the Mth kernel, t, u and v denote the height, width and depth of the filter. In the activation layer, $f(x)$ is a pixel-wise nonlinear function, i.e., rectified linear unit (ReLU) in this case [13].

$$f(x) = \max(0, x) \tag{3}$$

In the pooling layer, $f(x)$ is a layer-wise nonlinear down-sampling function aiming at reducing progressively the size of the feature representation. A fully connected layer is considered to be a type of convolutional layer whose convolutional kernel has the size of 1 * 1. The softmax is often added to the last fully connected layer to compute the probabilities of a given input $I_i$ for given output classes [9].

## 2 Literature Review

This section is intended to provide a comprehensive survey of recent studies on applying deep learning for early cancer detection, cancer diagnosis and prognosis.

In recent years, a bunch of papers has been published about the application of deep learning in cancer detection and diagnosis. In [14], Forslid et al. have used CNN on oral cancer dataset. Two different CNN models were used, namely ResNet and VGG. Results were very good as compared to other image diagnosis techniques. In [15], Litjens et al. have presented prostate cancer identification in biopsy specimens and breast cancer metastasis detection in sentinel lymph nodes using deep learning and concluded that deep learning improves the accuracy of prostate cancer diagnosis and breast cancer staging. Authors employed CNN. In [16], Albayrak et al. developed a deep learning based feature extraction algorithm to detect mitosis in breast histopathological images. In the proposed algorithm, the CNN model was used to extract features which were used to train a support vector machine (SVM) for the detection of mitosis. In [17], Folmsbee et al. explored methods for efficiently training convolutional neural networks for tissue classification using active learning (AL). Author employed comparison between random learning and active learning with CNN to identify seven tissue classes (stroma, lymphocytes, tumor, mucosa, keratin pearls, blood and background/adipose). The author found that the active learning strategy provides an average 3.26% greater performance than random learning for a given training set size. In [18], Krizhevsky et al. used AlexNet to construct a CNN model to classify benign or malignant tumors from the breast histopathological images. In [19], Iyer et al. have used the pre-trained model, VGG19 to extract information from the pathological images specific for lung cancer. A model consisting of deep convolution network-based image classification has been proposed for predictions on mutations in genes signature information of lung cancer among Indian populations. In [20], El-Sherbiny et al. proposed—Brain/Lung/Breast (BLB) automated detection system. It precisely predicts the occurrence of cancer and segments the expected region of tumor/cancer in MRI/CT scan/mammography images. This system proposes different classification techniques including support vector machine (SVM), ExtraTrees and convolutional neural network (CNN). CNN performed exceptionally well in the detection of lung cancer. In [21], Chen et al. proposed a deep cascade network for mitosis detection in breast histology slides. They first trained a fully connected network model to extract mitosis candidates from the whole histology slides and then fine-tuned a CaffeNet model for the classification of mitosis. Three networks with different configurations of fully connected layers were trained, and the scores were averaged to generate the final output. In [22], Arvidsson et al. compared two different techniques; by training the networks using color augmentation and by using digital stain separation using an auto-encoder. The author achieved accuracies of 95% for classification of benign versus malignant tissue and 81% for Gleason grading for data from the same site as the training data. In [23], Albarqouni et al. explored deep CNN in a biomedical context, a multiscale CNN architecture, was developed with an aggregation layer which was introduced after the softmax layer to aggregate the prediction results with the annotation results from multiple participation. In [24], Pratiher et al. have deployed deep learning framework to classify elastic scattering spectra of biological tissues into normal and cancerous ones. Authors experimented to show the superiority of the convolutional neural network extracted deep features over classical handcrafted biomarkers. The proposed method employs elastic scattering spectra of the tissues as input to CNN. In [25], Wichakam et al. proposed a combined system consisting of CNN and SVM for mass detection on digital mammograms. CNN was used on mammographic patches to get the high-level feature representation of the image. This high-level feature set was used as input to SVM for classification of mammograms. In [26], Xu et al. proposed a stacked sparse auto-encoder based algorithm to classify nuclei in breast cancer histopathology. In [27], Xia et al. proposed a system of tumor classification by pre-training a CNN from samples of different tissue types in histopathological images, then fine-tuning the obtained pre-trained on a particular tissue type. CNN showed improvement over training from scratch with limited data. In [28], Khoshdeli et al. applied convolutional neural networks for grading of the tumor and decomposing tumor architecture from H&E stained histology sections of the kidney. In [29], Xu et al. proposed deep convolutional neural network activation features to perform classification, segmentation and visualization in large-scale tissue histopathology images. Authors used a pre-trained ImageNet network on features extracted by CNN.

After going through the extensive literature survey, we found that the classification of dysplasia at the cellular level has not been done by any researcher so far using deep learning. Since malignancy of cells starts with dysplasia, we need to detect dysplasia and classify different stages of dysplasia in order to check whether cells are going to be malignant or not. We have undertaken this research work after identifying the research gap discussed above and also to early detect any malignancies at the cellular level. The methodology of our research work is described in detail in the next section.

## 3 Materials and Methodology

In this study, we have taken data of 52 patients suffering from oral dysplasia. Two H&E stained histopathological samples are taken from each patient. The data is taken from Indira Gandhi Govt. Dental College and Hospital, Jammu, India. Based on CIN classification, these data samples were classified into four different classes, namely normal, mild dysplasia, moderate dysplasia and severe dysplasia. The images of H&E slides were captured with the slides were observed and analyzed under a multi-headed microscope (Nikon Eclipse E400 DS-F12) having a digital camera mounted over it and connected and configured with an attached computer. All the images were clicked at 40X. From these images, individual cells were cropped out manually. These individual cell images were then subjected to semi-automatic image processing task which included brightness and contrast enhancement. Then, we did data augmentation on these images, which included rotation and translation. All the images were resized to 128 * 128 pixels.

Some original images and final dataset images from the dataset are shown in Fig. 3a and b, respectively:

In the next step, we created a dataset from these images by labeling each image appropriately. A specialist doctor from the department of oral pathology did all the labeling. Any image which was doubtful was excluded from the dataset. Once

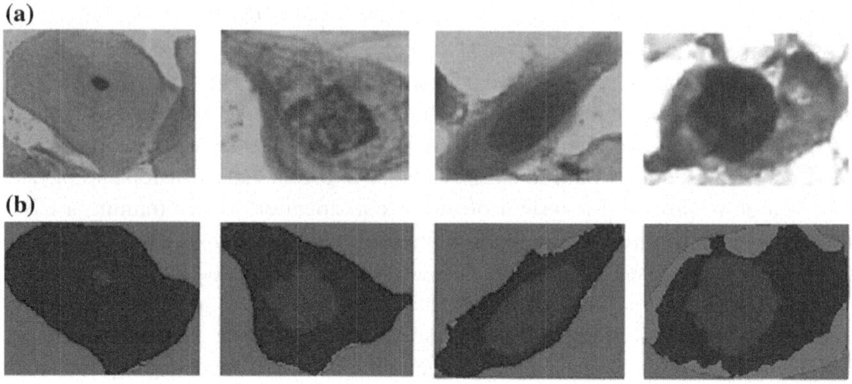


**Fig. 3** **a** Some images of original cells extracted from H&E stained microscopic slides, **b** images of the above cells after pre-processing and segmentation

the original images dataset was created we implied further pre-processing on it by segmenting cells and background in images so as to exclude any unnecessary details. As can be seen in Fig. 3b, the segmented cells are shown. Dark blue color signifies cytoplasm of the cell, light blue color inside dark blue portion signifies nucleus of the cell and the red part signifies background. Gray part can also be seen in some cell images, but it does not signify anything, and it is just blank space so it can be ignored. There are a total of 2557 images in our dataset. A total of 716 images belonging to a normal class, 614 belonging to mild dysplasia, 588 belonging to moderate dysplasia and 639 belonging to severe dysplasia.

CNN was created and trained from scratch. Python language along with the Tensorflow and Keras deep learning libraries were used to build the CNN model. The architecture of CNN used is given in Fig. 4.

CNN was created and trained from scratch. Python language along with the Tensorflow and Keras deep learning libraries were used to build the CNN model.

All layers and their internal structures are given in Fig. 4. We have added two dropout layers, one before flattening layer and one after first fully connected layer to minimize overfitting.

Final layer being output layer which uses softmax to generate four outputs since our problem is four-class classification problems. It gives four probabilities as outputs for four classes. Dataset was divided into training set and testing set in a 70:30 ratio, respectively. The training set contained 1790 images, and testing set contained 767 images. The images in output are converted into grayscale images.

| Layer (type) | Output Shape | Param # |
|---|---|---|
| input_1 (InputLayer) | (None, 128, 128, 1) | 0 |
| conv2d_1 (Conv2D) | (None, 128, 128, 32) | 320 |
| max_pooling2d_1 (MaxPooling2 | (None, 43, 43, 32) | 0 |
| conv2d_2 (Conv2D) | (None, 43, 43, 64) | 18496 |
| max_pooling2d_2 (MaxPooling2 | (None, 15, 15, 64) | 0 |
| conv2d_3 (Conv2D) | (None, 15, 15, 80) | 46160 |
| max_pooling2d_3 (MaxPooling2 | (None, 5, 5, 80) | 0 |
| dropout_1 (Dropout) | (None, 5, 5, 80) | 0 |
| flatten_1 (Flatten) | (None, 2000) | 0 |
| dense_1 (Dense) | (None, 512) | 1024512 |
| dropout_2 (Dropout) | (None, 512) | 0 |
| dense_2 (Dense) | (None, 4) | 2052 |

**Fig. 4** CNN model layer-wise architecture

## 4 Results and Discussion

The classification of the dysplastic cells was carried out by CNN into four different classes.

When CNN was trained on training data it gave accuracy of 94.6% and in testing, it gave an accuracy of 90.22%. Our CNN was trained for a total of 50 epochs. The confusion matrix is given in Fig. 5.

Other observations like precision, recall, F1-score for each class and the average of all classes are given in Table 2. The formulas for precision, recall and F1-score are given as:

$$\text{Precision} = \frac{\text{TP}}{\text{TP} + \text{FP}} \tag{4}$$

$$\text{Recall} = \frac{\text{TP}}{\text{TP} + \text{FN}} \tag{5}$$

$$\text{F1Score} = \frac{2 * (\text{Recall} * \text{Precision})}{(\text{Recall} + \text{Precision})} \tag{6}$$

where TP are true positives, TN is true negatives, FP is false positives and FN are false negatives.

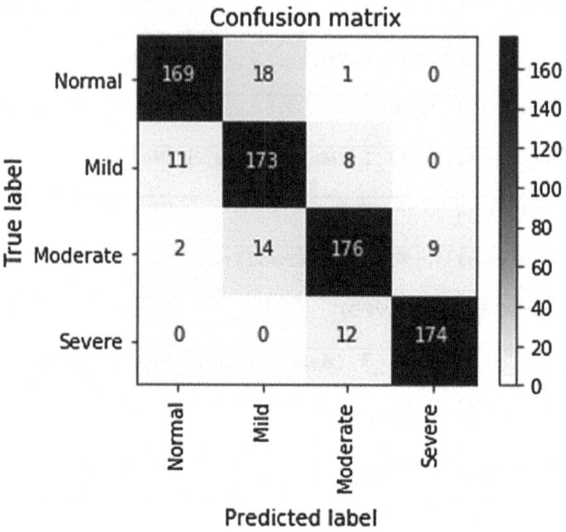


**Fig. 5** Confusion matrix for testing of 767 oral epithelial cell images

**Table 2** Calculated precision, recall and F1-score values from the confusion matrix

| Class label | Precision | Recall | F1-score |
|---|---|---|---|
| Normal | 0.93 | 0.90 | 0.91 |
| Mild | 0.84 | 0.90 | 0.87 |
| Moderate | 0.89 | 0.88 | 0.88 |
| Severe | 0.95 | 0.94 | 0.94 |
| Average | 0.90 | 0.90 | 0.90 |

Some of the images classified by our CNN are given in Fig. 6.

One important point worth discussing here is the pre-processing of image data. While working on image data especially medical image data, pre-processing plays an important role in improving the accuracy of the machine learning model. When initially images were given to CNN to train, it gave poor training accuracy (73% approx.). After pre-processing, it produced better training accuracy (94.6%). Also, one more thing worth discussing that we came across our experiment was that with multiclass classification problem, the training and testing accuracies of CNN were not satisfactory particularly for medical image data where accuracy is the most important of all things under consideration.

## 5 Conclusion

Our proposed deep learning system although not in real time but produced promising results. Although, our proposed deep learning system has not achieved higher

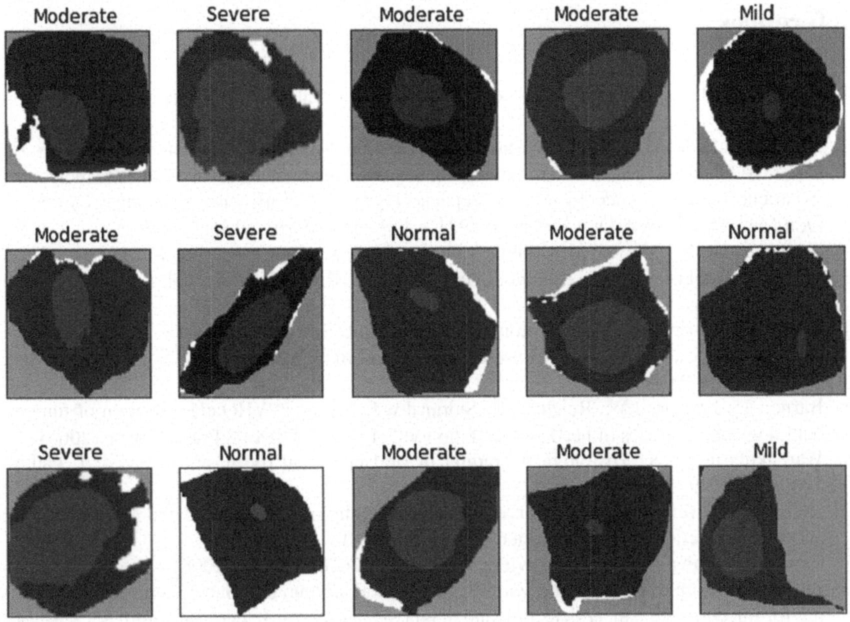


**Fig. 6** Some predicted images along with the labels

accuracies as compared to the experienced specialist concerned oral pathologist, yet it has produced satisfactory accurate results.

## 6 Future Scope

In the future, we will increase the dataset size to input more and more training data so that our proposed deep learning model will learn more efficiently. Also, we will fine-tune our deep learning model so that it can achieve more accuracy in predicting the true label. At present, our deep learning model is able to classify four different classes. In the future, we would like to include three more classes and hence making it classify seven different classes.

**Acknowledgements** I would like to thank the faculty of the Department of Oral Pathology and Department of Oral Medicine and Radiology from the Institute of Indira Gandhi Govt. Dental College and Hospital for their timely support in undertaking this research work.

## References


1. Sugerman, P.B., Savage, N.W.: Current concepts in oral cancer. Aust. Dent. J. **44**(3), 147–156 (1999)
2. Ariji, Y., Fukuda, M., Kise, Y., Nozawa, M., Yanashita, Y., Fujita, H., Katsumata, A., Ariji, E.: Contrast-enhanced computed tomography image assessment of cervical lymph node metastasis in patients with oral cancer by using a deep learning system of artificial intelligence. Oral Surg. Oral Med. Oral Pathol. Oral Radiol. J. S2212–4403(18)31201 (2018)
3. Gnananandar, G., Masthan, K., Babu, A.N., Jacobina, J.: Histomorphometric analysis of oral epithelial dysplasia and oral squamous cell carcinoma. Biomed. Pharmacol. J. 10(1): 395–400 (2017)
4. Fischer, D.J., Epstein, J.B., Morton, T.H., Schwartz, S.M.: Interobserver reliability in the histopathologic diagnosis of oral pre-malignant and malignant lesions. J. Oral Pathol. Med. 33(2): 65–70 (2004)
5. Barnes, L., Eveson, J.W., Reichart, P., Sidransky, D.: Book, "WHO classification of tumors: pathology and genetics of head and neck tumors". Lyon CEDEX 08 Press, France (2005)
6. Warnakulasuriya, S.: Histological grading of oral epithelial dysplasia: revisited. J. Pathol. **194**(3), 294–297 (2001)
7. Xiao, H.D., Bifulco, C.B.: Cellular and Molecular Pathology, Oral, Head and Neck Oncology and Reconstructive Surgery. Elsevier, pp 57–78 (2018)
8. Fass, L.: Imaging and cancer: a review. Mol. Oncol. **2**(2), 115–152 (2008)
9. Zilong, Hu, Tang, Jinshan, Wang, Ziming, Zhang, Kai, Zhang, Lin, Sun, Qingling: Deep learning for image-based cancer detection and diagnosis—a survey. Pattern Recognit. **83**, 134–149 (2018)
10. Doi, K.: Computer-aided diagnosis in medical imaging: historical review, current status and future potential. Computer. Med. Imagin. Gr. **31**(4), 198–211 (2007)
11. Intro to deep learning. https://medium.com/intro-to-artificial-intelligence/deep-learning-series-1-intro-to-deep-learning-abb1780ee20, Last accessed 2019/01/14
12. LeCun, Y., Bengio, Y., Hinton, G.: Deep learning. Nature **521**, 436–444 (2015)
13. Xu, B., Wang, N., Chen, T., Li, M.: Empirical evaluation of rectified activations in convolutional network. arXiv preprint arXiv:1505.00853 (2015)
14. Forslid, G., Wieslander, H., Bengtsson, E., Wählby, C., Hirsch, J.-M., Stark, C.R., Sadanandan, S.K.: Deep convolutional neural networks for detecting cellular changes due to malignancy. In: IEEE International Conference on Computer Vision Workshops (ICCVW), Venice, pp. 82–89 (2017)
15. Litjens, G., Sánchez, C.I., Timofeeva, N., Hermsen, M., Nagtegaal, I., Kovacs, I., van de Kaa, C.H., Bult, P., van Ginneken, B., van der Laak, J.: Deep learning as a tool for increased accuracy and efficiency of Histopathological diagnosis. In: Scientific Reports, Vol. 6, pp. 262–286 (2016)
16. Albayrak, A., Bilgin, G.: Mitosis detection using convolutional neural network based features. In: IEEE 17th International Symposium on Computational Intelligence and Informatics (CINTI), pp. 335–340 (2017)
17. Folmsbee, J.: Active deep learning: Improved training efficiency of convolutional neural networks for issue classification in oral cavity cancer. In: 2018 IEEE 15th International Symposium on Biomedical Imaging, Washington, DC, pp. 770–773 (2018)
18. Krizhevsky, A., Sutskever, I., Hinton, G.E.: Imagenet classification with deep convolutional neural networks. In: Advances in Neural Information Processing Systems, pp. 1097–1105 (2012)
19. Iyer, A., Hima Vyshnavi A.M., Namboori Krishnan, P.K.: Deep convolution network-based prediction model for medical diagnosis of lung cancer—a deep pharmacogenomic approach: deep diagnosis for lung cancer. In: Second International Conference on Advances in Electronics, Computers and Communications, pp. 1–4 (2018)
20. El-Sherbiny, B., Nabil, N., Emad, Y., Ayman, N., Mohiy, T., Raouf, A.A.: BLB (Brain/Lung cancer detection and segmentation and Breast Dense calculation). In: First International Workshop on Deep and Representation Learning (IWDRL), Cairo, pp. 41–47 (2018)

21. Chen, H., Dou, Q., Wang, X., Qin, J., Heng, P.-A.: Mitosis Detection in Breast Cancer Histology Images via Deep Cascaded Networks. AAAI, pp. 1160–1166 (2016)
22. Arvidsson, I., Overgaard, N.C., Marginean, F.-E., Krzyzanowska, A., Bjartell, A., Åström, K., Heyden, A.: Generalization of prostate cancer classification for multiple sites using deep learning. In: IEEE 15th International Symposium on Biomedical Imaging, Washington, DC, pp. 191–194 (2018)
23. Albarqouni, S., Baur, C., Achilles, F., Belagiannis, V., Demirci, S., Navab, N.: Aggnet: deep learning from crowds for mitosis detection in breast cancer histology images. IEEE Trans. Med. Imagin. **35**, 1313–1321 (2016)
24. Pratiher, S., Bhattacharya, S., Mukhopadhyay, S., Ghosh, N., Pasupuleti, G., Panigrahi, P.K.: Deep features using a convolutional neural network for early-stage cancer detection. In: Proceedings Volume 10679 Optics, Photonics, and Digital Technologies for Imaging Applications, Vol. 1067902 (2018)
25. Wichakam, I., Vateekul, P.: Combining deep convolutional networks and SVMs for mass detection on digital mammograms. In: 8th International Conference on Knowledge and Smart Technology (KST), IEEE, pp. 239–244 (2016)
26. Xu, J., Xiang, L., Hang, R., Wu, J.: Stacked sparse autoencoder (SSAE) based framework for nuclei patch classification on breast cancer histopathology. 2014 IEEE 11th International Symposium on Biomedical Imaging, IEEE, pp. 999–1002 (2014)
27. Xia, T., Kumar, A., Feng, D., Kim, J.: Patch-level tumor classification in digital histopathology images with domain adapted deep learning. In: 40th Annual International Conference of the IEEE Engineering in Medicine and Biology Society (EMBC), Honolulu, HI, pp. 644–647 (2018)
28. Khoshdeli, M., Borowsky, A., Parvin, B.: Deep learning models differentiate tumor grades from H&E stained histology sections. In: 40th Annual International Conference of the IEEE Engineering in Medicine and Biology Society (EMBC), Honolulu, HI, pp. 620–623 (2018)
29. Xu, Y., Jia, Z., Wang, L.B., Ai, Y., Zhang, F., Lai, M., Chang, E.I.: Large scale tissue histopathology image classification, segmentation, and visualization via deep convolutional activation features. BMC Bioinf. **18**(1), 281 (2016)

# Internet of Things-Based Hardware and Software for Smart Agriculture: A Review

**Brij Bhushan Sharma and Nagesh Kumar**

**Abstract** Smart agriculture with IoT (Internet of things) gives rise to agribusiness and may fulfill the demand of food in the world. Agribusiness will become more productive when IoT is introduced. As the farmers are becoming more aware to IoT, the possibility of introducing new technologies is gaining momentum in agribusiness. The IoT is highly efficient, accessible, persistent and of exposed nature. The farmers may use sensors for monitoring the crops, soil and also analyze the crop production. This will give boost up to precision agriculture. The information about farms like temperature, moisture level and soil mineral level, pH value of soil and air quality can be collected remotely by using IoT on smartphones or computer systems. Today's IoT techniques and devices are so advanced and are mainly application dependent providing smart systems to be deployed in specific areas. In this paper, a review of IoT devices and protocols is presented and it also throws light on issues arise during the implementation of IoT in agriculture.

**Keywords** IoT · Agriculture · Agribusiness · Smart systems · Sensors

## 1 Introduction

Agribusiness is considered as the premise of life for the species of humans as it is a principle wellspring of sustenance grains and other crude materials. It expects fundamental occupation in the improvement of country's economy. It additionally gives plentiful work chances to the general population of the nation for the improvement of the financial state of the nation and it is very important to develop the agrarian segment, lamentably several agriculturists still utilizing the old traditional methods of the agriculture techniques for cultivating, which results in low yielding of harvests of food grains or crops. In case of industries, where ever the robotization automation

B. B. Sharma (✉) · N. Kumar
School of Electrical and Computer Science, Shoolini University of Biotechnology and Management Sciences, Solan, Himachal Pradesh, India
e-mail: brijbhushan@shooliniuniversity.com

N. Kumar
e-mail: nageshkumar@shooliniuniversity.com



P. K. Singh et al. (eds.), *Proceedings of ICRIC 2019*, Lecture Notes in Electrical Engineering 597, https://doi.org/10.1007/978-3-030-29407-6_13

has been introduced and persons are provided with the programmed hardware the success has been achieved. So we need to execute current science and innovation technologies in the agribusiness to increase the yield among the imperative things that may go to the rancher's advantage is the way to control the utilized of common sources and regular habitat which horticultures rely upon along these lines, this issue has caught farmer's enthusiasm to execute agro-ecological remote checking strategy in their horticulture business [1].

We need to enhance the versatility and convenience of the new keen homestead framework by beating the issue of utilization confinements of wired gadgets in agribusiness by utilizing remote correspondence module. In any case, since the current innovation dimension of the remote correspondence framework experiences a power-lack issue, the advancement of the low-power remote correspondence module is being enacted, and with the improvement of the battery innovation, the change speed from the wired framework to the remote framework. As per industrialization, the utilization of automation and innovation to agribusiness is endeavoring from the need of large scale of manufacturing of nourishment. Computerized or automated agriculture started with the IT products and with collection of crop growth [2]. The collection of data regarding crops can play a major role in computerized agriculture so the collection of data from the crop a sensor is required which can fetch data from the environment and on the basis of that data a proper system can be created a server is also required in computerized or automated agribusiness to store the environmental and growth parameter data of the particular crop so the farmer can adjust the required parameter for the proper cultivation of the crop [3].

Since all the smart farming involves a green house or a nursery with the help of combination of electrical and electronic gadgets an equation of a proper crop growth, crop drying, and storage can be satisfied these gadgets includes: microcontroller, fans, radiators, air-conditioning, air systems, temperature maintenance and monitoring devices, etc., in these kind of systems a wired network can be a serious issue or disadvantage so a complete IoT, WSN operated system is required to avoid the confusion regarding wires or complexity of the system [4]. IoT term speaks to a general idea for the capacity of system gadgets to detect and gather information from our general surroundings and after that share that information over the Internet where it tends to be handled and used for different intriguing purposes. The IoT is included savvy machines interfacing and speaking with different machines, items, situations and frameworks. Presently multi day's each individual are associated with one another utilizing heaps of correspondence way. Where the most famous correspondence way is Web so in another word we can say Web which interface people groups.

In this paper, the review of current technologies in smart agriculture will be studied and various IoT devices and their comparisons is presented. The rest of the paper composed of background study, current techniques and issues and conclusions.

## 2 Background Study

This section is particularly focused on the smart agriculture publications [4]. Proposed that the sensors and microcontrollers of each of the three nodes are effectively interfaced with raspberry pi and remote correspondence is accomplished between different nodes. All perceptions and test tests demonstrates that venture is a finished answer for field exercises, water system issues, and capacity issues utilizing remote-controlled robot, brilliant water system framework and a savvy distribution center administration framework individually. Usage of such a framework in the field can enhance the yield of the harvests and generally speaking generation.

In [5] authors have tried to show SmartFarmNet, a spearheading exertion in building a versatile sensor information securing, investigation, and perception stage for brilliant cultivating applications, in light of the Web of Things. They displayed the engineering structure of the stage that intends to help basically any IoT gadgets, permit fast ingestion and perception of IoT information utilizing zero-programming exertion (do-it-without anyone else's help standards), and give a virtual research facility condition to representation and sharing of study information. The proposed system utilizes an interesting and novel ongoing measurable examination approach that empowers close ongoing reactions to client questions (approving the stage's capacity to scale so as to deal with high-speed information streams). Through assessment utilizing genuine cultivating information.

In [6] the authors' provided a comparison table of techniques used in smart farming, they did comparison in between seven techniques from which they carried out the results of those techniques and also issues present in these techniques and they have concluded that water being a valuable asset must be used efficiently. Agriculture is one of those territories which devours parcel of water. Water system to the ranch is a tedious procedure and must be done on opportune premise. As pointed through this work an auto water system framework estimating the dampness content, moistness, as temperature, rain location and the water level. Later reaping the abundance of water from the development field and reused back to the tank. The created framework additionally cautions the agriculturist with ringer when there is intruder (human/creature) into the farm. Thus, the proposed framework bargains about the water system framework in savvy way utilizing Internet of things (IoT) which tackled the ebb and flow issues identified with cultivating, for example, by decreasing human endeavors, wastage of water and refreshing the rancher about the live state of the field on the cell phone.

In [7] authors provided a complete survey step by step on big data in explored few techniques and stated that the crop management is possible by giving required measure of water, manures and all updates about product development can be taken from anyplace to wherever. Machine-to-machine architecture can be associated and all meteorological data is taken to furthermore, arrangement can be connected and executed. Future work is to give security framework to the field too with regards to the yields. The unregistered individual or any creature ought to be perceived and implication ought to be sent to the agriculturist's versatile.

In [8] the content gives a comparative study on IoT's and its applications in smart farming, in which they concluded that data gathered from different examines the work can be additionally reached out in two different ways. Reliability and scalability can be improved and the open-source programming languages such as R and python could be used as a program, the improvement of brilliant irrigation framework could be executed in other manors, for example, citrus edits and breaking down the execution. The informational index can be as yet expanded to enhance the exactness of the framework in confirmation conspire further complexities of the convention are diminished without bargaining security highlights. The whole work can be even converged with distributed computing environment. From the past work, a portion of the new choices can be made in products. There are sensors which can do astonishing things in the horticulture. The nation needs great horticulture and it could be made still savvy. The informational collection is kept up for each savvy work in farming and can be utilized for further reference. Utilizing drone with all the climate and temperature data the sort of yield which must be planted in agribusiness can be found. Which trim suits to which condition, those verifiable data can be found what is more, send to agrarian specialists. With that information, he can plant new yields. Additionally, if the field has the ability to develop by spreading the seeds. It can likewise be robotized. Another gadget might be concocted and made to spread the seeds crosswise over fields dependent on soil type data. What's more, if the atmosphere is transformed it can likewise be hinted through canny frameworks so that some extraordinary seeds can be spreaded. Huge information assumes an extraordinary job in keeping up the dataset for climate information, soil type qualities, in view of the information gathered the seeds can be tossed by rural specialists or by automaton-like gadget to shower the seeds. Another critical challenge is that the exploration has demonstrated that the kind of compost can be distinguished for a specific soil.

The authors in [9] are also focused to smart agriculture, in which she provided a survey on smart agriculture using IoT's, in this, she proposed that the framework empowers powerful soil, water, dampness, gatecrashers recognition parameters has been checking and refreshing utilizing IoT. This empowers powerful soil upkeep furthermore, siphon controlling component. This defeats the manual activities required to screen and keep up the horticultural cultivates in both programmed and manual modes.

In [10] a review on Internet of things-based application smart agricultural system, authors stated that they had studied about the techniques based on GPRS and concluded that there is need of those kind of framework, which is ongoing information analyzer which can send information around the globe utilizing the idea of IoT. In the future, we can deal with WiFi, Lora-based innovation.

In [11] authors proposed that IoT can be a highlighted area for researchers. Horizon 2020, the EU programme for research and advancement incorporates a work program in 2016–2017, which centers around maintainable sustenance security which will be a vehicle to advance nourishment security and wellbeing by acclimatizing innovation into nourishment store network.

In [12] authors presented that the PLC and ZigBee gadget with RF usefulness could control all the above-said farming procedure like seeding, water system, furrowing and preparing. Water system framework could be accomplished consequently with the assistance of PLC and level sensor. For this, the field could be isolated into a few sections. For seeding process, photoelectric sensor and transport ought to be utilized. The development of plastic plate was constrained by the transport engine. For furrowing process, the apply autonomy guideline could be included. For the control of tractor adapt engine could be utilized. At that point, ultrasonic sensor and photoelectric sensor additionally utilized for furrowing process. Because of this mechanization of rural process the expense for keeping up the field additionally diminished. Blending of fertilization in a right proportion will be finished utilizing PLC. At that point, it is changed over into fluid frame by providing water. At that point, it will be blended and send through the pipe or sprayer. For this procedure, single-stage enlistment engine and level sensor will be utilized.

In [13] the authors have proposed framework, which is a smart farm wired/remote framework with hub—server—database and outer control structure. So as to take care of the power issue and space confinement of the wired framework, which is a restriction of the existing keen homestead, we executed a correspondence strategy utilizing Bluetooth and LPWAN module as low-power module. Also, by applying a straightforward answer for the bit misfortune that can happen in remote correspondence, the inter-module message trade has been effectively performed also, by applying an institutionalized message trade strategy that can be connected as a current-related specialized field, the plausibility of extending the innovation application in the IoT field can be affirmed.

## 3 Comparison of Current Techniques and Issues

The following table presents a comparison of techniques. Major fields and techniques that can be involved in smart agriculture business are big data, evapotranspiration, wireless sensor network, cyber-physical system and cloud computing with IoT's (Table 1).

## 4 Conclusion

Smart agribusiness may be the future of agriculture industry and IoT will play a major role in development. The IoT will give a rise to precision agriculture and also make it more accurate and proficient. The applications of IoT in agribusiness include the monitoring of conditions like moisture, pH level, mineral level and humidity, etc. IoT will operate on all of these conditions and the automation will provide the balancing methods of these parameters. This will make a farmer's work easy and he also can operate the IoT devices from home. IoT will reduce the cost of doing

**Table 1** Comparisons and issues of smart agriculture techniques

| Sr. No. | Techniques | Issues |
|---|---|---|
| 1. | (IoT), sensors, cloud computing, mobile computing, big-data analysis [2] | Soil nutrient sensors are not used |
| 2. | Bus concept, ZigBee protocols based on IEEE 802.15.4, hybrid network [3] | Not energy saving and data fusion, directions are left for future research |
| 3. | Raspberry pi, arduino microcontrollers, xbee modules [6] | If failure of any particular part or device is not informed and has to be tested manually |
| 4. | ZigBee with Wings [7] | The technique can achieve convenient wireless connection only within a short distance |
| 5. | WSN with GSM technology [8] | Provides only precision values that is not accurate and is not cost efficient |
| 6. | WSUs and a WIU, based on microcontroller, ZigBee, and GPRS technologies [11] | The investment in electric power supply is expensive |

farming by reducing the wastage of fertilizers, pesticides and other resources. There is a need of developing low-cost sensors and devices which can be used by every farmer to take out good crops from their fields. Also there is a requirement of new protocols and algorithms which can be used specifically in agriculture. The future if agribusiness will be bright if everyone to use IoT and sensor systems. But for making it possible it is necessary to educate farmers about the tools and techniques.

## References


1. Al-Fuqaha, A., Guizani, M., Mohammadi, M., Aledhari, M., Ayyash, M.: Internet of things: a survey on enabling technologies, protocols, and applications. IEEE Commun. Surv. Tutorials **17**(4), 2347–2376 (2015)
2. Ivanov, S., Bhargava, K., Donnelly, W.: Precision farming: sensor analytics. IEEE Intell. Syst. **30**(4), 76–80 (2015)
3. Chetan Dwarkani M., Ganesh Ram R., Jagannathan S., Priyatharshini, R.: Smart farming system using sensors for agricultural task automation. In: IEEE Technological Innovation in ICT for Agriculture and Rural Development (TIAR), pp. 49–53. IEEE, Chennai (2015)
4. Gondchawar, N., Kawitkar, R.S.: IoT based smart agriculture. Int. J. Adv. Res. Comput. Commun. Eng. **5**(6), 177–181 (2016)
5. Jayaraman, P.P., Yavari, A., Georgakopoulos, D., Morshed, A., Zaslavsky, A.: Internet of things platform for smart farming: experiences and lessons learnt. Sensors **16**(11), 1884 (2016)
6. Keerthana, B., Nivetha, P., Boomika, M., Mathivathani, M., Niranjana, A.: IoT based smart security and monitoring devices for agriculture. Int. J. Future Revolution Comput. Sci. Commun. Eng. **4**(2), 342–345 (2018)
7. Biradar, H.B., Shabadi, L.:. Review on IOT based multidisciplinary models for smart farming. In: 2nd IEEE International Conference on Recent Trends in Electronics, Information & Communication Technology (RTEICT), pp. 1923–1926. IEEE (2017)

8. Srilakshmi, A., Rakkini, J., Sekar, K.R., Manikandan, R.: A comparative study on Internet of Things (IoT) and its applications in smart agriculture. Pharmacognosy J. **10**(2), (2018)
9. Veena, S., Mahesh, K., Rajesh, M., Salmon, S.: The survey on smart agriculture using IoT. Int. J. Innovative Res. Eng. Manage. **5**(2), 63–66 (2018)
10. Navinay, M.K., Gedam, R.: A review paper on internet of things based application smart agricultural system. Int. J. Latest Eng. Manage. Res. **2**(4), 69–71 (2017)
11. Nukala, R., Panduru, K., Shields, A., Riordan, D., Doody, P., Walsh, J.: Internet of Things: a review from 'farm to fork'. In: 27th Irish conference on Signals and Systems Conference (ISSC), pp. 1–6. IEEE, (2016)
12. Dhivya, A., Infanta, J., Chakrapani, K.: Automated agricultural process using PLC and ZigBee. J. Artif. Intell. **5**(4), 170–177 (2012)
13. Yoon, C., Huh, M., Kang, S. G., Park, J., Lee, C.: Implement smart farm with IoT technology. In: 20th International Conference on Advanced Communication Technology (ICACT), pp. 749–752. IEEE (2018)

# Smart Approach for Real-Time Gender Prediction of European School's Principal Using Machine Learning

**Yatish Bathla, Chaman Verma and Neerendra Kumar**

**Abstract** Supervised machine learning is used to solve the binary classification problem on four datasets of European Survey of Schools: Information and Communication Technology (ICT) in Education (known as ESSIE) which is supported by European Union (EU). To predict the gender of the principal based on their response for the ICT questionnaire, the authors applied four supervised machine learning algorithms (sequential minimal optimization (SMO), multilayer perception (ANN), random forest (RF), and logistic regression (LR) on ISCED-1, ISCED-2, ISCED-3A, and ISCED-3B level of schools. The survey was conducted by the European Union in the academic year 2011–2012. The datasets have total 2933 instances\ & 164 attributes considered for the ISCED-1 level, 2914 instances\ & 164 attributes for the ISCED-2 level, 2203 instances\ & 164 attributes for the ISCED-3A level and 1820 instances\ & 164 attributes for the ISCED-3B level. On the one hand, SMO classifier outperformed others at ISCED-3A level and on the other hand, LR outperformed others at ISCED-1, ISCED-2, and ISCED-3B. Further, real-time prediction and automatic process of the datasets are done by introducing the concepts of the web server. The server communicates with the European Union web server and displays the results in the form of web application. This smart approach saves the data process and interaction time of humans as well as represents the processed data of the Weka efficiently.

**Keywords** Supervised machine learning · Classification · Real time · Sensitivity · Principal gender prediction

---

Y. Bathla · N. Kumar
Doctoral School of Applied Informatics and Applied Mathematics, Óbuda University, Budapest, Hungary
e-mail: yatish.bathla@phd.uni-obuda.hu

N. Kumar
e-mail: neerendra.kumar@phd.uni-obuda.hu

C. Verma (✉)
Department of Media and Educational Informatics, Eötvös Loránd University, Budapest, Hungary
e-mail: chaman@inf.elte.hu

N. Kumar
Department of Computer Science and IT, Central University of Jammu, Jammu, India



P. K. Singh et al. (eds.), *Proceedings of ICRIC 2019*, Lecture Notes in Electrical Engineering 597, https://doi.org/10.1007/978-3-030-29407-6_14

## 1 Introduction

Machine learning (ML) has been using in the various sectors like computer vision, text and speech recognition, spam filter on the email, online shopping and viewing recommendation, credit card fraud detection, and many more. In 2011, European Commission has been conducted a survey over 190,000 filled questionnaires from students, teachers, and principals in 27 European Union (EU) countries to analysis the Information and Communication Technology (ICT) in ISCED level-1 (primary level of education), ISCED level-2 (lower secondary level of education), and ISCED level-3 (upper secondary level of education), distinguishing level-3A academic and level-3B vocational [1]. Here, the authors considered principals' dataset belongs to all the ISCED levels to predict the gender of European schools' principal based on the survey response. In this research work, only response from the ICT questionnaire was considered. The machine learning (ML) is applications of artificial intelligence (AI) research that can access data and use it learn for themselves. It makes intelligent decisions based on data [2]. The supervised learning assumes that training examples are labeled by class labels with classified and predictive modeling is a most popular one as it forecasts a target and dependent attribute based on the value of other attributes [3]. In ML, Bayes networks, artificial neural networks and support vector machines, and different knowledge representation models are used for the classification and regression problems. Here, various classifiers are used for learning decision trees and rules are used to support decision-making methods [4]. The SVM is a supervised learning model that represents the binary classification for linear and nonlinear versions [5]. It optimally separates the data into two categories and performs the classification by constructing the $N$-dimensional hyperplane [6]. Random forest (RF) is a combination of tree predictors so that each tree depends on the values of a random vector with the same distribution for all trees in the forest that sampled independently [7]. The multilayer perception or artificial neural network (ANN) is simple mathematical models for the distribution over $X$ or both $X$ and $Y$, but also intimately associated with a learning algorithm [8]. The binary logistic regression (LR) is suitable for a binary classification problem. Previously, students' demographic features such as residence state and gender were predicted using the supervised machine learning classifiers [9–12]. Also, an experimental study was conducted to predict the age group of the university's students [13].

In this research work, four supervised machine learning algorithms, i.e., sequential minimal optimization (SMO), multilayer perception, random forest (RF), and binary logistic regression (LR) are applied by using the ML software (Weka version 3.8.3.) to predict the principals' gender of based on the ICT questionnaire. Further, the authors introduced concept of the web server in the real-time prediction of principals' gender of schools toward ICT. In the real-time systems, the absolute deadline for task begins with time zero and the relative deadline is with respect to the task released time. According to the paper [14], real-time tasks are produced due to the occurrence of either internal or external events. Also, real-time predictive models of European school students' nationality to monitor the online ICT access and ICT-based activities were suggested by paper [15]. Also, the prediction of the Indian and Hungarian university student's attitude in the real time was conducted in the paper [16]. With the proposed web server concept, all the procedures are now real time and automatic. This smart approach saves the human interaction time in terms of extracting, processing, and representing the dataset obtained from the survey. The concept of the web application was proposed in the paper [17]. This approach could be applicable to other areas of AI like robot navigation with obstacle avoidance in unknown environment [18]. The present study is conducted to predict the gender of European school's principal in real time using the web server.

The present study is categorized into five major sections. Section 2 explains techniques to preprocess, train-test, classify, and validation. Section 3 explores the experimental results at various levels of schools ISCED. Here, knowledge flow environment is used for generating the ROC curve and comparing the results based on the genders. Section 4 discusses the smart approach to process the data automatically and evaluate the Weka results using the web server. Section 5 concludes the essence of the present study with a significant recommendation for future work.

## 2 Methods and Techniques

### *2.1 Dataset*

Four secondary datasets of European school's principal have been downloaded from the European Union (EU) website. There is a various level of school's division (ISCED level-1 (primary level of education), ISCED level-2 (lower secondary level of education), and ISCED level-3 (upper secondary level of education), distinguishing level-3A academic and level-3B vocational) participated in this survey. In this research work, only questionnaire specific from the Information and Communication Technology (ICT) is considered. The survey is divided into seven parts: device used in the schools (desktop computer, laptop whether internet access, digital reader, mobile phone, interactive whiteboard, data projector), number of computers and interactive whiteboard (computer laboratories, classrooms, school library, other locations in school), Internet access(broadband speed, main means), ICT maintenance (school-school staff, school-external company contracted by the school, school-external unit arranged by educational authorities, school-other), Available at school (homepage or website, school email addresses, local and wireless LAN, virtual learning environment), professional development of teachers (introductory internet use, advanced courses on applications and internet use, equipment-specific training, pedagogical use of ICT, learning applications, multimedia, participation in peer learning communities), ICT coordinator (school, available full time, rewarded for function, supporting pedagogical ICT use), ICT teaching problematic (not enough (computers,

**Table 1** ESSIE datasets of European's principal

| Level | Initial instance | Initial attribute | Missing value | N/A value | Missing gender |
|---|---|---|---|---|---|
| ISCED 1 | 3019 | 208 | 38,828 | 77,163 | 86 |
| ISCED 2 | 2995 | 208 | 34,460 | 73,226 | 81 |
| ISCED 3A | 2248 | 208 | 36,971 | 60,217 | 45 |
| ISCED 3B | 1873 | 208 | 35,713 | 50,551 | 53 |

laptops, internet-connected computers, interactive whiteboard), insufficient bandwidth, teachers lacking skills), lacking support (technical, pedagogical), lack (teaching material, national language content, integrating ICT in curriculum, pedagogical models), (school time organization), (school space organization), and so on. The authors have trained and tested four different datasets (ESSIE) using *K*-fold with four classifiers. It has been collected by the European Commission Information Society and Media Directorate-General which is also available online [1] in which ISCED levels dataset has been explained below in Table 1.

## 2.2 Preprocessing

Before use dataset, it is essential to improve data quality [19]. There are a few numbers of techniques used for data preprocessing [20] as aggregation, sampling, dimension reduction, variable transformation, and dealing with missing values. To make quality and stabilized dataset, Weka version 3.8.1 tool is applied to four different files belongs to different ISCED levels. The *Missing Value* filter counted 38,828 missing values in the ISCED-1 level dataset, 34,460 in ISCED-2, 36,971 in ISCED-3A, and 35,713 in ISCED-3B level. Now, there is a need to handle the missing data values. The missing values are handled with the *Replace Missing Value* filter which replaces missing values with mean and mode values of the whole dataset. The re-scaling of the dataset to the range of 0–1 is achieved using the *Normalize* filter normalized all data except target attribute gender. In the four datasets, questions numbered from SC01Q01A to SC06Q01, from SC23Q01A to SC23Q05E, and from SC25Q01 to SC21att are removed by using the self-reduction method as they are indexed and mean values. Hence, after applied self-reduction, 164 attributes with 2933 instances are selected from ISCED-1, 164 attributes with 2914 instances are selected from ISCED-2, 164 attributes with 2203 instances are selected from ISCED-3A, and 164 attributes with 1820 instances are selected from ISCED-3B. The processed dataset has been explained below in Table 2.

The gender attribute has three classes 1-female, 2-male, *X*-misplaced. Hence, 265 instances belong to *X* category are removed manually from all the ISCED levels. We considered the gender as a class variable and it is converted to nominal from numeric. The ESSIE survey used the mixed approach of data scale such as nominal,

**Table 2** Processed datasets of European's principal

| Level | Initial instance | Initial attribute |
|---|---|---|
| ISCED 1 | 2933 | 164 |
| ISCED 2 | 2914 | 164 |
| ISCED 3A | 2203 | 164 |
| ISCED 3B | 1820 | 164 |

ordinal, interval, and categorical, etc. The responses provided by schools' principals were numeric in nature.

## 2.3 *Testing and Validation*

The training and testing of the datasets are conducted using *K*-fold cross-validation method (CV) which is validation approach to sampling the dataset. In this, $k$ number of subset among which one is used as ($k$–1) test set and rest of sets shall be ($k$) train sets [10]. We applied *K*-fold CV method with fixed $k = 10$ folds.

## 2.4 *Knowledge Flow Environment*

To predict the gender, we used knowledge flow environment (KFE) which is a substitute for the Weka Explorer. The experimental layout of supervised machine learning with filters, classifiers, evaluators, and visualizers interactively on a 2D canvas is shown in Fig. 1. The extracted datasets are trained, tested, and validated using four classifier algorithms at 10-fold CV. To predict the gender of principal, we used four supervised machine learning classifiers such as sequential minimal optimization (SMO), multilayer perceptron(ANN), random forest (RF), and logistic regression (LR) in KFE.

## 2.5 *Performance Evaluation*

The authors have considered the following performance measures are used to justify the gender prediction:

a. Accuracy: The number of correct predictions of DA status from overall predictions.
b. Error: The number of incorrect predictions of DA status from the overall predictions.

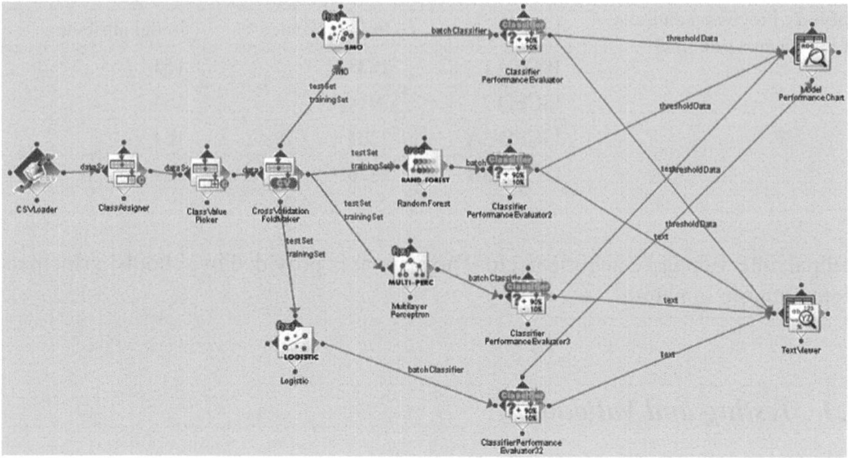


**Fig. 1** Knowledge flow environment for the gender prediction

c. Receiver operating characteristics curve (ROC): Receiver operating characteristics curve shows the true positive rate (sensitivity or TPR) on the $y$-axis and false positive rate (1-specificity or FPR) at $x$-axis at various cutoffs.
d. Area Under ROC (AUC): To show the accuracy of models' area under the curve of ROC is also appropriate.
e. Confusion matrix: It shows results in the form of matrix whereas rows present actual gender and column presents predicted gender.

# 3 Experiments Results, Analysis, and Evaluation

## *3.1 Confusion Matrices*

To produce significant confusion matrices machine learning classifier of Weka software played a vital role to evaluate results. The results of an experiment based on using 10-fold cross-validation for ISCED levels are represented in Tables 3, 4, 5, and 6. It is the combination of four confusion matrices generated by respective models.

a. At ISCED level-1 schools, the maximum number of correct prediction for female principals (1451) is provided by the RF with 164 attributes but it fails to correctly classify of male principals (242) as shown in Table 3. After feature extraction, LR outperformed the RF in the prediction of principal gender. At ISCED level-1 schools, LR with 164 attributes predicted correctly 1269 females out of total 2060 and predicted correctly 424 males out of 688. Therefore, LR performed well in the prediction of the gender of principals at ISCED level-1 schools.

**Table 3** Confusion matrices at 10-folds cross-validation for ISCED-I level

| Models | | SMO | | ANN | | RF | | LR | |
|---|---|---|---|---|---|---|---|---|---|
| Prediction | | Prediction | | Prediction | | Prediction | | Prediction | |
| Gender | | F | M | F | M | F | M | F | M |
| Actual | F | 1281 | 412 | 1115 | 578 | 1451 | 242 | 1269 | 424 |
| | M | 748 | 492 | 654 | 586 | 904 | 336 | 703 | 537 |

**Table 4** Confusion matrices at 10-folds cross-validation for ISCED-II level

| Models | | SMO | | ANN | | RF | | LR | |
|---|---|---|---|---|---|---|---|---|---|
| Prediction | | Prediction | | Prediction | | Prediction | | Prediction | |
| Gender | | F | M | F | M | F | M | F | M |
| Actual | F | 903 | 658 | 858 | 703 | 1126 | 435 | 1046 | 515 |
| | M | 581 | 772 | 617 | 736 | 757 | 596 | 671 | 682 |

**Table 5** Confusion matrices at 10-folds cross-validation for ISCED-III A level

| Models | | SMO | | ANN | | RF | | LR | |
|---|---|---|---|---|---|---|---|---|---|
| Prediction | | Prediction | | Prediction | | Prediction | | Prediction | |
| Gender | | F | M | F | M | F | M | F | M |
| Actual | F | 415 | 505 | 446 | 474 | 291 | 629 | 418 | 502 |
| | M | 298 | 985 | 489 | 794 | 203 | 1080 | 339 | 944 |

**Table 6** Confusion matrices at 10-folds cross-validation for ISCED-III B level

| Models | | SMO | | ANN | | RF | | LR | |
|---|---|---|---|---|---|---|---|---|---|
| Prediction | | Prediction | | Prediction | | Prediction | | Prediction | |
| Gender | | F | M | F | M | F | M | F | M |
| Actual | F | 693 | 331 | 658 | 366 | 780 | 244 | 683 | 341 |
| | M | 447 | 349 | 437 | 359 | 526 | 270 | 416 | 380 |

b. At ISCED level-2 schools, the maximum number of correct predictions for female principals (1126) is provided by the RF with 164 attributes but it fails to correctly classify of male principals (435) as shown in Table 4. After feature extraction, LR performs little better than the RF in the prediction of principal gender. At ISCED level-2 schools, LR with 164 attributes predicted correctly 1046 females out of total 1764 and predicted correctly 515 males out of 869. Therefore, LR performed slightly better in the prediction of the gender of principals at ISCED level-II schools.
c. At ISCED level-3A schools, the maximum number of correct prediction for female principals (418) is provided by the LR with 164 attributes and also it provides correct prediction of male principals (502) as shown in Table 5. After feature extraction, SMO outperformed the RF in the prediction of principal gender. At ISCED level-3A schools, SMO with 164 attributes predicted correctly 415 females out of total 654 and predicted correctly 505 males out of 795. Therefore, SMO performed excellently in the prediction of the gender of principals at ISCED level-III A schools.
d. At ISCED level-3B schools, the maximum number of correct prediction for female principals (780) is provided by the RF with 164 attributes but it fails to correctly classify of male principals (244) as shown in Table 6. After feature extraction, LR performed slightly better than the RF in the prediction of principal gender. At ISCED level-3B schools, LR with 164 attributes predicted correctly 683 females out of total 1170 and predicted correctly 341 males out of 584. Therefore, LR performed satisfactorily in the prediction of the gender of principals at ISCED level-3B schools.

## 3.2 *Evaluation Matrices and ROC's*

To produce significant evaluation matrices, machine learning classifier of Weka software played a vital role to evaluate the performance of the gender prediction. The results of an experiment based on using 10-fold cross-validation for ISCED levels are represented in Tables 7, 8, 9, and 10. It can measure the performance of the features responsible for gender prediction.

**Table 7** Evaluation matrices for ISCED level-I

| Models | SMO | ANN | RF | LR |
|---|---|---|---|---|
| AUC | 0.618 | 0.655 | 0.681 | 0.683 |
| ROC | 0.577 | 0.597 | 0.624 | 0.628 |
| Accuracy (%) | 60.5 | 58 | 61 | 61.6 |
| Error (%) | 40 | 42 | 39 | 38.4 |

Table 8 Evaluation matrices for ISCED level-2

| Models | SMO | ANN | RF | LR |
|---|---|---|---|---|
| AUC | 0.578 | 0.602 | 0.640 | 0.637 |
| ROC | 0.575 | 0.571 | 0.617 | 0.621 |
| Accuracy (%) | 57.5 | 54.7 | 59.1 | 59.3 |
| Error (%) | 42.5 | 45.3 | 40.9 | 40.7 |

Table 9 Evaluation matrices for ISCED level-3A

| Models | SMO | ANN | RF | LR |
|---|---|---|---|---|
| AUC | 0.492 | 0.493 | 0.548 | 0.536 |
| ROC | 0.609 | 0.579 | 0.632 | 0.624 |
| Accuracy (%) | 63.5 | 56.3 | 62.2 | 61.8 |
| Error (%) | 36.5 | 43.7 | 37.8 | 38.2 |

Table 10 Evaluation matrices for ISCED level-3B

| Models | SMO | ANN | RF | LR |
|---|---|---|---|---|
| AUC | 0.593 | 0.495 | 0.513 | 0.499 |
| ROC | 0.558 | 0.572 | 0.594 | 0.590 |
| Accuracy (%) | 57.3 | 55.9 | 57.7 | 58.4 |
| Error (%) | 42.7 | 44.1 | 42.3 | 41.6 |

a. Data from Table 7 maximum accuracy 61.6% is achieved by LR classifier with 164 attributes to predict the gender of teachers at ISCED level-1 schools. The maximum area under the curve (AUC) is found 0.683 by LR with 164 attributes to prove the significant relevance of overall accuracy of the model for prediction. Again, receiver operating characteristics (ROC) curve is found 0.628 by LR which is above the benchmark of ROC curve as shown in Fig. 2. Further, LR has low misclassification error 38.4%.
b. Data from Table 8 maximum accuracy 59.3% is achieved by LR classifier with 164 attributes to predict the gender of teachers at ISCED level-2 schools. The maximum area under the curve (AUC) is found 0.637 by LR with 164 attributes to prove the significant relevance of overall accuracy of the model for prediction. Again, receiver operating characteristics (ROC) curve is found 0.621 by LR which is above the benchmark of ROC curve in Fig. 3. Further, LR has low misclassification error 40.7%.
c. Data from Table 9 maximum accuracy 63.5% is achieved by SMO classifier with 164 attributes to predict the gender of teachers at ISCED level-3A schools. The maximum area under the curve (AUC) is found 0.548 by RF with 164 attributes to prove the significant relevance of overall accuracy of the model for prediction. Again, receiver operating characteristics (ROC) curve is found 0.632 by RF

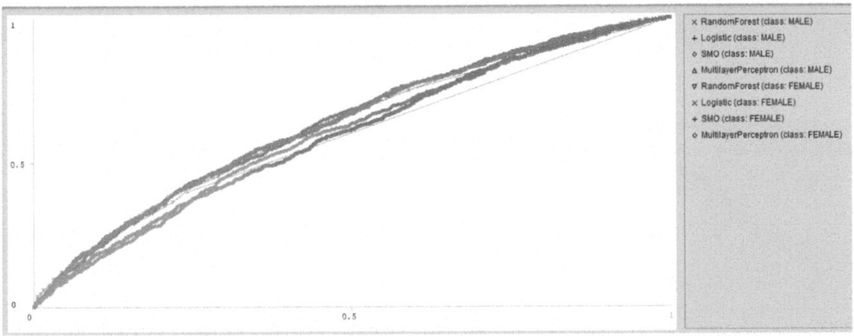


**Fig. 2** ROC of gender prediction at 10-Fold at ISCED level-I (primary level of education)

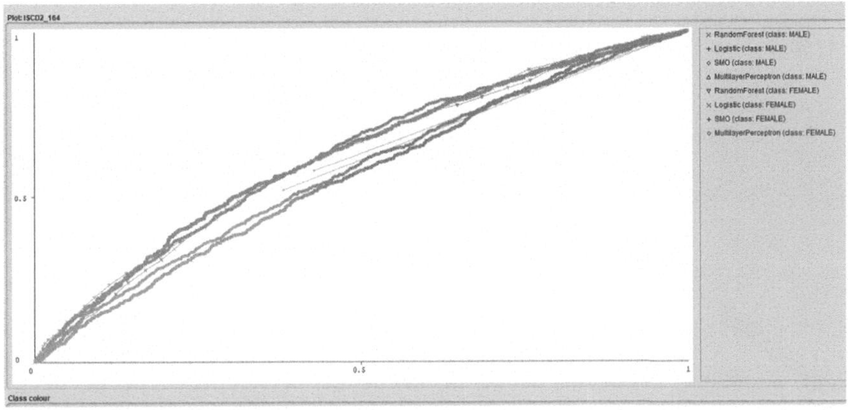


**Fig. 3** ROC of gender prediction at 10-Fold at ISCED level-2 (lower secondary level of education)

which is above the benchmark of ROC curve in Fig. 4. Further, SMO has low misclassification error 40.7%.

d. Data from Table 10 maximum accuracy 58.4% is achieved by LR classifier with 164 attributes to predict the gender of teachers at ISCED level-3B schools. The maximum area under the curve (AUC) is found 0.593 by SMO with 164 attributes to prove the significant relevance of overall accuracy of the model for prediction. Again, receiver operating characteristics (ROC) curve is found 0.594 by RF which is above the benchmark of ROC curve in Fig. 5. Further, LR has low misclassification error 41.6%.

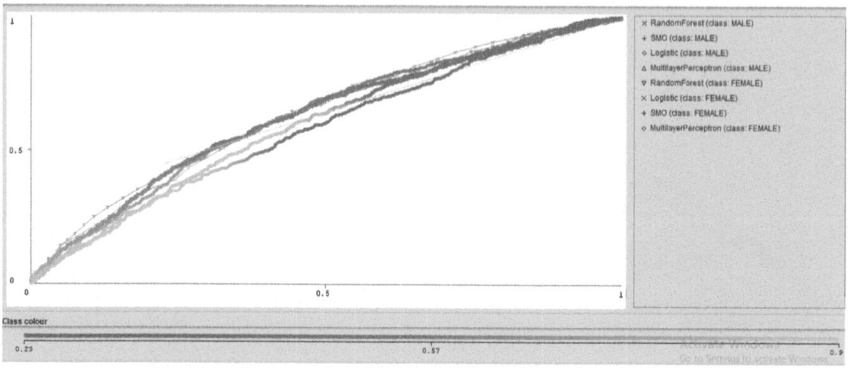


**Fig. 4** ROC of gender prediction at 10-Fold at ISCED level-3A (academic senior level of education)

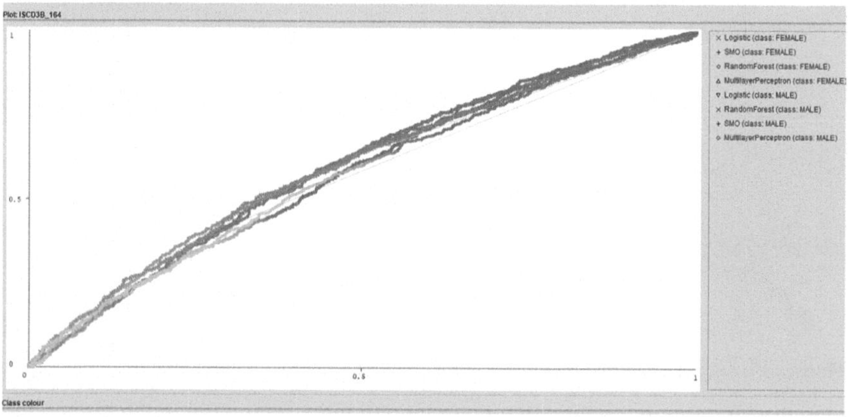


**Fig. 5** ROC of gender prediction at 10-Fold at ISCED level-3B (vocational senior level of education)

## 4 Web Server for Real-Time Prediction

All the operations during the gender prediction process require a dedicated human to download data from the EU website, processed the data manually, compile the data, and generate the results. It is time-consuming as well as there are more chances of errors. Therefore, the authors proposed the evaluation web server as shown in Fig. 6. It has an application named as SmartApp and evaluation database. It can extract the data from EU web server, processed the data with algorithm, and finally interact with the Weka software. All the results of European's principal gender prediction are displayed by the SmartApp and data is stored in the evaluation database. SmartApp is a web-based application could be stored in the HTML, JSP format.

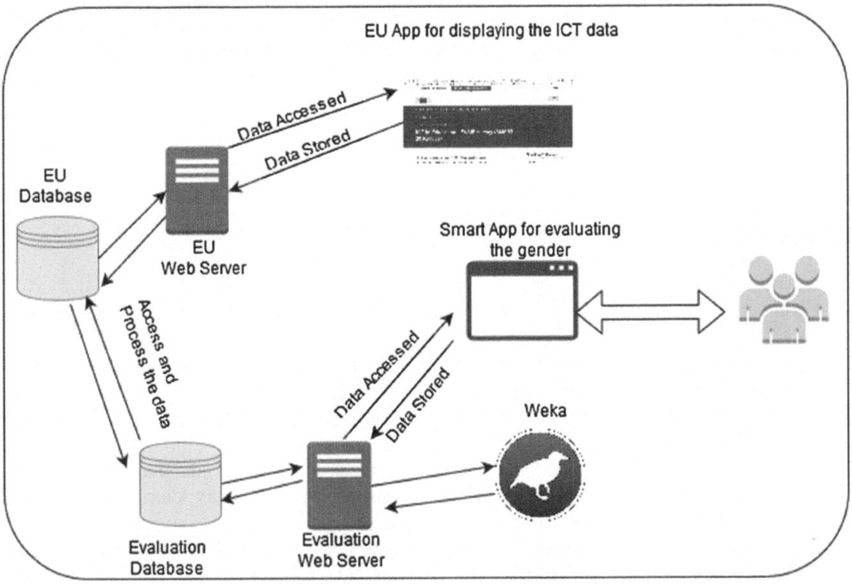


**Fig. 6** User interaction from the smart application

The evaluation database could be postgresql, MongoDB or Microsoft Azure SQL. This is the smart approach as it saves the time, predicts gender on the real time, and efficient user interaction through the SmartApp.

## 5 Conclusion

This paper has stabilized the attributes of European Commission datasets to predict the gender of principals at all the levels of education such as ISCED level-1, ISCED level-2, ISCED level-3A, and ISCED level-3B using machine learning. During the experimental study, LR has been proven as best approach that trained ISCED level-1, ISCED level-2 dataset, and ISCED level-3B dataset with *k*-fold cross-validation to predict the gender of the European principals. The maximum accuracy is achieved with 164 attributes by LR (61.6%) as compare to RF (61%) to predict principal gender at ISCED level-1. Again, LR classifier obtained the highest accuracy (59.3%) as compare to RF (59.1%) to predict principal gender at ISCED level-2. LR classifier also obtained the highest accuracy (58.4%) as compare to RF (57.7%) to predict principal gender at ISCED level-3B. But, SMO classifier obtained the highest accuracy (63.5\%) as compare to RF (62.2\%) to predict principal gender at ISCED level-3A. Finally, evaluation web server saves the time and represents the data smartly. The server is on the conceptual level.

**Acknowledgements** The authors are thankful to the European Commission to provide ESSIE dataset online. Also, second author's project is also sponsored by the Hungarian Government and Co-financed by the European Social Fund under the project "Talent Management in Autonomous Vehicle Control Technologies (EFOP-3.6.3-VEKOP-16-2017-00001)."

## References

1. European Commission.: https://ec.europa.eu/digital-single-market/news/ict-education-essie-survey-smart-20100039 (2011)
2. Meera Gandhi, G.: Machine learning approach for attack prediction and classification using supervised learning algorithms. Int. J. Comput. Sci. Commun. **1**(2), 247–250 (2010)
3. Furnkranz, J., Gamberger, D., Lavrac, N.: Foundations of rule learning, pp. 1–17. Springer Science & Business Media (2012)
4. Clerc, M.: The Swarm and the Queen: towards a Deterministic and adaptive particle swarm optimization. In: Proceedings of the IEEE Congress on Evolutionary Computation (CEC), pp. 1951–1957. IEEE, (1999)
5. Deshmukh, B., Patil, A., Pawar, B.: Comparison of classification algorithms using weka on various datasets. Int. J. Comput. Sci. Info. Technol. **4**(2), 85–90 (2011)
6. Cheu, R.L., Srinivasan, D., Tian E.: Support vector machine models for freeway incident detection. In: Intelligent Transportation Systems, Proceedings. vol. 1, pp. 238–243. IEEE (2003)
7. Kumar, S.R.: Comparative analysis of weka data mining algorithm random forest, random tree and lad tree for classification of indigenous news data. Int. J. Emerg. Technol. Adv. Eng. **5**(1), 507–517 (2015)
8. Minsky, M., Papert, S.: Perceptrons: an introduction to computational geometry. MIT press (2017)
9. Verma, C., Ahmad, S., Stoffová, V., Illés, Z., Dahiya, S.: Gender prediction of the European school's teachers using machine learning: preliminary results. In: International Advance Computing Conference, pp. 213–220. IEEE, India (2018)
10. Verma, C., Ahmad, S., Stoffová, V., Illés, Z.: Forecasting residence state of indian student based on responses towards information and communication technology awareness: a primarily outcomes using machine learning. In: International Conference on Innovations in Engineering, Technology and Sciences, In. Press., IEEE, India (2018)
11. Verma, C., Stoffová, V., Illés, Z., Dahiya, S.: Binary logistic regression classifying the gender of student towards computer learning in European schools. In: The 11th Conference of Ph.D. Students in Computer Science, pp. 45. Szeged University, Hungary (2018)
12. Verma, C., Stoffová, V., Illés, Z.: An ensemble approach to identifying the student gender towards information and communication technology awareness in european schools using machine learning. Int. J. Eng. Technol. **7**, 3392–3396 (2018)
13. Verma, C., Stoffová, V., Illés, Z.: Age group predictive models for the real time prediction of the university students using machine learning: preliminary results. In: International Conference on Electrical, Computer and Communication, In. Press., IEEE, India (2019)
14. Verma, C., Stoffová, V., Illés, Z.: Rate-Monotonic versus early deadline first scheduling: a review. In: International Conference on education technology and Computer Science in Building Better Future, pp. 188–193. University of Technology and Humanities, Poland (2018)
15. Verma, C., Ahmad, S., Stoffová, V., Illés, Z. Singh, M.: National identity predictive models for the real time prediction of european schools students: preliminary results. In: International Conference on Automation, Computational and Technology Management, In. Press., IEEE, London (2019)

16. Verma, C., Illés, Z., Stoffová, V.: Attitude prediction towards ict and mobile technology for the real-time: an experimental study using machine learning. In: The 15th International Scientific Conference eLearning and Software for Education, In. Press., university Politehnica of Bucharest, Romania (2019)
17. Bathla, Y., Albkree, M.: Web application for 3D software comparison considering PLM system. Hadmérnök **13**(1), 302–313 (2018)
18. Kumar, N., Vamossy, Z.: Robot navigation with obstacle avoidance in unknown environment. Int. J. Eng. Technol. **7**(4), 2410–2417 (2018)
19. Han, J., Kamber, M., Pei, J.: Data mining (Third Edition). In: The Morgan Kaufmann Series in Data Management Systems, pp. 1–38 (2012)
20. Hongbo, D.: Data mining techniques and applications: an introduction. In: Cengage Learning, pp. 21–97. (2010)

# GPU and CUDA in Hard Computing Approaches: Analytical Review

**Hardik Singh, Raavi Sai Venkat, Sweta Swagatika and Sanjay Saxena**

**Abstract** Hard computing, i.e., conventional computing, requires an exactly expressed analytical model. At the time of implementation of hard computing techniques, each time exact results are guaranteed. The fundamental premises and standards behind hard computing are precision, certainty, and rigor. The main problem of hard computing is the time consumption of different algorithms such as kriging interpolation algorithm, Smith–Waterman algorithm, LZW compression, and many more. GPU provides an efficient way to do massively parallel operations such as taking the square root of all values in a region of memory quickly. The operational frequency of GPU is slower than CPU but has more cores than the CPU. This is the main reason why the above algorithms perform better on GPU architecture compared to the CPU. CUDA-based implementation of several hard computing techniques on GPU gives more significant results in terms of time. This article provides an overview of the current literature of CUDA and GPU-based implementation of several hard computing techniques.

**Keywords** Graphics processing unit (GPU) · Computed unified device architecture (CUDA) · Hard computing techniques · Computational approaches

H. Singh · R. S. Venkat · S. Swagatika · S. Saxena (✉)
Department of Computer Science & Engineering, IIIT Bhubaneswar, Bhubaneswar, Odisha, India
e-mail: sanjay@iiit-bh.ac.in

H. Singh
e-mail: Id-B516057@iiit-bh.ac.in

R. S. Venkat
e-mail: B116040@iiit-bh.ac.in

S. Swagatika
e-mail: B516047@iiit-bh.ac.in



P. K. Singh et al. (eds.), *Proceedings of ICRIC 2019*, Lecture Notes in Electrical Engineering 597, https://doi.org/10.1007/978-3-030-29407-6_15

# 1 Introduction

In 1996, L.A. Zade (LAZ) introduced the term hard computing. According to LAZ, we term the computing as hard computing, if the precision result is guaranteed. The control action is unambiguous. And control action is formally defined (i.e., with mathematical model or algorithm). A graphics processing unit (GPU) is an unichip processor mainly used to increase the performance of video and graphics. Unified memory technology is supported by Nvidia which helps in memory management in GPU-based applications [1]. This architecture does not require the need for duplicating the information between the host system, GPU memory, and main memory. The memory shared between CPU, GPU, and host machine is known as a unified memory. Depending on where the threads are running whether on the host machine or GPU side, unified memory is automatically managed. GPU has more memory bandwidth and over thousands of hardware thread context with over hundreds of parallel compute threads making it suitable for highly parallel computation-intensive workloads. There is a 2.5–3.0 times increase in the GPU's performance every year compared to CPU's performance. CUDA developed by Nvidia provides an extensive programming framework that is suitable for GPU [2]. In CUDA, there is a hierarchy of threads, blocks, and grids, and each thread is executed in a SIMT fashion. There is a virtual mapping of threads to a random number of Streaming Multiprocessors (SMs) with the help of wraps. This article provides a brief comprehensive analysis of different hard computing techniques based on GPU and CUDA by considering current kinds of literature [3]. The organization of the paper is as follows: Sect. 2 describes GPU and CUDA. Different hard computing techniques are described in Sect. 3. Section 4 gives a brief introduction of a comprehensive analysis of different algorithms implemented on GPU and CUDA. The conclusion is given in Sect. 5. Valuable references are given in Sect. 6.

# 2 GPU and CUDS

The GPU's propelled abilities were utilized basically for 3D game rendering. Today, GPU's capabilities are mainly used to speed up computational loads in the fields of oil-gas exploration, scientific research, and financial modeling. GPUs are highly optimized for taking large data and performing the same operation quickly, while PC microprocessor has the characteristics to skip all over the place [4]. A CPU can work on a variety of different calculations, while a GPU (the brawn) concentrates on a particular task with all its computing power. CPU cores are mainly designed for sequential serial processing. CPU can greatly enhance the performance of a single task within a particular job, although it has wide-ranged tasks. In GPU, there are over thousands of smaller and effective cores capable of handling multiple processes simultaneously. They are about 50–100 times faster in the field that requires parallel processing such as machine learning and big data analysis. GPU's CUDA provides

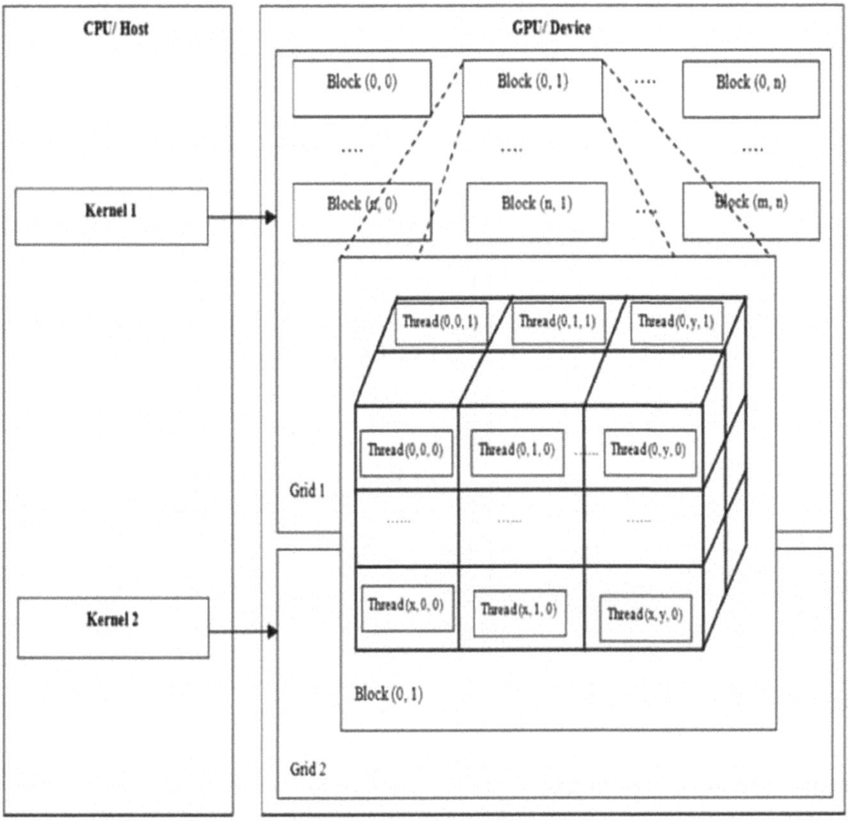


**Fig. 1** CUDA architecture [35]

a programming framework for Nvidia's GPU. Figure 1 gives the block diagram that represents the model of CUDA architecture, and Fig. 2 gives the architecture of the CUDA flow model.

## 3 Hard Computing Applications

Hard computing is based on numerical analysis, binary logic crisp systems, and crisp software. It is deterministic, requires exact input data, and is strictly sequential. Applications of hard computing include solving numerical problems (e.g., roots of the polynomial), searching and sorting techniques, and computational geometry problems (shortest tour in a graph, finding closest pair of points given a set of points, etc.). In general, any problem which can be solved within a fixed number of steps or *P*-class problem is solved using hard computing. Furthermore, applications of hard

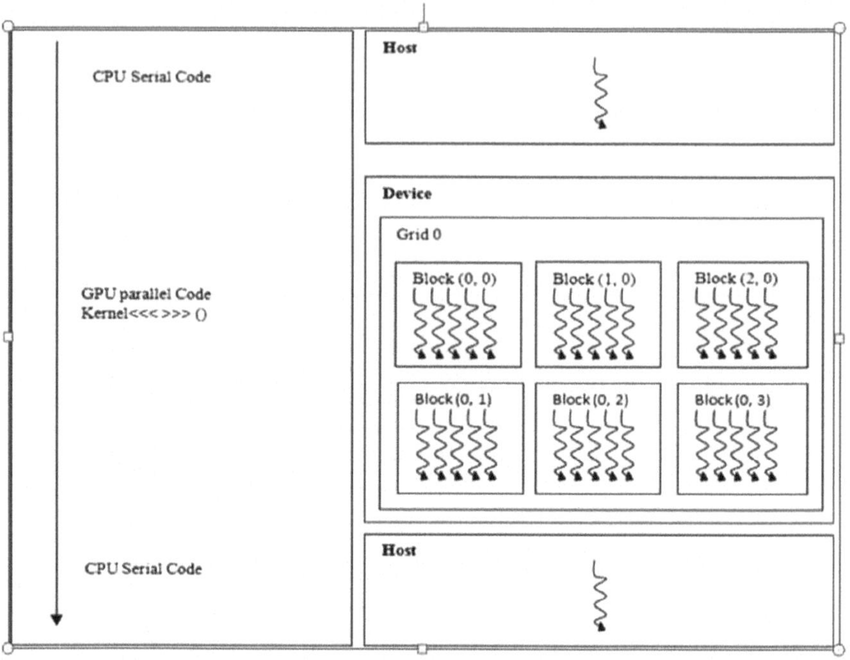


**Fig. 2** CUDA flow model [35]

computing can be seen in graph theory, deterministic finite automatons, scheduling problems, and other compute-intensive problems that have fixed input and can achieve complete accuracy and exclusiveness in the output.

# 4 Comprehensive Analysis of Different Algorithms Implemented on GPU

## *4.1 Description*

The similarity in genetic sequences has been an imperative piece of genomic research. To efficiently find IRs in DNA sequences, hashing is used. We can locate the IR by hashing each frame of $F$ bases in the forward direction and accessing the location of its downstream match.

#### 4.1.1 Results

Using the CUDA implementation, imperfect IRs can be found 18× faster compared to other imperfect IR extraction methods [5].

#### 4.1.2 Type of GPU

Nvidia Tesla.
K40 m, 12 GB VRAM.

### *4.2 Description*

The subset-sum problem comes under the category of NP-complete problems. This problem can be parallelized into three different architectures, i.e., a 240-core NVIDIA FX 5800 graphics processing unit (GPU), a 16-processor IBM ×3755 shared memory machine, and a 128-processor Cray Extreme Multi-threading (XMT) massively multi-threaded machine [6].

#### 4.2.1 Results

NVIDIA GPGPU is best suitable for small subset-sum problem sizes; it is observed that for large problems, the Cray XMT performs well, and for medium problems, the IBM ×3755 performs well.

#### 4.2.2 Type of GPU

Nvidia FX 5800, 4 GB.

### *4.3 Description*

With the help of register communication, parallel merge sort for the highly parallel systems do not get affected by oversegmentation while previous comparison-based sort do get affected. Following steps are there in this process:

(1) Eight elements are sorted in each thread making use of maximum register bandwidth.
(2) Binary then-linear searching approach.

(3) Using a register window and moving shared memory, oversegmentation is avoided [7].

#### 4.3.1 Results

Compared to thrust sort, efficient parallel merge sort is found to be 70% faster than non-stable optimized key-value sort, two times and three times faster than their key-value implementation and key-only implementation.

#### 4.3.2 Type of GPU

Nvidia GTX 580.

### *4.4 Description*

Edit distance plays a crucial role in natural language translations, virus scanners, security kernels, etc. The baseline algorithm used is Needleman–Wunsch alignment [8].

#### 4.4.1 Results

By careful management of memory usage and control flow divergence, this algorithm provides a 9.3× speedup over an efficient multi-threaded, CPU-based implementation.

#### 4.4.2 Type of GPU

NVIDIA GeForce GTX 275 GPU.

### *4.5 Description*

Computation is transformed into bit-wise operation and post-processing step; the LCS algorithm is optimized for one-to-many problems [9].

### 4.5.1 Results

Using the bit-vector operations, based on a independent computation of each one-to-one LCS comparison, inter-task parallelism is adopted. Multi-GPU approach made possible to get TeraCUPS performance in the estimation of the length of MLCS problem which is first for the LCS algorithm.

### 4.5.2 Type of GPU

NVIDIA M2090.
Fermi GPU.

## *4.6 Description*

Cryptography is a technique mainly used to meet safety requirements like authentication andprivacy. The SSL and TLS provide a secure connection on the Internet, and 70% of the time is devoured in executing the handshake phase by the RSA algorithm. This time can be significantly reduced by using a parallelized RSA algorithm instead of sequential [10].

### 4.6.1 Results

Runtime decreases by a significant amount using *P*-1 algorithm implementation. Runtime further decreases with an increase in test size.

### 4.6.2 Type of GPU

Nvidia GPU.

## *4.7 Description*

Simplex algorithm is mostly used for solving Linear Programming (LP) problems. The LP problem is used to find the optimal solution of a given problem [11].

#### 4.7.1 Results

GPU implementation is 24.5 faster than the best solution available.

#### 4.7.2 Type of GPU

NVIDIA Tesla C2050.

### *4.8 Description*

To count the number of cycles in a graph with approximate algorithm comes under the category of NP-complete problem using CUDA, based on backtracking with the lookahead technique. This algorithm is based on the assumption that all vertices start with vertex *s* and are numbered. The algorithm repeats the operation for all vertices of the graph [12].

#### 4.8.1 Results

Using the GPU thread-based algorithm to detect cycle significantly reduces the time required for computation compared to CPU.

#### 4.8.2 Type of GPU

NVIDIA 9500 GS.

### *4.9 Description*

In this work, MD simulation is used for simulating the large-sized system of liquid metal atoms. This simulation uses fine-grained spatial method. This method reduces the computation time of neighbor list update and interaction force calculation within the simulation model which decreases the overall computation time [13].

#### 4.9.1 Results

GPU implementation is 1.5 to 2 times faster than CPU 16-cores clusters and is nine to 11 times faster than sequential implementation.

#### 4.9.2 Type of GPU

NVIDIA 480 GTX, 580GTX and M2050.

### *4.10 Description*

In this algorithm, an adaptive mesh generation technique is used to minimize the number of particles present in each cell. A strategy named cell-based strategy is used to perform computations on each cell separately from other cells over a single time-step and also supplies a group of threads for each cell to perform the computations for that corresponding cell [14].

#### 4.10.1 Results

The code executed on a single CPU is 20 times slower than a single 448 core GPU.

#### 4.10.2 Type of GPU

NVIDIA Tesla C2050.

### *4.11 Description*

In this work, authors have thorough rCUDA, a progressive context offering remote acceleration of CUDA that permits a lessening on the numeral of accelerators in a cluster, providing improved litheness [15].

### *4.12 Description*

In this work, parallel computing approach is used to counter the failures of the progressive approach, i.e., time taken for the multiple sequence arrangements (NP-hard) which is exponentially high for the progressive approach of the large instances. For this, they introduced a GPU version of ClustalW v2.0.11 called CUDA ClustalW v1.0, which reduces the time taken exponentially [16].

#### 4.12.1 Results

ClustalW is 33 times slower than CUDA ClustalW.

#### 4.12.2 Type of GPU

NVIDIA Tesla C2050.

### 4.13 Description

MKP is an NP-hard problem that has more than one constraint in comparison with the knapsack problem and is solved by an algorithm based on Ant Colony Optimisation (ACO). The functions used to construct or modify solutions are probabilistic functions that are calculated through the trails left by the ants to communicate with other ants [17].

#### 4.13.1 Results

The time of execution and quality of solution are balanced according to the required conditions. For the best solution, the longer is the time required.

### 4.14 Description

For the simulation of elastically deformable objects, the author in this work used a multigrid approach. For large deformations, co-rotated strain formulation is used for simulating it. The effective method of solution for solving partial differential equations (PDE) on finite element discretizations implements modern work on multigrid models [18].

#### 4.14.1 Results

GPU implementation is 27 times faster compared to CPU implementation of a single-core CPU and four times faster than when implemented on eight CPU cores.

#### 4.14.2 Type of GPU

NVIDIA GeForce GTX 480.

## *4.15 Description*

Kriging algorithms are a cluster of interpolation methods and are used in many of the geological applications. The computation time of the algorithm is reduced by the help of some advanced mathematical functions, matrix-vector, and matrix-matrix multiplication. Time-consuming loops are reconstructed by minimizing memory transfer [19].

### 4.15.1 Results

Sequential program implementation is 18 times slower than CUDA implementation. OpenMP version on Quad cores CPU is 6.32 slower than CUDA.

### 4.15.2 Type of GPU

NVIDIA GeForce GTX 470.
NVIDIA GeForce 9800GT.

## *4.16 Description*

The numerical integration of SDE is a great tool for analysis of a different variety of problems. Generally, analytical methods cannot solve stochastic differential equations of practical value, and so direct numerical methods are used to obtain the solutions [20].

### 4.16.1 Results

GPU implementation is 675 times faster than CPU.

### 4.16.2 Type of GPU

NVIDIA Tesla 1060C.

## *4.17 Description*

Due to the exponential increase in data size, parallel computing is one of the best computing solutions available. For low volume data sets, we use GPU on-chip registers to minimize time for the data retrieval. For huge data volume data sets, we use an algorithm which uses matrix multiplication and on-chip GPU memory to high-computational speed [21].

### 4.17.1 Results

GPU-based algorithms are 3 to 8 times faster than the best-reported solutions.

### 4.17.2 Type of GPU

NVIDIA GTX 280.

## *4.18 Description*

The simulation model used for simulating accidental release on a local scale is the stochastic Lagrangian particle model. Stochastic Lagrangian particle model which is used as a benchmark for decision-making strategies on environmental studies is implemented on CUDA for better performance [22].

### 4.18.1 Results

GPU implementation is 80–120 faster than CPU implementation.

### 4.18.2 Type of GPU

NVIDIA GeForce 8800.
NVIDIA GeForce 8800 GTX.

## *4.19 Description*

Knapsack problem comes under the category of NP-hard problem which is solved here using the firefly algorithm based on firefly flashing. The main idea of the firefly algorithm is taken from the optical relationship among fireflies, and it is used because it is successful and inexpensive [23].

#### 4.19.1 Results

When the number of fireflies are less, GPU execution time is greater than CPU execution time, and when the number of fireflies are more, GPU execution time is lesser than the CPU execution time. Speedup observed is 320 times faster as compared to CPU when CUDA is implemented.

#### 4.19.2 Type of GPU

NVIDIA GeForce 9500 GT.

### *4.20 Description*

Harmony search is used to find the most optimal solution (here best harmony) by melody improvisation. Here, harmony search is used to solve the Traveling Salesman Problem which can be applied to find the most efficient route for data to travel between various modes [24].

#### 4.20.1 Results

The observed speedup achieved is significant for harmony search, parallel implementation using GPU. Specifically, for problems of larger size, the speedup is 32 times in comparison to problems of smaller size.

#### 4.20.2 Type of GPU

NVIDIA Fermi C2050 server.

### *4.21 Description*

This paper focuses on speeding up the Lempel–Ziv–Welch compression of TIFF images using CUDA. In LZW compression, an input string characters are converted into a string of codes with the help of a code table containing the mapping from strings to codes. The paper talks about the inverse of LZW compression, i.e., the LZW decompression [25].

#### 4.21.1 Results

The GPU implementation of this problem using CUDA is observed to be faster by three times as compared to the CPU implementation.

#### 4.21.2 Type of GPU

NVIDIA GeForce GTX 980.

### *4.22 Description*

The small scope approximation (SSA) is a method to evaluate the electromagnetic scattering from rough surfaces. To overcome the problem of bandwidth bottleneck, data transmission is reduced. To reduce the memory usage, the large spectrum of the surface is divided into further parts, and then calculations are done [26].

#### 4.22.1 Results

In this particular case, it is observed that GPU implementation of the problem is weaker than the CPU implementation, but the SSA methods based on CUDA executed on a GPU having 336 cores are observed to have a much faster speed of calculation as compared to the serial SSA method.

#### 4.22.2 Type of GPU

NVIDIA GeForce GTX 460.

### *4.23 Description*

One of the successful forms of optimization observed is the Simulated Annealing method. By using the algorithm, a global optimum can be found amongst the presence of local optima of huge numbers. It is used in operational research and IC design. The paper focuses on an appropriate number of iterations for SA and its parallelization [27].

#### 4.23.1 Results

Compared to the serial SA, the parallelization of SA done using CUDA improves the results by 27%. The best speedup is observed up to 14.84 as compared to the sequential version.

#### 4.23.2 Type of GPU

NVIDIA GeForce GTX 680.

### *4.24 Description*

The Smith–Waterman algorithm is categorized as a dynamic programming algorithm that creates an implicit array. In the array created, every cell represents a sub-problem in the alignment problem. The algorithm is used to find similar regions among the two strings of nucleic acid sequences or protein sequences [28].

#### 4.24.1 Results

Performance using CUDA is 14.5 GCUPS. It is approximate twice the performance observed on the CPU and more than 50% faster than the Sony Playstation implementation.

#### 4.24.2 Type of GPU

NVIDIA 9800 GX2.

### *4.25 Description*

The FDTD (finite difference time domain) method is today's most well-known technique for the solution of electromagnetic problems. It is used for solving Maxwell's equations in complex geometries. Here, it is used in the field of biomedical engineering to simulate an electromagnetic field using a human model [29].

#### 4.25.1 Results

The implementation of GPU using CUDA in this problem makes it 29 times efficient than the CPU.

#### 4.25.2 Type of GPU

NVIDIA Tesla C1060.

### *4.26 Description*

The models that are used for the canonical problem of dispatching, their computation time fundamentally rely upon the problem scale factor. Huge RAM data is required by the complicated parallel algorithm of the NP-hard problem of dispatching [30].

#### 4.26.1 Results

Computation time can be reduced using FPGA as a coprocessor, it has parallel computation.

#### 4.26.2 Type of GPU

Nvidia GeForce 680.

### *4.27 Description*

It is very difficult to find the roots of higher degree polynomial. We have a direct method only for degree 1 or 2. In the Durand–Kerner root method, the solution is refined till it converges. Parallel implementation of this algorithm produces better results [31].

#### 4.27.1 Results

Using the parallel implementation of this method, a polynomial size of 48,000 can be executed in 373.944 s, whereas the sequential version uses 4510 s. Therefore, parallel version of algorithm execution time is ten times faster than sequential.

#### 4.27.2 Type of GPU

Nvidia Tesla C2070.

### *4.28 Description*

The single source shortest path (SSSP) problem is used to calculate the path that is shortest and its corresponding distances from a source to other points. Here, Dijkstra's algorithm helps in solving the problem based on the principle of relaxation in which firstly an approximate distance is calculated and that is gradually replaced by more accurate calculations until we reach the correct distance calculation [32].

#### 4.28.1 Results

With respect to the CPU version, the GPU implementation using CUDA in this problem shows speedups from $13\times$ to $220\times$, and a performance gain up to 17% is recorded.

#### 4.28.2 Type of GPU

NVIDIA GeForce GTX 680.

### *4.29 Description*

String sorting falls in a general sorting category which comprises of integer sorting, multi-key sorting, and variable-length key sorting. Here, the challenging part is variable key length sorting. Here, the discussed algorithms are burstsort, multi-key quicksort, and MSD radix sort [33].

#### 4.29.1 Results

GPU implementation of radix sort by removing singleton segments has a speedup of more than ten times compared to other string sorting approaches. Banerjee et al. hybrid implementation (CPU + GPU) merge sort is observed to be faster by 20% for keys of fixed length and faster by 24% for keys of variable length compared to Davidson et al.

#### 4.29.2 Type of GPU

NVIDIA GPU.

### *4.30 Description*

The interacting multiple model (IMM) algorithm makes use of a lot of models for interaction through the mixing of state for tracking a target navigating through an arbitrary path [34].

#### 4.30.1 Results

There is a significant improvement in the performance of the algorithm based on GPU compared to the traditional CPU implementation. Using the GPU to solve the problem also makes the filtering error relatively small.

## 5 Conclusion

After going through several current kinds of literature, we have found that GPU and CUDA play a very vital role in several techniques or methods for faster computing. Almost speedup obtained in each method discussed above gives substantial outcomes because GPU provides multiple streaming processors to solve a problem. For example, several authors have implemented NP-complete problems using GPUs. Knapsack problem which is the example of NP-hard problem is implemented using the firefly algorithm based on firefly flashing. The main idea of the firefly algorithm is taken from the optical relationship among fireflies, and it is used because it is successful and inexpensive. Other algorithms such as kriging interpolation algorithm, *k*-means algorithms, and many more give significant results on GPU. Further, it is recommended that the use of CUDA is optional to use GPU. However, several different software provides the inbuilt functionality for the use of GPU.

## References

1. Saxena, S., Sharma, S., Sharma, N.: Parallel image processing techniques, benefits and limitations. Res. J. Appl. Sci. Eng. Technol. **12**(2), 223–238 (2016). https://doi.org/10.19026/rjaset.12.2324
2. Saxena, S., Sharma, N., Sharma, S., Singh, S., Verma, A.: An automated system for atlas based multiple organ segmentation of abdominal CT images. British J. Math. Comput. Sci. **12**(1), 1–14 (2016)

3. Huzmiev, I.K., Chipirov, Z.A.: Energy consumption powered by graphics processing units (GPU) in response to the number of operating computing unit. In:2016 2nd International Conference on Industrial Engineering, Applications and Manufacturing (ICIEAM) (2016)
4. Anuranjeeta, A., Saxena, S., Shukla, K., Sharma, S.: Cellular image segmentation using morphological operators and extraction of features for quantitative measurement. Biosci. Biotechnol. Res. Asia **13**(2), 1101–1112 (2016). https://doi.org/10.13005/bbra/2139
5. Baskett, W., Spencer, M., Shyu, C.R.: Efficient GPU-accelerated extraction of imperfect inverted repeats from DNA sequence. In: IEEE International Conference on Bioinformatics and Biomedicine (BIBM) (2017)
6. Bokhari, S.S.: Parallel solution of the subset-sum problem: an empirical study. In: Concurrency and Computation: Practice & Experience archive, vol. 24 issue 18 (2012)
7. Davidson, A., Tarjan, D.,Garland, M., Owens, J.D.: sEfficient parallel merge sort for fixed and variable length key. Innovative Parallel Computing (InPar) (2012)
8. Farivar, R., Kharbanda, H., Venkataraman, S., Campbell, R.H.: An algorithm for fast edit distance computation on GPUs. In: 2012 Innovative Parallel Computing, InPar (2012)
9. Ozsoy, A., Chauhan, A.,Swany, M.: Achieving teracups on longest common subsequence problem using GPGPUs. In: International Conference on Parallel and Distributed Systems (2013)
10. Vargas, M.P.P., Rodriguez, R.A.A., Parra, O.J.S.: Algorithm for the optimization of RSA based on parallelization over GPU SSL/TLS protocol. In: 2017 IEEE International Conference on Smart Cloud (SmartCloud) (2017)
11. Lalami, M.E., Baz, D.E., Boyer, V.: Multi GPU implementation of the simplex algorithm. In: 2011 IEEE International Conference on High Performance Computing and Communications (2011)
12. Mahdi, F., Safar, M.,Mahdi, K.: An algorithm for detecting cycles in undirected graphs using CUDA Technology. In: Digital Information and Communication Technology and Its Applications. DICTAP. Communications in Computer and Information Science, vol. 167. Springer, Berlin, Heidelberg (2011)
13. Jie, L., Li, K., Shi, L., Liu, R., Mei, J.: Accelerating solidification process simulation for large-sized system of liquid metal atoms using GPU with CUDA. J. Comput. Phys. **257**, 521–535 (2014)
14. Goldsworthy, M.J.: Monte Carlo algorithm for real gas flows. Comput Fluids (2014). https://doi.org/10.1016/j.compfluid.2014.01.033
15. Pena, A.J., Reano, C., Silla, F., Mayo, R., Orti, E.S.Q., Duato, J.: A complete and efficient CUDA-sharing solution for HPC clusters. Parallel Comput. **40**(10), 574–588 (2014). https://doi.org/10.1016/j.parco.2014.09.011
16. Hung, C.L., Lin, Y.S., Lin, C.Y., Chung, Y.C., Chung, Y.F.: CUDA ClustalW: an efficient parallel algorithm for progressive multiple sequence alignment on Multi-GPUs. Comput. Biol. Chem. **58**, 62–68 (2015)
17. Fringler, H., Caceres, E.N., Mongelli, N., Song S.W.: A CUDA based solution to the multi-dimensional knapsack problem using the ant colony optimization. Procedia Comput. Sci. **29**, 84–94 (2014)
18. Dick, C., Georgii, J., Westermann, R.: A real-time multigrid finite hexahedra method for elasticity simulation using CUDA. Simul. Model. Pract. Theory **19**(2), 801–816 (2011)
19. Cheng, T.: Accelerating universal Kriging interpolation algorithm using CUDA-enabled GPU. Comput. Geosci. **54**, 178–183 (2013)
20. Januszewski, M., Kostur, M.: Accelerating numerical solution of stochastic differential equations with CUDA. Comput. Phys. Commun. **181**(1), 183–188 (2010)
21. Li, Y., Zhao, K., Chu, X., Liu, J.: Speeding up k-Means algorithm by GPUs. J. Comput. Syst. Sci. **79**(2), 216–229 (2013)
22. Molnar, F., Szakaly, T., Mesazaros, R., Lagzi, I.: Air pollution modeling using a graphics processing unit with CUDA. Comput. Phys. Commun. **181**(1) 105–112 (2010)
23. Hajarian, M., Shahbahrami, A., Hoseini, F.: A parallel solution for the 0–1 knapsack problem using firefly algorithm. In: 2016 1st Conference on Swarm Intelligence and Evolutionary Computation (CSIEC) (2016)

24. Chain Le K., Abdullah, R.: Parallel strategies for harmony search on CUDA to solve traveling salesman problem. In: 2015 5th Conference on IT Convergence and Security (ICITCS) (2015)
25. Funasaka, S., Nakano, K., Ito, Y.: Fast LZW compression using a GPU. In: 2015 Third International Symposium on Computing and Networking (CANDAR) (2015)
26. Jiang, W.-Q., Zhang, M.. Wei, P.B., Nie, D.: Spectral decomposition modeling method and its application to EM scattering calculation of large rough surface with SSA method. In: IEEE Journal of Selected Topics in Applied Earth Observations and Remote Sensing, vol. 8, issue 4 (2015)
27. Wei, K.C., Wu, C.C., Yu, H.L.: Mapping the simulated annealing algorithm on to CUDA GPUs. In: 2015 10th International Conference on Intelligent Systems and Knowledge Engineering (ISKE) (2015)
28. Ligowski, L., Rudnicki, W.: An efficient implementation of smith-waterman algorithm on GPU using CUDA, for massively parallel scanning of sequence databases. In: 2009 IEEE International Conference on Parallel & Distributed Processing (2009)
29. Nagaoka, T., Watanabe, S.: A GPU-based calculation using the three-dimensional FDTD method for electromagnetic field analysis. In: 2010 Annual International Conference of the IEEE Engineering in Medicine and Biology (2010)
30. Fedosenko, Y.S., Reznikov, M.B., Plekhov, A.S., Chakirov, R., Houlden, N.: Scaling analysis of solving algorithms for canonical problem of dispatching in the context of dynamic programming. In: 2017 Internet Technologies and Applications (ITA) (2017)
31. Ghidouche, K., Couturier, R., Sider, A.: A parallel implementation of the durand-kerner algorithm for polynomial root-finding on GPU. In: 2014 International Conference on Advanced Networking Distributed Systems and Applications (2014)
32. Arranz, H.O., Torres, Y., Llanos, D.R., Escribano, A.G., A new GPU based approach to the shortest path problem. In: 2013 International Conference on High Performance Computing & Simulation (HPCS) (2013)
33. Neelima, B., Narayan, A.S., Prabhu, R.G.: String sorting on multi and many-threaded architectures: a comparative study. In: 2014 International Conference on High Performance Computing and Applications (ICHPCA) (2014)
34. Xin, Z., Shen, X.: A new approach to accelerate IMM algorithm of maneuvering target tracking based on CUDA. In: 2012 IEEE International Workshop on Electromagnetics: applications and Student Innovation Competition (2012)
35. Kalaiselvi, T., Sriramakrishnan, P., Somasundaram, K.: Survey of using GPU CUDA programming model in medical image analysis. **9**, 133–144 (2017)

# IoT-Based Home Automation with Smart Fan and AC Using NodeMCU

**Raj Desai, Abhishek Gandhi, Smita Agrawal, Preeti Kathiria and Parita Oza**

**Abstract** In today's world of the twenty-first century, the Internet of Things (IoT) has emerged exponentially. Many applications are developed in these fields of automation. When it comes to home automation, this concept can be integrated to make it smarter. It makes it easier to access and monitor different home appliances. This paper shows how IoT can be used for smart home automation using NodeMCU and android mobile application. The main focus of the paper is, how the sensor nodes collect the data and pass it to the mobile devices to perform some action as per the user's commands and provide support to IoT-based mutual controlling between fan and AC based on threshold temperature value.

**Keywords** Internet of Things (IoT) · Home automation · NodeMCU · Arduino Uno · Android · Smart fan and AC

## 1 Introduction

In this fast working world where the people are busy, it is difficult to manage and control home appliances manually. This work reduces their time and works as they can control the home by a single click on their smart devices [1]. It is like a blessing for elderly and disabled persons also. It gives remote access to the member of the house by which they can control the appliances. As energy is a very crucial factor

---

R. Desai · A. Gandhi (✉) · S. Agrawal · P. Kathiria · P. Oza
Department of CE, Institute of Technology, Nirma University, Ahmedabad, India
e-mail: 16mca015@nirmauni.ac.in

R. Desai
e-mail: 16mca012@nirmauni.ac.in

S. Agrawal
e-mail: smita.agrawal@nirmauni.ac.in

P. Kathiria
e-mail: preeti.kathiria@nirmauni.ac.in

P. Oza
e-mail: parita.oza@nirmauni.ac.in



P. K. Singh et al. (eds.), *Proceedings of ICRIC 2019*, Lecture Notes in Electrical Engineering 597, https://doi.org/10.1007/978-3-030-29407-6_16

in life, energy cost is saved by using smart sensors like motion sensor detects the motion in the house and it works accordingly with intelligence.

Internet of Things (IoT) has grown rapidly in the last decade; many industries have adopted IoT as their major services. IoT deals with intelligent objects which are connected with sensors and collect data and can communicate with surrounding people using different communication technologies. The goal of IoT is to control and manage objects around us in the more intelligent and useful manner [2] which improves the lifestyle of the humans by providing security, safety, and entertainment as a result of cost-effective living.

Statista has stepped into market analytics to build bridges between different sources, methodologies and survey periods and delivers consistent data points to the most relevant topics [3]. In Fig. 1, Statista provides revenue generated through the smart home market and expected to generate until 2021 based on historical data. Experts have estimated that the revenue of the home automation market will reach $12.81 billion by 2030.

Home automation system can be developed using a single controller which has the ability to instruct and control various appliances such as lights, fans, AC [4], power plugs, sensors, and security systems [5]. The greatest advantage of home automation is that through smart devices like smartphone, laptop, PCs, or tablets [6], and it managed and controlled various home appliances. Home automation system uses various wireless technologies [7] such as Bluetooth [8], ZigBee [9, 10], Wi-Fi [11], GSM [12] to communicate with microcontrollers. There are many IoT-based home appliances in which all the appliances can be controlled from their smart devices anywhere at any time as represented in Fig. 2. Here, the proposed IoT-based home

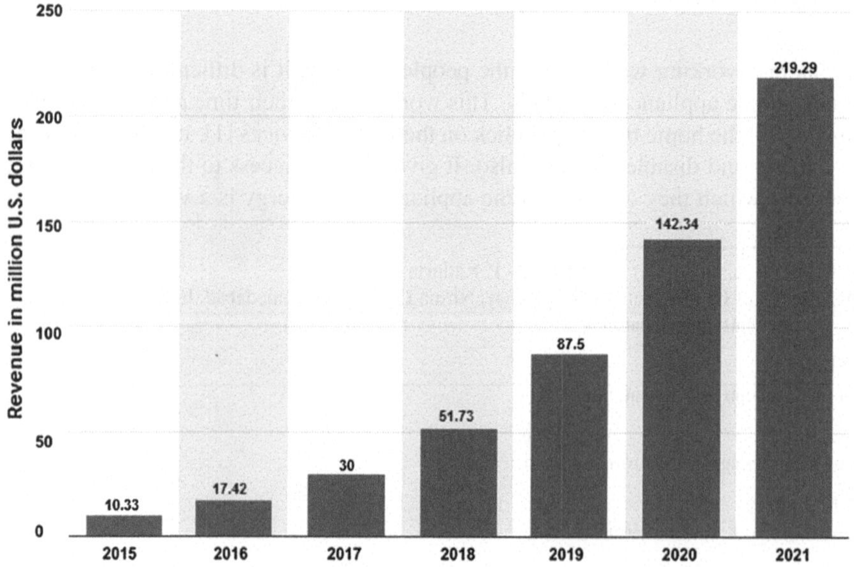


**Fig. 1** Revenue in smart home market [3]

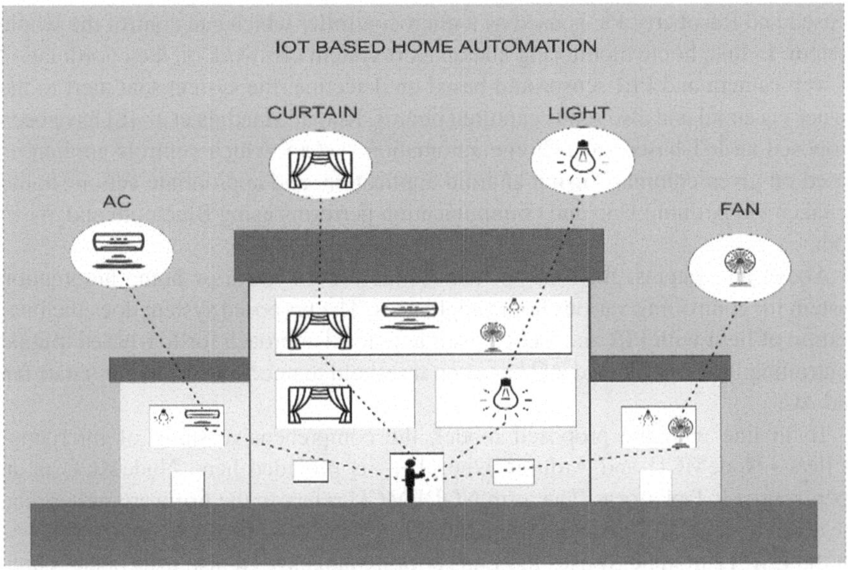


**Fig. 2** Concept of IoT-based home automation

automation system with integration of devices like a smart fan and AC (i.e., mutual controlling between fan and AC) for maintaining the room temperature which leads to saving power consumption.

The rest of this paper is organized as follows. Section 2—related work. Section 3—proposed model and its implementation. Section 4—challenges and research opportunity. Lastly, Section 5—concludes the paper.

## 2 Related Work

For home automation, appliances need to connect with sensor nodes and internet using microcontroller or microprocessor which operates through smart devices-based application. Momataz Soliman et al. [9] have been designed IoT enabled smart home using Arduino, where implementation is done using ZigBee technology in which their proposed model coordinate all devices like AC, light, and fan with Arduino UNO separately. So, that one by one operate all appliances (actuator) but devices did not communicate with each other. Sudhir Yadav et al. [13] have been proposed an IoT-based smart home security system in which model uses different types of sensors like PIR sensor, gas sensor, RFID, ultrasonic sensor, fingerprint sensor, water flow sensor, and so on for measuring required data and based on event occurrence, the system gives appropriate alert to the user via email or SMS. Tanwar et al. [14] have been proposed IoT-based home monitoring and alert system in which web camera 2.0, PIR sensor is used and Raspberry Pie is used as a microcontroller which can control the whole system. In this, home monitoring and an alert system can work on the coordination of web camera and PIR sensor and based on detection, the system sent alert to the owner via email and also sends captured photos. Kumar Mandula et al. [8] have been proposed an IoT-based smart home automation system, which controls appliances based on given command using android application, and appropriate action should be taken via Arduino Uno and communication performs using Bluetooth and Wi-Fi both.

Above all systems, focused on home monitoring system or home automation system for controlling various home appliances. The proposed system does the integration of light with PIR and LDR sensor and novel approach for IoT-based mutual controlling between fan and AC based on threshold temperature value for smart fan and AC.

To in line with the proposed model, the comprehensive study of microcontrollers—NodeMCU and Arduino Uno [15]—is provided here. NodeMCU is an open-source IoT platform. The term NODEMCU refers to the firmware rather than the development kit. It includes Esp8266 Wi-Fi socket and hardware which is based on the ESP 12 module. It uses the Lua scripting language. It also uses open-source projects like Lua-cjson, spiffs, etc. It uses Esp8266-12E as Wi-Fi module, similar to Esp8266 12 which has extra 6 GPIO's. NodeMCU uses the micro USB port for power, programming as well as debugging which has 2 $\times$ 2.54 mm 15-pin header with access to GPIO's, SPI, UART, ADC, and power pins. It also has reset and flash buttons. Power need to NodeMCU is 5 V via micro USB port. Dimensions of NodeMCU are 49 $\times$ 24.5 $\times$ 13 mm.

The Arduino Uno is a microcontroller board based on the ATmega328p. It has six analog inputs, 14 digital input/output pins, a 16 MHz ceramic resonator, a power jack, a USB connection, a reset button, and an ICSP header. Power it with an AC-to-DC adapter or connect it to a computer with a USB cable to get it started. It has SRAM of 2 kb EEPROM of 1 kb. Arduino can be powered through USB or by an external power supply. The output power is of two types 5 and 3.3 V. Arduino provides serial communication which is available on digital pins 0(Rx) and 1(Tx). Dimensions of Arduino are 2.7 in $\times$ 2.1 in [68.6 mm $\times$ 53.3 mm]. It is not directly connected with Wi-Fi instead of external communication chips like Esp8266 for Wi-Fi module have to be used.

For better understanding, the comparison between Arduino Uno and NodeMCU based on different parameters is provided in Table 1. Based on given comparative study and consideration of speed, storage, and other working aspects in the proposed model, NodeMCU is used for interfacing with sensor nodes and smart devices with a Wi-Fi module. System memory and clock speed of esp8266 is more than Arduino as well as flash memory that is also greater than Arduino of esp8266.

**Table 1** Comparison Arduino Uno and NodeMCU

| Parameters | Arduino Uno | NodeMCU |
|---|---|---|
| Processor | ATMega328p | – |
| System memory | 2 kb | <45 kb |
| Clock speed | 16 MHz | 26–52 MHz |
| Input voltage | 7–12 V | 5 V |
| Operating voltage | 5 V | 3.3 V |
| Flash memory | 32 kb | Up to 128 MB |
| Communication supported | IEEE 802.11 b/g/n IEEE 802.15.4 433RF BLE 4.0 via Shield | IEEE 802.11 b/g/n |
| Development environments | Arduino IDE | Arduino Ide, Lua Loader |
| Programming Languages | C, C++ | C, C++, Lua |
| I/O Connectivity | SPI, I2C, UART, GPIO | UART, GPIO |

## 3 Proposed System and Implementation

The proposed system consists of different components which are used based on different needs. These components can be classified based on the functionality of the component. Table 2 comprises of different categories of components based on sensor/hardware name, image of the component, its mechanism, and purpose to use.

The proposed system's layered architecture provides the appropriate solution in such a way that in future if one component of any layer is replaced or added, then no need to upgrade the whole system [13].

For example, right now data transmission in layer-2 is possible through Wi-Fi module, if required, it can be replaced or added with Bluetooth module then updation is required in that respective layer only. Figure 3 graphically represents the layered architecture of the proposed system and provides information related to component(s) exist in each layer for better understanding. For proposed layered architecture, the working of each layer is discussed herewith:

**Layer 1**: External hardware and sensors: According to the change in the environment, sensors handover data to the microcontroller.
**Layer 2**: Microcontroller and internet: Data provided by sensors is analyzed and then actions are taken according to it. The device by which data is transferred from microcontroller to a web server.
**Layer 3**: Web server: Command that is given by the user is transferred to the microcontroller using the web server.
**Layer 4**: Application: Actual user interface for the user interaction to execute commands.

### Proposed System Implementation

The proposed system implemented for android-based smart devices developed in Android Studio. When the application starts, the starting activity displays multiple

**Table 2** Different sensors and hardware specification

| Sensor/hardware name | Image of sensor/hardware | Mechanism | Major use |
|---|---|---|---|
| DTH 11 Temperature Sensor |  | Humidity measurement | Digital temperature and humidity sensor |
| PIR Sensor |  | Passive infrared | Motion detection, living moving object detection |
| LDR Sensor |  | Light detector | The resistance that changes with the light intensity that falls upon it |
| IR Transmitter |  | Receive infrared signal | In receives, infrared signals received from the remote control |
| Stepper Motor |  | Converts digital pulses into mechanical shaft rotation | Speed control |
| Esp8266 12e |  | Internet connectivity and micro controlling | It connects microcontroller with internet through Wi-Fi and also used as a controller |

buttons to the user for controlling different appliances. When the user selects any button, an appropriate command is passed to the web server, web server formulates appropriate JSON object corresponding to the command given by the user, which then passes the JSON object to Esp8266. Esp8266 than parses the JSON object and gives the data to NodeMCU which then performs suitable actions on to the appliances. The overall architecture of the proposed approach with the functions of all modules shown in Fig. 4.

For understanding the interaction between different components (i.e., objects/nodes) and state diagrams are represented in Fig. 5. The state diagram generally provides the abstract description of the behavior of the system. In Fig. 5, it shows the changes in the state as per the given command. Initially, the command is given to the server which in turn passes to the NodeMCU. Different sensors are used for the smart working of appliances like PIR sensor is used to detect motion, DHT11 sensor is for humidity and temperature information and so on.

Complete working android application module of the proposed system is shown in Fig. 6a, b. Complete working hardware module of the proposed system is shown in Fig. 6c, d.

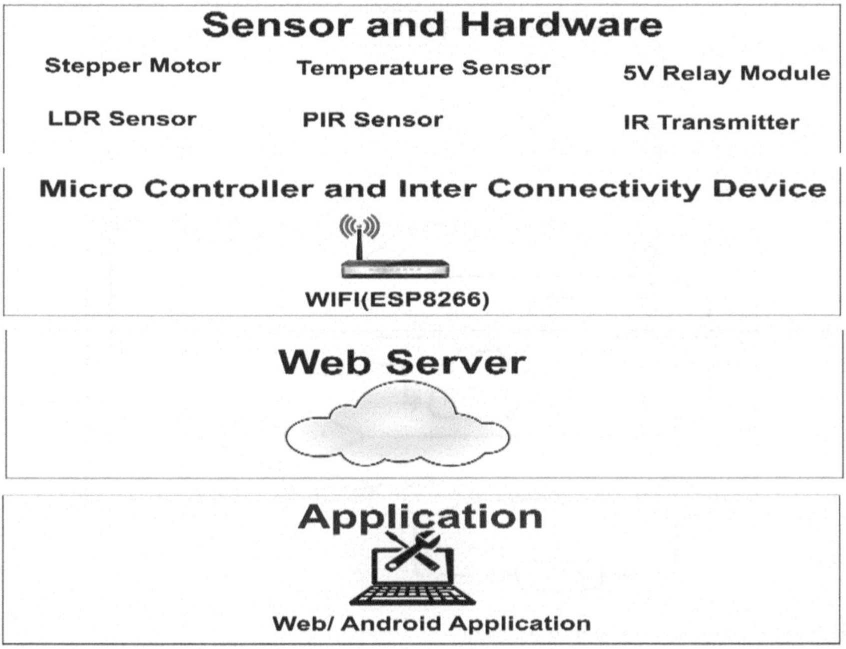


**Fig. 3** Layered architecture of the proposed system

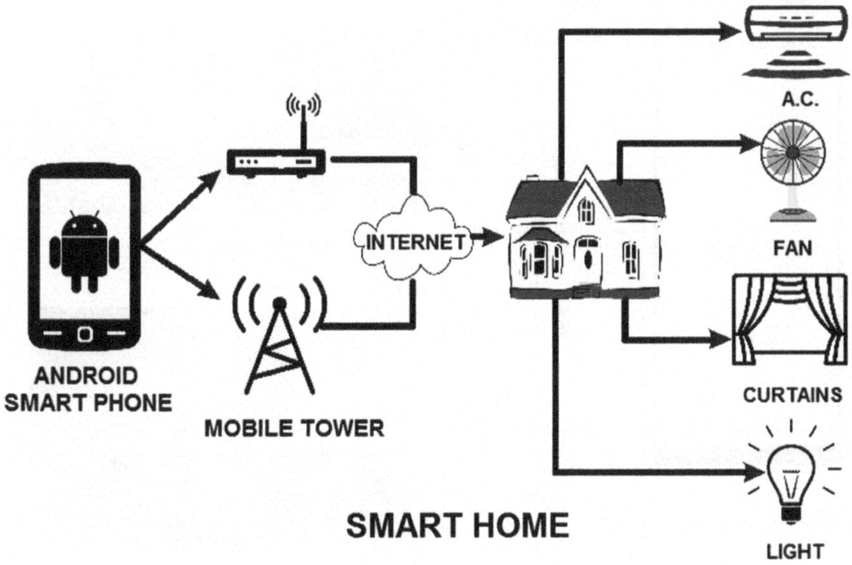


**Fig. 4** System architecture of the proposed system

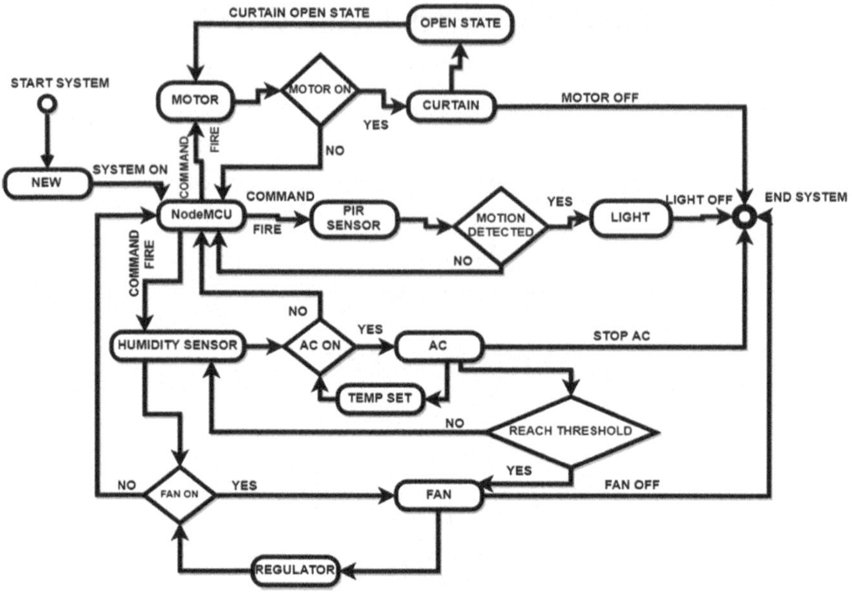


Fig. 5 State diagram of the proposed system

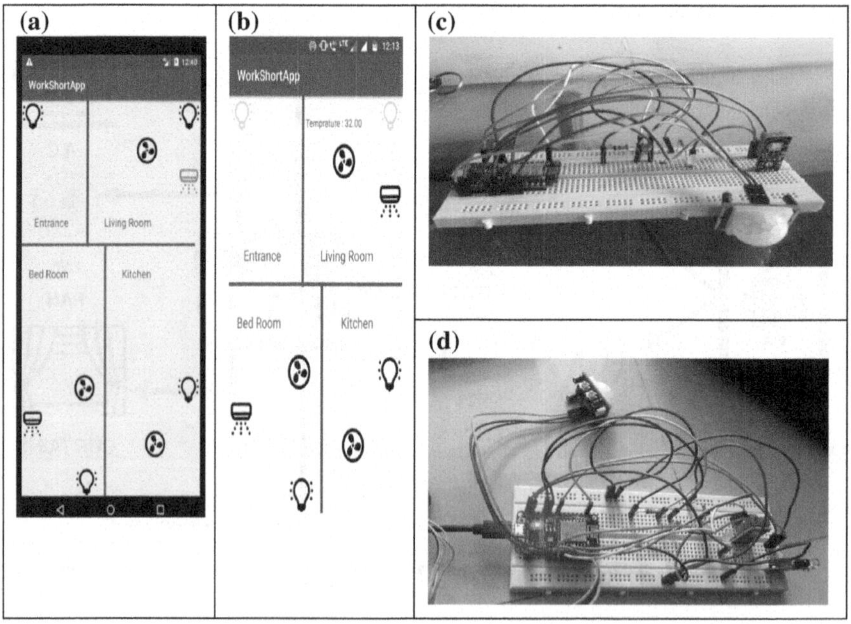


**Fig. 6** **a** working module in an android device. **b** working module in an android device. **c** working hardware module. **d** working hardware module

Figure 6a, b provides an android application which communicates with hardware module as shown in Fig. 6c, d, respectively. Here, all the sensors are connected to NodeMCU and the appliances are managed through it. The real-time sensor data is passed onto NodeMCU. The data is then passed to the server where it is stored in JSON object, and it is then accessed from mobile. When the respective event occurred from mobile, the required command is passed as a JSON object to NodeMCU and afterward, it parsed to activate the appropriate pin(s).

## 4 Challenges and Research Opportunities

One of the primary challenges is to integrate the sensors and applications with embedded devices. Another main challenge is about the security and privacy for connected devices. As the deployed sensors give a large amount of data proper data handling, storing and security are the main concern.

### *4.1 Big Data*

A large number of data is being collected from sensors. It comes in the form of 3 V that is variety, velocity, and volume. This data cannot be simply stored or processed in central RDBMS system for that NoSQL database also be used [16, 17]. IoT big data analytics [18] can be used to find unseen patterns, trends, examine and reveal new information [19].

### *4.2 Distributed Computing*

To deal with the big data distributed approach is more efficient than the centralized data storage system. Rather than computing on cloud servers, computing chips are installed with sensors for computing [20].

### *4.3 Privacy and Security*

No specific architecture is there for IoT, there are different layers in it and attacks can be done on any of them such as denial of service attack, unauthorized access of tags, code injection, or tag cloning [21].

### *4.4 Edge Computing*

Storage and computing data on to cloud nowadays is a too lengthy process, instead, edge computing can be used by which local client machines can be effectively utilized and can provide timely and intelligent services [22, 23].

## 5 Conclusion

In this paper, the prototype of IoT-based home automation system is presented. It is shown that different appliances can be controlled through the task already defined. By connecting the web services and sensors, appliances controlling will become easier and more human-friendly. This paper also presents the proposed model for smart home automation using IOT integration with web services, different sensors, and microcontroller. The user can remotely control the home appliances through an android phone. The result of the proposed system has met our expectations, the sensors and the app work on real-time scenarios. The proposed IoT-based home automation system with integration of devices like a smart fan and AC (i.e., mutual controlling between fan and AC) for maintaining room temperature leads to saving power consumption. It may be concluded that the proposed system will be friendly and useful to all the generations as well as for easy controlling of home appliances. This paper also confirms the advantage of NodeMCU over Arduino in terms of cost, connectivity, and many more parameters.

## References

1. Baviskar, J., et al.: Real-time monitoring and control system for the greenhouse based on 802.15. 4 wireless sensor network. In: Fourth International Conference on Communication Systems and Network Technologies (CSNT). IEEE 2014
2. https://en.wikipedia.org/wiki/Internet_of_things/
3. https://www.statista.com/statistics/644418/smart-home-automation-revenue-in-india/
4. Kathiria, P., Vadgama, B., Shah, D., Ashishdeep, A., Agrawal, S.: IoT Based approach for controlling electrical peripheral devices of auditorium. Int. J. Adv. Res. Comput. Sci. **8**(5). (2017)
5. Felix, C., Raglend, I.J.: Home automation using GSM. In: Signal Processing, Communication, Computing and Networking Technologies (ICSCCN), 2011 International Conference on. IEEE (2011)
6. Al-Ali, A.-R., Al-Rousan, M.: Java-based home automation system. IEEE Trans. Consum. Electron. **50**(2), 498–504 (2004)
7. Oza, P., Sharma, P.: Optimized data aggregation protocol in WSN for automation of water sprinklers. Energy **1400**, 1600 (1800)
8. Mandula, K., et al.: Mobile-based home automation using the internet of things (IoT). In: International Conference on. IEEE Control, Instrumentation, Communication and Computational Technologies (ICCICCT) (2015)

9. Gill, K., et al.: A zigbee-based home automation system. IEEE Trans Consumer Electron. **55**(2) (2009)
10. Han, D-M., Lim, J.-H.: Design and implementation of smart home energy management systems based on zigbee. IEEE Trans. Consumer Electron. **56**(3) (2010)
11. Soliman, M., et al.: Smart home: Integrating the internet of things with web services and cloud computing. In: IEEE 5th International Conference on Cloud Computing Technology and Science (CloudCom), vol. 2. IEEE (2013)
12. Wang, M., et al.: An IoT-based appliance control system for smart homes. In: Fourth International Conference on Intelligent Control and Information Processing (ICICIP). IEEE (2013)
13. Yadav, S., Verma, J., Agrawal, S.: SUTRON: IoT-based industrial/home security and automation system to compete the smarter world. Int. J. Appl. Res. Info. Technol. Comput. **8**(2), 193–198 (2017)
14. Tanwar, S., Patel, P., Patel, K., Tyagi, S., Kumar, N., Obaidat, M.S.: An advanced Internet of Thing based security alert system for the smart home. In: 2017 International Conference on Computer, Information and Telecommunication Systems (CITS), pp. 25–29. IEEE (2017)
15. https://www.arduino.cc/
16. Agrawal, S., Verma, J.P., Mahidhariya, B., Patel, N., Patel, A.: Survey on Mongodb: an open-source document database. Database **1**(2), 4 (2015)
17. Agrawal, S., Patel, A.: A study on graph storage database of NoSQl. Int. J. Soft Comput. Artif. Intell. Appl. (IJSCAI) 5(1) (2016)
18. Verma, J.P., et al.: Big data analytics: Challenges and applications for text, audio, video, and social media data. Int. J. Soft Comput. Artif. Intell. Appl. (IJSCAI) 5(1) (2016)
19. https://www.whizlabs.com/blog/iot-and-big-data/
20. Chien, S.Y., Chan, W.K., Tseng, Y.H., Lee, C.H., Somayazulu, V.S., Chen, Y.K.: Distributed computing in IoT: system-on-a-chip for smart cameras as an example. In: The 20th Asia and South Pacific Design Automation Conference, pp. 130–135. IEEE (2015)
21. Ren, Z., Liu, X., Ye, R., Zhang, T.: Security and privacy on the internet of things. In: 2017 7th IEEE International Conference on Electronics Information and Emergency Communication (ICEIEC), pp. 140–144. IEEE (2017)
22. Ren, J., Pan, Y., Goscinski, A., Beyah, R.A.: Edge computing for the internet of things. IEEE Netw. **32**(1), 6–7 (2018)
23. Tanwar, S., Kumari, A., Tyagi, S., Kumar, N.: Verification and validation techniques for streaming big data analytics in the internet of things environment. IET Networks (2018)

# Sampling Approaches for Imbalanced Data Classification Problem in Machine Learning

**Shivani Tyagi and Sangeeta Mittal**

**Abstract** Real-world datasets in many domains like medical, intrusion detection, fraud transactions and bioinformatics are highly imbalanced. In classification problems, imbalanced datasets negatively affect the accuracy of class predictions. This skewness can be handled either by oversampling minority class examples or by undersampling majority class. In this work, popular methods of both categories have been evaluated for their capability of improving the imbalanced ratio of five highly imbalanced datasets from different application domains. Effect of balancing on classification results has been also investigated. It has been observed that adaptive synthetic oversampling approach can best improve the imbalance ratio as well as classification results. However, undersampling approaches gave better overall performance on all datasets.

**Keywords** Imbalanced dataset · Machine learning · Resampling · Undersampling · Oversampling

## 1 Introduction

Real-world datasets like medical data of cancerous and non-cancerous cases, credit card fraud data, intrusion detection datasets and sensor-based events reporting are inherently imbalanced. In an imbalanced dataset with binary-valued classes, one class of data far outnumbers the other class which is generally more important to study. Credit card transactional datasets, for example, have a typical ratio of 10,000:1 or more for legitimate transactions to fraudulent ones. The problem of imbalanced datasets in machine learning is an important research issue [1]. A dataset is imbalanced when out of two possible classes of outcomes, occurrence of one is very rare. The performance of chosen classification model depends upon quality and quantity

S. Tyagi (✉) · S. Mittal
Department of Computer Science and Engineering, Jaypee Institute of Information Technology Noida, Noida, U.P, India
e-mail: shivani.tg94@yahoo.com

S. Mittal
e-mail: sangeeta.mittal@jiit.ac.in

 
P. K. Singh et al. (eds.), *Proceedings of ICRIC 2019*, Lecture Notes in Electrical Engineering 597, https://doi.org/10.1007/978-3-030-29407-6_17

of training dataset. More the volume of training dataset, more is the expected accuracy of classification [2]. In imbalanced datasets with two-valued classes, accuracy may not show the clear picture of classification results. For example, if a dataset has 99% negative and 1% positive class values, then accuracy can reach up to 99% even if the classifier does not classify even a single instance of positive class correctly. In many applications like disease detection and intrusion detection, it is more important to correctly predict the minority class. That is, non-identification of a fraudulent application will be more expensive on part of company than suspecting a normal application to a credible one [3]. The actual assessment can be obtained by analysing other parameters like balanced accuracy, negative predictive values and positive predictive values.

In machine learning, class imbalance issue can be addressed by either tuning the dataset itself or by changing learning methods of underlying algorithm. At the algorithm level, the results of classifier are tuned by approaches like adaboosting, bagging and cost-sensitive learning [4]. With respect to data-level-based imbalance handling, occurrence of both classes is equated algorithmically to improve imbalance ratio. In this paper, we outline and evaluate various data-level imbalance handling methods for their capability of improving imbalanced ratio of five highly imbalanced datasets of different application domains.

The paper has been organized in four more sections. Section 2 describes the imbalance data classification and related problems. In Sect. 3, various resampling methods for class imbalance have been studied briefly. Results obtained on various metrics of comparison have been presented and discussed in Sect. 4. Section 5 includes the paper conclusion.

## 2 Resampling Methods for Class Imbalance

Data-level methods dealing with imbalanced datasets aim to obtain balanced distribution by performing some modification or alteration in number of majority and minority class instances [5]. This process to reduce imbalance by changing number of instances of each class type is known as resampling. Resampling can be done in two ways, namely undersampling and oversampling.

### *2.1 Undersampling*

When large number of instances are available in the original dataset, then undersampling can be an effective means to improve imbalance. Undersampling does balancing by decreasing the more frequently occurring class. Redundancy in data instances is used to choose samples to be removed and create a balance in classes to train classification models. Here, several undersampling techniques implemented in imbalance package of R language have been evaluated [6]. The techniques are summarized as follows:

**CNN—Condensed Nearest Neighbour**: In nearest neighbour classification, test instances are mapped to the label of instance of the training data to which it is most similar. In *K*-nearest neighbour (K-NN), most frequently occurring label among '*K*'-nearest neighbours of test case is assigned. However, K-NN is very slow, and as for every test instance, it checks all the training instances for similarity. CNN has been designed to speed up K-NN by reducing instances to be checked [7]. Outliers are identified and removed for obtaining minimal set of instances that were correctly classified by 1-NN. These instances become the absorbed points and are used in K-NN. In this work, CNN has been used for data undersampling majority data in imbalanced datasets.

**TL—Tomek Links**: In a dataset, a pair of instances are said to have a Tomek link if each of them are nearest neighbour of other while belonging to different class. Such pairs are generally boundary cases or noise and thus likely to be misclassified. Therefore, these instances can be removed without affecting the quality of dataset.

- The Tomek link function searches for the instances that are tomek links using 1-NN for the given dataset.
- To reduce imbalance, majority class examples involved in Tomek links are removed [8].

**OSS—One-Sided Selection**: To solve the problem of imbalanced datasets, the authors in [9] proposed OSS approach.

- In OSS, synthetic samples are generated by using the approach of Tomek links followed by the CNN method.
- Firstly, any overlapping, borderline and/or noisy instances are removed.
- The links of majority class are discarded until for every instance its nearest neighbour from remaining dataset belongs to its own class [10].
- Then, the instances obtained are processed using CNN.

**ENN—Edited Nearest Neighbour**: ENN is also based on K-NN classification. It extends OSS by considering three nearest neighbours of each instance of frequently occurring class. An instance is removed if its class is different from at least two of its three nearest neighbours [11].

**NCL—Neighbourhood Cleaning Rule**: NCL is similar to ENN as it also looks for inconsistencies in classification of three nearest neighbours. However, NCL has been initially designed for cleaning the dataset based on neighbour similarity property [12]. It works by looking for neighbours of rarer class and removes any opposite class label if found in this set.

## 2.2 Oversampling

The process of oversampling is used when the available dataset is imbalanced but small. These methods balance the dataset by creating synthetic samples of minority class.

**SMOTE—Synthetic Minority OverSampling Technique**: It is a powerful sampling method in which the minority class is oversampled by creating synthetic samples [13]. Matching rows according to K-NN are randomly chosen, and their convex combinations are prepared to obtain new samples. SMOTE has been used for generating artificial fraud transactions for creating balanced credit card datasets [14, 15].

**MWMOTE—Majority Weighted Minority Oversampling Technique**: It is a modification over SMOTE technique and works better when the quality of input data is low [16].

- First ranks the minority class samples according to their ease of being distinguishable from other class.
- Similar samples of only those minority class instances that are easily distinguishable are generated by using clustering methods [17].

**PDFOS—Probability Density Function Estimation-Based Oversampling**: PDFOS generates synthetic data by approximating a Gaussian multivariate distribution [16, 18]. Probability density function of existing samples of minority class is estimated and extrapolated to generate matching instances. Details about this method can be obtained from [18].

**RWO—Random Walk Oversampling**: Random walk oversampling generates synthetic samples for minority class while preserving its variance and mean [19]. RWO function is also a part of imbalance package in *R* [16]. The process involved in RWO oversampling technique is as follows:

- RWO generates synthetic samples by randomly walking through real data and creates samples with similar dispersion.
- The weighted information of less frequent samples helps in the generation of final synthetic sample by using clustering approach in order to make the range of generated samples similar to some of the rarer class cluster member.

**ADASYN—Adaptive Synthetic Sampling**: ADASYN approaches for imbalanced learning are a two-class imbalanced classification technique. ADASYN approach improves classification learning in two ways:

- Decreasing the bias, if any, brought in by the distribution of classes;
- By improving learning by shifting the classification decision boundaries.

ADASYN is an oversampling function available in *R*'s library SMOTE family package [14]. A more detailed description of approach can be seen in [20].

**WRACOG—Wrapper for Rapidly Converging Gibbs Algorithm**: WRACOG [24] is an enhancement of RACOG algorithm and a wrapper-oriented classification technique.

- It first generates a synthetic sample focusing on increasing the minority class by approximating probability distribution.
- This process is repeated until further improvement.
- The samples with the highest probability of getting misclassified get selected by WRACOG model.

WRACOG algorithm uses the process of Gibbs algorithm for data sampling while improving many aspects related to the usage of Gibbs method of sampling [21].

# 3 Performance Evaluation

## *3.1 Datasets Used*

We used publicly available, processed real datasets for evaluation:

**Dataset-1**: **CREDIT CARD FRAUD DETECTION DATASET**

The dataset has total of 284,807 transactions. It was collected by European cardholders and analysed during a research collaboration by Andrea Dal Pozzolo and his peers [22].

**Dataset-2**: **PUBCHEM BIOASSAY DATASET**

The dataset has a total of 467,892 records of primary and confirmatory bioassay data from the different types of screening including bioactivity outcomes and potency performed using HTS technology. The main class attributes are labelled as active and inactive compounds for drug like small molecules properties present in bioassay of a patient. The dataset is publicly available and provided by Pubchem [23].

**Dataset-3**: **YEAST DATASET**

It is about applying several bio-statistical tests on yeast bacteria to detect protein. It is available for public usage at [24]. We used yeast dataset with endoplasmic reticulum lumen (ERL) as minority class and combined the other classes as majority class.

**Dataset-4**: **SATIMAGE DATASET**

A fraction of original data of this dataset has been used here [25]. We used damp as minority class instances and others as majority class instances.

**Table 1** Dataset class distribution

| Dataset | Majority class | Minority class | Imbalance ratio | No. of features |
|---|---|---|---|---|
| DS 1: Credit card data (September 2013) | 284,315 | 492 | 0.0017 | 28 |
| DS 2: PubChem bioassay data (2011) | 466,438 | 1454 | 0.0031 | 145 |
| DS 3: Yeast sata (1996) | 1479 | 5 | 0.0033 | 8 |
| DS 4: Satimage data (1993) | 5809 | 626 | 0.1067 | 36 |
| DS 5: Diabetes data (2015) | 500 | 268 | 0.53 | 8 |

**Dataset-5**: **DIABETES DATASET**

It is a dataset of Indian women heritage and records their vital parameters to establish reasons of diabetes by taking samples of effected and non-effected females' data [26].

Table 1 represents the composition of all the datasets used along with their imbalance ratio. Clearly, the datasets are extremely imbalanced. Even a null classifier would predict over 99% accuracy.

## 3.2 *Metrics*

For evaluating resampling techniques, the following metric is considered.

**Imbalance Ratio**: For a given two-class dataset, the imbalance ratio is computed as follows:

$$\text{imbalance Ratio} = \#\text{instances}_{\text{minority}} / \#\text{instances}_{\text{majority}} \tag{1}$$

The chosen algorithms assume the issue as imbalanced classification problem. We have considered the confusion matrix given in Table 2 for evaluating metrics. However, classical metrics of accuracy and confusion matrix will not be able to capture the actual fraud identification rate due to skewness in instances of each class. Thus, metrics that balance the detection of both classes have been considered.

**Positive Predictive Value or Precision**: PPV is a measure of correct positive results among all positive predictions. The positive predictive value (PPV) is defined

**Table 2** Confusion matrix for evaluating classification

| Predicted | Actual | |
|---|---|---|
| | Normal | Fraud |
| Normal | True negatives (TN) | False negatives (FN) |
| Fraud | False positives (FP) | True positives (TP) |

as follows:

$$\mathrm{PPV} = \mathrm{TP}/(\mathrm{TP} + \mathrm{FP}) \quad (2)$$

**Negative Predictive Value**: NPV is the proportion of correctly identified negative values (0 here) among all negative predicted ones. The negative predictive value is defined as follows:

$$\mathrm{NPV} = \mathrm{TN}/(\mathrm{TN} + \mathrm{FN}) \quad (3)$$

**Specificity** measures ratio of actual negatives that are correctly identified.

$$\mathrm{Specificity} = \mathrm{TN}/(\mathrm{TN} + \mathrm{FP}) \quad (4)$$

**Sensitivity** computes percentage of actual positives that are correctly identified.

$$\mathrm{Sensitivity} = \mathrm{TP}/(\mathrm{TP} + \mathrm{FP}) \quad (5)$$

**Balanced Accuracy** is average detection rate obtained on either class.

$$\mathrm{Balanced\ Accuracy} = (\mathrm{Sensitivity} + \mathrm{Specifcity})/2 \quad (6)$$

**Prevalence** is the term used for how often the condition yes actually occurs.

$$\mathrm{Prevalence} = (\mathrm{TP} + \mathrm{FN})/(\mathrm{TP} + \mathrm{FP} + \mathrm{FN} + \mathrm{TN}) \quad (7)$$

### *3.3 Machine Learning Algorithms Used*

The following popular machine learning algorithms are used for evaluating the classifier results after applying and before applying the resampling techniques on dataset:

**K-Nearest Neighbours**: K-NN classifiers are supervised learning methods that classify new instances based on various similarity measures [7].
**Neural Networks (NN)**: For fraud detection, the commonly used method is NNs. NNs recognize related patterns and predict values based on associative memory of patterns it learnt [8].
**Support Vector Machine (SVM)**: SVM is used for supervised learning where the data is linearly classified and analysed [6, 8].

# 4 Results

This study used five datasets to run through different sampling and classification. Results by applying undersampling and oversampling techniques are shown in Tables 3 and 4.

Table 5 represents the results of performance evaluation of machine learning algorithms over chosen metrics before applying resampling to the dataset. There are few NaNs in the table where the classifier could not detect even a single true positive or true negative for the result set.

The best resampling techniques from each category were selected that is NCL ADASYN, and the results of performance evaluation of ML classifier on the dataset obtained by NCL and ADASYN techniques are shown in Tables 6, 7 and 8.

Figures 1 and 2 show the results of balanced accuracy of ML classifiers for NCL and ADASYN resampled data.

## *4.1 Discussion of Results*

From the result in Sect. 3.4, the performance is interpreted as follows:

**Table 3** Imbalance ratio of undersampling techniques

| Techniques used | Imbalance ratio | | | | |
|---|---|---|---|---|---|
| | DS 1 | DS 2 | DS 3 | DS 4 | DS 5 |
| CNN | 0.2046 | 0.2033 | 0.0568 | 0.1226 | 0.53 |
| OSS | 0.2340 | 0.2331 | 0.1815 | 0.1137 | 0.712 |
| NCL | 0.2739 | 0.2739 | 0.0.2083 | 0.2511 | 0.75 |
| ENN | 0.2041 | 0.2041 | 0.0033 | 0.108 | 0.557 |
| TL | 0.2324 | 0.2324 | 0.0033 | 0.1107 | 0.6802 |

**Table 4** Imbalance ratio of oversampling techniques

| Techniques used | Imbalance ratio | | | | |
|---|---|---|---|---|---|
| | DS 1 | DS 2 | DS 3 | DS 4 | DS 5 |
| PDFOS | 0.0052 | 0.0049 | 0.099 | 0.215 | 0.3943 |
| MWMOTE | 0.0020 | 0.0018 | 0.5986 | 0.6341 | 0.696 |
| SMOTE | 0.75 | 0.71 | 0.754 | 0.74 | 0.75 |
| RWO | 0.0020 | 0.0018 | 0.698 | 0.715 | 0.736 |
| ADASYN | 0.9999 | 0.9987 | 0.9899 | 0.9799 | 0.955 |
| WRACOG | 0.0017 | 0.00119 | 0.819 | 0.8599 | 0.866 |

**Table 5** Performance evaluation of machine learning algorithms before resampling

| Classifier metrics used | | Dataset used | | | | |
|---|---|---|---|---|---|---|
| | | DS 1 | DS 2 | DS 3 | DS 4 | DS 5 |
| Accuracy | SVM | 0.986 | 0.978 | 0.985 | 0.997 | 0.998 |
| | K-NN | 0.998 | 0.985 | 0.991 | 0. 967 | 0.997 |
| | NN | 0.99825 | 0.989 | 0.971 | 0.986 | 0.992 |
| PPV | SVM | NaN | NaN | 0.015 | 0.056 | NaN |
| | K-NN | NaN | NaN | 0.58 | 0.807 | 0.851 |
| | NN | 0.99 | 0.891 | 0.805 | 0.916 | 0.956 |
| NPV | SVM | NaN | NaN | 0 | NaN | 0.564 |
| | K-NN | 0.99 | 0.891 | 0.853 | 0.9020 | 0.928 |
| | NN | NaN | NaN | 0.003 | NaN | 0.149 |
| Sensitivity | SVM | 0.93 | 0.897 | 0.896 | 0.746 | 0.864 |
| | K-NN | 0 | 0 | 0.210 | 0.164 | 0.128 |
| | NN | 1 | 0.976 | 0.895 | 0.901 | 0.945 |
| Specificity | SVM | 0.92 | 0.872 | 0.81 | 0.89 | 0.823 |
| | K-NN | 1 | 0.98 | 0.971 | 0.996 | 0.99 |
| | NN | 0 | 0.001 | 0.002 | 0 | 0 |
| Prevalence | SVM | NaN | NaN | 0.261 | 0.5946 | 0.698 |
| | K-NN | 0.001 | 0 | 0.091 | 0.566 | 0.771 |
| | NN | 0.99 | 0.89 | 0.981 | 0.996 | 0.998 |

### Testing the Effectiveness for Balancing Dataset

In undersampling techniques, NCL showed improved imbalance ratio results for all five datasets. In oversampling techniques, ADASYN gave up to 0.99 imbalance ratio. From Tables 3 and 4, out of all the resampling techniques, NCL in undersampling and ADASYN in oversampling showed best overall results. Therefore, we further evaluated these two for checking performance of selected machine learning algorithms.

### Performance Evaluation of ML Algorithms

It can be seen from the results in Table 5 that none of the classification mechanisms is best in all respects. Therefore, the results have been scrutinized and interpreted with respect to performance question relevant. There were few NaNs values where the classifier could not detect even a single true positive or true negative value. We evaluated these classifiers again after applying NCL and ADASYN resampling techniques to get improved performance.

**Table 6** Performance evaluation of machine learning algorithms after applying NCL

| Classifier values | | Dataset used | | | | |
|---|---|---|---|---|---|---|
| | | DS 1 | DS 2 | DS 3 | DS 4 | DS 5 |
| Metrics | Classifier techniques | After applying NCL | | | | |
| Accuracy | SVM | 1 | 1 | 1 | 1 | 1 |
| | K-NN | 1 | 1 | 1 | 1 | 1 |
| | NN | 1 | 1 | 1 | 1 | 1 |
| PPV | SVM | 1 | 1 | 1 | 1 | 1 |
| | K-NN | 1 | 1 | 1 | 1 | 1 |
| | NN | 1 | 1 | 1 | 1 | 1 |
| NPV | SVM | 1 | 1 | 1 | 1 | 1 |
| | K-NN | 1 | 1 | 1 | 1 | 1 |
| | NN | 1 | 1 | 1 | 1 | 1 |
| Sensitivity | SVM | 1 | 1 | 1 | 1 | 1 |
| | K-NN | 1 | 1 | 1 | 1 | 1 |
| | NN | 1 | 1 | 1 | 1 | 1 |
| Specificity | SVM | 1 | 1 | 1 | 1 | 1 |
| | K-NN | 1 | 1 | 1 | 1 | 1 |
| | NN | 1 | 1 | 1 | 1 | 1 |
| Prevalence | SVM | 0.78 | 0.79 | 0.80 | 0.79 | 0.69 |
| | K-NN | 0.69 | 0.65 | 0.65 | 0.79 | 0.61 |
| | NN | 0.81 | 0.89 | 0.72 | 0.78 | 0.81 |

**Performance Evaluation of Classifier After Applying Resampling Techniques**

From Tables 6 and 7, we can clearly see that the results have been improved a lot for all ML algorithms. K-NN and NN showed similar results for the chosen metrics. But, SVM scored best among all the classifiers used as well as overall also in performance.

**Testing Overall Effectiveness of Classification**

Balanced accuracy is used as a measure to detect classification effectiveness in imbalanced datasets. Out of all chosen methods, SVM classifier showed overall perfect results for both NCL undersampled data and for ADASYN oversampled data.

The resampling techniques helped to reduce imbalance ratio of the datasets and furthermore removed NaN values for classifiers by improving the skewness of imbalanced datasets for better classification results.

**Table 7** Performance evaluation of machine learning algorithms after applying ADASYN

| Classifier values | | Dataset used | | | | |
|---|---|---|---|---|---|---|
| | | DS 1 | DS 2 | DS 3 | DS 4 | DS 5 |
| Metrics | Classifier techniques | After applying ADASYN | | | | |
| Accuracy | SVM | 0.92 | 0.89 | 0.90 | 0.98 | 0.89 |
| | K-NN | 0.93 | 0.81 | 0.89 | 0.90 | 0.87 |
| | NN | 0.91 | 0.83 | 0.95 | 0.96 | 0.87 |
| PPV | SVM | 0.91 | 0.98 | 0.96 | 0.99 | 0.98 |
| | K-NN | 0.94 | 0.97 | 0.98 | 0.97 | 0.98 |
| | NN | 0.89 | 0.97 | 0.98 | 0.95 | 0.98 |
| NPV | SVM | 0.91 | 0.97 | 0.99 | 0.99 | 0.82 |
| | K-NN | 0.94 | 0.95 | 0.84 | 0.80 | 0.79 |
| | NN | 0.89 | 0.96 | 0.81 | 0.89 | 0.81 |
| Sensitivity | SVM | 0.93 | 0.95 | 0.98 | 0.98 | 0.99 |
| | K-NN | 0.93 | 0.94 | 0.91 | 0.95 | 0.99 |
| | NN | 0.91 | 0.95 | 0.91 | 0.91 | 0.99 |
| Specificity | SVM | 0.87 | 0.92 | 0.77 | 0.81 | 0.73 |
| | K-NN | 0.85 | 0.89 | 0.69 | 0.78 | 0.79 |
| | NN | 0.71 | 0.90 | 0.69 | 0.61 | 0.80 |
| Prevalence | SVM | 0.93 | 0.81 | 0.81 | 0.80 | 0.79 |
| | K-NN | 0.89 | 0.71 | 0.72 | 0.63 | 0.66 |
| | NN | 0.91 | 0.81 | 0.61 | 0.95 | 0.65 |

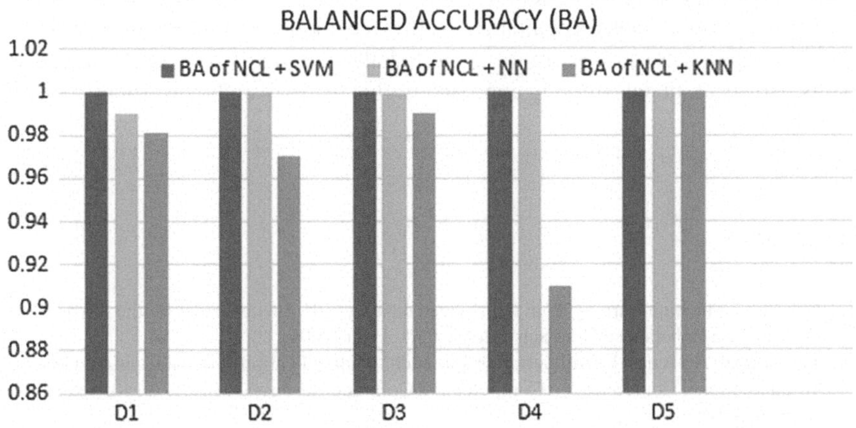


**Fig. 1** Balanced accuracy after applying NCL

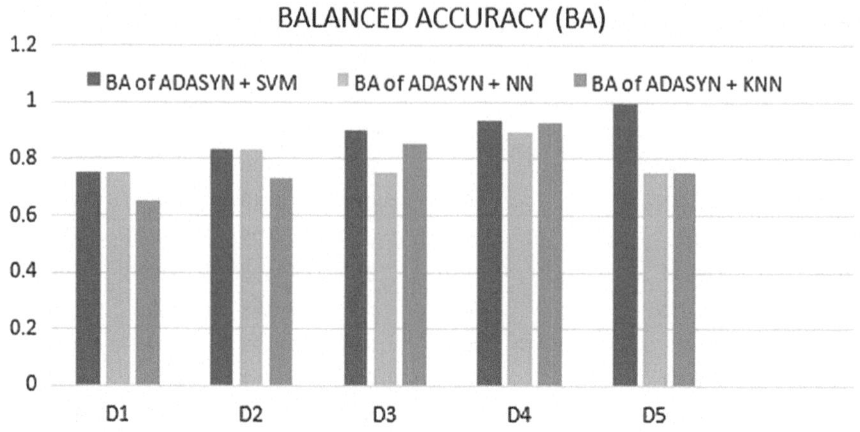


**Fig. 2** Balanced accuracy after applying ADASYN

## 5 Conclusion

In this paper, a number of resampling techniques to improve the imbalance ratio of highly imbalanced datasets has been evaluated. Imbalance data classification is a peculiar classification problem due to very high imbalance in instances of minority and majority class instances in data as examples. Popular machine learning algorithms SVM, K-NN and NN were evaluated on different metrics. It is concluded that the effect of balancing techniques resulted improved performance of classifier. Undersampling approaches handle the dataset skewness in better ways, and hence, perform well over all metrics compared to other techniques. It has been observed that ADASYN approach can best improve the imbalance ratio as well as classification results. However, undersampling approaches give better overall performance on all datasets.

## References


1. Provost, F.: Machine learning from imbalanced data sets 101. Invited paper for the AAAI2000. Workshop on Imbalanced Data Sets, Menlo Park, CA (2000)
2. Hanskunatai, A.: A new hybrid sampling approach for classification of imbalanced datasets. In: 3rd International Conference on Computer and Communication Systems (ICCCS), pp. 67–71. Nagoya (2018)
3. Liu, X., Wu, J., Zhou, Z.: Exploratory undersampling for class-imbalance learning. In: in IEEE Transactions on Systems, Man, and Cybernetics, Part B (Cybernetics), vol. 39, no. 2, pp. 539–550 (2009)
4. Zhang, J., Cui, X., Li, J., Wang, R.: Imbalanced classification of mental workload using a cost-sensitive majority weighted minority oversampling strategy. Cogn. Technol. Work (2017)
5. Practical Guide to deal with Imbalanced Classification Problems in R: https://www.analyticsvidhya.com/blog/2016/03/practical-guide-deal-imbalanced-classification-problems

6. Package unbalanced Documentation in R: https://cran.r-project.org/web/packages/unbalanced/unbalanced.pdf
7. Introduction to k-Nearest Neighbour Classification and Condensed Nearest Neighbour Data Reduction: http://www.math.le.ac.uk/people/ag153/homepage/KNN/OliverKNNTalk.pdf
8. Tomek, I.: Two modifications of cnn. IEEE Trans. Syst. Man Cybern. **6**, 769–772 (1976)
9. Kubat, M., Matwin, S., et al.: Addressing the curse of imbalanced training sets: one-sided selection. In: Machine Learning-International Workshop Then Conference, pp. 179–186. Morgan Kaufmann Publishers, Inc (1997)
10. Oversampling: https://en.wikipedia.org/wiki/Oversampling/
11. Wilson.D.: Asymptotic properties of nearest neighbor rules using edited data. In: IEEE Transactions on Systems, Man and Cybernetics, pp. 408–421 (1972)
12. Laurikkala, J.: Improving identification of difficult small classes by balancing class distribution. In: Artificial Intelligence in Medicine, pp. 63–66 (2001)
13. Padmaja T.M., Dhulipalla N., Krishna P.R., Bapi R.S., Laha A.: An unbalanced data classification model using hybrid sampling technique for fraud detection. In: Lecture Notes in Computer Science, vol. 4815. Springer, Berlin, Heidelberg (2007)
14. Smotefamily Package Documetation in R: https://cran.r-project.org/web/packages/smotefamily/smotefamily.pdf
15. Chawla, N.V., et al.: SMOTE: synthetic minority over-sampling technique. arXiv preprint arXiv:1106.1813 (2011)
16. Imabalnce Package Documetation in R: https://cran.r-project.org/web/packages/imbalance/imbalance.pdf
17. Barua, S., Islam, M.M., Yao, X., Murase, K.: Mwmote-majority weighted minority oversampling technique for imbalanced data set learning. IEEE Trans. Knowl. Data Eng. **26**(2), 405425
18. Gao, M., Hong, X., Chen, S., Harris, C.J., Khalaf, E.: PDFOS: Pdf estimation based oversampling for imbalanced two-class problems. Neurocomputing **138** (2014)
19. Zhang, H., Li, M.: Rwo-sampling: a random walk over-sampling approach to imbalanced data classification. Info. Fusion **20**, 99116
20. He, H., Bai, Y., Garcia, E., Li, S.: ADASYN: Adaptive synthetic sampling approach for imbalanced learning. In: Proceedings of IJCNN (IEEE World Congress on Computational Intelligence). IEEE International Joint Conference. pp. 1322–1328 (2009)
21. Das, B., Krishnan, N.C., Cook, D.J.: RACOG and wRACOG: two probabilistic oversampling techniques. IEEE Trans. Knowl. Data Eng. **27**(1), 222234
22. Newman, D.J., Asuncion, A.: UCI machine learning repository. Transformed datasets are available at http://www.ulb.ac.be/di/map/adalpozz/imbalanced-datasets.zip(2007)
23. PbChem Bioassay Data, UCI machine learning repository datasets are available at: http://archive.ics.uci.edu/ml/datasets/PbChem+Bioassay+Data
24. Yeast DataSet, UCI machine learning repository datasets are available at: http://archive.ics.uci.edu/ml/datasets/Yeast?ref=datanews.io
25. Statlog, UCI machine learning repository datasets are available at: https://archive.ics.uci.edu/ml/datasets/Statlog+(Landsat+Satellite)
26. Pima Indians Diabetes Database: https://www.kaggle.com/uciml/pima-indians-diabetes-database

# Sentiment Analysis and Mood Detection on an Android Platform Using Machine Learning Integrated with Internet of Things

**Diksha Kushawaha, Debalina De, Vandana Mohindru and Anuj Kumar Gupta**

**Abstract** Mental health is considered as one of the most sensitive topics of research and it is highly affected by an individual's mood and sentiments. Social media has been proven to be one of the major catalysts in deterioration and fickleness of one's mind. In this paper, we present an android application called "moody buddy" ingratiated with a heartbeat analyzing hardware kit which would detect and analyze the moods and emotions of an individual very close to accuracy. Mood recognition and sentiment analysis is a vast and complex area of research. Moreover, monitoring human emotions is found out to be one of the technically challenging aspects. So, in order to achieve the quality output of our research and testing work, we have taken help from artificial intelligence and Internet of Things domain. Here, we have considered the activity of the user on his/her social networking as a starting point of our research work. The concept of logistic regression is used in our software. In order to solidify our idea more, we are adding a hardware component which would monitor the heartbeat of the person and its modulation. In case of any abnormality examined in the heart rate, the questionnaire appears again. At the end, a cumulative of the hardware component's results and software component's would help us analyze and detect the current mood of the individual to very close to high accuracy value.

**Keywords** Mood and sentiment analysis · Artificial intelligence · Internet of Things · Heart rate monitoring · IBM Watson · Bag of words

---

D. Kushawaha · D. De · V. Mohindru (✉) · A. K. Gupta
Department of Computer Science and Engineering, Chandigarh Group of Colleges - College of Engineering, Mohali, Punjab, India
e-mail: vandanamohindru@gmail.com

D. Kushawaha
e-mail: dikshakushwaha99@gmail.com

D. De
e-mail: debalina322000@gmail.com

A. K. Gupta
e-mail: anuj21@hotmail.com



P. K. Singh et al. (eds.), *Proceedings of ICRIC 2019*, Lecture Notes in Electrical Engineering 597, https://doi.org/10.1007/978-3-030-29407-6_18

## 1 Introduction

According to a study conducted by WHO for the National Care Of Medical Health (NCMH), it observes that at least 6.5% [1] of the Indian population suffers from several mental ailments some of which are noted down to have weird and absurd causes as well. The average suicide rate in India is almost 10.9 for every lakh people and the majority of people who commit suicide lies between the age bracket of below 44 [1]. Coincidentally, another major report has been published on how India surpasses all the other developing countries in its excessive usage of social networking sites. Now, it is not that social networking is all about communication purposes, the youth today expresses far more intimate information. Now according to the policies of this kind of social networking sites, they cannot manipulate the contents of any users unless and until it is beyond their terms and condition. The study says that social networking sites are filled with extreme instigating posts and visually stimulating sources, these results show the causes behind self-destruction in teenagers. Looking at others flashy lifestyle stories over Instagram feed or just watching over another magazine like life statues, there is a constant realization of one not being able to achieve such a glorious life. This is a very serious mental condition which means a never-ending obsession of achieving the unreal world. At severe cases, one might encounter doubt in oneself, a never-ending loop of self-realization and under-estimation. Because of giving too many hours in front of the gadget will leads to lower back problems, eye problem, and neck pain. This is also proven that it will cause illness. It also leads to the problem of physical illness when you are spending time sitting around and chat with people on Facebook or any other social media. Spending too much time on social media causes physical illness like blood pressure problem, a certain type of cancers or nutrition problems and many other which is not beneficial for the wellness of the people. Using too much internet mainly social media will cause depression and ADHD (i.e., an excessive eagerness to seek attention from others by doing unusual deeds), impulsive disorder, problems with proper mental functioning, paranoia, and fore-loneliness. The social media platform can be used to look back at the different memories. However, it can also distort the way in which you remember certain tidbits from your life. Another crucial problem due to these social media is the lack of understanding between close ones and with oneself as well. The report says that most of the teenagers suffer from this condition where one fails to understand one's mood properly or does not understand how to react to certain emotions.

From the Fig. 1, we find that most prominent age groups affected by mental disorders are between the age groups of 12–18 and the middle-aged peoples. Middle age crisis is considered as one of the traumatic situations for such mental disorders as noted. As per the statistics, it is observed that 50% of the patients used the antidepressant-like SSRIs. In the age group of 60 years, the most constant mixture of Benzodiazepines, and SSRIs whereas in the middle (30–44) age group of patient uses the mixture of SSRIs and Benzodiazepines. And the most frequent mixture in the group of youngest age were SSRIs and the combination of SSRIs and Benzodiazepines [1].

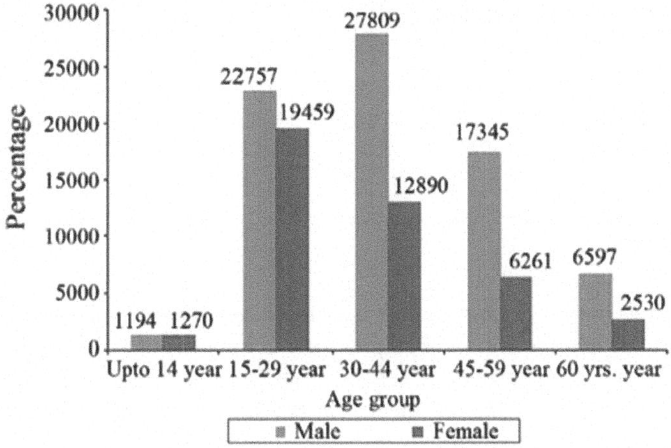


**Fig. 1** The rate of depression in different age groups

The college years are also difficult. Sometimes teenagers or college-goers suffer from ragging, frustration, and depressions. Moreover, certain students get this weird obsession of showing off themselves by creating something which is beyond their scope. Hence, creating an illusion of something is beyond their capability. Sometimes, they sought help from the social media platform to portray themselves something which they wish for. In this attempt, they often get so caught up that they forget the difference between virtual and reality. Most of the cases of mental disorders have been noted due to social media actions. Excessive social media interactions also lead to social media addiction, which on sudden abstain can create huge trauma in the user's mental health.

Now, since everything is digitized these days so why do not we use a gadget to predict our mood, analyze our emotions, and help us to cheer up as well. This was the prime motivation behind our project. All though we have been discussing the cons of social networking sites, so for analyzing we are using the social networking sites only, it means using it in a much positive manner. As for analyzing any kind of data and train it with the help of machine learning, the software requires to collect its own data set. So, all the data will be withdrawn from the user's social networking sites and their activities on them. Collecting the data and training it accordingly is one of the basic yet essential steps of machine learning and it depends upon the social media entirely. Once the data has been created, "bag of words" gets formed. Then testing is done by the entire software itself. In an overview, we modified the usage of social networking sites in solving a problem which has been created by itself only [2, 3].

# 2 Related Technologies Used

In the same fields, quite a handful of technologies has been developed which are stated as follows.

## *2.1 Twiggle*

Twiggle is a technology company that uses machine learning and natural language processing to develop search technologies for e-commerce sites [1]. In March 2017, Twiggle announced the debut of its Semantic API, its inaugural product, which gives online retailers the ability to enhance their existing search capabilities by adding semantic understanding to their existing search engine [4].

## *2.2 The North Face*

The North Face is considered as one of the biggest e-commerce retailer sites, which provides a holistic approach to the customer to buy stuff from their sites. The North Face is known to use IBM Watson's AI technology as a way of virtually communicating with its users. For example, based on the recent search history of the user, it makes its very own assumptions and via a continual learning process, it suggests the best item required by the user. Like if the user is asking for buying a jacket, then by virtue of the IBM Watson, it asks the specifications and then it shows the results.

## *2.3 Virtual Assistance*

### 2.3.1 Google Now

Google Now is another product in association with Google Feed as shown in Fig. 2. Just like the name suggests, it works as a virtual assistant, serving all the actions and activities which are required by the user. It is installed with natural language processor which understands the verbal commands of the user and accordingly it places its action. For example, if the user says play music, then it will automatically apply its algorithm and play the needed song which the user is demanding [5, 6].

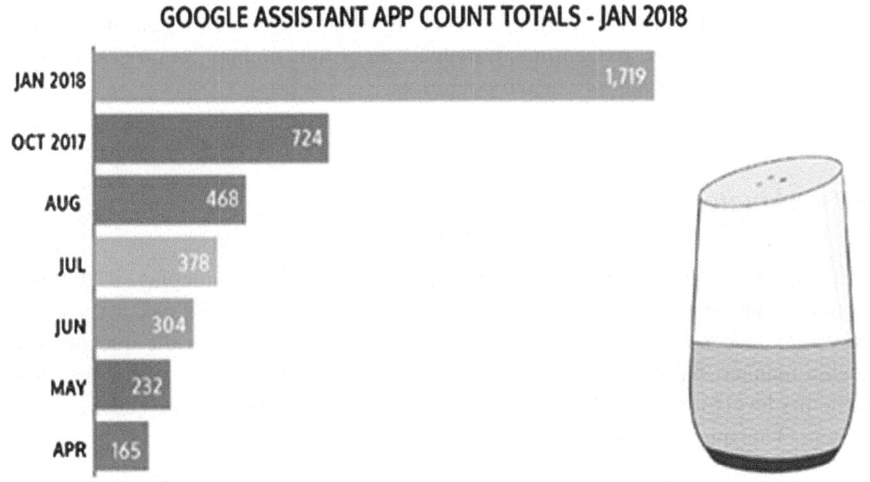


**Fig. 2** Google assistant application usage in respective years

### 2.3.2 Alexa

Amazon Alexa is another famous virtual assistant [7] as shown in Fig. 3. Just like Google assistant, it has also got certain features like voice recognition, sentiment recognition kind of responses. Alexa shows a very unique system of machine learning concept, which is neural networks. Just like the human neural systems, it also

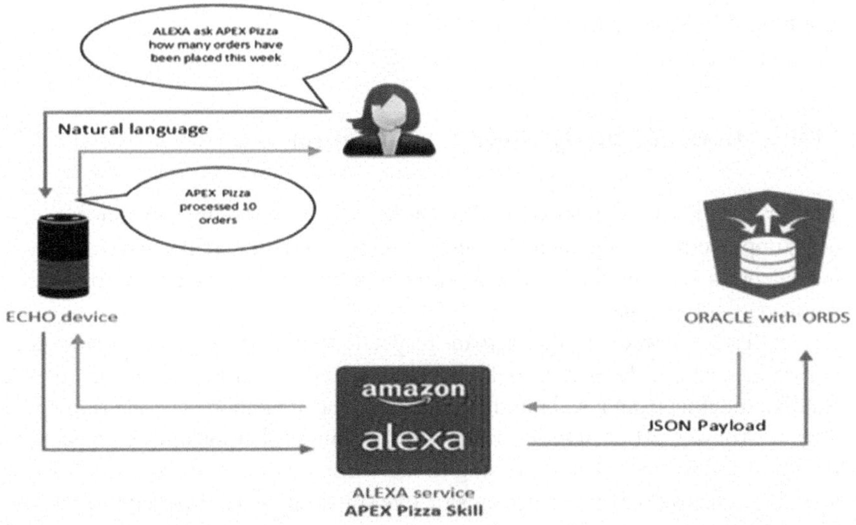


**Fig. 3** Working of Alexa

provides the same organic feelings of it being used. Alexa shows certain features of sentiment analysis. Its sentiment analysis algorithm mainly comprises of such algorithms involving that of voice recognition.

### *2.4 Akinator*

Akinator is an online game and mobile app [8, 9]. During gameplay, it attempts to determine what fictional or real-life "character" which would ask a set of 12 questions. It has got features of artificial intelligence, which sets the questions in such a manner that based on the attempts and the experience level of the user, the difficulty level increases. The Akinator is known for its highly accurate questionnaire system. This questionnaire gives a highly accurate result [10–13].

### *2.5 Chatbots*

A chatbot is nothing but a textual platform which is aided with artificial intelligence properties. Chatbot means chatting with robots. These kinds of applications are solely designed to check how the robots would function when they actually deal with humans. It has been designed for the prime motive of customer interaction purpose. Some chatbots use sophisticated natural language processing systems, but work in a much-simplified manner, where it breaks down all the necessary keywords present in natural language from the user and creates its own particular patterns and keeps on learning from experiences [14].

## 3 Objectives of Moody Buddy Application

- To decrease the rate of depression: One of the biggest pros of this invention would be an improvement in the mental health of the country. India ranks first in terms of most depressed countries. So, it is a prime time for some development to be done at this particular arena.
- To manifest a wholesome solution to the respective moods detected by the software: For an instance, if the person is feeling low or sad (according to the calculated results), timeline from Facebook full of vivid colorful memories would pop up on the screen or suggestion of the users favorite song might also pop up as a suggestion. Different actions are taken according to the current mood of the person. Similar ways if the person is angry or in a hyperactive situation, motivational quotes would pop up or motivational videos will start playing automatically.
- To invent a breakthrough technology in the field of mental health care: Talking about mental health, not much has been contributed by the field of technologies. So

for the very first time, such technology will evolve which would intelligently detect one's mood with a very high range of precision even if one fails to understand how to react to certain emotionally charged situations.
- To create a great child mental health monitoring device for all the parents: These days almost all the teenagers within the age limit of 13–19 are connected with some kind of android exposure. Teenagers are an age group which is popular for their tantrums and unpredictable mood swings. Sometimes, the parents fail to understand the root cause or rather the mood of their offsprings. Also in this twenty-first century where most of the parents are working, there always lies a gap in the attachment level which is required. Hence, the child sometimes fails to express his or her emotions freely. Using this gadget, working parents could also track the moods of their child and make assumptions what might have gone throughout the day in their child's life and take the required steps to do the needful.
- To decrease the rate of suicides occurring in our country for the reason being poor mental health and negligence: If the underlying problem of depression gets solved, then there is no question of one's ending up life [2].
- To provide a holistic approach to solve a problem: A combined effort of both physical and virtual help is used up to improvise the current situations.
- To incorporate it with all the existing technologies present and give it a high edge makeover.

## 4 The Algorithm of Moody Buddy Application

The proposed application has mainly four components

- Heart rate monitoring device
- Data analysis
- A questionnaire using IBM Watson
- Depending on the result, corresponding action will be performed.

### *4.1 Heart Rate Monitoring Device*

The user can use the device as BANDS to wear which will record the heart rate of the user for collecting data every minute as shown in Fig. 4. If the heart rate is higher than the normal limit for a prolonged period that is more than 15 min, then two assumptions can be made:

- The user is angry or tensed.
- The user is excited.

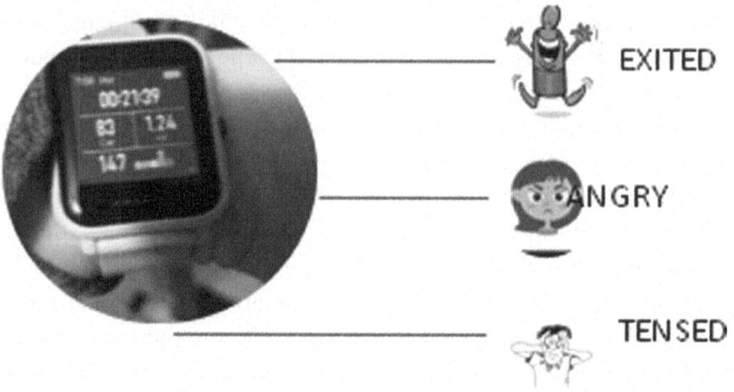


**Fig. 4** Working heartbeat monitoring device

According to the assumption, IBM Watson is used to get accurate results. If the heart rate is less than the normal rate, then there might be two possibilities

- The person might be depressed.
- He is suffering from an ailment.

Whenever such abnormality will take place again, the notification will pop up asking for permission to IBM Watson questionnaires which predict the current mood of the user [15].

## *4.2 Data Analysis*

At first, our application will be built "A BAG OF WORDS" where all the words in the English dictionary will be recorded and segmented into three categories with the help of "SENTIMENTAL ANALYSIS"

***POSITIVE, NEUTRAL, and NEGATIVE***

*For example*: If the user posts something on Twitter that contains words like good, beautiful, happy, etc., these keywords will be counted to be POSITIVE [16, 17]. The essence of this functionality is taken from the concept of sentiment analysis, wherein the algorithm we use natural language processing (NPL) to classify the documents in the same segregation of keywords [18, 19].

***POSITIVE, NEUTRAL, and NEGATIVE***

Anything which will be posted by the user in the social networking site or be it uploading any pictures will be analyzed by our algorithms. Here, we will train our data sets first. For that, we will need some previously present data sets of social media activities which can be brought down by D libraries in machine learning (python) libraries. The data set will be then applied with a number of algorithms and equations which are noted as follows. If our data is in the form of the image, we use "image processing" in order to analyze that data; in this, we apply "principal component analysis" algorithm [20].

Principal component analysis can be described as signal processing and transformation of a given set in which $n$ number of input vectors (variables) with the same length $K$ formed in the $n$-dimensional vector

$$x = [x_1, x_2, x_3, \ldots, x_n]^{\mathrm{T}}$$

into a vector $y$ according to

$$y = A(x - m_x) \tag{1}$$

These enable to form a simple formula (1) but it is necessary to keep in the mind that each row of the vector $x$ consists of $K$ values belonging to one input. The vector $m_x$ in Eq. (1) is the vector of mean values of all input variables defined by the relation

$$E\{x\} = 1/k \sum_{k=1}^{k} x_k \tag{2}$$

Matrix $A$ in Eq. (1) is determined by the covariance matrix $C_x$.

Rows in the $A$ matrix are formed using the eigenvectors $E$ of $C_x$ ordered according to corresponding eigenvalues in descending order. The evaluation of the $C_x$ matrix given by the relation.

$$C_x = E\left\{(x - m_x)(x - m_x)^{\mathrm{T}}\right\} = 1/k \sum_{k=1}^{k} x_k {x_k}^{\mathrm{T}} - m_x {m_x}^{\mathrm{T}} \tag{3}$$

As the vector $x$ of input variables is $n$-dimensional, then the size of $C_x$ is $n \times n$. The elements $C_x$ $(i, i)$ lying in its main diagonal are called as variances

$$C_x(i, i) = E\left\{(x_i - m_i)^2\right\} \tag{4}$$

Of $x$ and the other values $C_x(i, j)$ determine the covariance between input variables $x_i, x_j$.

$$C_x(i, j) = E\left\{(x_i - m_i)(x_j - m_j)\right\} \tag{5}$$

Between input variables $x_i, x_j$ . The rows of $A$ in Eq. (1) are orthonormal so the inversion of PCA is possible according to the relation

$$x = A^{\mathrm{T}} y + m_x \tag{6}$$

The Eq. (1) defines the kernel of PCA which has some other unique properties resulting from the matrix theory which can be used in the signal and image processing to achieve various goals.

**PCA Use for Image Compression**

Reducing the data volume is a pretty common task which is a part of the image processing algorithm. The principles of image compressing are shown in algorithms stated in Eqs. (1), (2), and (4). Algorithms based on the image color reduction are mostly lossy appearing; however, the outcomes which come out are acceptable in the application of most of the cases. One of the most common image processing algorithms is the image transformation from color to the gray-level (intensity) image $I$. Its implementation is usually based on the weighted sum of three color components $R$, $G$, $B$ according to the relation

$$I = w3B + w2G + w1R \tag{7}$$

The $R$, $G$, and $B$ (Red–Blue–Green) matrices contain image color components; the weights w are determined with all the possibilities of human perception [16]. The PCA method observes the problem in some another way. Equation 6 gives the idea where the matrix $A$ is replaced by the matrix $A$ in which only one highest (instead of $n$) eigenvalues are used for its forming. The vector $\hat{x}$ of reconstructed variables is given by the relation

$$\hat{x} = ATky + xm \tag{8}$$

True-color images of size $N*M$ are generally saved in the three-dimensional matrix $P$ with size $N*M*3$ that means the information about the intensity of color components is stored in the three given planes. The vector of input variables $x$ in Eq. (1) can be formed as the $n = 3$-dimensional vector of each component of the color.

Forming three 1-dimensional vectors $x_1, x_2, x_3$ from each plane $P$ $(M, N, i)$ with the length of $M{\cdot}N$ can be more helpful for better programming and understanding. The covariance matrix $C_x$ and corresponding matrix $A$ are then solved and then 3-dimensional reconstructed vector $\hat{x}$ according to Eq. (8) may be called as the first, the second, and the third component of the given image. According to the matrix theory, it implies that the image formed by reconstruction with the matrix $A1$ (only the first—largest eigenvalue was used for its definition) contains the maximum number of information so that the image will have the maximum contrast. Following are the properties could be significant in the following image processing.

If the data is in the form of "text," then the whole text is broken down into sentences and sentences are broken down into words and we use "bag of words" in which keywords (value is assigned to words whether positive, negative, or neutral) are stored and categorized into three ways

1. Words which show the positive attitude are categorized into "positive."

2. Words which show negative attitude are categorized into "negative."
3. If the sentence contains both positive and negative keywords, then that sentence is categorized into "neutral."

## *4.3 IBM Watson Questioning Answering System Proposed in Natural Language Processing*

One more technology we are using to analyze more accurate mood of the user, after getting the result from the "data analysis" and "heart rate monitoring device," we combine the results of both methods and depending upon that "questionnaire" is pop-up on the screen of the user, and user gives the answers in one word whether "yes" or "no" [16]. This questionnaire uses a simple algorithm as used in online game "Akinator" (described above).

Depending upon the answers given by the user, we can easily analyze the mood of the user with "high accuracy."

- IBM Watson plays a vital role in our application; this is one of the major factors through which we can easily get more accurate results.
- This special technology can easily analyze the mood with the unstructured content of the user.
- The basic working principle of IBM Watson is to parse keywords in a clue while searching for related terms as responded by the user.
- Watson gives instantaneously frequent reaction which is consistent; it generates the corresponding response and is immune to human players' psychological tactics, such as jumping between sections on every clue.
- Mountains of structured and unstructured (About 80%) content are over the Cloud, IBM Watson struggle to understand every day.

## *4.4 Once the Questionnaire Does Its Work and We Get a Calculated Result, Provisions Are Made of What to Be Done for Each and Every Emotion*

- If a person is found out to be sad (i.e., low heart rate and a greater percentage of neutral and negative keywords used in 24 h), then timeline from Facebook or the user's favorite song will pop up. Or even clips of funny videos or most searched videos by the user will come up as suggestions. Since the android application will have the access to the phone book as well, top five frequently contact people will also lined up on the screen in case the user wants to talk to someone.

- If the person is unwell (i.e., low heart rate and not much social media activity throughout the day), then again top five frequently contacted person's number will pop up along with a local doctor's phone number.
- If the person is angry (i.e., showing very high heart rates and greater percentage of neutral and negative keywords used in 24 h), an online audio or video to calm down will pop up on the screen along with five frequently contact number and suggestions of motivational and funny videos will be shown on the user's screen.
- If the person is tensed (i.e., showing very high heart rates and a greater percentage of neutral keywords used in 24 h), then five frequently contact number and suggestions of motivational and funny videos will be shown on the user's screen.
- If the person is happy (i.e., showing very high heart rates and a greater percentage of positive keywords used in 24 h), a message would pop up that "congratulations" for being motivated and happy throughout the day.

Figure 5 explains the overall working of the proposed moody buddy application.

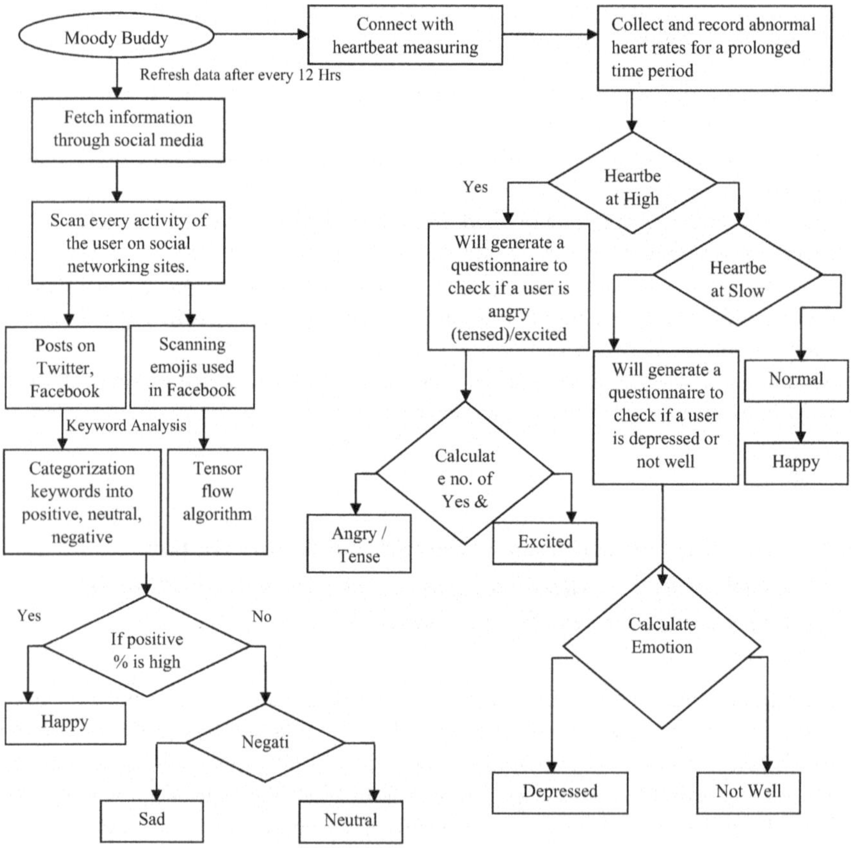


**Fig. 5** Flow chart of moody buddy application

## 5 Comparisons with the Related Technologies

Considering Table 1, the comparison chart we find that there has been no such considerate work done in the field of mood detection and sentiment analysis. Big brands like Google have implemented quite a few algorithms for sentiment analysis. However, in terms of detecting the mood of the user that to very close to accuracy, it is a long way to go.

**Table 1** Comparison between related technologies and moody buddy application

| Existing state of art | Technologies used in the art | Year in which was discovered | Application | Drawbacks |
|---|---|---|---|---|
| Twiggle | Artificial intelligence, deep learning, natural language processing | 2014 | It finds its application in the e-commerce area by enriching your product catalog with standardized structured data, product attribution, and automatic clustering to enable one and their customers to find products in a snap | Clearly, it has nothing no scope to detect on the mood detection part. Its work is sorely concentrated on keyword analysis and experiential learning |
| Chatbots | Neural networks, deep learning, natural language processing, data mining | 2014 | It finds its application as an artificial conversational tool which is used in many other software industries like Facebook, Google, etc. This kind of technology is also used in toys like Barbie | It works like a simple machine which works learns under the continuous learning experiences to provide the user with an all-round experience, however, no possibilities of sentiment analysis are shown |

(continued)

**Table 1** (continued)

| Existing state of art | Technologies used in the art | Year in which was discovered | Application | Drawbacks |
|---|---|---|---|---|
| Google assistant | Artificial intelligence, natural language processing, android (4.1 jelly beans), big data analysis, data mining | 2012 | Acts as a virtual assistant to the user<br>It also learns from experiences | Though it shows some part of sentiment analysis like wishing one during one's birthdays or options like playing games or complimenting however it fails in mood detection |
| North face | Artificial intelligence in the platform of IBM Watson, natural language processing | 1966 | It mainly finds its application in e-commerce | No such scope for sentiment analysis or mood detection |
| Alexa | Artificial intelligence, natural language processing, IoT | 2014 | It finds its application in home automation systems, a virtual assistant | No such scope for sentiment analysis or mood detection |
| Akinator | Artificial intelligence | 2007 | It is a questionnaire game | It is just an entertainment game. No such scope for sentiment analysis or mood detection |

## 6 Conclusion

From the above discussion, we can conclude that a breakthrough solution in the path of mental health is being made. Taking social media inputs as the starting points, we have formed a solution to take over the depression which has been caused by the social media itself. The way in which we have made the culprit of the entire depressing component as a key ingredient of this project is a true stunner. We are not stressing on the fact that all these will reduce the usage of social media. Rather on seeing the results on their social networking activity even getting the preprogrammed solutions which are curated especially for each and every moods and emotion of the user, to some level the user's mood will be elevated. Now in this twenty-first century where we are always attached to our cell phones, it feels good to know that our gadgets could intelligently take care of ourselves. This application will also help to cherish the mood of the user to a certain level of happiness. According to Neurological science, inspirational quotes stimulating funny videos are said to instill very firm optimistic vibes within the users. So the videos, links we would be using would really impact the mental health of the user. It would create a small amount of timelapse for the user to get rid of all the hardships and encroaching realities. Also in our solution list, we provide with frequently contacted numbers. It says that sharing thoughts or pain helps to calm down one. All in all, we have devised a technology which would not only help us to detect the current mood of the user (very close to accuracy) and if the according to the moods, well-tested solutions will be provided. This project aims at finding a solution to the problems which seem fairly challenging and intriguing to us, by creating a solution from the source which is the prime responsible key behind all the problems.

## References


1. Grover, S., Dutt, A., Avasthi, A.: An overview of Indian research in depression. Indian J. Psychiatry **52**(Suppl1), S178 (2010)
2. https://www.verywellmind.com/depression-statistics-everyone-should-know-4159056
3. Byrne, G.J., Pachana, N.A.: Anxiety and depression in the elderly: do we know any more? Curr. Opin. Psychiatry **23**(6), 504–509 (2010)
4. Lev, I., Shalom, E., Avidor, A.: Hierarchic model and natural language analyzer. U.S. Patent Application 15/275,620, filed March 29, 2018
5. https://www.techrepublic.com/article/google-assistant-the-smart-persons-guide/
6. Castro, D., New, J.: The Promise of Artificial Intelligence. Center for Data Innovation (2016)
7. Chung, H., Iorga, M., Voas, J., Lee, S.: Alexa, can i trust you? Computer **50**(9), 100–104 (2017)
8. https://www.academia.edu/6233778/Report_about_Akinator
9. Zięba, M., Tomczak, J.M., Brzostowski, K.: Selecting right questions with restricted Boltzmann machines. In: Progress in Systems Engineering, pp. 227–232. Springer, Cham (2015)
10. Golzadeh, H., Ekárt, A., Faria, D.R., Buckingham, C.D., Manso, L.J.: Emotion recognition using spatiotemporal features from facial expression landmarks
11. Connor Adams Sheets, December: How Does Akinator Work? Behind the Genie that "Reads Your Mind" (2018)
12. Gonsior, B., Sosnowski, S., Buß, M., Wollherr, D., Kühnlenz, K.: An emotional adaption approach to increase helpfulness towards a robot. In: 2012 IEEE/RSJ International Conference on Intelligent Robots and Systems, pp. 2429–2436. IEEE (2012)
13. Kühnlenz, B., Sosnowski, S., Buß, M., Wollherr, D., Kühnlenz, K., Buss, M.: Increasing helpfulness towards a robot by emotional adaption to the user. Int. J. Soc. Robot. **5**(4), 457–476 (2013)
14. Toxtli, C., Monroy-Hernández, A., Cranshaw, J.: Understanding chatbot-mediated task management. In: Proceedings of the CHI Conference on Human Factors in Computing Systems (CHI '18), Paper 58, 6 pages. ACM, New York, NY, USA (2018). https://doi.org/10.1145/3173574.3173632
15. https://healthdocbox.com/Sleep_Disorders/70907190-How-to-analyze-stress-from-heart-rate-heart-rate-variability-a-review-of-physiology.html
16. Chen, Y., Argentinis, J.D.E., Weber, G.: IBM Watson: how cognitive computing can be applied to big data challenges in life sciences research. Clin. Ther. **38**(4), 688–701 (2016)

17. Müller, O., Junglas, I., Debortoli, S., vom Brocke, J.: Using text analytics to derive customer service management benefits from unstructured data. MIS Q. Executive **15**(4), 243–258 (2016)
18. http://ataspinar.com/2016/01/21/sentiment-analysis-with-bag-of-words/
19. Nowak, J., Taspinar, A., Scherer, R.: LSTM recurrent neural networks for short text and sentiment classification. In: International Conference on Artificial Intelligence and Soft Computing, pp. 553–562. Springer, Cham (2017)
20. Ke, H., Shaoping, M.: Text categorization based on. Concept indexing and principal component analysis. In: 2002 IEEE Region 10 Conference on Computers, Communications, Control, and Power Engineering. TENCOM'02 Proceedings, vol. 1, pp. 51–56. IEEE (2002)

# Predictive Strength of Selected Classification Algorithms for Diagnosis of Liver Disease

**Prateek Singh, Deepak Chahal and Latika Kharb**

**Abstract** Human liver is believed to be the largest gland in human body. Weight of healthy human livers is around 1.2–1.5 kg and contributes approximately 3.3% of total body weight. Liver breaks down nutrients from our daily diet into substances which are less toxic to our body such as ammonia to a much less toxic substance called urea. According to data released by National Institute of Nutrition (NIN), Hyderabad suggests that food we consume are lesser nutritious then we used to consume in last three decades. According to data published by WHO in 2017, deaths due to liver disease reached 259,749 or 2.95% of total deaths which makes liver disease one of leading cause of death in India. With the power of machine learning and data science, we can provide better information to doctors so that they can start treatment of disease at its initial stage of disease. This paper investigates the performance of logistic regression, K-nearest neighbor algorithm, decision tree and support vector machine algorithm on liver reports of Indian patients. Dataset of Indian patients is collected from UCI repository. Some patient whose age exceeds 79 years is listed as of age 80 years. Algorithms are evaluated on the basis of: (i) recall, (ii) precision, (iii) F1-score and (iv) support.

**Keywords** National institute of nutrition · Machine learning · Data science · Logistic regression · K-nearest neighbor · Decision tree and support vector machine · UCI repository

## 1 Introduction

Patients with liver disease have been continuously increasing because of [1] poor quality food, excessive consumption of alcohol [1], consuming fatty acids that cause obesity which results non-alcoholic fatty liver disease (NAFLD), induction of medication that induced liver problems. Liver diseases are increasing day by day that

P. Singh · D. Chahal · L. Kharb (✉)
Jagan Institute of Management Studies, Sector-5, Rohini, Delhi, India
e-mail: latika.kharb@jimsindia.org



P. K. Singh et al. (eds.), *Proceedings of ICRIC 2019*, Lecture Notes in Electrical Engineering 597, https://doi.org/10.1007/978-3-030-29407-6_19

ranks at 10th leading cause of death in India. Indian Cuisines are known or extensive use of spices they use spices to add flavors. Spicy food items are not good for our liver functionality consumption of hot chilies which produce heat in our body are hard to digest for our liver whereas there are some good spices like turmeric and ginger that help our liver to recover from various diseases. Functioning of liver disease is hard to discover at early stage as liver function normally even if it is partially damaged. Liver diseases can be diagnosed by analyzing patient's blood enzymes. Moreover, with the development of machine learning algorithms patients can diagnose their liver diseases at its initial stage. Through usage of classification models, one can check for the occurrence or absence of liver disease.

In this paper, five classification algorithms [2] were measured to compare their performance based on data of liver patients namely, support vector machines, Naive

| Index | Age | Gender | Total_Bilirubin | Direct_Bilirubin | Alkaline_Phosphotase | Alamine_Aminotransferase | Aspartate_Aminotransferase | Total_Protiens | Albumin | Albumin_and_Globulin_Ratio | Datas |
|---|---|---|---|---|---|---|---|---|---|---|---|
| 0 | 65 | Female | 0.7 | 0.1 | 187 | 16 | 18 | 6.8 | 3.3 | 0.9 | 1 |
| 1 | 62 | Male | 10.9 | 5.5 | 699 | 64 | 100 | 7.5 | 3.2 | 0.74 | 1 |
| 2 | 62 | Male | 7.3 | 4.1 | 490 | 60 | 68 | 7 | 3.3 | 0.89 | 1 |
| 3 | 58 | Male | 1 | 0.4 | 182 | 14 | 20 | 6.8 | 3.4 | 1 | 1 |
| 4 | 72 | Male | 3.9 | 2 | 195 | 27 | 59 | 7.3 | 2.4 | 0.4 | 1 |
| 5 | 46 | Male | 1.8 | 0.7 | 208 | 19 | 14 | 7.6 | 4.4 | 1.3 | 1 |
| 6 | 26 | Female | 0.9 | 0.2 | 154 | 16 | 12 | 7 | 3.5 | 1 | 1 |
| 7 | 29 | Female | 0.9 | 0.3 | 202 | 14 | 11 | 6.7 | 3.6 | 1.1 | 1 |
| 8 | 17 | Male | 0.9 | 0.3 | 202 | 22 | 19 | 7.4 | 4.1 | 1.2 | 2 |
| 9 | 55 | Male | 0.7 | 0.2 | 290 | 53 | 58 | 6.8 | 3.4 | 1 | 1 |

**Fig. 1** Sample of dataset from UC Irvine Machine Learning Repository

```
In [6]: sns.countplot(data=dataset, x = 'Dataset', label='Count')
Out[6]: <matplotlib.axes._subplots.AxesSubplot at 0xd58adb0>
```

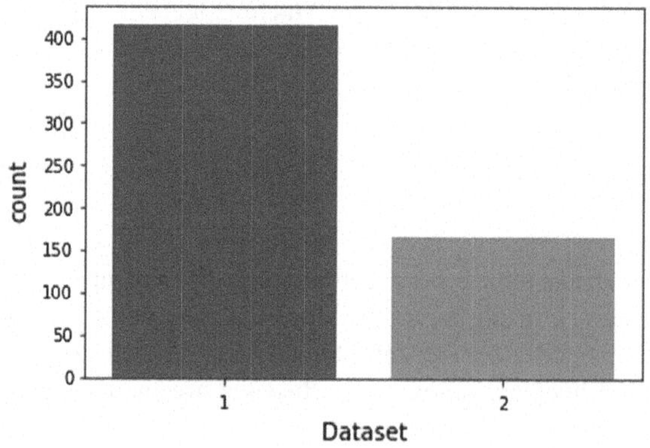


**Fig. 2** Liver patients are represented with 1 and non-liver patients are denoted by 2

Bayes classification, K-nearest neighbor, decision tree classification and logistic regression. The patient's records are collected from the UC Irvine Machine Learning Repository (also called as UCI repository). Dataset contains patient's records from Andhra Pradesh state in India, it contains 584 records, four hundred and sixteen (416) patients have liver ailments and one hundred and sixty-seven (167) patients have healthful liver. The dataset column is a class label dividing groups into liver patients (1) or not (2) (Figs. 1 and 2).

## 2 Data Visualization and Data Analysis

See Figs. 3, 4 and 5.

## 3 Proposed Technique

For detecting liver disease patients, proposed techniques are used in this paper. The comparisons are made for different machine learning algorithms viz., logistic regression algorithm, support vector machine algorithm, Naive Bayes classification algorithm [2], K-nearest neighbor algorithm and decision tree to find out which algorithm suits best and can categorize patients suffering from liver disease; cross-validation of data is done through *k*-folds cross-validation model. Figure 6 shows the architectural diagram for representing the overall system framework.

```
In [3]: dataset.info()

<class 'pandas.core.frame.DataFrame'>
RangeIndex: 583 entries, 0 to 582
Data columns (total 11 columns):
Age                           583 non-null int64
Gender                        583 non-null object
Total_Bilirubin               583 non-null float64
Direct_Bilirubin              583 non-null float64
Alkaline_Phosphotase          583 non-null int64
Alamine_Aminotransferase      583 non-null int64
Aspartate_Aminotransferase    583 non-null int64
Total_Protiens                583 non-null float64
Albumin                       583 non-null float64
Albumin_and_Globulin_Ratio    579 non-null float64
Dataset                       583 non-null int64
dtypes: float64(5), int64(5), object(1)
memory usage: 47.9+ KB
```

**Fig. 3** Only gender is non-numeric value in our dataset. There are 10 features and one output

```
In [8]: sns.countplot(data=dataset, x = 'Gender', label='Count')
```

```
Out[8]: <matplotlib.axes._subplots.AxesSubplot at 0xf7a3190>
```

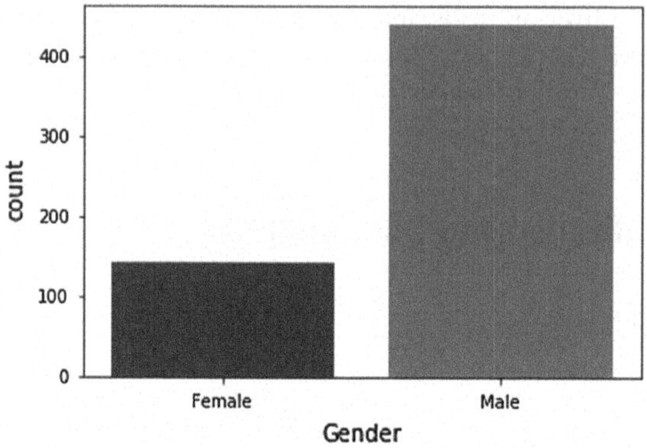


**Fig. 4** Number of males and females in dataset

```
In [9]: GndA = sns.FacetGrid(dataset, col="Dataset", row="Gender", margin_titles=True)
        GndA.map(plt.hist, "Age", color="red")
        plt.subplots_adjust(top=0.9)
        GndA.fig.suptitle('Disease by Gender and Age');
```

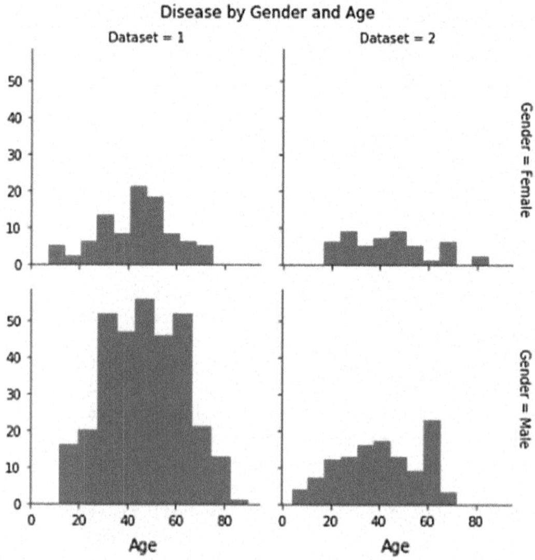


**Fig. 5** Female having age 45–55 are more prone to have liver diseases, whereas males having age 35–65 are highly prone to have liver disease

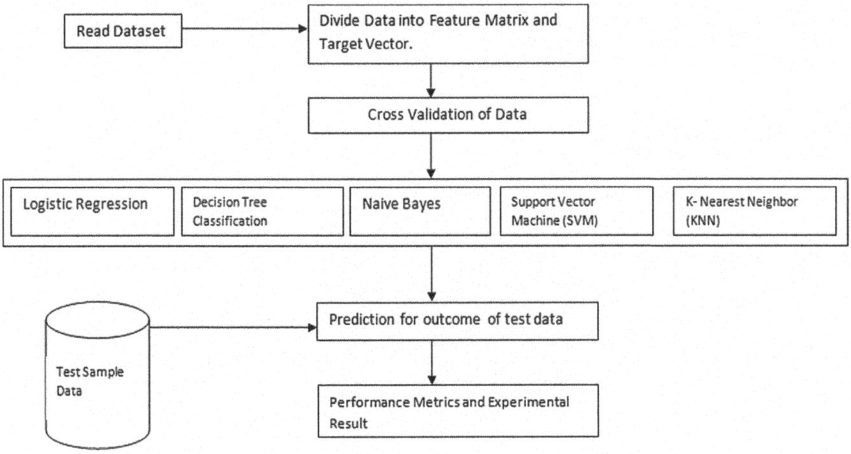


**Fig. 6** Architectural diagram for system framework

## *3.1 Cross-Validation Algorithm*

Cross-validation is similar to training and test split but it is applied on more subsets. It means we split our data into $K$ subsets and prepare our model on $K - 1$ subset. In this paper, we have applied $K$-folds cross-validations, dataset are divided into $k$ subsets and the last subset is considered as test data. We then average the model against each of the folds and then test it against our test set. In this paper, we have used 15 folds that are divided into 15 each and from these parts, 14 parts are the training sets and the last part is taken as test set. It ensures that each piece of dataset is helpful for training and testing and therefore helps to shrink the trial bias (Fig. 7).

## *3.2 Logistic Regression Algorithm*

The logistic regression algorithm is highly used to predict binary values in a given set of variables (True/False, 0/1). To represent categorical variables, dummy variables are used; in this paper, we used OneHotEncoder to convert categorical values to dummy encoding. To predict which class a particular data belongs, threshold can be set. According to this threshold, the obtained estimated property is classified. Let say predicted value $\geq 0.5$, then classify patient has liver disease.

$$Y = b_0 + b_1 x_1$$

$$P = \frac{1}{1 + e^{-y}} \quad \text{(Sigmoid Function)}$$

$$\ln\left(\frac{p}{1 - p}\right) = b_0 + b_1 x_1 \quad \text{(Logistic Regression Formula)}$$

```
In [14]: from sklearn.preprocessing import LabelEncoder,OneHotEncoder
         label_encoder = LabelEncoder()
         X[:,1] = label_encoder.fit_transform(X[:,1])
         one_hot_encoder = OneHotEncoder(categorical_features = [1])

         from sklearn.model_selection import KFold
         kf = KFold(n_splits=15)
         kf.get_n_splits(X)
         print(kf)
         for train_index, test_index in kf.split(X):
             print("train:", train_index, "TEST:", test_index)
             X_train, X_test = X[train_index], X[test_index]
             y_train, y_test = Y[train_index], Y[test_index]

         KFold(n_splits=15, random_state=None, shuffle=False)
         train: [ 39  40  41  42  43  44  45  46  47  48  49  50  51  52  53  54  55  56
           57  58  59  60  61  62  63  64  65  66  67  68  69  70  71  72  73  74
           75  76  77  78  79  80  81  82  83  84  85  86  87  88  89  90  91  92
           93  94  95  96  97  98  99 100 101 102 103 104 105 106 107 108 109 110
          111 112 113 114 115 116 117 118 119 120 121 122 123 124 125 126 127 128
          129 130 131 132 133 134 135 136 137 138 139 140 141 142 143 144 145 146
          147 148 149 150 151 152 153 154 155 156 157 158 159 160 161 162 163 164
          165 166 167 168 169 170 171 172 173 174 175 176 177 178 179 180 181 182
          183 184 185 186 187 188 189 190 191 192 193 194 195 196 197 198 199 200
          201 202 203 204 205 206 207 208 209 210 211 212 213 214 215 216 217 218
          219 220 221 222 223 224 225 226 227 228 229 230 231 232 233 234 235 236
          237 238 239 240 241 242 243 244 245 246 247 248 249 250 251 252 253 254
          255 256 257 258 259 260 261 262 263 264 265 266 267 268 269 270 271 272
          273 274 275 276 277 278 279 280 281 282 283 284 285 286 287 288 289 290
          291 292 293 294 295 296 297 298 299 300 301 302 303 304 305 306 307 308
          309 310 311 312 313 314 315 316 317 318 319 320 321 322 323 324 325 326
          327 328 329 330 331 332 333 334 335 336 337 338 339 340 341 342 343 344
          345 346 347 348 349 350 351 352 353 354 355 356 357 358 359 360 361 362
          363 364 365 366 367 368 369 370 371 372 373 374 375 376 377 378 379 380
```

**Fig. 7** Label encoding and *K*-folds cross-validation

Cost function used in logistic regression is different from cost function we use in linear regression this is because linear regression uses squared errors as cost function. If we use it in logistic regression, then it will show non-convex function parameter. Gradient Descent will only converge global minimum only if our function is convex in nature.

$$\text{Cost}(h\theta(x), Y = -\log(h\theta(x)) \quad \text{if } y = 1$$

$$-\log(1 - h\theta(x)) \quad \text{if } y = 0$$

The gradient of cost function with respect to each component of parameter vector [3] is:

$$\frac{\partial}{\partial \theta_j} = (y - h_\theta(x))x_j$$

Once we acquire gradient, our next step is to revise the fresh set of parameter vector using [3] (Fig. 8)

```
In [17]: from sklearn.linear_model import LogisticRegression
         regressor = LogisticRegression(random_state = 0)
         regressor.fit(X_train,y_train)
         pred_LR = regressor.predict(X_test)

         print(pred_LR)

         [1 1 1 1 1 1 1 1 1 1 1 1 1 1 1 1 1 1 1 1 1 1 1 1 1 1 1 1 1 1 1 1 1 1 1 1 1
          2]
```

**Fig. 8** Predictions made by logistic regression where pred_LR represents predictions

$$\theta_j := \theta_j + \alpha\left(y^i - h_\theta\left(x^i\right)\right)x_j^i \quad (\propto = \text{step towards direction of gradient}).$$

## 3.3 *Support Vector Machine Algorithm (SVM)*

SVM algorithm is a popular choice for linear and non-linear problems. SVM try to locate largest margin between points on either side of decision line [4]. It is less prone to overfitting and by the use of appropriate kernel, they are considered to work well even if data is not linearly separable [5]. Support vector tries to find hyperplane that separate the outcome of classes with intend of finding greatest distance to the neighboring point of two classes. Hyperplanes are decision boundaries that help to classify the data points. Given labeled training data, the algorithm [6] yields a ideal hyperplane that helps to classify new precedent. In two dimensional spaces, this hyperplane is a line separating a plane in two sections where each class lays in either side.

To maximize the margin between the data points and the hyperplane, loss function helps to maximize the margin is hinge loss.

$$c(x, y, f(x)) = \begin{cases} 0, & \text{if } y * f(x) \geq 1 \\ 1 - y * f(x), & \text{else} \end{cases}$$

$$c(x, y, f(x)) = (1 - y * f(x)) + \text{cost is } 0$$

if the predicted value and actual value have same sign, if different signs are there then we need to calculate the loss value.

$$\min_w \lambda ||w||^2 + \sum_{i=1}^{n} 1 - y_i \langle x_i, w \rangle)_+$$

Partial derivates work with respect to weights to find the gradients. With the help of gradient, we can update our weights (Fig. 9).

```
In [20]: from sklearn.svm import SVC
         classifier_SVM =  SVC(kernel = 'rbf',random_state = 0, shrinking = True)
         classifier_SVM.fit(X_train,y_train)
         pred_SVM = classifier_SVM.predict(X_test)

In [21]: pred_SVM

Out[21]: array([1, 1, 1, 1, 1, 1, 1, 1, 1, 1, 1, 1, 1, 1, 1, 1, 1, 1, 1, 1, 1, 1,
                1, 1, 1, 1, 1, 1, 1, 1, 1, 1, 1, 1, 1, 1, 1, 1], dtype=int64)
```

**Fig. 9** Predictions made by support vector machine where pred_SVM represents predictions

$$\frac{\delta}{\delta w_k}\lambda||w||^2 = 2\lambda w_k$$

$$\frac{\delta}{\delta w_k}(1 - y_i\langle x_i, w\rangle)_+ = \begin{cases} 0, & \text{if } y_i\langle x_i, w\rangle \geq 1 \\ -y_i x_{ik}, & \text{else} \end{cases}$$

## 3.4 Decision Tree Classification

As decision trees work for categorical as well as continuous input and output variables, we will first split our data according to type of situations and scenarios till we reach our optimal homogenous sets of data, in other words making data less impure. Impurity is traces of one class division into another. To measure impurity in data, we use term Entropy. It can be calculated with the help of probability [7].

$$H = -\sum p(x)\log p(x)$$

If we have multiple features in dataset. Which feature we should select for Division. If we divide classes into multiple branches, the information gain for given node is defined as [7] (Figs. 10 and 11):

$$\text{Information Gain}(n) = \text{Entopy}(x) - \left(\left[\text{weighted average}\right] * \text{entopy}(\text{children for feature})\right)$$

## 3.5 K-Nearest Neighbors

It is often used when model training should be quick, but predictions are typically slower. It categorizes a new data record by comparing it with analogous records from training set [3]. If dataset record consists set of numbers say $n$, we can find distance

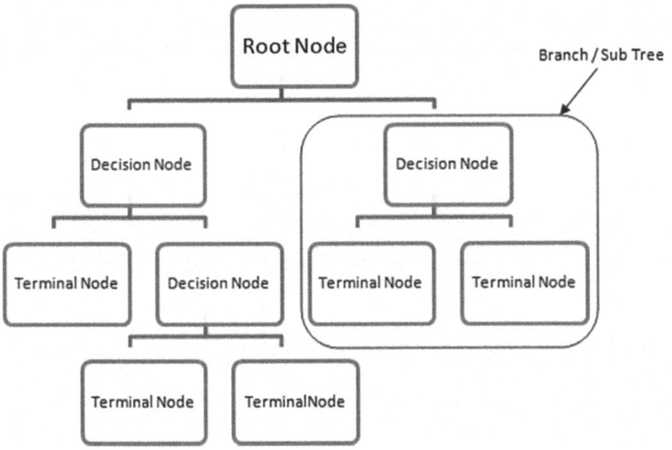


**Fig. 10** Decision tree classification

```
In [22]: from sklearn.tree import DecisionTreeClassifier
         Classifier_DCT = DecisionTreeClassifier(criterion = 'entropy', random_state = 0)
         Classifier_DCT.fit(X_train, y_train)
         pred_DCT = Classifier_DCT.predict(X_test)

In [23]: pred_DCT

Out[23]: array([1, 1, 1, 1, 1, 1, 2, 2, 1, 1, 1, 2, 1, 1, 1, 1, 1, 1, 2, 2, 1, 1,
                1, 1, 1, 1, 1, 1, 1, 1, 1, 1, 2, 1, 1, 1, 2, 1], dtype=int64)
```

**Fig. 11** Predictions made by decision tree classification where pred_DCT represents predictions

between records with distance formula

$$d = \sqrt{n_1^2 + n_2^2 \ldots n_n^2}$$

$$\sqrt{\sum_{i=1}^{n} (q_i - p_i)^2}$$

When algorithm is making prediction for new data point, it finds the nearest known record and assigns it to new record. This would be 1 nearest neighbor classifier [4]. The number of neighbor we consider for classifying the new data point is according to the $k$ value we provide to our algorithm, and it picks the class that is most common among the neighbors. Training phase is faster in KNN, because we index known records for fast distance calculations to new data point.

The K-nearest neighbor algorithm firstly chooses the number $K$ of neighbors; then takes the nearest neighbor; then counts number of data points in category, and

```
In [24]: from sklearn.neighbors import KNeighborsClassifier
         classifier = KNeighborsClassifier(n_neighbors  = 2)
         classifier.fit(X_train, y_train)
         pred_KNN = classifier.predict(X_test)

In [25]: pred_KNN

Out[25]: array([1, 1, 1, 1, 1, 1, 2, 1, 1, 1, 1, 1, 1, 1, 1, 1, 1, 1, 1, 2, 2, 1,
                1, 1, 1, 1, 1, 2, 1, 1, 1, 1, 1, 1, 1, 1, 1, 1], dtype=int64)
```

**Fig. 12** Predictions made by K-nearest neighbor where pred_KNN represents predictions

finally assigns the new data point to the category [8] where you counted the most neighbors (Fig. 12).

### *3.6 Naive Bayes Algorithm*

Naive Bayes algorithm based on Bayes Theorem works on supposition that outcome of one aspect of a class is independent of values [1, 9] in other attributes; this supposition is called class conditional independence [1, 8]. It is mostly helpful for extremely big datasets. Bayes theorem provides method of calculating posterior probability [10, 11]:

$$P(c|x) = \frac{P(x|c)P(c)}{P(x)}$$

$$P(c|X) = P(x_1|c) \times P(x_2|c) \times \cdots \times P(x_n|c) \times P(c)$$

$P(c|X)$ Posterior probability of class ($c$, target) given predictor ($x$, attributes).
$P(c)$ Prior probability of class.
$P(x|c)$ Likelihood probability of predictor given class.
$P(x)$ Prior probability of predictor.

It would be computational expensive to compute $P(X|C_i)$, to reduce computation in evaluation naïve assumption of class computation independence is made. It presumes that values of attributes are conditionally independent of each other, the class labels of tuples, i.e., that there are no dependence relationship between the attributes.

$$P(c_i|x_0, \ldots, x_n) \propto P(x_0, \ldots, x_n|c_i)P(c_i)$$

$$\propto P(c_i)\prod_{j=1}^{n} P\left(x_j|c_i\right)$$

In case where features are continuous in nature, they may follow Gaussian distribution (Fig. 13).

```
In [26]: from sklearn.naive_bayes import GaussianNB
         classifier = GaussianNB()
         classifier.fit(X_train, y_train)
         pred_NB = classifier.predict(X_test)

         pred_NB

Out[26]: array([2, 1, 1, 1, 1, 2, 2, 2, 1, 2, 1, 2, 2, 1, 1, 1, 1, 1, 2, 2, 1, 2,
                2, 1, 1, 1, 2, 1, 1, 1, 1, 1, 1, 2, 2, 2, 2, 2], dtype=int64)
```

**Fig. 13** Predictions made by Naïve Bayes classification where pred_NB represents predictions

**Table 1** Confusion matrix structure

| Predicted values | Actual values | |
|---|---|---|
| | Positive (1) | Negative (0) |
| Positive (1) | True Positive | False Positive |
| Negative (0) | False Negative | True Negative |

## 4 Performance Metrics and Experimental Result

The basic performance measures derived from the confusion matrix. Confusion matrix is a table often used to describe performance of classification model on the basis of test data [12]. Confusion matrix is also known as error matrix. Various measures such as precision, recall, f1-score and support can be derived from confusion matrix (Tables 1 and 2; Figs. 14, 15, 16, 17, 18, 19, 20, 21, 22, 23 and 24).

Accuracy: Accuracy is the ratio of number of correct prediction divided by number of total predictions (Table 3; Figs. 25, 26, 27, 28, 29 and 30).

$$\frac{\text{True Positive} + \text{True Negative}}{\text{True positive} + \text{True Negtaive} + \text{False positive} + \text{False Negtaive}}$$

## 5 Conclusion

In this paper, various machine learning models were used to detect liver disease in patients. Precision, recall, F-Score, support, micro-average, macro-average, weighted-average and accuracy score are used to evaluate the performance for proposed system. The accuracy of logistic regression, K-nearest neighbor, support vector machine, Naive Bayes and decision tree classification is 86, 84, 84, 71 and 76, respectively.

**Table 2** Understanding concepts behind

| Precision | Recall | *F*1-score |
|---|---|---|
| Precision signifies the proximity of two or more measurements with each other. Thus, precision is inversely linked to standard error. If standard error is huge, sample estimates are fewer precise. True positives are data points that are predicted as positive by model and are actually positive (correct predictions) and the false positive are the data points predicted as positive but actually labeled as negative | Capacity of a model to find out the important cases inside a dataset. Exact meaning of recall is the number of true positives divided by number of true positives plus number of false negative [13]. False negative are the that are predicted as negative but they are actually positive (incorrect prediction) recall = $\frac{\text{true positive}}{\text{true positive}+\text{false positive}}$ | *F*1-score can be interpreted as a weighted mean of precision and recall, where fl-score achieve its best value at 1 and worst case at 0. The general commitment of precision and recall in fl-score is equivalent. It is required to seek balance between precision and recall. *F*1-score is also known, as *F*-Score or *F*-measure [3] $F1 = 2 \times \frac{(\text{precision} \times \text{recall})}{(\text{precision}+\text{recall})}$ |
| **Micro-average** | **Macro-average** | **Weighted-average** |
| Micro-average aggregates to contribution of all classes to calculate average metric. If we are using multi-class classification, micro-average is preferable because there can be class imbalance, i.e., there can be many more examples in one class than of other class. It calculates the metrics by counting the total true positives, false negative and false positives | Macro-average computes the metric independently of each class and then returns the average. In other words, it calculates the metrics of each class and computes their unweighted mean. Unlike micro-average, it treats all classes equally | Each data point contributes in equal portion in final mean; some data points contribute more "weights" than others. If weights of all data points equal, then weighted mean will be equal to arithmetic mean. Calculate metrics and their average weight using support (the number of true instances for each label). This account macro for label imbalance, thus it can result in a *F*-Score that is not among precision and recall |

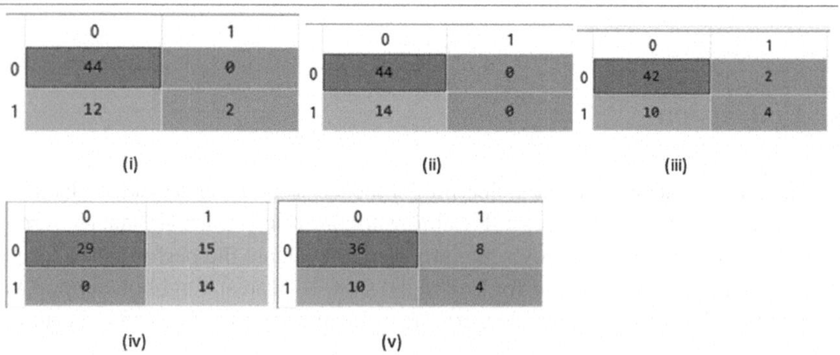


**Fig. 14** Confusion matrix for: (i) logistic regression, (ii) support vector machine, (iii) K-nearest neighbor, (iv) Naïve Bayes, (v) decision tree classification

Fig. 15 Graphical report for support vector machine

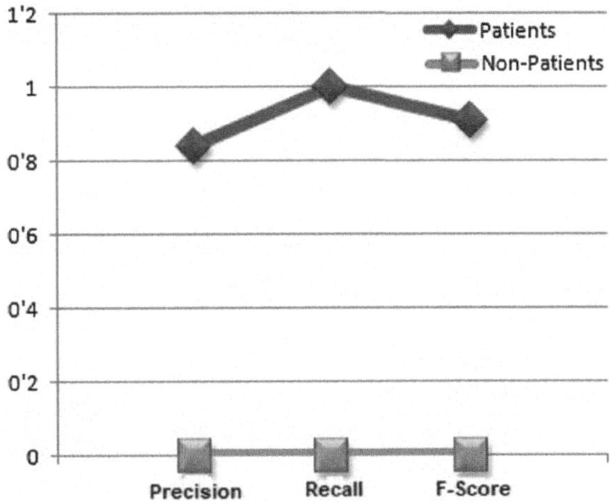


```
In [34]: print(classification_report(y_test, pred_SVM, target_names=target_names))
                    precision    recall  f1-score   support

    Liver Patients       0.84      1.00      0.91        32
Non-Liver Patients       0.00      0.00      0.00         6

         micro avg       0.84      0.84      0.84        38
         macro avg       0.42      0.50      0.46        38
      weighted avg       0.71      0.84      0.77        38
```

**Fig. 16** Classification report for support vector machine

**Fig. 17** Graphical report for Naïve Bayes

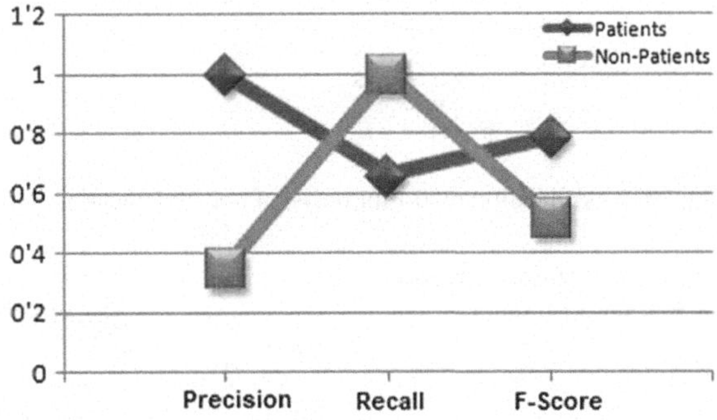

```
In [35]: print(classification_report(y_test, pred_NB, target_names=target_names))
                    precision    recall  f1-score   support

    Liver Patients       1.00      0.66      0.79        32
Non-Liver Patients       0.35      1.00      0.52         6

         micro avg       0.71      0.71      0.71        38
         macro avg       0.68      0.83      0.66        38
      weighted avg       0.90      0.71      0.75        38
```

**Fig. 18** Classification report for Naïve Bayes

**Fig. 19** Graphical report for decision tree classification

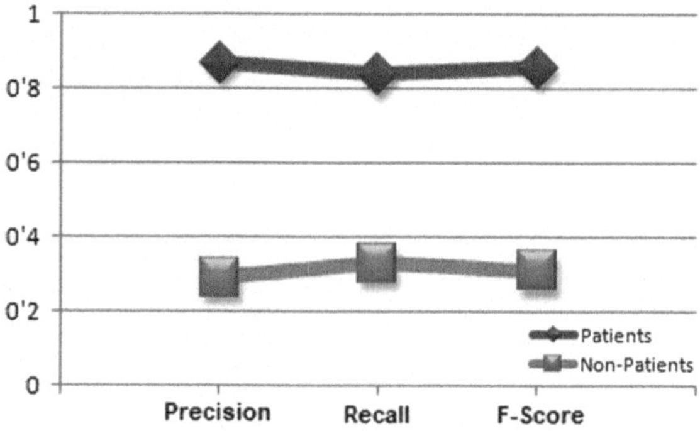


```
In [36]: print(classification_report(y_test, pred_DCT, target_names=target_names))
                    precision    recall  f1-score   support

    Liver Patients       0.87      0.84      0.86        32
Non-Liver Patients       0.29      0.33      0.31         6

         micro avg       0.76      0.76      0.76        38
         macro avg       0.58      0.59      0.58        38
      weighted avg       0.78      0.76      0.77        38
```

**Fig. 20** Classification report for decision tree classification

**Fig. 21** Classification report for K-nearest neighbor

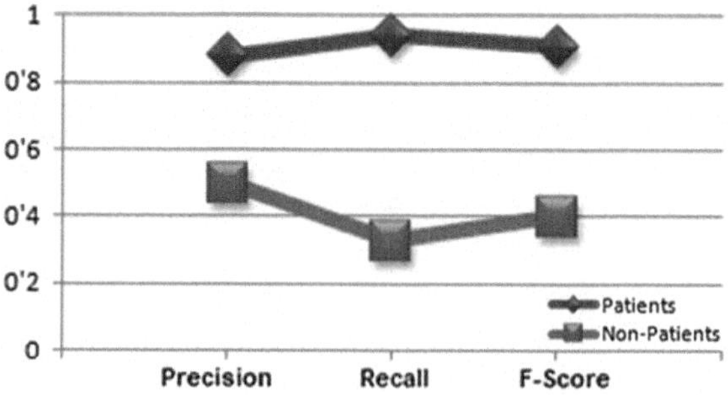

```
In [37]: print(classification_report(y_test, pred_KNN, target_names=target_names))

                    precision    recall  f1-score   support

    Liver Patients       0.88      0.94      0.91        32
Non-Liver Patients       0.50      0.33      0.40         6

         micro avg       0.84      0.84      0.84        38
         macro avg       0.69      0.64      0.65        38
      weighted avg       0.82      0.84      0.83        38
```

**Fig. 22** Classification report for K-nearest neighbor

**Fig. 23** Graphical report for logistic regression

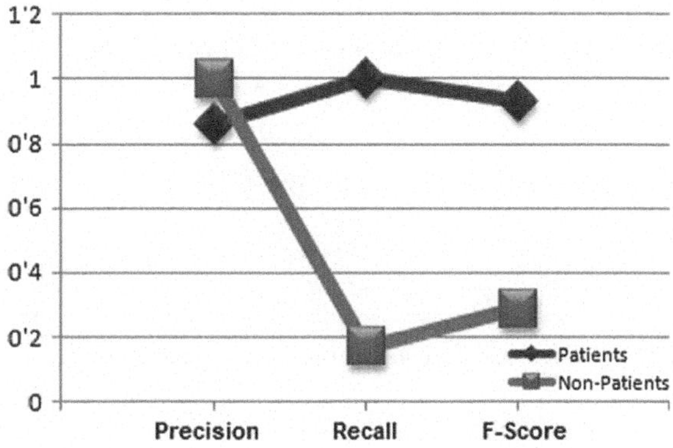


```
In [33]: from sklearn.metrics import classification_report
         target_names = ['Liver Patients', 'Non-Liver Patients']
         print(classification_report(y_test, pred_LR, target_names=target_names))

                    precision    recall  f1-score   support

    Liver Patients       0.86      1.00      0.93        32
Non-Liver Patients       1.00      0.17      0.29         6

         micro avg       0.87      0.87      0.87        38
         macro avg       0.93      0.58      0.61        38
      weighted avg       0.89      0.87      0.83        38
```

**Fig. 24** Classification report for logistic regression

**Table 3** Accuracy score for different models

| Models | Logistic regression | K-nearest neighbor | Support vector machine | Decision tree classification | Naive Bayes |
|---|---|---|---|---|---|
| Accuracy score | 86.84 | 84.21 | 84.21 | 76.31 | 71.05 |

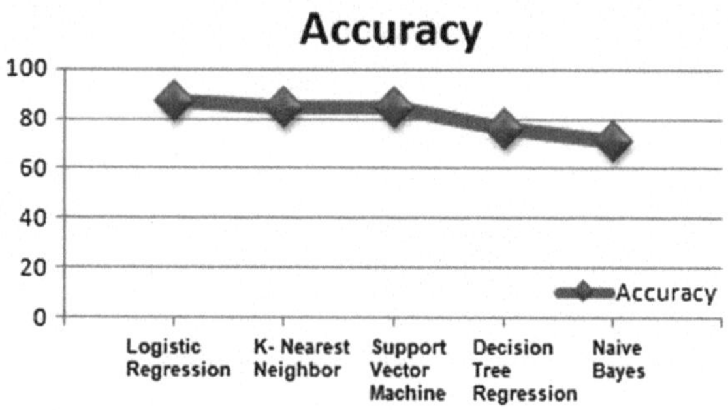


**Fig. 25** Graphical accuracy report

```
In [38]: acc_NB = accuracy_score(y_test, pred_NB)*100

In [39]: acc_NB

Out[39]: 71.05263157894737
```

**Fig. 26** Accuracy score for Naïve Bayes

```
In [40]: acc_DCT = accuracy_score(y_test, pred_DCT) * 100
         acc_DCT

Out[40]: 76.31578947368422
```

**Fig. 27** Accuracy score for decision tree classification

```
In [41]: acc_KNN = accuracy_score(y_test, pred_KNN) * 100
         acc_KNN

Out[41]: 84.21052631578947
```

**Fig. 28** Accuracy score for K-nearest neighbor

```
In [42]: acc_LR = accuracy_score(y_test, pred_LR) * 100
         acc_LR

Out[42]: 86.8421052631579
```

**Fig. 29** Accuracy score for logistic regression

```
In [43]: acc_SVM = accuracy_score(y_test, pred_SVM) * 100
         acc_SVM

Out[43]: 84.21052631578947
```

**Fig. 30** Accuracy score for support vector machine

## References

1. Ramana, B.V., Babu, M.S.P., Venkateswarlu, N.B.: A critical study of selected classification algorithms for liver disease diagnosis. Int. J. Database Manage. Syst. (IJDMS) **3**(2) (2011)
2. Ombabi, A.H., Lazzez, O., Ouarda, W., Alimi, A.M.: Deep learning framework based on word2vec and cnn for users interests classification. In: 2017 Sudan Conference on Computer Science and Information Technology (SCCSIT), pp. 1–7, IEEE (2017)
3. Sarkar, D., Bali, R., Sharma, T.: Practical Machine Learning with Python: A Problem-Solver's Guide to Building Real-World Intelligent Systems. Springer Nature, New York (2018)
4. Betrouni, N., Delval, A., Chaton, L., Defebvre, L., Duits, A., Moonen, A., Leentjens, A.F., Dujardin, K.: Electro encephalography based machine learning for cognitive profiling in Parkinson's disease: preliminary results. Mov. Disord. **34**, 210–217 (2019)
5. Syaifudin, Y.W., Puspitasari, D.: Twitter data mining for sentiment analysis on peoples feedback against government public policy. MATTER: Int. J. Sci. Technol. **3**(1), 110–122 (2017)
6. Gaikwad, G., Joshi, D.J.: Multiclass mood classification on Twitter using lexicon dictionary and machine learning algorithms. In: 2016 International Conference on Inventive Computation Technologies (ICICT), Coimbatore, pp. 1–6 (2016)
7. https://medium.com/machine-learning-101/chapter-3-decision-trees-theory-e7398adac567
8. Haque, M.R., Islam, M.M., Iqbal, H., Reza, M.S., Hasan, M.K.: Performance evaluation of random forests and artificial neural networks for the classification of liver disorder. In: 2018 International Conference on Computer, Communication, Chemical, Material and Electronic Engineering (IC4ME2), pp. 1–5. IEEE (2018)
9. Murty, M.N., Devi, V.S.: Bayes classifier. In: Pattern Recognition. Undergraduate Topics in Computer Science. Springer, London (2011)
10. https://www.journalbinet.com/uploads/2/1/0/0/21005390/38_jstei_analysis_of_classification_algorithms_for_liver_disease_diagnosis.pdf
11. Chou, S.Y., Trappey, A.: Advanced Concurrent Engineering. https://doi.org/10.1007/978-1-84882-762-2
12. Yassin, F.M., Lazzez, O., Ouarda, W., Alimi, A.M.: Travel user interest discovery from visual shared data in social networks. In: 2017 Sudan Conference on Computer Science and Information Technology (SCCSIT), Elnihood, pp. 1–7 (2017)
13. Arun Raj Kumar, P., Selvakumar, S.: Distributed denial of service attack detection using an ensemble of neural classifier. Comput. Commun. **34**(11), 1328–1341 (2011). https://doi.org/10.1016/j.comcom.2011.01.012

# A Review of Applications, Approaches, and Challenges in Internet of Things (IoT)

**Anil Sharma and Renu Sharma**

**Abstract** Internet of things "IoT" is now an integral part of our life. Its widespread usage is almost in every application starting from smart home, smart cities, smart farming, remote monitoring, industrial automation, and transport and the list is cumulative. Although this is very useful technology, many challenges are hampering its growth. The crucial challenges faced in the development of IoT are: security, privacy, constrained resources (as first layer in IoT is of sensors and they are quite constrained as far as resources are concerned), interoperability, and integration. Tremendous work has been done in these directions but much more is needed to overcome these barriers. Based on the literature review, this paper provides comprehensive information on IoT, its applications, approaches, and various challenges faced by existing approaches used in smart devices, which in turn can be helpful in finding future prospects.

**Keywords** Web/Internet of things · Smart health · Smart farming · Security thermostats · Smart meters

## 1 Introduction

Internet and its applications have become an integral part of today's life. The term "Internet of things" (IoT) is a combination of two words "Internet" and "things," where Internet is World Wide Web which is based on TCP/IP protocol, while things are distinguishable entities, so semantically IoT is a system of connected objects having exclusive address which is based on typical communication protocols [1, 2]. The IoT is anything that can be a part of network and can be communicated using different types of sensors [1, 3]. IoT refers to a system in which sensors can gather data from a network and thereafter share that data over the Internet, where it can be utilized for various interesting reasons [1, 4]. Additionally, this is inter-networking

A. Sharma · R. Sharma (✉)
Lovely Professional University, Phagwara, Kapurthala (Jalandhar), Punjab, India
e-mail: renusharma1978@yahoo.com

A. Sharma
e-mail: anil.19656@lpu.co.in



P. K. Singh et al. (eds.), *Proceedings of ICRIC 2019*, Lecture Notes in Electrical Engineering 597, https://doi.org/10.1007/978-3-030-29407-6_20

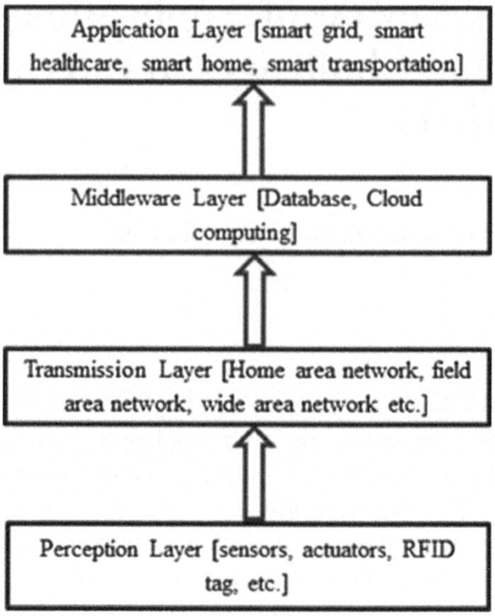


**Fig. 1** Layered architecture [7]

of physical objects, vehicles (moreover implied as "relate gadgets" and "smart Gadgets") assemblies and individual things presented with tools, programming, sensors, actuators, by using framework that approves these articles to collect and operate data [5]. For illustration, an object, in the IoT, has the capacity to deal with exchange information in system like in smart home all the sensors are deployed to give information and by using Web, a man-made request that can be entrusted an IP address and furnished with the capability to replace information [6]. A smart architecture is based upon layers as depicted in Fig. 1.

IoT typically involves many wireless sensor networks (WSNs) and for identification, radio-frequency identification devices (RFIDs) can be used. A WSN comprises smart sensors that can connect through a medium like radio communication. RFID devices are easy to use. They mainly comprise two parts: (1) integrated circuits (ICs) with some computational abilities and (2) a transmitter for communication [8, 9]. In the 1970s, structures for checking meters remotely on the power-driven framework, telephone lines were used [9]. During the last 20 years progression in remote sensing, permitted "Machine-to-Machine" (M2M undertaking and assembling answers for hardware observing and operation to turn into wide-ranging appreciation). Most of these untimely M2M suggestions were depended on close function built framework and exclusive on manufacturing particular principles instead of Internet Protocol (IP)-based framework and Internet principles [10]. To associate objects through IP to the Web is not a first thought. The starting Internet enabled gadget is an IP-linked toaster with Internet connectivity could be used was highlighted in 1990 at an Internet consultation [11], early Internet enabled devices like a coke machine at the University of Carnegie Mellon in the USA [12]. With this early development, a strong area of innovative works into brilliant entity networking helped to make the establishment for today's Internet of things [13].

Furthermore, like the quantity of Web-attached strategy grows, evaluation of activity they produce is predictable to go up fundamentally. For example, Cisco predicts that movement of Internet delivered through non-PC items will climb from 40% in 2014 to only below 70% in 2019 [14]. Cisco similarly anticipated that the amount of M2M affiliations (incorporating into current, home, social insurance, auto, and other IoT objects) will rise from 24% of each and related device in 2014 to 43% in 2019 [10, 15]. Moreover, the IoT is a growing matter of mechanical, communal, and monetary importance. Shopper things, solid merchandise, autos and trucks, present-day and effectiveness parts, sensors, as well as other ordinary articles are being merged with Internet accessibility and viable data informative capacities that assurance to change the means of effort, survive, and play [16]. It is estimated that the effects of IoT on the Internet and financial system are extraordinary. With several associations, billions of IoT devices and an overall monetary impact of extra $11 trillion by 2025 are projected [15, 16].

# 2 Application Areas of Internet of Things

## *2.1 Smart Home*

The smart home is the main area of application of IoT. From the Amazon Echo to the Nest Thermostat, there are numerous things accessible that customers can be in command of their voices to create their lives more associated than in previous times. The keen home machines can be partitioned into two classes: (1) Non-schedulable machines, for example, lights, PCs, and TVs, depend on manual control to finish an undertaking and are required just when the clients are home [17]. (2) Schedulable home machines, which can be planned for ideal operation, are exchanged on/off whenever, e.g., clothes washer, cooling system, press, water warmer, and EVs. The machines which can finish an undertaking with no manual control, for example, ventilation and cooling system and water warmer are schedulable [18].

## *2.2 Smart City*

A smart city provides an intelligent way to deal with components such as transport, power, health, smart buildings, and smart atmosphere. Moreover, smart city basically improves the living standards of citizens by connecting every object to the Internet. Smart activity controller, keen stopping smart wastage, smart water supply, etc. are few of the instances [19].

### 2.3 *Wearable*

In IoT, watches are no longer just for time. The Apple Watch and various smart watches have changed our wrists into phone holsters by engaging elements education, phone calls, etc. and that is just the tip of the ice sheet. Likewise, devices Fitbit and Jawbone have changed the well-being scene by giving people more data about their workouts [19].

### 2.4 *Smart Farming*

The use of IoT can really help in the farming sector. As sensors can help to monitor need of manure, pesticides, and irrigation, effective utilization can be done on the resources. Innovation has crossed obstacles by utilizing remote monitoring, remote organizing, and portability to efficiently usage of the resources. This can really help in rural advancement [20].

### 2.5 *Machine-to-Machine*

Machine-to-machine (M2M) interchanges, similarly called MTC for "Machine-Type-Communication" by the Third Generation Partnership Project (3GPP) [21], allude to correspondences among little and economical gadgets wherever no or minimal human intervention is required [21, 22]. There have been around 5 billion M2M gadgets associated with the remote systems, and their number is going to be 50 billion before this decade. These gadgets are heterogeneous, running from remote sensors with lesser requirements [23, 24].

## 3 Literature Review

In [25], authors have described that little devices can be fitted with computational power. Authors also described the utility of bionetworks in this area. In [26], authors have proposed a system based on sensors to give safety for elderly people. Sensors are used widely at floor, bathtub, toilet, and shower [1]. In this, big data has been used. The main focus is to provide safety to aged people in the bathroom through IoT. In [27], authors have explained prediction of an event in an IoT-based environment. Bayesian event prediction model has been used. A prototype system has been used to forecast outbound flight delay events after studying inbound flight delays [1]. In [28], a system is proposed based on IoT, which is using big data analysis to provide a solution to the water supply of USA which is actually based on 17,000 different systems. Authors have suggested for one system based upon WSNs which can efficiently manage the water supply with keeping in mind of existing system and providing solution for the existing problems like legitimate usage of water, concerns of pressure and flow, and water losses. In [29], authors have developed an interaction system between a human and machine. To make a more humanized interaction, various aspects of interaction have been dealt like analysis of various existing emotional models, developing new models, recognition of emotions through text. In this system based upon smart home technology, after text analysis, a more personalized, Chinese language-based interaction system has been proposed. In [30], authors have proposed routing protocol to deal emergency fast routing. Authors have explained the need of fast routing because IoT devices are kept on increasing and in case of disaster or emergency reduced response time is must to decrease the effect. Authors have described emergency response IoT-based global information decision (ERGID). The main aim is to have reduced response time. Authors have demonstrated the above protocol on STM32W108 sensor nodes. In [31], author presents a framework for semantic interoperability between different formats of documents. In an IoT-based environment, information is being produced in abundance and that is too in different formats. Authors have suggested common dictionary "CONEX" for same group of semantic document and a "semantic interference algorithm" (SIA) has been proposed. Proposed framework has been implemented on flight booking. In [32], authors have presented a context-aware parking system. In this by using technology, a new parking system is given which is capable to use cloud and sensors to have efficient parking. This [32] paper uses semantic interoperability technique. A semantic inference algorithm has been proposed to infer a document. This can really help to provide interoperability. If one sees into the problem areas of IoT, they will find that it is mainly comprised of privacy, security, energy efficiency, and interoperability. In Table 1, literature of these four categories has been analyzed intensely, which can help researchers to have possible future prospects.

**Table 1** Literature review and findings

| Terms | Literature review | Techniques | Strength | Limitation | Future context | References |
|---|---|---|---|---|---|---|
| 1. Privacy | In 2019 by Arbia Riahi Sfar et al. | Markovian game-based solution to protect privacy | Integration of IoT context and privacy in global system | Based upon one model | Long-term player payoffs | [33] |
| | In 2016 by Lukas Malina et al. | Information privacy k-anonymity, homomorphic encryption, aggregate marks, ring mark, quality-based marks, and their point of view in IoT | Proof knowledge, employ public key encryption | The small RAM memory for these privacy solutions | Need a temper proof module to produce large size of signature | [34] |
| | In 2016 Ioannis Chatzigiannakis et al. | Privacy by design approach, zero knowledge protocol, elliptic curve cryptography (ECC) approach | It is quite effective in privacy preserving and can be useful for other smart city applications, e.g., intelligent buildings | It is just a public key cryptography having lesser key parameters | To expansion of protocols which can prove the relations for encrypted values without telling them | [35] |
| | In 2013 Du kun kun et al. | Software platform of management system is used for human interaction | It is based on text emotion recognition and human interaction | It can be easily tracked and no middleware is used to secure personal information | An encryption key or protocol is needed for hiding the personal data | [29] |
| 2. Security | In 2016 victor Chang et al. | An integrated security approach CCAF multilayered security | It can discover and block viruses and trojans | Missing of real-life examples | Need to use different prototypes for better security solutions | [36] |

(continued)

**Table 1** (continued)

| Terms | Literature review | Techniques | Strength | Limitation | Future context | References |
|---|---|---|---|---|---|---|
| | In 2015 by FLAUZAC Olivier et al. | An architecture with multiple SDN controller and scalable with multiple SDN domains in networks for security purposes in IoT devices | (a) Guarantee of independence of each domain in failure of network (b) Security of entire network to prevent attacks | Not suitable in real environment | Need to extend SDN domain and investigate security mechanism. Test the system in larger scale | [37] |
| | In 2015 Jiafu wan et al. | In this multilayered context, aware architecture and context aware vehicular security approaches are highlighted | It can be adapted in user's situation and can be incorporated in security services | It is not suitable for proper security or authentication | Need to improve vehicle safety and develop proactive context aware system | [32] |
| | In 2013 Rodrigo Roman et al. | Distributed architecture and centralized system are sued in this paper | Access control, trust managements for verifying data | Authentication issue is not tackled | Network security protocols can be developed for fault tolerance | [38] |
| 3. Energy efficient | In 2018 Rolf Arne Kjellby et al. | Self-sustainable through energy harvesting | Lesser energy is required | Not possible if source of light is not there | Can be applied to many applications | [39] |
| | In 2016 Min Wei et al. | The demand response (DR) approach and IoT energy management (EMS, EMA) platform for common information model | It allowed to devices implemented with less time, effort, cost, and complexity | It does not support (DR) the real-time energy price | Need to implement and extend genuine rechargeable batteries, sun-based board, wind turbines, and waste warmth recuperation | [40] |

(continued)

**Table 1** (continued)

| Terms | Literature review | Techniques | Strength | Limitation | Future context | References |
|---|---|---|---|---|---|---|
| | In 2016, T. Qiu et al. | In this, a new protocol ERGID proposed and DIM, RECP mechanisms are used | It ensures the routing selection for each node, lower chances of packet delay | This technique is low in power consumption in big scale | In future, a trade-off can be used between E2E delay and energy expenditure | [30] |
| | In 2016 by Jau-yang change | Distributed clustering computing technique for sensing node is used | Suitable where more IoT devices are employed and it can help in reducing energy consumption | Fails in large node connectivity | It is suggested to extend the lifetime to power | [40] |
| | In 2015 Emad Felemban et al. | MMSPEED for remote sensor systems to give benefit separation and probabilistic QoS ensures in the auspiciousness and unwavering quality areas | Useful for probabilistic multipath, efficient, reliable | Low in measurement of power consumption | Need to improve power consumption in real network | [41] |
| | In 2015 by Tie Qiu et al. | The STET approach in view of crossing tree arrangement for WSNs. It consolidates two sorts of time synchronization SRP and RRP for energy-efficient | The energy consumption is low | Time synchronization accuracy not improved as compare to existing approaches | To enhance time synchronization exactness with low vitality utilization | [42] |

(continued)

**Table 1** (continued)

| Terms | Literature review | Techniques | Strength | Limitation | Future context | References |
|---|---|---|---|---|---|---|
| | In 2013 By M. Aissani et al. | Real-time routing protocols DPD, NEP and SNGF speed protocols are used | Low cost, end-to-end delay and better relay speed | Waste energy of nodes and overload the network link | Need for mechanism of delay packet and void tolerant | [43] |
| 4. Interoperability | In 2018 B. Kang et al. | Intelligent gateway for interoperability, CoAP | Self-configurable | Variety of sensors are hard to integrate | To include heterogeneous device | [44] |
| | In 2017 by Shuo Yang et al. | A framework for semantic interoperability. It is based on documents | Improve the information fusion across different information system, improve the reusability of document | Not based on real Web-based cases | Need to expand DocLang to more information systems | [45] |
| | In 2017 by M. Ganzha et al. | Various ontologies of health, logistics, and transportation have been discussed based upon agents, middleware | Semantic interoperability has been tried | Full functionality is pending | More work can be done to achieve semantic interoperability at larger domains | [46] |
| | In 2016 by J. Stassner et al. | Semantic interoperability, translation layer | An effective translation layer can be very helpful for interoperability | Main cause of interoperability problem is due to lack of standardization | Translation layer covering more number of devices | [47] |
| | In 2016 by C. T. Ovengalt et al. | Virtual object and CEP technique are used for incompatible devices | Context broker | Complex in large volume data and | New simulation environment can be developed | [48] |

# 4 Challenges

## *4.1 Privacy*

From the literature review, it is found that privacy is the main issue in IoT. In much appearance including a layer of encryption will deal with most protection issues. However, in IoT, the genuine esteem lies in the examples of correspondence and the metadata that is never scrambled, including the planning and repetition of information trade [49]. For example, by just observing the information activity that is going all through your home, a hoodlum can remotely make sense of your living examples and the times your home might be empty, and in this manner plan to burglarize your home without coming near your home. In order to realize opportunities, strategies have to develop individual privacy preference over a broad scale of expectations [37, 50].

## *4.2 Security*

It is found that security is the most challenging issue of IoT worldwide. The IoT is based on a wide range of semiconductor advancements, including power administration gadgets, sensors, and microchips [40]. Execution and security necessities differ significantly starting with one application then onto the next. The more noteworthy the volume of touchy information we exchange over the IoT, the more noteworthy the danger of information and wholesale fraud, gadget control, information misrepresentation, IP robbery, and considerably server/arrange control [37, 48].

## *4.3 Energy Efficient*

From the findings given above, it is noticeable that time organization is necessary for the utilization of broad-scale wireless sensor networks (WSNs). Nevertheless, numerous procedures of time synchronization encounter the evil impacts of high correspondence overheads while looking for high exactness [42]. Such overheads cause basic lifetime shrinkage of WSNs since incessant information data correspondence expends much vitality which is confined in every sensor at the center point. The vitality utilization expanded quickly with the improvement of WSNs thickness [29]. It is found that the battery reinforcement or long existence of sensors or gadgets is the enormous challenge.

### 4.4 Interoperability

The challenge of interoperability is facing worldwide in the area of Internet of things. The Web of things is a network between individuals, procedures, and things and the communication between heterogeneous devices. In IoT industry companies developing load of devices every day with different operating systems, different platforms, and different programmability, these dissimilar devices cannot communicate with each other so this is the problem of interoperability [8]. Intelligent gateways can really help to solve this problem [44].

## 5 Conclusion and Future Work

Authors have enlightened the introduction of IoT and its applications areas. Literature survey of this paper highlights the different areas and their respective challenges, which can be used as a ready reckoner to pick a particular area to pursue further research. Implementation of IoT is facing challenge in terms of security, privacy, energy efficiency, and interoperability. Comprehensive information is given to get an insight into these different parameters. A lot of work is already done but much more is still need to be done in all these above-discussed directions. Growth rate of smart devices is tremendous. IoT will definitely be premier technology in coming years. The paper contains the real-life examples and applications of Internet of things. In future work, the various limitations and challenges in smart devices are needed to tackle and the area of IoT should be explored with new techniques. Further research can be done to overcome the problem of interoperability that is major challenge of Internet of things of the hour.

## References


1. Kaur, K., Sharma, A.: Interoperability among Internet of Things (IoT) components using model-driven architecture approach. In: Fong, S., Akashe, S., Mahalle, P. (eds.) Information and Communication Technology for Competitive Strategies, ICTCS-2017, pp. 519–534. Springer, Singapore (2019)
2. Homepage of postspaces Internet of Things. http://www.postscapes.com/internet-of-things-definition/. Last accessed 2018/09/18
3. Muztoba, M., et al.: Robust communication with IoT devices using wearable brain machine interfaces. In: International Conference on Computer-Aided Design, pp. 200–207, IEEE/ACM, Austin (2015)
4. Ding, G., Wang, L., Wu, Q.: Big data analytics in future Internet of Things. arXiv preprint arXiv:1311.4112 (2013)
5. Webpage of Wikipedia. https://en.wikipedia.org/wiki/Internet_of_things. Last accessed 2018/09/18
6. Webpage of Internet of Things agenda. http://internetofthingsagenda.techtarget.com/definition/Internet-of-Things-IoT. Last accessed 2018/09/18

7. Webpage of researchgate. https://www.researchgate.net/figure/The-five-layer-IoT-architecture-114_fig7_309606980. Last accessed 2018/10/17
8. Stojkoska, B.L.R., Trivodaliev, K.V.: A review of Internet of Things for smart home: challenges and solutions. J. Clean. Prod. **140**, 1454–1464 (2016)
9. Webpage of Wikipedia. https://en.wikipedia.org/wiki/Machine_to_machine. Last accessed 2018/12/20
10. Webpage of automation world. https://www.automationworld.com/cloud-computing/know-difference-between-iot-and-m2m. Last accessed 2018/12/20
11. Webpage of living internet. http://www.livinginternet.com/i/ia_myths_toast.html. Last accessed 2016/12/18
12. Webpage of cs. https://www.cs.cmu.edu/~coke/history_long.txt. Last accessed 2016/09/18
13. Webpage of tools IETF. https://tools.ietf.org/html/rfc7452. Last accessed 2016/09/18
14. Rose, K., Eldridge, S., Chapin, L.: The internet of things: an overview. In: The Internet Society, pp. 1–50 (2015)
15. Shelkikar, R.P., Wagh, N.S.: Review paper based on women tracking device using concept of "Internet of Things". Int. J. Appl. Innov. Eng. Manage. **5**, 63–73 (2016)
16. Perera, C., Liu, H., Jayawardena, S., Chen, M.: A survey on internet of things from industrial market perspective. IEEE Access **2**, 1660–1679 (2014)
17. Wilson, C., Hargreaves, T., Hauxwell-Baldwin, R.: Benefits and risks of smart home technologies. Energy Policy **10**, 72–83 (2017)
18. Mitchell, S., Villa, N., Stewart-Weeks, M., Lange, A.: The internet of everything for cities. Connecting People, Process, Data, and Things to Improve the 'Livability' of Cities and Communities. Retrieved from https://www.cisco.com/c/dam/en_us/about/ac79/docs/ps/motm/IoE-Smart-City_PoV.pdf. Cisco (2013)
19. Webpage of Internet of Things, http://internetofthingswiki.com/iot-applicationsexamples/541/#Smart-Cities. Last accessed 2018/09/18
20. Mohanraj, I., Ashokumar, K., Naren, J.: Field monitoring and automation using IOT in agriculture domain. Procedia Comput. Sci. **93**, 931–939 (2016)
21. Barki, A., Abdullah, A., Gharout, S., Traoré, J.: M2M security: challenges and solutions. IEEE Commun. Surv. Tutorials **8**(2), 1241–1254 (2016)
22. Biral, A., Centenaro, M., Zanella, A., Vangelista, L., Zorzi, M.: The challenges of M2M massive access in wireless cellular networks. Digit. Commun. Netw. **1**(1), 1–19 (2015)
23. Istepanian, R.S., Hu, S., Philip, N.Y., Sungoor, A.: The potential of Internet of m-health Things "m-IoT" for non-invasive glucose level sensing. In: 2011 Annual International Conference Engineering in Medicine and Biology Society, EMBC, pp. 5264–5266, IEEE (2011)
24. Bonomi, F., Milito, R., Zhu, J., Addepalli, S.: Fog computing and its role in the internet of things. In: Proceedings of the First Edition of the MCC Workshop on Mobile Cloud Computing, pp. 13–16, ACM (2012)
25. Gill, A.Q., Phennel, N., Lane, D., Phung, V.L.: IoT-enabled emergency information supply chain architecture for elderly people: the Australian context. Inf. Syst. **58**, 75–86 (2016)
26. Koo, D.D., Lee, J.J., Sebastiani, A., Kim, J.: An internet-of-things (IoT) system development and implementation for bathroom safety enhancement. Procedia Eng. **145**, 396–403 (2016)
27. Karakostas, B.: Event prediction in an IoT environment using naïve bayesian models. Procedia Comput. Sci. **83**, 11–17 (2015)
28. Koo, D., Piratla, K., Matthews, C.J.: Towards sustainable water supply: schematic development of big data collection using Internet of Things (IoT). Procedia Eng. **118**, 489–497 (2015)
29. Du, K.K., Wang, Z.L., Hong, M.: Human machine interactive system on smart home of IoT. J. China Univ. Posts Telecommun. **20**, 96–99 (2013)
30. Qiu, T., Xia, Y. Lv, F., Chen, N., Wan, J., Tolba, A.: ERGID: an efficient routing protocol for emergency response Internet of Things. J. Netw. Comput. Appl. **72**, 104–112 (2016)
31. Yang, S., Guo, J., Wei, R.: Semantic interoperability with heterogeneous information systems on the internet through automatic tabular document exchange. Inf. Syst. **69**, 195–217 (2017)
32. Wan, J., Zhang, D., Zhao, S., Yang, L., Lloret, J.: Context-aware vehicular cyber-physical systems with cloud support: architecture, challenges, and solutions. IEEE Commun. Mag. **52**(8), 106–113 (2014)

33. Safar, A.R., Challal, Y., Moyal, P., Natalizio, E.: A game theoretic approach for privacy preserving model in IoT-based transportation. IEEE Trans. Intell. Transp. Syst. 1–10 (2019)
34. Malina, L., Hajny, J., Fujdiak, R., Hosek, J.: On perspective of security and privacy—preserving solutions in the internet of things. Comput. Netw. **102**, 83–95 (2016)
35. Chatzigiannakis, I., Vitaletti, A., Pyrgelis, A.: A privacy—preserving smart parking system using an IoT elliptic curve based security platform. Comput. Commun. 165–177 (2016)
36. Chang, V., Ramachandran, M.: Towards achieving data security with the cloud computing adoption framework. IEEE Trans. Serv. Comput. **9**, 138–151 (2016)
37. Olivier, F., Carlos, G., Florent, N.: New security architecture for IoT network. Procedia Comput. Sci. **52**, 1028–1033 (2015)
38. Roman, R., Zhou, J., Lopez, J.: On the features and challenges of security and privacy in distributed internet of things. Comput. Netw. **57**, 2266–2279 (2017)
39. Kjellby, R.A., Johnsrud, T.E., Loetveit, S.E., Cenkeramaddi, L.R., Hamid, M., Lozano, B.B.: Self-powered IoT devices for indoor applications. In: 31st International Conference on VLSI Design and 17th International Conference on Embedded systems (VLSID) (2018)
40. Chang, J.Y.: A distributed cluster computing energy-efficient routing scheme for internet of things systems. Wireless Pers. Commun. **82**(2), 757–776 (2015)
41. Felemban, E., Lee, C., Ekici, E.: MMSPEED: multipath, Multi-SPEED protocol for QoS guarantee of reliability and timeliness in wireless sensor networks. IEEE Trans. Mob. Comput. **5**, 738–754 (2006)
42. Qiu, T., Chi, L., Guo, W., Zhang, Y.: STETS: a novel energy-efficient time synchronization scheme based on embedded networking devices. Microprocess. Microsyst. **39**, 1285–1295 (2015)
43. Aissani, M., Bouzand, S., Fareb, A., Laidoui, M.A.: EA-SPEED: energy-aware real-time routing protocol for wireless sensor networks. Int. J. Inf. Commun. Technol. **5**, 22–44 (2013)
44. Kang, B., Choo, H.: An experimental study of a reliable IoT gateway. ICT Express **4**(3), 130–133 (2018)
45. Yang, S., Guo, J., Wei, R.: Semantic interoperability with heterogeneous information systems on the internet through tabular document exchange. Inf. Syst. **69**, 195–217 (2017)
46. Ganzha, M., Paprzycki, M., Pawlowski, W., Szmeja, P., Wasielewska, K.: Semantic interoperability in the Internet of Things; an overview from IINTER-IoT perspective. J. Netw. Comput. Appl. **81**, 111–124 (2017)
47. Strassner, J., Diabl, W.W.: A semantic interoperability architecture for internet of things data sharing and computing. In: IEEE 3rd World Forum on Internet of Things (WF-IoT), pp. 609–614 (2016)
48. Ovengalt, C.T., Djouani, K., Kurien, A.M., Chibani, A.: Towards a scalable and optimised context broker for seamless interoperability within IoT systems. Procedia Comput. Sci. **94**, 91–96 (2016)
49. Park, S., Kim, Y., Chang, H.: An empirical study on security expert ecosystem in the future IoT service environment. Comput. Electr. Eng. **52**, 199–207 (2016)
50. Almeida, V.A.F., Doneda, D., Monteiro, M.: Governance challenges for the Internet of Things. IEEE Internet Comput. **19**(4), 56–59 (2015)

# Intellegent Networking

# Web Search Personalization Using Semantic Similarity Measure

**Sunny Sharma and Vijay Rana**

**Abstract** Web search personalization is the process of providing personalized results to the user for his query. In this paper, we present a relevance model to personalize search results which is based on query personalization. The user query is directly matched to the keywords of the user profile, and the original query is altered according to the keywords which is more likely similar or related according to the similarity measure. By finding the similarity between the user original query and user profile, a linear combination of preference space is generated at run-time to determine more accurately which pages are truly the most important with respect to the modified query. A heuristic algorithm is used to maintain the user profile based on the ongoing behavior. Our experiments prove that retrieving the search results based on query modification is effective in providing the personalized results to the user.

**Keywords** Web search personalization · Query modification · Semantic similarity · Semantic annotations

## 1 Introduction

Web search personalization is the process of providing the personalized search results to the user for his/her query by taking the advantage of knowledge acquired through the Web usage mining. Different clients have diverse interests. Thus, for some users, the initial document may be related, but for others not. Thus, we study Web personalization [2]. We have projected an easy way for personalized search based on long-term user queries that matches or outperforms the state-of-the-art for this task. Current search engines are not personalized up to the mark: returned search results based on the keywords entered by the user. From a set of documents, those items that

S. Sharma (✉)
Department of Computer Applications, Chandigarh University, Mohali, Punjab, India
e-mail: sunny202658@gmail.com

V. Rana
Department of Computer Applications, SBBS University, Khiala, Punjab, India



P. K. Singh et al. (eds.), *Proceedings of ICRIC 2019*, Lecture Notes in Electrical Engineering 597, https://doi.org/10.1007/978-3-030-29407-6_21

best match the query are returned to the user. Personality of the user is not considered when processing a query [14]. Personalized Web search, on the other hand, requires the specific profile for each user into the process of finding the considered documents [19] for a search query by mounting the amount of a priori input information offered to the search algorithms. In this personalization process, our main focus on the query modification.

For the purpose of query personalization, user preferences are provided in the form of a user profile. The goal is to locate which set of preferences should be considered in the context of a given query and personalize the query accordingly. Our approach to modify the query is based on similarity score between query and user profile data. The user profile contains information like previous searched queries, clicked links, meta tags [12], and semantic annotation-based Web documents [23]. Our system is composed of different modules and algorithms: A simple interface consisting of a search box, Web user identification techniques, intuitive algorithms for preprocessing of Web user query and for finding the semantic similarity between the current user query and the past queries. The user profile is also continuously being updated while user interacts with the systems. And at the final, the results are being provided to the user after discovering and analyzing the patterns. The patterns are primarily based on the semantic similarity score and data mining techniques [26]. By using accessible software package, an analyst can find the similarity distance, of words and phrases [15]. But somehow, these methods are still insufficient for natural language processing and information extraction in the light of ambiguous keywords. At the end, we present that the proposed work retrieves overall better results than direct approaches such as the KNN method in terms of accuracy.

In the next section, we present some novel work to personalize and describe its theoretical backgrounds. Section 3 describes the detailed methodology with terminologies and the process of personalization. Section 4 outlines the process and implementation for similarity, and using such a system in practice, the experiments in this section analyze the feasibility and usefulness of our approach, followed by the evaluation of the experimental work and the conclusion. In the conclusion section, we briefly summarize our work and outlining our future plan.

## 2 Related Work

The author in [22] performs Web personalization. It mainly focuses on the re-ranking of the results that are already ranked by the search engine for the different users. They have personalized (re-ranked) on the basis of the clicks and dwell time of the user for a Web page. The paper begins with probabilistic graph model to identify the relationship between user, user's query and the ranked documents. Thereafter, the author of paper [25] states that using the collected user profile directly for the purpose of personalization is not a good practice because it can deteriorate the Web search results. The author starts the paper with a new embedding approach to make user profile. The manuscript is focused on learning user profile for search personalization.

To handle these problems, they represent user profile using a user embedding and two projection matrix which are based on LDA-based vector embeddings of the submitted queries and the documents.

In this paper [17], the author presented a new embedding model to model the 3-way relationship to re-rank the documents for personalization. The 3-way relationship is described as user, query and documents which is represented as three-column matrix with three embedding vectors. The general architecture of the proposed model includes three crucial steps. In the first step, they extract feature maps from the embedding triple vectors. In the second step, they do perform the routing algorithm to only one final capsule which is the composition of three vectors. Finally, the length of the vector output of this final capsule is a score measured and is used to calculate the objective function. The paper [3] presents Personalized Social Document Representation (PerSaDoR) collected in bookmarking system which is based on the social information. The PSDR is expected to provide social representations for each document in order to rank the retrieved documents. The system is valuated with large data sets which show that it increases linearly as the data sets grow, so eventually, it can be employed to large data sets [8, 11].

However after all, the author in [10] presents still there is dearth of personalization which creates problems for the users as well as for the search engine to perceive whether the results presented to the user fit well or not? As the user behavior presents only false signals. So, in response, they collected user explicit response.

The other problems that may occur are: A Web user seeking for the information on given term would return search results having the specified term but would fall short to get back the Web document that is described by its synonymy term [1]; data sets of the information are not interlinked with one another which make mining even more complicated to manage and produce the identical set of outcomes in the equivalent way for all the clients [13].

We, in this paper, purely depend on the semantic similarity between the user query and the user profile. That similarity helps to change the query into the personalized query which matched to annotation-based documents for further expanding.

## 3 Methodology

Figure 1 depicts picture of our proposed model for search personalization. The user query is required to be preprocessed before taking it into the approach. The preprocessed query is matched with user profile for semantic similarity. After analyzing it with the user profile, the query is altered to personalized query which is passed to third-party search engines. The results from these search engines are parsed in order to extract meta information. Ontologies are used to describe the Web documents. The fetched documents are further ranked and provided to the users. A user profile can be learnt automatically from the past history of the user. The user profile is being updated implicitly while the user is interacting with the provided results. Implementation results show that our technique to Web personalization is efficient

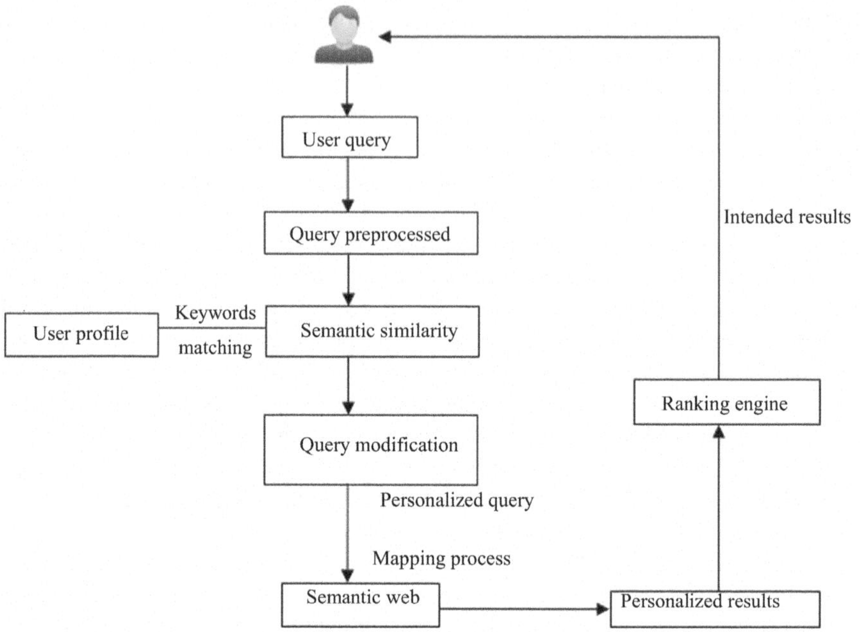


**Fig. 1** Overall working of personalization

and effective. A number of ontology-based search tools have been developed [4, 6, 7, 13, 20, 27].

Earlier work (OntoBroker, WebKB, Corese, SHOE OntoSeek) mainly focused on ontology language for knowledge representation. In this manuscript, we focus on information retrieval.

## 3.1 User Identification

Web usage mining is the process of collecting and analyzing the Web usage logs, and it is important to preprocess the Weblogs before actually using it. One of the key issues in preprocessing [13] is Web user identification to identify each unique user on the Web before personalizing the contents to the user. User can be well identified by three approaches [21] IP address, Cookies and User Identification.

## 3.2 Query Preprocessing

It is better to preprocess the query before passing it to the algorithms for finding the similarity. Web search engine plays a vital role for the user in retrieving the intended information from the Web. Nevertheless in many cases, search engines have some difficulties in retrieving the information. These difficulties can be come due to the ambiguous nature of the user query. So, in due course, it is important for the search engines to preprocess the user query before taking it in the regards. The preprocessing includes removal of stop words, spelling correction, stemming and tokenization. A well-defined preprocess approach is stated in Algorithm 1. In our system, we exploit supervised learning, can also be described as the corpus-based approach, and suitable machine learning-based classifiers such as Decision Tree (D-Tree) and K-Nearest Neighbor (KNN) are applied to a manually annotated data set.

**Algorithm 1: Effective Preprocessing for User Query**

```
Input: input query.
Output: concept/ tokens.
parse (): designed library for parsing the sentence,
stem (): designed library for stemming the keywords,
sw[]: an array contains all the stopwords.

1. Import sw[], stemmer, dic[], k=0
2. Input t = query
3. arr[] = parse(t)
4. for i= 1 to size of arr[]
        for j= 1 to size of sw[]
         if words[i] == sw[j]
         Eliminate words[i]
         end if
      end for
   end for
5. arr2[] = meaningful query without stopwords
6. for i= 1 to size of arr2[]
        if arr[i] == compound word
        stemmer.stem(arr[i])
        end if
 end for
7. arr3[] = stemmed words
8. for i= 1 to size of arr3[]
        for j = 1 of dic[]
        compute min distance(arr3[], dic[j])
        end for
        k++
        arr4[k] = arr3[i]
 end for
9. for i= 1 to size of arr[]
        print arr[i]
    end for
10. End
```

## 3.3 User Profile

In our system, the ontology-based user profile is formed implicitly while the users interact with the system. The client profile is basically reference ontology in which every concept has associated terms showing the client's apparent enthusiasm for that concept. It stores user's hits with meta tags generated after analyzing the query and the document itself. The user profile is constructed by analyzing the browsing behavior of the user which includes user past queries, hits, specifically the content, length and time spent on each Web page they visit. The Web documents, the user visits, are implicitly classified into the concepts with meta tags contained in the reference ontology. User feedback is not required.

## 3.4 Semantic Similarity

We start by describing a general description of our similarity-based personalized search and the intuitions behind it. It provides the most personalized results with respect to the user query according to the context in which it occurs. It computes the similarity between user query and the user profile. User profiling is a vital part of Web personalization. A user profile is a set of personal data includes past queries, hits with meta tags associated with a specific user. Integrating tagging information can assist to advance personalization and retrieval techniques. The basic plan of the tag mark extension is to store the search query, e.g., "mobile phones," in memory when the user evaluates search results. Whenever he finds a relevant Web document and clicks on it, our system will automatically convert the search query to tags and automatically store to the user profile with the clicked link. Tag marking is a well-situated way to enhance the search and personalization. Before starting any Web project, it is imperative to understand the target.

The user query is directly matched to the keywords of the user profile, and the original query is altered according to the keywords which is more likely similar or related according to the similarity measure. The similarity measure [5] among words is computed either directly using the terms in WordNet or the associated meanings of terms those defined in WordNet, respectively. Probability model and page rank algorithms [9] have been used to resolve the issues relating to re-ranking. Probability model is based on probability of relevant and non-relevant results while PageRank computes the backlinks of Web pages. Both these algorithms address the solely problems of ranking not of personalization, therefore, the motivation to propose context similarity-based Web personalization. The working algorithm for finding semantic similarity is given in Algorithm 2, and Fig. 2 outlines the algorithm.

**Fig. 2** Similarity between user query and user profile

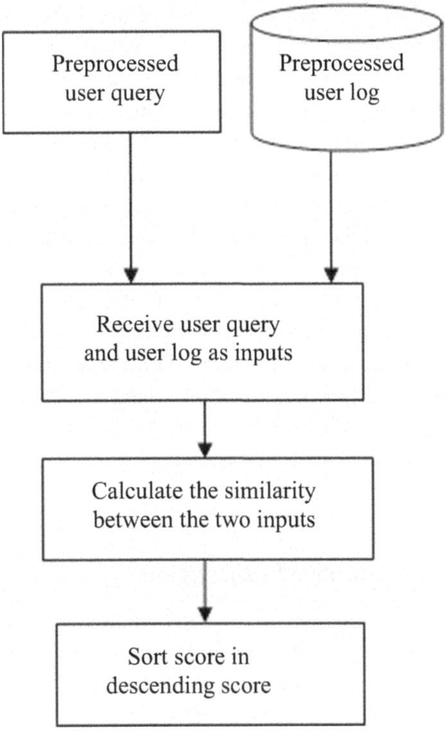


## Algorithm 2: Semantic Similarity

***Input:*** *keywords*
***Output:*** *similarity distance (user profile, user query)*
*Len (xi,xj): the length of the shortest path from synset xi to synset xj in WordNet.*
*depth (ci): the length of the path to synset ci from the global root entity and depth(root)=1*
*lso(xi,xj): the lowest common subsumer of x1 and x2*

```
1.  Import WordNet (wn), user logs as array ul[]
2.  Declare the array score[], sw[]
3.  sw[]=wn.getSynset(query)
4.  for i=1 to size of sw
        for j=1 to size of ul
            score[]=(2*depth (lso (sw[i],[j])) / (len (sw[i],ul[j]) + 2 * depth (sw[i],ul[j])))
         j++
        end for
      i++
    end for
5.  Sort score in descending order.
6.  End.
```

The user query is preprocessed before passing for semantic similarity. Various algorithms have been designed and used here as removing stop words, spelling correction, stemming and tokenization which has been described above. The user query after preprocessing is transferred to the log files where the identification of the user is identified and the data accessed by the user is determined from the database. The similarity match is performed between the users input query and the tags by the semantic similarity algorithms. If the similarity found between the user input and the accessed data is some extent similar, then the user receives the result by exploiting some patterns even the data can be provided from the local cache. There is no need of accessing, again and again, the same inputs. The data of the cookies is stored in the database of the log files. So, fetching the data from the database if the probability of the similar data is high, then the desired result is given to the user, taking the data from the log files database. That similarity helps to change the current query into the personalized query which matched to annotation-based documents for further expanding illustrated below.

### *3.5 Query Modification*

Query personalization is an optimization problem. Query q given by the user u, the intention is to find the parts of the profile of u, when combined with q, will retrieve the personalized results (Fig. 3).

- **Preference space**: Given a query q is modeled to the user profile p, a set of preference space is produced from p which is related to the original query q.
- **Parameter estimation**: It generates estimations on the set of personalized queries generated by integrating preferences of a subset of p into the original query q.
- **Personalized query construction**: This is the module used for generating the actual personalized query.

**Method**

We consider two methods into account: (i) similarity between $t$ and $t_j$, and (ii) similarity between $t_j$ and user profile, where query term: $t$; related term: $t_j$. The user

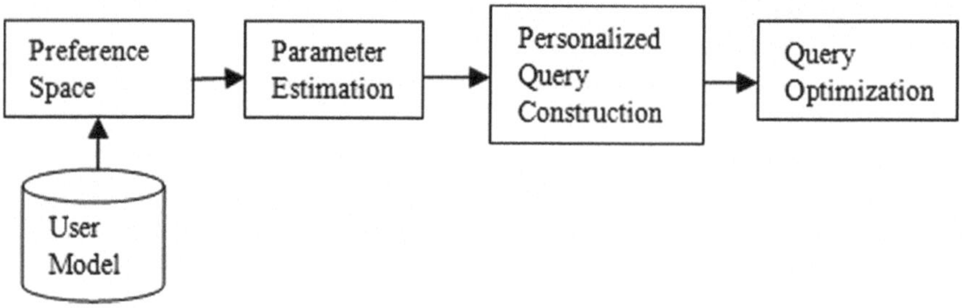


**Fig. 3** Personalized query construction

profile is represented as a unit vector: $p_u = \{w_{t1}, w_{t2}, \ldots, w_{tn}\}$, where $w_{tj}$ is user term frequency.

After computation of these two similarities, a merge operation is required to obtain a final ranking value that indicates the similarity of $t_j$ with $t$ with respect to the user u. We obtain a final ranking for indicating the similarity in terms of Weighted Borda-Fuse as rank $U_t(t_i)$.

**Algorithm 3: Query Modification**

*U : a user; Q: a query.*
*$P_u$ [m] ← extract profile of u*
*for all $t_i$ € Q do*
*l ← list of neighbor of ti*
*for all $t_j$ € l do*
*$t_j$.value ← Rank U $t_i$ ($t_j$)*
*sort l on the basis of $t_j$ value and consider only top k terms in l,*
*make a logical OR between $t_i$ and all terms of l,*
*Update Q'*

## 3.6 *Ontology-Based Retrieval*

Ontology is defined as "Explicit specification of conceptualization." In general, it refers to the action or process of framing a concept or idea of something with the use of explicit knowledge. On the Web, It aims to solve the limitations of keyword-based search engines. The system takes input query as SPARQL query. The query is executed using a list of semantic entities that satisfy the query, with the above formal instance, and the annotated documents are retrieved, re-ranked and presented to the user in order to satisfy his/her needs. The documents can be manually annotated or by scanning the whole documents and finding the most frequently used keywords on the page which can be used to annotate the Web documents. To make the use of domain knowledge in data mining process or any other sources, the first step is to make the model of domain knowledge in such a way that machine can analyze and process it. Ontology is the best way to make it remarkable. OWL has been an accepted standard language for defining the ontology. Our framework of ontology-based information retrieval can be viewed as an upgradation of simple keyword-based search retrieval techniques by developing ontology knowledge base instead of keyword-based index. The general outline of semantic retrieval process is shown in Fig. 4. Our system consists of a usual interface which initially takes a user query which is further analyzed to formal Resource Description Query Language (RDQL) query. This query could be generated from the knowledge used as ontology, as in [18]. The RDQL query is analyzed next to the knowledge domain, which fetches a list of results that satisfy the user query. A query matching system [16] is used to match

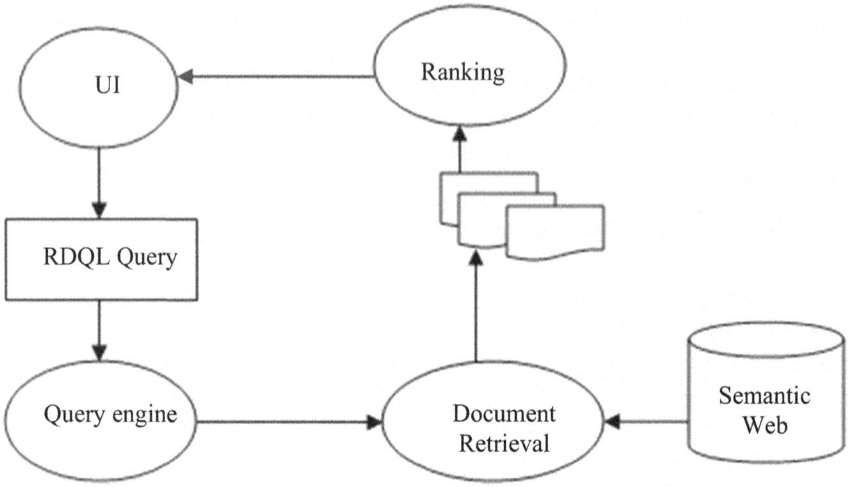


**Fig. 4** Explicit view of ontology-based information retrieval

the query to the annotated Web documents. Finally, the annotated documents with these instances are fetched, re-ranked according to the user behavior and presented to the user.

It is more effective to retrieve the information from the Web when having proper domain knowledge. The domain knowledge can be implemented on the Web using ontology. Ontology can be seen as an explicit specification of conceptualization (classification of objects with words and examples) [24]. All the conceptual knowledge is stored in the ontology level which further transformed to ontology library using OWL. When a Web user interacts with an agent with various data parameters, the agent begins to explore the ontology to discover the entire feasible nodes associated with the user's specified parameters. Thereafter, this intuitive becomes reasonable because all the data sets are well defined in semantic Web and interlinked with one another. We design a Web of interlinked information for our work, where we analyze the information by following semantic Web approach. In this study, we reflect on e-commerce data which is downloaded from the eBay, Amazon and other platforms. We wish to explore some manually made ontology. For conciseness, categories like mobile phone, camera, laptop and tablet-related information are measured. We encode generic information using OWL. The semantic Web technologies, XML [25] and RDF, are used. XML: XML is an extensible markup language which is more than the HTML. It provides the facility of syntax creation for content structure within the Web documents. RDF: It stands for Resource Description Framework. RDF is an easy language which is used to describe the resources and the relationships between the different resources.

## 4 Experiments and Results

The modified query is forwarded to third-party search engines, and the results retrieved are parsed in order to extract information by scanning the whole individual document and looking for the most occurring keywords which can be taken as meta tags. Ontologies are used to define different keywords.

The entire method of Web search personalization works as follows:

1. The user makes a query of his choice on the search engine,
2. The user's profile is generated on the server side (the user's computing device),
3. The query goes to the similarity module for finding the similarity with annotations (tag marking) in user profile, and the query is altered to the personalized query,
4. With respect to the altered query, the documents are returned from third parties,
5. The personalized documents are re-ranked and presented to the user.

The personalized results are hereby shown in Fig. 5 for the query "best 4G phones". The user query is transferred to the user profile model for developing the patterns. The generated pattern is further exploited to the annotation-based documents for generating personalized results. The results are worth showing in Fig. 5 that the user is much interested in these phones based on his past history. The user profile is updated every time the user interacts with the system.

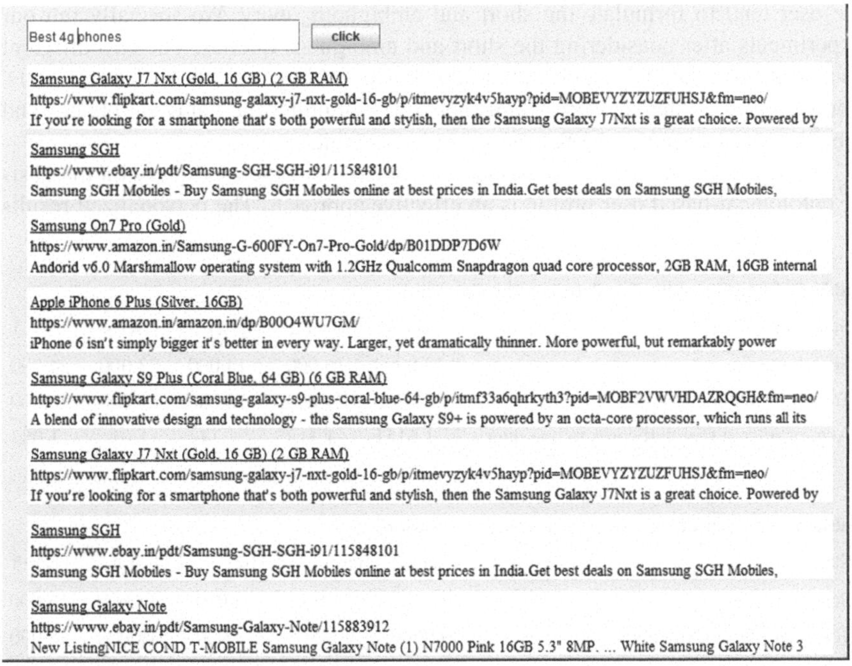


**Fig. 5** Personalized results

The system has been designed on Eclipse, an essential tool for any java developer and implemented on open-source general-purpose scripting language java, including, XML, SQL and Workbench. The modules are being implemented with knowledge-based technique and integrated with WordNet 3.0, one of the popular databases for the NLP domain.

## 5 Discussion and Evaluation of Experimental Results

### Data Set

In our experiments, five data sets are collected from five different users in the two sessions, and each user is supposed to submit number of queries. Each user is assigned a different computer. For each query, the user indentified the set of relevant pages. For each query, at most ten pages are evaluated by the user as relevant or irrelevant. Each query with relevant documents comprises a search record in the user profile. The statistics of the data set is shown in Tables 1 and 2.

While using search engine, users usually formulate the ambiguous queries to the search engine. The search engine always presents the results to the user, however, failed to understand the user's search of real intention. On the other hand, Web search personalization provides the accurate results as intended by the users even the user tend to formulate the short and ambiguous query. We specially ran our experiments after considering the short and ambiguous queries. The two different users can formulate the exact same query, however, they want the different results. For example, for the keyword "Samsung" one can prefer the Samsung phones and the other one can prefer the Samsung camera or tablets.

Our experimental results show that reformulation of the user query on the basis of ontological-based user profile is an effective approach. The personalized results

**Table 1** Statistics of the five data sets (first session)

| Statistics | User 1 | User 2 | User 3 | User 4 | User 5 |
|---|---|---|---|---|---|
| Search queries | 13.0000 | 16.000 | 14.000 | 23.000 | 24.000 |
| Total relevant documents | 21.0000 | 21.0000 | 16.000 | 21.000 | 26.000 |
| Avg. of related search records to one query | 1.6151 | 1.3125 | 1.142 | 1.61 | 1.083 |

**Table 2** Statistics of the five data sets (second session)

| Statistics | User 1 | User 2 | User 3 | User 4 | User 5 |
|---|---|---|---|---|---|
| Search queries | 17.0000 | 11.0000 | 18.0000 | 21.0000 | 16.000 |
| Total relevant documents | 39.0000 | 31.0000 | 31.0000 | 51.0000 | 38.000 |
| Avg. of related search records to one query | 2.2941 | 2.8181 | 1.7222 | 2.4285 | 2.375 |

are hereby shown in Fig. 5 for the query "best 4G phones" after modifying it to the Q = "Samsung + phones + 4g" based on user history and query formulation.

In order to measure the effectiveness of our method proposed in this paper, the precision is measured. The precision is fraction of fetched or retrieved documents that are relevant. The recall which is also used in many cases is the fraction of relevant documents that were retrieved.

$$\text{Precision} = \text{TP}/(\text{TP} + \text{FP})$$

i.e., (number of True Positive (correctly fetched documents))/(Total number of documents retrieved):

$$\text{Recall} = \text{TP}/(\text{TP} + \text{FN})$$

i.e., (Number of True Positive (correctly fetched pages))/(Total number of relevant documents retrieved):

$F1$ is also used for finding the conservation average: $F1 = (2{\cdot}P{\cdot}R)/(P + R)$.

We computed precision as the fraction of predicted documents for query that agree with preferences obtained from the human. We retrieve the results from the different e-commerce platforms: eBay, Amazon and Flipkart, which are further re-ranked and provided to the user. The user finds his relevant results with a very general query "best 4g phones"; however, he is interested in Samsung and Apple phones. In the initial stages when the system does not have enough information about the behavior of the user, it becomes tiresome task to provide the results with good accuracy. The precision is measured at the three different sessions. The first session is when the user formulates the query first time and the system has no knowledge about the user intention; the second session is when the user formulates the same search query and gets much better recommendation due to the browsing behavior collected at the first stage. And similarly, precision is measured for the third and last stage which is considered as the most personalized results based on the data set collected in the first and second sessions.

We compared our results to the eBay and Amazon search engines. The same query is submitted three times to these search engines. The personalized results are considered at third time which retrieved a large number of personalized documents, but only first ten results were considered to limit the study. The mean precision of the different search engines for the queries is shown in Fig. 6. Comparing the mean precision, our model has high precision followed by the Amazon, while ebay received the lowest precision. Note that each study used a different set of query logs, so results cannot be directly compared.

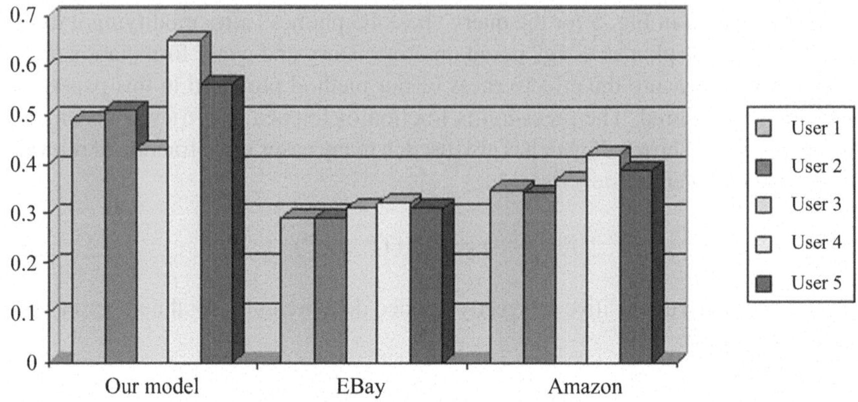


**Fig. 6** Precision at personalized level

# 6 Conclusion

In this paper, we present our system that takes user query as input and offers the personalized results after correlating the user query with the usage logs. We have proposed a new simple and novel approach for personalized search based on long-term behavioral signals that match or outperform the state-of-the-art for this task. This research study can assist other researchers, developers and other stake holders in building successful e-commerce business. However, this research work is in progress. In the future, we devise to validate the findings in this paper by carrying out a user study to verify the efficiency of the ethics identified.

# References

1. Arenas, M., Ugarte, M.: Designing a query language for RDF: marrying open and closed worlds. ACM Trans. Database Syst. (2017). https://doi.org/10.1145/3129247
2. Berkovsky, S., Freyne, J.: Web personalization and recommender systems. In: Proceedings of the 21th ACM SIGKDD International Conference on Knowledge Discovery and Data Mining, pp. 2307–2308, Aug 10–13, ACM, Sydney, NSW, Australia (2015). https://doi.org/10.1145/2783258.2789995
3. Bouadjenek, M.R., Hacid, H., Bouzeghoub, M., Vakali, A.: Persador: personalized social document representation for improving web search. Inf. Sci. **369**, 614–633 (2016). https://doi.org/10.1016/j.ins.2016.07.046
4. Corby, O., Dieng-Kuntz, R., Faron-Zucker, C.: Querying the semantic web with corese search engine. In: Proceedings of the 16th European Conference on Artificial Intelligence, pp. 705–709, Aug 22–27, IOS Press, Valencia, Spain (2004). https://dl.acm.org/citation.cfm?id=3000150
5. Gleich, D.F.: PageRank beyond the web. SIAM Rev. **57**, 321–363 (2015). https://doi.org/10.1137/140976649

6. Guha, R., McCool, R., Miller, E.: Semantic search. In: Proceedings of the 12th International Conference on World Wide Web, pp. 700–709, May 20–24, ACM, Budapest, Hungary (2003). https://doi.org/10.1145/775152.775250
7. Heflin, J., Hendler, J.: Searching the web with SHOE. In: Proceedings of the AAAI Workshop on AI for Web Search (WS' 00), pp. 35–40. AAAI Press, California (2000)
8. Ho, S.Y., Bodoff, D.: The effects of web personalization on user attitude and behavior: an integration of the elaboration likelihood model and consumer search theory. MIS Q. **38**, 497–520 (2004). https://doi.org/10.25300/misq/2014/38.2.08
9. Jiang, Y., Wang, X., Zheng, H.T.: A semantic similarity measure based on information distance for ontology alignment. Inform. Sci. **278**, 76–87 (2014). https://doi.org/10.1016/j.ins.2014.03.021
10. Kim, J.J.: Ontology construction and its applications with text mining techniques. Inform. Sci. Soc. **33**, 20–23 (2015)
11. Kim, J.Y., Teevan, J., Craswell, N.: Explicit in situ user feedback for web search results. In: Proceedings of the 39th International ACM SIGIR Conference on Research and Development in Information Retrieval, pp. 829–832, July 17–21, ACM, Pisa, Italy (2016). https://doi.org/10.1145/2911451.2914754
12. Lopes, P., Roy, B.: Dynamic recommendation system using web usage mining for E-commerce users. Proc. Comput. Sci. **45**, 60–69 (2015). https://doi.org/10.1016/j.procs.2015.03.086
13. Lopez, V., Pasin, M., Motta, E.: AquaLog: An ontology-portable question answering system for the semantic web. In: Proceedings of the 2nd European Conference on the Semantic Web: Research and Applications, pp. 546–562, May 29–June 01, Springer, Heraklion, Greece (2005). https://doi.org/10.1007/11431053_37
14. Lv, Y., Fuxman, A., Chandra, A., Wu, Z.: U.S. Patent Application No. 14/268,953 (2015)
15. Mikolov, T., Le, Q.V., Sutskever, I.: Exploiting similarities among languages for machine translation. arXiv preprint arXiv:1309.4168 (2013)
16. Musgrove, T.A., Walsh, R.H.: U.S. Patent No. 9,400,838. U.S. Patent and Trademark Office, Washington, DC (2016)
17. Nguyen, D.Q., Vu, T., Nguyen, T.D., Phung, D.: A capsule network-based embedding model for search personalization. arXiv preprint arXiv:1804.04266 (2018)
18. Rawassizadeh, R., Dobbins, C., Nourizadeh, M., Ghamchili, Z., Pazzani, M.: A natural language query interface for searching personal information on smartwatches. In: Proceedings of the IEEE International Conference on Pervasive Computing and Communications Workshops, pp. 679–684, Mar 13–17, IEEE Xplore Press, Kona, HI, USA (2017). https://doi.org/10.1109/percomw.2017.7917645
19. Ricci, F., Rokach, L., Shapira, B.: Recommender systems: introduction and challenges. In: Ricci, F., Rokach, L., Shapira, B. (eds.) Recommender Systems Handbook, pp. 1–34. Springer, Boston, MA (2015)
20. Rocha, C., Schwabe, D., de Aragao, M.: A hybrid approach for searching in the semantic web. In: Proceedings of the 13th International World Wide Web Conference, pp. 374–383, May 17–20, ACM, USA (2004). https://doi.org/10.1145/988672.988723
21. Saxena, N., Katiyar, V., Husain, M.: A personalized web search model using user identification technique. Int. J. Multi. Res. Mod. Educ. **1**, 215–217 (2015). http://rdmodernresearch.org/wp-content/uploads/2015/09/29.pdf
22. Smyth, B., Coyle, M., Briggs, P., McNally, K., O'Mahony, M.P.: Collaboration, reputation and recommender systems in social web search. In: Ricci, F., Rokach, L., Shapira, B. (eds.) Recommender Systems Handbook, pp. 569–608. Springer, Boston, MA (2015)
23. Vandenbussche, P.Y., Atemezing, G.A., Poveda-Villalón, M., Vatant, B.: Linked Open Vocabularies (LOV): a gateway to reusable semantic vocabularies on the web. Seman. Web **8**, 437–452 (2017). https://doi.org/10.3233/SW-160213
24. Vigneshwari, S., Aramudhan, M.: Social information retrieval based on semantic annotation and hashing upon the multiple ontologies. Ind. J. Sci. Technol. **8**, 103–107 (2015). https://doi.org/10.17485/ijst/2015/v8i2/57771

25. Vu, T., Nguyen, D.Q., Johnson, M., Song, D., Willis, A.: Search personalization with embeddings. In: Proceedings of the 39th European Conference on Information Retrieval, pp. 598–604, Apr 8–13, Springer, Aberdeen, UK (2017). https://doi.org/10.1007/978-3-319-56608-5_54
26. Witten, I.H., Frank, E., Hall, M.A., Pal, C.J.: Data Mining: Practical Machine Learning Tools and Techniques, 4th edn. Elsevier Science, p. 654. ISBN-10: 0128043571 (2016)
27. Zhang, L., Yu, Y., Zhou, J., Lin, C., Yang, Y.: An enhanced model for searching in semantic portals. In: Proceedings of the 14th International Conference on World Wide Web, pp. 453–462, May 10–14, ACM, Chiba, Japan (2005). https://doi.org/10.1145/1060745.1060812

# Efficient Data Transmission in Wireless Sensor Networks

**Brahm Prakash Dahiya, Shaveta Rani and Paramjeet Singh**

**Abstract** Wireless sensor networks (WSNs) are collection of sensor nodes. The main goal of wireless sensor nodes is to sense the medium and process the data to specified location. WSNs suffer with many technical challenges such as node deployment, battery power, self-configuration, slow convergence, and packets drop during the data transmission. This paper focuses on efficient data transmission, quick convergence and optimizes lifetime of WSNs using hybrid artificial bee colony with salp (HABCS). This paper also generates the different frame formats such as hello message, bond message, route request message, route reply message, and data packets. The proposed approach is implemented using network simulator-2 and it is more energy-efficient than other existing classical and SI protocols.

**Keywords** SI · ACOPSO · Hybrid artificial bee colony with salp

## 1 Hybrid Artificial Bee Colony with Salp Meta-Heuristic

### *1.1 Standard ABC*

Artificial bee colony (ABC) algorithm was introduced to optimize multi-objectives function. It is popular because, it uses less computational energy because it works in less number of phases. Its work is based on three groups of bees: Employed, onlooker and scout bees [1]. Employed bee works in exploitation phase. It shares information with onlooker bee of path and reachability of food. Onlooker bee will select a food

B. P. Dahiya (✉)
I. K. G. Punjab Technical University, Kapurthala, Punjab, India
e-mail: brahmprakashdahiya@gmail.com

S. Rani · P. Singh
Giani Zail Singh Campus College of Engineering and Technology, Bathinda, India
e-mail: param2009@yahoo.com

P. Singh
e-mail: garg_shavy@yahoo.com

 
P. K. Singh et al. (eds.), *Proceedings of ICRIC 2019*, Lecture Notes
in Electrical Engineering 597, https://doi.org/10.1007/978-3-030-29407-6_22

sources based on information supplied by employed bee and fitness function. In the last, scout bees work in exploration phase.

They search food source randomly and update status again as an employed bee. Other hand, if food source is fully exploited then the employed bees and onlookers bees change the phase exploitation to exploration phase and become scout bees. The solution is to move globally called exploration process. Exploration identifies the most suitable area in the search space and avoids local solution. The solution is to move locally is called exploitation process. Standard artificial bee colony (ABC) optimization has faced many problems related to slow convergence area, narrow exploited in employed bee phase and not proper balancing between exploitation and exploration phases [1] (Fig. 1).

Therefore, hybrid artificial bee colony with salp (HABCS) meta-heuristic approach is proposed with high convergence and increase balancing rate between exploitation and exploration phases based on salp behaviour in employed bee phase. HABCS approach is combination of two families: Standard ABC and salp swarm, respectively. Salp is family of Salpidae. Slap chain is designed using leader and follower. It has two sets: Leader and follower, respectively. Leader is head swarm and follower follows leader. Salp swarm approach is represented as [2] (Fig. 2):

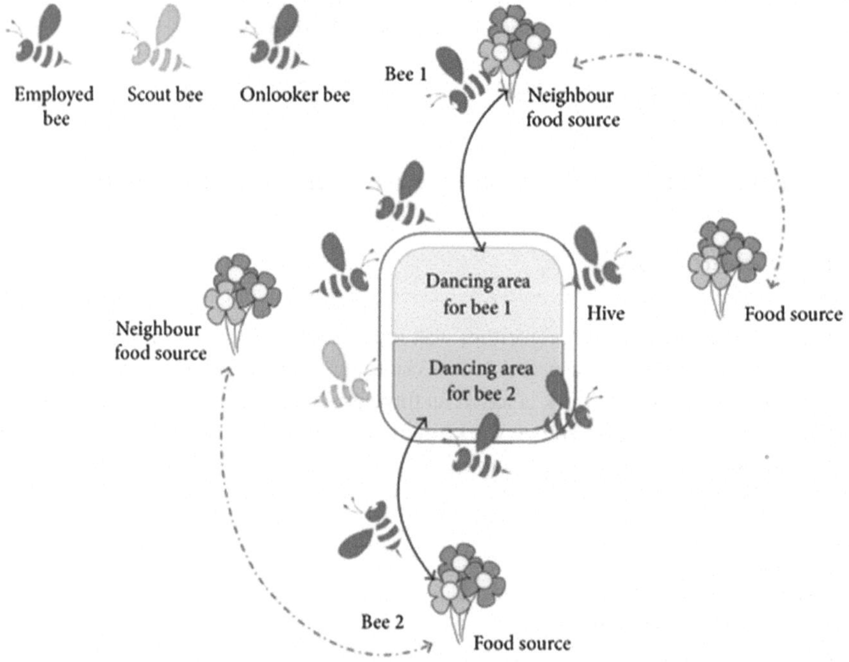


**Fig. 1** Artificial bee colony working cycle

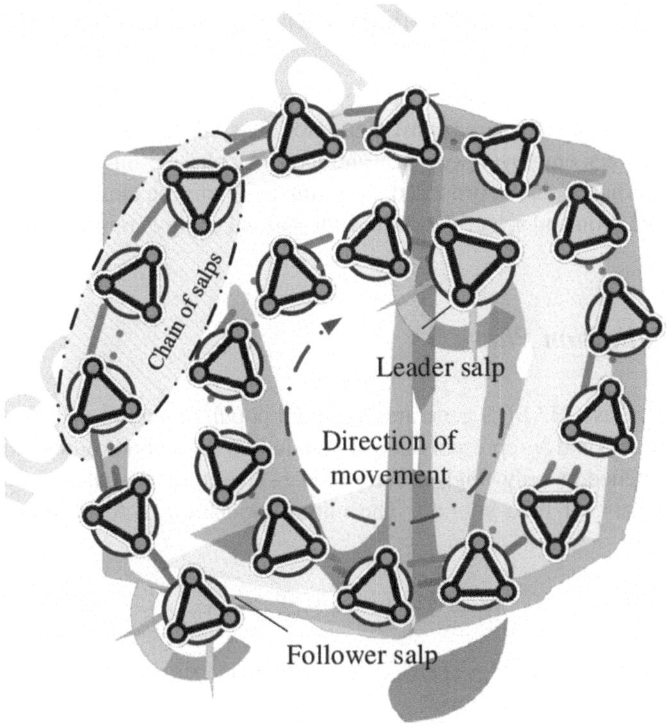


**Fig. 2** Salp swarm working

$$P_{1j} = \begin{cases} f_j + C_1\big((u_j - l_j)C_2 + l_j\big) & C_3 \geq 0 \\ f_j - C_1\big((u_j - l_j)C_2 + l_j\big) & C_3 < 0 \end{cases} \tag{1}$$

where, $P_{1j}$ represents the position of first leader. $f_j$ represents the location of food source and $j$th represents dimension of food source. Upper and lower bounds are represented by $u_j$ and $l_j$, respectively. $C_1$, $C_2$, and $C_3$ are random coefficients. $C_1$ indicates balancing between exploration and exploitation phases. The coefficient $C_1$ is defined as:

$$C_1 = 2\mathrm{e}^{-\left(\frac{4y}{Y}\right)^2} \tag{2}$$

According to Eq. 2, $y$ presents the current cycle and $Y$ defines maximum no. of cycles.

## 2 Hybrid Artificial Bee Colony with Salp Meta-Heuristic

This section proposes hybrid artificial bee colony with salp (HABCS) meta-heuristic approach for fast and large area convergence to solve ABC problem. It also improves balancing between employer bee phase and scout bee phase using the salp coefficient function $C_1$. Therefore, the proposed algorithm performs the cluster head selection, cluster formulation and select the paths with optimal energy consumption.

### 2.1 Initialization Phase

In the initial phase, HABCS generates an initial population number (PN), here $P$ is initial population and $N$ represents size of population. SN is food source. Single food source is searched by only single employed bee. Therefore, (Equal no. of employed bee = Equal no. of food source). The initial population of food source (SN) is represented as:

$$X_{ij} = X_{l_j} + r(0, 1)\left(X_{u_j} - X_{l_j}\right) \tag{3}$$

where, $i \in (1, 2, \ldots \text{SN})$ and $j \in (1, 2, \ldots D)$. $D$ indicates dimensional vector. $X_{l_j}$ and $X_{u_j}$ define lower and upper limits.

### 2.2 Improved Employed Bee Phase

Improved employed bee phase replaces existing food source location by new one using the following equation:

$$W_{ij} = \left\{X_{ij} + \left(\emptyset_{ij}\left(X_{ij} - X_{kj}\right)\right) + \begin{matrix} f_j + C_1\left(\left(u_j - l_j\right)C_2 + l_j\right)C_3 \\ f_j - C_1\left(\left(u_j - l_j\right)C_2 + l_j\right)C_3 \end{matrix}\right. \tag{4}$$

where $C_1 = 2\mathrm{e}^{-\left(\frac{4y}{Y}\right)^2}$.

According to Eq. 4, select the high fitness value from all food sources. $W_{ij}$ is obtained after comparison with $X_{ij}$ only if $W_{ij}$ is better as compared to $X_{ij}$. If $W_{ij}$ is not better than $X_{ij}$ then it will be exploited with existing solution $X_{ij}$. This phase improves search ability using local exploitation. Therefore, it performs balance between exploitation and exploration phase using the $C_1$ salp coefficient. Thus, HABCS is resolved dimensional problem.

### 2.3 Improved Onlooker Bee Phase

The salp with $C_1$ coefficient has a better outcome and it is used in the whole updating process. When salp does not perform balancing, then salp phase will be ended, and in this way, onlooker phase of standard *ABC* algorithm will be started.

$$P_i = \frac{f_i}{\sum_{i=1}^{SN} f_i} \quad (\text{if } r(0,1) \leq P_i) \tag{5}$$

$$Y_{ij} \begin{cases} W_{ij} & (\text{if } r(0,1) \leq P_i) \\ X_{\text{optimal},j} & \text{otherwise} \end{cases}$$

$$X_{ij} \begin{cases} Y_{ij} & (\text{if } f(Y_{ij}) \leq f(X_{ij})) \\ X_{\text{optimal},j} & \text{otherwise} \end{cases}$$

$$Y_{ij} = \left\{ X_{ij} + (\emptyset_{ij}(X_{ij} - X_{kj})) + \begin{matrix} f_j + C_1((u_j - l_j)C_2 + l_j)C_3 \\ f_j - C_1((u_j - l_j)C_2 + l_j)C_3 \end{matrix} \right. \tag{6}$$

Onlooker bees activate to enhance the solution of employed bee using Eq. 4. A greedy selection procedure is applied on $X_{ij}$ and $Y_{ij}$ and selected the best one. If fitness value of $Y_{ij}$ is best one as compared to $X_{ij}$ then save the details of new one and delete details of old one. In the parallel way, salp is activated and it performs balancing in exploitation and exploration phases. Otherwise, the previous position is reserved in memory.

### 2.4 Improved Scout Bee Phase

In standard *ABC*, the scout bee phase generates random solution. It is suffered with pick convergence rates due to multiple cycles. The modify scout bee phase covers the whole search space with $C_1$ coefficient to keep up the balancing with employed phase and shares the information fast. It plays an important role in fast convergence and improves diversity in the population.

$$X_{ij} = (X_{l_j} + r(0,1)(X_{u_j} - X_{l_j}))C_1 \tag{7}$$

## 3 Efficient Data Transmission Using HABCS Meta-Heuristic

This section proposes optimal cluster head selection, cluster formulation, and optimal path selection from source to destination using HABCS meta-heuristic approach. It is working in three cycles, i.e., HABCS-Cluster initialization phase, HABCS-Route Searching Phase, and HABCS data transmission phase.

## *3.1 HABCS-Initialization Phase*

HABCS-Initialization phase starts after sensor nodes randomly deployed over homogeneous regions. It performs CHs selection and formulation. The working of each phase will consist of number of steps which are to be carried out in sequence as follows:

**Step 1: Exchange hello message** The first step allows the hello message local flooding, which helps all SNs to be aware of the current status of their nearby neighbors. In second step, the choosy flooding of hello messages from BS enables each SN to compute its shortest distance (squared Euclidean distance) to the BS. The hello message acts as a control message, communicated between nodes during the HABCS-Initialization phase and contains preliminary information about the network components. The frame format of hello message is shown in Fig. 3.

Where the fields of the packet are specified as follows:

- Sequence no of frame (SEQ): It has a unique sequence number of 2 byte each for all the event communications, which get incremented when a new message is created.
- Type of message (MSG): This field is of 1 byte each and tells about the type of message (data or control) in the packet.
- Source Identification (SID): This 1-byte field provides the source identity.
- Node Type (NODE): This field denotes the type of node (BS/CH/SN) and is of 1 byte each.
- Hop-Count (H-COUNT): It counts the number of hops for a packet from its source to destination and of 2 byte in size.
- Previous Node Identification (PID): This 1-byte field contains the identity of previous node, which forwarded the message at last hop.
- Energy of Previous Node (E-PID): The 2 byte-field contains the residual energy of previous node, which forwarded the message at last hop.

When the hello message arrives for the first interval time, all SN will perform updating in neighbor node table, which it maintains along with the values of PID and

| 2byte | 1byte | 1byte | 1byte | 2byte | 1byte | 2byte |
|---|---|---|---|---|---|---|
| **SEQ** | **MSG** | **SID** | **NODE** | **H-Count** | **PID** | **EPID** |

**Fig. 3** Frame format of hello message

EPID. Further, the SN confirms if the NODE is set to be BS. At final stage, the hello message from the BS is again broadcast with the H-COUNT, PID and E-PID updated. This choosy hello messages flooding from BS helps each SN to acknowledge the status of the BS and assists to compute the shortest hop distance to each SN. At the completion of step 1, each SN will have the neighboring node table updated with distance and remaining energy of each node.

**Step 2: Cluster Head Selection** Optimal cluster head selection is must to improve the lifespan of WSNs and reduce energy consumption. CH is acting as gathering data from the various radio nodes and performs data aggregation to handle the huge amount of data. The optimal cluster head selection optimizes WSNs performance. Therefore, this section proposes the HABCS meta-heuristic approach for optimal cluster head selection with following steps that are represented as:

a. *Fitness function notation*:

This notation creates a fitness function to analyse food source of population. The cluster head selection using HABCS approach adopts the following objectives such as cluster head selection based on highest remaining energy, selection of minimum distance between cluster head and base station, and selection of minimum distance between existing cluster head and newly elected cluster head.

$$f_i \leftrightarrow \mathrm{MAX}\left(E_{rmn_i}\right) \tag{8}$$

$$f_i \leftrightarrow \frac{1}{\min\left(D_{bs,\mathrm{CH}_i}(\max) + D_{\mathrm{CH}_i,\mathrm{CH}_l}(\max)\right)} \tag{9}$$

Merged Eqs. 8 and 9 represented as:

$$f_i \leftrightarrow Q\frac{\mathrm{MAX}\left(E_{rmn_i}\right)}{\min\left(D_{bs,\mathrm{CH}_i}(\max) + D_{\mathrm{CH}_i,\mathrm{CH}_l}(\max)\right)} \tag{10}$$

where $Q = 1$ is proportion constant.

$$f_i \leftrightarrow \frac{MAX\left(E_{rmn_i}\right)}{\min\left(D_{bs,\mathrm{CH}_i}(\max) + D_{\mathrm{CH}_i,\mathrm{CH}_l}(\max)\right)} \tag{11}$$

Equation 11 is used to drive fitness value of food source. Therefore, optimal cluster head selection is performed by using HABCS meta-heuristic approach. After every cycle, CH will be replaced by best one based on highest fitness value. After execution of these phases, optimal CHs are obtained for each round which exhibit optimal attributes among a finite set of SNs. These CHs will now act as single point of contact for event communication and data aggregation.

**Step 3: Broadcast CHs list to SNs** After selection of CHs, the BS will broadcast the list of CHs to all the SNs in the network. This is a type of flooding message which will

| 2byte | 1byte | 1byte | 1byte | 2byte | 1byte | 1byte |
|---|---|---|---|---|---|---|
| SEQ | MSG | SID | NODE | H-Count | CH-ID | CL-ID |

**Fig. 4** Frame format of bond message

make SNs aware about the current status of other nodes updated as CHs. Whenever list of CHs will be updated after each round, it will be rebroadcasted among SNs in the network. Further, CHs receiving this message will update their node type as CH.

**Step 4: Bond Message Generation and Cluster Formulation** Now cluster formulation is the most important phase in WSNs. This section shows the creation of clusters. In the initial stage, every cluster head sends bond message to the nearest sensor nodes. After receiving bond messages from cluster heads, every sensor node shows interest in cluster formulation. The frame format of bond message is shown in Fig. 4.

SEQ, MSG, SID, NODE, H-Count have discussed in frame format of hello message. Where the other fields of the packet are specified as follows:

- Cluster Head Identification (CH-ID): The 1-byte field contains the CH identity.
- Cluster Identification (CL-ID): The 1-byte field contains the cluster identity.

But at the time sensor nodes are confused because these nodes have already got the message to formulate cluster from other cluster head also. Therefore, to avoid this confusion, weight ($w_i$) is assigned to each sensor node based on distance. Thus, optimal cluster is obtained at the end of this process (Table 1).

$$W_i \leftrightarrow CD_{\mathrm{CH}_p, rn_i}(\mathrm{Max}) \tag{12}$$

**Table 1** Cluster formulation

| Sensor nodes ($rn_i$) | Cluster heads ($\mathrm{CH}_p$) | Cluster head weight ($u_i$) | Sensor node weight ($w_i$) | $W_i = \lvert u_i * w_i \rvert$ | CH chosen |
|---|---|---|---|---|---|
| $n_1$ | $CH_1$, $CH_2$, $CH_3$ | 3 | 0.89 | 3 | $CH_3$ |
| $n_2$ | $CH_1$, $CH_3$ | 2 | 0.78 | 1 | $CH_1$ |
| $n_3$ | $CH_1$, $CH_3$ | 2 | 0.98 | 1 | $CH_1$ |
| $n_4$ | $CH_1$, $CH_2$, $CH_3$ | 3 | 0.56 | 3 | $CH_3$ |
| $n_5$ | $CH_1$, $CH_3$ | 2 | 0.34 | 1 | $CH_1$ |
| $n_6$ | $CH_2$, $CH_3$ | 2 | 0.87 | 2 | $CH_2$ |
| $n_{20}$ | $CH_3$, $CH_4$ | 2 | 0.32 | 1 | $CH_1$ |

## 3.2 HABCS-Route Searching Phase

The next challenge is to find the shortest route to improve energy efficiency of WSNs. This section helps in selecting the shortest route selection using HABCS meta-heuristic approach. Optimal route selection from sensor nodes to base station is effective for energy efficiency of WSNs. It selects optimal route based on the least hop count from sensor nodes to base station and it consumes less energy during transmission.

**Step 1: Initiate Route-Request (R-REQ)** In the beginning, the SN initiates one Route-Request (R-REQ) message to each nearby neighbour node with a unique route ID, to explore multiple disjoint routes form source to final destination as shown in Fig. 5.

Not all R-REQ messages will arrive at the BS and few of them will be dropped out by the intermediate nodes in order to avoid common routes that go through same SNs. Frame format of R-REQ message is as in Fig. 6.

SEQ, MSG, SID, PID, E-PID have discussed in frame format of hello message. The fields of the R-REQ message are specified as:

- Target Identification (TID): This 1-byte field provides the target or destination identity.

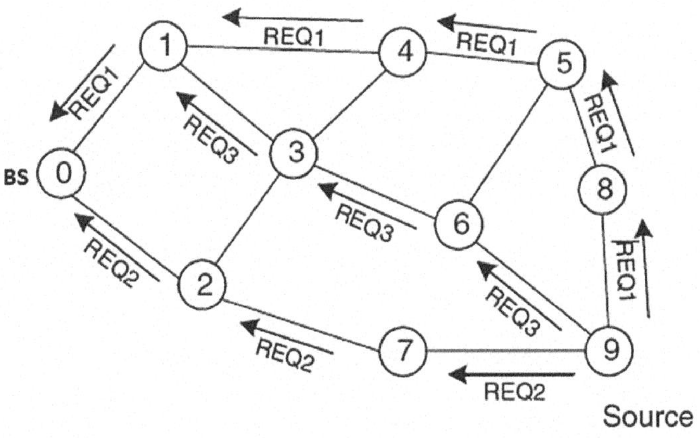


**Fig. 5** Route request cycle

| 2byte | 1byte | 1byte | 1byte | 1byte | 2byte | 1byte | 1byte |
|---|---|---|---|---|---|---|---|
| SEQ | MSG | SID | TID | RID | R-Cost | PID | E-PID |

**Fig. 6** Frame format of R-REQ message

- Route Identification (RID): A 1-byte unique route ID is being assigned to every route discovered through R-REQ message.
- Route Cost (R-COST): It contains the accumulated route cost associated with every route ID consumes 2 bytes.

After receiving the R-REQ message, an intermediate SN checks SID and TID fields, respectively. If the TID represents BS that is in neighbor node table and performed route table updating. The R-REQ message is sent to the BS directly with updated PID and E-PID. Otherwise, the SN has to select one of the neighbor's nodes to forward the R-REQ message until BS is located. The selection of the neighbor node to forward the R-REQ message will depend upon two factors. In first condition, it should be joined the same pair of BS and source nodes and it should have been selected same path. Secondly, the link cost of the selected neighbor has to be minimum among all the available neighbor nodes. Further, if no one of the neighbors fulfill these terms and conditions, the R-REQ message will basically be dropped out from the network. In this way, multiple disjoint routes will be obtained from SNs to BS with minimum route cost.

**Step 2: Select optimal Route** Now to obtain an optimal route from available set of multiple disjoint routes is a multimodal optimization problem, which can be solved by incorporating HABCS meta-heuristic. The procedure for selecting optimal route is as follows:

a. *Fitness Function evaluation*:

In this section, the fitness of different path and maximum fitness value will be evaluated by HABCS approach. This approach is based on objectives such as sensor nodes collected data sent to base station with the least number of hop count and energy utilization during transmission data from sensor node to base station. These objectives are achieved to find minimum distance based on less number of hop count that is represented as:

$$f_i \leftrightarrow \mathrm{Min}\left\{HC_{rn_i,rn_{N+1}}(\mathrm{Max})\right\} \tag{13}$$

$$f_i \leftrightarrow \mathrm{Min}\left\{D_{(rn_i,bs)}(\mathrm{Max}) + D_{(\mathrm{CH}_l,bs)}(\mathrm{Max})\right\} \tag{14}$$

Fitness value is evaluated after merging Eqs. 13 and 14:

$$f_i = Q\mathrm{Min}\left\{HC_{rn_i,rn_{N+1}}(\mathrm{Max}) + \left(D_{(rn_i,bs)}(\mathrm{Max}) + D_{(\mathrm{CH}_l,bs)}(\mathrm{Max})\right)\right\} \tag{15}$$

where $Q$ is proportional constant, here $Q = 1$.

$$f_i = \mathrm{Min}\left\{HC_{rn_i,rn_{N+1}}(\mathrm{Max}) + \left(D_{(rn_i,bs)}(\mathrm{Max}) + D_{(\mathrm{CH}_l,bs)}(\mathrm{Max})\right)\right\} \tag{16}$$

Therefore, optimal route $\mathrm{Route}_{\mathrm{optimal}}$ will be obtained to transmit data from source to destination.

| 2byte | 1byte | 1byte | 1byte | 1byte | 2byte | 1byte | 1byte |
|---|---|---|---|---|---|---|---|
| SEQ | MSG | SID | TID | RID | D-Rate | PID | E-PID |

**Fig. 7** Frame format of R-REP message

**Step 3: Initiate Route Reply (R-REP) Message** After selection of optimal route, BS communicates the route reply (R-REP) messages to the source node through this optimal route to assign data rate for transmitting data packets. The frame format of R-REP message is shown as in Fig. 7.

SEQ, MSG, SID, PID, E-PID have discussed in frame format of hello message. The fields of the R-REP message are specified as follows:

- Target Identification (TID): This 1-byte field provides the target or destination identity.
- Route Identification (RID): A 1-byte unique route ID is being assigned to every route discovered through R-REQ message.
- Data Rate (D-RATE): It represents the actual data rate assigned for the optimal route and maximum of 2 byte in size.

When an in-between SN gets the R-REP message, it examines its routing table entry that corresponds to SID, TID, and RID values. After that it forwards the message to the next hop after updating the fields PID and E-PID. But if a source node gets the R-REP message, it first searches the entry related to TID and RID from its routing table, then update it with the data transmission rate assigned in the R-REP message.

## *3.3 ABCS-Data Transmission Phase*

After selection of optimal routes, the source node starts transmission of data packets with the allocated data rates on selected optimal path through data packets; the detailed procedure is explained in following step:

**Step 1: Initiate Data Transmission**
After data rate assignment to optimal route, a (TDMA) time division multiple access schedule sends data to reduce collision chances. Data is collected from different sources and redundancy is eliminated. Then the aggregated data is sent to the base station. The frame format of DATA packet is shown as in Fig. 8.

SEQ, MSG, SID, PID, E-PID have discussed in frame format of hello message. The fields of the DATA packet are specified as follows:

- Target Identification (TID): This 1-byte field provides the target or destination identity.

| 2byte | 1byte | 1byte | 1byte | 1byte | 2byte | 4byte | 40byte | 1byte | 1byte |
|---|---|---|---|---|---|---|---|---|---|
| SEQ | MSG | SID | TID | RID | PKT-ID | T-STAMP | PAYLOAD | PID | E-PID |

**Fig. 8** Frame format of DATA packet

- Route Identification (RID): A 1-byte unique route ID is being assigned to every route discovered through R-REQ message.
- Data packet Identification (PKT-ID): It contains a unique ID for data packet which is transmitted and of 2-byte size.
- Time Stamp (T-STAMP): It carries the time stamp. It is time duration of data message that is created at the source node.
- Data Payload (PAYLOAD): It contains the actual data which is transmitted form SID to TID with maximum size of 40 bytes.

In all steps, the SN can obtain the next hop by using its routing table and the information placed in the data message that are SID, TID, and RID. At the BS, the values of PID and E-PID will be updated each time a DATA packet arrives which helps to monitor the status of multiple routes being used time to time. Further, to identify the failure route, the BS also performs monitoring of delay of data packets on very route. When the delay is crossed the already defined threshold, the BS pre-assumes that the route is destroyed and initiate R-REQ message to re-discover the optimal routes again.

## 4 Performance Evaluation and Result Analysis

The proposed HABCS meta-heuristic approach is implemented using network simulator-2 (NS-2) tool. NS-2 evaluates the performance of the proposed approach. It is open source software and installed using Kubuntu that is open source operating system. HABCS meta-heuristic technique performance is evaluated with other classical routing and swarm intelligence algorithms. The wireless sensor network is represented by 50, 100, 150, 200, 250, and 300 sensors nodes, respectively There sensor nodes are deployed in 200 * 200 $m^2$ square area. Initial energy is 10,000 J, 20,000 J, 34,950 J, 48,000 J, 55,500 J, and 60,000 J, respectively, depends on different WSNs scenarios. Simulator is defined a multi-path routes between the sensor nodes and cluster head (CH). It is also performed single path routing between cluster head and base station (BS) to send data packets. The proposed HABCS algorithm results are compared with results of LEACH [3], ESO [3], HBO [4], ACO [5], ABC [6], PSO [7], and hybrid ant colony optimization and particle swarm optimization (ACOPSO) [8]; these are based on performance metrics such as remaining energy, consumed energy, network lifetime, and end to end delay.

Figure 9 shows remaining energy with the same energy in Joules. Remaining energy is also called residual energy. It is remaining energy after completion of each round in the wireless sensor network. Figure 9 shows that proposed algorithm delivers better result and the highest remaining energy ratio as compared to existing classical and SI protocols on the same energy level. Therefore, proposed algorithm is more energy-efficient than the existing classical and SI protocols.

Figure 10 shows that proposed algorithm delivers better result and the consumed energy ratio as compared to existing classical and SI protocols on the same energy level. Therefore, proposed algorithm is more energy-efficient than the existing classical and SI protocols.

Figure 11 shows performance of proposed algorithm with existing algorithms LEACH, ESO, HBO, ACO, PSO, ABC, and ACOPSO, respectively, in term of end to end delay using the same energy level. End to end delay means total time taken to transfer the data over the network. Figure 11 shows the proposed algorithm delivers better results in term on end to end delay as compared with LEACH, ESO, HBO,

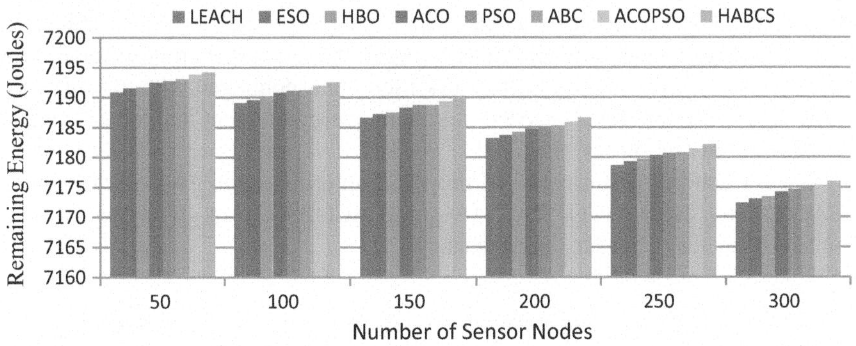


**Fig. 9** Remaining energy with the same energy level in Joule

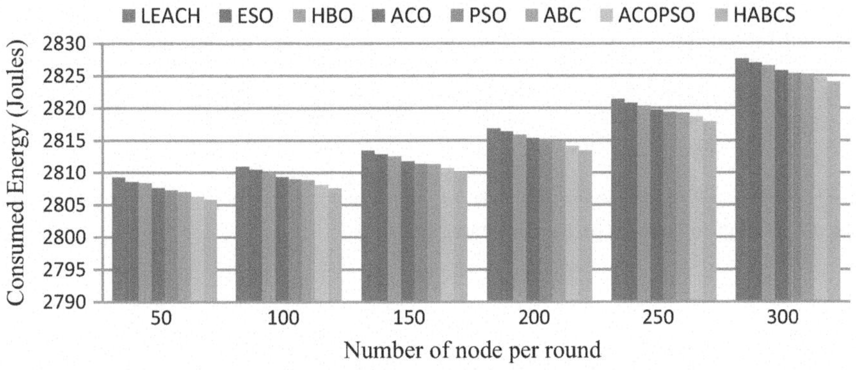


**Fig. 10** Consumed energy with the same energy level in Joule

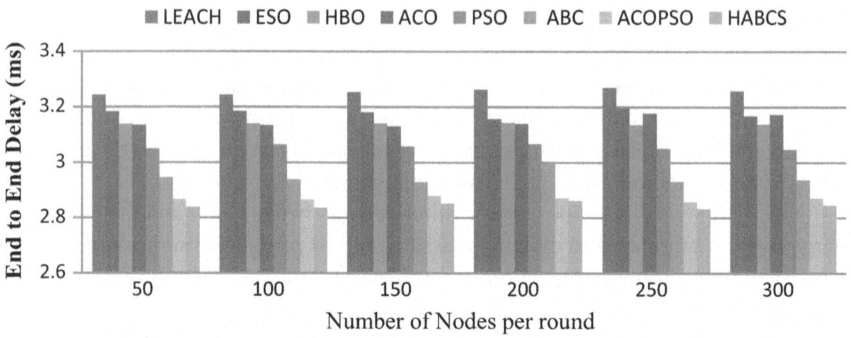


**Fig. 11** End to end delay with the same energy level

ACO, PSO, ABC, and ACOPSO. Therefore, proposed algorithm is more energy-efficient than other existing classical and SI algorithms.

Figure 12 shows performance of proposed algorithm with existing algorithms LEACH, ESO, HBO, ACO, PSO, ABC, and ACOPSO respectively in term on network lifetime using the same energy level. Figure 12 shows that proposed algorithm delivers better results in term on network lifetime as compared with existing protocols. Therefore, proposed algorithm is more energy-efficient than other existing classical and SI algorithms.

Figure 13 shows performance of proposed algorithm with existing algorithms LEACH, ESO, HBO, ACO, PSO, ABC, and ACOPSO, respectively, in term on number of rounds (LND) last node dead.

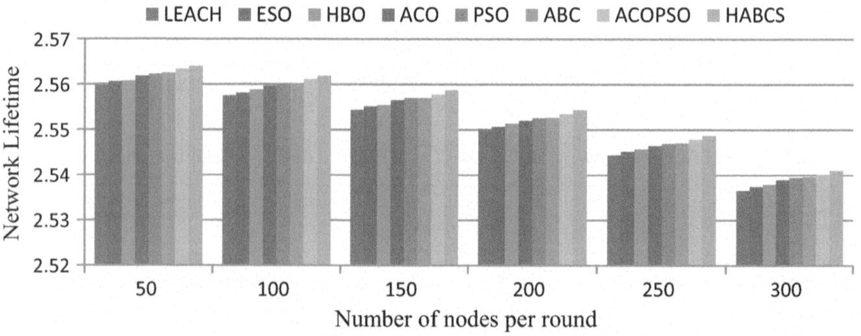


**Fig. 12** Lifetime of network on the same energy level

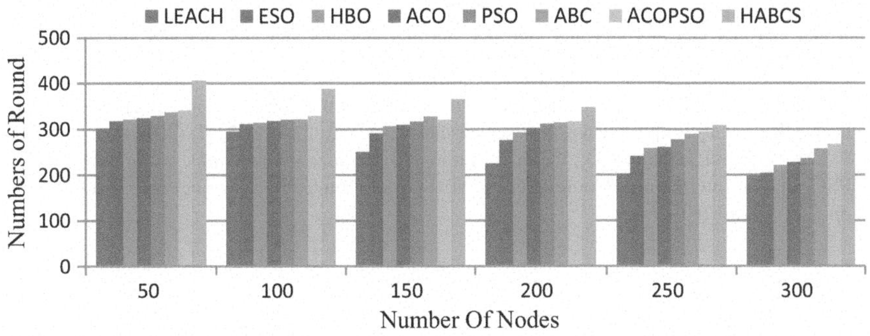


**Fig. 13** Lifetime of network on the same energy level

## 5 Conclusion and Future Work

This paper is presented novel algorithms to improve lifetime of wireless sensor networks. Here, a hybrid artificial bee colony with salp (HABCS) meta-heuristic is proposed with high convergence and increase balancing rate between exploitation and exploration phase. HABCS approach is a combination of two families: Standard ABC and salp swarm, respectively; wireless sensor networks (WSNs) are collection of sensor nodes. This paper is generated the different frame formats such as hello message, bond message, route request message, route reply message, and data packets. This paper is performed efficient data transmission, quick convergence by using hybrid artificial bee colony with salp (HABCS). In future work the proposed algorithms can be applied in mobile network.

**Acknowledgements** The authors acknowledge I. K. Gujral Punjab Technical University, Kapurthala for providing research facilities.

## References

1. Karaboga, D., Okdem, S., Ozturk, C.: Cluster based wireless sensor network routing using artificial bee colony algorithm. Wirel. Netw. **18**(7), 847–860 (2012)
2. Mirjalili, S., Amir, H.G., Mirjalili, S.Z., Saremi, S., Faris, H., Mirjalili, S.M.: Salp swarm algorithm: a bio-inspired optimizer for engineering design problems. Adv. Eng. Soft. **114**, 163–191 (2017)
3. Chandramouli, H., Desai, S.C., Jagadeesh, K.S., Dhruve, K.D.: Enhancing network lifetime in wireless sensor networks adopting elephant swarm optimization. Glob. J. Comput. Sci. Technol.: Netw. Web Secur. **13**(4), 0975–4350 (2013)
4. Selvi, M., Nandhini, C., Thangaramya, K., Kulothungan, K., Kannan, A.: HBO based clustering and energy optimized routing algorithm for WSN. In: Eighth International Conference on Advanced Computing (ICoAC), Chennai, pp. 89–92 (2016)
5. Gupta V., Sharma S.K.: Cluster head selection using modified ACO. In: Fourth International Conference on Soft Computing for Problem Solving, pp. 11–20. Springer (2015)

6. Ahmad, T., Haque, M., Khan, A.M.: An energy-efficient cluster head selection using artificial bees colony optimization for wireless sensor networks. In: Advances in Nature-Inspired Computing and Applications, pp. 189–203. Springer (2017)
7. Rao, P.C.S., Jana, P.K., Banka, H.: A particle swarm optimization based energy efficient cluster head selection algorithm for wireless sensor networks. Wirel. Netw. **23**(7) (2016)
8. Pal, D., Verma, P., Gautam, D., Indait, P.: Improved optimization technique using hybrid ACO-PSO. In: 2nd International Conference on Next Generation Computing Technology, Dehradun, pp. 277–282 (2016)

# Lifetime Improvement in Wireless Sensor Networks Using Hybrid Grasshopper Meta-Heuristic

**Brahm Prakash Dahiya, Shaveta Rani and Paramjeet Singh**

**Abstract** The energy efficiency and lifetime of wireless sensor networks (WSNs) are the more focusing points. The WSNs faced many challenges during the data transmission. Node deployment, leader selection, and optimal route selection are challenges that affect the energy level and lifetime of WSNs. Many existing techniques have been proposed to node deployment, cluster leader and optimal route selection. But, all existing techniques have not given satisfactory results in the network energy optimization. Therefore, this paper presents hybrid artificial grasshopper optimization algorithm (HAGOA). It is an inherited behavior of artificial grasshopper optimization and artificial bee colony variance. The proposed algorithm will place sensor nodes using artificial grasshopper optimization technique. These sensor nodes may be static or dynamic that depends on the network scenario. The cluster head selection and optimal route selection will perform using artificial bee colony variance. It also performs balancing between exploration and exploitation phases in the given search space. This algorithm is a combination of two families: Artificial grasshopper and ABC Variance, respectively. It compares with existing classical and swarm intelligence (SI) protocols in the terms of remaining energy, sensor node lifetime, consumed energy, end to end delay and maximum number of rounds.

**Keywords** Wireless sensor networks · Swarm optimization · Ant colony optimization · Energy consumption · Artificial grasshopper

---

B. P. Dahiya (✉)
I. K. G. Punjab Technical University, Kapurthala, Punjab, India
e-mail: brahmprakashdahiya@gmail.com

S. Rani · P. Singh
Giani Zail Singh Campus College of Engineering and Technology, Bathinda, India
e-mail: param2009@yahoo.com

P. Singh
e-mail: garg_shavy@yahoo.com

 
P. K. Singh et al. (eds.), *Proceedings of ICRIC 2019*, Lecture Notes in Electrical Engineering 597, https://doi.org/10.1007/978-3-030-29407-6_23

# 1 Introduction

Wireless sensor networks (WSNs) are collection of small sensor nodes which have sensing, communication, and computational capabilities. The energy efficiency and lifetime of WSNs are more focusing points. WSNs have capability to sense, process, and transmission of the data from sensor nodes to base station. Battery life is important factor during the sensor nodes deployment in geographical area. In wild areas, battery charging is not easily possible. So, energy utilization is the main issue [1]. The network lifetime is utilized by the following factors such as node deployment, cluster leader selection, cluster formulation, and optimal path selection [2]. Wireless sensor networks (WSNs) are progressively used in the fields such as military, medicinal services, natural, organic, basic health observation, and state built forecasting [3–6].

A novel technique, namely hybrid artificial grasshopper optimization algorithm (HAGOA) is proposed for promising node deployment, cluster leader selection, cluster formulation, and optimal path selection. The rest of the paper is organized as follows: Section 2 presents the hybrid artificial grasshopper optimization algorithm. Artificial bee colony (ABC) variance is presented in Sect. 3. Energy optimization using HAGOA is discussed in Sect. 4. HAGOA execution is presented in Sect. 5. Section 6 introduces the performance evaluation and results. In the last, conclusion and future works are discussed in Sect. 7.

# 2 Proposed Hybrid Artificial Grasshopper Optimization Algorithm (HAGOA)

## 2.1 *Artificial Grasshopper Optimization (AGOA)*

Artificial grasshopper optimization (AGOA) is population-based meta-heuristic algorithm. It is a nature-inspired approach that performs balancing between exploitation and exploration phases. It starts working with in a group. The group is gathering of larva and adult grasshoppers, respectively. The adult phase describes the exploration search in the given search space, and larva phase describes the exploitation in given search area. Both phases work together and find promising global solution in given search space. Therefore, AGOA can be better performed with static and dynamic sensor node deployment in WSNs. It can enhance convergence area in given search space.

**Grasshopper is represented as**:

AGOA is a set of grasshoppers. $\%A$ indicates adult grasshoppers and remaining $\%L$ indicates larva grasshoppers. The management of group is performed by resistance based learning that assists in finding promising position target. The grouping of the most extreme wellness grasshopper is considered as the middle grouping and the position of target is identified with the position of selector using Eq. 1:

$$A_{T_{\mathrm{P}}} = \sum \frac{A_{\mathrm{sel}(n_1, n_2, n_3, n_4, \ldots n_{NV})}}{\text{Number of selector}} \quad 1 \le NV \le N \tag{1}$$

where $A_{T_{\mathrm{P}}}$ represents the position of target and $A_{\mathrm{sel}}$ indicates position of selector. Selector is larva grasshopper. During the selection process, the selector is elected randomly.

The position of target and mute selector is updated in the left, right, and central direction in the given search space by using Eqs. 2, 3, and 4, respectively. Mute selector is adult grasshopper.

$$A_{T_{\mathrm{p}'}} = \left(A_{T_{\mathrm{P}}} + \mathrm{rnd} * \mathrm{per}_i * \left(A_{T_{\mathrm{P}}} - A_{\mathrm{sel}}\right)\right) \tag{2}$$

where $A_{T_{\mathrm{p}'}}$ indicates the updated position of target. Per $i$ is % of enhancement in fitness of mute selector. The rnd is random function and generates number in range [0, 1] and where 0 and 1 indicates the lower and upper bound. The update in fitness value of mute selector is represented by using Eqs. 3 and 4, respectively:

$$A_{\mathrm{sel}'} = \begin{cases} \mathrm{rnd} * \left(2 * A_{T_{\mathrm{P}}} - A_{\mathrm{sel}}\right) + A_{T_{\mathrm{P}}} \ \mathrm{if}\left(2 * A_{T_{\mathrm{P}}} - A_{\mathrm{sel}}\right) < A_{T_{\mathrm{P}}} \\ \mathrm{rnd} * A_{T_{\mathrm{P}}} + \left(2 * A_{T_{\mathrm{P}}} - A_{\mathrm{sel}}\right) \ \mathrm{if}\left(2 * A_{T_{\mathrm{P}}} - A_{\mathrm{sel}}\right) > A_{T_{\mathrm{P}}} \end{cases} \tag{3}$$

$$A_{\mathrm{sel}'} = \begin{cases} \left(\mathrm{rnd} * A_{\mathrm{sel}} + A_{T_{\mathrm{P}}}\right) \ \mathrm{if}\left(A_{\mathrm{sel}} < A_{T_{\mathrm{P}}}\right) \\ \left(\mathrm{rnd} * A_{T_{\mathrm{P}}} + A_{\mathrm{sel}}\right) \ \mathrm{if}\left(A_{\mathrm{sel}} > A_{T_{\mathrm{P}}}\right) \end{cases} \tag{4}$$

The position of mute selector can be changed in the left or right direction using Eq. 3, otherwise Eq. 4 can be applied for center direction. Here, rnd is random number.

The interleaving process defines the success of mute selector if it improves best position in the last round RND:

$$\mathrm{succ}(k, \mathrm{RND}, W) \begin{cases} 1 \ \mathrm{BEST}_{\mathrm{RND},k,W} < \mathrm{BEST}_{\mathrm{RND}-1,k,W} \\ 0 \ \mathrm{BEST}_{\mathrm{RND},k,W} = \mathrm{BEST}_{\mathrm{RND}-1,k,W} \end{cases} \tag{5}$$

where $\mathrm{succ}(k, \mathrm{RND}, W)$ is the success of mute selector $k$ in group $W$ at round RND. $\mathrm{BEST}_{\mathrm{RND},k,W}$ is best position found by mute selector $k$ at round RND (last round). 1 indicates mute selector higher success rate with best position near the solution and 0 indicates lower success rate without significant enhancement. Therefore, using all the above equations, one grasshopper may beat another grasshopper. The adult grasshopper beats the other adult grasshoppers in the group. All defeated grasshoppers leave the group. Some grasshoppers die and other grasshoppers switch the group. The concept of the grasshopper algorithm can be implemented in WSNs for node deployment and cluster leader selector.

# 3 Artificial Bee Colony (ABC) Variance

Artificial bee colony with salp (HABCS) meta-heuristic approach is proposed with high convergence and increase balancing rate between exploitation and exploration phases based on salp behavior in employed bee phase, Onlooker bee phase, and Scout bee phase. Artificial bee colony variance is a combination of two families: Standard ABC and salp swarm, respectively. Standard ABC salp is family of salpidae. Slap chain is designed using leader and follower. It has two sets: leader and follower, respectively. Leader is head swarm and followers follow the leader. Salp swarm approach is represented as [7]:

$$P_{1j} = \begin{cases} f_j + C_1\big((u_j - l_j)C_2 + l_j\big)C_3 \geq 0 \\ f_j - C_1\big((u_j - l_j)C_2 + l_j\big)C_3 < 0 \end{cases} \tag{6}$$

where $P_{1j}$ represents the position of first leader. $f_j$ represents the location of food source, and $j$th represents dimension of food source. Upper and lower bounds are represented by $u_j$ and $l_j$, respectively. $C_1$, $C_2$, and $C_3$ are random coefficients. $C_1$ indicates balancing between exploration and exploitation phases. The coefficient $C_1$ is defined as:

$$C_1 = 2\mathrm{e}^{-\left(\frac{4y}{Y}\right)^2} \tag{7}$$

According to Eq. 7, $y$ presents the current cycle and $Y$ defines maximum no. of cycles. In this section, HABC variance meta-heuristic approach is discussed that performs the cluster head selection, cluster formulation and select the paths with optimal energy consumption.

## 3.1 *Initialization Phase*

In the initial phase, ABC variance generates an initial population number (PN), here $P$ is initial population and $N$ represents size of population. SN is food source. Single food source is searched by only single employed bee. Therefore, (Equal no of employed bee = Equal no of food source). The initial population of food source (SN) is represented as:

$$X_{ij} = X_{l_j} + r(0, 1)\big(X_{u_j} - X_{l_j}\big) \tag{8}$$

where $i \in (1, 2, \ldots \text{SN})$ and $j \in (1, 2, \ldots D)$. $D$ indicates dimensional vector. $X_{l_j}$ and $X_{u_j}$ define lower and upper limits.

### 3.2 Improved Employed Bee Phase

Improved employed bee phase replaces existing food source location by new one using the following equation:

$$W_{ij} = \left\{ X_{ij} + \left( \emptyset_{ij} \left( X_{ij} - X_{kj} \right) \right) + \begin{matrix} f_j + C_1\left( \left( u_j - l_j \right) C_2 + l_j \right) C_3 \\ f_j - C_1\left( \left( u_j - l_j \right) C_2 + l_j \right) C_3 \end{matrix} \right. \quad (9)$$

where $C_1 = 2e^{-\left(\frac{4v}{Y}\right)^2}$.

According to Eq. 9, select the high fitness value from all food sources. $W_{ij}$ is obtained after comparison with $X_{ij}$ only if $W_{ij}$ is better as compared to $X_{ij}$. If $W_{ij}$ is not better than $X_{ij}$ then it will be exploited with existing solution $X_{ij}$. This phase improves search ability using local exploitation. Therefore, it performs balance between exploitation and exploration phase using the $C_1$ salp coefficient. Thus, HABCS is resolved dimensional problem.

### 3.3 Improved Onlooker Bee Phase

The salp with $C_1$ coefficient has a better outcome and it is used in the whole updating process. When salp does not perform balancing then salp phase will be ended and in this way, onlooker phase of standard ABC algorithm will be started.

$$P_i = \frac{f_i}{\sum_{i=1}^{SN} f_i} \quad (\text{if } r(0,1) \le P_i) \quad (10)$$

$$Y_{ij} \begin{cases} W_{ij} & (\text{if } r(0,1) \le P_i) \\ X_{\text{optimal},j} & \text{otherwise} \end{cases}$$

$$X_{ij} \begin{cases} Y_{ij} & \left(\text{if } f\left(Y_{ij}\right) \le f\left(X_{ij}\right)\right) \\ X_{\text{optimal},j} & \text{otherwise} \end{cases}$$

$$Y_{ij} = \left\{ X_{ij} + \left( \emptyset_{ij} \left( X_{ij} - X_{kj} \right) \right) + \begin{matrix} f_j + C_1\left( \left( u_j - l_j \right) C_2 + l_j \right) C_3 \\ f_j - C_1\left( \left( u_j - l_j \right) C_2 + l_j \right) C_3 \end{matrix} \right. \quad (11)$$

Onlooker bees are activated to enhance the solution of employed bee using Eq. 11. A greedy selection procedure is applied on $X_{ij}$ and $Y_{ij}$ and selected the best one. If fitness value of $Y_{ij}$ is best one as compared to $X_{ij}$ then save the details of new one and delete details of old one. In the parallel way, salp is activated and it performs balancing in exploited and exploration phase. Otherwise, the previous position is reserved in memory.

### 3.4 Improved Scout Bee Phase

In standard ABC, the scout bee phase generates random solution. It is suffered with pick convergence rates due to multiple cycles. The modify scout bee phase covers the whole search space with $C_1$ coefficient to maintain the balancing with employed phase and shares the information fast. It plays an important role in fast convergence and improves diversity in the population.

$$X_{ij} = \left(X_{l_j} + r(0,1)\left(X_{u_j} - X_{l_j}\right)\right)C_1 \tag{12}$$

## 4 Energy Optimization Using HAGOA

### 4.1 Fitness Function Notation for Node Deployment

The artificial grasshopper optimization algorithm is applied on WSNs to optimize energy. Here, numbers of sensor nodes (SNs) are treated to be the number of larva grasshoppers, and mobile nodes (MNs) are treated to be the number of adult grasshoppers. The concept of attacking by grasshoppers derives when data has to be transferred by the sensor nodes. In this section, HAGOA performs static and dynamic node deployment, cluster formulation, and optimal cluster head selection.

$$\text{SNs} = \{rn_1, rn_2, rn_3, rn_4, \ldots rn_N\} \quad 1 \le i \le N \tag{13}$$

Here, SNs indicates set of static sensor nodes.

$$\text{MNs} = \{m_1, m_2, m_3, m_4, \ldots m_N\} \quad 1 \le i \le N \tag{14}$$

Here, MNs indicates set of dynamic sensor nodes.

The total area covered in WSNs and numbers of nodes in given search area are represented as:

$$\text{MIN}_{\text{no}_{\text{nodes}}} = \frac{A_r}{\pi \cdot \text{ran}^2} \tag{15}$$

where ran represents the communication range or sensing range and $\text{MIN}_{\text{no}_{\text{nodes}}}$ indicates the least number of sensor nodes that are needed to make connected network during the data transmission. $A_r$ indicates the total area to be covered.

$$A_r = \left(X_i - X_j\right) * \left(Y_i - Y_j\right) \tag{16}$$

The distance between $rn_i$, and $rn_j$ located in position $(X_i, X_j)$ and $(Y_i, Y_j)$ is calculated as:

$$d(rn_i, rn_j) = \sqrt{(X_i - X_j)^2 + (Y_i - Y_j)^2} \tag{17}$$

Sensing range between sensor node $rn_i$ and $rn_j$ is $r_s$ radius. If $d(rn_i, rn_j) \leq r_s$ is said to be connected nodes.

$$\mathrm{Cov(SN)} = \sum \begin{matrix} 1, d(rn_i, t) \leq r_s \\ 0, d(rn_i, t) > r_s \end{matrix} \tag{18}$$

where $d$ is the distance between sensor node $(rn_i)$ and target node $(t)$. The coverage (SN) function equal to one means target is sensed or covered. Zero indicates target that is not sensed or covered.

$$f_1(M_i) = \sum \begin{matrix} f_1(M_i) + 1,\ d(M_i, t) \leq r_s \text{ and } t! \in \{\mathrm{SN}, \mathrm{MN}/M_i\} \\ f_1(M_i) \text{ Otherwise} \end{matrix} \tag{19}$$

where $f_1(M_i)$ indicates mobile sensor node fitness. Here, target $(t)$ does not belong to static nodes (SN) and exempted $(M_i)$ other mobile sensor nodes (MN).

$$\text{Fitness Ratio} = \left(\frac{f_1(M_i)}{\sum t}\right)\% \tag{20}$$

where fitness ratio defines total number of targets can be covered or sensed by each mobile sensor node $(M_i)$.

TCov defines the total coverage of static node and mobile sensor node.

$$\mathrm{TCov} = \mathrm{Cov(SN)} + \sum_{i=1}^{m} f_1(\mathrm{MN}) \tag{21}$$

Derived the fitness function to identity the week node and optimize the position of mobile sensor nodes:

$$f_1 = \frac{\beta_1}{\mathrm{Dis}} + \beta_2 * \mathrm{Del} + \beta_3 * \mathrm{Drop}_{\mathrm{ratio}} \tag{22}$$

$f_1$ is the fitness function to calculate week node, $\beta_1$, $\beta_2$ and $\beta_3$ are constants such as $\beta_1 + \beta_2 + \beta_3 = 1$. Week node means connected node but exploited with week coverage, high delay, and high drop ratio. $\mathrm{Drop}_{\mathrm{ratio}}$ indicates the drop ratio of packets. Dis indicates distance between sensor node and previous sensor node.

### 4.2 Fitness Function Notation for Cluster Head Selection

This notation creates a fitness function to analyze food source of population. The cluster head selection using ABC variance that adopts the following objectives such as cluster head selection based on highest remaining energy, selection of minimum distance between cluster head and base station and selection of minimum distance between existing cluster head and newly elected cluster head.

$$f_i = \frac{\mathrm{MAX}\left(E_{rmn_i}\right)}{\min\left(D_{bs,\mathrm{CH}_i}(\max) + D_{\mathrm{CH}_i,\mathrm{CH}_l}(\max)\right)} \tag{23}$$

Equation 23 is used to drive fitness value of food source.

### 4.3 Fitness Function Evaluation for Optimal Path Selection

In this section, the fitness of different path and maximum fitness value will be evaluated by ABC variance. The objective is achieved to find minimum distance based on less number of hop count that is represented as:

$$f_i = \mathrm{Min}\left\{HC_{rn_i,rn_{N+1}}(\mathrm{Max}) + \left(D_{(rn_i,bs)}(\mathrm{Max}) + D_{(\mathrm{CH}_l,bs)}(\mathrm{Max})\right)\right\} \tag{24}$$

Therefore, this derivation can be applied to find fitness value in every round.

## 5 HAGOA Execution

### 5.1 HAGOA Initialization Phase

It performs sensor node deployment and setup the base station position. It is also deployed some dynamic node for the backup purpose in emergency time. After the deployment of static nodes, the minimum number of mobile sensor nodes are placed using Eq. 15. The fitness function helps to calculate maximum number of sensed target by each mobile sensor nodes only when these target nodes are un-sensed or uncovered by static and other mobile nodes. It also reduces the overlapping chances in the coverage area. Figure 1 shows the working of HAGOA.

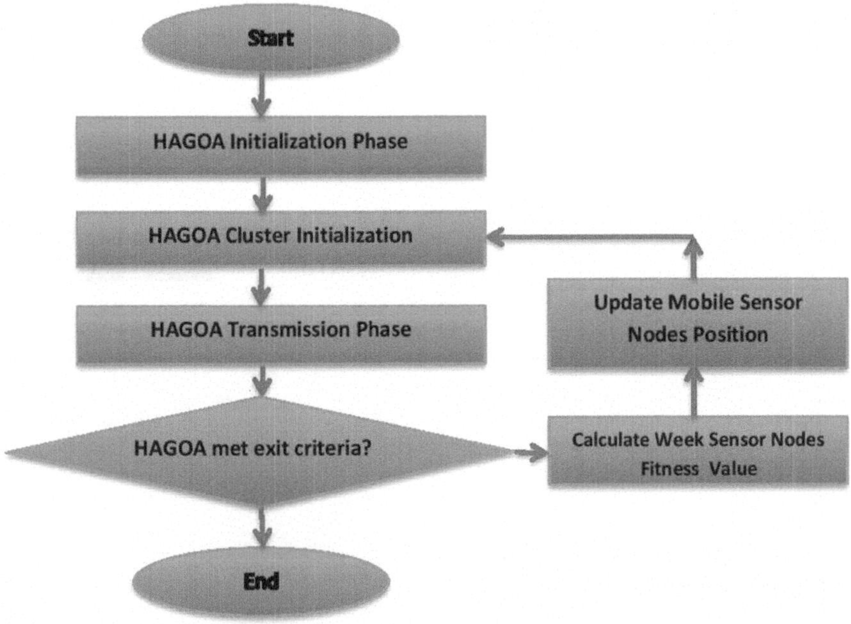


**Fig. 1** Flow chart AGOA meta-heuristic algorithm in WSNs

## 5.2 *HAGOA Cluster Phase*

It performs cluster formulation and cluster leader selection. Optimal cluster head selection is must to improve the lifespan of WSNs and reduce energy consumption. Cluster leader is acting as gathering data from the various radio nodes and performs data aggregation to handle the huge amount of data. The optimal cluster leader selection optimizes WSNs performance. Therefore, this section proposes the HAGOA meta-heuristic approach for optimal cluster leader selection with following steps that is represented as:

**Cluster Leader Selection Algorithm**:

*Input and Parameters Setup:*

PN ← It indicated population number
Mi ← Maximum Round (R)
D ← It indicates vector dimension
SN ← It indicated sources of food
$X_{l_j}$ ← It indicates lower limit of elements
$X_{u_j}$ ← It indicates upper limit of elements

**Parameters Initialization** →

```
$ ← Parameter Controller
P_i ← Probability of selected food source
C1 ← Balancing Coefficient
Output:
CH_l ← X_optimal,j
{
start:
R ← 0                                              Initialization Bee Phase
for (i=1,2,.......................................,SN) do        →
Produce r(0,1)
If r ≤ $ then
Produce X_ij ϵ (−1,0)
{
else
Produce X_ij ϵ (0,1)
}
end if
Evaluate optimal solution and fitness fi (X_ij)
Update C1 co-efficient
R++
end for
for (i=1,2,.......................................,SN) do
Produce W_ij by using equation no. 9.        Improved Employed Bee Phase
Evaluate optimal solution and fitness fi (W_ij)        →
R++
if fi (X_ij) < fi (W_ij) then
X_ij ← W_ij
fi (X_ij) ← fi (W_ij)
else
Previous existence solution is stored in memory
end if
Update C1 co-efficient
end for
if R==Mi then
Stored optimized solution in memory and W_ij is optimal solution and exit.
CH_l ← X_optimal j
Update C_1 co-efficient
end if
end for
```

**for (i=1,2,……………………………………,SN) do** $(if\ r(0,1) \leq P_i)$
then
$Y_{ij} \leftarrow W_{ij}$ ***Improved Onlooker Bee Phase***
else
Previous existence solution is stored in memory
end if
Evaluate optimal solution and fitness fi ($Y_{ij}$) and fi ($X_{optimal,j}$)
If fi ($Y_{ij}$) $\leq$ fi ($X_{optimal,j}$) then
$X_{ij} \leftarrow Y_{ij}$
if fi ($Y_{ij}$) > fi ($X_{optimal,j}$) then
$X_{optimal,j} \leftarrow Y_{ij}$
fi ($X_{optimal,j}$) $\leftarrow$ fi ($Y_{ij}$)
end if
end if
if solution is unsatisfactory then replace with new solution generated by using equation no. 13.
R++ ***Improved Scout Bee Phase***
end for
if R== $M_i$ then
$CH_l \leftarrow X_{optimal,j}$
end if
}

After execution of this algorithm, optimal CHs are obtained for each round which exhibit optimal attributes among a finite set of SNs. These CHs will now act as single point of contact for event communication and data aggregation.

## 5.3 *HAGOA Transmission Phase*

It selects the data rate. It is also performed the optimal path selection basis on least hop. This section helps in selecting the shortest route selection using HAGOA meta-heuristic approach. Optimal route selection from sensor nodes to base station is effective for energy efficiency of WSNs. It selects optimal route based on the least hop count from sensor nodes to base station and it consumes less energy during transmission.

**Optimal Route Selection Algorithm**:

*Input and Parameters Setup:*

PN ← It indicated population number
Mi ← Maximum Round (R)
D ← It indicates vector dimension
SN ← It indicates sources of food
$X_{l_j}$ ← It indicates lower limit of elements
$X_{u_j}$ ← It indicates upper limit of elements
$ ← Parameter Controller
$P_i$ ← Probability of selected food source
C1 ← Balancing Coefficient

**Parameters Initialization**

Output:
$Route_{HC}$ ← $X_{optimal,j}$
{
start:
R ← 0
for (i=1,2,………………………………………,SN) do
Produce $r(0,1)$
If r ≤ $ then
Produce $X_{ij} \in (-1,0)$
{
else
Produce $X_{ij} \in (0,1)$
}
end if
Evaluate optimal solution and fitness fi ($X_{ij}$)
Update C1 co-efficient
R++
end for

**Initialization Bee Phase**

for (i=1,2,………………………………………,SN) do
Produce $W_{ij}$ by using equation no. 9
Evaluate optimal solution and fitness fi ($W_{ij}$)
R++
if fi ($X_{ij}$) < fi ($W_{ij}$) then
$X_{ij}$ ← $W_{ij}$
fi ($X_{ij}$) ← fi ($W_{ij}$)
else

**Improved Employed Bee Phase**

Previous existence solution is stored in memory
end if
Update C1 co - efficient
end for
if R==Mi then
Stored optimized solution in memory and $W_{ij}$ is optimal solution and exit.
$Route_{HC} \leftarrow X_{optimal,j}$
Update C1 co -efficient
end if
end for
for (i=1,2,……………………………………,SN) do
$(if\ r(0,1) \leq P_i)$ then Improved Onlooker Bee Phase
$Y_{ij} \leftarrow W_{ij}$
else
Previous existence solution is stored in memory
end if
Evaluate optimal solution and fitness fi $(Y_{ij})$ and fi $(X_{optimal,j})$
If fi $(Y_{ij}) \leq$ fi $(X_{optimal,j})$ then
$X_{ij} \leftarrow Y_{ij}$
if fi $(Y_{ij}) >$ fi $(X_{optimal,j})$ then
$X_{optimal,j} \leftarrow Y_{ij}$
fi $(X_{optimal,j}) \leftarrow$ fi $(Y_{ij})$
end if
end if
if solution is unsatisfactory then replace with new solution generated by using equation no 13 **Improved Scout Bee Phase**
R++
end for
if R== Mi then
$Route_{optimal} \leftarrow X_{optimal,j}$
end if
}

Therefore, optimal route $Route_{optimal}$ will be obtained to transmit data from source to destination. After the completion of first round it checks the energy level. If energy is left for execution, it performs monitoring of whole network and find out the week nodes in the network using fitness function. If week nodes will be existed in the network then HAGOA will locate mobile sensor node between the week node and connected sensor nodes basis on best positioning fitness function and helps to avoid local optima and hidden node problem. It improves the coverage and network lifetime.

## 6 Performance Evaluation and Result Analysis

The proposed HAGOA meta-heuristic algorithm is implemented using network simulator-2 tool. The wireless sensor network is represented by 50, 100, 150, 200, 250, and 300 sensors nodes, respectively. These sensor nodes are deployed in 200 * 200 $m^2$ areas. Initial energy is 10,000 J. Sensor node coverage range is varying from 20 to 90 m. The proposed algorithm is defined a multipath route between the sensor nodes and cluster head (CH). It is also performed single hop routing between cluster head and base station (BS) for transmission of data packets. The proposed HAGOA algorithm results are compared with results of AGOA [8], LEACH [7], ACO [9], PSO [10], ABC [11], hybrid ant colony optimization and particle swarm optimization (ACOPSO) [12] and hybrid artificial bee colony with salp (HABCS) [13] and they are based on the performance metrics such as remaining energy, consumed energy, sensor node lifetime, and end to end delay.

Figure 2 shows that proposed algorithm delivers better result and the highest remaining energy ratio as compared to existing classical and SI protocols on the same energy level.

Figure 3 shows the proposed algorithm delivers better results in term of end to end delay as compared with LEACH, ACO, PSO, ABC, ACOPSO, HABCS, and

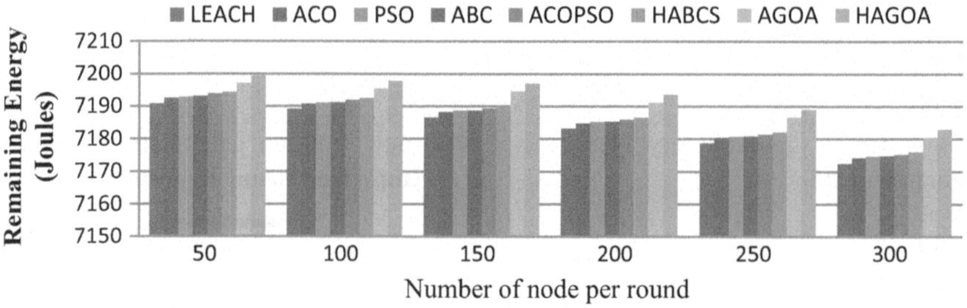


**Fig. 2** Remaining energy

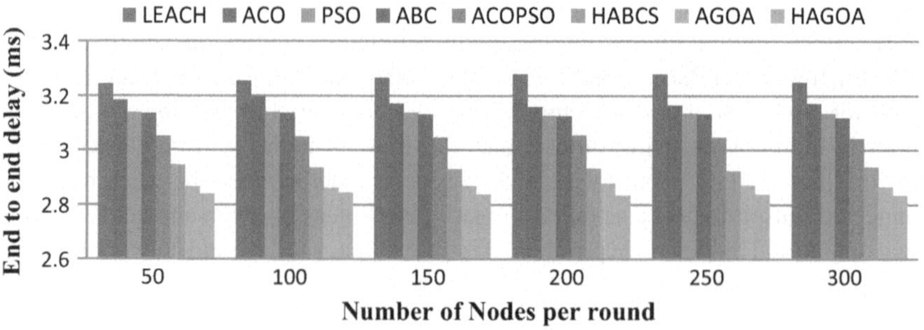


**Fig. 3** End to end delay

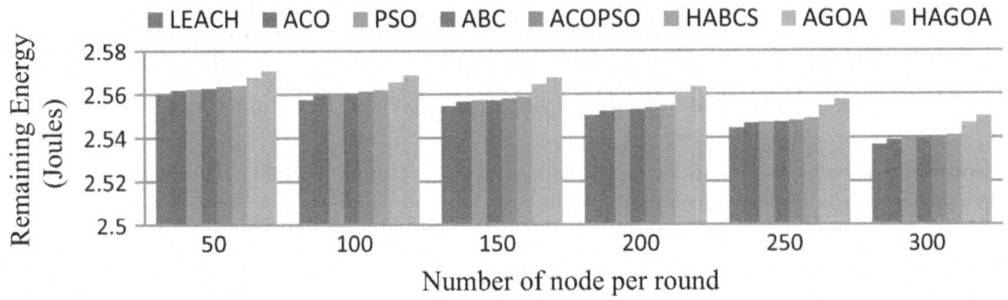


**Fig. 4** Lifespan of sensor node on same energy level

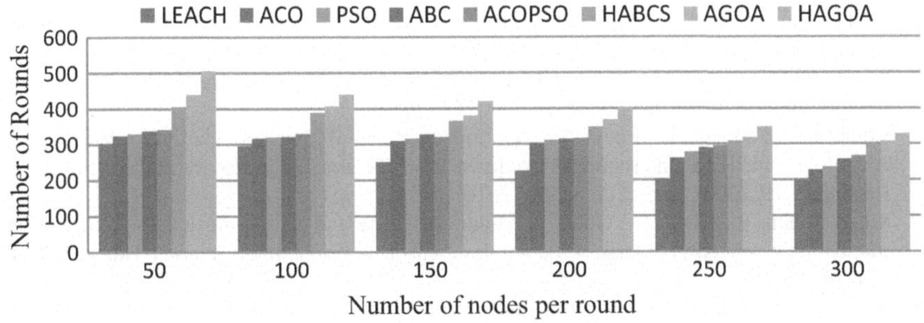


**Fig. 5** Network lifespan

AGOA. Figure 4 shows the proposed algorithm delivers better results in term of network lifetime as compared with existing protocols.

Figure 5 shows performance of proposed algorithm with existing algorithms LEACH, ACO, PSO, ABC, ACOPSO, HABCS, and AGOA, respectively, in term of number of rounds.

## 7 Conclusion and Future Work

In WSNs existing classical and SI algorithms are suffered with node deployment, cluster leader selection, and optimal route selection. Prefect node deployment means more energy saving during data transmission. Thus, this paper HAGOA meta-heuristic algorithm is performed promising node deployment, cluster leader selection, and optimal route selection. It is implemented using NS-2 simulator. The efficiency of proposed algorithm has already evaluated with benchmarks function using MATLAB in previous work. The proposed HAGOA has given best results in terms of remaining energy, consumed energy, network lifetime, and end to end delay as compared to LEACH, ACO, PSO, ABC, ACOPSO, HABCS, and AGOA, respectively.

Therefore, proposed algorithm is more energy efficient than other existing classical and SI algorithms. The future of this work can be considered to inherit the behavior of another swarm to optimized energy and increase the lifetime of WSNs.

**Acknowledgements** The authors acknowledge I. K. Gujral Punjab Technical University, Kapurthala for providing research facilities.

## References

1. Jia, D., Zhu, H., Zou, S., Hu, P.: Dynamic cluster head selection method for wireless sensor network. IEEE Sens. J. **16**(8), 2746–2754 (2016)
2. Nayak, P., Devulapalli, A.: A fuzzy logic-based clustering algorithm for WSN to extend the network lifetime. IEEE Sens. J. **16**(1), 137–144 (2016)
3. Suganya, S., Padmaja, D., Latha, Y.S.: Optimization and lifetime anticipation for WSN using fuzzy logic. In: IEEE International Conference on Electrical, Electronics and Optimization Techniques (ICEEOT), pp. 2353–2360 (2016)
4. Xu, Z., Chen, L., Chen, C., Guan, X.: Joint clustering and routing design for reliable and efficient data collection in large-scale wireless sensor networks. IEEE Internet Things J. **3**(4), 520–532 (2016)
5. Weijie, Lu: Donglin Bai Energy efficient distributed lifetime optimizing scheme for wireless sensor networks. Trans. Tianjin Univ. **22**(1), 11–18 (2016)
6. Mihaela, I.C., Eduardo, M., Margarita, S.J., Julio, R.B., Javier, R., Antonio, J.C.: Energy efficiency and quality of data reconstruction through data coupled clustering for self-organized large-scale WSNs. IEEE Sens. J. **16**(12), 5010–5020 (2016)
7. Khediri, S.E.L., Nejah, N., Anne, W., Abdennaceur, K.: A new approach for clustering in wireless sensors networks based on LEACH. Procedia Comput. Sci. **32**, 1180–1185 (2014)
8. Saremi, S., Mirjalili, S., Lewis, A.: Grasshopper optimization algorithm: theory and application. Adv. Eng. Softw. **105**, 30–47 (2017). ISSN 0965-9978
9. Gupta, V., Kumar, S.S.: Cluster head selection using modified ACO. In: Fourth International Conference on Soft Computing for Problem Solving, pp. 11–20, Springer (2015)
10. Rao, S., Jana, P.K., Haider, B.: A particle swarm optimization based energy efficient cluster head selection algorithm for wireless sensor networks. Wirel. Netw. **23**(7), 2005–2020 (2015)
11. Ahmad, T., Haque, M., Khan, A.M.: An energy-efficient cluster head selection using artificial bees colony optimization for wireless sensor networks. In: Advances in Nature-Inspired Computing and Applications, pp. 189–203 (2017)
12. Pal, D., Verma, P., Gautam, D., Indait, P.: Improved optimization technique using hybrid ACO-PSO. In: 2nd International Conference on Next Generation Computing Technology, Dehradun, pp. 277–282 (2016)
13. Dahiya, B.P., Rani, S., Singh, P.: Hybrid artificial bee colony with Salp (HABCS) meta-heuristic approach for energy optimization in wireless sensor networks. Sci. Technol. J. (Ciência e Técnica Vitivinícola J.) **33**(12), 73–97 (2018). ISSN: 0254-0223, E-ISSN: 2416-3953

# Routing Topologies and Architecture in Cognitive Radio Vehicular Ad hoc Networks

**Priya Bakshi, Prabhat Thakur and Payal Patial**

**Abstract** With the advancement in the wireless communication, there has been an immense growth in the number of vehicles on the road. The unpredictable nature of vehicular ad hoc network (VANET) due to random increase and decrease of the nodes/users (vehicle) on the roads is a challenging issue. Moreover, the increase in the number of nodes creates the problem of spectrum scarcity due to shortage of licensed spectrum for vehicular services. In order to solve the issue of spectrum scarcity, the cognitive radio network (CRN) has been developed which exploits the unlicensed spectrum for communication without affecting the licensed communication that is using interference avoidance. The CRNs are more vulnerable to the security and the privacy of the networks because the transmission parameters required for communication avoid the interference with the licensed and unlicensed users. Moreover, the safety message among the vehicles ensures the safety of the vehicles in the cognitive radio vehicular ad hoc network (CR–VANET) and also manages the sharing of the licensed/primary and the unlicensed/secondary users in the network. The routing and network topologies are a challenging issue due to mobility of vehicles. Therefore, in this review paper, we present the applications and various routing topologies for CR–VANET.

**Keywords** Cognitive radio · Radio spectrum · Routing topologies · Vehicular ad hoc networks · V-to-V communication

---

P. Bakshi (✉) · P. Patial
Department of Electronics and Communication Engineering, Chandigarh University, Mohali, Punjab, India
e-mail: priyabakshi34@gmail.com

P. Thakur
Department of Electrical and Electronics Engineering Sciences, University of Johannesburg, Auckland Park Kingsway Campus, Johannesburg 2006, South Africa

 
P. K. Singh et al. (eds.), *Proceedings of ICRIC 2019*, Lecture Notes in Electrical Engineering 597, https://doi.org/10.1007/978-3-030-29407-6_24

# 1 Introduction

Increase in number of vehicles on the road has conveyed main emphasis on enhancing safety of vehicles. With this demand, we are observing arise in expansion of the novel presentations and facilities for VANET situations. The most common instances are:

(i) Safety of roads
(ii) Collision avoidance
(iii) Traffic management
(iv) Vehicle-to-vehicle communication (V2V) [1].

For VANET, all the vehicles/nodes need communication with each other (V2V)) and other roadside units (vehicle-to-infrastructure (V2I)). The radio spectrum is the compulsory resource for communication and the allocated spectrum bands for VANET by the Federal Communications Commission (FCC) and European Telecommunications Standards Institute (ETSI) are 75 and 30 MHz of spectrum in 5.9 GHz band [2, 3]. However, in future with increase in vehicles, these allocated bands will be insufficient for communication; therefore, we need some platforms which can fulfill this demand of spectrum bands. The cognitive radio network is an emerging solution to overcome the problem of spectrum shortage. The main idea of CRN is to share similar bandwidth with secondary/unlicensed users (SUs) without causing any interference to primary/licensed users (PUs). Cognitive radio equipment detects the available channel and allows SUs to occupy the bandwidth. The CR includes spectrum knowledge representation which vigorously changes routing topology protocol in order to satisfy requirement for more proficiency [4].

The use of CR in the VANET can improve the issue of spectrum shortage. Therefore, CR–VANET is an emerging technology that can support VANET applications. Vehicular ad hoc networks defined as unstructured ad hoc network shifting on the road. The vehicle can communicate with each other or with roadside infrastructure to increase the safety of roads. The main challenges for cognitive radio network (CRN) deal between nodes in VANET in high-mobile network under dynamic channel situations. In addition to this, the unpredictable behavior of VANET, data security, node mobility, and priority assignment are the main challenges [4]. There are some challenges in developing CR–VANET. The open problems in CRN development network like control channel (CC), joint spectrum sensing (JSS), and cognitive implementation architecture (CIA).

# 2 Overview of Cognitive Radio and Its Architecture

Spectrum users are categorized as licensed users, and unlicensed users or cognitive radio. Licensed users legally operate in a specific frequency band while cognitive users are not granted with any specific frequency to transmit and receive the data. Cognitive radio provides the spectrum to unlicensed users for communication purpose [5–7]. Cognitive radio network users identify the spectrum holes, these holes

**Fig. 1** Cognitive radio for vehicular ad hoc networks (CRV) architecture (noncentric) [10]

are defined as a provisionally nonutilized spectrum that can be accessed by CUs. If a spectrum band is available, then cognitive radio (CR) fully utilizes the channel, although primary users are there.

The major applications of Cognitive Radio are:

(i) White space and regulation [8]
(ii) Smart grids [9]
(iii) Wireless sensor network (WSN)
(iv) Public security and medicinal network
(v) Internet of Things
(vi) Vehicular networks [10]

Cognitive radio in VANET is one of the major submissions of CR. Each vehicle in a geographical area can interconnect with each other straightly via some communication infrastructure.

The VANET architecture of cognitive radio vehicles based with on-board units (OBUs) and infrastructure services. The network architecture for cognitive radio are as follows:

(i) Incomplete infrastructure support [10].
(ii) Complete structure provision [10].

The architecture is defined as without infrastructure support or noncentric construction. The detecting could be communal among the network. Organization support nonappearance minimizes the geographical attention of communication performed in Fig. 1. The repeaters and routers are connected through main highway with fixed infrastructure support and low complexity. The range of data is fixed by minimum infrastructure, thus data are conveyed only to the vehicle in the range of services installed above in Fig. 2.

## 3 Cognitive Radio Vehicular Ad hoc Network

Currently, all automakers are capitalizing to stretch infotainment explanations and new substitutions to passengers and car drivers. Geographies and facilities tend to

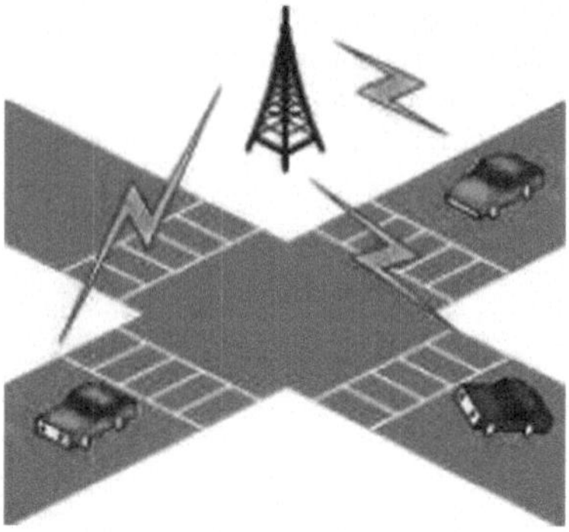

**Fig. 2** Limited structure cognitive radio architecture [10]

overload the obtainable range in automotive background. In cars, high suffering of the internet is determined in high traffic roads. Cognitive radio can be introduced to vehicle communication.

Cognitive radio enhances the throughput and also CR–VANETs enable more users to function in high user-friendly scenarios. Transceiver with reconfigurable software-defined radio (SDR) equipments is being included to vehicles. The operating parameters can be dramatically changed via software, which reduces data flexibility and operations of dissimilar bands [10]. So, hardware limitations are optimized in the development of novel equipments. The spectrum sensing (SS) is fundamental in cognitive radio networks as well as in CR–VANET. SS means to detect the presence of license users or secondary users in a specific frequency band more correctly; this reduces the practice of spectrum holes resourcefully.

Spectrum sensing (SS) methods are categorized as:

(i) Per-vehicle sensing.
(ii) Spectrum database (DB) methods.
(iii) Cooperation.

In per-vehicle sensing, each car achieves the range sensing independently and separately from the others. SS is performed with traditional spectrum sensing strategies as energy detection (ED), matched filter detection (MFD), cyclostationary detection (CD), and others [11]. However, despite the fact that individual car can achieve its own sensing with no record of any infrastructure support, the accuracy of the sensing decreases in situations of obstructions in Fig. 3 [11].

## 4 Literature Review

Uddin et al. [12] proposed the cognitive radio (CR) enabled vehicular ad hoc networks mainly for the multiagent relied on traffic management system (TMS). Due to the continuous interest of researchers in the concepts of CR, the demand of CR-based vehicular ad hoc networks was flourishing. It played out a major role in the

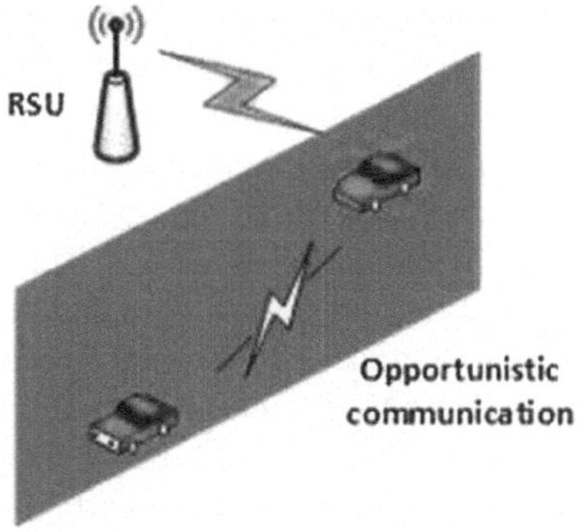


**Fig. 3** Adaptable communication in VANET [10]

communication of sensor nodes with high efficiency and reliability. In VANET, the sensor nodes are the moving vehicles that continuously changed the topology of the network. In the present research work, a cognitive radio-based network was proposed mainly for the vehicles that used the multiagent. A skeleton was introduced mainly for the learning purposes and for the decision making. The present technique had two basic phases. The first one was the dynamic and the second was the semi-dynamic. The dynamic was the communication of vehicle to vehicle without accessing any other device for the data exchange, whereas the same dynamic was the data communication of vehicle with roadside units (RSU). The communication was taking place by accessing a wireless link with the use of cognitive radio. Subsequently, a cluster formation approach was utilized to acquire the more accuracy in the data transmission. The experiment was proven better in terms of the throughput, consumption of energy, delay considered while data packets were sent, and the overhead of the network. Elgaml et al. [13] recommended the low delay and high throughput CR–VANET. Because of the advancement of the technology, the wireless sensor networks were utilized for various purposes in most of the applications and CR–VANET was one of them. It exploited CR for the access of unused channels by vehicles under the radio region. Therefore, CR–VANET was not just passed through the common challenges of cognitive ratio mainly the spectrum sensing. But it also faced some difficult issues related to the mobility of the sensor nodes in the network. The current research was described the low delay and high throughput CR region method for the better working criteria of CR–VANETs that had the tendency to manage with the general stands of VANET as IEEE802.11p. The experiment depicted that the current technique was easily declined the time consumption and the throughput was flourishing. Ahmed et al. [14] described the cooperative spectrum sensing, particularly for the CR–VANET with the detailed description and the research challenges of CR. These days, VANETs are being more interested in the field of intelligent traffic management and it enhanced in several ways to serve the people with better performance while data are transmitted from one node to the other. In VANET, the sensor nodes are the moving vehicles. It was supportive to various kinds of applications, mainly for the safety purposes, traffic efficiency, and even for the entertainment requirements. It makes the driving more comfortable with a lot of facilities. The cognitive radio simply increased the bandwidth mainly for the communication of moving vehicles. The main purpose of using cooperative spectrum sensing (CSS) was to exploit the temporary and spatial diversity to search out quickly the primary users of the data. The current research introduced the detailed description of the CSS approach mainly for the CR–VANETs. The major advantages of the technique were discussed in the literature review section. The other objective of the research was to clarify the common research challenges and future direction of the current research approach. JalilPiran et al. [15] proposed the fuzzy-based sensor fusion approach mainly for the CR–VANET. Generally, in the field of wireless sensor networks (WSN). The sensor fusion was introduced to integrate it with the gathered data via sensors basically to give access for the unified interpretation. The biggest characteristic of the sensor fusion was the high-level data in the statistical and in the definitive ways. It was difficult to acquire by using a single sensor. The current research initialized a dynamic sensor fusion method which was fully dependent upon the fuzzy theory. For the present approach, input sensor readings and final output were considered. The sensor nodes in CR–VANET were concatenated with the diverse sensors. Along with this, the crash severity was utilized as the consequent variable. The procedure and the fusion were obtained via using the fuzzy logic criteria. The results obtained from the present research approach were introduced the applicable system mainly to decline the causality rate of the vehicles. Eze et al. [16] give the explanation about CR technology which was utilized with VANET. The present status, issues, and research trends were described in detail. The vehicular networks and the cognitive radio networks both were being a fascinating advancement of the technology. The applications of cognitive radio were utilized for the intelligent vehicles and to sort out the common issues occurred because of the scarce spectrum. Current research gave the description of cognitive radio advancements which set a goal to enhance the efficiency in the vehicle communication. The description given in the research was introduced the dynamic technique and the major issues that were linked to the cognitive radio. Additionally, the research work also identified the different issues which were faced by the design and the creation of cognitive radio vehicular ad hoc networks. Table 1 defined the basic comparison of TCR, VANET, and CR–VANET, respectively.

**Table 1** Comparison of traditional cognitive radio (TCR), VANET, and CRV [17, 18]

| | TCR | VANET | CRV |
|---|---|---|---|
| Range | ~30 km | Few km | Few km |
| Mobility characteristics | Stationary | High mobility can, exceeds 100 km per hour | High mobility can exceed 100 km per hour |
| Topology | Centralized | V2I, V2V, and V2P | Both integration path |
| Performance | Throughput delay | Transmission delay and capacity | Transmission capacity and delay |

# 5 CR–VANET for Topology and Routing Protocols

CR–VANET has the characteristics of high mobility and rapid dynamic topology and data reliability for road safety. Designing appropriate media access control (MAC) protocol for CR–VANET is the main task. According to the review paper [4], there is still no domination structure of the CR–VANET protocol.

Moreover, in order to provide a normal view of Cognitive Radio Vehicular Ad hoc Network (CRVANET) MAC-LAYER, we summarize some essential articles [19–22] and present a preliminary structure of the CR–VANET protocol (MAC). The slot design of CR–VANET network system is shown in Fig. 4.

Routing in CR–V faces unique challenges when compared with mobile ad hoc networks (MANETs). In VANETs, the routing topologies are expected to accept the vehicle mobility of the node and the state of the channel. In CR–V, allotted route shall be more rapidly changed according to the event of license users and frequency in the spectrum pool [24]. In most of these topologies routing methods, the route is designed during the route-discovery and it can be modified when messages are missed or novel PU activity is detected. Algorithms suffer from performance degradation, when the spectrum state or spectrum availability, vehicle node positions modify speedily than the rate of route-updates. Various routing methods for VANETs have been implemented in [23]. Moreover, most of them are focused on short-range communication in ad hoc networks to aid intelligent transport system (ITS) in urban areas. The distance between binary vehicle nodes is much smaller than broadcast range. Furthermore, none of them measured the adaptable range access unlicensed industrial–scientific–medical (ISM) band to obtain more capacity. At current, the research on routing schemes for CR–VANET is really rare.

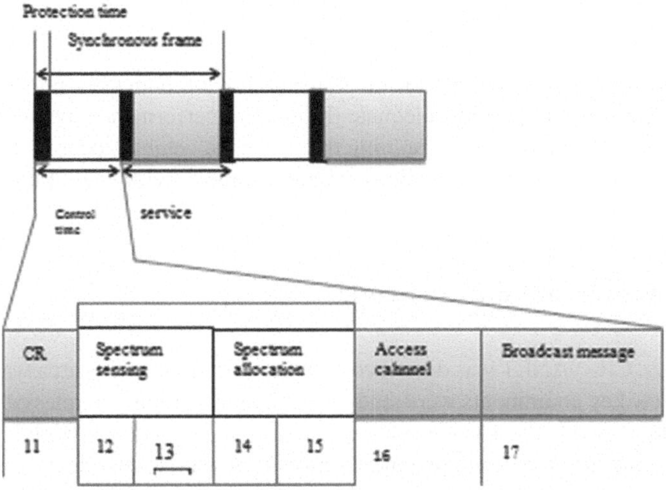


**Fig. 4** Slot design of CR–VANET [23]

**Table 2** Comparison between dissimilar routing protocols [25]

| Methods | Advantages | Disadvantages |
|---|---|---|
| PCTC | Most effective and reliable topology than other, low rerouting frequencies and delay | Requires local link knowledge consequence in huge message overhead |
| SABE | Overhead for choosing delay vehicle nodes is substantially reduced | Robustness of proposed algorithm shall be enhanced |
| Co-Cast | Channel overlap in frequency | Overhead for communication might be a problem |

A. *Prediction-based Cognitive Topology Control (PCTC) Routing*

It is a distributive prediction-based technology control method over the cognitive capability of routing CR–V. It is a middleware like cross-layer module residing. It utilizes cognitive connection data availability forecast, which is conscious of distortion to primary users and predicts the available duration connections. Based on connection predicts, it considers the dynamic modification of the topology and builds enhancing effective methods, which is objective in reducing routing frequency and enhancing end-to-end (E2E) delay network performance like as throughput and delay.

B. *Spectrum Aware Beacon Less Geographical Routing (SABE)*

The main objective of spectrum aware beacon is that the routing rules as well as the resource allocation policy is made by receivers on a per-packet and hop basis, so that the SABE protocol can be effective in spectrum dynamic. A CRV broadcasts advancing request packet and adds its data available resources and location.

C. *Cognitive Multicasting uses Orthogonal Frequency Division Multiplexing (OFDM)*

It is a cognitive multicast routing protocol inspired by on-demand multicast routing protocol. Its attempt to provide adequate throughput performance by choosing and using idle ISM channels. After searching the channels, vehicle nodes collaborate to form a multicast tree using and ad hoc demand distance vector [23] (Table 2).

## 6 Conclusion and Future Scope

In this paper, CR–VANET was described and the work done in this area was categorized and few key attainments were studied. CR is a developing wireless announcement topology (WAT) that can spread the vehicle announcement networks characteristics. CR motivates the development and growth in vehicle-to-vehicle, vehicle-to-infrastructure, and vehicle-to-peer communications. Though, there are still structural problems to combine the cognitive radio vehicles. The main problems and challenges in cognitive radio vehicles have been studied. Various high-lighted issues SS, MPs of observation, interference, end-to-end delay, and security system are the main concepts in CR–VANET. The future scope will improve the lack of data-adaptive approach and optimization of the recent CR–VANET varieties.

## References


1. Hartenstein, H., Laberteaux, K. (eds.): VANET: Vehicular Applications and Inter-Networking Technologies, vol. 1. Wiley (2009)
2. Kenney, J.: Dedicated short-range communications (DSRC) standards in the united states. Proc. IEEE **99**(7), 1162–1182 (2011)
3. ETSI-Intelligent Transport: Intelligent Transport Systems. http://www.etsi.org/technologies-clusters/technologies/intelligent-transport. Accessed 11 Mar 2014
4. Lim, J.M.Y., Chang, Y.C., Alias, M.Y., Loo, J.: Cognitive radio network in vehicular ad hoc network (VANET): a survey. Cogent. Eng. **3**(1), 1191114 (2016)
5. Thakur, P., Singh, G., Satashia, S.N.: Spectrum sharing in cognitive radio communication system using power constraints: a technical review. Perspect. Sci. **8**, 651–653 (2016)
6. Bharti, B., Thakur, P., Singh, G.: Framework for spectrum sharing in cognitive radio network for military applications. IEEE Potentials, Sept. 2017 (accepted). https://doi.org/10.1109/mpot.2017.2751656
7. Thakur, P., Kumar, A., Pandit, S., Singh, G., Satashia, S.N.: Advanced frame structures for hybrid spectrum accessing strategy in cognitive radio communication system. IEEE Commun. Lett. **21**(2), 410–413 (2017)
8. Sum, S., Villardi, G.P., Rahman, R.A., Baykas, T., Tran, H.N., Lan, Z., Sun, C., Alemseged, Y., Wang, J., Song, C., Pyo, C.W., Filin, S., Harada, H.: Cognitive communication in TV white spaces: an overview of regulations, standards, and technology. IEEE Commun. Mag. **51**(7), 138–145 (2013)
9. Bu, S., Yu, F.R.: Green cognitive mobile networks with small cells for multimedia communications in the smart grid environment. IEEE Trans. Veh. Technol. **63**(5), 2115–2126 (2014)
10. Carvalho, F., Lopes, W., Alencar, M., Filho, J.: Cognitive vehicular networks: an overview. Procedia Comput. Sci. **65**, 107–114 (2015)
11. Tandra, R., Mishra, S.M., Sahai, A.: What is a spectrum hole and what does it take to recognize one? Proc. IEEE **97**(5), 824–848 (2009)
12. Uddin, S.M., Mansoor, N., Hossain, S.: Cognitive radio enabled VANET for multi-agent based intelligent traffic management system. In: Proceedings of the First International Conference on Advanced Information and Communication Technology (ICAICT-16) (2016)
13. Elgaml, N., Khattab, A., Mourad, H.A.: Towards low-delay and high-throughput cognitive radio vehicular networks. ICT Express **3**(4), 183–187 (2017)
14. Ahmed, A.A., Alkheir, A.A., Said, D., Mouftah, H.T.: Cooperative spectrum sensing for cognitive radio vehicular ad hoc networks: an overview and open research issues. In: 2016 IEEE Canadian Conference on Electrical and Computer Engineering (CCECE), pp. 1–4. IEEE (2016)
15. JalilPiran, M., Ali, A., Suh, D.Y.: Fuzzy-based sensor fusion for cognitive radio-based vehicular ad hoc and sensor networks. Math. Probl. Eng. **2015**, 1–9 (2015)
16. Eze, J., Zhang, S., Liu, E., Eze, E.: Cognitive radio technology assisted vehicular ad-hoc networks (VANETs): current status, challenges, and research trends. In: 2017 23rd International Conference on Automation and Computing (ICAC), pp. 1–6. IEEE (2017)
17. Rubinstein, M.G., Abdesslem, F.B., de Amorim, M.D., Cavalcanti, S.R., Alves, R.D.S., Costa, L., Duarte, O., Campista, M.: Measuring the capacity of in-car to in-car vehicular networks. IEEE Commun. Mag. **47**(11), 128–136 (2009)
18. Felice, M.D., Mohammady, R.D., Chowdhury, K.R., Bononi, L.: Smart radios for smart vehicles: cognitive vehicular networks. IEEE Veh. Technol. Mag. **7**(2), 26–33 (2012)

19. Islam, M., Koh, C., Oh, S., Qing, X., Lai, Y., Wang, C., Liang, Y.C., Toh, B.E., Chin, F., Tan, G.L., Toh, W.: Spectrum survey in Singapore: occupancy measurements and analyses. In: 3rd International Conference on Cognitive Radio Oriented Wireless Networks and Communications, Luxemborg, pp. 1–7 (2008)
20. Tsukamoto, K., Matsuoka, S., Altintas, O., Tsuru, M., Oie, Y.: Distributed channel coordination in cognitive wireless vehicle-to-vehicle communications. In: Proceedings of International Conference on Wireless Access in Vehicular Environment (WAVE 2008), Dearborn, MI, USA (2008)
21. Zhongding, L., Chin, F.: A reliable and power efficient beacon structure for cognitive radio systems. IEEE Trans. Broadcast. **54**(2), 182–187 (2008)
22. Haykin, S.: Cognitive radio: brain-empowered wireless communications. IEEE J. Sel. Areas Commun. **23**(2), 201–220 (2005)
23. Kim, J., Krunz, M.: Spectrum-aware beaconless geographical routing protocol for cognitive radio enabled vehicular networks. Mob. Netw. Appl. **18**(6), 854–866 (2013)
24. Guan, Q., Yu, F.R., Jiang, S.: Topology control and routing in cognitive radio mobile ad hoc networks. In: Cognitive Radio Mobile Ad Hoc Networks, pp. 209–225. Springer, New York, (2011)
25. Kim, W., Oh, S.Y., Gerla, M., Park, J.S.: COCAST: multicast mobile ad hoc networks using cognitive radio. In: MILCOM 2009–2009 IEEE Military Communications Conference, pp. 1–7 (2009)

# Parameter Optimization Using PSO for Neural Network-Based Short-Term PV Power Forecasting in Indian Electricity Market

**Harendra Kumar Yadav, Yash Pal and M. M. Tripathi**

**Abstract** Because of developing concern to environmental changes, renewable energy is being looked as a key alternative to the conventional sources. Photovoltaic power is a green and an abundant renewable energy source. Photovoltaic power is dependent upon the solar irradiation which is highly intermittent in nature. So the precise forecasting of PV power is necessary to improve the operation of an electrical grid with the distributed energy resources. This paper proposes a novel hybrid model by combining particle swarm optimization (PSO) and the feed-forward neural network (FFNN) together. Proposed hybrid model is applied to forecast the PV power in Indian electricity market. One-year data consisting of hourly PV power generation, direct radiation, diffused radiation, and ambient temperature from the Indian energy market has been used for PV power forecasting. The developed model is applied for one-week ahead PV power forecasting for winter, summer, rainy, and autumn season, respectively. The performance of the proposed hybrid model outmatches and compared with some recently reported model.

**Keywords** Solar irradiation · Autocorrelation · Neural network · PSO and PV power forecasting

## 1 Introduction

The Due to the rising temperature, economic issues, climate change, and carbon emission the electricity market structure of the world is adopting new strategies to be relevant. As a result, renewable energy is emerging as the possible energy option of

H. K. Yadav (✉)
SREE, NIT Kurukshetra, Kurukshetra, Haryana, India
e-mail: nitkkrharendra@gmail.com

Y. Pal
Department of Electrical Engineering, NIT Kurukshetra, Kurukshetra, Haryana, India
e-mail: yash_pal1971@yahoo.com

M. M. Tripathi
Department of Electrical Engineering, DTU, New Delhi, India
e-mail: mmmtripathi@gmail.com



P. K. Singh et al. (eds.), *Proceedings of ICRIC 2019*, Lecture Notes in Electrical Engineering 597, https://doi.org/10.1007/978-3-030-29407-6_25

the future. The solar photovoltaic (PV) power is attracting more people in the world as an alternative solution to the future power problems. The growth rate of the PV power installations is increasing rapidity due to the continuous reduction in the cost of the solar PV panels. India has the tremendous scope of the solar power generation due to the geographical location of the country.

From last two decades, the advancement in smart grid technologies has made it possible to integrate the solar PV generation with the power grid. The smart metering infrastructure encourages the consumer to become a prosumer. The prosumers are allowed to consume the electricity, produce energy and are also allowed to inject extra power into the smart grid as per the rules laid down by electricity regulatory commission. In many countries, governments are promoting the distributed power generation and providing subsidy also on the installation of small photovoltaic rooftop power generation plant.

However, the PV power is intermittent and difficult to dispatch, due to the varying nature of solar irradiation and it may cause a power imbalance. Power distribution centers are facing difficulty to dispatch the power to the consumers, due to this power imbalance. Solar PV power forecasting based on the limited geographical as well as previous year data may be the key solution to this problem and may enhance the effective use of solar PV power generation. Solar PV power forecasting can reduce the impact of uncertainty in solar PV power generation, resulting in maintenance of power quality, improvement in system reliability and enhancement in solar PV power generation penetration into the grid [1].

Many researchers have developed different models for the PV power forecasting. It is observed that generally two approaches are adopted for the PV power forecasting, indirect [2–8] and direct approach [9–16]. In the direct approach researchers are forecasting solar PV power output directly and in indirect approach researchers are first forecasting solar irradiation from the available metrological data, followed by the solar PV power forecasting with the help of solar irradiation forecasting.

Many statistical and artificial intelligence techniques are used for PV forecasting. Persistence model is a benchmark model for the entire developed model for both types of forecasting. ARIMA [2], wavelet analysis [3], wavelet-ARIMA, support vector regression (SVR) [4], spatio-temporal [5], artificial neural network (ANN) [6], and hybrid ANN-based methods [8] is used for the indirect forecasting. In the direct forecasting methods, support vector regression (SVR) [9–14], ANN [15], fuzzy logic [16], ANFIS [17], and other many hybrid methods [18–24] are also used. These methods have some advantage and disadvantage over each other. All the above-cited models are developed for the specific time horizon forecasting, mostly for the one-day and one-week ahead forecasting.

ANN method is adopted for all type of forecasting by the researchers, for linear as well as nonlinear modeling. ANN model requires setting the model parameters and networking topology. The results of ANN models are highly correlated with the model parameters. As a result, the performance obtained from the ANN method is not up to the desired level and it requires improvement in the approach. In this paper, a hybrid model combining feed-forward neural network and particle swarm optimization is used for PV power forecasting. PSO is used for optimizing FFNN parameters, bias, and weights.

The one-year hourly solar PV power generation data obtained from Kolkata, India located at latitude 25.64°, longitude 79.45°, for the year 2014 is used for training and testing purpose of the proposed model. This one-year data of one-hour resolution are divided into four groups January to March, April to June, July to September, and October to December which corresponds to the winter, summer, rainy, and autumn seasons in India, respectively. Again data of each group is divided into two groups for training and testing purpose. Simulation results are quite encouraging as mean absolute percentage error, root mean square error and mean absolute error have been calculated to evaluate performance of the developed model. All the error parameters have shown that developed model is very accurate in short-term PV power forecasting for each season.

Rest of the paper is arranged in the following manner. Section 2 discusses the correlation of solar PV power with the input variables. Section 3 discusses the proposed approach. Section 4 presents the working of a proposed hybrid model of solar PV forecasting and Sect. 5 discusses various performance evaluation matrices. Results and discussions are presented in Sect. 6. Section 7 presents the conclusion.

## 2 Factors Influencing the Solar PV Power Output

Solar direct radiation, diffuse radiation, and atmospheric temperature are used as an input variable to developed forecasting model. This section shows graphical and Pearson correlation coefficient $r$ to study correlation between PV power output w.r.t. solar direct radiation, solar diffuse radiation, and atmospheric temperature.

### *2.1 Solar Radiation Intensity*

As we know that the photovoltaic power is highly dependent upon the solar irradiation. The solar irradiation is the combination of direct radiation and diffuse radiation. Seven days data from June 1, 2014 to June 7, 2014 were used to draw a curve of PV power output versus solar direct radiation and PV power output versus solar diffuse radiation as shown in Figs. 1 and 2, respectively, to show the correlation between them.

Figure 1 shows that PV power output is highly correlated with direct radiation. So we can say that the dependency of PV power is very high on direct radiation and therefore it is taken as one of the main input variables that affects PV power output. Similarly, Fig. 2 shows relation between PV power output and solar diffuse radiation and it shows a positive relationship between them. But it shows less correlation as compared to the solar direct radiation, so the dependency of the PV power is less on the solar direct radiation, and it is also considered as an important input factor.

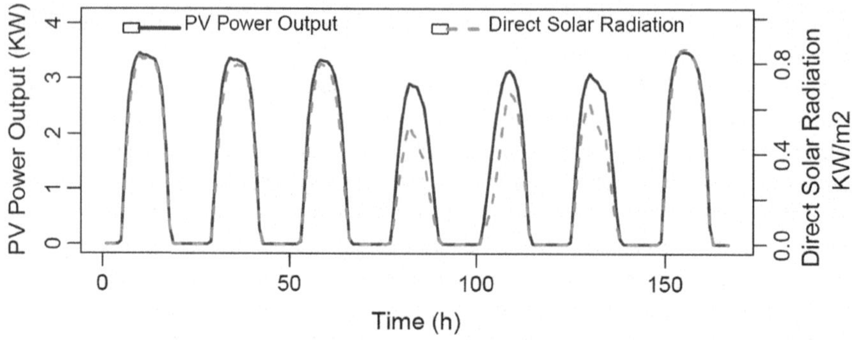


**Fig. 1** Curve of PV power output and direct solar radiation

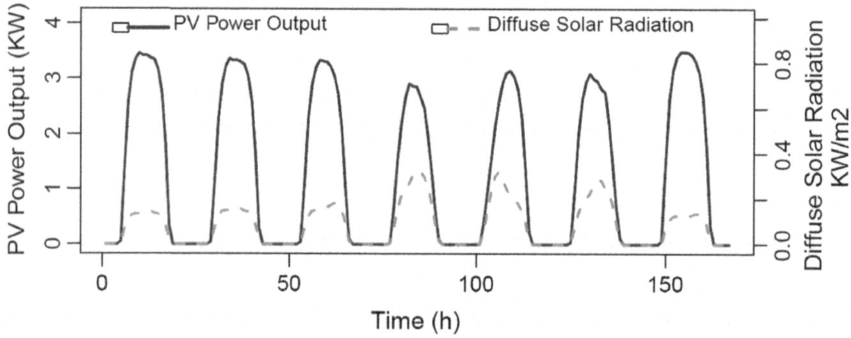


**Fig. 2** Curve of PV powers output and diffuse solar radiation

## 2.2 *Temperature*

PV power conversion efficiency is dependent upon the atmospheric temperature. PV cell voltage and current both are dependent on the atmospheric temperature. As the temperature is increased, the PV cell voltage is reduced and PV cell current is increased, but the reduction in voltage is more as compared to increase in current.

So according to the maximum power tracking the overall power output from PV cell is reduced as temperature increases and the conversion efficiency is reduced. In order to analyze effect of atmospheric temperature on PV power output, data of seven days from June 1, 2014 to June 7, 2014 were used to draw a curve of PV power output versus temperature as shown in Fig. 3. The occurrence of the local maximum point of the temperature and PV power slightly differ from each other as shown in Fig. 3 and shows as lightly positive correlation between the PV power and temperature.

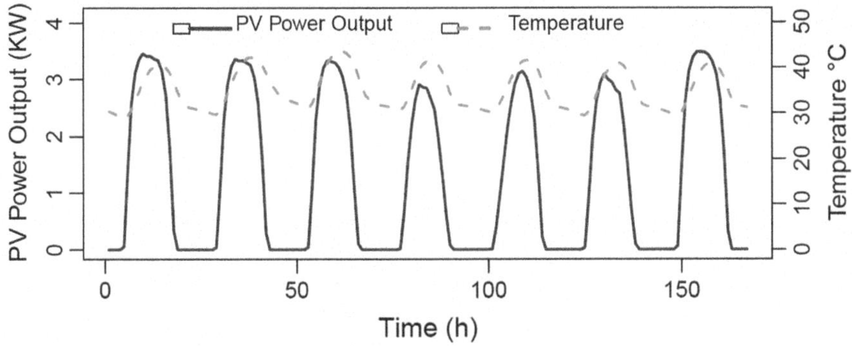


**Fig. 3** Curve of PV power output and atmospheric temperature

## *2.3 Pearson Correlation Coefficient*

Pearson coefficient *r* is used to study correlation between input variables with PV power output. Pearson coefficient *r* is calculated by the following Eq. (1) [18].

$$r = \frac{\sum AB - \frac{\sum A \sum B}{N}}{\sqrt{\left(\sum A^2 - \frac{\sum A^2}{N}\right) - \left(\sum B^2 - \frac{\sum B^2}{N}\right)}} \quad (1)$$

where *A* is PV power output, *B* is the different input variables, and *N* is the number of samples. Table 1 shows degree of correlation coefficient and their respective value of *r*.

The data of seven days of four seasons from March 1, 2014 to March 7, 2014, June 1, 2014 to June 7, 2014, September 1, 2014 to September 7, 2014, and December 1, 2014 to December 7, 2014 were collected and analyzed. Table 2 shows the correlation between PV power output and input variables such as solar direct radiation, solar diffuse radiation, and atmospheric temperature. From Table 2, it is clear that solar direct radiation shows strongly linear autocorrelation with the PV power with autocorrelation coefficient 0.993, 0.993, 0.993, and 0.987 of March, June, September, and December, respectively. Solar diffuse radiation shows slightly less autocorrelation with the PV power with autocorrelation coefficient 0.867, 0.897, 0.899, and

**Table 1** Degree of correlation coefficient

| S. No. | *r* | Degree of correlation |
|---|---|---|
| 1. | −1 | Completely negative |
| 2. | <0 | Negative |
| 3. | 0 | Uncorrelated |
| 4. | >0 | Positive |
| 5. | +1 | Completely positive |

**Table 2** Correlation coefficients of input variable w.r.t. PV power output

| S. No. | Season | Influence factor | Correlation coefficient ($r$) |
|---|---|---|---|
| 1. | Winter | Direct solar irradiation | 0.993 |
| | | Diffuse solar irradiation | 0.867 |
| | | Atmospheric temperature | 0.605 |
| 2. | Summer | Direct solar irradiation | 0.993 |
| | | Diffuse solar irradiation | 0.897 |
| | | Atmospheric temperature | 0.684 |
| 3. | Rainy | Direct solar irradiation | 0.993 |
| | | Diffuse solar radiation | 0.899 |
| | | Atmospheric temperature | 0.684 |
| 4. | Autumn | Direct solar irradiation | 0.987 |
| | | Diffuse solar radiation | 0.861 |
| | | Atmospheric temperature | 0.709 |

0.861 of March, June, September, and December, respectively. The autocorrelation coefficient of atmospheric temperature is moderate with the PV power with autocorrelation coefficient 0.605, 0.684, 0.684, and 0.709 of March, June, September, and December, respectively.

## 3 Proposed Approach

There are several approaches which have been developed for the PV power forecasting on the various time scale. In this paper, a fusion of feed-forward neural network and optimization algorithm PSO is used for developing the forecasting model. Feed-forward parameters are tuned by the optimization algorithm PSO.

## 3.1 Feed-Forward Artificial Neural Network

Generalized three-layer architecture of feed-forward neural network is shown in Fig. 4. It consists of three layers input layer, hidden layer, and an output layer. In the literature, it is found that a number of hidden layers are highly correlated with the output. In many cases, it is found as number of hidden layers increase output of neural networks model increased and so the accuracy of the model. At the same time model complexity and computational time also increases which is undesirable. In certain applications, as a number of hidden layer increases output of the network deviates from the actual result. In this paper, neural network has been optimized with an appropriate number of hidden layers with the help of hit and trial method.

The proposed neural network has been trained using particle swarm optimization (PSO) algorithm so that mean square error (MSE) becomes minimal for given set of input-output parameters as given by the following Eq. 2 [20].

$$f(w_t, \beta) = \sum_{i=1}^{N} [y_i - f(x_i, w_t, \beta)]^2 \tag{2}$$

where $x_i$ and $y_i$ are the input and target set, respectively, $w_t$ is the weighted value of input link, $\beta$ represents threshold correlation coefficient value, and $N$ is a number of hidden neurons in each layer.

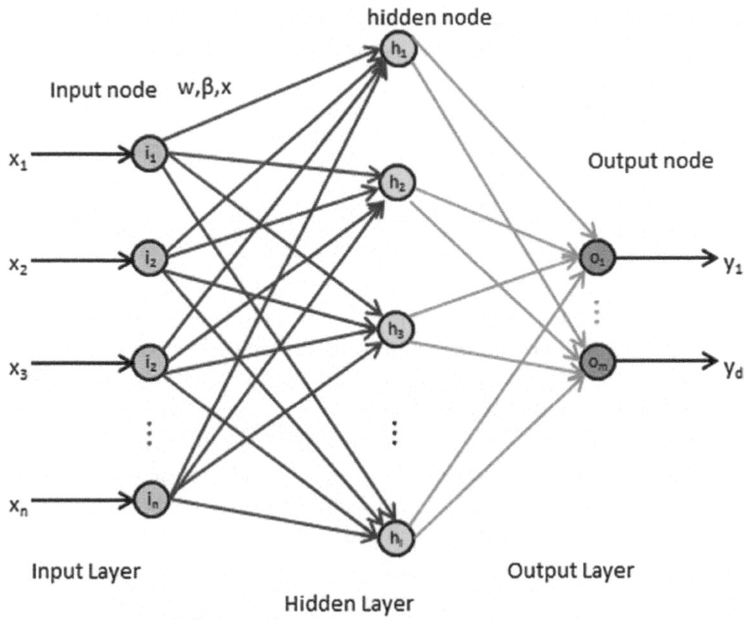


**Fig. 4** Architecture of feed-forward artificial neural network

With the help of PSO, the new value of wt and $\beta$ are updated with a substantial increase in weight ($\delta_{\text{wt}}$) and correlation ($\delta_{\beta}$) coefficient of approximation parameters, as given in Eqs. 3 and 4 after iteration, respectively.

$$w_t = w_t + \delta_{\text{wt}} \tag{3}$$

$$\beta = \beta + \delta_{\beta} \tag{4}$$

Corresponding values of ($\delta_{\text{wt}}$) and ($\delta_{\beta}$) could be evaluated by approximating input functions by their linearization as given by the following Eq. 5.

$$f\left(x_i, w_t + \delta_{\text{wt}}, \beta + \delta_{\beta}\right) \approx f(x_i, w_t, \beta) + J_i\left(\delta_{\text{wt}}, \delta_{\beta}\right) \tag{5}$$

The updating process of wt and $\beta$ for evaluating the function $f(\text{wt}, \beta)$ is continued till the desired value of a function or a maximum number of iteration is not achieved.

### 3.2 *Particle Swarm Optimization*

Particle swarm optimization is a search-based stochastic optimization algorithm proposed in 1995 by Kennedy and Eberhart [22]. PSO is encouraged by social and competitive activities of swarms to fulfill their need. The algorithm is supervised by personal; overall experience and present movement of swarms to decide their next position in search space [17]. If initial population of PSO has M individuals, each individual is being treated as a volume-less particles in *N*-dimensional space, the particle moves according to Eqs. 6 and 7 below.

$$v_i(t+1) = \omega \times v_i(t) + c_1 r_1\left(x_{\text{Pbest}_i} - x_i(t)\right) + c_2 r_2 (x_{\text{Gbest}} - x_i(t)) \tag{6}$$

$$x_i(t+1) = x_i(t) + v_i(t+1) \tag{7}$$

where $\omega$ is inertia factor, $c_1$ and $c_2$ are positive acceleration coefficients and $r_1$ and $r_2$ are two different random numbers uniformly distributed between [0 and 1].

Figure 5 shows the search mechanism of the optimization technique PSO using velocity update Eq. 6 and position update Eq. 7 as given above.

In this paper, values of $c_1$ and $c_2$ are extracted for a different number of hidden layers of feed-forward neural network by using hit and trial method for minimum forecasting error. IWA (inertia weight approach) is used during optimization process and inertia weight $\omega$ is modified by Eq. 8 [14, 17].

$$\omega = \omega\text{max} - \frac{(\omega\text{max} - \omega_{\text{min}})}{\text{max iteration}} \times \text{current iteration} \tag{8}$$

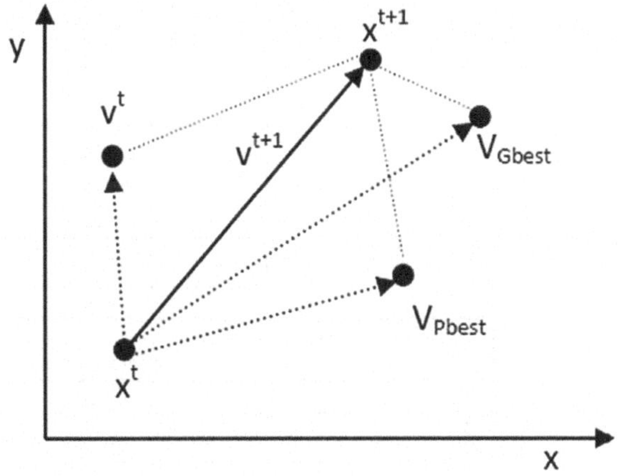


**Fig. 5** PSO search mechanism for updating the position of the particle

where $\omega_{max}$ and $\omega_{min}$ are final and initial inertia weight, respectively. Max iteration is the highest number of iteration and present iteration is a current number of iteration.

## 4 Working of Proposed Forecasting Model

A hybrid PV power forecasting method is proposed here that combines feed-forward artificial neural network (FFANN) and PSO. The flowchart in Fig. 6 shows detail working on the proposed hybrid method.

### *4.1 Data Summary*

In this paper, one-year data from January 1, 2014 to December 31, 2014 is obtained from Kolkata, India located at latitude 25.64°, longitude 79.45°, and data is divided into four data groups according to the season in India. Again every season data is divided into two parts for training and testing purpose of developed models. In each season, first two-month data is used for training purpose and next one-week data of next month is used for testing purpose [25].

### *4.2 Pre-processing Stage*

In pre-processing stage, all data are scaled between ranges of 0 and 1 using Eq. 9 as training algorithm give more accurate result when all data are scaled in same range [8].

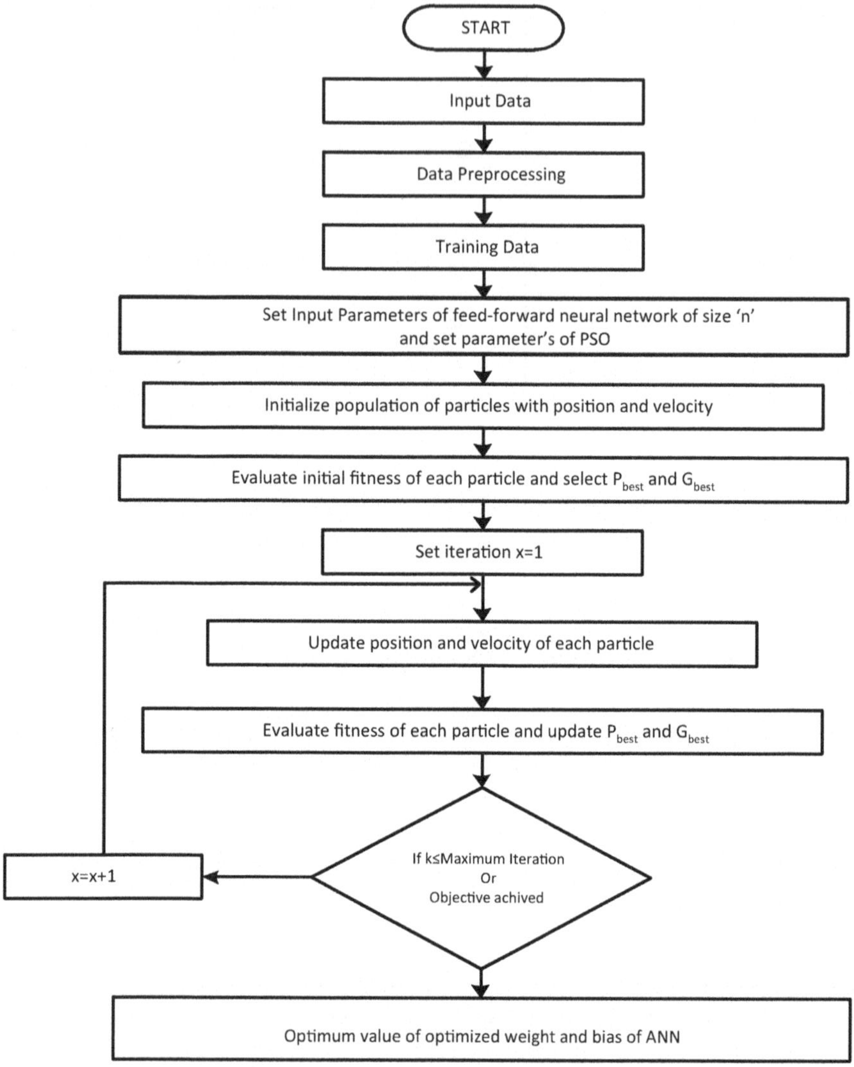


**Fig. 6** Flowchart of methodology used to develop forecasting model

$$x_{\text{scaled}} = y_{\min} + (y_{\max} - y_{\min}) \times \frac{x - x_{\min}}{x_{\max} - x_{\min}} \tag{9}$$

where $y_{\max}$ and $y_{\min}$ being 1 and 0.1, respectively, and $x_{\max}$ and $x_{\min}$ is maximum and minimum value of corresponding parameters, and $x_{\text{scaled}}$ is normalized value. The data was rescaled back to original dimensions after the network has been processed.

### 4.3 Data Preparation

As stated above, one-year data of solar radiation (direct and diffuse radiation), temperature and PV power are taken for testing the model. Model is tested for Indian context data of Kolkata region. In this paper, one-year data of one-hour resolution are divided into four groups January–March, April–June, July–September, and October–December according to season winter, summer, rainy, and autumn, respectively. Again each group data is divided into two sets, i.e., training set and testing set. Training and testing set data are divided such that first two-month data are used for training of model and one-week data are used for testing.

### 4.4 Training Stage

Once data are normalized and divided into groups, two-month data of each group are used to train ANN-PSO hybrid model, for forecasting of 1-week ahead, PV power of the corresponding group which is not considered in training dataset. During the training process of the model, model parameters are adjusted according to inputs and output data given for training. Feed-forward artificial neural network parameters are optimized and updated with help of particle swarm optimization algorithm. Weights and bias of feed-forward neural network are being updated during the iteration process. The parameter related to PSO algorithm is given in Table 3.

The training process stops whenever the iteration will reach maximum value or goal of training error is achieved. Summary of the workflow of the feed-forward artificial neural network is given below.

- Number of input: 4 (direct and diffuse radiation, temperature, and PV power)
- Number of hidden layers: 5
- Number of input layer: 3
- Number of output layer: 1
- Comparison function: MAPE, MAE, and RSME
- Data distribution: Two months (one-hour resolution) for training and one-week data of the next month is used for testing of the corresponding group.

**Table 3** Parameters of PSO [17]

| Parameters | Value |
|---|---|
| Size of particles | 30 |
| Total count of iterations | 2500 |
| $C_1$ | 1–2.5 |
| $C_2$ | 2.5–3 |
| Initial inertia weight $\omega_{max}$ | 0.9 |
| Initial inertia weight $\omega_{min}$ | 0.4 |

### 4.5 Forecasting Stage

Developed ANN-PSO hybrid model is used to forecast PV power for one-week time horizon of different four seasons. Direct and diffuse solar radiation and temperature are used as an input variable and PV power as a target for the developed model.

## 5 Performance Evaluation Index

There is no single indicator that can evaluate the performance of forecasting models. Based on the literature, there are many performance evaluation methods which are used by researchers to evaluate the performance of forecasting models.

Following three evaluation indexes, root mean squared error (RSME), mean absolute percentage error (MAPE), and mean absolute error (MAE) is used to check feasibility and ability of proposed PV power forecasting model as given by Eqs. 10, 11, 12, and 13, respectively [1, 17].

- Root mean square error

$$\mathrm{RMSE} = \sqrt{\frac{1}{N}\sum_{j=1}^{N}\left(X_{mj} - X_{cj}\right)^2} \tag{10}$$

- Mean absolute percentage error

$$\mathrm{MAPE} = \frac{100}{N}\sum_{j=1}^{N}\frac{\left(X_{mj} - X_{cj}\right)}{\overline{X}_m} \tag{11}$$

$$\overline{X}_m = \frac{1}{N}\sum_{j=1}^{N} X_m \tag{12}$$

- Mean absolute error

$$\mathrm{MAE} = \frac{1}{N}\sum_{j=1}^{N}\left|\left(X_{mj} - X_{cj}\right)\right| \tag{13}$$

where $X_{mj}$ is $j$th measured value, $X_{cj}$ is $j$th calculated value, and $N$ is the number of measurement taken. $\overline{X}_m$ is the average PV power of the forecasting period. Average

PV power is used in Eq. 11 to keep away from adverse outcome of PV power close to zero [14].

## 6 Result and Discussion

As mentioned above that the one-year data from January 2014 to December 2014 is obtained from Kolkata, India for the purpose of training and testing of the developed PSO tuned FFNN model. A code in MATLAB and R environment is developed for training and testing purpose of the model. The one-year data is divided into four groups according to the season, January–March (winter), April–June (summer), July–September (rainy), and October–December (autumn), respectively. From each group, first two-months data are used for training purpose and one-week data of last month is used for testing purpose.

Figures 7, 8, 9, and 10 show the one-week ahead forecasted and actual solar PV power for the months of March 2014, June 2014, September 2014, and December

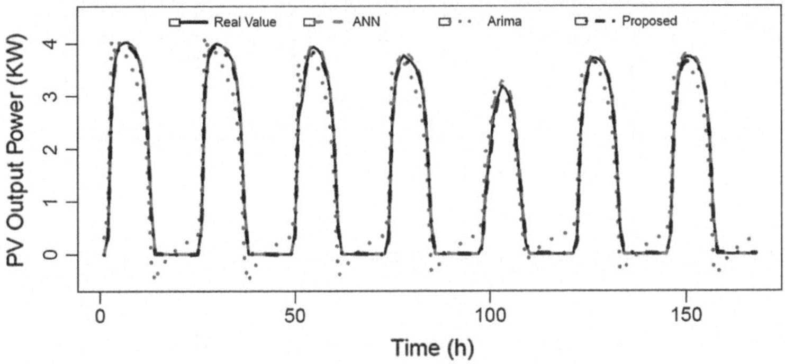


**Fig. 7** Forecasted result of the March 1st week

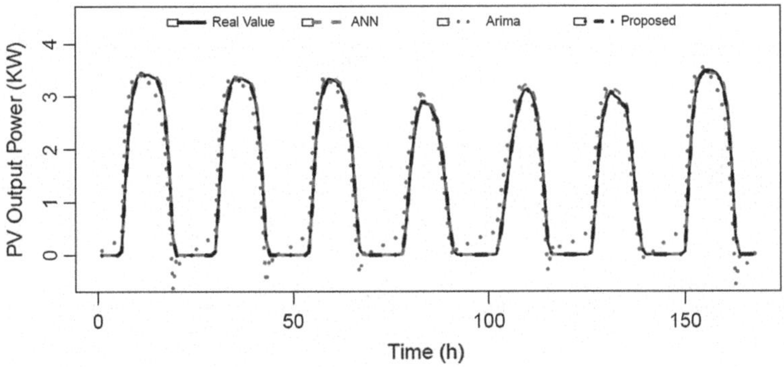


**Fig. 8** Forecasted result of the June 1st week

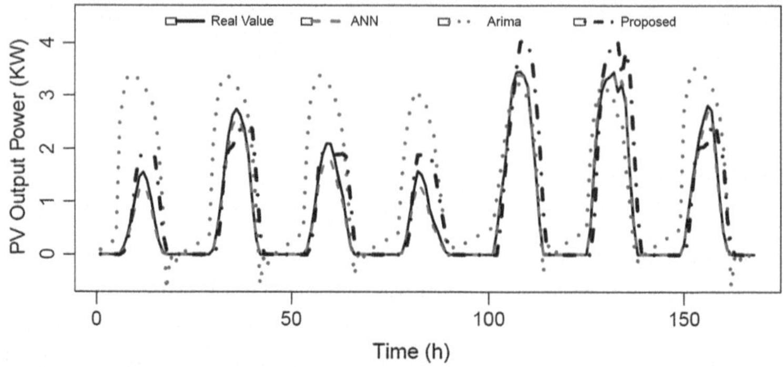


**Fig. 9** Forecasted result of the September 1st week

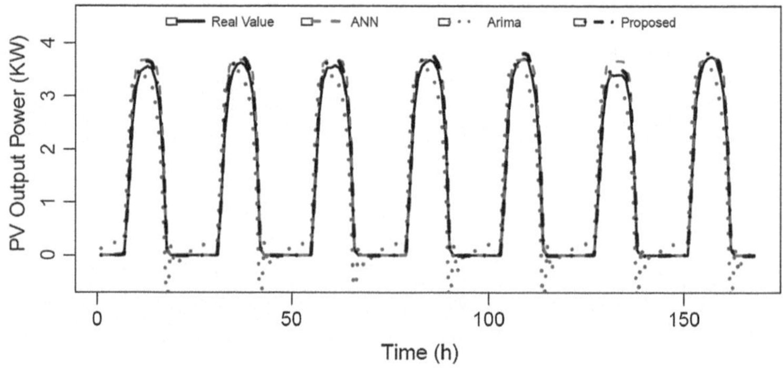


**Fig. 10** Forecasted result of the December 1st week

2014, respectively. The mean absolute percentage error (MAPE) has been shown for all the four months in Figs. 11, 12, 13, and 14, respectively. It is evident from these plots that the proposed model is able to forecast the week ahead solar PV power with least error. The average MAPE of one week for the months of March 2014,

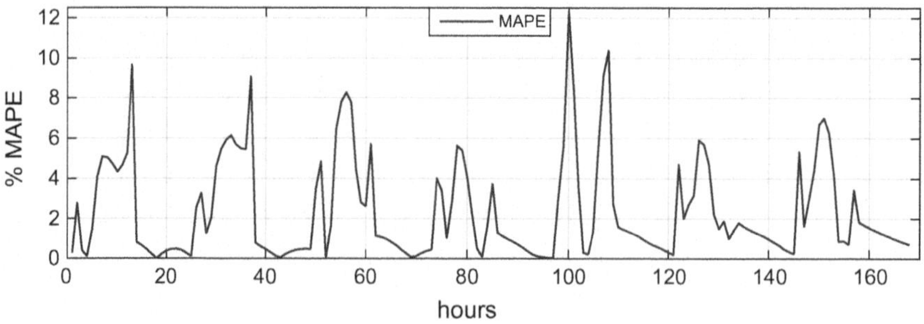


**Fig. 11** One week absolute % MAPE error of March 1st week

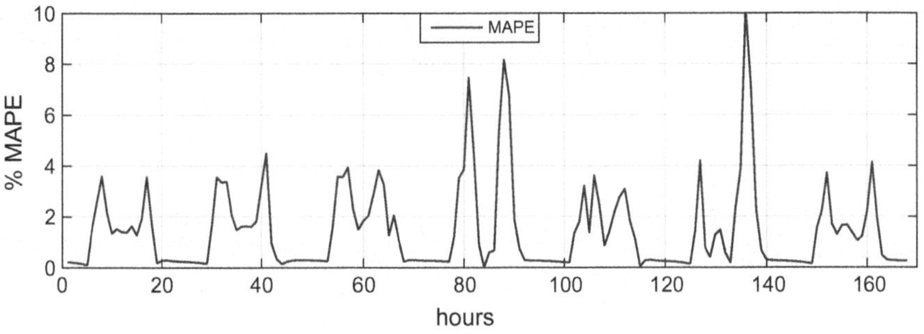


**Fig. 12** One week absolute % MAPE error of June 1st week

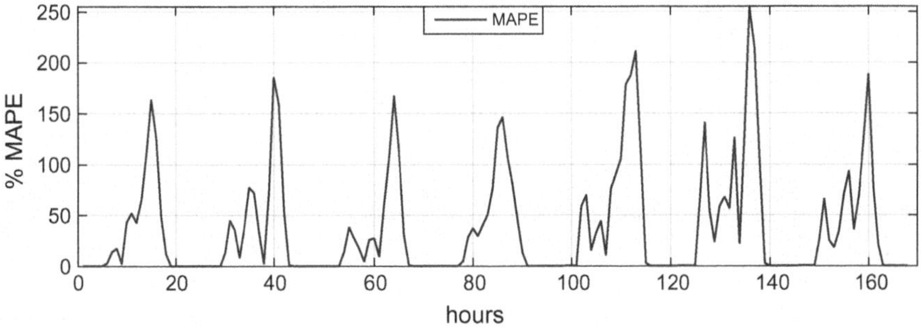


**Fig. 13** One week absolute % MAPE error of September 1st week

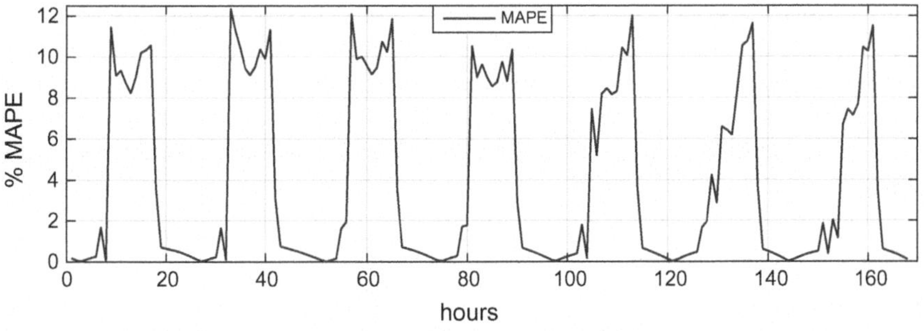


**Fig. 14** One week absolute % MAPE error of December 1st week

June 2014, September 2014, and December 2014 are 2.36, 1.45, 29.18, and 3.80%, respectively.

In the month of June forecasted results is best with minimum error as compared to the other seasons because during these periods solar irradiation is consistent. The result of the September month shows that forecasted results are poor as in these periods the solar irradiation is very random due to the rainy season, humidity in the

**Table 4** Summarized result of all four seasons

| Season | Evaluation matrix | Forecasting methods | | |
|---|---|---|---|---|
| | | ARIMA | ANN | Proposed approach |
| Winter | MAPE | 24.699 | 3.019 | **2.365** |
| | MAE | 0.344 | 0.042 | **0.033** |
| | RMSE | 0.434 | 0.066 | **0.048** |
| Summer | MAPE | 24.585 | 3.960 | **1.454** |
| | MAE | 0.315 | 0.051 | **0.019** |
| | RMSE | 0.404 | 0.076 | **0.029** |
| Rainy | MAPE | 24.429 | **7.488** | 29.180 |
| | MAE | 0.315 | **0.056** | 0.217 |
| | RMSE | 0.407 | **0.097** | 0.341 |
| Autumn | MAPE | 26.326 | 7.847 | **3.798** |
| | MAE | 0.328 | 0.098 | **0.047** |
| | RMSE | 0.468 | 0.166 | **0.072** |

air, and random temperature variation. Also during rainy season, the solar irritation is highly variable due to clouds. In December, forecasted result is comparable with the summer season.

Along with MAPE, the MAE and RMSE have also been calculated to evaluate the accuracy of the proposed model. Table 4 presents the summarized MAPE, RMSE, and MAE error results of all the four seasons for the ARIMA, ANN, and proposed PSO tuned FFNN model for week ahead solar PV power forecasting. All the three-season errors are very good for the proposed model for each season except rainy season where the ANN has performed better. However, the summer season (sunny days) shows better solar PV power forecasting result than other seasons as the error is very less in this period. The forecasting results indicate that all four season results are good and able to perform PV forecasting with an acceptable error, except rainy season.

## 7 Conclusion

Although renewable power, especially solar PV power generation has become a reality of today, its integration to the grid is a challenge which can be tackled by better forecasting of solar PV power generation in advance. This paper presents the development of a new hybrid model combining PSO and feed-forward neural network (FFNN) for prediction of solar photovoltaic (PV) power generation in the Indian electricity market. The developed model is applied for one-week ahead solar PV forecasting using the data of the year 2014 from Kolkata, India. The results are presented for four different seasons namely winter, summer, rainy, and autumn. The results show that the developed model is able to make week ahead forecast of solar PV power with minimum possible error as evident from MAPE, RMSE, and MAE results. The MAPE for winter, summer, rainy, and autumn season's are 2.36, 1.45, 29.17, and 3.79%, respectively. All error parameters are best for the summer season and poorest for the rainy season. The proposed hybrid FFNN-PSO model emerges as a right tool for forecasting of solar PV power generation which can be further applied to other renewable energy markets.

## References

1. Yadav, H.K., Pal, Y., Tripathi, M.M.: Photovoltaic power forecasting methods in smart power grid. In: Annual IEEE India Conference (INDICON), pp. 1–6 (2015)
2. Lauret, P., Voyant, C., Soubdhan, T., David, M., Poggi, P.: A benchmarking of machine learning techniques for solar radiation forecasting in an insular context. Sol. Energy **112**, 446–457 (2015)
3. Zhang, P., Takano, H., Murastra, J.: Daily solar radiation prediction based on wavelet analysis. In: Proceedings of SICE Annual Conference, Tokyo, Japan, pp. 13–18 (2011)
4. Lyu, L., Kantardzic, M., Arabmakki, E.: Solar irradiance forecasting by using wavelet based denoising. In: IEEE Conference, 978-1-4799-4509-2/14/$31.00 ©2014 IEEE
5. Tascikaraoglu, A., Sanandaji, B.M., Chicco, G., Cocina, V., Spertino, F., Erdinc, O., Paterakis, N.G., Catalao, J.P.S.: Compressive spatio-temporal forecasting of meteorological quantities and photovoltaic power. IEEE Trans. Sustain. Energy **7**(3), 1293–1304 (2016)
6. Ashraf, I., Chandra, A.: Artificial neural network based models for forecasting electricity generation of grid connected solar PV power plant. Int. J. Global Energy Issues **21**(1/2), 119–130 (2004)
7. Tripathi, M.M., Singh, S.N., Upadhyay, K.G.: Price forecasting in competitive electricity markets: an analysis. In: Proceedings of International Conference on Energy Engineering (ICEE-2009), Puducherry, India, 7–9 January 2009, paper no. EEE4214
8. Upadhyay, K.G., Tripathi, M.M., Singh, S.N.: An approach to short term load forecasting using market price signal. In: International Conference on Distribution (CIRED 2007), Vienna, Austria, May 2007, paper 0487, pp. 21–24
9. Tanaka, K., Uchida, K., Ogimi, K., Goya, T., Yona, A., Senjyu, T., Funabashi, T., Hwan Kim, C.: Optimal operation by controllable loads based on smart grid topology considering insolation forecasted error. IEEE Trans. Smart Grid **2**(3), 438–444 (2011)
10. Mellit, A., Pavan, A.M.: A 24-h forecast of solar irradiance using artificial neural network: application for performance prediction of a grid-connected PV plant at Trieste, Italy. Solar Energy **84**, 807–821 (2010)
11. Yang, H.T., Huang, C.M., Huang, Y.C., Pai, Y.-S.: A weather based hybrid method for 1- day ahead hourly forecasting of PV power output. IEEE Trans. Sustain. Energy **5**(3), 917–926 (2014)
12. Saini, L.M., Aggarwal, S.K., Kumar, A.: Parameter optimisation using genetic algorithm for support vector machine-based price-forecasting model in national electricity market. IET Gener. Transm. Distrib. **4**(1), 36–49 (2010)
13. Shi, J., Lee, W.J., Liu, Y., Yang, Y., Wang, P.: Forecasting power output of photovoltaic system based on weather classification and support vector machine. IEEE Trans. Ind. Appl. **48**(3), 1064–1069 (2012)
14. Baharin, K.A., Rahman, H.A., Hassan, M.Y., Gan, C.K.: Short-term forecasting of solar photovoltaic output power for tropical climate using ground-based measurement data. J. Renew. Sustain. Energy **8**, 053701-1–053701-19 (2016)
15. Catalao, J.P.S., Mariano, S.J.P.S., Mendes, V.M.F., Ferreira, L.A.F.M.: Short-term electricity prices forecasting in a competitive market: a neural network approach. Electric Power Syst. Res. **77**(10), 1297–1304 (2007)

16. Yona, A., Senjyu, T., Funabashi, T., Kim, C.H.: Determination method of insolation prediction with fuzzy and applying neural network for long term ahead PV power output correction. IEEE Trans. Sustain. Energy **4**(2), 527–533 (2013)
17. Catalao, J.P.S., Pousinho, H.M.I., Mendes, V.M.F.: Hybrid wavelet-PSO-ANFIS approach for short-term electricity prices forecasting. IEEE Trans. Power Syst. **26**(1), 137–144 (2010)
18. Bizzarri, F., Bongiorno, M., Brambilla, A., Gruosso, G., Gajani, G.S.: Model of photovoltaic power plants for performance analysis and production forecast. IEEE Trans. Sustain. Energy **4**(2), 278–285 (2013)
19. Zhifeng, Z., Jianjun, T., Tianjin, Z., Linlin, Z.: PV power short-term forecasting model based on the data gathered from monitoring network. China Commun. Sig. Process. (2) (2014)
20. Rukhaiyar, S., Alam, M.N., Samadhiya, N.K.: A PSO-ANN hybrid model for predicting factor of safety of slope. Int. J. Geotech. Eng. https://doi.org/10.1080/19386362.2017.1305652
21. Li, L.L., Cheng, P., Lin, H.C., Dong, H.: Short-term output power forecasting of photovoltaic systems based on the deep belief net. Adv. Mech. Eng. **9**(9), 1–13 (2017)
22. Kennedy, J., Eberhart, R.: Particle swarm optimization. In: Proceedings of IEEE International Conference of Neural Networks, vol. 4, pp. 1942–1948 (1995)
23. Heo, J.S., Lee, K.Y., Garduno-Ramirez, R.: Multiobjective control of power plants using particle swarm optimization techniques. IEEE Trans. Energy Convers. **21**(2), 552–561 (2006)
24. Shahidehpour, M., Yamin, H., Li, Z.: Market Operations in Electric Power Systems: Forecasting, Scheduling and Risk Management. Wiley, New York (2002)
25. Pfenninger, S., Staffell, I.: Long-term patterns of European PV output using 30 years of validated hourly reanalysis and satellite data. Energy **114**, 1251–1265 (2016). https://doi.org/10.1016/j.energy.2016.08.060

# Exploring the Effects of Sybil Attack on Pure Ad Hoc Deployment of VANET

**Nishtha and Manu Sood**

**Abstract** A type of ad hoc network formed among moving vehicles that come in one another's radio transmission range is called Vehicular Ad Hoc Network (VANET). VANET may be deployed using three architectures: pure ad hoc, WLAN and hybrid. Vehicle to Vehicle (V2V) and Vehicle to Infrastructure (V2I) are the two modes in VANET used for communication among vehicles. These features not only distinguish a VANET from other ad hoc networks but also make these networks more exposed to attacks and increase their complexity. These networks being the primary mechanism for communication in VANETs, appropriate and timely delivery of information is of prime importance. Due to the existence of various vehicular traffic scenarios, a single category of routing protocols is not sufficient for the VANETs. Therefore, various categories of routing protocols have been tailored to meet specific kinds of routing requirements in this framework. Most significant ones are ad hoc/topology-based, position-based, geocast-based, cluster-based, broadcast-based protocols. There exist a number of attacks that apart from affecting various other parameters also affect the routing protocols in these VANETs. Most of these attacks may be launched in all the above-mentioned three architectures. But, one of the most dangerous attacks is the Sybil attack that may be initiated in pure ad hoc deployment of VANET where vehicles communicate with one another in one to one manner using carry forward approach. In this paper, with the help of illustrative example for each of the five categories of routing protocols, we show how Sybil attack affects these protocols. Considering the ad hoc scenarios in VANETs, we also discuss a prevention mechanism for the Sybil attack briefly.

**Keywords** Vehicular ad hoc network (VANET) · Global positioning system (GPS) · Vehicle to vehicle (V2V) communication · Vehicle to infrastructure (V2I) communication · Pure ad hoc architecture

Nishtha (✉) · M. Sood
Department of Computer Science, Himachal Pradesh University, Shimla, India
e-mail: nishtha16@yahoo.com

M. Sood
e-mail: soodm_67@yahoo.com



P. K. Singh et al. (eds.), *Proceedings of ICRIC 2019*, Lecture Notes in Electrical Engineering 597, https://doi.org/10.1007/978-3-030-29407-6_26

## 1 Introduction

Vehicular ad hoc network (VANET) is a wireless communication network that is created among vehicles moving on the road for the purpose of communications [1]. One of the main reasons behind the deployment of VANET is to share real-time traffic information while some other reasons being to make entertainment-based and user-related services such as internet access, e-commerce, business-related services, etc. accessible to vehicles involved in intra-vehicular communication (IVC) [2, 3]. Every vehicle participating in VANET is equipped with On-Board Unit (OBU) that contains processors, memory, Global Positioning System (GPS) unit, sensors as well as user interfaces for connection establishment with other vehicles in the VANET [1, 4]. Two modes of communication in VANET are (a) Vehicle to Vehicle (V2V) communication in which vehicles communicate in a peer to peer manner with each other, and (b) Vehicle to Infrastructure (V2I) communication where vehicles communicate through Road Side Units (RSUs) [1, 3–7]. There exist three different architectures for deploying VANETs. One of these is the pure ad hoc architecture in which vehicles communicate only with one another in V2V mode. In this architecture, a packet forwarded from the source vehicle to the destination vehicle moves using multi-hop approach and each in-between vehicle also acts as a router. The second type of architecture that can be deployed is pure cellular in which the RSU behaves as an access point by forming wired connections with other RSUs and wireless connection with vehicles. In this architecture, RSU acts as a centralized infrastructure for a group of vehicles in the specified region. The third type is the hybrid architecture in which both ad hoc and pure cellular architectures simultaneously exists [8].

Two aspects of VANET that significantly distinguish these networks from other types of ad hoc networks are the architecture and broadcasting of beacons. Vehicles in VANET that are one hop apart periodically transmit beacons among each other. These beacons are the packets containing information about the vehicle's ID as well as its velocity and every vehicle broadcasts these beacons at regular intervals of time. Other neighboring vehicles store this information in a tabular form [9]. Therefore, a VANET helps in increasing the capability and safety of any transportation system. But, security of VANET is of utmost importance as many attacks exist to degrade its performance or sometimes totally demolishes this network [10]. Among these, one of the most hazardous attacks is Sybil attack. In Sybil attack, by forging or creating fake identities, a malicious vehicle creates an illusion that a few extra vehicles are present in the network although these extra vehicles are virtual, they do not exist in reality. The malicious vehicle is also known as Sybil attacker and the virtual vehicle(s) created by it through fake identities are known as Sybil vehicle(s) [3, 5, 10–12]. The presence of these fake vehicles affect real traffic scenario and also affect voting, misbehavior detection, fair resource allocation, data aggregation and routing in VANET [5, 13, 14]. Sybil attack can effectively disrupt the functioning of the network only in the peer to peer networks. The existence of centralized infrastructure makes Sybil attack easily detectable [5, 13, 14]. Out of the three architectures, in pure cellular and hybrid architectures where a centralized infrastructure in the form of an RSU is always present, Sybil attack is easily detectable. Therefore, the likelihood of Sybil attack in pure ad hoc architecture is the highest where no centralized infrastructure is present and all vehicles communicate in a peer to peer manner with one another.

In order to utilize the full functionality of VANET, the presence of RSU is compulsory. The RSUs are deployed in most of the scenarios such as cities, urban areas, highways, etc. But in spite of this, the specific situations in which the RSUs are not deployed are (a) Interior parts of a country, (b) Hostile/Disputed locations, (c) War-field/Enemy-area towards which group of army vehicles are moving, and (d) Services from RSU are temporarily withdrawn due to natural calamities or some other reason.

There are few situations where vehicles communicate with one another in pure ad hoc manner and only these are the particular situations where a malicious vehicle by entering in VANET may launch a Sybil attack. But, real-time traffic information sharing is the prerequisite in VANET and that too with minimum or zero time delay. In order to meet the communication requirements of three different architectures present in VANET, diverse categories of routing protocols have been framed. These categories include ad hoc/topology-based, position-based, geocast-based, cluster-based and broadcast-based routing protocols [8, 15–18].

Vehicles in VANET are recognized by their Internet Protocol (IP) addresses and Media Access Control (MAC) addresses and a malicious vehicle may easily spoof these addresses [12]. Majority of the routing protocols such as position-based, geocast-based protocols, etc. in VANET utilize GPS information provided by OBU for making routing-related judgments [16]. But this GPS information is also liable to be spoofed or jammed by a malicious vehicle [19]. Therefore, because of the vehicle's ID spoofing, spoofing and jamming of GPS signals, a high probability of Sybil attack in pure ad hoc VANETs always exist.

The rest of the paper is organized as follows. In Sect. 2 of this paper, the impact of the Sybil attack on one of the significant ad hoc/topology-based protocols, i.e. Dynamic Source Routing (DSR) protocol is discussed with the help of an illustrative diagram. Section 3 presents the influence of Sybil attack on one of the significant position-based protocols, i.e. Anchor-based Street and Traffic-Aware Routing (A-Star) protocol with the help of another illustrative diagram. Section 4 highlights impact of Sybil attack on one of the significant geocast-based protocols, i.e. Mobicast supported with another illustrative diagram Sect. 5 depicts the effects of the Sybil attack on one of the significant cluster-based protocol, i.e. Cluster-Based Routing (CBR) protocol and on one of the significant broadcast-based protocol, i.e. Density-aware reliable broadcasting protocol (DECA) in Sect. 6 again with the help of appropriate illustrations. Further, a mechanism for prevention of Sybil attack is discussed in Sect. 7, followed by the conclusion in Sect. 8.

# 2 Ad Hoc/Topology-Based Routing Protocols

These routing protocols are specially deployed for pure ad hoc networks. In these protocols packet from a source to destination is forwarded using the carry-forwarding approach in a multi-hop manner. In these protocols, the next intermediate vehicle used to forward packet is chosen on the basis of network topology information [8, 15–18].

## *2.1 Dynamic Source Routing (DSR) Protocol*

DSR protocol comes under reactive routing category of protocols in which the route is established only when it is required. DSR does not use periodic beacons but dynamically determine the route on the basis of cached information. This cached information is accumulated by each vehicle. This information is regarding source routes that every vehicle has learned in the network. When a packet from a source vehicle is to be forwarded to a destination which is not in the transmission range of the source vehicle then the packet is forwarded using multi-hopping. In place of using a routing table in order to establish a routing path, DSR uses dynamic source routing. This is a routing technique in which a source vehicle that needs to forward packets to the destination vehicle establishes the entire route depending on the prior information it has stored and accordingly set up a route up to the destination. While forwarding the packet, the source vehicle encapsulates this information in the packet's header. Therefore, whenever a source vehicle has to forward a packet to the destination, it searches in the route cache for a route to the destination. If a route is present in the route cache then this route is established to forward the packets, otherwise, the route is searched using route discovery protocol. In route discovery, the source vehicle broadcasts a route request (RREQ) packets to all vehicles in its transmission range and on discovering a route from an intermediate vehicle the source vehicle receives back a route reply (RREP) packets. These intermediate vehicles present in the network again rebroadcast the RREQ packets. In this way, the route request broadcasted in the network ultimately reaches the destination vehicle and the destination vehicle finally replies back to the source vehicle by sending the RREP packet. An intermediate vehicle may utilize its own route cache and may not employ route cache information it has received from another vehicle. In a multi-hop forwarding while transmitting packets from an intermediate vehicle to another vehicle the route maintenance mechanism checks that every hop in the specified path is working at that time or not. In case a transmission error occurs, then the intermediate vehicle sends back an error packet containing addresses of both the intermediate vehicles between which the error occurred. The source vehicle on receiving the error packets removes the intermediate vehicle for which the error occurred. It also destroys all paths involving the said vehicle from its source cache [15, 16, 20].

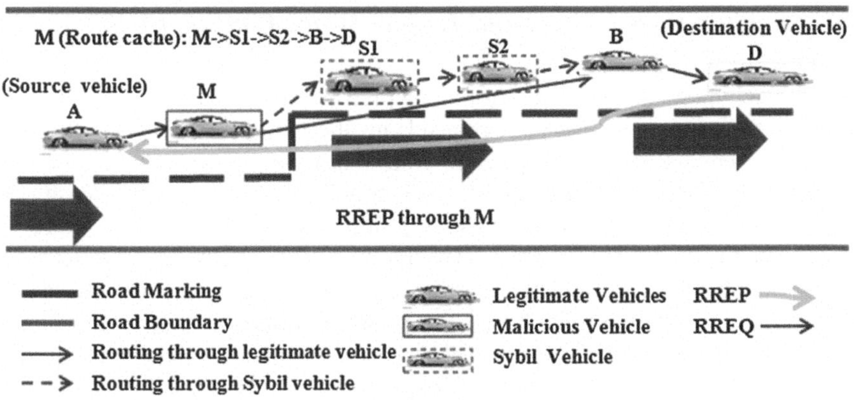


**Fig. 1** Sybil attack on dynamic source routing (DSR) protocol

Consider a VANET topology with six vehicles where a source vehicle *A* needs to forward a packet to destination vehicle *D* as shown in Fig. 1. Using the DSR protocol, in order to establish a path among the source and the destination, vehicle *A* searches its route cache. If there does not exist such a route in its route cache, then the specific path is discovered using the route discovery protocol. Suppose vehicle *M* is a malicious vehicle that has managed to enter in the network and has launched two Sybil vehicles: *S*1 and *S*2. Suppose in the route cache of vehicle *A* no route to vehicle *D* exists. Therefore, the route is discovered using the route discovery mechanism. Hence, the source vehicle *A* broadcasts a route request (RREQ) packet. Thus, the route request (RREQ) packet reaches the malicious vehicle *M* that has a path to the destination vehicle *D*. As a result, the path discovered from *A* to *D* is through *M* and is $A \rightarrow M \rightarrow S1 \rightarrow S2 \rightarrow B \rightarrow D$. Therefore, the RREP packet is sent back to the source vehicle *A*. But, in the meantime, the malicious vehicle destroys the Sybil vehicle *S*2. As a result, an error packet is sent back to the source vehicle *A*. Consequently, vehicle *A* destroys all paths that involve vehicle *S*2 from its route cache. In this way, the presence of Sybil vehicle disrupts the DSR protocol by increasing routing overhead, communication delay as well as over utilizes network resources.

## 3 Position-Based Routing Protocols

These are one of the most significant protocols in VANET. These protocols do not require routing tables and GPS information is used to locate vehicles. The GPS information is further utilized for making forwarding decisions between source and destination [8, 15–18].

### 3.1 Anchor-Based Street and Traffic-Aware Routing (A-Star)

One of the important position-based routing protocols is A-Star that is specifically designed for urban areas. The basic principle involved in this protocol is to forward packet through specific anchor paths in cities that have the maximum number of vehicles. Each road is being assigned a weight and roads with a sparse density of vehicles are provided with more weights and vice versa. Therefore, in this protocol, those anchor paths are preferred that are having minimum weight and as a result have the maximum number of vehicles. In order to analyze which of the road has maximum vehicles at any instance of time, this protocol utilizes bus route information. Therefore, to ensure packet delivery roads that have maximum vehicles are preferred as anchor paths. Each vehicle is provided with local street map information, and anchor paths are calculated by means of Dijkstra's least weight algorithm. Apart from using anchor paths, this protocol works on greedy forwarding strategy [8, 21].

Consider a VANET topology with nineteen vehicles in which intermediate vehicle *B* receives a packet from source vehicle that is to be forwarded to destination vehicle *D* as shown in Fig. 2. Using A-Star routing protocol, the packet from *B* is to be forwarded to a vehicle by calculating the anchor paths among roads {*R*1, *R*2, *R*3, *R*4}. Assume that in road *R*3, a malicious vehicle *M* has managed to enter and has placed a Sybil vehicle *S* in the network. Using the A-Star protocol, minimum weight

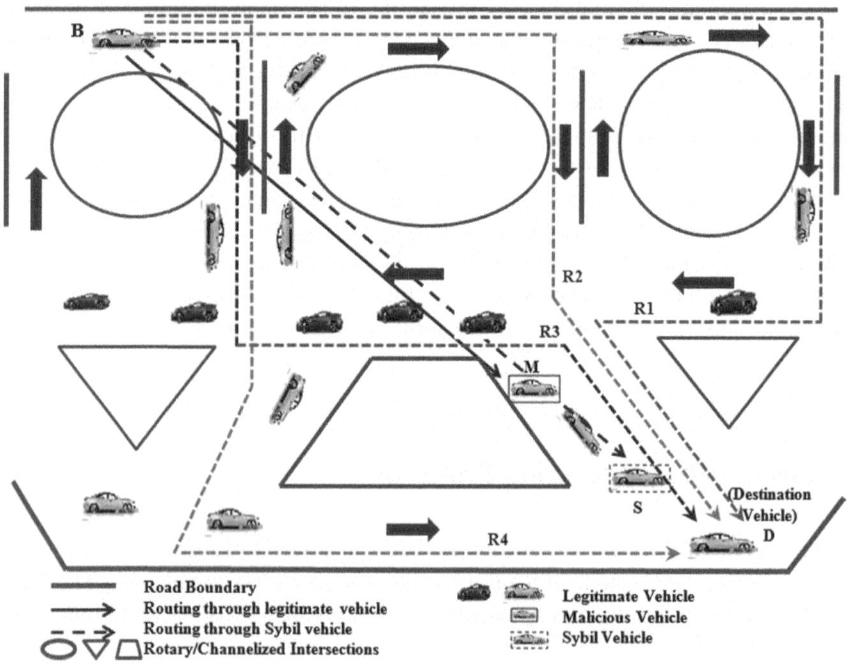


**Fig. 2** Sybil attack on anchor-based street and traffic-aware routing (A-Star)

is with road *R*3. As a result, road *R*3 is selected as it is having the maximum number of vehicles and is also the shortest to reach the destination vehicle *D*. Therefore, the packet is forwarded along the road *R*3 to vehicle *S* because it is the next hop of vehicle *B* and is nearest to destination among all neighbors of *B*. But, in fact, the packet is forwarded to the vehicle *M*, the physical vehicle of Sybil vehicle *S*. It shows that A-Star protocol is affected by the placement of a Sybil vehicle in the network.

## 4 Geocast-Based Routing Protocols

These protocols are used in situations where a message is to be delivered to all the vehicles moving in a specified geographic area known as Zone of Relevance (ZOR) [8, 15–18].

### *4.1 Mobicast*

One of the important protocols in the geocast-based category of protocols is mobicast. This protocol helps in prevention of temporal network fragmentation problem. Consequently, mobicast messages are distributed effectively to each and every vehicle in the specified Zone of Relevance (ZOR) within a suitable time. Unlike other geocast protocols mobicast considers time as an important issue. In mobicast, whenever an event occurs the first vehicle that notifies this event is considered responsible to broadcast this event-related message to all other vehicles in the ZOR. This is achieved in this protocol by dynamically estimating the exact ZOF (zone of forwarding). ZORt is divided into four quarters with the center of the region to be the same as the location of the vehicle that is held accountable for the broadcasting of event-based message. To accomplish this, mobicast protocol defines ZOAt (zone of approaching), ZORt (zone of relevance) and ZOFt (zone of forwarding). The protocol assumes that irrespective of time every vehicle disseminate the message to other vehicles even if these vehicles are not moving in the particular ZOR [22].

Consider a VANET topology with eighteen vehicles. Assume that vehicle *M* is a malicious vehicle that has managed to enter in the network and has launched a Sybil vehicle *S*. Let vehicle *B* met with an accident on the highway as shown in Fig. 3. The first vehicle that reveals this event is vehicle *S* moving behind vehicle *B*. The dotted circle outside *S* denotes its communication range. Using this protocol, with vehicle *S* as its origin, its communication range is divided into four quadrants: ZOR1, ZOR2, ZOR3, and ZOR4. Using mobicast protocol, vehicle *S* becomes responsible for broadcasting the event-related messages in the network. But in reality, the physical location of Sybil vehicle *S* is the vehicle *M*. As a result, being a Sybil vehicle *S* may broadcast fake messages to disturb the traffic in VANET. In this protocol, if the Sybil vehicle launched by a malicious vehicle is not selected as a vehicle responsible to

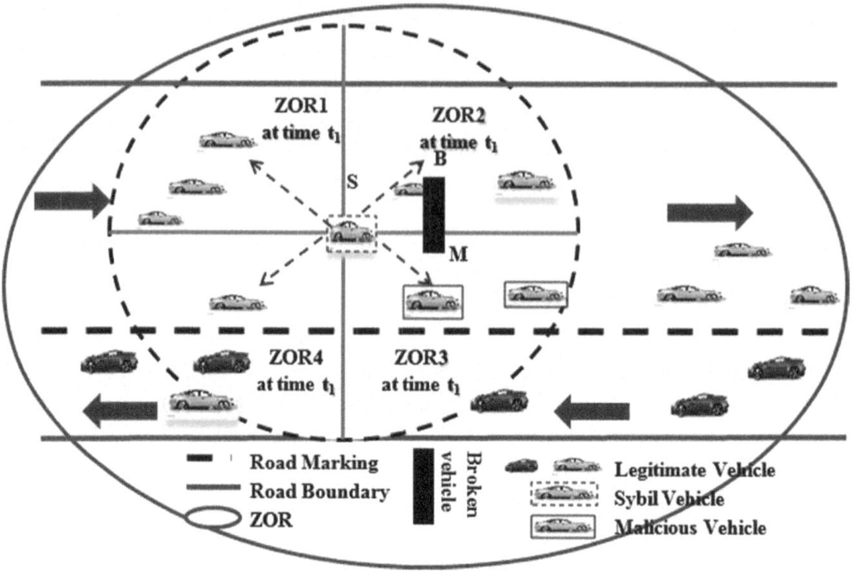


**Fig. 3** Sybil attack on mobicast [22]

rebroadcast the message, even then the addition of Sybil vehicle(s) results in extra utilization of network resources, consequently disrupting the mobicast protocol.

## 5 Cluster-Based Routing Protocol

Cluster-based routing protocols are characterized by the formation of a number of clusters with one member in each cluster elected as a cluster head (CH). The cluster head selection in these routing protocols is done using different strategies and it is accountable for all management related functions in the cluster [8, 15–18].

### *5.1 Cluster-Based Routing (CBR) Protocol*

In this protocol, the geographical area is bifurcated into segments of equal size known as grids. Each grid is similar to a cluster having a cluster header and a unique ID. Maps are used to divide the geographic area into grids. The data packet is forwarded through these grids by the cluster header. As we are considering V2V routing only, therefore, we assume that RSU is not placed in this geographic area. Even if an RSU is there we assume that the RSU is not in the working state. In this protocol, the cluster header is selected on the basis of its location. Hence, a vehicle that is nearer to the center of a grid is selected as a cluster header [23].

The cluster header selection is made by broadcasting LEAD, APPLY, and LEAVE messages. The LEAD message is broadcasted by the cluster header, APPLY message is broadcasted by any vehicle to become a cluster header and LEAVE message is transmitted by the cluster header when it is about to leave the grid. When a vehicle in the grid does not receive a LEAD message from the cluster header within a specific time interval it then broadcasts a APPLY message specifying coordinates of its grid and its own position. If the cluster header receives this APPLY message then it sends a reply with a LEAD message to this vehicle. If any other vehicle that is not the cluster header receives APPLY message from any other vehicle, then this vehicle checks that if it is nearer to the grid center. If this vehicle is nearer to the grid then it sends an APPLY message specifying its grid coordinates and its own position to the other vehicle. But, if a vehicle present in the grid for a specific amount of time does not receive a LEAD or an APPLY message then this vehicle itself becomes the cluster header. When a cluster header is about to leave the grid it broadcasts a LEAVE message and then accordingly the new cluster header is selected. Each cluster header in a grid maintains a cluster header table that keeps its own location, the location of all vehicles present in the grid as well as the location of all grids existing in its neighborhood. This protocol in order to forward packets from source to destination uses optimal neighbor cluster header strategy. As the forwarded packets contain destination location, the header in the grid always identifies where the destination is located. Therefore, whenever a packet that is to be forwarded to a specified destination comes in a grid, the cluster header in the grid consults its cluster header table to find its neighbor grids. As the cluster header always makes out the direction where the destination exists. With this direction, it compares the angle formed by this grid center with each of its neighbors' grid center. It then selects the particular grid to forward the packet which forms the least angle with the direction to the destination. But, the cluster header forward the packet to a specific grid computed on the basis of optimal neighbor cluster header selection. Using this strategy, the packet is only forwarded if the grid header in the neighboring grid has a higher priority than the grid header from where the packet is to be forwarded [16, 17, 23].

Consider a VANET topology with nine vehicles arranged in VANET. The geographic area is divided into grids $G1$, $G2$, $G3$, $G4$, $G5$, and $G6$. The grid $G1 = \{A, B\}$ is having two vehicles, $G3 = \{G\}$ having one vehicle, $G5 = \{C, M, S, E\}$ four vehicles, $G6 = \{D, F\}$ two vehicles and Grid $G2$ and $G4$ are empty grids without any vehicle at a particular instance of time as shown in Fig. 4. Let us assume that the source vehicle $A$ present in grid $G1$ needs to send a packet to destination vehicle $D$ in grid $G6$. Let a malicious vehicle $M$ enters the grid $G4$ and is able to produce one Sybil vehicles $S$. Suppose malicious vehicle $M$ is able to place $S$ as a center header in $G4$. Using CBR protocol, the packet from $A$ is to be forwarded to grid $G4$ as it forms the least angle with the direction towards the destination. But in fact, the packet is forwarded to vehicle $M$, the physical location of $S$ thereby, disturbing the CBR protocol.

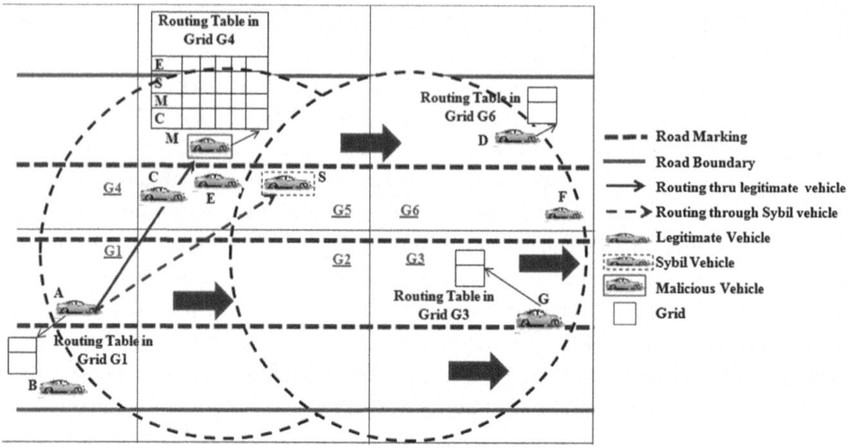


**Fig. 4** Sybil attack on cluster-based routing (CBR) protocol

## 6 Broadcast-Based Routing Protocols

Single-hop messages called beacons in VANET are the main criterion on which these protocols make their routing decisions. These beacon messages are flooded periodically in the network to acquire information regarding neighboring vehicles that are at a distance of one-hop [8, 15–18].

### *6.1 Density-Aware Reliable Broadcasting Protocol (DECA)*

One of the routing protocols in the broadcast-based category is Density-Aware Reliable Broadcasting (DECA). It is tailored for dense traffic areas such as urban areas and also for sparse areas such as highways. This protocol does not utilize GPS information but only employ information gathered through beacon from one-hop neighbors. In this protocol whenever a vehicle needs to broadcast a message among all the one-hop neighbors it searches for a neighbor vehicle that has the highest density of local information in order to further rebroadcast the messages. As a result, this vehicle is made responsible for rebroadcasting the message. On receiving a broadcast message every vehicle checks whether it is the selected vehicle to rebroadcast the messages. If a vehicle is the selected vehicle for rebroadcasting then this vehicle broadcasts the message otherwise it stores the message. If vehicles do not receive a rebroadcast message from a selected vehicle within a specified time, all these vehicles start rebroadcasting the message. With the help of periodic beacons, vehicles identify their neighbors. Each vehicle further identifies which of its neighboring vehicles have not got the message and it then again rebroadcast the message for those neighbor vehicles [15, 18, 24].

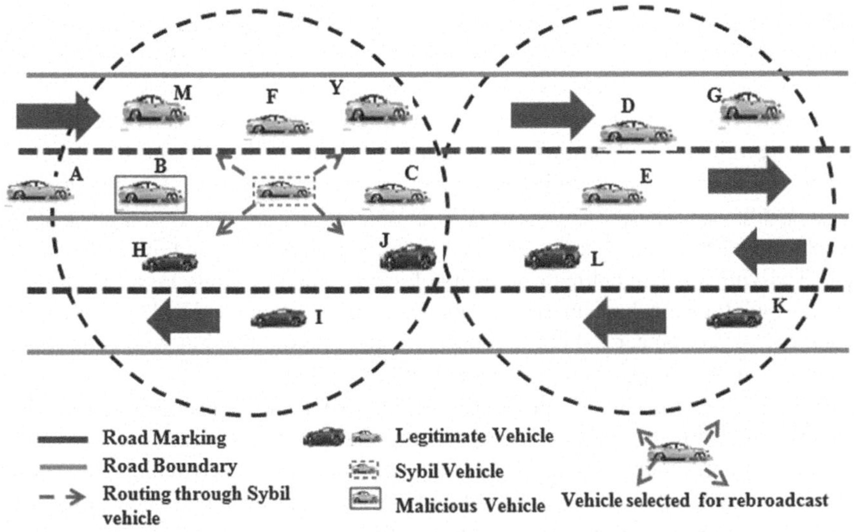


**Fig. 5** Sybil attack on density-aware reliable broadcasting (DECA) protocol

Consider a VANET topology with fifteen vehicles. In this topology, ten vehicles $\{A, B, C, F, H, I, J, M, S, Y\}$ are at a distance of one hop and five vehicles $\{D, E, G, L, K\}$ are also one hop apart with one another at an instance of time. Let a malicious vehicle *M* enters in the VANET and creates a Sybil vehicle *S* as represented in Fig. 5. This protocol does not consider GPS information and only relies on periodic beacon messages. Vehicle *M* and *S* make their presence recognized by sending periodic beacons. Let a vehicle *A* needs to broadcast a message. Thus, it selects a vehicle among all its one-hop neighbor vehicles having the highest density of local information. *S* is a vehicle that has the highest density of local information and is selected as a vehicle to rebroadcast the message. But, *S* being a Sybil vehicle it may broadcast fake messages in the network. Hence, this will disrupt DECA routing protocol with the dissemination of fake messages in the network. The situation will become worst if the malicious vehicle is able to produce more and more Sybil vehicles in the network, therefore, it becomes impossible for legitimate vehicles to analyze the validity of messages. As a result, all the legitimate vehicles in the network are under the influence of fake beacons transmitted by Sybil vehicle(s) consequently, disrupting the DECA protocol.

## 7 A Mechanism for Prevention of Sybil Attack in Pure Ad Hoc VANET

It has been discussed above that all categories of routing protocols designed for pure ad hoc VANET are prone to Sybil attack. We have analyzed various scenarios where

VANETs operate in an ad hoc manner using V2V communication among the moving vehicles. Most of these scenarios can be interior parts of a country, some hostile areas, disputed locations, war-fields, situations arising from natural calamities, etc. In all these situations, especially hostile areas, war-fields are the scenarios where security of moving vehicles is of utmost importance since the soldiers move in a troop of vehicles so as to cooperate with one another in a number of ways. All categories of routing protocols designed for pure ad hoc VANETs are defenseless against Sybil attack. Therefore, due to vulnerabilities present in VANET, enemies may easily launch a Sybil attack, making the security of these vehicles a major issue. Therefore, we here look for a mechanism to fight this challenge. As explained by Douceur, Sybil attack may easily be carried out in a peer-to-peer environment but with centralized management, we can easily detect such an attack [14].

Thus, the placement of a centralized authority may lessen or prevent such attacks. To overcome Sybil attack one way could be the placement of dedicated vehicles after a number of legitimate vehicles in the troop [25]. Each dedicated vehicle present in VANET is responsible for observation and management of neighbor vehicles that fall in its radio range in the same manner as the cluster head controls its members in a cluster. All these dedicated vehicles in the group collaborate with one another for data sharing and management related information. Therefore, at any instance of time, each of the dedicated vehicles present in the network has information about entire VANET.

Consider a troop of fifteen vehicles moving toward a war-field as shown in Fig. 6. As depicted, three of the vehicles *A*, *B* and *C* are dedicated vehicles having their respective radio ranges specified as a cluster. The only function of these dedicated

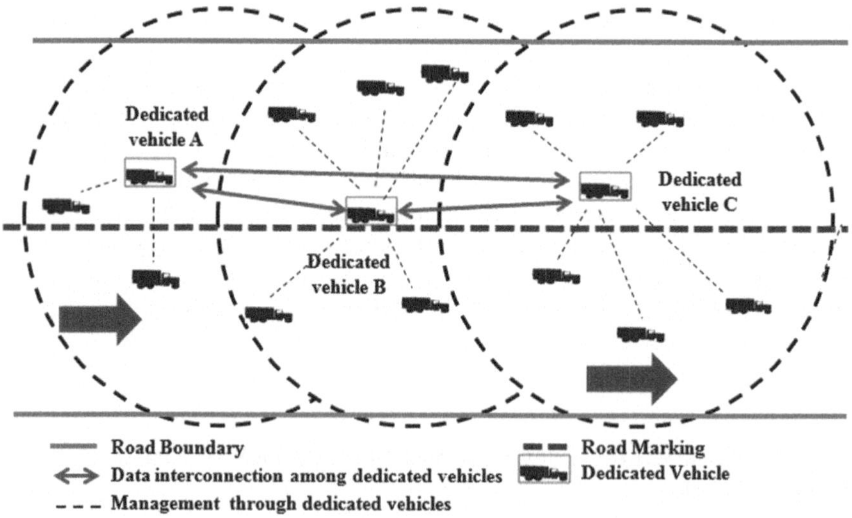


**Fig. 6** Prevention of Sybil attack by placing dedicated vehicles to observe and manage its neighboring vehicles

vehicles is to control, monitor and store management related information of vehicles falling in their clusters. Trust establishment among vehicles in VANET is one of the most promising techniques for detection of Sybil attack. In this technique, the reliability of vehicle is computed on the basis of its behavior with other neighboring vehicles that are at a distance of one hop apart in the network. For example, (a) a vehicle that is working together with other vehicles in the network for message exchange will add to its trust value; whereas (b) a vehicle that adds or deletes packets, drops few or all packets, forward the same packets repeatedly, etc., its trust value is accordingly decreased. Therefore, depending on specified metrics each vehicle calculates the trust value of every vehicle in their cluster and forwards this information to its respective dedicated vehicle. A minimum trust value is declared by the dedicated vehicle. If the trust value of any vehicle is less than the specified minimum trust value then these vehicles are kept in the category of suspected malicious vehicle. Therefore, by computing the trust value for each vehicle, the dedicated vehicles $A$, $B$ and $C$ judges the malicious behavior of vehicle(s) present in their respective cluster. Consequently, the vehicles that are in the category of a malicious vehicle are forbidden to take part in the network [26]. Dedicated vehicles present in the network also interchange the data-related information with one another.

As a result, a single dedicated vehicle is capable of viewing a complete picture of VANET. Hence, these dedicated vehicles thus acting as a centralized infrastructure help in preventing Sybil attack in such scenarios at any instance of time.

## 8 Conclusion

In this paper, it has been shown that for pure ad hoc deployment of VANET, each protocol in every category of routing protocols is prone to Sybil attack. Sybil attack may disrupt these routing protocols through any of these actions: (a) non-delivery of forwarded packets, (b) forwarding wrong packets, (c) delay in forwarding packets, (d) disrupting the voting strategies such as cluster head selection in case of cluster-based routing protocols, (e) inability to detect misbehaving vehicle etc. This results in deteriorating the functionality and quality of service in VANET, making such networks even unsuitable for use at times.

One of the most appropriate methods for prevention of an attack in such a VANET is to design the routing protocols in a structured way that delivers packets to destination vehicle(s) in minimum time, have optimal efficiency for routing of packets and with minimum consumption of network resources. These protocols must include mechanisms for detection and prevention of malicious vehicle(s). Moreover, the GPS information used in most of the routing protocols in VANET must be secured by having an inbuilt mechanism to detect and overcome spoofing and jamming of GPS signals in each and every vehicle taking part in VANET. In this context, we have explored the idea of using a group of dedicated vehicles in preventing the attacks such as the Sybil attack and have suggested that the combination of dedicated vehicles in the manner shown can prevent the VANET from getting affected by the Sybil attack. One such scenario is when soldiers are either approaching the war-fields or moving towards hostile locations in groups of vehicles with deployment of ad hoc VANET for communication. Using dedicated vehicles and the trust value associated with each vehicle falling in their radio ranges, the malicious vehicle(s) can be identified before it can launch the Sybil attack. In our future work, we would like to explore the implementation of the proposed mechanism for the detection, removal, and prevention of the Sybil attack in such pure ad hoc VANETs.

## References


1. Singh, K.D., Rawat, P., Bonnin, J.: Cognitive radio for vehicular ad hoc networks (CR-VANETs): approaches and challenges. EURASIP J. Wirel. Commun. Netw. **49**, 1–22 (2014). https://doi.org/10.1186/1687-1499-2014-49
2. Durresi, M., Durresi, A., Baroll, L.: Emergency broadcast protocol for inter-vehicle communications. In: 11th International Conference on Parallel and Distributed Systems Fukuoka Japan (2005). https://doi.org/10.1109/icpads.2005.147
3. Jin, D,. Song, J.S.: A traffic flow theory aided physical measurement-based sybil nodes detection mechanism in vehicular ad-hoc network. In: IEEE/ACIS 13th International Conference on Computer and Information Science (ICIS), pp. 281–286 (2014). https://doi.org/10.1109/icis.2014.6912147
4. Liang, W., Li, Z., Zhang, H., Wang, S.: Bie R Vehicular ad hoc networks: architectures, research issues, challenges and trends. IJ Distrib. Sens. Netw. **11**(8), 1–11 (2015). https://doi.org/10.1155/2015/745303
5. Sharma, A.K., Saroj, S.K., Chauhan, S.K., Saini, S.K.: Sybil attack prevention and detection in vehicular ad-hoc network. In: International Conference on Computing, Communication and Automation Noida, pp. 594–599 (2016). https://doi.org/10.1109/ccaa.2016.7813790
6. Eze, E.C., Zhang, S.J., Liu, E.J., Eze, J.C.: Advances in vehicular ad-hoc networks (VANETs) challenges and road-map for future development. IJ Autom. Comput. **13**(1), 1–18 (2016). https://doi.org/10.1007/s11633-015-0913-y
7. Nassar, L., Jundi, A., Golestan, K., Sattar, F., Karray, F., Kamel, M., Boumaiza, S.: Vehicular ad-hoc networks(VANETs): capabilities, challenges in context-aware processing and communication gateway. In: international conference on autonomous and intelligent systems. In: Kamel, M., Karray, F., Hagras, H. (eds.) Autonomous and Intelligent Systems. AIS 2012. Lecture Notes in Computer Science, Springer, Berlin, Heidelberg, vol. 7326, pp. 42–49 (2012). https://doi.org/10.1007/978-3-642-31368-4_6
8. Li, F., Wang, Y.: Routing in vehicular ad hoc networks: a survey. IEEE Veh. Technol. Mag. **2**(2), 12–22 (2007). https://doi.org/10.1109/mvt.2007.912927)
9. Papadimitratos, P., Fortelle, A.D.L., Evenssen, K., Brignolo, R., Cosenza, S.: Vehicular communication systems: enabling technologies, applications, and future outlook on intelligent transportation. IEEE Commun. Mag. **47**(11), 84–95 (2009). https://doi.org/10.1109/mcom.2009.5307471
10. Al-Kahtan, M.S.: Survey on security attacks in vehicular ad hoc networks (VANETs). In: 6th International Conference on Signal Processing and Communication Systems. Gold Coast, QLD, pp. 1–9 (2012). https://doi.org/10.1109/icspcs.2012.6507953
11. Vasudeva, A., Sood, M.: Survey on sybil attack defense mechanisms in wireless ad hoc networks. J. Netw. Comput. Appl. **120**, 78–118 (2018). https://doi.org/10.1016/j.jnca.2018.07.006
12. La, V.H., Cavalli, A.: Security attacks and solutions in vehicular ad hoc networks: a survey. IJ Adhoc Netw. Syst. **4**(2), 1–20 (2014). https://doi.org/10.5121/ijans.2014.4201

13. Newsome, J., Shi, E., Song, D., Perrig, A.: The sybil attack in sensor networks: analysis and defenses. In: 3rd International Symposium on Information Processing in Sensor Networks. Berkeley, CA, USA. pp. 259–268 (2004). https://doi.org/10.1109/ipsn.2004.239019
14. Douceur, J.R.: The sybil attack. In: Druschel, P., Kaashoek, M.F., Rowstron, A.I.T. (eds.) First International Workshop on Peer-to-Peer Systems. Springer-Verlag, London, UK. pp. 251–260 (2002)
15. Vats, J., Tejpal, G.: Sharma S Routing protocols and attacks in vehicular ad hoc networks (VANETs): a review. IJ Adv. Res. Comput. Sci. Softw. Eng. **6**(12), 185–198 (2016)
16. Kakarla, J., Sathya, S.S., Laxmi, B.G., Babu, B.R.: A survey on routing protocols and its issues in VANET. IJ Comput. Appl. **28**(4), 38–44 (2011). https://doi.org/10.5120/3373-4663
17. Rehman, S., Khan, M.A., Zia, T., Zheng, L.: Vehicular ad hoc networks (VANETs)—an overview and challenges. J. Wirel. Netw. Commun. **3**(3), 29–38 (2013). https://doi.org/10.5923/j.jwnc.20130303.02
18. Nagaraj, U., Dhamal, P.: Broadcasting routing protocols in VANET. Netw. Compl. Syst. **1**(2), 13–19 (2011)
19. Mitch, R.H., Dougherty, R.C., Psiaki, M.L., Powell, S.P., O'Hanlon, B.W., Bhatti, J.A.: Humphreys TE signal characteristics of civil GPS jammers. In: 24th International Technical Meeting of the Satellite Division of the Institute of Navigation. Portland, OR, pp. 1907–1919 (2011)
20. Johnson, D.B., Maltz, D.A.: Dynamic source routing in ad hoc wireless networks. In: Imielinski, T., Korth, H. (eds.) Mobile Computing, The Kluwer International Series in Engineering and Computer Science. Springer, Boston, MA, vol. 353, pp. 153–181 (1996). https://doi.org/10.1007/978-0-585-29603-6_5
21. Seet, B.C., Liu, G., Lee, B., Foh, C., Wong, K., Lee, K.: A-star: a mobile ad hoc routing strategy for metropolis vehicular communications. In: Mitrou, N., Kontovasilis, K., Rouskas, G.N., Illiadis, I., Merakos, L. (e.ds) Networking 2004, LNCS, Springer, Berlin, Heidelberg, vol. 3042, pp. 989–999 (2004). https://doi.org/10.1007/978-3-540-24693-0_81
22. Chen, Y., Lin, Y., Lee, S.: A mobicast routing protocol for vehicular ad-hoc networks. ACM/Springer Mobile Netw. Appl. **15**(1), 20–35 (2010). https://doi.org/10.1007/s11036-009-0176-3
23. Luo, Y., Zhang, W., Hu, Y.: A new cluster based routing protocol for VANET. In: Second International Conference on Networks Security, Wireless Communications and Trusted Computing, pp. 176–180 (2010). https://doi.org/10.1109/nswctc.2010.48
24. Nakom, N.N., Rojviboonchai, K.: DECA: density-aware reliable broadcasting in vehicular ad hoc networks. In: The ECTI International Conference on Electrical Engineering/Electronics, Computer, Telecommunications and Information Technology. Chiang Mai, Thailand. IEEE Xplore, pp. 598–602 (2010)
25. Piro, C., Shields, C., Levine, B.N.: Detecting sybil attack in mobile ad hoc networks. In: Securecomm and Workshops, Baltimore, MD. IEEE Xplore, pp. 1–11 (2006). https://doi.org/10.1109/seccomw.2006.359558
26. Singh, R., Singh, J., Singh, R.: A novel sybil attack detection technique for wireless sensor networks. J. Adv. Comput. Sci. Technol. **10**, 185–202 (2017)

# Analysis and Design of WDM Optical OFDM System with Coherent Detection Using Different Channel Spacing

**Sakshi Sharma, Davinder Parkash and Sukhpreet Singh**

**Abstract** In this paper, WDM optical OFDM system model with coherent detection has been proposed for different channel spacing. Coherent optical OFDM system is mostly used in various applications because of its various advantages such as high spectral efficiency and flexibility. Four different frequency channels with same power are used and their effects on the output in term of Q factor, minimum BER is analyzed using optisystem software. The proposed OFDM system provide high Q-factor with minimum bit error rate (BER) making system more efficient. At a channel spacing of 100 GHz the system provides the best result.

**Keywords** Orthogonal frequency division multiplexing · Coherent detection · Bit error rate and quality factor

## 1 Introduction

With the rapid growth in the communication at a global level, transmission of data at higher rate is in demand these days. Fiber optics communication is a technology that is used to transmit signals like data, video or voice [1]. The ongoing communication systems are generally based on the transmission through the microwave, radio or the coaxial cables, etc. that has its own restrictions about the bandwidth and faces high amount of loss during the transmission that makes these traditional systems outdated for meeting the existing and future demands of the bandwidth. On the other hand Optical fiber has the provision of large bandwidth and has fewer losses for the transmission that make the optical fiber suitable for the future demands of audio, video, and combined data transmission. This is the reason optical fiber is taken as a

S. Sharma (✉) · D. Parkash · S. Singh
Department of Electronics and Communication, Chandigarh University, Gharuan, Mohali, Punjab, India
e-mail: sakshi.ss708@gmail.com

D. Parkash
e-mail: drdavinderchechi@gmail.com

S. Singh
e-mail: sukhpreet.ece@cumail.in



P. K. Singh et al. (eds.), *Proceedings of ICRIC 2019*, Lecture Notes in Electrical Engineering 597, https://doi.org/10.1007/978-3-030-29407-6_27

reliable way of communication media in the recent telecommunication engineering [2].

In a traditional communication system, the high-speed data transmission was provided to the user by using wired network structure, specifically optical fiber, whereas the mobile-based communication only supports the wireless network infrastructure. It has been seen that the wireless network faces several issues such as degradation in system performance [3].

Figure 1 shows the block diagram of the OFDM system. Orthogonal Frequency Division Multiplexing (OFDM) modulation technique uses multi-carrier with different subcarriers which are orthogonal to each. OFDM is a method of digital signal modulation in which a single data stream is split across several separate narrowband channels at different frequencies to reduce interference and crosstalk. Nowadays the OFDM is applied in several wireless applications. It is utilized in various 3G, 4G, WLAN standards; Digital Video Broadcasting (DVB) [4]. Coherent optical OFDM system is mostly used in various applications because of its various advantages such as high spectral efficiency and flexibility [5]. OFDM is also used to remove channel dispersion. It is widely used in systems because of its ease of signal processing due to less complexity of efficient algorithm of FFT/IFFT [6]. Both multiplexing and modulation are attained using inverse fast Fourier transform (IFFT) [6].

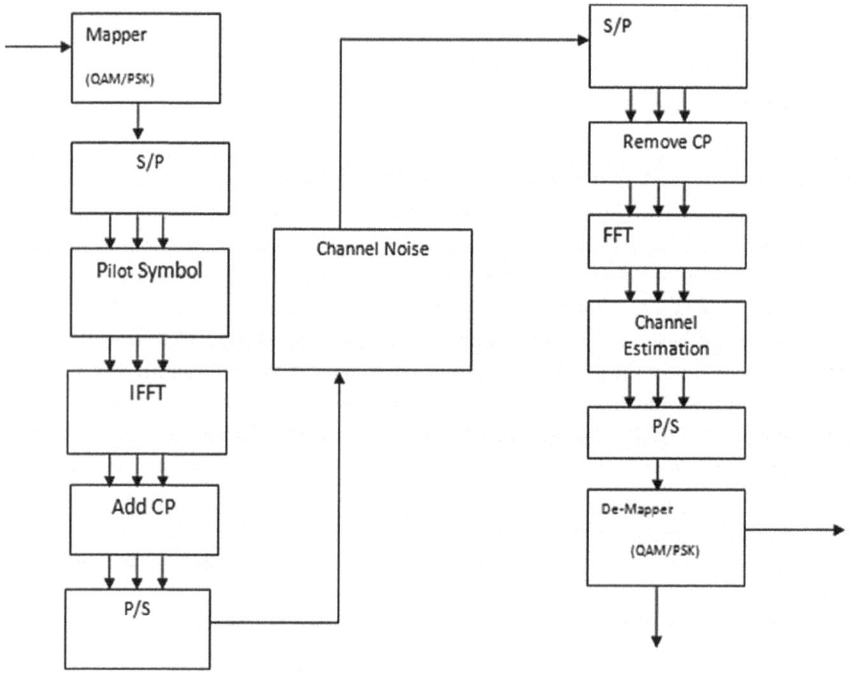


**Fig. 1** Block diagram of OFDM [4]

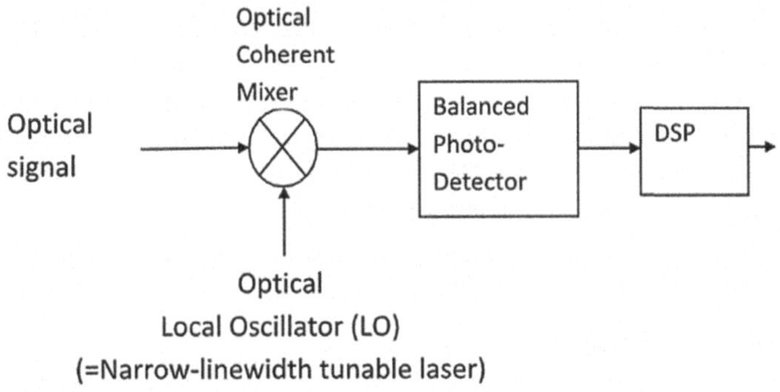


**Fig. 2** Optical coherent detector [7]

**Optical detection methods**:

- Non-coherent detection
- Differentially coherent detection
- Hybrid of non-coherent and differentially coherent detection
- Coherent detection.

**Coherent detection**

Among all the mentioned detection methods, coherent detection is the most efficient as presented in Fig. 2. Coherent optical OFDM (CO-OFDM) is considered as the technology of coming generation in the field of optical communication. In this technique, receiver determines the decision variables with respect to amplitude and phase information. This mechanism provides highest amount of flexibility in modulation formats, where coding of information is employed both in amplitude and phase. Coherent detection demands the receiver to contain information regarding the carrier phase when the Local Oscillator (LO) performs demodulation and behaves as absolute phase reference. Conventionally, Phase-Locked Loop (PLL) is employed to obtain carrier synchronization [7].

Optical systems can employ both the optical as well as electrical PLL. Utilization of Electrical PLL is found to be more beneficial than optical in case of duplex systems because the trans-receiver can employ a single laser as TX and LO together.

In this paper, the author developed a mechanism for downstream communication in OFDM-PON system by using various lengths of fiber at remote node for various m-QAM mapped OFDM signals over various data rates. The proposed scheme was implemented by using the coherent detection at the user end. The analysis of proposed work was done on various subcarriers [8]. OFDM was a desired modulation arrangement that currently attained a lot of consideration in the fiber-optic network. The fundamental preferred standpoint of optical OFDM was that it can adapt to practically unlimited measure of ISI. In fast optical transmission frameworks, ISI was caused for example by chromatic scattering and it was severe issue in long-term frameworks whose bit rate was higher [9].

This research work presents a performance analysis of coherent WDM optical OFDM system for different channel spacing. In this system, four different channel frequency having power −4 dBm are used to study their effect on Q factor and bit error rate at different channel spacing of 50, 100, and 150 GHz. Section 2 describes the design and simulation of WDM optical OFDM system, Sect. 3 discusses the simulation result, and Sect. 4 concludes the whole work.

## 2 Design and Simulation

Figure 3 shows the block diagram of OFDM system with coherent detection consists of two parts transmitter as well as receiver with transmission channel. Transmitter parts consists of PRBS (Pseudo Random Bit Sequence) generator, QAM generator, WDM multiplexer 4 × 1, OFDM modulator, Quadrature modulator, NRZ pulse generator, CW laser source, MZM (Mach Zehnder Modulator), optical filter and receiver part consists of photodetector, Quadrature demodulator, QAM decoder, WDM demultiplexer 1 × 4, OFDM demodulator, BER analyzer. PRBS generator is a binary sequence generator which is obtained using deterministic algorithm, difficult to predict and used in telecommunication, encryption, simulation, etc. NRZ pulse generator is a binary sequence generator in which one is represented by positive pulse and zero is represented by negative pulse. Mach Zehnder modulator is employed for controlling the varying amplitude of optical signal. Laser is an optical source which is used for generating a coherent beam of light. WDM multiplexer is used to combine multiple wavelengths into single wavelength optical signal. OFDM

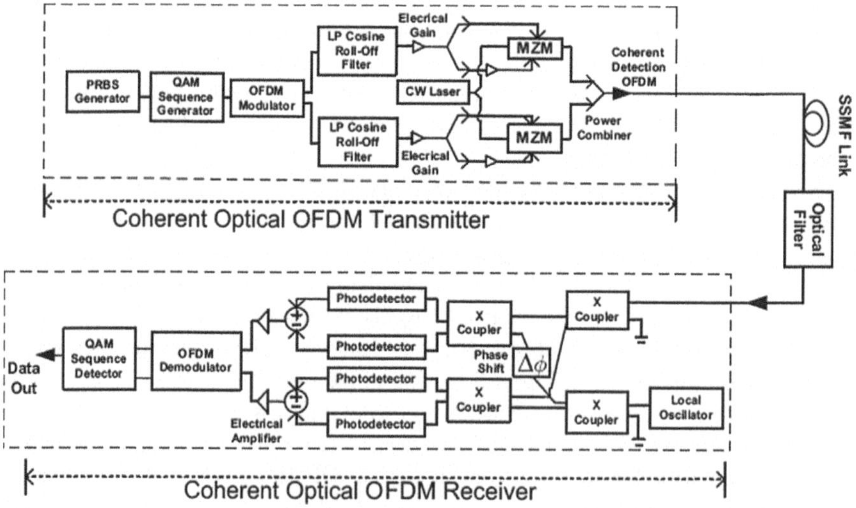


**Fig. 3** Block diagram of WDM OFDM system based on coherent detection [8]

modulator uses input data in different format like BPSK, QAM, etc. Optical amplifier is used to amplify the optical signal. Semiconductor optical amplifier (SOA), Raman amplifier, etc. are the examples of optical amplifier. WDM demultiplexer is used to split single wavelength into multiple wavelength at receiver side. OFDM demodulator demodulates an OFDM input signal by using an FFT operation. Photodetector is responsible for optical to electrical conversion. Different types of photodetector are PIN, Avalanche photodiode (APD). This simulation utilizes three visualizes in particular optical power meter, optical range analyzer and BER analyzer. Optical power meters give the power both in dBm and Watts. BER analyzer naturally computes the BER value, Q factor and show eye graph.

The Simulink model for Coherent WDM Optical OFDM mechanism is presented in Figs. 4, 5 and 6. First of all, there are four transmitters are utilized that are of 10 Gbps, after that by utilizing the WDM mechanism these four signals of 10 Gbps are multiplexed. Then, for one-span 60 km transmission fiber, an optical amplifier mechanism is utilized. Moreover, bit-error-rate is additionally determined when the equalizer is introduced so as to demonstrate the basic enhancement when the equalizer is utilized in the framework. Afterward, for every condition, the constellation diagrams are additionally analyzed, at a value of attained power of just about −4 dBm. The noteworthy enhancement would be noticed from the transmitted framework with equalizer, particularly when the fiber length is expanded, from those constellation diagrams (Table 1).

## 3 Results and Discussion

In this section the performance of downstream coherent detection m-QAM-mapped OFDM is analyzed for various parameters like Constellation Diagram, BER, Q-factor, etc. for different frequency spacing of 50, 100 and 150 GHz.

The Electrical constellation of receiving coherent detection m-QAM-mapped OFDM downstream signal (512 subcarriers) at ONU is shown in Fig. 7. The red section in this graph depicts the obtained signal whereas the blue section signifies the occurrence of noise in the signal because of transmission by feeder fiber. Because the constellation diagrams are clear therefore performance obtained for downstream coherent detection OFDM is better.

Figures 8, 9 and 10 Shows amplitude variation on the $y$-axis that ranges from 0 to 1 whereas time variation on the $x$-axis that ranges from 0 to 1. The minimum value of BER is 0.0029295 which is less than 50 and 150 GHz. The maximum value of Q-factor for proposed model is 9.22647 at spacing frequency of 100 GHz which is higher than other spacing frequency at 50 and 150 GHz.

Table 2 shows the comparison of the different parameters like Q-factor, BER and eye diagram of OFDM system model with coherent detection. The different values of Q-factor depict that at higher q-factor the noise in the system is minimum and the BER is also minimum which makes the system more reliable. It has been observed from the tables that OFDM system better results at spacing of 100 GHz.

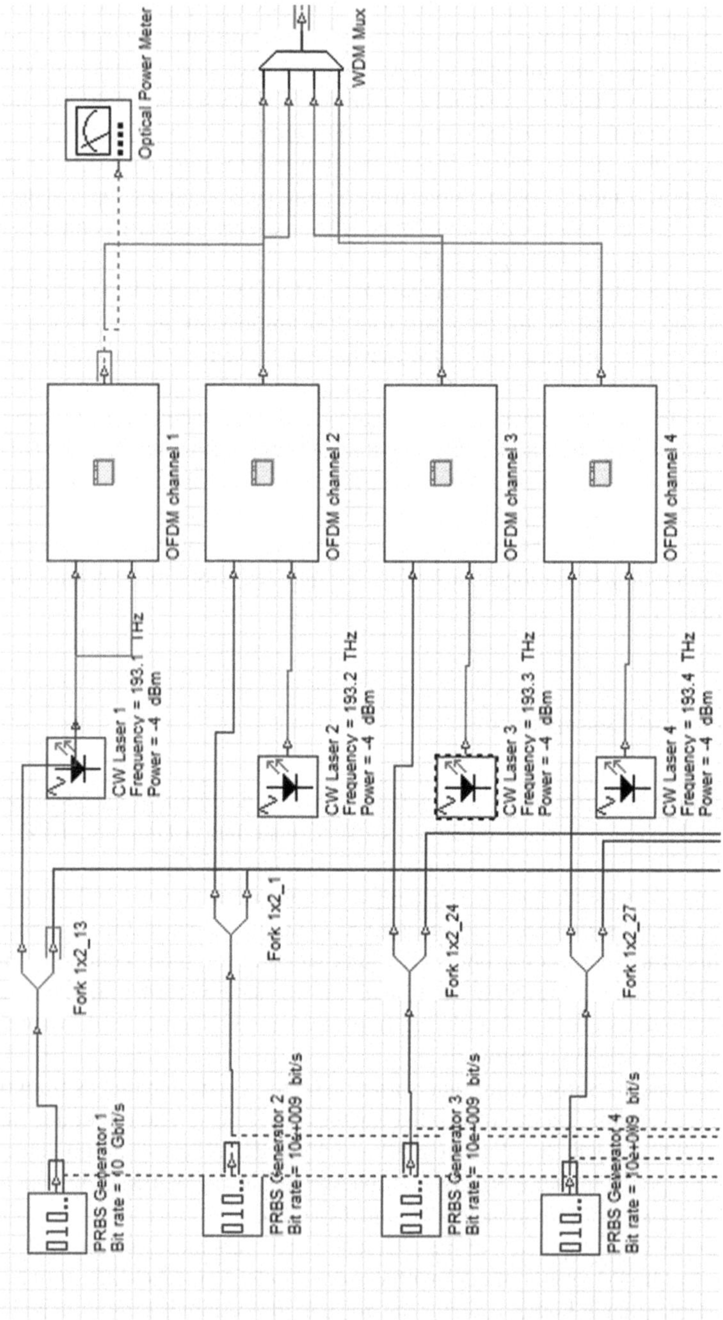


**Fig. 4** Optical OFDM transmitter

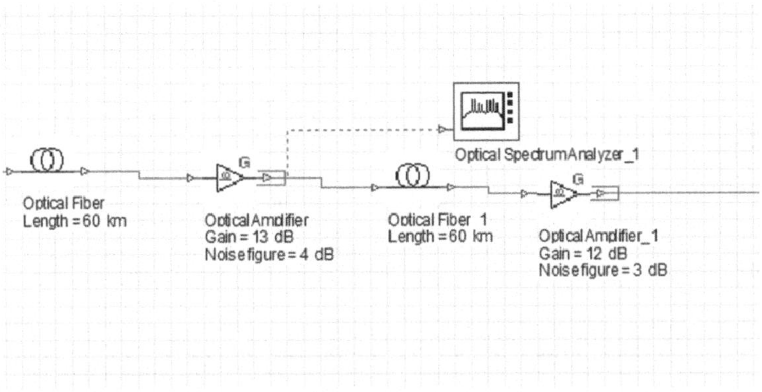


**Fig. 5** Optical transmission link

## 4 Conclusion

The proposed OFDM system model with coherent detection is used to investigate the results for different values of BER and Q-factor at different spacing between the LASER frequencies. At a frequency of 193.2 THz a minimum value of BER i.e. 0.0029295 and high-quality factor i.e. 9.22647 is achieved using optical amplifier having channel spacing 100 GHz. The proposed system provides high-quality factor which reduced the noise which makes it more efficient system. The simulation results have demonstrated that the results obtained for channel spacing of 100 GHz are better as compare to results obtained for channel spacing of 150 GHz as well as for 50 GHz.

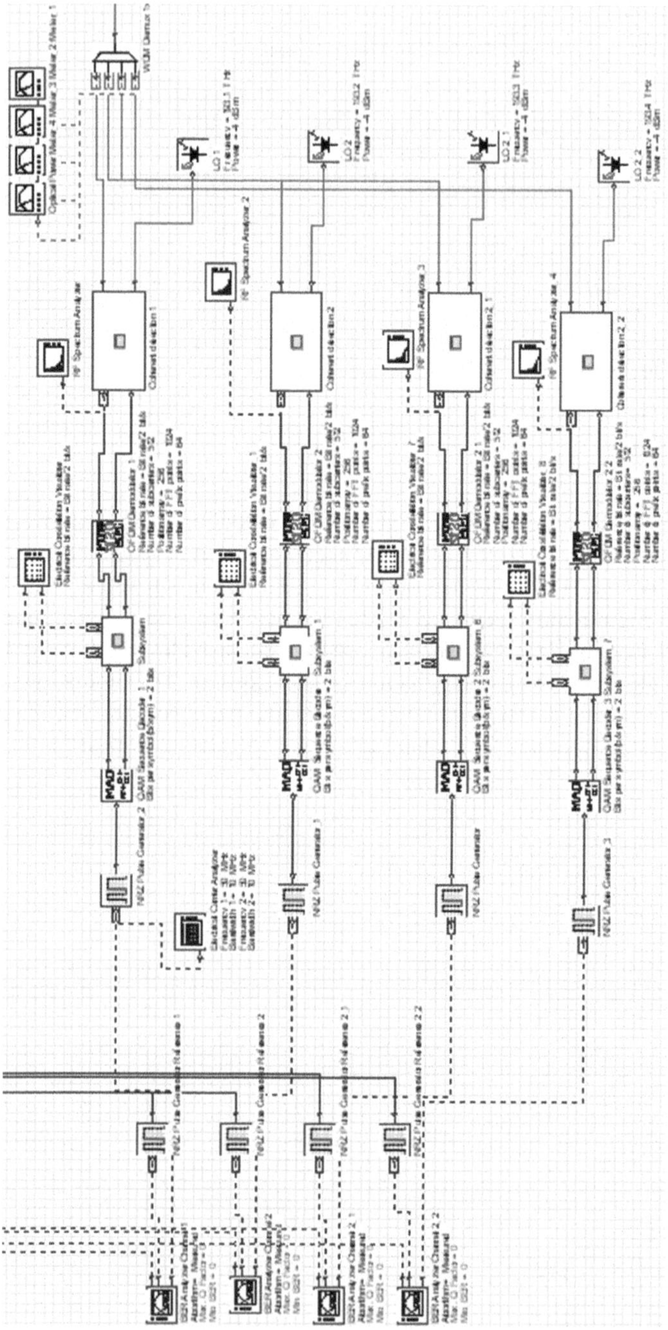

**Fig. 6** Optical OFDM receiver

**Table 1** Design parameter specification

| Parameters | Values (unit) |
|---|---|
| Channel power | −4 dBm |
| Bit rate | 10 Gbps |
| Attenuation loss | 0.2 dB/km |
| Optical fiber length | 60 km |

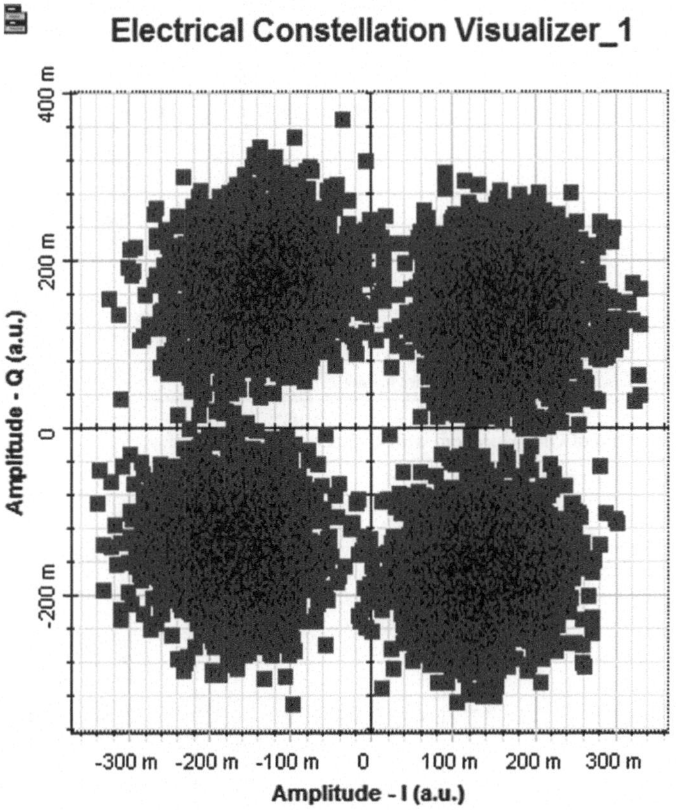


**Fig. 7** Constellation diagram of downstream coherent detection OFDM signal

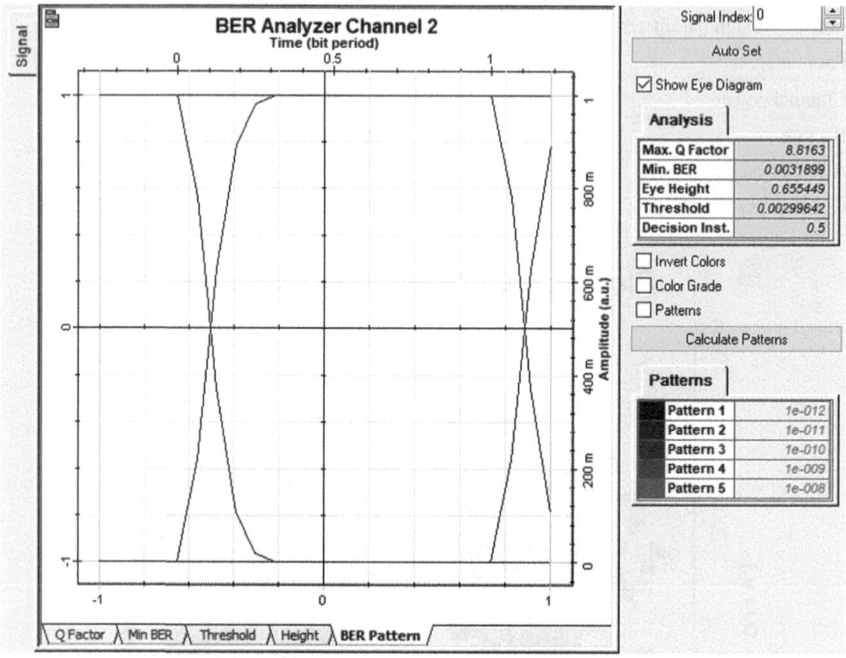


**Fig. 8** Eye diagram of OFDM system with channel spacing 50 GHz

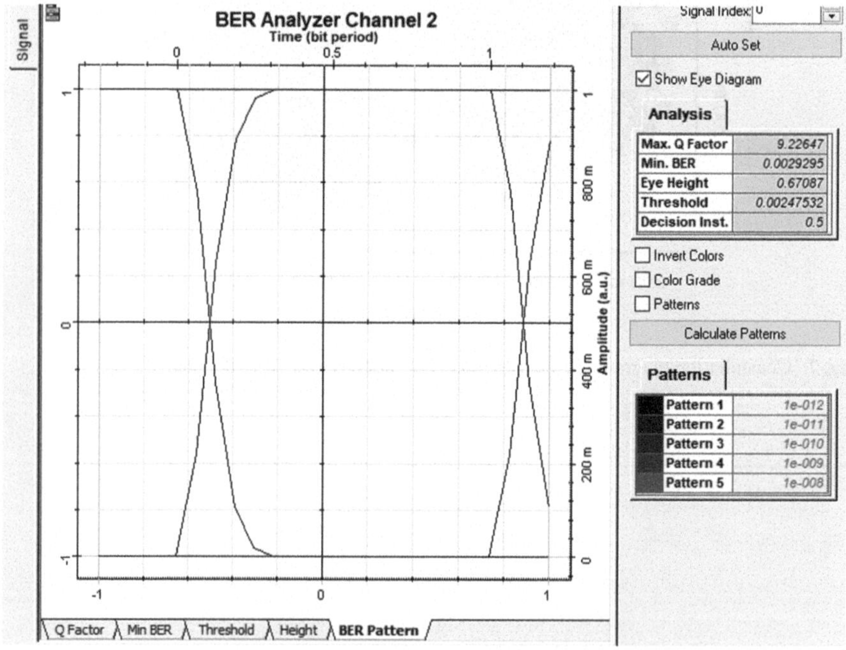


**Fig. 9** Eye diagram of OFDM system with channel spacing 100 GHz

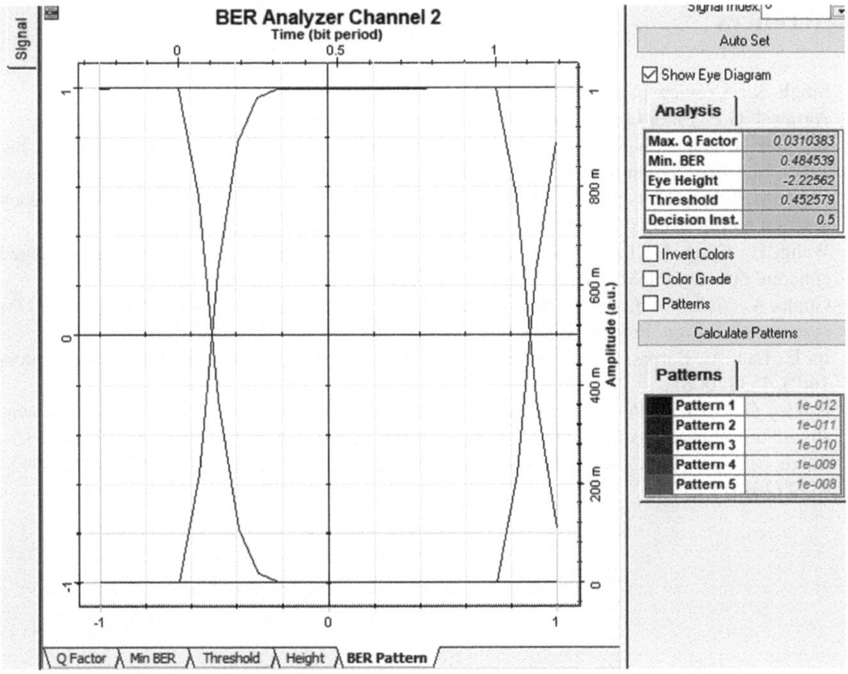


**Fig. 10** Eye diagram of OFDM system with channel spacing 150 GHz

**Table 2** Parameter analysis for different spacing frequency of 50, 100 and 150 GHz

| Parameters | 50 GHz | 100 GHz | 150 GHz |
|---|---|---|---|
| Max. Q factor | 8.8163 | 9.22647 | 0.0310383 |
| Min. BER | 0.0031899 | 0.0029295 | 0.484539 |
| Eye height | 0.655449 | 0.67087 | −2.22562 |
| Threshold | 0.00299642 | 0.00247532 | 0.452579 |
| Decision inst. | 0.5 | 0.5 | 0.5 |

## References


1. Singh, K.: A review paper on: optical fiber communication **3**(8) (2017)
2. Agrawal, G.P.: Fiber Optics Communication System, 3rd edn. Wiley, New York (2002)
3. Mittal, P., Chauhan, N.S., Gaurav, A.: Coherent detection optical OFDM system (IJSRET). Int. J. Sci. Res. Eng. Technol. **4**(4), 342 (2015)
4. Khan, M., Iqbal, S., Asghar, W.: A review paper on: the PAPR analysis of orthogonal frequency division multiplexing (OFDM). Int. J. Mobile Netw. Commun. Telem. **4**(1), 1–13 (2014)
5. Wang, H., Kong, D., Li, Y., Wu, J., Lin, J.: Performance evaluation of (D)APSK modulated coherent optical OFDM system. Optical Fiber Technol. **19**(3), 242–249 (2013)
6. Gupta, A., Malhotra, J.: On the optimization of channel spacing in the hybrid WDM-COOFDM system. Int. J. Sign. Process. Image Process. Patt. Recogn **9**(7), 353–360 (2016)
7. Ip, E., Lau, A., Barros, D., Kahn, J.: Coherent detection in optical fiber systems. Opt. Express **16**(2), 753 (2008)
8. Pandey, G., Goel, A.: Performance analysis of long-reach coherent detection OFDM-PON downstream transmission using m-QAM-mapped OFDM signal. J. Opt. Commun. **38**(4), 46 (2016)
9. Singh, M., Kaur, K.: Coherent detection WDM optical OFDM system. IJARCCE **2**(12), 4793–4797 (2013)

# Design and Investigation of Multiple TX/RX FSO Systems Under Different Weather Conditions

**Shubham Mahajan, Davinder Parkash and Harjeevan Singh**

**Abstract** In the proposed model, effect of number of transmitters and receivers on performance of proposed free-space optical (FSO) communication system is simulated as well as analyzed under various weather conditions like clear, haze and fog. FSO communication provides a strong and efficient method for transmission and reception of information through free channels because of its higher data transfer capacity and inbuilt ability of security. The effect of environmental parameters has been investigated for range of 1 km. The effect of attenuation increases in bad weather conditions affecting the performance of FSO system while designing a system the weather conditions have to be taken care. The effect is to beat the impact of fog attenuation on FSO system by assessment and execution through simulated results of the executed model with respect to Q factor, height of eye, power received and bit error rate. The proposed multiple TX/RX FSO system shows better results as compared to the 1 TX/1 RX system.

**Keywords** FSO · Optical fiber · Optical wireless communication · Q-factor and weather conditions

## 1 Introduction

FSO communication system is an optical communication innovation that utilizes line of sight (LOS) way propagating in free space for transmission of information between two points for communication. FSO basically is "fiberless optics" transmission. "Free-space" in this relates to air, space, vacuum or remote. The FSO correspondence depends on the optical communication with contrast to system ignoring

S. Mahajan (✉) · D. Parkash · H. Singh
Department of Electronics and Communication, Chandigarh University, Gharuan, Mohali, Punjab, India
e-mail: mahajanshubham2232579@gmail.com

D. Parkash
e-mail: drdavinderchechi@gmail.com

H. Singh
e-mail: harjeevansingh.ece@cumail.in



P. K. Singh et al. (eds.), *Proceedings of ICRIC 2019*, Lecture Notes in Electrical Engineering 597, https://doi.org/10.1007/978-3-030-29407-6_28

utilizing solid, for example, fiber optical link of transmission. For the most part an FSO system is embraced when a physical association is impractical but a practicable arrangement because of staggering expenses or different analysis. It is additionally embraced to satisfy the necessity of higher transfer speed so as to deal with the information transmission of the system [1]. Various services such as voice, data, text, video, etc. are used to transmit data from one end to other. Laser beam radiate through arbitrary medium, for example, climate or beneath the surface of water has been concentrated widely for a long time. This innovation has a few points of interest over customary radio frequency (RF) or microwave communication outstanding to its expanded power effectiveness, low mass and space demand, higher oriented (i.e. bigger antenna gain), and prize free very higher data transfer capacity [2]. FSO Technology is the progressive remote communication which furnishes high transfer speed with high information rate of transmission over high connection extend. The real test of FSO innovation is climatic depletion that hampers information transmission of the system by expanding bit error rate. Travelling wave semiconductor optical amplifier is useful devices that can take reduce the impact of fog in FSO system [1].

FSO communication is relied upon to very important job in 5G communication systems. FSO link offers a promising option in contrast to the traditional fiber optic links used for backhoes connects because of the simplicity of sending, fast setup time, and low cost. Optical fiber communication and FSO use the similar infrared (IR) wavelengths of light with high rate of data transmission. FSO system requires less power, less cost with higher privacy and high date rates. These kinds of connections are appropriate for 1–2 Gbps with distances up to 1–5 km [3–6]. In addition, it tends to be utilized as a choice for expanding range of fiberless communication. Despite this other advantages of FSO are no requirement of licensed band, no radiation effects, easy to install [7]. It uses the idea of transmitting data at high speed using optical beam from TX end to RX end in free space. Due to which clear line of sight (LOS) is needed to be maintained between both ends for effective communication, LOS offers various advantages both to service providers and telecommunication.

Fiber optics is imperative for worldwide broadband network as with the accelerating number of users there is demand of wide broadband network that can provide less delay in transmission. In any case, since this innovation exclusively utilizes the air as transmission mode, defenselessness towards weather conditions is unavoidable. These aggravations will essentially influence FSO transmission efficiency. The environmental turbulences affect the power received along with decreasing the system performance.

Figure 1 shows the Block diagram of FSO system. In this system, the information is sent from transmitter to receiver through space. Information is moved on to modulator where optical carriers (LASER/LED) are used to fed transmitter and modulated carrier signal with information signal is moved on to atmospheric channel. Information is received and detected at receiver end and sent to destination using photo-detector, filter and demodulator. In this work, we have used multiple TX/RX i.e. number of laser beams in FSO system for analyzing communication link performance. The numerous TX/RX used to analyze the performance and efficiency

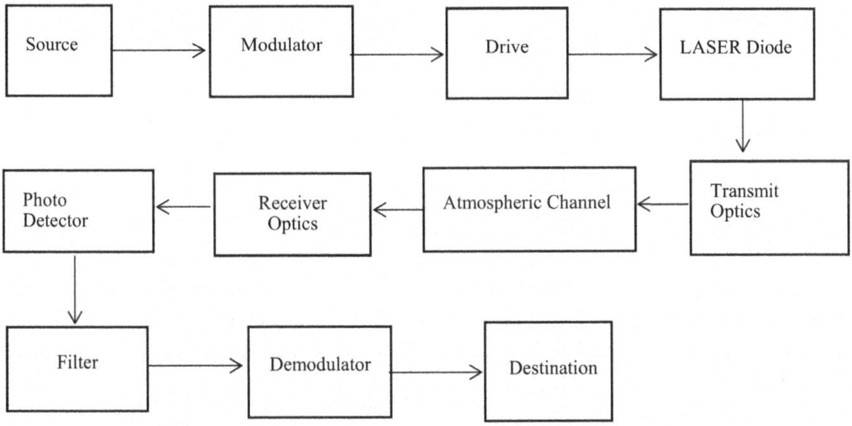


**Fig. 1** Block diagram of FSO communication system

of FSO system under different weather conditions. The simulation is done by using optical system simulator named OptiSystem 13.0 by Optiwave.

# 2 FSO Challenges

## *2.1 Absorption and Scattering Loss*

In the process of propagation through earth's atmosphere the laser beam has to interact with particles and gas molecules present in atmosphere. Absorption and scattering loss is major reason for loss in atmosphere channel [8–10]. Attenuation in atmosphere is represented as below:

$$\alpha = \mathrm{e}^{-\sigma d} \tag{1}$$

distance of measurement is denoted by $d$, coefficient attenuation particular to length per unit is denoted by $\sigma$.

$$\sigma \cong \frac{3.912}{V}\left(\frac{\lambda}{550}\right)^{-q} \tag{2}$$

Visibility is denoted by $V$ and is measured in km, wavelength is denoted by $\lambda$ and measured in nm, diffusing particles distribution is of size $q$ attained by Kim and Kruse [11, 12].

### 2.2 Fog

It results in atmospheric attenuation leading to absorption and scattering losses. It will result in high attenuation 350 dB/km in dense fog conditions [12].

$$\alpha_{\mathrm{Fog}} = \frac{10 \log V\%}{V\ (\mathrm{km})}\left(\frac{\lambda}{\lambda o}\right)^{-q} \tag{3}$$

Visibility is denoted by $V$, measured in km, $V\%$ is percentage of air drop, wavelength is denoted by $\lambda$ in nm, range of visibility is denoted by $\lambda_o$ (550 nm), and diffusing particles distribution is of size $q$. Now by Kruse model [12] parameter of $q$ is defined as

$$q = \begin{cases} 1.6 & \text{if } V > 50\,\mathrm{km}; \\ 1.3 & \text{if } 6\,\mathrm{km} < V < 50\,\mathrm{km}; \\ 0.585V & \text{if } V < 6\,\mathrm{km}. \end{cases} \tag{4}$$

Kim [13], has defined $q$ parameter as below:

$$q = \begin{cases} 1.6 & V > 50\,\mathrm{km} \\ 1.3 & 6\,\mathrm{km} < V < 50\,\mathrm{km} \\ 0.16V + 0.34 & 1\,\mathrm{km} < V < 6\,\mathrm{km} \\ V - 0.5 & 0.5\,\mathrm{km} < V < 1\,\mathrm{km} \\ 0 & V < 0.5\,\mathrm{km} \end{cases} \tag{5}$$

From Eq. (2), we get to know that for higher wavelengths there is less attenuation.

### 2.3 Rain

Its brunt is not more than fog. Wavelength of FSO is smaller than drops of rain. It has been suggested that 2.5 mm/h. corresponds to light rain and 25 mm/h corresponds to heavy rain [14, 15]. According to Carbonneau [16]

$$\text{Attenuation of rain} = 1.076 * R_3^2 \tag{6}$$

### 2.4 *Snow*

Attenuation because of snow is more than rain and not as much as fog as size of snow particles is in the middle of fog and rain. Attenuation because of snow is 30–350 dB/km [17, 18].

## 3 Simulation and Design

The proposed model is classified into three number of parts named as transmitters Tx, propagation channels and receivers Rx as shown in Fig. 2. Transmitter comprises of information source, NRZ modulator, and continuous wave (CW) laser. The data source is a pseudo-random binary source (PRBS) generator, which delivers a chain of numbers that is required for transmission. The result of the PRBS is chain of bits including a combination of 1's (ON) and 0's (OFF). Data in the subsystem is encoded using NRZ encoding framework where "1" is characterized by vital bit and "0" is characterized through a substitute bit. The simulation model represented in Fig. 3 has PRBS generator which produced information at data rate of 10 Gbps and light is regulated utilizing Mach Zehnder Modulator (MZM). The laser source operates at power of 10 dBm and wavelength of 1550 nm. The gain of the amplifier is set to 10 dB. Signal is then specifically sent to the recipient through FSO channel.

Attenuation of three different weather conditions clear, fog and haze are 0.44 dB/km, 6.8 dB/km and 8.8 dB/km, respectively. Initially, the weakening quality is set as 0.44 dB/km (clear). The optical signals from the FSO channel are collected by photodetector (APD). Low pass Bessel filter is used to utilize the channel-free from noise. This simulation utilizes three visualizers' in particular optical power meter, optical range analyzer and BER analyzer. Optical power meter gives the power both

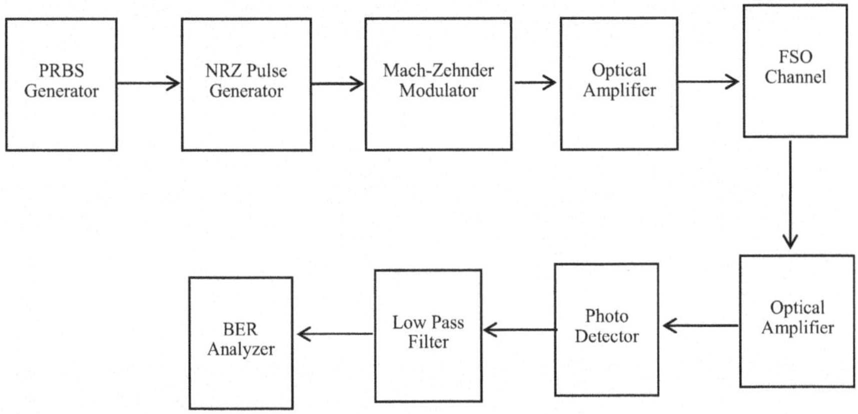


**Fig. 2** Block diagram of 1TX/1RX FSO system

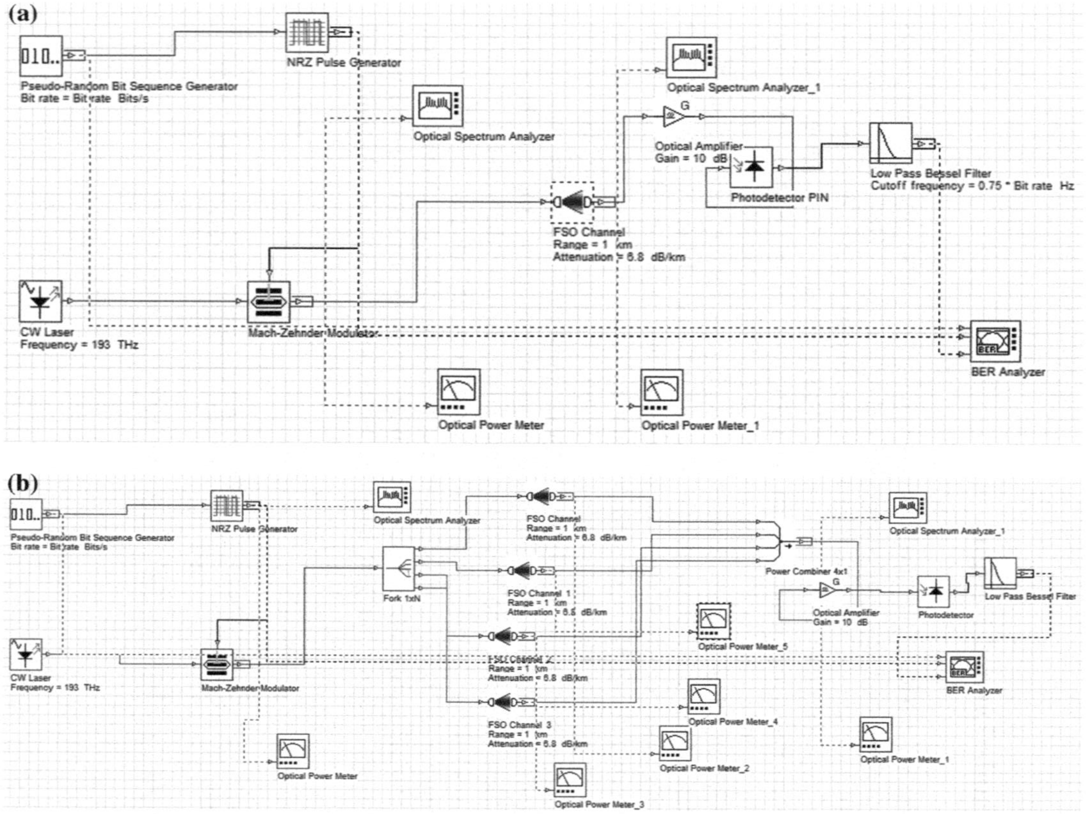


**Fig. 3** **a** Simulation model of 1 TX/1RX FSO systems. **b** Simulation model of 4 TX/4 RX FSO systems

**Table 1** FSO system design parameters

| Parameters | Value |
|---|---|
| Transmission rate | 10 Gbps |
| Transmitted power | 10 dBm |
| Wavelength | 1550 nm |
| Modulation format | NRZ |
| Range | 1 km |
| Gain | 10 dBm |
| LASER frequency | 193.1 THz |
| Clear weather condition | 0.44 dB/km |
| Fog weather condition | 6.8 dB/km |
| Haze weather condition | 8.8 dB/km |

in dBm and Watts. BER analyzer naturally computes the BER value, Q factor and eye diagram. The proposed FSO System Design Parameters are presented in Table 1.

The numerous Tx/Rx FSO system simulation model as presented in Fig. 3b under different weather condition. Further despite using a single FSO channel, the proposed model has four channels. Power splitter is implemented for replicating the input signal to all four channels.

## 4 Results and Discussions

The proposed FSO interface system is used to examine the performance of the system with respect to Q-factor, BER, overall eye diagram and received power. The proposed FSO system has implemented 1 TX/1 RX and 4 TX/4 RX by taking bit rate of 10 Gbps, power equal to 10 dB and range of 1 km. Both the systems are examined under the three different weather conditions clear, fog and haze. Figure 4a eye diagram of 1 TX/1RX with BER value of 2.6884e-138, Q-factor of 25.0033 and eye height of 7.57637e-005 Fig. 4b shows the eye diagram of 4 TX/4 RX with eye height of 7.86847e-005, BER of 5.33472e-224 and Q-factor 31.9287 of under clear weather conditions. Figure 5a shows eye diagram of 1 TX/1 RX with BER value of 1.78819e-014, Q-factor of 7.57478 and eye height of 1.19173e-005 Fig. 5b shows eye diagram of 4 TX/4 RX with eye height of 6.8778e-005, BER of 9.09062e102 and Q-factor 21.3848 of under fog weather conditions. Figure 6a eye diagram of 1 TX/1 RX with BER value of 3.29806e-007, Q-factor of 4.97248 and eye height of 4.90184e-006. Figure 6b shows the eye diagram of 4 TX/4 RX with eye height of 4.03178e-005, BER of 2.8413e-050 and Q-factor 14.8631 of under haze weather conditions. Results are represented in tabular form in Tables 2, 3 and 4.

(a)

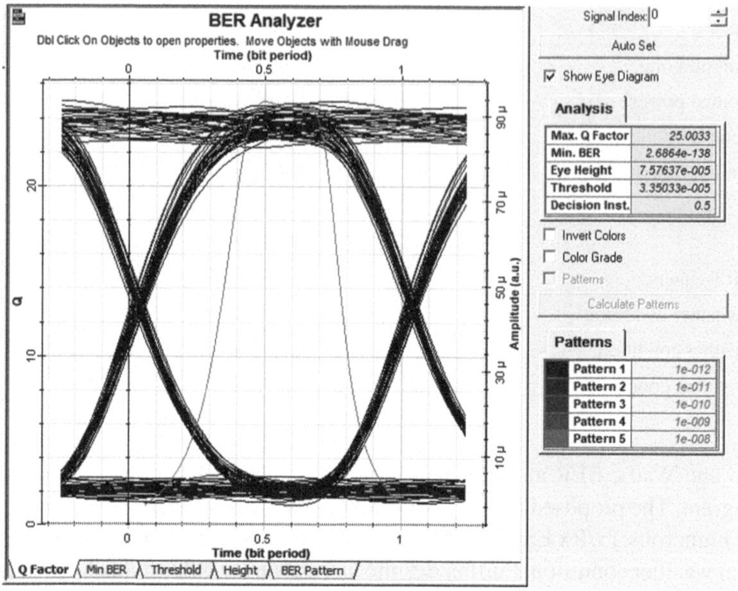


(b)

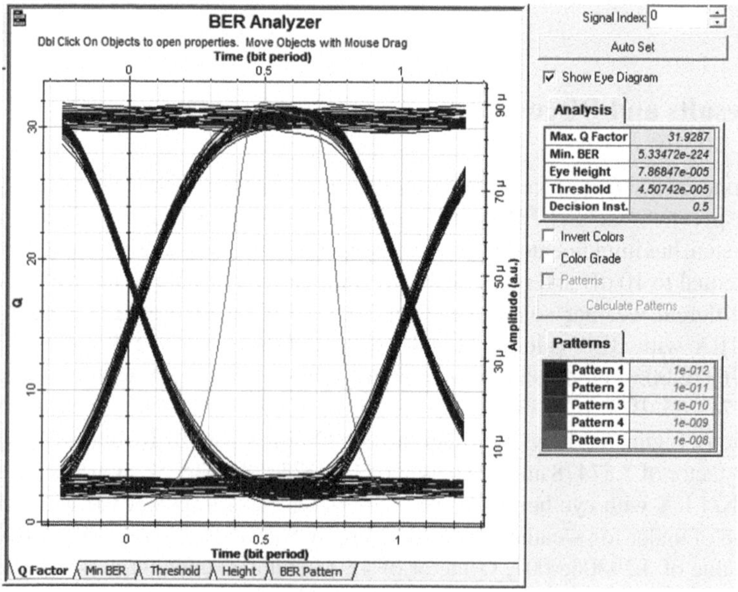


**Fig. 4** Eye diagram of 1 TX/1 RX. **b** Eye diagram of 4 TX/4 RX under clear weather condition

(a)

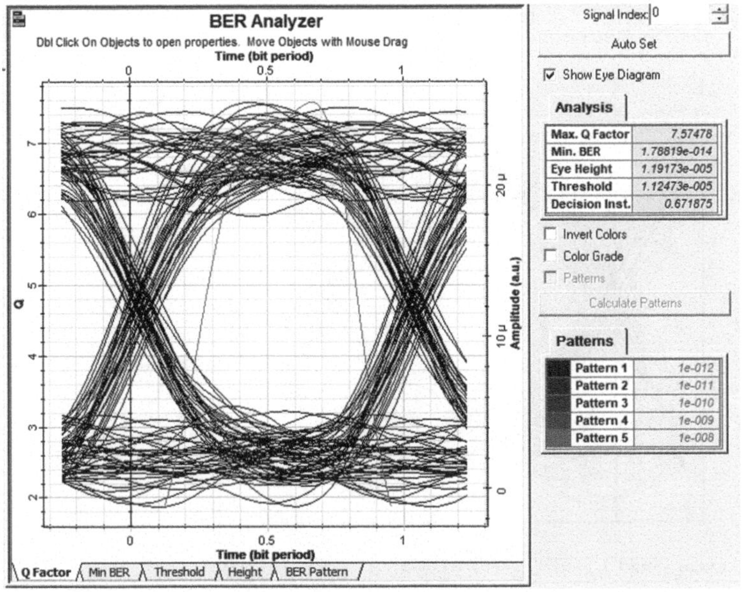


(b)

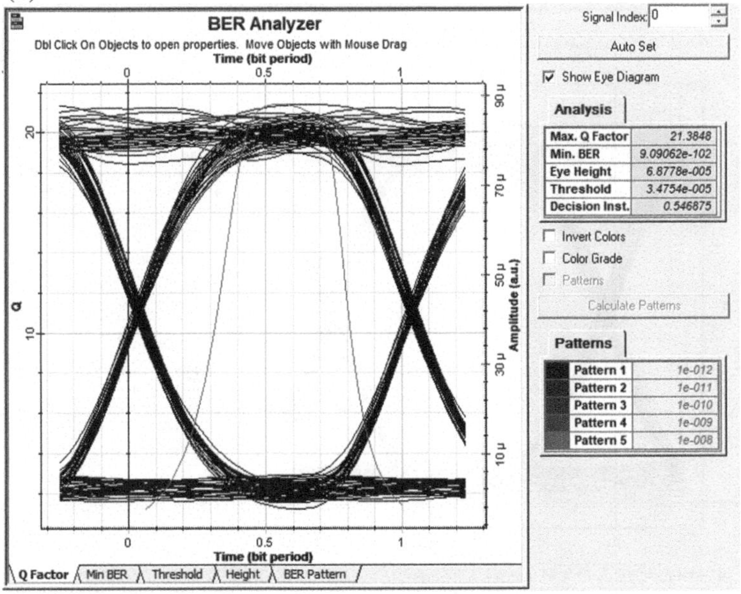


**Fig. 5** Eye diagram of 1TX/1 RX. **b** Eye diagram of 4 TX/4 RX under fog weather conditions

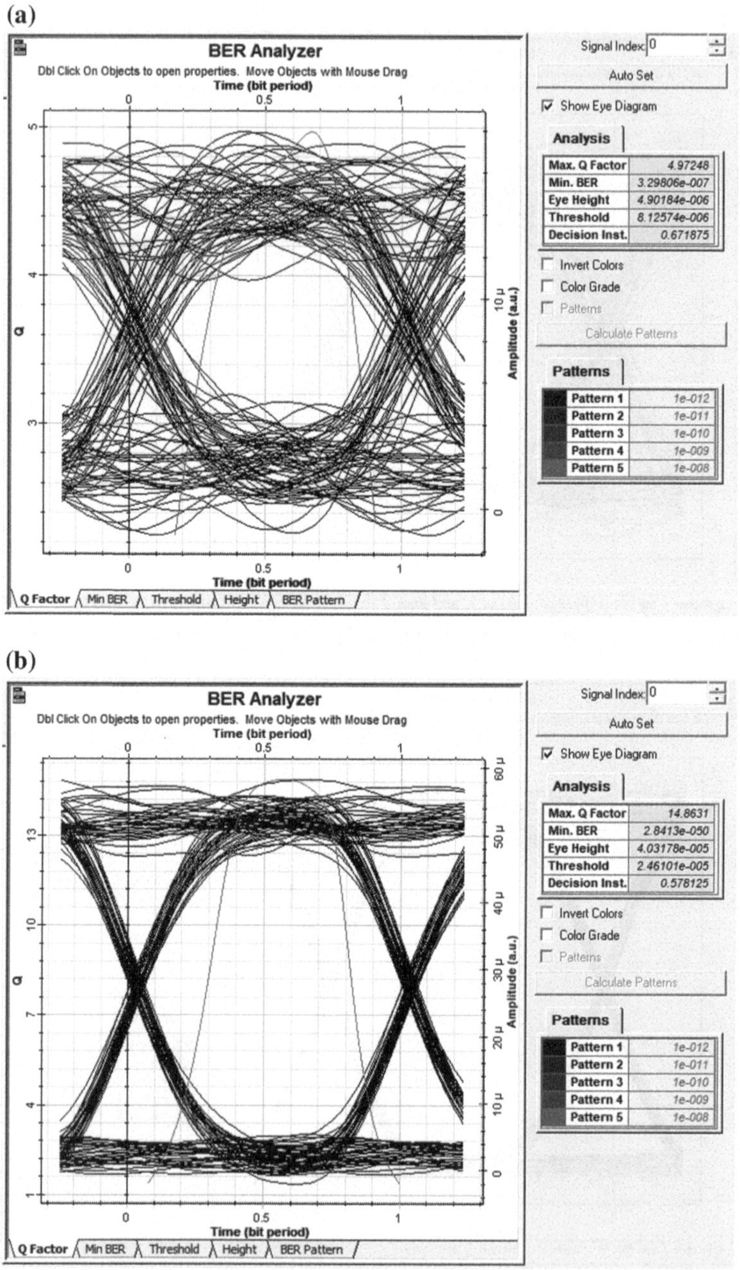


**Fig. 6** **a** Eye diagram of 1 TX/1 RX. **b** Eye diagram of 4 TX/4 RX under haze weather conditions

**Table 2** FSO system results under the effect of clear weather conditions

| TX/RX combination | Q-factor | Bit error rate | Eye height | Power received |
|---|---|---|---|---|
| 1 TX/1 RX | 25.0033 | 2.6884e–138 | 7.57637e–005 | −23.873 dBm |
| 4 TX/4 RX | 31.9287 | 5.33472e–224 | 7.86847e–005 | −17.782 dBm |

**Table 3** FSO System results under the effect of fog weather conditions

| TX/RX combination | Q-factor | Bit error rate | Eye height | Power received |
|---|---|---|---|---|
| 1 TX/1 RX | 7.57478 | 1.78819e–014 | 1.19173e–005 | −30.232 dBm |
| 4 TX/4 RX | 21.3848 | 9.09062e–102 | 6.8778e–005 | −24.142 dBm |

**Table 4** FSO System results under the effect of haze weather conditions

| TX/RX combination | Q-factor | Bit error rate | Eye height | Power received |
|---|---|---|---|---|
| 1 TX/1 RX | 4.97248 | 3.29806e–007 | 4.90184e–006 | −32.232 dBm |
| 4 TX/4 RX | 14.8631 | 2.8413e–050 | 4.03178e–005 | −26.142 dBm |

## 5 Conclusion

The proposed FSO system has been simulated and analyzed under various weather conditions for 1 TX/1 RX and 4 TX/4 RX at 1550 nm wavelength. The effect of environmental parameters has been investigated for range of 1 km. The proposed FSO interface system is used to examine the performance of the system with respect to Q-factor, BER, overall eye diagram and received power. The proposed FSO system has implemented 1 TX/1 RX and 4 TX/4 RX by taking bit rate of 10 Gbps and with 10 dB power. It has been concluded that the FSO system with 4 TX/4 RX is more efficient than 1 TX/1 RX FSO system. The proposed FSO system provides an efficient link for the transmission of data under dense weather conditions for higher data rates.

## References

1. Hossain, F., Afroze, Z.: Eliminating the effect of fog attenuation on FSO link by multiple TX/RX system with travelling wave semiconductor optical amplifier. In: 2nd International Conference on Advances in Electrical Engineering (ICAEE 2013), Dhaka, Bangladesh, 19–21 Dec 2013
2. Kaushal, H., Kaddoum, G., Jain, V.K., Kar, S.: Experimental investigation of optimum beam size for FSO uplink. Optics Commun. **400**, 106–114 (2017)
3. Bloom, S., Korevaar, E., Schuster, J., Willebrand, H.: Understanding the performance of free space optics. J. Opt. Netw. **2**, 178–200 (2003)
4. Kaushal, H., Kaddoum, G.: Optical communication in space: challenges and mitigation techniques. IEEE Commun Surveys Tutorials **19**, 57–96 (2017)
5. Ijaz, M., Harris, G., Adebisi, B., Rajbhandari, S., Popoola, W.: Experimental study of the beam wander mitigation in free space optical communications using single input multiple output

system (PIMRC). IEEE 28th Ann. Int. Symp. Pers. Indoor Mobile Radio Commun. **978**(1), 1–5 (2017)
6. Dev, K., Nebuloni, R., Capsoni, C., Fiser, O., Brazda, V.: Estimation of optical attenuation in reduced visibility conditions in different environments across free space optics link. 10th Eur. Conf. Antennas Propag. **11**(12), 1708–1713 (2017)
7. Rashid, F.U., Semakuwa, S.K.: Performance analysis of free space optical communication under the effect of rain in Arusha Region, Tanzania. Int J Eng Res Technol. **3**, 1523 (2014)
8. Long, R.K.: Atmospheric attenuation of ruby lasers. Proc. IEEE. **51**(5), 859–860 (1963)
9. Langer, R.M.: Effects of atmospheric water vapour on near infrared transmission at sea level. In: Report on Signals Corps Contract DA-36–039-SC-723351, J.R.M. Bege Co., Arlington, Mass (1957)
10. Sizum, M., de Fornel, M., Nabousi, M.A.I.: Propagation of optical and infrared waves in the atmosphere. Proc. SPIE **43**(02), 319–329 (2003)
11. Kim, I.I., Achour, M.: Free-space links address the last-mile problem, vol. 37 (2001)
12. Kruse, P.W. et al.: Elements of Infrared Technology: Generation, Transmission and Detection. Wiley, New York (1962)
13. Kim, I., Mc Arthur, B., Korevaar, E.: Comparison of laser beam propagation at 785 and 1550 nm in fog and haze for optical wireless communication. Proc. SPIE **4214**, 26–37 (2001)
14. Suriza, A.Z., Rafiqul, I.M., Wajdi, A.K., Naji, A.W.: Proposed parameters of specific rain attenuation prediction for free space optics link operating in tropical region. J Atmosp. Solar Terres. Phys. **94**, 93–99 (2013)
15. Vavoulas, A., Sandalidis, H.G., Varoutas, D.: Weather effects on FSO network connectivity. J. Opt. Comm. Net. **4**(10), 734–740 (2012)
16. Carbonneau, T.H., Wisely, D.R.: Opportunities and challenges for optical wireless; the competitive advantage of free space telecommunications links in today's crowded market place. SPIE Conf. Optic. Wirel. Commun. **3232**, 119–128 (1998)
17. Kim, I.I., Korevaar, E.: Availability of free space optic (FSO) and hybrid FSO/RF systems. Light pointe Tech report. http://www.opticalaccess.com (2019)
18. Kaushal, H., Kaddoum, G.: Optical communication in space: challenges and mitigation techniques. Commun Surv Tutorials https://doi.org/10.1109/comst.2603518.ieee (2016)

# Dynamic Distance Based Lifetime Enhancement Scheme for HWSN

**Sumit Kumar Gupta, Sachin Kumar, Sudhanshu Tyagi and Sudeep Tanwar**

**Abstract** Nowadays Wireless Sensor Network (WSN) is popular research field for new scholars. This technique has huge potential in various fields like energy efficiency, data gathering, network security, etc. The bitty sensor nodes are formed a wireless sensor network which has capabilities of communication, transmission and sensing with certain limitations. Limited capability of energy or battery life which attracts many scholars to do the research and find appropriate solutions. Cluster-based heterogeneous WSN (HWSN) is a possible solution to improve the network life time. Keeping in the view of above, this is the key area we have taken in this paper to improve the network life time with the use of less battery consumption. Here, we proposed "Dynamic Distance based Lifetime Enhancement scheme for Heterogeneous WSN". In the proposed scheme cluster head is selected by dynamic hopping for data transmission to base station. By the selection of dynamics hopping transmission, network life time improved significantly as compared to pre-existing techniques like LEACH, SEECP protocol.

**Keywords** WSN · Clustering · Companion node · Dynamic radius · Dynamic hopping

S. K. Gupta
Department of ECE, SRMS College of Engineering, Technology and Research, Bareilly, India
e-mail: sumikg@gmail.com

Department of ECE, Amity School of Engineering & Technology, Lucknow Campus, Uttar Pradesh, India

S. Kumar
Department of ECE, Amity School of Engineering and Technology, Lucknow Campus, Uttar Pradesh, India
e-mail: skumar3@lko.amity.edu

S. Tyagi (✉)
Department of ECE, Thapar Institute of Engineering & Technology, Patiala, Punjab, India
e-mail: s.tyagi@thapar.edu

S. Tanwar
Department of Computer Science and Engineering, Institute of Technology, Nirma University, Ahmedabad, Gujarat, India
e-mail: sudeep149@rediffmail.com



P. K. Singh et al. (eds.), *Proceedings of ICRIC 2019*, Lecture Notes in Electrical Engineering 597, https://doi.org/10.1007/978-3-030-29407-6_29

# 1 Introduction

WSN is still challenging topic for researchers in different areas like energy efficiency, clustering, network security, data gathering and node heterogeneity [1–4]. Many researchers are working in the field of energy efficiency, how can be improved the network life span so that for a long time, we can collect the sensed data like temperature, humidity, etc. In the network, there is no such element in which we can increase the life time except the sensor node. The sensor node has the different functional unit like sensing, power management, signal processing and transmission unit. In all these units, the power unit is most limited resource which we cannot replace once we deployed because the placement of sensor node is entirely random. There are many applications like surveillance and military area where the replacement of battery is impossible that's why researchers are always trying to increase the battery life of SN and to decrease the distance between SN and BS. To accomplice this task, Researchers have proposed different routing protocol that can prevent to deplete the energy of sensor node.

There are many areas like environment monitoring in which temperature, humidity, pollution control, etc., military operations like surveillance, vehicle monitoring system, earthquake monitoring, target tracking and surveillance system, patients monitoring system, where WSNs are playing a vital role [14]. All these applications and many more can be performed by a group of sensor nodes with some essential parameter like energy, security, data gathering, etc. Although there are many more issues discussed in literature, we are discussing heterogeneity parameter in WSNs as they are playing significant role in real-time scenario.

Cluster is defined as a group of similar members or like objects which have same property [3]. Here the similar object is sensor node and the same property is distance. Thus Cluster is formed using distance criteria of closer sensor node. In the literature, we have lots of clustering based protocols. Clustering-based protocol [6] and threshold-based protocol [15, 16] are widely used protocol which is using to enhance the network life time using less energy of the sensor node. To improve network life, researchers are using node heterogeneity in which sensor nodes have different energy levels in the network [7, 8]. Low Energy Adaptive Clustering Hierarchy (LEACH) is the very basic protocol for clustering which shows enhancement of lifetime and stability of the network [9]. In clustering, few sensor nodes formed the cluster and one of them worked as Cluster Head (CH) and other sensor node worked as Cluster Member (CM).

## *1.1 Motivation*

Although many researchers have done a lot of work in literature with above motivation [6, 9, 15, 16]. But a lot of potentials is still available in the field of clustering and node heterogeneity of sensor nodes Which motivate us to do research work in this area. So our objectives are

- To study of the different protocols related to clustering and node heterogeneity.
- To enhance the lifetime of network using heterogeneity of sensor nodes.

This paper is organized as follows: A brief summary of related work is provided in Sect. 2. The proposed approach is discussed in Sect. 3. The performance evaluation and analysis are discussed in Sect. 4. Conclusion is summarized in Sect. 5 with possible future scope.

## 2 Literature Review

In the literature review, Clustering and thresholding technique are the most popular technique in the routing protocol. LEACH is most popular clustering protocol in which clustering is done on the probability basis and one of the sensor node is elected as CH and rest are the cluster member (CM) [6]. CM selection is based on the distance from the base station and cluster head. This protocol is divided into two phases, one is setup phase and the second is steady-state phase. In the setup phase, cluster formation is done and steady-state phase transmission of data from sensor node to base station is done via cluster head. LEACH protocol is the very first paper which is based on homogeneity on the basis of sensor nodes. Stable Election Protocol (SEP) is another very basic research paper based heterogeneity of sensor nodes which improves the energy efficiency of the whole network [7]. EHE-LEACH, i.e. Enhanced Heterogeneous LEACH is another protocol based on direct communication and clustering is done after certain radius of region which shows better energy efficiency [9] compared to [7].

Kumar et al. [5] proposed Enhanced Threshold Sensitive Stable Election Protocol (ETSSEP) which shows better result in comparison of [7] about the network lifetime and stability. This protocol is based on dynamically changing cluster head selection probability. It elects CHs on the basis of remaining energies of sensor nodes and minimum number of clusters per round of the communication. Mittal et al. [11] proposed Distance-based Residual Energy Efficient Stable Election Protocol (DRESEP) for wireless sensor network. This is an event-driven protocol and a reactive algorithm. It uses dual-hop communication between CH and BS to balance the load. This protocol improved the utilization of energy and network lifetime significantly. Benkirane et al. [12] proposed Distance-based Stable Election Protocol (DB-SEP) for heterogeneous WSNs. In this paper, authors select the CHs on the basis of initial energy of SN and distances between SNs and the BS. The result shows DB-SEP increases the network lifetime as compared to SEP. Kumar et al. [13] proposed Multihop Communication Routing (MCR) protocol for heterogeneous WSNs. In this paper, authors select the CHs on the basis of weighted probabilities among different types of sensor nodes. In this protocol, the CHs send the data using the optimal multi-hop path to the BS. Hence MCR is the multi-hop version of EEHC.

TEEN i.e. Threshold-sensitive Energy Efficient Network and TSEP, i.e. Threshold sensitive Stable Election Protocol are the threshold-based protocol in which the sensed data is transmitted to the base station only when the particular threshold has reached [15, 16]. Such protocols are used Soft Threshold (ST) and Hard Threshold (HT) to avoid redundancy of data and to transmit the data to CH respectively. Zhang et al. [18] proposed dynamic heterogeneity of sensor node in terms of energy. They proposed Adaptive Clustering algorithm for Dynamic Heterogeneous wireless sensor networks (ACDHs) in which they combined static heterogeneity and dynamic heterogeneity. They added a certain percentage of sensor node in the network when the operations of the network are evolved. In this protocol, CH is selected on the basis of initial energy of the SN, remaining energy of SN and average energy of network. Singh et al. [19] has proposed multilevel heterogeneous network model which is based on HEED protocol. In the proposed MLHEED (Multi-Level HEED) protocol, the heterogeneity is divided into *n*th-level. This protocol has two parameters as primary and secondary parameter. The primary parameter decides the heterogeneity level using the secondary parameters. At every level of heterogeneity, the numbers of nodes are determined by the secondary parameter. The MLHEED protocol has implemented fuzzy in which four variables are used for deciding the cluster head like residual energy, node density, average energy and distance between BS and SNs.

Tyagi et al. [20] have proposed a Lifetime Extended Multi-Levels Heterogeneous routing protocol for wireless sensor networks (LE-MHR). In this protocol, authors have proposed k levels horizontal energy heterogeneity which improves network lifetime at a significant level. Rizk et al. [21] have proposed stable and energy-efficient clustering protocol for heterogeneous WSNs. In this paper network has divided into two cluster in which one cluster has only Normal Nodes (NN) which sense the data and transmit to another cluster. Another cluster has Advance Nodes (AN) which collected the data and transmit to base station. In this protocol advance nodes make sure have enough energy and maintain minimum energy with respect to normal nodes. Zaatouri et al. [22] present a comparative analysis among different protocols of homogeneous and heterogeneous environments. In homogeneous environment, all sensor nodes have the same technical specification like initial energy, bandwidth, etc. while in heterogeneous environment, sensor nodes have different technical specification like initial energy, bandwidth, etc. In heterogeneous environment, some sensor nodes have different initial energy than others. In this paper, researchers are classified a number of protocols and maintained comparative chart on the basis of the node's typology and mobility aspects. Aznoli et al. [23] reviewed the deployment mechanism of the sensor node which has been used in WSN systematically. Sensor node can be deployed on the basis of deterministic and non-deterministic deployment techniques. They have reviewed a number of publications from the year 2004 to 2016 and produced a number of open issues which needs to be addressed. Table 1 shows the comparison of different protocols with proposed approach.

**Table 1** Comparison of pre-existing protocols

| Protocol | CH selection | LS | ST | Merit | Demerit |
|---|---|---|---|---|---|
| LEACH [6] | Random | α | – | LT increased | Unbalance load |
| SEP [7] | WEP | α | α | LT better | Stable region is poor |
| EHE-LEACH [9] | WEP | α | α | LT and ST improved | – |
| MCR [13] | WEP | α | α | LT improved | Poor stability |
| TEEN [15] | Random | α | – | LT improved | Throughput reduced |
| TSEP [16] | WEP | α | α | LT and ST improved | – |
| ACDN [18] | WEP | α | – | LT improved | – |
| MLHEED [19] | WEP | α | – | LT improved | – |
| SEECP [17] | Fixed | α | α | LT improved | SN exhausted early |
| Proposed approach | Fixed | α | α | LT and ST improved | – |

LS-Life span, LT-Lifetime, ST-Stability, SN-Sensor node

# 3 Proposed Approach

Here, we discussed the energy model used in this approach. Then we discussed selection of CH and CMs of cluster. We discussed for selection criteria of dynamic region and dynamic hopping of transmission. In the proposed protocol, we select the CH in deterministic manner but the selection of dual-hop communication is based on node density which select the radius of circle through which outer circle CH will send the data to the inner circle CH that has the higher energy. The proposed protocol is a reactive, energy-efficient, load-balanced and stable protocol for HWSN. In the proposed protocol, CH selection is deterministic approach which is based on leftover energy of each sensor node [17]. Each simulation round of the proposed protocol is divided into the two phases i.e. setup and steady-state phase. During the ow of paper, we use following symbols as elaborated in Table 2.

## *3.1 Assumptions*

In addition to proposed protocol, we have made some following assumptions-

- It is assumed that the network is heterogeneous i.e. SNs have different level of energy.
- It is assumed that all the SNs and BS are stationary after the deployment.
- It is assumed that sensor nodes have bidirectional capability of transmission.
- It is assumed that those SNs are at the boundary of the network can communicate without any problem.
- It is assumed that Received Signal Strength(RSS) is calculated with help of distance.

**Table 2** Symbols and abbreviation

| Symbol abbreviation | |
|---|---|
| $N$ | Total number of nodes |
| $M * M$ | Size of the network |
| $E_{DA}$ | Data aggregation energy |
| $K$ | packet size |
| $_{mp}$ | Radio amplifier energy for multipath |
| $_{fs}$ | Radio amplifier energy for free space path |
| $E_n$ | Initial energy of normal nodes |
| $E_i$ | Initial energy of intermediate nodes |
| $E_a$ | Initial energy of advance nodes |
| $\alpha$ | Incremental energy of intermediate nodes |
| $\beta$ | Incremental energy of advance nodes |
| $a$ | Percentage of normal nodes |
| $b$ | Percentage of intermediate nodes |
| $c$ | Percentage of advanced nodes |

- It is assumed that every SN have the capability for data aggregation by which multiple data packets can compress into single data packet.

## 3.2 *Problem Formulation*

Recently Mittal et al. [17] proposed distance-based routing protocol which is energy-efficient and stable protocol for WSN. CH determination on the basis of residual energy of the alive SNs. It is deterministic protocol which eliminates the uncertainties of CH election. Each round of the SEECP is bifurcated in two phases as setup and steady-state phase as proposed in LEACH [6], The advantage of SEECP are as follows:

- SEECP choose the criteria to select CH is deterministic which eliminates the uncertainty of number of CH.
- SEECP is based on residual energy of each node which improves the certainty of CH.
- SEECP is randomly distributed which does not require inclusive information of the network.
- In SEECP, CH communicate with the BS using two-hop communication which improves the energy efficiency of the network.
- Although the above-mentioned advantages improve the network life it may be vulnerable to the following limitations.
- Although SEECP has improved the life time of first node to die but reduce the stability of the overall network life which depends on the difference between last

node die and first node die. This is the main crucial parameter of any network because if the first node will die in early stage, the remaining nodes must sustain in the network for the longer life so that we can get the information from the network like monitoring of the forest area, it is least bothered if one node die early stage but we must get the information for longer time.

- In the SEECP, there is no such valid justification for the selection of radius.
- In SEECP, once the selection of CH is done that node remains CH until its energy is higher which impose to die faster which may reduce the efficiency of the network.

## 3.3 *Energy Model*

In this section, we cover the energy model used in the proposed protocol. The basic energy model is LEACH protocol and it is homogeneous environment based model [6]. The Received Signal Strength (RSS) is the criteria for selection of cluster members. The selection of CH is random and sensor node are randomly distributed. But SEECP is heterogeneous environment based model [17]. It shows the three level of heterogeneity. SEECP have three types of sensor nodes like normal node, intermediate node and advance node. The selection of CH is deterministic and sensor node are randomly distributed. It has fixed 5% selection of nodes to be CH.

In this paper, we have used radio model same in [6] in Fig. 1. According to this radio model, the transmitting energy is given by as in Eq. (1)

$$E_{Tx}(k, d) = k * E_{\text{elec}} + k * \varepsilon_{\text{amp}} * k * d^n \tag{1}$$

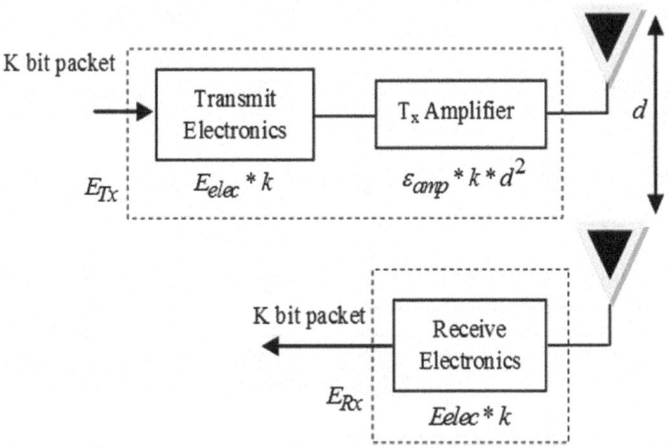


**Fig. 1** Energy model [10]

where $n$ (2 or 4) is path-loss component and amplification factor $\varepsilon_{\text{amp}}$ ($\varepsilon_{fs}$ or $\varepsilon_{\text{mp}}$) are defined as free path and multipath environment respectively.

$$E_{Tx}(k,d) = \begin{cases} k * E_{\text{elec}} + k * \varepsilon_{\text{fs}} * d^2, \ (d < d_0) \\ k * E_{\text{elec}} + k * \varepsilon_{\text{fs}} * d^4, \ (d \geq d_0) \end{cases} \quad (2)$$

The above set of Eqs. (2) is the dissipation energy per bit of transmitter circuitry and $d_0 = \sqrt{\frac{\varepsilon_{\text{fs}}}{\varepsilon_{\text{mp}}}}$ The receiving energy is

$$E_{Rx}(k) = E_{\text{elce}} * k \quad (3)$$

The proposed protocol has three-level heterogeneous environment as shown in Fig. 2 in which level one has only normal nodes which are a times of $N$ and has $E_n$ energy. Level two has intermediate nodes which are $b$ times of $N$ and has $E_i = \alpha E_n$ energy. Level three has advance nodes and has $E_a = \beta E_n$ energy. So initial energy of all nodes is given by Eq. (4).

$$E_{\text{total}} = E_n + \alpha * E_n + \beta * E_n \quad (4)$$

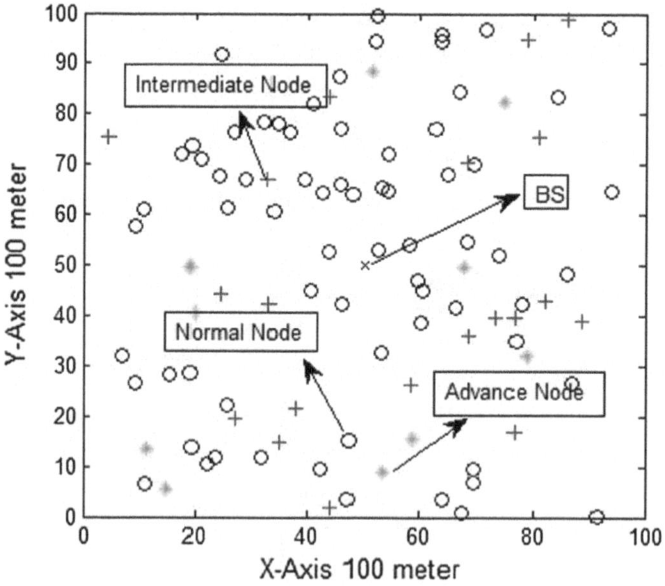


**Fig. 2** Random node deployment in the network

### 3.4 Setup and Steady-State Phase

The selection of setup and steady-state phase is same as in [6] but the CH selection is on the basis of remaining energy of SNs as in [17]. In the first round, BS will select only 5% CHs randomly as in [17] but in the following rounds, the node that has higher remaining energy will elect as CH [17]. In the second phase, every cluster member transmits its sensed data to the respective CH only when the event triggered i.e. $H(T) \leq C(V)$ where C(V) is the current value of the sensed parameter and the hard threshold value is given by $H(T)$. The re-transmission is only possible when the condition $C(V) - S(V) \geq S(T)$ is satisfied where $S(V)$ is the sensed attribute and the soft threshold value is given by $S(T)$ [15]. The aggregated data is sent by CHs to BS either using dynamic hop or single-hop route depending on the distance among CH, BS and radius of dynamics region.

### 3.5 Dynamic Region Selection

The selection of dynamic region is based on node density of the network. The node density will remain same for any type of network. It may be square region, rectangular region or circular region as in our case it is circular region as in [17]. So we choose the square network in which we choose dynamic circular region on the basis of node density.

$$\frac{N_1}{M^2} = \frac{N_2}{\pi R_d^2} \tag{5}$$

$$\Rightarrow R_d^2 = \frac{N_2}{N_1}\left(\frac{M^2}{\pi}\right) \tag{6}$$

$$\Rightarrow R_d = \sqrt{\frac{N_2}{N_1}\left(\frac{M^2}{\pi}\right)} \tag{7}$$

where $M$ is the dimension of square network, $R_d$ is the dynamic radius of circular region, $N_1$ is the total number of SNs in square network and $N_2$ is total number of SNs in dynamic circle. BS is situated in the center of circle i.e. center of the square network and center of circle coincide. Therefore, as we increase the number of node inside the circle as radius of the circle will increase.

### 3.6 Dynamic Hoping Selection Process

The selection of dynamic hoping is based on fixed 5% CH whenever in the inner region, there is 3 or 4 CHs than only the sensed data is transferred via dual hope communication otherwise data is transferred by single-hop communication in both region (inner and outer). In such way the load balancing can be done efficiently. In this protocol for distant CH, dynamic hop communication exists if the $d_{\text{CH-BS}}$ is greater than dynamic radius $R_d$ and inner region has 3 or 4 CH otherwise single-hop communication will exist and CH will send the data directly to BS for both inner and outer regions. In dual-hop communication, distant CH will look another CH which is known as companion node within dynamic region that has higher energy to transmit the data to BS.

### 3.7 Algorithm of Proposed Approach

In this paper, we use the following algorithm for the proposed approach. The selection of CH is fixed for 5% after that CM are elected which send the sensed data to CH as per threshold value. CH aggregates the data and sends the data to BS using single or dual-hop transmission. The selection of hopping is based as per the algorithm.

## 4 Performance Evaluation

### 4.1 Performance Metrics

**Network lifetime**: This is most wildly used metric for the analysis of the sensor network. Network lifetime is defined as the total time during which the sensor nodes will transmit the data of the dedicated task to the base station where the time is defined in terms of number of round. Sensor nodes will transmit the data till they are alive which means they have energy if all the dedicated energy is drained out, the sensor node is considered as dead. Hence the network lifetime is the difference between Last Node Alive (LNA) and the starting of the network.

**Algorithm 1 Pseudo code for Proposed Protocol**

1. Initialize the network parameters
2. Identifies the normal, intermediate and advance Nodes
3. BS selects5%CH at first Round
4. CH broadcast message for CM
5. CH broadcast TDMA Schedule for CM
6. CM checks C(V)
7. if C(V) $\geq$ H(T) then

8. Test = C(V)-S(V)<br>
9. else go to 6<br>
10. end if<br>
11. if test ≥ S(T) then<br>
12. Data send to CH<br>
13. else go to 6<br>
14. end if<br>
15. CH aggregates data<br>
16. CH checks distance CH to BS<br>
17. If $dCH2BS < Rd$<br>
18. Data transmit to BS using single hop transmission<br>
19. Else if ($d_{CH2BS} > R_d$) and (*more CH within* $R_d$)<br>
20. Data send to BS using Dual hop transmission<br>
21. Else<br>
22. Data send to BS using Single hop transmission<br>
23. Check higher energy SN for CH<br>
24. Select 5% CH for next Round<br>
25. Go to 4

**Stability**: The another most wildly used metric is Network Stability. Stability of the network is defined differently by different researchers. We considered stability in terms of first Node Die (FND) i.e. this is the time duration from start of operation until first sensor node that has exhausted its total energy. After that nodes start to die as they exhausted their energy. So the network is stable till all the sensor nodes are alive, after that, they start die out and decrease the stability of the network.

**Throughput**: This is another performance metric which shows the network quality. This is defined by different researchers differently but the overall meaning will remain same. Throughput is defined as the total number of bits per second which are transmitted to base station.

The performance of proposed protocol is evaluated in terms of FND, HNA and LNA for the evaluation of network stability. The comparing protocols are implemented in our simulation environment by MATLAB. The 10 different iterations are performed on designed network which has 100 sensor nodes randomly distributed in square sensing field that have base station (BS) at the $x$-axis of 50 m and $y$-axis of 50 m. The characteristics of network are used for simulations as summarized in Table 3.

This analysis is done to analyze the effect of proposed protocol in heterogeneous environment only because in the homogeneous environment, it will perform better due to different SN of higher energy. Simulations are performed with different temperature ranging from 0° to 200 °F in different regions of the network. 50 °F is set as the hard threshold and 20 °F is set as the soft threshold values [17]. In the heterogeneous environment, 20% SNs are set as intermediate nodes and 10% SNs are set as advance nodes of the total nodes. The normal node has $E_0 = 0.1\ J$ as initial energy and the advance nodes and intermediate nodes are assigned 3 ($\beta = 3$) and 2 ($\alpha = 2$) times initial energy of normal node respectively.

**Table 3** Simulation parameters

| Parameter value | |
|---|---|
| $N$ | 100 |
| $M * M$ | $100 * 100\ m^2$ |
| BS | $x$-axis = 50 m, $y$-axis = 50 m |
| $E_{DA}$ | 5 nJ/bit/round |
| $K$ | 4000 bits |
| $mp$ | 0.0013 pJ/bit/$m^4$ |
| $fs$ | 10 pJ/bit/$m^2$ |
| $E_n$ | 0.10 J |
| $E_i$ | $\alpha$ times of $E_n$ |
| $E_a$ | $\beta$ times of $E_n$ |
| $\alpha$ | 2 |
| $\beta$ | 3 |
| $a$ | 70% of N |
| $b$ | 20% N |
| $c$ | 10% of N |

Figure 3 shows that different dynamic region that has different node density. As we increase the node density correspondingly dynamic radius will increase. Initial when the node density was low there is no significant improvement in FND of the proposed protocol but as we reached to 20 node density, the radius becomes 25.24 m that shows the significant improvement that means we have achieved better result at low radius as compared to [17]. Figure 4 shows dynamic radius of different node density. As we increase the node density, the dynamic radius will increase.

Figure 5 shows the average result of LEACH, EHE-LEACH, SEECP and proposed protocol with different radius of different node density under 10 iterations of heterogeneous environments. The CH selection process is same as [17] which has 5% of alive sensor nodes in every round. The selection of CH is based on the residual energy of the SNs. In every round CH transfer the data to BS either via direct or single-hop communication using higher energy CH in the calculated radius of circle. This radius is calculated on the basis of node density. As the node density increases, the radius proportionally increases. Both the protocols are reactive type protocols because the transmission is possible when the conditions are met. Sensor nodes sensed the data continuously but can transmit only when the threshold condition is met.

Figure 5a–c show the stability of different routing protocols. No doubt LEACH performances least stability other than analyzed protocols because of LEACH is homogeneous routing protocol in which all SNs have equal energy. Other protocols have heterogeneous routing protocols.

Figure 5a shows better result other than analyzed protocols in terms of FND. Since FND shows the inactiveness of the first SN i.e. it will no longer with the network

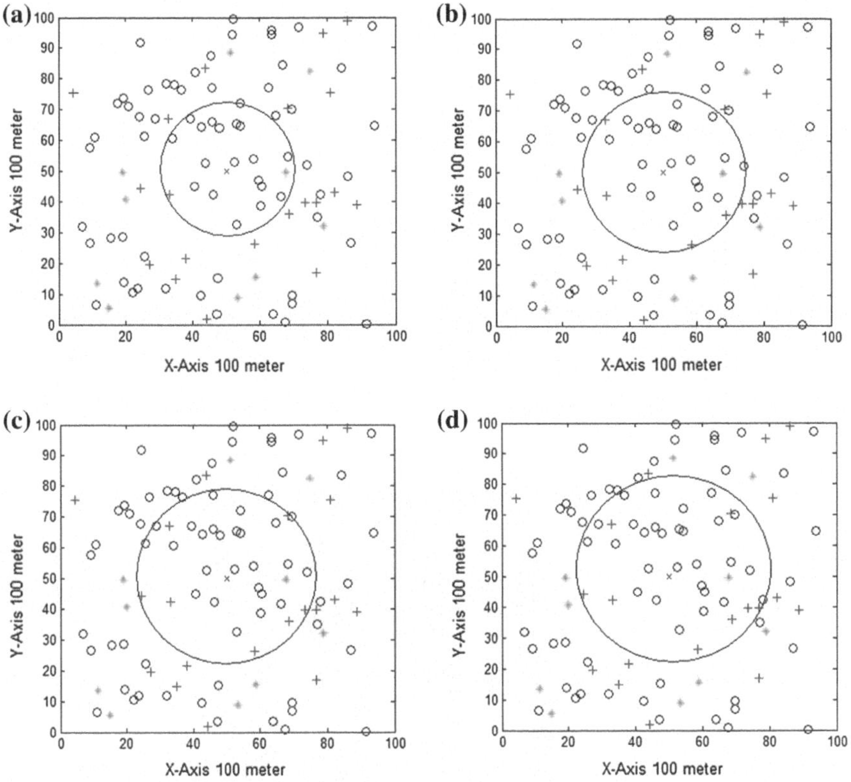


**Fig. 3** Scenarios of **a** dynamic radius of 20 SNs, **b** dynamic radius of 25 SNs, **c** dynamic radius of 30 SNs and **d** dynamic radius of 35 SNs

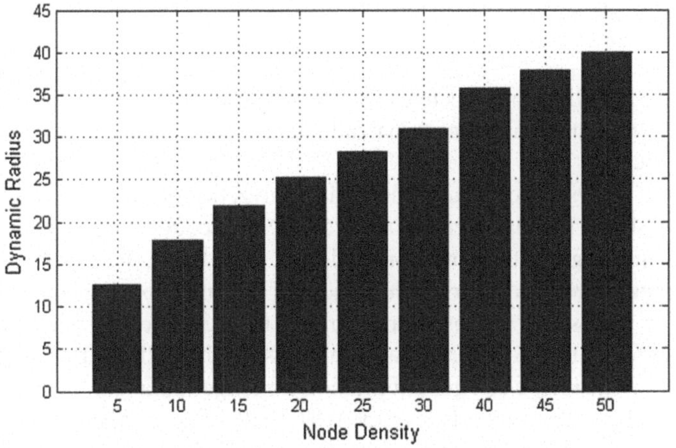


**Fig. 4** Different node density versus dynamic radius

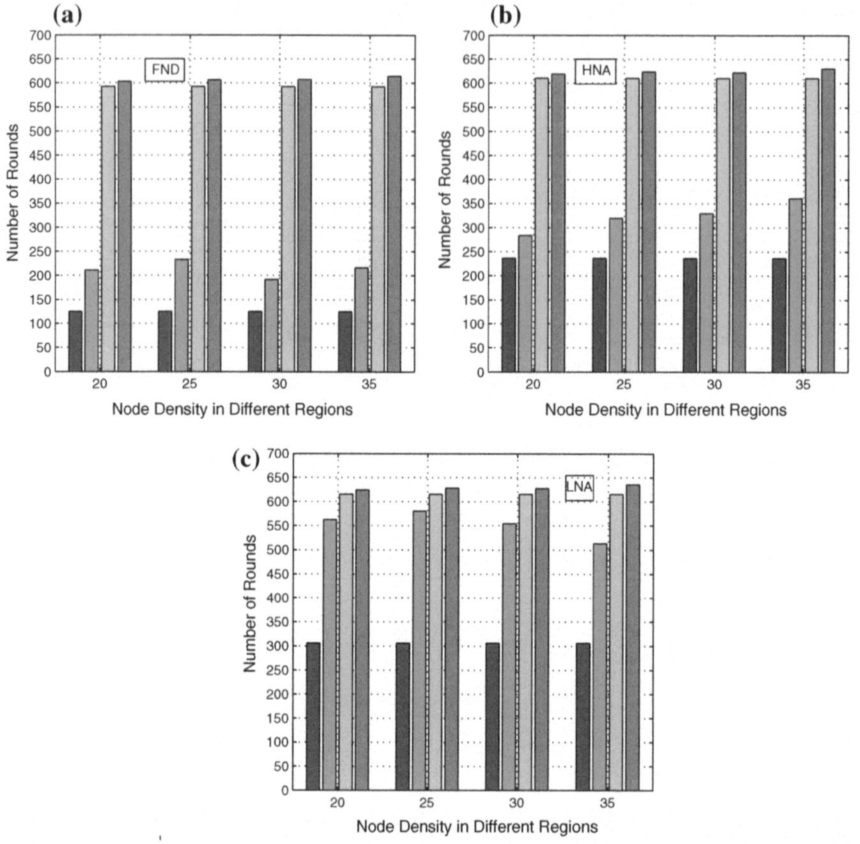


**Fig. 5** Scenarios of **a** FND at different node density, **b** HNA at different node density, and **c** LNA at different node density

and can't send the data to respective CH. The proposed protocol shows better result because of single-hop communication and data is transmitted only when triggered valued is reached.

Figure 5b shows better result other than analyzed protocols in terms of HNA. Since HNA shows the inactiveness of the 50% SNs i.e. 50% sensor node has died and they are no longer with the network and can't send the data to respective CH. The proposed protocol shows better result because of dynamic hopping communication.

Figure 5c shows that proposed protocol has not better results other than EHE-LEACH but better than SEECP protocol in terms of LNA. EHE-LEACH shows the better result; it means we can get information for longer duration.

Figure 6a shows the comparison of LEACH, EHE-LEACH, SEECP and proposed protocol for the throughput analysis. Throughput shows that how many number of bits are transmitted to base station from the cluster head. The comparison figure shows that proposed protocol outperforms as compared to other protocols. This shows

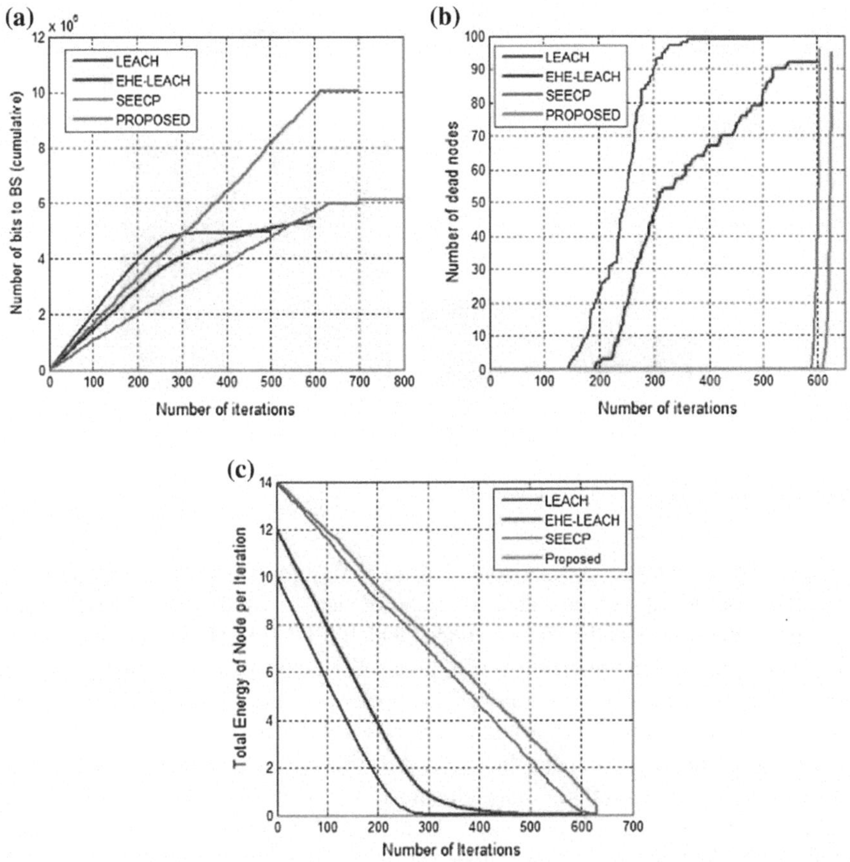


**Fig. 6** **a** Comparison of throughput of different protocols, **b** number of dead nodes of different protocols, and **c** total energy of node per round

that proposed protocol transmits more numbers of bits to base station which can analyze for the intended purpose. The proposed protocol may transmit large number of data in the network because of dynamic selection of transmission of data to BS via either dual-hop or single-hop communication. When 3 or more CHs are within the dynamic radius, transmission is done using dual-hop otherwise transmission is done via single-hop whether CH is in dynamic radius or out of dynamic radius.

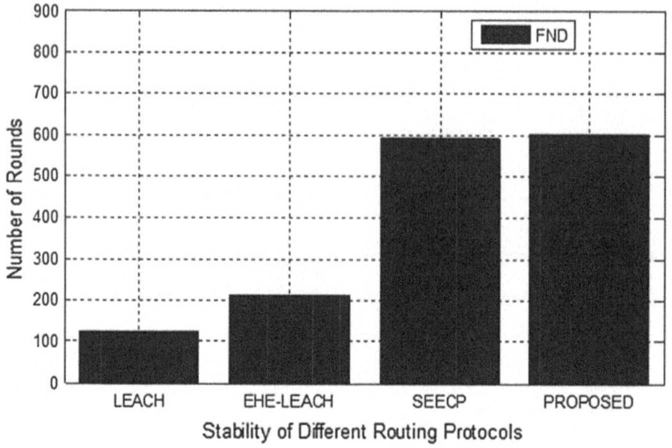


**Fig. 7** Stability of different routing protocol

Figure 6b shows the comparison of dead nodes for LEACH, EHE-LEACH, SEECP and proposed protocol. This graph shows that proposed protocol have longer life time of the sensor nodes. Sensor nodes which use LEACH protocol have least life time because of homogeneous sensor nodes. Figure 6b also shows that proposed protocol has high network stability. This happens because of heterogeneity of the sensor nodes but the proposed protocol performs well because of selection of CH in every round among higher energy sensor nodes. Figure 6c shows the total energy of sensor nodes in a given round. This is the most important metric of the network. As soon as the energy of the SN exhausted, the node will die and will not send any data to CH or BS. As long as the sensor node sustain in the network, it will have energy and transmit the data.

Figure 7 shows the stability of LEACH, EHE-LEACH, SEECP and proposed protocol. Stability performance metric shows the FND of different protocols which shows first node die in the network. As soon as the first node die in the network, the network became unstable. The proposed protocol outperforms with respect to other protocol. This happens because the selection of higher residual energy of CH and dynamic selection of transmission hopping. Therefore, always higher energy CH will select and balance energy load among the sensor node.

## 5 Conclusion

There are many designing issues in routing protocols in WSN like energy efficiency, load balancing and stability and these problems can be solved by different routing algorithm. This paper describes threshold-based routing protocol that is suitable for time-based critical events. In this proposed protocol, residual energy is the parameter of CH selection. This protocol selects the dynamic radius based on node density which shows better result than [17]. The deterministic CH selection approaches improve load balancing among the SNs and provide balanced energy consumption using dynamic hopping. This proposed protocol is outperforming over SEECP because of dynamic region and dynamic hopping of communication The mobility of BS and SNs are another crucial area of research in wireless sensor network which is untouched in this proposed protocol. Hence the mobility of SN and BS may be considered as future scope for designing the protocols in the applications of real-time scenario of wireless sensor network.

## References


1. Afsar, M.M., Tayarani-N, M.: Clustering in sensor networks: a literature survey. J. Netw. Comput. Appl. **46**, 198226 (2014)
2. Pantazis, N.A., Nikolidakis, S.A., Vergados, D.D.: Energy-efficient routing protocols in wireless sensor networks: a survey. IEEE Commun. Surv. Tutorials **15**(2), 551591 (2013)
3. Rostami, A.S., Badkoobe, M., Mohanna, F., Keshavarz, H., Hosseinabadi, A.A.R., Sangaiah, A.K.: Survey on clustering in heterogeneous and homogeneous wireless sensor networks. J. Supercomput. **7**(4), 277–323 (2018)
4. Memon, I., Hussain, I., Akhtar, R., Chen, G.: Enhanced privacy and authentication: an efficient and secure anonymous communication for location based service using asymmetric cryptography scheme. Wirel. Pers. Commun. **84**, 1487–1508 (2015)
5. Kumar, S., Verma, S.K., Kumar, A.: Enhanced threshold sensitive stable election protocol for heterogeneous wireless sensor networks. Wirel. Pers. Commun. **85**, 16 (2015)
6. Heinzelman, W.B., Chandrakasan, A., Balakrishnan, H.: Energy efficient communication protocol for wireless micro sensor networks. In: Proceedings of the 33rd Annual Hawaii International Conference on System Sciences (HICSS-33), p. 223. IEEE (2000)
7. Smaragdakis, G., Matta, I., Bestavros, A.: SEP: a stable election protocol for clustered heterogeneous wireless sensor networks. In: Proceedings of the International Workshop on SANPA (2004)
8. Kumar, D., Aseri, T.C., Patel, R.: B-EEHC-energy efficient heterogeneous clustered scheme for wireless sensor networks. Comput. Commun. **32**, 662667 (2009)
9. Tyagi, S., Kumar, N.: A systematic review on clustering and routing techniques based upon LEACH protocol for wireless sensor networks. J. Netw. Comput. Appl. **36**(2), 623–645 (2013)
10. Tyagi, S., Gupta, S.K., Tanwar, S., Kumar, N.: EHE-LEACH: enhanced heterogeneous LEACH protocol for lifetime enhancement of wireless SNs, 978-12013 IEEE (2013)
11. Mittal, N., Singh, U.: Distance-based residual energy efficient stable election protocol for WSNs. Arab. J. Sci. Eng. **40**, 1637–1646 (2015)
12. Benkirane, S., Benihssane, A., Hasnaoui, M.L., Laghdir, M.: Distance based stable election protocol (DB-SEP) for heterogeneous wireless sensor network. Int. J. Comput. Appl. **58**(16), 915 (2012)
13. Kumar, D., Aseri, T.C., Patel, R.B.: Multi-hop communication routing (MCR) protocol for heterogeneous wireless sensor networks. Int. J. Inf. Technol. Commun. Converg. **1**(2), 130–145 (2011)
14. Tanwar, S., Kumar, N., Rodrigues, J.J.P.C.: A systematic review on heterogeneous routing protocols for wireless sensor network. J. Netw. Comput. Appl. **53**, 39–56 (2015)
15. Manjeshwar, A., Agrawal, D. P.: TEEN: a routing protocol for enhanced efficiency in wireless sensor networks. In: Proceedings of International Parallel and Distributed Processing Symposium (IPDPS'01) workshops, pp. 2009–2015. San Francisco, CA, USA (2001)

16. Kashaf, A., Javaid, N., Khan, Z.A., Khan, I.A.: TSEP: threshold-sensitive stable election protocol for WSNs. In: Proceedings of 10th International Conference on Frontiers of Information Technology, pp. 164168. IEEE, Islamabad (2012)
17. Mittal, N., Singh, U., Sohi Singh, B.: SEECP: a stable energy efficient clustering protocol for wireless sensor networks. Wirel. Netw. **23**, 18091821 (2017)
18. Zhang, J., Chen, J.: ACDH: An adaptive clustering algorithm for dynamic heterogeneous wireless sensor networks. Wirel. Netw. (2017)
19. Singh, S., Chand, S., Kumar, B.: Multilevel heterogeneous network model for wireless sensor networks. Telecomm. Syst. **64**, 259277 (2017)
20. Tyagi, S., Tanwar, S., Gupta, S.K., Kumar, N., Rodrigues, J.J.P.C.: A lifetime extended multi-levels heterogeneous routing protocol for wireless sensor networks. Telecommun. Syst. (2015)
21. Rizk, R., Farouk, F., Zaki, F.W.: Towards energy efficiency and stability in heterogeneous wireless sensor networks. Wirel. Pers. Commun. **96**, 43474365 (2017)
22. Zaatouri, I., Guilou, A.B., Alyaoui, N., Kachouri, A.: A comparative study of the energy efficient clustering protocols in heterogeneous and homogeneous wireless sensor networks. Wirel. Pers. Commun. **97**, 64536468 (2017)
23. Aznoli, F., Navimipour, N.J.: Deployment strategies in the wireless sensor networks: systematic literature review, classification, and current trends. Wirel. Pers. Commun. **95**, 819–846 (2017)

# On Security of Opportunistic Routing Protocol in Wireless Sensor Networks

**Archana Sharma and Yashwant Singh**

**Abstract** With the passage of time, Wireless Sensor Networks (WSNs) gain importance and have become one of the most fascinating areas of research in the past few years. Reliable and efficient data routing, that helps the data to reach its destination, remains the bottom line of research problem. Hence, various routing protocols are developed which are based on different parameters. But opportunistic routing has fascinated many researchers these days because of its broadcasting nature that makes it more efficient than the old routing methods. However, while routing in this sensor network the loss of data takes place because of its security lack. Therefore, security is also a challenging issue with the routing of WSN. This paper presents an overview of the WSN, its architecture, applications of WSN in different fields presented. An overview of the security aspect of routing in WSN is given. Finally, a comparison of different security methods of opportunistic routing is also presented.

**Keywords** Wireless sensor networks · Routing protocol · Opportunistic routing · Security

## 1 Introduction

A wireless sensor network can be defined as an assembly of different sensor nodes, clustered to form a collaborative network. Sensor networks include spatially distributed independent sensors that track physical and natural conditions at different locations, like temperature, pressure, motion, vibration, etc. and gather the data and pass this data to a main location regarded as a sink. The Wireless sensor network consists of the extensive amount of sensor nodes, ranging in numbers, from a few hundred to even thousands, and they are connected through wireless transmission. The Sensor nodes are typically known as motes. The sensor node typically has five

A. Sharma (✉) · Y. Singh
Department of Computer Science and Information Technology, Central University of Jammu, Samba, Jammu and Kashmir 181143, India
e-mail: archyshar1995@gmail.com

Y. Singh
e-mail: yash22222k1@gmail.com

 
P. K. Singh et al. (eds.), *Proceedings of ICRIC 2019*, Lecture Notes in Electrical Engineering 597, https://doi.org/10.1007/978-3-030-29407-6_30

key components (as appeared in Fig. 1): radio transceiver, microcontroller, memory, electronic sensing device and vital source, typically a battery [1]. In WSN, the sensor nodes are deployed in two types: organized and unorganized. In unorganized deployment, the nodes are spread in an abandoned environment in which no human intervention is allowed, whereas, the case in which all the nodes are arranged in a specific manner, is the organized deployment.

The Wireless Sensor Network is one of the developing technologies that is enabling the aggregation of information in different scenario ranging from environmental monitoring to various health and military applications. The applications of WSN are numerous, however, these are based on strategies like monitoring, controlling and tracking. Figure 2 below displays us some of the applications of WSN in different fields.

The process of selection of the best path for travelling of data packets to reach their destination is regarded as routing. If the motes and the sink have less distance

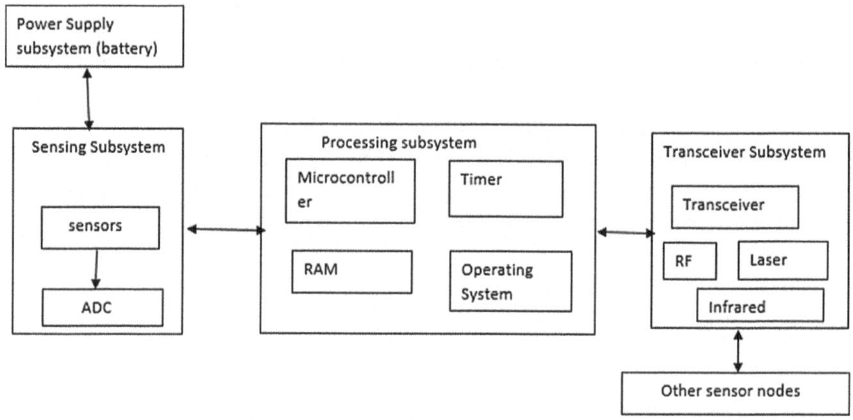


**Fig. 1** Architecture of wireless sensor network [1]

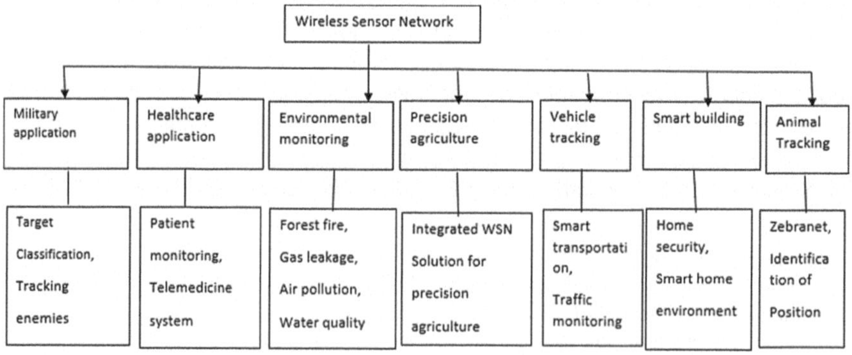


**Fig. 2** Applications of wireless sensor networks

between them, then it is a single-hop communication, but if the sensor nodes are placed far away from the sink node, then it is a multi-hop communication [2]. The Opportunistic Routing (OR) is an innovative routing methodology that uses features of broadcast communication [16]. It broadcasts the data packet so that numerous neighboring nodes can overhear it. Later, the node that has successfully received that broadcasted packet coordinate among one another and then selection of best node is done which is used for packet forwarding. Finally, the packet of data is again broadcasted and it is performed again and again until the packet attains its final destination. It has two main components: selection of candidates and coordination method. The candidate selection method helps in the creation of a list of nodes which are used for data forwarding. Coordination method helps in selection of the best node from the list, which can be used for reaching the destination. It increases reliability [14] and also has better performance in comparison to traditional routing methods.

The security is most important parameter in any field. Security in these networks is a very challenging task because there is no physical protection, an open and shared communication medium is used, nodes are deployed in a hostile environment, constraints on sensor nodes make WSNs more vulnerable to Denial of Service (DOS) attack, a distributed network architecture and many more. Security in WSN consists of network security, data security, source security. For safe and secure communication, some security necessities should be fulfilled by WSN. The Security requirements of WSN include confidentiality, authentication, integrity, availability, authorization, and non-repudiation. Some secondary security requirements are data freshness, source localization, and self-organization [20]. These requirements provide information of the sensor network.

OR protocols working is greatly affected by the existence of malicious nodes and invaders. Thus, a large number of security improvement methods are developed by the researchers, that is based on the specific characteristics of OR protocols. These security-related protocols are divided into classes: trust-management method, game theory method, and other interrelated protocols [25]. In the trust management process, every node of the network is assigned a distinct value known as the trust value. This trust value shows their level of trustworthiness. These trust values are determined based on their successful or unsuccessful communications with the other sensor nodes of network. Likewise, the game-theory-based method considered the cooperation between the nodes. In this method, network cooperation can be intended as a match that is based on certain rules. Other security-related protocol includes Anonymous Wireless-mesh Encoded Opportunistic Routing, A packet salvaging model for OR methods, Modeling and performance evaluation of security attacks for OR.

In this paper, Sect. 2 discusses the literature survey, Sect. 3 provides an analysis of various security protocols in opportunistic routing and the last section concludes the paper.

## 2 Literature Survey

In the current scenario, great amount of work has been done in the area of WSN and opportunistic routing. WSN is a network of small nodes of sensors, whose components encapsulate various processes like sensing, processing of data and communication. There are various factors like fault tolerance, scalability, production, costs, operating environment, sensor network topology, etc. The protocol stack of WSN consists of physical layer, data link layer, network layer, transport layer, application layer, power managing level, mobility managing level, and a task managing level [3].

The standards can be defined as the protocol and functions used by the sensor nodes to interface different kind of networks. Some of the standards used in WSN are IEEE802.15.4, ZigBee, WirelessHART, Wibree, etc. There are two types of configurations of sensor nodes. In one configuration, nodes directly communicate with the sink, while in second configuration nodes do not communicate directly with the sink because of the presence of large coverage area [4]. In [5] author presented that the design of WSN depends on its application. If there is any change in the application, then design of WSN is also changed. The topologies of network also change if any, physical changes occur in the WSN.

WSN has various applications in different fields. Some of the important applications of WSN are military and security applications, civil applications, environmental applications and ecological application [6]. There are some factors that affect practical applications of WSN. These factors are cost, energy consumption, miniaturization, location, performance, mobility and hardware security [7]. Rashid and Rehmani [8] performs a survey of deployment of WSN in the urban areas. The WSN in urban areas can be used in gully pot monitoring, disaster management, transportation applications and security applications.

The routing protocol finds out the best path between sensor nodes and sinks node and then transmits the data through this best path. Node distribution strategy, data reporting method, nature of the node, fault tolerance, network dynamics and node localizations are the various factors that affect the designs of routing protocol [9]. In [10] different routing protocols are formed for WSN on the basis of issues that are degrading the performance of WSN. The classification is based on five main key features: energy efficiency, operational model, routing objectives, network architecture, and route selection. On basis of energy efficiency, two protocols are given: Fixed and Adjustable. On the basis of operational model: path selection, QoS based, delivery model. The Routing objectives include non-real time, network lifetime and real-time. The network architecture-based routing includes Data-Centric, Hierarchical, Location-Based. The route selection includes Proactive, reactive and hybrid. Chahal and Singh Gill [11] compare pegasis protocol with leach protocol and found that pegasis protocol is a chain-based protocol improvement over leach. HEED Protocol increases lifetime by distributing energy consumption between nodes and also minimizes the overhead. Misra and Goswami [12] classify routing protocol in various categories on the basis of factors like path establishment, structure of network, originator of communication and operation of the protocol. The path establishment classifies protocol in the following types: proactive, reactive and hybrid protocols. The network structure includes the flat, hierarchical, location-based and direct protocol. The initiator of communication includes source and destination protocol. The protocol operation includes negotiation, QoS, coherent, query-based and multipath protocol.

The opportunistic routing is one of the most promising technologies because it uses broadcasting method of communication. Opportunistic Routing broadcasts the packets of data to the nearby nodes and after that the best node is chosen which is used to forward the packet. This cycle continues until packets reach its final destination [13]. The Opportunistic routing overcomes various issues of WSN like dreadful conditions of environment, shortage of predefined network infrastructure, etc. The allocation of resources is one of most challenging issues which is also overcome by OR. OR has upgraded reliability, efficiency and throughput of the networks [14]. It can be considered as an amalgamation of routing protocols and media-access control. The protocols like EFFORT and EEOR reduces the overall energy consumption of network and EQGOR protocol enhances quality of service parameter of WSN [15]. The main objective of opportunistic routing is overhearing and cooperation between the nodes. OR has better performances in comparison to traditional routing techniques [16]. Extremely Opportunistic Routing (ExOR) [17] is the first OR protocol and that has three phases: selection of the forwarding candidates, allowing transmissions, and making a decision to forward the packet or not. It makes a list of prioritized candidates. The main opportunistic routing protocols used are Energy Efficient Opportunistic Routing (EFFORT), Simple Opportunistic Adaptive Routing (SAOR), EEOR, Exclusive opportunistic routing (ExOR), Energy-Aware Opportunistic Routing (EAOR) [18].

There are various issues related to security are present in WSN. The major task for holding any security procedure in WSN is determined by sensor size, processing power, storage, and type of works performed by the sensors [19]. Modares et al. [20] classified the security requirements of WSNs into following types: Data Authentication, Data Confidentiality, Data Integrity, Availability, Data Freshness. The common security requirements are availability, confidentiality, integrity and authentication [21]. Some other requirements are also considered, they are known as secondary requirements. Those are source-localization, self-organization and data-freshness [22]. To overcome these security issues in WSN various secure protocols is used. In [23] a secure routing protocol based on a robust neighborhood monitoring system (NMS) is proposed. This protocol is energy-efficient. The protocol drops the packet if the malicious nodes are present in the routing path. It confirms a consistent node by node delivery of packets in the routing path. Another secure protocol is proposed known as Authenticated Routing for Ad hoc Networks (ARAN), that identifies as well as secures from any nasty assaults done by the outsiders and chums in a single explicit adhoc environment [24].

The security-related protocols of opportunistic routing into three categories: trust-based protocols, game-based protocols, and other protocols [25]. Trust-based is a renowned method applied for the protection against attacks and also for removing malicious nodes from the network. The Game-based methods try to convert the cooperation among the network nodes into a game. In this game, the players are the nodes that try to enhance their utility. It uses a mathematical model to solve and find the winner of a game.

The Cryptography can be used for security issue, but this will interrupt the functioning of WSN. Kumar and Singh [26] proposed a new energy-efficient and secure routing protocol known as trust and packet load balancing based opportunistic routing protocol (TPBOR). This protocol is based on Trust management mechanism. It is an energy-efficient protocol and secures the network by calculating the trust merits of nodes. In calculation of trust merit of nodes, least number of parameters are considered. The parameters like energy and packet delivery that appears to be necessary for a network are not taken. The trust and location-aware routing (TLAR) [27] protocol is a protocol used for commune location concerned routing in the WSN. TLAR protocol is robust in detecting and segregating the malicious nodes. This protocol fails to provide load balancing in the network. It is not applied for the real testbeds. An opportunistic routing protocol is known as Secure and Scalable Geographic Opportunistic Routing with received signal strength (SGOR) [28] provides us both scalability and security in WSNs. The proposed protocol has higher packet delivery rate. In this paper, a method for verifying the secure location of adhoc networks was not discussed. Rani [29] formed a new OR metric for wireless sensor network named as a trust aware energy-efficient OR metric for WSN. The proposed metric is resultant of trustworthiness and energy consumption of nodes. This metric prevents malicious activities and maintains data integrity. The proposed metric has higher security and has equal energy load distribution. The one-way delay rises with a minor increase in overhead computation. Trust-based Security Protocol (TSP) [30] which secures network of blackhole attack. In TSP, the trust is built on the quantity of messages that are transported successfully and three other factors also: Social Group value (SGV), Credits and Hop count. This protocol does not calculate the relative delivery probability of the nodes and also does not detect the node that provides forge information regarding delivery probability. Cao et al. [31] proposed a protocol named Trajectory-Driven Opportunistic Routing (TDOR). In this protocol, the source node calculates an immune path that should be followed to reach the destination and that path is used for delivery of message. This protocol provides efficient message delivery with lower overheads of routing. In [32] authors proposed a model that will benefit in improving the security of opportunistic routing protocols named it as A Novel Packet Salvaging Method. The proposed mechanism is based on the Discrete-Time Markov Chain and also uses node redundancy that is present in each hop of protocol. Darehshoorzadeh et al. [33] proposed a novel method that will aid in understanding the behavior of nasty nodes and also help in knowing their effects on wireless mesh networks. It uses a Discrete Time Markov Chain (DTMC) and can be applied for calculation of the packet drop ratio. In these methods, removing the malicious nodes from the network is difficult. A Potential Threat (PT) [34] based protocol is an opportunistic routing protocol. This protocol has been developed on the basis of a process that selects the features of the inward nodes on the basis of the factors like instant messages delivery, reputation of nodes and their history. The proposed protocol helps in securing the network against the malicious nodes. This protocol helps in the better exploitation of resources of networks, timely delivery of messages, reduces overhead and least amount of dropped packets. In this method, probabilistic metrics are not merged which calculate the nature of node.

The OR protocols rely on the fact that all nodes are benign and are eager to work together with each other. All nodes are however not benign, the attackers may affect some nodes and thus these nodes may act maliciously. These malicious nodes lead to wastage of network resources and moreover also affects the routing performance. In order to protect the network from these malicious acts, either nodes are to be secured from attacks or the communication is to be secured. For promoting cooperation of all the nodes in network, several other categories of protocols, like trust-based and game-based protocols are used.

## 3 Comparative Analysis of Security Methods of Opportunistic Routing in WSN

Security has been an important challenge and hence various attacks launched on the WSN. Table 1 shows these attacks, access level, the nature of attack, the protocol stack and their defenses [35, 36].

Because of wireless nature of WSN, an attacker can easily participate or observe the communication taking place in network and also due to lack of a proper security mechanism in many of routing protocols, it makes easier for an attacker to attack on these networks. Thus, the WSNs are susceptible to various kind of security attacks as shown in Table 1. Attackers may add some data bits to channel, replay the old packets or deploy some malicious node to attack the network. So, proper defense measures are to be used to protect the WSN from these security attacks.

A large number of security attacks are to be launched on the traditional routing protocols of WSN. Similar to that, various security attacks may be intended and initiated for OR-based security protocols [25]. Table 2 below lists these attacks and provide their description also.

The security attacks that can be initiated on the opportunistic routing is shown in Table 2. These attacks can be started by injecting misinformation into network, by using some false identity, by using some malicious nodes and by various other means. These attacks are launched by the attacker for decreasing the efficiency of the network, decreasing the routing performance and wasting resources of network. In order to protect the networks from these security attacks, various security methods are to be used.

In this section, the different security protocols of opportunistic routing are compared and the result is shown in the table below. Table 3 shows the name of various security attacks, their candidate selection and candidate coordination methods, attack models, security solution and also their advantages and disadvantages.

**Table 1** Different security attacks in WSN

| Name of attack | Nature of attack | Access level | Protocol stack | Defenses |
|---|---|---|---|---|
| Jamming [37] | Modification and interruption | Active | Physical and data link layers | Prioritization of messages, changing of mode, lowering of duty cycle, etc. |
| Tampering [37] | Modification and interception | Active | Physical layer | Hiding the nodes and tamper-proofing |
| Collision [37] | Modification and interruption | Active | Data link layer | Using error correcting codes |
| Traffic manipulation [35] | Modification and interruption | Active | Data link layer | By using methods like watch dog and back-off value |
| Exhaustion [37] | Modification and interruption | Active | Data link layer | Rate limitation |
| Unfairness [35] | Modification and interruption | Active | Data link layer | Small frames |
| Sybil attack [36] | Modification and interception | Active | Datalink layer, physical and network layers | Authentication |
| Sinkhole attack [36] | Fabrication, modification and interruption | Active | Physical layer and data link layer | Identity certificate having some unique information |
| Wormhole attack [36] | Modification and interception | Active | Data link layer and network layer | Packet leash |
| Hello flood [38] | Fabrication and modification | Active | Network layer | Authentication and packet leashes |
| Acknowledgement flooding [35] | Fabrication and modification | Active | Network layer | Bi-directional link verification |
| Traffic analysis [35] | Confidentiality and availability | Passive | Data link layer | False paths can be used to confuse an attacker |
| Flooding [35] | Interruption and fabrication | Active | Transport layer | Mystify the attacker |
| Overwhelming of sensors [35] | Interruption | Active | Application layer | Tuning of sensors |

**Table 2** Different security attacks on OR

| Name of attack | Description | Purpose |
|---|---|---|
| Impersonation [25] | Injects false information in the network or uses false identity | Attack the whole network |
| Black hole attack [25, 35] | By injecting false information in the network, malicious node tries to attack data packet | Decreases the performance of network |
| Worm hole attack [35] | An attacker attempts to trace packets of data at one network location and after that tunnel them to another location and their repetition is also done | Creates sinkholes in the network |
| Gray hole attack [35] | Some received packet of data are selected by the malicious node and then they are forward them | Making detection more difficult |
| Packet modification [25] | Malicious node modifies packet content and tries to insert false information | Wastage of resources |
| Routing looping [25] | Malicious node access and deploys the information of routing, this leads to rotation of packet in loop of nodes | The packet never reaches its destination |
| Sybil on-off [25] | A malicious node falsifies identities of various nodes, so that it can create multiple illegitimate identities in the network | Decreases efficiency of fault-tolerant schemes |
| Replay attack [25] | Repetitively reinserting packets of data into network | Misinformation is infused into routing tables |

From Table 3, it may be analyzed that various security methods can be used for enhancing the security of WSN. To defend against the various security attacks on the wireless networks, these protocols use methods like trust management and game theory-based methods. The trust management secures the nodes by invalidating the impact of malicious nodes present in the network. The game theory-based protocols facilitate association between the nodes. These security protocols are used not only to defend against the malicious nodes present in the networks but also to isolate them from the network. Therefore, these protocols protect the network from various kinds of security attacks and also increase its routing performance.

**Table 3** OR security methods

| Name | Candidate selection | Candidate coordination | Attack models | Security solution | Advantages | Disadvantages |
|---|---|---|---|---|---|---|
| TPBOR [26] | Relay selection criteria | Network coding | Black hole, gray hole | Trust-management | Reduces security overhead and thus increases the life of network | Consumption of energy of nodes on calculation of trust value |
| TLAR [27] | Direct and indirect observations | Geographic routing principle | Black hole, Packet modification attacks, Badmouth attack | Trust management | Provide solutions for routing process in WSN having resources constrained | Not applicable when multiple attacks are launched on network |
| Trust aware energy-efficient metric [29] | Calculation of forwarding ratio, acknowledgment impact, and energy consumption | Acknowledgment sincerity value | Energy cost model | Trust management | Prevention of malicious nodes and maintaining data integrity | End-to-end delay is high |
| SGOR [28] | Calculating trust value and combining it with distance | Timer-based | Location-spoofing, greyhole and blackhole attacks | RSS analysis, trust management | Scalability and securing network against various attacks | Probability of creation of attacks on the sink and also depends on transmitting power of nodes |
| TSP [30] | Social-Group Value (SGV) | Trust-distribution technique | Black hole attack | Trust management | Mitigate blackhole attack | Certain messages are sent to the malicious nodes |
| TDOR [31] | Trajectory computation | Message relaying | – | – | Efficient and reliable message delivery | Higher delivery latency |

(continued)

**Table 3** (continued)

| Name | Candidate selection | Candidate coordination | Attack models | Security solution | Advantages | Disadvantages |
|---|---|---|---|---|---|---|
| A novel packet salvaging model [32] | Depends on which OR protocol has been used | Depends on OR protocol | Black hole attack | Packet salvaging | No overheads in data packets | Isolation of malicious nodes is not possible |
| Modeling and performance evaluation of the security attacks for OR [33] | Depends on OR protocol | Depends on OR protocol | Black-hole attack | – | Applicable on almost every opportunistic routing protocol | Hard to invalidate the effects of malicious nodes |
| Potential Threat (PT) based security protocol [34] | Satisfaction of average time constraint | Frequency of authentic IDs | Black hole attack | Game theory method | Reduces overheard in message delivery and least dropping of packet | Can be applied only to black hole attack |

# 4 Conclusion

Wireless sensor network is a grouping of sensor nodes which have numerous practical applications. However, in the real world, there are many applications that become a challenging issue which needs to be overcome for efficient functioning of WSN. In this paper, the structure of Wireless Sensor Networks is described. Specifically, the main features and applications of the WSN are presented. Second, the brief summary of routing in WSN is presented. Third, the opportunistic routing in WSN is presented and various security protocols of opportunistic routing protocol are compared. Finally, the comparison of security attacks on WSN and security attacks on OR is given. In the last, the various security protocols of opportunistic routing are compared.

# References

1. Manshahia, M.S.: Wireless sensor network: a survey. Int. J. Sci. Eng. Res. **7**(4), 710–716 (2016)
2. Shabbir, N., Hassan, S.R.: Routing protocols for wireless sensor networks (WSNs). Wirel. Sens. Netw. Insights Innov. **4**, 22–37 (2017)
3. Akyildiz, I.F., Su, W., Sankarasubramaniam, Y., Cayirci, E.: Wireless sensor networks: a survey. Comm. ACM **38**(4), 393–422 (2002)
4. Sarammal, K., Roseline, R.A.: A review: wireless sensor networks and its application, platforms, standards and tools. Int. J. Comput. Trends Technol. **4**(8), 2905–2911 (2013)
5. Bhende, M., Wagh, S.J., Utpat, A.: A quick survey on wireless sensor networks. In: 2014 Fourth International Conference on Communication Systems and Network Technologies, CSNT, pp. 160–167 (2014)
6. Aldeer, M.M.N.: A summary survey on recent applications of wireless sensor networks. In: Proceeding—2013 IEEE Student Conference Research Development SCOReD 2013, pp. 485–490 (2015)
7. Zhang, S., Zhang, H.: A review of wireless sensor networks and its applications. IEEE International Conference on Automation Logistics ICAL, pp. 386–389 (2012)
8. Rashid, B., Rehmani, M.H.: Applications of wireless sensor networks for urban areas: a survey. J. Netw. Comput. Appl. **60**, 192–219 (2016)
9. Yang, X., Deng, D., Liu, M.: An overview of routing protocols on Wireless Sensor Network. In: Proceedings of 2015 4th International Conference Computer Science Network Technology ICCSNT 2015, pp. 1000–1003 (2016)
10. Abdullah, M., Ehsan, A.: Routing protocols for wireless sensor networks: classifications and challenges. J. Electron. Commun. Eng. Res. **2**, 5–15 (2014)
11. Chahal, S., Singh Gill, N.: Comparative study of various WSN routing protocols. Indian Journal Science Technology **9**, 1–6 (2016)
12. Misra, S., Goswami, S.: Network routing: fundamentals, applications, and emerging technologies, 1st edn. Wiley, Hoboken (2017)
13. Patel, T.: Opportunistic routing in wireless sensor networks : a review. In: 2015 IEEE International Advance Computing Conference (IACC), pp. 983–987 (2015)
14. Jadhav, P., Satao, P.R.: A survey on opportunistic routing protocols for wireless sensor networks. Proced. Comput. Sci. **79**, 603–609 (2016)
15. Mounika, M., Chinnaswamy, C.N.: Opportunistic routing protocols for wireless sensor networks: a survey. **7**, 928–931 (2016)

16. Saidi, H., Grete, D., Adnane, A.: Opportunistic routing in wireless sensors networks. In: ICCWCS'17 International Conference on Computer Wireless Communication Systems, pp. 1–5 (2017)
17. Biswas, S., Morris, R.: Opportunistic routing in multi-hop wireless networks. ACM SIGCOMM Comput. Commun. Rev. **34**, 69–74 (2004)
18. Sharma, M., Singh, Y., Kumar, N.: Opportunistic routing in wireless sensor networks: a comparative analysis. Opportun. Rout. Wirel. Sens. Netw. A Comp. Anal. **1**(6), 94 (2014)
19. Pathan, A.S.K., Lee, H.W., Hong., C.S.: Security in wireless sensor networks: issues and challenges. ICACT2006, pp. 1043–1048 (2006)
20. Modares, H., Salleh, R., Moravejosharieh, A.: Overview of security issues in wireless sensor networks. In: Proceedings of CIMSim 2011 3rd International Conference on Computational Intelligence Modelling Simulation, pp. 308–311 (2011)
21. Grover, J., Sharma, S.: Security issues in wireless sensor network-a review. In: 2016 5th International Conference on Reliability, Infocom Technologies and Optimization ICRITO 2016 Trends Future Directions, pp. 397–404 (2016)
22. Anwar, R.W., Bakhtiari, M., Zainal, A., Abdullah, A.H., Qureshi, K.N.: Security issues and attacks in wireless sensor network. World Appl. Sci. J. **30**(10), 1224–1227 (2014)
23. Sen, J., Ukil, A.: LNCS 6018—a secure routing protocol for wireless sensor networks, pp. 277–290 (2010)
24. Sanzgiri, K., Levine, B.N., Shields, C., Belding-Royer, E.M.: A secure routing protocol for ad hoc networks. In: 10th IEEE International Conference on Network Protocols (ICNP' 02), pp. 1–10 (2002)
25. Salehi, M., Boukerche, A.: Secure opportunistic routing protocols: methods, models, and classification. Wirel. Netw. **25**, 559 (2017)
26. Kumar, N., Singh, Y.: Trust and packet load balancing based secure opportunistic routing protocol for WSN. In: 4th IEEE International Conference on Signal Processing, Computing and Control ISPCC 2k17, pp. 463–467 (2017)
27. Vamsi, P.R., Kant, K.: Trust and location-aware routing protocol for wireless sensor networks. IETE J. Res. **2**, 634 (2016)
28. Lyu, C., et al.: SGOR: Secure and scalable geographic opportunistic routing with received signal strength in WSNs. Comput. Commun. **59**, 37–51 (2015)
29. Rani, S.: An energy efficient based opportunistic routing metric for wireless sensor networks. Int. J. Adv. Res. Comput. Sci. Softw. Eng. Res. **7**, 191–197 (2017)
30. Woungang, I.: Trust-based security protocol against blackhole attacks in opportunistic networks. 2013 IEEE 9th International Conference on Wireless and Mobile Computing, Networking and Communications, pp. 724–729 (2013)
31. Cao, Y., et al.: A trajectory-driven opportunistic routing protocol for VCPS. IEEE Trans. Aerosp. Electron. Syst. **9251**, 1–13 (2018)
32. Salehi, M., Boukerche, A.: A novel packet salvaging model to improve the security of opportunistic routing protocols. Comput. Net. **122**, 1–46 (2017)
33. Salehi, M., Boukerche, A., Darehshoorzadeh, A.: Modeling and performance evaluation of security attacks on opportunistic routing protocols for multihop wireless networks. Ad Hoc Netw. **50**, 1–43 (2016)
34. Chhabra, A., Vashishth, V., Sharma, D.K.: A game theory based secure model against black hole attacks in opportunistic networks. In: 2017 51st Annual Conference on Information Sciences and Systems (2017)
35. Singh, Y., Mohindru, V.: Wireless sensor network security attack a node clone attack. Lap Lambert (2017)
36. Raza, M.A., Raza, B., Aftab, A.: Comparative study of security attacks on wireless sensor networks. Int. J. Multidiscip. Sci. Eng. **5**, 16–22 (2014)
37. Wood, A.D., Stankovic, J.A.: Denial of service in sensor networks. Computer **35**, 54–62 (2002)
38. Pooja, M., Singh, V.: Security issues and sybil attack in wireless sensor networks. Int. J. P2P Netw. Trends Technol. **3**, 7 (2013)

# The Significance of Using NDN in MANET

**Mosleh Hamoud Al-Adhaileh, Farkhana Muchtar, Abdul Hanan Abdullah and Pradeep Kumar Singh**

**Abstract** Lately, there has been an increase in the trend among Mobile Ad hoc NETwork (MANET) researchers to use Named Data Networking (NDN) as a network stack solution in MANET. Thus, this paper presents an in-depth review of the potential uses of NDN in MANET environment. In addition, this paper also provides future research direction that could be undertaken on the subject.

**Keywords** Mobile ad hoc network · MANET · Named data networking · NDN

## 1 Introduction

Lately, there has been an increase in the trend among MANET researchers to use NDN as a network layer solution in MANET. This trend can be seen clearly through literatures of [1–7] review articles.

According to [5], this trend indicates that there is very wide potential to be explored on the use of NDN in wireless ad hoc networks. By replacing end to end dedicated link sessions and IP address as endpoint identity with named data retrieval, NDN has the great potential to be the solution for MANET environment. However, proper procedures and approaches first need to be realized before NDN can be used in wireless multi-hop networking environments such as MANET.

M. H. Al-Adhaileh
Deanship of E-Learning and Distance Education, King Faisal University, Al-Ahsa, Saudi Arabia
e-mail: madaileh@kfu.edu.sa

F. Muchtar · A. H. Abdullah
School of Computing, Faculty Engineering, Universiti Teknologi Malaysia, 81310 Skudai, Johor, Malaysia
e-mail: farkhana2@gmail.com

A. H. Abdullah
e-mail: hanan@ut.my

P. K. Singh (✉)
Department of CSE and IT, JayPee University of IT, Waknaghat, Solan, Himachal Pradesh 17334, India
e-mail: pradeep_84cs@yahoo.com



P. K. Singh et al. (eds.), *Proceedings of ICRIC 2019*, Lecture Notes
in Electrical Engineering 597, https://doi.org/10.1007/978-3-030-29407-6_31

The history of Named Data Networking (NDN) is closely related to Content-Centric Network (CCN) introduced by Van Jacobson under the supervision of Paolo Alto Research Center (PARC) in 2006 [8]. NDN is part of Future Internet Architecture Program and receives grant allocation from U.S. National Science Foundation. In its early stages, NDN researchers used CCN implementation called CCNx for their research needs. Subsequently, NDN research communities argued that they require their own NDN implementation and forked CCNx version 0.7.2 source code and created NDNx version 0.1 in August 2013 to enable NDN researchers to further develop their NDN implementation according to their research objectives. This is because NDN research communities have different objectives than CCN as both are under different research fund despite sharing the same architecture and philosophy regarding how future Internet architecture should be designed.

NDN and CCN are categorized as Information-Centric Networking (ICN) even though ICN begin to be used in 2010, inspired by Van Jacobson's lecture in the 2006 Google Tech Talk titled "A New Way to Look at Networking". Although there are other network protocols in the same category of ICN such as DONA, TRIAD and Pursuit but NDN and CCN are the only ICN implementation that does not completely rely on Internet Protocol (IP) stack and capable of being on its own without IP technology.

At present, there are copious amount of articles that focus on discussing NDN-based MANET. This is due to the increase of the trend among MANET researchers to use NDN as a network stack solution in MANET. Hence, the aim of this paper is to presents an in-depth review of the potential uses of NDN in MANET environment. In addition, this paper also provides future research direction that could be undertaken on the subject.

The paper is organized as follows. Section 1 will provide an introductory outline. Section 2 elaborates on the history and evolution of NDN and CCN. Section 3 will provide the introduction of NDN. Section 4 discusses the advantages of NDN for MANET solution and Sect. 5 presents the final conclusions.

## 2 History and Evolution of NDN and CCN

NDN and CCN are categorized as Information Centric Networking (ICN) even though ICN begin to be used in 2010, inspired by Van Jacobson's lecture in the 2006 Google Tech Talk titled "A New Way to Look at Networking". Although there are other network protocols in the same category of ICN such as DONA, TRIAD and Pursuit but NDN and CCN are the only ICN implementation that does not completely rely on Internet Protocol (IP) stack and capable of being on its own without IP technology.

We laterally use NDN name to represent both NDN and CCN (before CCNx 1.0) to simplify explanation and avoid confusions on the selection of names, whether it is ICN, CCN or NDN. This is because CCN and NDN are based on the same network architecture and shares similar philosophy in its previous work although

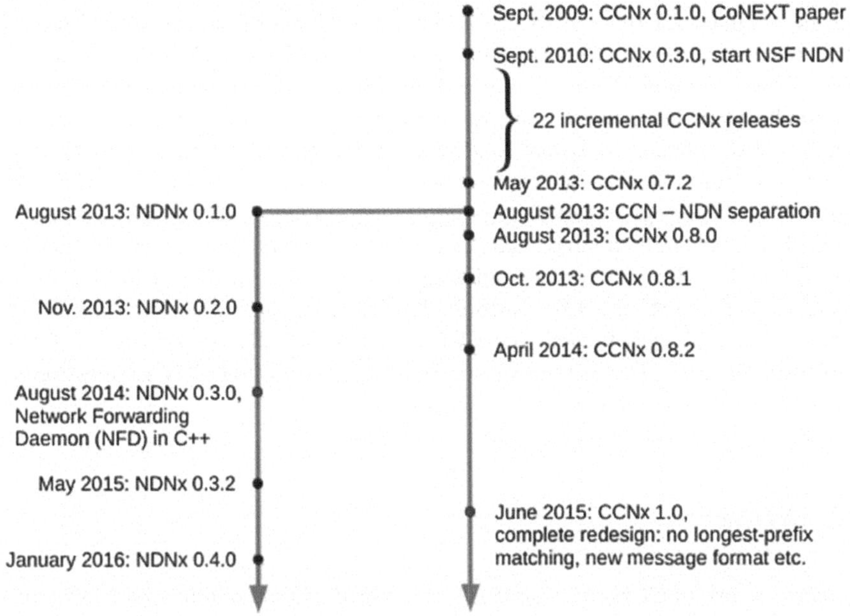


**Fig. 1** Evolution of CCN and NDN projects

technical implementation differs beginning with NDNx version 0.1. Although ICN is too general, it does have its own history before CCN and NDN even existed. To facilitate readers understanding, Fig. 1 illustrates the difference between CCN and NDN in terms of implementation.

## *2.1 CCNx*

PARC developed CCN implementation called CCNx as a step forward towards the actual realization of a much concrete CCN architecture to assist researchers and developers identify issues that arise in the implementation of CCN as the new Internet architecture as well as seeking new potentials that can be featured into CCN to make CCN implementation more practical and ready for use in real-life situations. CCNx was released as an open-source project under the GNU GPL License to attract interest of other parties such researchers from the academic world and industry to participate in the development of CCN.

CCNx 1.0 was developed from scratch based on the experience they gained when developing CCNx 0.x. Source code for CCNx 1.0 is no longer released as an open-source project, but instead shared with communities based on two different license versions namely, free evaluation license for academic institutions and commercial evaluation license for commercial organizations. Binary format of CCNx 1.0 can still be obtained for free but its use is limited to Ubuntu Linux platform and not available in other OS platforms like Windows OS and Mac OS. However, in January 2016, PARC released CCNx 1.0 source code publicly in github.com with simplified version licensing term (http://www.parc.com/news-release/111/parc-offers-content-centric-networkingccnx-software-to-advance-next-generation-internet.html) [9].

There are some architectural changes between CCNx 0.x and CCNx 1.0. The main difference is the design of forwarding engine and different packet formats especially packet headers (combination of static and optional headers) because CCNx 1.0 uses nested TLV based packet formats rather than the use of XML binary in CCNx 0.x. Interest matching in CCNx 1.0 is performed with exact matching which is, in contrast, to name prefixes matching method in CCNx 0.x and it is still being practiced in NDN implementation [9]. There are lots of other changes added to CCNx 1.0 however, it cannot be covered completely in this section.

## 2.2 NDNx

NDNx is a fork of CCNx version 0.7.2 with added additions following NDN Consortium research interest such as ndn-tlv, NDN link protocol (NDNLP), NDN link state routing protocol based on OSPF (OSPFN), NDN Common Client Libraries with TLV support and ndnSIM for simulation purposes. Currently, CCNx still maintains the use of binary xml for network packet format as opposed to NDNx that has shifted to Type-Length-Value (TLV) format that is more flexible and efficient. NDNx 0.1 was released in August 2013 extending up to version 0.3 alpha 1 in February 2014 being the last NDNx implementation based on CCNX implementation before it was replaced with NDN Platform, as NDN come into being the new implementation developed from scratch under NDN Next Phase project (NDN-NP).

## 2.3 CCNNDN-NP (NDN Next Phase)

NDN Next Phase or also known as NDN-NP is a continuation of NDN implementation development undertaken by NDN Consortium based on their own experience using CCNx implementation and NDNx implementation before deciding to develop their own NDN implementation from scratch called the NDN Platform. NDN Platform version 0.3 was released in August 2014 consisting of Named Forwarding Daemon (NFD) version 0.2.0, ndn-cxx library version 0.2.0 and also some NDN Common Client libraries suite (NDN-CCL) version 0.3 such as NDN-CCP version 0.5 (C++), PyNDN2 version 2 beta 1 (Python), NDN- JS version 0.5 (JavaScript) and jNDN version 0.1 (Java).

NDN-NP gave focus on NDN implementation that is researchers and developers friendly with a simpler modular design that is easy to modify. NDN Platform makes it easier for researchers and developers to add new features into NDN such as better forwarding strategy, separated routing protocol implementation, robust security practice, easier NDN application development through ndn-cxx and NDN-CCL libraries as well as easier and accurate NDN simulation platforms (ndnSim2).

## 3 Introduction of NDN

Named Data Networking is a research project that receives research grant from National Science Foundation (NSF) since 2010 until now, 2016. NDN Consortium is headed by Lixia Zhang, a UCLA Professor and Van Jacobson, a UCLA adjunct professor with the focus on designing and developing alternative network protocols to replace TCP/IP as part of the next generation Internet research effort under NSF.

Eight universities joined for NDN Next Phase, funded by NSF for 2014–2016, led by UCLA (University of California, Los Angeles) with collaborations from Cisco, Qualcomm, Comcast, Verisign, Panasonic, Intel, Huawei, Alcatel-Lucent and Orange as representatives from the communications industry in developing and exploring NDN-related ideas to ensure NDN is a practical solution for future generation Internet.

During its early stages, NDN uses CCN implementation developed by PARC (Paolo Alto Research Center, own by Xerox) called CCNx until the second phase of NDN-NP (NDN Next Phase) where they have already developed their own NDN implementation as a result from their own experience obtained while using CCNx for their own research purposes. This is because NDN community has many different focuses other than PARC despite sharing the same research base.

At the same time, they found that old CCNx implementation (before CCNx 1.0) was not modular and makes it was difficult for researchers to develop new ideas or components to meet NDN research interest community. Additionally, NDN developed on their own by NDN consortium enables them to use new TLV (Type, Length, Value) format network packet instead of xml binary used in CCNX 0.7.x previously. This new network packet format is called NDN-TLV.

Therefore, in developing NDN-NP, NDN community themselves has developed their own NDN implementation beginning with the development of NFD (Named Forwarding Daemon) and several other NDN libraries such as NDN-CPP, NDN-CXX, NDN.js and JNDN as well as several other NDN support tools to enable NDN community to conduct better NDN research and development especially for the purpose of adapting NDN in real-world of industries.

### *3.1 NDN Hourglass Architecture*

Like TCP/IP, NDN also uses the concept of hourglass architecture in multiple layers of network stack. TCP/IP multiple-layer design is practical, whereby even though the design of TCP/IP network packet is permanent but the technology that supports

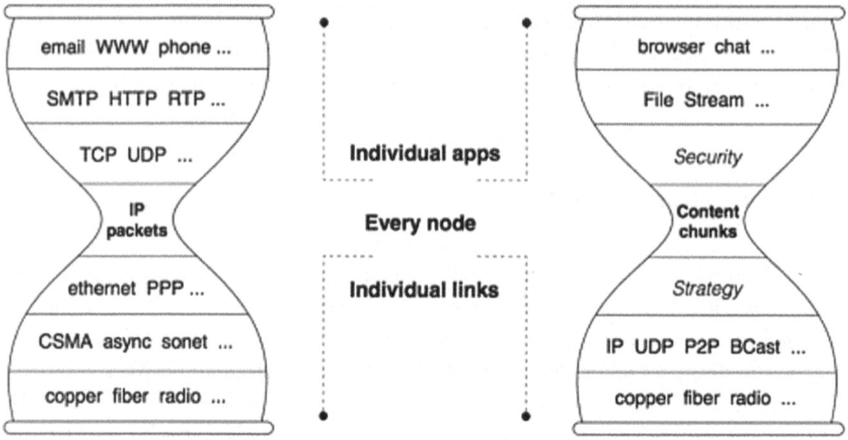


**Fig. 2** Internet and NDN hourglass architecture [10]

the TCP/IP network layer whether it is overlay technology (layer 4 and above) or underlay technology (layer 2 and below) are various. As example, for transport layer (layer 4), various transport protocol options can be used over TCP/IP network layer such as TCP, UDP, ICMP and so on. Meanwhile, TCP/IP network layer can be implemented in various forms of physical networks such as ethernet, wifi, cellular network, telephone network, fiber optic network and so on.

TCP/IP network packet is the basic design for TCP/IP stack where each network packet represents communication session binding between one end point to the other end point and each end point is identified with ip address. Therefore, the unique identity of each TCP/IP network packet is practically a combination of source ip address, destination ip address and sequence identity of packet.

In NDN however, NDN network packet (content chunck) is the basis of the NDN network stack (Fig. 2).

## *3.2 NDN Network Packet Format*

In NDN, there are two types of network packets, namely interest packets and packet data (also known as content packet). Unlike TCP/IP, network layer communication is two-way, initiated by content requester (or known as content consumer) by sending interest packets to content producer (also known as content source) based on routing tables, FIB tables such as TCP/IP network Packet forwarding. The difference is that, FIB table in NDN stores content name prefix along with network face but FIB table in TCP/IP stores end point IP address. Further information on FIB table in NDN can be found in next section.

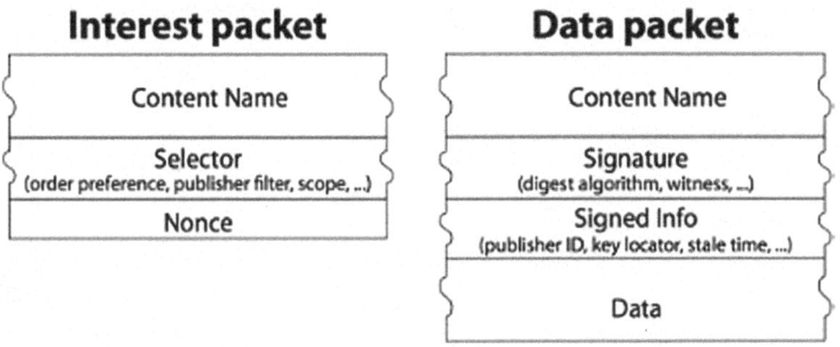


**Fig. 3** NDN packet structure [10]

When content requester receives interest packet, the requested data packet forwarded by content provider goes through reversed path of interest packet where reversed path of interest packet is stored in the PIT table (routing table for data packet) (Fig. 3).

## *3.3 NDN Operation*

To perform interest and content packet forwarding, NDN routing mechanism uses three data structures namely, Pending Interest Table (PIT), Forwarding Information Base (FIB) and Content Store (CS) as shown in Fig. 4. Forwarding strategy component determines when and where interest packet forwarding is performed. PIT is the routing table that stores interest information that has been forwarded but has not yet been met. Each entry record in PIT contains the desired content name prefix and information on which network face interest is received for that content.

In NDN, network communication input/output occurs via network face, a superset to network interface concept used in the TCP/IP network stack. NDN network face can refer to network interface, local connection or application that is available on the NDN forwarder.

When the NDN forwarder receives interest packet, forwarding strategy component will first check with content store to determine if the desired content by interest packet is already available in content store or not. If yes, forwarding strategy proceed to forward content packet of desired content from content store to the network face used by the interest packet. Otherwise, the desired content name prefix information will be stored in PIT as well the as network face id that specifies the origin of interest packet.

Next, forwarding strategy component will refer to FIB table to perform next hop interest forwarding up to the point where interest packet reaches the content provider (or NDN forwarder that stores the content in content store). FIB entry consists of available content name prefix and a list of network face that can be passed by in order to obtain the content. If the content name prefix that is required by interest

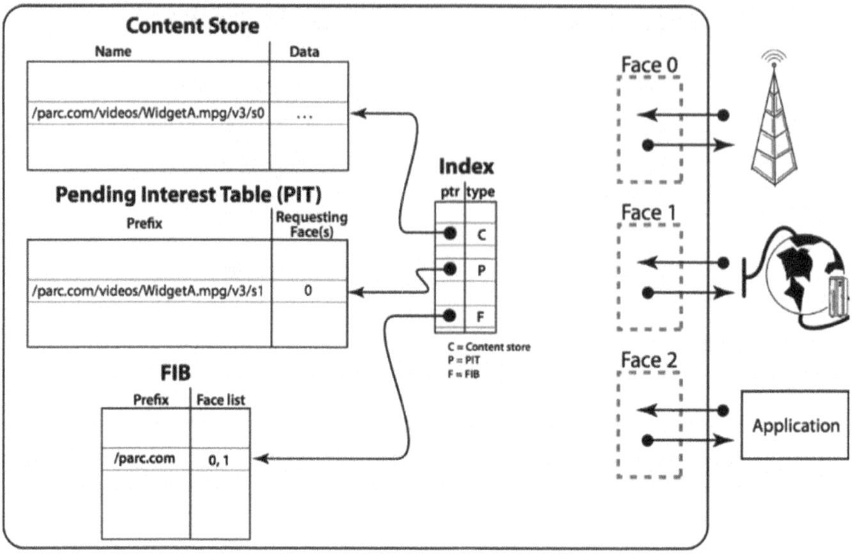


**Fig. 4** NDN node [10]

and accepted by NDN forwarder is not found in the FIB table, interest will then be dropped automatically. If available, then the interest will be recorded in PIT and forwarded through recorded network face in FIB until the desired content is found.

Content packet forwarding, on the other hand, is performed by forwarding strategy component after the content provider has received interest packet of matching content through similar network path taken by interest packet of the desired content but in the opposite direction. PIT entries are used by forwarding strategy component to determine which network face interest of content originates from and uses the same network face to deliver the desired packet content.

In addition to content provider, another NDN forwarder will store the content packet in content store before forwarding it to next-hop NDN forwarder (depending on content store strategy, subset of forwarding strategy). After content packet is forwarded to next hop NDN forwarder PIT entry that refers to content packet will be deleted. If interest packet for the same content packet is received again (possibly from other content requester) forwarding strategy component will retrieve content packet from content store (if it is still available) to fulfil that interest packet without having to be forwarded to next hop NDN forwarder or even content provider.

Content store serves as network cache to enable multiple requests for the same content can be performed efficiently without having to repeatedly retrieve the original location of the content. Instead, NDN forwarders can act as a local copy storage of desired content, especially popular content (multiple content requester). Due to this, broadcast or multicast communication can be performed more efficiently in NDN than TCP/IP network stack approach.

# 4 NDN Advantages for MANET Solution

There are two main features that makes NDN very suitable for MANET namely: (1) Identification is based on data, not host of data (2) Content is secured by itself and can be cached anyway in network [11, 12].

According to authors in [12], NDN does not require point-to-point routing information as being practiced in host-based MANET because data is searched based on the identity of the data itself rather than the host that stores the data. Without the need to create dedicated link session between source and destination, NDN provides a more practical and reliable facility for dynamic network requirements such as MANET.

More interestingly, security features are part of the NDN standard architecture where security is placed on the data itself instead of on the host of data or even container of content (e.g. connection). Default security features in NDN enable MANET implementation to be much more secure without the need for additional complex security features such as host-centric MANET [3, 4, 13].

Additionally, content in NDN can be obtained from more than just one host and does not necessarily be taken from the original source of content because the content is not uniquely tied to the host. Separation of content and host allows content to be stored in network cache or better known as content store in network. Content-store allows data to be acquired locally thus increasing the availability of the data itself and reduces the duration required to perform content delivery to content consumer [1, 14–17].

From Name-based content routing, network caching and adaptive forwarding features in NDN provides NDN some advantages as the network stack solution in MANET namely.

## *4.1 Suitable for Wireless Communication*

NDN does not require a dedicated end-to-end link connection between source and destination but instead uses receiver-driven approach where the data is sent asynchronously in the network because host identifier is not required for content dissemination purposes [8, 10]. Because of this, NDN based MANET can exploit wireless broadcasting method without having the need to set destination nodes thus making multi-hop forwarding to be performed with more flexibility and simplicity [9].

Asynchronous communication in NDN is compatible with connection-less nature of wireless broadcasting, therefore, enables smart-broadcast routing to be fully utilized for content routing purposes in NDN based MANET and this is the most popular approach used in existing NDN based MANET [3].

If unicast communication is used for content request and content delivery purposes, each unique face of forwarder in every individual node represents unique unicast connection link between each node in the neighbourhood. Unlike host-centric

MANET, it is not necessary to have unique unicast connection to every neighbour nodes in NDN based MANET. Contrary to host-centric MANET, unique unicast connection in NDN based MANET need not be obtained through exclusive neighbourhood discovery mechanism but instead, it is dynamically created through content discovery and path discovery that is performed as smart discovery similar to that conducted in Airdrop [38] and Dynamic Unicast approach [19].

In addition, network caching through content-store within NDN allows content consumers to obtain requested content from the nearest source thus increasing the reliability of wireless communication. Content-store also helps to reduce network packet redundancy and collision of broadcast communication especially in multi-hop environment [20].

## *4.2 Node Mobility Support*

NDN has the advantage in terms of mobility support because it does not use point-to-point communication model, therefore, allowing mobile node to communicate with other nodes based on content's identity instead of content's host identity [8, 10]. Content consumer communicates based on desired content and does not need the maintain dedicated routing path to the content provider [11]. Therefore, NDN can cope with unstable network conditions such as intermittent, broken link of connection and dynamic changes of network topology in MANET [1].

NDN naturally support content consumer mobility because content consumer only needs to resend interest of unsatisfied content segment based on FIB entries reference to other intermediate nodes that acts as new forwarding relay [1, 9, 21–23]. Copy of required content segment with shortest path obtainable in content-store allows the content segment retransmission to be performed without the need for content provider itself shortens the required content delivery distance [22].

However, content provider mobility and intermediate node mobility support are not easily performed as in content consumer mobility in NDN based MANET because it requires extra mechanisms to update latest FIB entries in neighbour nodes to ensure interest forwarding will not be sent to distant locations [9, 23–27]. Nonetheless, content store in NDN is able to reduce problem posed by node mobility in MANET by increasing the availability of desired content and reducing the dependence on the content provider for data packet retransmission [25]. Additionally, NDN based MANET can use broadcasting method that easily can perform content rediscovery if there are changes to content topology due to node mobility [26], as being practiced in existing NDN based MANET solution such as LFBL [20, 28], E-CHANET [29] and Airdrop [38].

## 4.3 No Exclusive Control Mechanism Is Required

Host-centric MANET inherited end-to-end communication from infrastructured networking such as TCP/IP. In the end-to-end communication, dedicated link session is required to allow one end to communicate with the other end smoothly as each communication is based on unique host identity (such as IP address) rather than content (such as data, files, service, Function or user). Therefore, host-centric MANET relies heavily on exclusive control mechanism in dynamic routing protocol because accurate and updated routing information is greatly required to establish a reliable and stable dedicated link session [3].

Instead, NDN does not require network topology information to perform data dissemination because interest and data forwarding depended only on available network faces either forwarding face is a broadcast channel or dedicated unicast link to neighbour nodes [20]. Therefore, NDN-based MANET does not require dedicated control mechanism to update routing table of multi-hop paths to represent dynamic network topology information [29]. Without exclusive control mechanism and control message traffic, NDN based MANET can save network bandwidth usage and battery power consumption.

Furthermore, NDN based MANET also does not require decentralized dynamic IP address allocation mechanism or also known as auto-configuration mechanism to assign IP addresses to each mobile node in MANET [29]. Auto-configuration protocol is used in Host Centric MANET whether it is for setting IP address for new mobile node joining the network or mobile node disconnected from the network and returns the link to other mobile nodes in the network.

To date, there is still no robust and effective auto-configuration protocols to be used in MANET particularly one that can ensure the uniqueness of network address of each node. Further, some auto-configuration solution such as Duplicate Address Detection (DAD) causes broadcast storm problems and extra overhead [30]. Omitting the need for auto-configuration mechanisms in NDN based MANET can overcome auto-configuration problems for MANET and salvages energy usage caused by auto-configuration protocol.

## 4.4 Better and Simpler Routing Approach

Unlike forwarding mechanisms in TCP/IP, forwarding mechanism in NDN is smart and adaptive. In fact, communication in NDN can even be done without the need for routing mechanism which is the opposite in TCP/IP. NDN forwarder has forwarding plane that work two ways whereby, content request is performed by sending interest packet of desired content and the content is then sent back in the form of data packet using similar route but in the opposite direction. Therefore, content routing can be performed though/using forwarding strategies in forwarding plane without the need for exclusive routing protocol as being practiced in the host-centric MANET [31].

NDN also does not require complete network topology information when routing is performed because communication between mobile nodes is not conducted as end-to-end conversation. Instead, each node in the network communicate with each other based on identity (name prefix) of the requested data [20].

Hop-to-hop communication is performed by referring to the network face that has the potential to retrieve the requested data and this reference is obtained from the FIB table that can be updated either through proactive convergence or through reactive content discovery. Interest packets that refer to the same desired content will not be forwarded repeatedly, but instead it is only recorded in PIT entries and if interest packet for the same content originated from the same face, it will automatically drop the interest packet. Hop-to-hop content delivery is performed based on 'breadcom trail' method that refers to PIT table as to determine which network face does the interest packet originate from about or content request till it reaches content consumer. If there is new content request for the same content, the requested content will be retrieved directly from content store and need not be re-requested from content provider again [3].

Besides that, NDN allows multipath forwarding to enable interest and data packet to be sent using more than one path unlike host-centric network such as TCP/IP that practiced single shortest path routing (quote). Multipath forwarding has two main advantages if it is used in MANET namely internal load balancing advantages on network traffic routing and also adaptive forwarding that is, if broken link occurs due to node mobility, current intermediate node can provide NACK message to the previous intermediate node to resend the failed network packet via an alternative route. Adaptive forwarding through multipath routes allows broken link recovery to be performed without the need to wait for global routing convergence to be performed [19, 32].

Network cache through content- store in NDN forwarder also provide many advantages to be used in MANET especially in terms of content availability where content consumers are able to obtain the desired content through local copy placed in content store. Content-store shortens the distance to obtain the content thus indirectly reduces the amount of time required to perform content routing [33]. Content providers also benefit from content-store feature in NDN because the same content does not need to be sent repeatedly to more than one content consumer, hence making NDN based MANET to be very compatible with multicast communication method in reducing MANET's network traffic [1, 3, 20, 33]. Content-store feature can also indirectly reduces negative impacts due to node mobility and other causes that often result in changes to network topology because, when content delivery fails to be performed, broken link occur but the same content still can be recovered from content-store without the need to do retransmission from content provider [31].

## 4.5 Better Security Support

Security is internally backed up in NDN compared to TCP/IP and other host-centric networks. Unlike security practices in host-centric networking, NDN uses data-centric security approach whereby that security is placed on the data itself rather than on the host of data or container of data.

When a particular content needs to be sent in NDN, that content will be broken into segments or fragments of contents and each of the segmented content will be encrypted and forwarded to content consumer as data packet together with its digital signature in the packet header that is generated using content producer private key. To enable consumer content to re-integrate all the segments and convert it back to its original content form, public key from content producer is required [10, 34–36]. The policies on how content producers distribute and share public key of encrypted content are open to any methods and approaches and this issue is also among the many topics of research that are increasingly gaining attention among researchers [36].

Encryption on data level in NDN enables content privacy and data access control to be performed easily without much dependency on third-party security implementation. Content privacy and data access control can be determined based on digital signature approach used as well as public-key policy practiced in the network. As an easy example, content providers can decide who is able to accept the available content through public-key sharing policy where only content consumer that has the public key for digital signature can receive the content owned by that particular content provider. Since content naming prefix is in the form of hierarchies, data access policy can be performed easily based on name prefix hierarchy without involving any complex mechanisms as practiced in host-centric networking [10, 34, 35].

Since NDN uses the hierarchical name prefix for content identity, NDN does not require explicit mechanism to provide names to content such as in DNS or auto-configuration mechanism for data naming, therefore, liberating NDN free from name hijacking attacks or also known as DNS hijacking. Other Information Centric Networks such as DONA and Pursuit do not have this advantage because they are using flat name prefix for content identity [36].

In addition to that, NDN also has the advantage/have some advantages in terms of security compared to host-centric networking because its adaptive forwarding strategy can detect anomalies generated by intruders. In fact, NDN forwarding strategy also prevent occurrences of DDOS attacks because content request is based on name prefix instead of host address and furthermore, same interest packet will not be sent repeatedly by the forwarder but rather, it is only included in PIT entries if it originates from different network face and if it is from the same network face, the forwarder simply drops the interest packet [36, 37].

### 4.6 Energy Efficient

In NDN, interest aggregation is performed efficiently because the same interest packet will not be "forwarded" repeatedly but contrarily stored in PIT table or dropped if it comes from the same network face. Further, if the data packet has already been forwarded by one of each relay, the data packet is then stored in content-store that acts as network cache and it is directly retrieved from content-store instead of the original location of content provider. NDN also uses anycast data forwarding method whereby, if the content being requested is coming from more than just one content consumer, it will not be sent repeatedly by content provider but instead the same content will be retrieved from local copy stored in content-store or provided in the form of replicated repository [18, 38].

Interest aggregation, in-network caching and anycast forwarding in NDN forwarding strategy make NDN more energy efficient than any other network solutions [38]. Additionally, energy conservation is achieved in NDN through the reduction of network packet redundancy using those mentioned features and energy conservation is critical in mobile networks such as MANET [19].

Energy efficiency of MANET can also be achieved through NDN based MANET because there is no exclusive control mechanism used to perform global network convergence as practiced in host-centric MANET and at the same time there is also no extra network traffic when control message needs to be sent repeatedly to ensure that routing information is up-to-date and accurate [18, 19].

In terms of content delivery recovery due to node mobility in MANET, NDN possesses superior energy-efficient mechanism because content provider does not need to retransmit data packet to be sent to content consumers. This is because intermediate node can perform data packet forwarding retry through broadcasting method or adaptive forwarding method whereby data packet will be retransmitted through an alternative path. Without the need to use complex mechanism for broken link recovery and retransmission from content providers to consumer content, energy can be effectively be conserved through NDN based MANET solution [19].

## 5 Conclusion

From the explanations provided in this article, it is opined that this paper has met its intended objectives, which was to discuss the potentials of using NDN for MANET. This project is still ongoing. In future research undertakings, the focus will be narrow down to revealing the potential of using NDN in improving energy efficiency of MANET only, i.e., as the following:

i. comparison of the energy consumption pattern of the NDN based content routing mechanism with the HCN based routing protocol (OLSRd, Babeld, Batman-adv and BMX6) by measuring two-parameter metrics relating to energy efficiency; energy consumption and file transfer completion time.

ii. comparison of the energy consumption pattern of the NDN based content routing mechanism with the HCN based routing protocol when in use with different network traffic content. Said content will include video on demand (VOD) and live streaming video, which due to its unique nature will have differing needs for the file transfer network traffic.
iii. comparison of the energy consumption pattern of the NDN based content routing mechanism with the HCN based routing protocol with regards to node mobility parameters and network topology changes in the testbed.

All the garnered comparison results above is to suggest that the NDN based WAHN outperforms or vice versa to the other routing protocols in terms of energy efficiency. Furthermore, the results obtained from future work propose above shall serve as a stepping stone to better understand the real potential of the NDN based content routing mechanism and how it can be used as a solution to the issue of energy consumption in MANET.

## References


1. Amadeo, M., Campolo, C., Molinaro, A., Ruggeri, G.: Content-centric wireless networking: a survey. Comput. Netw. **72**, 1–13 (2013)
2. Amadeo, M., Campolo, C., Molinaro, A.: Forwarding strategies in named data wireless ad hoc networks: design and evaluation. J. Netw. Comput. Appl. **50**, 148–158 (2015)
3. Liu, X., Li, Z., Yang, P., Dong, Y.: Information-centric mobile ad hoc networks and content routing: a survey. Ad Hoc Netw. **58**, 255–268 (2017)
4. Saxena, D., Raychoudhury, V., Suri, N., Becker, C., Cao, J.: Named data networking: a survey. Comput. Sci. Rev. **19**, 15–55 (2016)
5. Conti, M., Boldrini, C., Kanhere, S.S., Mingozzi, E., Pagani, E., Ruiz, P.M., Younis, M.: From MANET to people-centric networking: milestones and open research challenges. Comput. Commun. **71**, 1–21 (2015)
6. Lee, U., Rimac, I, Hilt, V.: Greening the internet with content-centric networking. In: Proceedings of the 1st International Conference on Energy-Efficient Computing and Networking (e-Energy'10), pp. 179–182. ACM, New York (2010)
7. Braun, T., Trinh, T.A.: Energy efficiency issues in information-centric networking. In: Pierson, J.-M., Costa, G.D., Dittmann, L. (eds.) Energy efficiency in large scale distributed systems, LNCS, vol. 8046, pp. 271–278. Springer, Heidelberg (2013)
8. Jacobson, V., Smetters, D.K. Thornton, J.D., Plass, M.F, Briggs, N.H., Braynard, R.L.: Networking named content. In: Proceedings of the 5th International Conference on Emerging Networking Experiments and Technologies (CoNEXT'09). ACM, New York (2009)
9. Anastasiades, C.: Information-centric communication in mobile and wireless networks. Dissertation of Doctor of Philosophy, University of Bern, Bern, Switzerland (2016)
10. Zhang, L., Afanasyev, A., Burke, J., Jacobson, V., Claffy, K., Crowley, P., Pa-padopoulos, C., Wang, L., Zhang, B.: Named data networking. SIGCOMM Comput. Commun. Rev. **44**(3), 66–73 (2014)
11. Zhu, Z., Afanasyev, A., Zhang, L.: A new perspective on mobility support. Named-Data Networking Project, Technical Report (2013)
12. Zhu, Z.: Support mobile and distributed applications with named data networking. Ph.D. Thesis, University of California Los Angeles (2013)

13. Yi, C., Abraham, J., Afanasyev, A., Wang, L., Zhang, B., Zhang, L.: On the role of routing in named data networking. In: Proceedings of the 1st International Conference on Information-centric Networking (ICN'14), pp. 27–36. ACM, New York, NY, USA (2014)
14. Zeng, Y., Hong, X.: A caching strategy in mobile ad hoc named data network. In: 2011 6th International ICST Conference on Communications and Networking in China (CHINACOM), pp. 805–809 (2011)
15. Sourlas, V., Flegkas, P., Tassiulas, L.: Cache-aware routing in information-centric networks, In: 2013 IFIP/IEEE International Symposium on Integrated Network Management (IM 2013), pp. 582–588 (2013)
16. Dalaly, S.M.S.R.: A content caching strategy for named data networking. Ph.D. Thesis, National University of Singapore (2014)
17. Zhang, G., Li, Y., Lin, T.: Caching in information centric networking: a survey. Comput. Netw. **57**(16), 3128–3141 (2013)
18. Amadeo, M., Campolo, C., Iera, A., and Molinaro, A.: Named data networking for IoT: an architectural perspective. In: 2014 European Conference on Networks and Communications (EuCNC 2014), pp. 1–5 (2014)
19. Anastasiades, C., Weber, J., Braun, T.: Dynamic Unicast: Information-centric multi-hop routing for mobile ad-hoc networks. Comput. Netw. **107**(2), 208–219 (2016)
20. Meisel, M., Pappas, V., Zhang, L.: Ad hoc networking via named data. In: Proceedings of the Fifth ACM International Workshop on Mobility in the Evolving Internet Architecture (MobiArch'10), pp. 3–8. ACM, New York (2010)
21. Tyson, G., Sastry, N., Cuevas, R., Rimac, I., Mauthe, A.: A survey of mobility in information-centric networks. Commun. ACM **56**(12), 90–98 (2013)
22. Jiang, X., Bi, J., Wang, Y.: What benefits does NDN have in supporting mobility In: 2014 IEEE Symposium on Computers and Communication (ISCC), pp. 1–6 (2014)
23. Zhang, Y., Afanasyev, A., Burke, J., Zhang, L.: A survey of mobility support in named data networking. In: 2016 IEEE Conference on Computer Communications Workshops (INFOCOM WKSHPS), pp. 83–88 (2016)
24. Wang, L., Afanasyev, A., Kuntz, R., Vuyyuru, R., Wakikawa, R., Zhang, L.: Rapid traffic information dissemination using named data. In: Proceedings of the 1st ACM Workshop on Emerging Name-Oriented Mobile Networking Design—Architecture, Algorithms, and Applications (NoM'12), pp. 7–12. ACM, New York (2012)
25. Azgin, A., Ravindran, R., Wang, G.: Mobility study for named data networking in wireless access networks. In: 2014 IEEE International Conference on Communications (ICC), pp. 3252–3257 (2014)
26. Azgin, A., Ravindran, R., Wang, G.: A scalable mobility-centric architecture for named data networking. arXiv:1406.7049 (2014)
27. Azgin, A., Ravindran, R., Chakraborti, A., Wang, G.-Q.: Seamless producer mobility as a service in information centric networks. In: Proceedings of the 3rd ACM Conference on Information-Centric Networking (ACM-ICN'16), pp. 243–248. ACM, New York (2016)
28. Meisel, M., Pappas, V., Zhang, L.: Listen first, broadcast later: topology-agnostic forwarding under high dynamics. In: Annual Conference of International Technology Alliance in Network and Information Science, pp. 1–8 (2010)
29. Amadeo, M., Molinaro, A., Ruggeri, G.: E-CHANET: routing, forwarding and transport in information-centric multihop wireless networks. Comput. Commun. **36**(7), 792–803 (2013)
30. Kumar, H.: Issues and trends in auto configuration of IP address in MANET. Int. J. Comput. Commun. Control **3**, 353–357 (2008)
31. Yi, C.: Adaptive forwarding in named data networking. Ph.D. Thesis, The University of Arizona, Arizona (2014)
32. Yi, C., Afanasyev, A., Wang, L., Zhang, B., Zhang, L.: Adaptive forwarding in named data networking. SIGCOMM Comput. Commun. Rev. **42**(3), 62–67 (2012)
33. Ho, C. Y., Tseng, C.C.: A case study of cache performance in ICN-various combinations of transmission behavior and cache replacement mechanism. In: 2015 17th International Conference on Advanced Communication Technology (ICACT), pp. 323–328 (2015)

34. Hamdane, B., Serhrouchni, A., Fadlallah, A., Fatmi, S.G.E.: Named-data security scheme for named data networking. In: 2012 Third International Conference on the Network of the Future (NOF), pp. 1–6 (2012)
35. Tourani, R., Mick, T., Misra, S., Panwar, G.: Security, privacy, and access control in information-centric networking: a survey. arXiv:1603.03409 [cs]ArXiv: 1603.03409
36. Lutz, R.: Security and privacy in future internet architectures—benefits and challenges of content centric networks. arXiv:1601.01278 [cs]ArXiv:1601.01278
37. Yi, C., Afanasyev, A., Moiseenko, I., Wang, L., Zhang, B., Zhang, L.: A case for stateful forwarding plane. Comput. Commun. **36**(7), 779–791 (2013)
38. Kim, D., Kim, J.-H., Moon, C., Choi, J., Yeom, I.: Efficient content delivery in mobile ad-hoc networks using CCN. Ad Hoc Netw. **36**(1), 81–99 (2016)

# Using NDN in Improving Energy Efficiency of MANET

**Farkhana Muchtar, Pradeep Kumar Singh, Raaid Alubady, Ili Najaa Aimi Mohd Nordin, Radzi Ambar, Mohd Najwadi Yusoff, Deris Stiawan and Mosleh Hamoud Al-Adhaileh**

**Abstract** This paper seeks to extol the virtues of named data networking (NDN), as an alternative to host-centric networking (HCN), for its prominent features that can be taken advantage of to significantly reduce energy consumption demands in a mobile

F. Muchtar
Faculty Engineering, School of Computing, Universiti Teknologi Malaysia, 81310 Skudai, Johor, Malaysia
e-mail: farkhana2@gmail.com

P. K. Singh (✉)
Department of CSE and IT, JayPee University of IT, Waknaghat, Solan, Himachal Pradesh 17334, India
e-mail: pradeep_84cs@yahoo.com

R. Alubady
Information Technology Department, College of Information Technology, University of Babylon, Hillah, Iraq
e-mail: alubadyraaid@itnet.uobabylon.edu.iq

I. N. A. M. Nordin
Faculty of Engineering Technology, Universiti Tun Hussein Onn Malaysia, 84600 Pagoh, Malaysia
e-mail: ilinajaa@uthm.edu.my

R. Ambar
Faculty of Electrical and Electronic Engineering, Universiti Tun Hussein Onn Malaysia, 86400 Batu Pahat, Malaysia
e-mail: aradzi@uthm.my

M. N. Yusoff
School of Computer Sciences, Universiti Sains Malaysia, 11800 Gelugor, Pulau Pinang, Malaysia
e-mail: najwadi@usm.my

D. Stiawan
Faculty of Computer Science, Universitas Sriwijaya, Jalan Srijaya Negara, Palembang 30139, Indonesia
e-mail: deris@unsri.ac.id

M. H. Al-Adhaileh
Deanship of E-learning and Distance Education, King Faisal University, Al-Ahsa, Kingdom of Saudi Arabia
e-mail: madaileh@kfu.edu.sa



P. K. Singh et al. (eds.), *Proceedings of ICRIC 2019*, Lecture Notes in Electrical Engineering 597, https://doi.org/10.1007/978-3-030-29407-6_32

ad hoc network (MANET) implementation. Therefore, a NDN-based content routing mechanism was compared with two types of HCN routing protocol implementations in this study: OLSR and Batman-adv. The experimental results obtained from this research provide early evidence that NDN can increase the energy efficiency of MANET compared to the use of HCN primarily TCP/IP on the network stack solution for MANET. Of particular note would be NDN-based content routing's viability as a solution for energy consumption issues that plague wireless multi-hop ad hoc networks. Last but not least, this paper also provides the future research direction that could be undertaken on the subject.

**Keywords** Mobile ad hoc network · MANET testbed · Named data networking · NDN · Energy efficiency · Host-centric · HCN · OLSR and Batman-adv

## 1 Introduction

Energy is a limited resource on any mobile nodes in mobile ad hoc network (MANET) because the energy source only comes from the battery which only store limited amount of energy supply, just to enable the mobile devices to be used even if it is not connected to a fixed power supply such as a mesh router in a wireless mesh network. Furthermore, the mobile device in the MANET not only serves as a consumer, but also a router or relay to network traffic from other mobile devices to form a multi-hop wireless network [1, 2]. Because of that, energy conservation issue in MANET is one of the critical issues and needs to be addressed effectively, given that the lifespan of a mobile device in a MANET must be maximized as good as possible. This is because the lifespan of mobile devices in MANET influenced directly the lifespan of the network itself [3–6].

Until now, after several decades host-centric MANET exists, and there is still no evidence indicating that the increase in energy efficiency of the host-centric MANET can be done effectively. The approach used in the host-centric MANET to tackle dynamic network topology with additional mechanisms is still less efficient even though various new better methods have been proposed [7–10].

Circa 2000 a new paradigm appeared to replace host-centric, known as NDN. NDN has vast potential to improve the network energy efficiency of future Internet applications [11, 12].

This paper has been divided into six main sections as follows: Sect. 1 provides the introduction to the paper including the background information and problem addressed. Section 2 critically evaluates the related work carried out in this field. Section 3 discusses the advantages of NDN architecture that makes NDN more suitable to be used as a network stack solution in MANET compared to HCN, e.g., TCP/IP. Section 4 describes the research methodology used in this study. Section 5 provides a preliminary result obtained from real-world testbed-based experiment conducted in this research. And finally, Sect. 6 concludes the paper and presents our future work.

## 2 Related Work

This section takes an in-depth look at the related work carried out by other researchers and published in journals, conference proceedings, and technical reports. A critical analysis is carried out on the available literature with a view of understanding those works as well as identifying the potential issue that can be explored as a research topic.

Energy-efficient strategies have been researched for decades in IP-based MANET, and various solutions have been proposed as a result. In [13, 14], several types of link costs have been proposed based on the expected power consumption. They are used for the calculation of classical shortest path algorithms in order to find the optimal end-to-end path. In [15–17], the authors present the algorithm to find the minimum transmission power that preserves network connectivity. Since lower transmission power introduces a smaller transmission range, this approach effectively minimizes the effect of interference as well as the energy consumption for transmission itself. In [18, 19], the energy consumption is minimized by making nodes sleep, provided that network connectivity is preserved. Since low power is dissipated in sleep mode, they spare energy by turning off the unnecessary nodes for a certain scheduled period. The schemes in [20–22] minimize the number of message transmissions by information aggregation, optimized flooding and controlling the frequency of control messages.

Despite being in existence for several decades, there is still currently no evidence to indicate that energy efficiency of a host-centric MANET can be done more effectively. The approach used in a host-centric MANET to tackle dynamic network topology with additional mechanisms is still inefficient, though various new improved methods have been proposed.

In a host-centric MANET, an exclusive control mechanism is required to ensure that the link session for end-to-end connection between two nodes is always updated reactively or proactively. Any additional control mechanisms and broadcasted control messages will further increase energy consumption in a host-centric MANET solution.

Also, the network routing update process is done more frequently in an infrastructure network as the network topology in MANET is dynamic due to node mobility and wireless signal interference. A host-centric MANET requires an exclusive routing protocol to enable network routing updates to be dynamic, with this exclusivity costing further energy consumption in turn.

Additionally, the IP address assignment on each mobile node also requires an auto-configuration mechanism. This need to ensure conflict-free IP address assignment further increases the energy consumption within a host-centric MANET.

From the past research mentioned above, what makes this research similar to all of them is the same focus the researchers have, i.e., improving energy efficiency in MANET. On the other hand, what makes our research differ from all the previous research is in terms of the proposed solution, where we suggest our solution by using alternative solution of HCN paradigm, which is by using NDN paradigm, which offers a prominent feature that can be taken advantage of in order to significantly reduce the energy consumption in MANET.

# 3 Justification of Using NDN for MANET

While the TCP/IP network or HCN paradigm has been proven to serve the Internet effectively for nearly five decades, HCN still remains a stumbling block for MANET. After obtaining clearer insights on the limitations and disadvantages of host-centric networking-based MANET (HCN-based MANET) from this section, attention is now focused on other available network architecture, like NDN, that can be utilized as an alternative to improve energy efficiency in MANET. In justifying this proposed solution, a series of past research that highlighted the advantages of NDN, especially in terms of energy efficiency, is presented in this section.

According to authors in [23], there is vast potential to be explored by using NDN in wireless ad hoc networks. By replacing end-to-end dedicated link sessions and switching IP address as endpoint identity with named data retrieval, NDN has the potential to be ideal for the MANET environment. However, proper procedures and approaches first need to be realized before NDN can be used in wireless multi-hop networking environments like MANET.

NDN has vast potential to improve the network energy efficiency of future internet applications [11, 12]. In fact, experiments conducted by authors in [11] show that NDN-enabled routers produced far better results in terms of energy efficiency compared to CDNs and P2P networks.

Energy efficiency in MANET is improved through the use of NDN by the following reasons:

i. In NDN, interest aggregation is performed efficiently because the same interest packet will not be 'forwarded' repeatedly, but is instead stored in a Pending Interest Table (PIT) or dropped if it comes from the same network face. If the data packet has already been forwarded by one of each relay, the data packet is then stored in a content store that acts as network cache and it is directly retrieved from the content store, instead of the original location of the content provider. NDN also uses anycast data forwarding method, whereby if the requested content comes from multiple content consumers, it will not be sent repeatedly by content provider, but the same content will instead be obtained from a local copy stored in a content store or provided in the form of replicated repository [24, 25].
ii. Interest aggregation, in-network caching, and anycast forwarding in NDN forwarding strategy make NDN more energy efficient than any other network solutions [24]. Energy conservation is achieved in NDN through the reduction of network packet redundancy as energy conservation is critical in mobile networks like MANET [26].
iii. Energy efficiency in a NDN-based MANET can also be achieved because there is no exclusive control mechanism to perform global network convergence, as practiced in a HCN-based MANET. Additionally, there is no extra network traffic when control messages need to be sent multiple times to ensure that routing information is up-to-date and accurate [25, 26].
iv. In terms of content delivery recovery due to node mobility in MANET, NDN uses superior energy-efficient mechanisms as the content provider does not need

to retransmit the data packet to be sent back to the content consumers. This is because intermediate nodes can perform data packet forwarding and retry using broadcasting methods or adaptive forwarding methods. This means data packets will be retransmitted using an alternative path. Without the need to use complex mechanisms for broken link recovery and retransmission from content providers to content consumers, energy can be effectively conserved by using a NDN-based MANET solution [26].

From recognition given in the literature on the advantages of using NDN in improving energy efficiency, hence, we believe that NDN would be very suitable to be used in MANET in improving the energy efficiency.

## 4 Research Methodology

The methodology used in this study has been adapted from the design science research methodology (DSRM) [27]. Figure 1 shows the research framework adopted in this study where the links between the stages in the DSRM, methods adopted in each stage, and the main outcome (deliverable) at the completion of each stage. The vertical downward arrows show the main process flow while the horizontal arrows indicate the methods adopted at each stage and the expected outcomes of these methods.

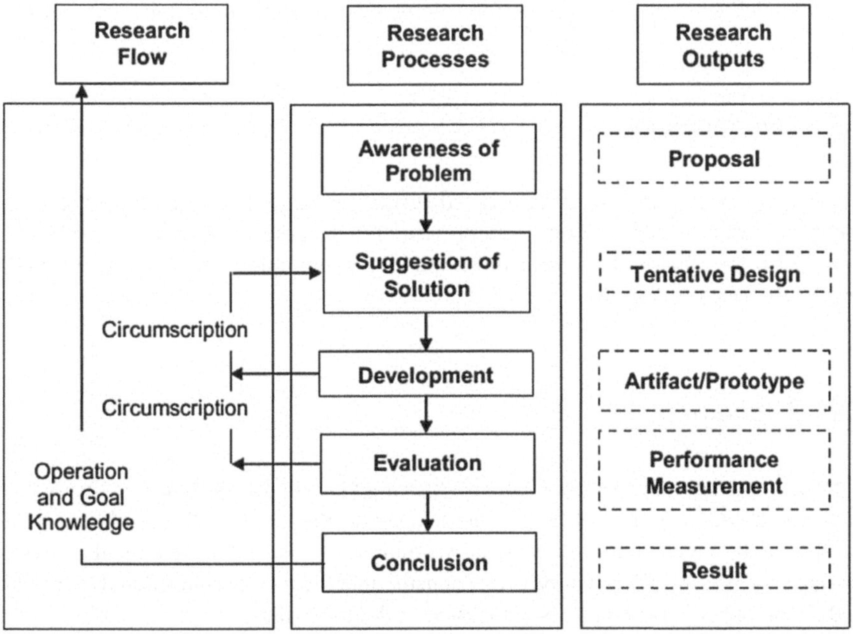


**Fig. 1** Design research approach [27]

The proposed methodology consists of five main stages. They are namely (i) awareness of the problem, (ii) suggestion, (iii) development, (iv) evaluation, and (v) conclusion. The details about the work carried out at each stage are explained below.

### 4.1 Awareness of the Problem

The first step in conducting this research would be to fully understand the energy efficiency issue of the MANET and to figure out the crucial factor that affects it. Then, the relation between energy efficiency of the MANET and the HCN paradigm shall be analyzed to determine whether this issue is suitable to be explored. The main outcome of this stage would be the research gap that needs to be filled at the end of this research.

### 4.2 Suggestion

Throughout this stage, a detailed research plan based on the research gap identified in Stage 1 would be put forward. The research plan would include many important points including research focus and motivation, description of research problem along with the research questions, relevant areas to be consulted, research approach and final deliverables and contributions along with the limitations and future work.

### 4.3 Development

The heart of the research work is conveyed during this stage. The development stage is the implementation stage of the concept created in the preceding stage. Two types of MANET implementation has been developed for this research, HCN-based MANET and NDN-based MANET.

### 4.4 Evaluation

During this stage, an extensive set of experiments will be carried out. From this analysis, a conclusion shall be prepared to relate the result with the objective of this research, which is to improve energy efficiency of MANET. Statistical analysis methods such as ANOVA will be used to ensure that the result is interpreted correctly and a suitable conclusion can be generated at the final stage.

### *4.5 Conclusion*

Finally, the performance measures collected at the previous stage will be plotted and tabulated depending on the type of data for easy visualization and comparison. Based on the results of the comparison of results, conclusions will be derived so as to which architecture performs better under which conditions.

## 5 Preliminary Result

In this research, we use the testbed method to compare energy efficiency of NDN-based MANET with HCN-based MANET implementation available. In comparison, we chose to use the two most popular and active open-source dynamic routing protocols in the mesh network community, OLSR and Batman-adv. While we use self-learning NDN to represent dynamic content routing of NDN based MANET.

### *5.1 Testbed Setup*

The focus of this preliminary study is that we first want to see the difference in energy consumption between NDN-based MANET with two HCN-based MANETs (OLSR and Batman-adv) when file transfers are done in a multi-hop wireless ad hoc network environment. No mobility node is created as the focus is more on observation of routing behavior within the multi-hop ad hoc network itself (Fig. 2).

We chose the multicast scenario which is the same content in one producer node retrieved by multiple consumer nodes as shown in Fig. 3.

This scenario is selected to see how far NDN advantage over HCN is the same content will not be sent repeatedly if via the same intermediate node and in-network caching mechanism using the content store in NFD to reduce energy consumption in each node in multi-hop ad hoc network.

Banana Pro single-board computer (SBC) is chosen as a mobile device in this testbed because the process of data gathering especially energy consumption is much easier to do than other mobile devices such as smartphones and tablets (Fig. 4).

This is because Banana Pro has a serial terminal that allows remote access to be done in serial communication and does not interfere with energy consumption readings rather than wired Ethernet networks that significantly affect energy consumption readings when experiments are performed. At the same time, Banana Pro is not included with unnecessary sensors and devices including display, and hence, it energy consumption reading is more focused on wireless network communication activities.

Energy consumption measurement is done at node 4 because all network traffic that happens when the experiment is executed will be via node 4. Monsoon power

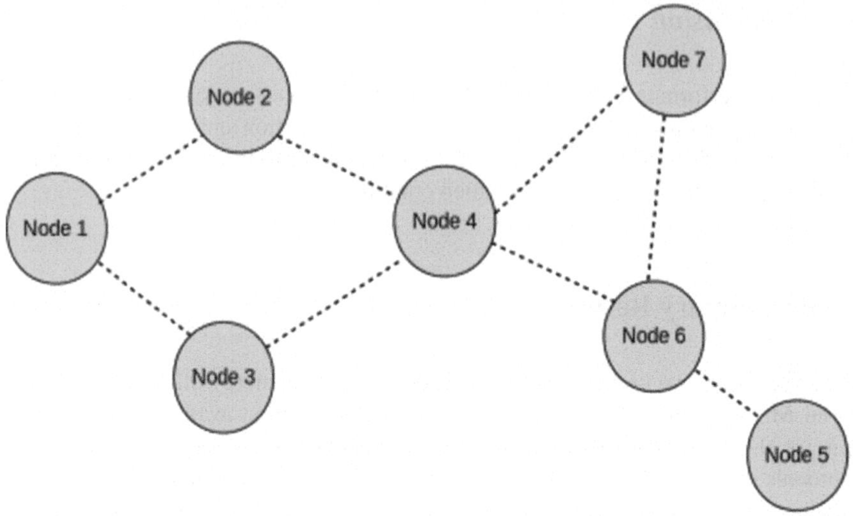


**Fig. 2** Network topology of testbed

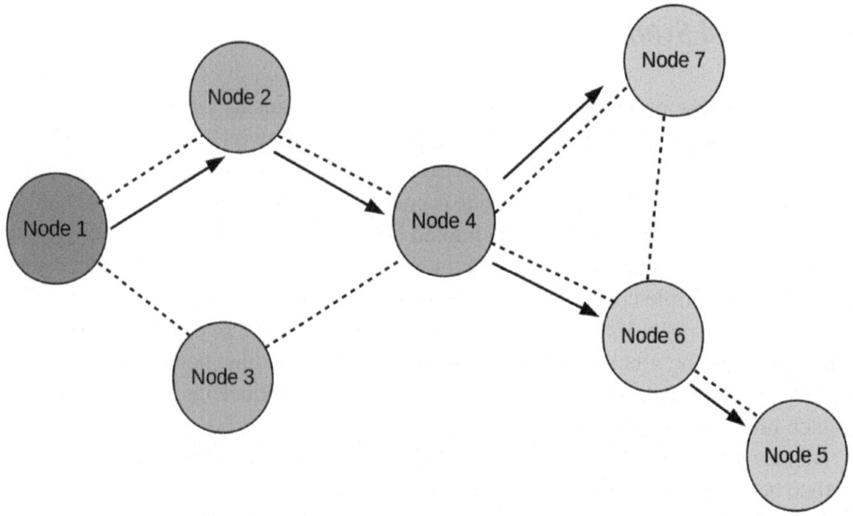


**Fig. 3** Experiment scenario, single content, single producer, multiple consumer

**Fig. 4** Banana Pro single-board computer

monitor is used to capture power consumption of node 4, and calculations will be made later to find out the value of energy consumption of node 4 during the experiment (see Fig. 5).

In addition, node 4 is controlled in experiments using a serial console approach compared to other nodes to avoid energy consumption reading disruption if using a wired Ethernet network.

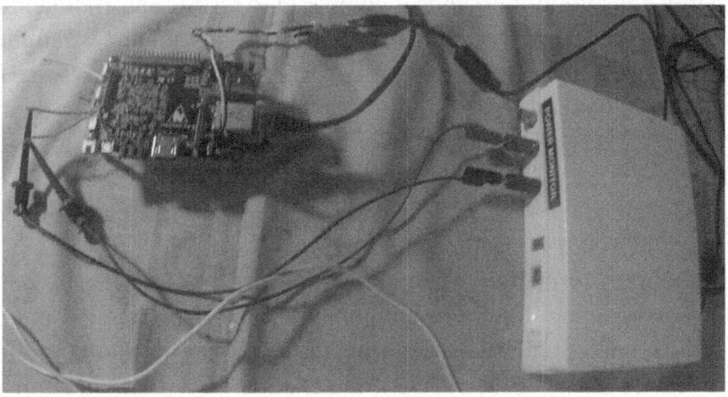

**Fig. 5** Energy consumption measurement using monsoon power monitor

**Table 1** Experiment result

| Dynamic routing solution | Energy consumption (J) | File transfer completion time (min:s) |
|---|---|---|
| NDN | 104.20 | 01:13.31 |
| OLSR | 228.25 | 02:30.75 |
| Batman-adv | 225.04 | 02:41.04 |

## *5.2 Experiment Result*

Prior to comparison, first experiment results were tested first using the one-way ANOVA significance test to ensure that the three values of experiment result (completion time and energy consumption) of the three different dynamic routing solutions were significantly different.

Table 1 shows the experiment result obtained in our preliminary test. Energy consumption reading is derived from node 4, and file transfer completion time is obtained on average from each consumer node (nodes 5, 6, and 7).

From a one-way ANOVA, test shows the mean of energy consumption result within all groups significantly different in each group ($p = 1.369\text{E}{-}64 < \alpha = 0.05$). Mean of NDN result is significantly different from the mean of OLSR result ($p = 6.5958\text{E}{-}13 < \alpha = 0.05$) and mean of Batman-adv result ($p = 7.1021\text{E}{-}13 < \alpha = 0.05$).

Therefore, comparisons can be made to energy consumption results for NDN, OLSR, and Batman-adv. It is to be noted here, file transfer completion time is only used as a reference that NDN does not sacrifice the performance of file transfer in reducing energy consumption in node 4 compared to OLSR and Batman-adv.

Figure 6 shows a comparison of energy consumption of node 4 when file transfer is done using NDN, OLSR, and Batman-adv. File transfers made using NDN are much lower than OLSR and Batman-adv. Even node 4 consumed energy node when using NDN (104.20 J) is less than half when using OLSR (228.25 J) and Batman-adv (225.04 J).

The results obtained prove that NDN-based MANET is more energy efficient than two HCN-based MANETs: OLSR and Batman-adv.

Low consumption of energy when using NDN versus OLSR and Batman-adv may be known to be due to file transfer completion time as shown in Fig. 7. The use of NDN allows completion time of file transfer to be shorter that is 1 min 13.31 s compared to OLSR around 2 min 30.75 s and Batman-adv which is 2 min 41.04 s.

NDN allows multiple content for single content based MANET single request can be made more efficient in terms of energy consumption and robustness compared to HCN based MANET, and this is what make NDN is more ideal to be a solution for MANET network stack compared to HCN.

This is evidenced by the experiment result we obtained in our preliminary study through energy consumption reading at node 4 as a relay for all network traffic in the experiment.

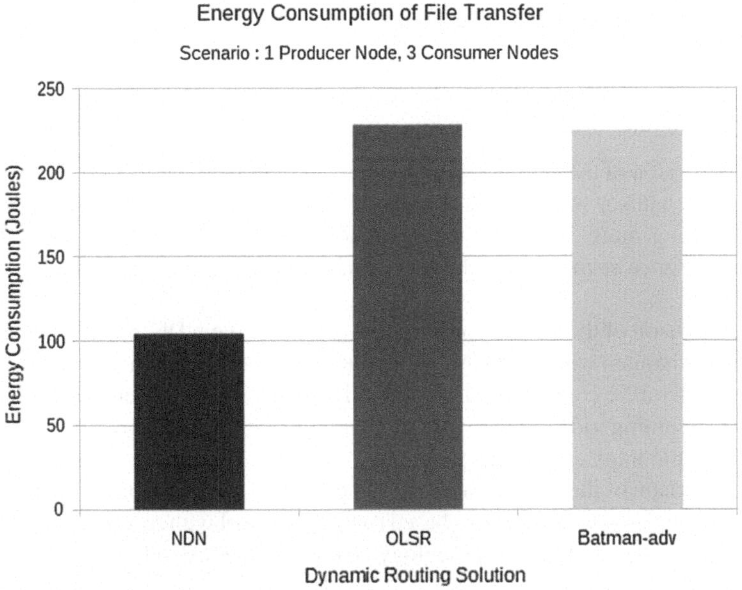


**Fig. 6** Energy consumption comparison during file transfer

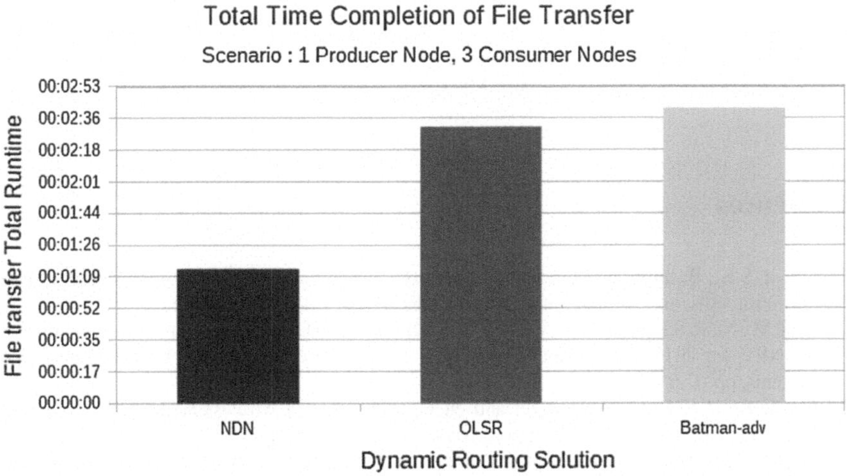


**Fig. 7** File transfer completion time comparison

## 6 Conclusion and Future Work

This paper discussed only surface level of our research work. This project is still ongoing. In future research undertakings, the focus will be as the following:

i. Comparison of the energy consumption pattern of the NDN-based content routing mechanism with the HCN-based routing protocol (OLSRd and Babel) by measuring more parameter metrics relating to energy efficiency and network performance such as throughput or goodput, delay, jitter, packet loss rate, and latency.
ii. Comparison of the energy consumption pattern of the NDN-based content routing mechanism with the HCN-based routing protocol when in use with different network traffic content. Said content will include video on-demand (VOD) and live streaming video, which due to its unique nature will have differing needs for the file transfer network traffic.
iii. Comparison of the energy consumption pattern of the NDN-based content routing mechanism with the HCN-based routing protocol with regard to node mobility parameters and network topology changes in the testbed.

All the garnered propose comparison results above is to suggest that the NDN based MANET outperforms or vice versa compared to HCN based MANET in terms of energy efficiency. Furthermore, the results obtained from the future work propose above shall serve as a stepping stone to better understand the real potential of the NDN-based content routing mechanism and how it can be used as a solution to the issue of energy consumption in MANET.

## References

1. Sarkar, S.K., Basavaraju, T.G., Puttamadappa, C.: Ad Hoc Mobile Wireless Networks: Principles, Protocols, and Applications, 2nd edn. CRC Press, Boca Raton (2013)
2. Patil, M., Naik, S.R.R., Nikam, V.B., Joshi, K.K.: Extended ECDSR protocol for energy efficient MANET. In: 2015 International Conference on Advanced Computing and Communication Systems, pp. 1–6 (2015)
3. Shpungin, H., Li, Z.: Throughput and energy efficiency in wireless ad hoc networks with Gaussian channels. IEEE/ACM Trans. Netw. **20**(1), 15–28 (2012)
4. Jumira, O., Zeadally, S., Jumira, O., Zeadally, S.: Energy Efficiency in Wireless Ad Hoc Networks. Wiley, London (2012)
5. Ray, N.K., Turuk A.K.: Energy conservation issues and challenges in MANETs. In: Lakhtaria, K.I. (eds.) Technological Advancements and Applications in Mobile Ad-Hoc Networks: Research Trends, pp. 291–318. IGI Global (2012)
6. Yitayal, E., Pierson, J.M., Ejigu, D.: A balanced battery usage routing protocol to maximize network lifetime of MANET based on AODV. In: Balandin, S., Andreev, S., Koucheryavy, Y. (eds.) Internet of Things, Smart Spaces, and Next Generation Networks and Systems 2014. LNCS, vol. 8638, pp. 266–279. Springer, Cham (2014)
7. Holland, G., Vaidya, N.: Analysis of TCP performance over mobile ad hoc networks. In: Proceedings of the 5th International Conference on Mobile Computing and Networking (MobiCom'99), pp. 219–230. ACM, New York (1999)

8. Lambrou, T.P., Panayiotou, C.G: A survey on routing techniques supporting mobility in sensor networks. In: Fifth International Conference on Mobile Ad-hoc and Sensor Networks, (MSN'09), pp. 78–85 (2009)
9. Meisel, M., Pappas, V., Zhang, L.: Ad hoc networking via named data. In: Proceedings of the Fifth International Workshop on Mobility in the Evolving Internet Architecture (MobiArch'10), pp. 3–8. ACM, New York (2010)
10. Han, H., Wu, M., Hu, Q., Wang, N.: Best route, error broadcast: a content-centric forwarding protocol for MANETs. In: 2014 IEEE 80th Vehicular Technology Conference (VTC Fall 2014), pp. 1–5 (2014)
11. Lee, U., Rimac, I., Hilt, V: Greening the internet with content-centric networking. In: Proceedings of the 1st International Conference on Energy Efficient Computing and Networking (e-Energy'10), pp. 179–182. ACM, New York (2010)
12. Braun, T., Trinh, T.A: Energy efficiency issues in information-centric networking. In: Pierson, J.-M., Costa, G.D., Dittmann, L. (eds.) Energy Efficiency in Large Scale Distributed Systems 2013. LNCS, vol. 8046, pp. 271–278. Springer, Berlin (2013)
13. Senouci, S.-M., Pujolle, G.: Energy efficient routing in wireless ad hoc networks. In: Proceedings of IEEE International Conference on Communications (ICC 2004), vol. 7, pp. 4057–4061 (2004)
14. Kwon, S., Shroff, N.B.: Energy-efficient interference-based routing for multi-hop wireless networks. In: IEEE International Conference on Computer Communications (INFOCOM 2006), pp. 1–12 (2006)
15. Ingelrest, F., Simplot-Ryl, D., Stojmenovic, I.: Optimal transmission radius for energy efficient broadcasting protocols in ad hoc and sensor networks. IEEE Trans. Parallel Distrib. Syst. **17**(6), 536–547 (2006)
16. Cardei, M., Wu, J., Yang, S.: Topology control in ad hoc wireless networks with hitch-hiking. In: IEEE Communications Society Conference on Sensor and Ad Hoc Communications and Networks (SECON 2004), pp. 480–488 (2004)
17. Lin, S., Zhang, J., Zhou, G., Gu, L., He, T., Stankovic, J.A.: ATPC: adaptive transmission power control for wireless sensor networks. In: Proceeding of the 4th International Conference on Embedded Networked Sensor Systems (SenSys 2006). ACM, New York (2006)
18. Carle, J., Simplot-Ryl, D.: Energy-efficient area monitoring for sensor networks. Computer **37**(2), 40–46 (2004)
19. Cardei, M., Thai, M., Li, Y., Wu, W.: Energy-efficient target coverage in wireless sensor networks. In: IEEE International Conference on Computer Communications (INFOCOM 2005), vol. 3, pp. 1976–1984 (2005)
20. Xia, D., Vlajic, N.: Near-optimal node clustering in wireless sensor networks for environment monitoring. In: IEEE International Conference on Advanced Information Networking and Applications (AINA 2007), pp. 632–641 (2007)
21. Ingelrest, F., Simplot-Ryl, D., Stojmenovic, I.: Smaller connected dominating sets in ad hoc and sensor networks based on coverage by two-hop neighbors. In: 2007 2nd International Conference on Communication System Software and Middleware, pp. 1–8 (2007)
22. Han, B., Fu, H., Li, L., Jia, W.: Efficient construction of connected dominating set in wireless ad hoc networks. In: Proceedings of the 2004 IEEE International Conference on Mobile Ad-hoc and Sensor Systems (MASS 2004), pp. 570–572 (2004)
23. Conti, M., Boldrini, C., Kanhere, S.S., Mingozzi, E., Pagani, E., Ruiz, P.M., Younis, M.: From MANET to people-centric networking: milestones and open research challenges. Comput. Commun. **71**, 1–21 (2015)
24. Kim, D., Kim, J., Moon, C., Choi, J., Yeom, I.: Efficient content delivery in mobile ad-hoc networks using CCN. Ad Hoc Netw. **36**(Part 1), 81–99 (2016)
25. Amadeo, M., Campolo, C., Iera, A., Molinaro, A.: Named data networking for IoT: an architectural perspective. In: 2014 European Conference on Networks and Communications (EuCNC 2014), pp. 1–5 (2014)

26. Anastasiades, C., Weber, J., Braun, T.: Dynamic unicast: information centric multi-hop routing for mobile ad-hoc networks. Comput. Netw. **107**(Part 2), 208–219 (2016)
27. Vaishnavi, V.K., Kuechler, W.: Design Science Research Methods and Patterns: Innovating Information and Communication Technology, 2nd edn. CRC Press Inc, Boca Raton, FL, USA (2015)

# Image Processing and Computer Vision

# Fingerprint Biometric Template Security Schemes: Attacks and Countermeasures

**Reza Mehmood and Arvind Selwal**

**Abstract** Biometrics is one of the most promising technologies for providing secure authentication in modern computing applications and eradicates the issues associated with the traditional authentication systems. Almost 50% of the security infrastructure comprises fingerprint biometric. As the market share of fingerprints is increasing tremendously, its security is becoming a challenge for research community. In this paper, a brief review of different fingerprint template security schemes has been presented. Moreover, various masquerade attacks on fingerprint template have been studied and their countermeasures are presented. A comparative analysis of different template security schemes based on different performance metrics like FAR, FRR, and EER is also provided. It was seen that the methods employed for fingerprint may not work for other biometric traits like iris, face, etc., because of their difference in dimensions of templates. This paper allows to find the research gaps in the existing template security algorithms and suggests further development in the field of biometric template protection.

**Keywords** Biometrics · Fingerprint · Template security · Feature vector

## 1 Introduction

Biometrics is the science of identifying an individual from its physical and behavioral characteristics. Biometrics is employed in almost every place for authentication purposes to ensure better security and eradicate the issues that were associated with the traditional authentication systems. May it be an MNC or an airport, biometrics plays a significant role in making the authentication process fast and more secure. We cannot deny the importance of biometric system in marking the attendance of students and even employees, thus preventing the chance of buddy punching. All

R. Mehmood (✉) · A. Selwal
Department of Computer Science and Information Technology, Central University of Jammu, Jammu and Kashmir 181143, India
e-mail: rzmhmd26@gmail.com

A. Selwal
e-mail: arvind.cuj@gmail.com



P. K. Singh et al. (eds.), *Proceedings of ICRIC 2019*, Lecture Notes in Electrical Engineering 597, https://doi.org/10.1007/978-3-030-29407-6_33

biometric systems work on biometric traits. The biometric traits used to identify an individual can be a fingerprint, hand geometry, face, iris, gait recognition, etc. The fingerprint is the most broadly used biometric trait among all other traits in a biometric system. Fingerprint recognition technology has rapidly grown in these past years due to its good performance and low cost of acquisition devises. A fingerprint consists of different patterns of dark and bright regions which are called ridges and valleys, respectively. Ridge feature point is further divided into local ridge pattern (ending, bifurcation) and global ridge pattern (loop, delta, and whorl). The features of a fingerprint pattern at the local level are called minutia points. Ridge ending is a point where the ridges end suddenly and ridge bifurcation is a point where the ridge gets divided into two or more ridges. Ridge ending and ridge bifurcation are commonly used minutia types as all other types are combinations of these two (Fig. 1).

A fingerprint biometric system involves two steps: first is the enrollment step, and second is the verification step. During enrollment, fingerprint of an individual is acquired using a sensor and the quality of image captured is enhanced before further processing. After that, features are extracted from the improved image and stowed

**Fig. 1** Global and local feature points of a sample fingerprint [1]

in a secure place in the form of a template. The features used by most of fingerprint systems are generally called minutia which is the representation of a fingerprint having bifurcation and termination. Feature extraction process involves estimating the ridge orientation ($\theta$) at a ridge point ($x$, $y$) followed by binarization and minutia extraction. The database stores the template in the form of a matrix of size $n \times 3$ with ($x$, $y$, $\theta$) values. During verification, the stored template in the database is compared to the query template from the new user using a matching algorithm. If it is a match, then the user is accepted, otherwise rejected [2]. The different stages of a fingerprint biometric system are shown in Fig. 2 (adapted from [2]).

The performance of a biometric system can be identified by its percentage of accepting the genuine users and rejecting the imposters. The proportion of imposters that the biometric system will incorrectly accept is defined as the false accept rate (FAR) of that system as given in Eq. 1.

$$\text{FAR} = \frac{\text{Number of imposters accepted}}{\text{Total number of imposters}} \tag{1}$$

The proportion of genuine users that a system will incorrectly reject is defined as the false reject rate (FRR) as shown in Eq. 2.

$$\text{FRR} = \frac{\text{Number of genuine users rejected}}{\text{Total number of users}} \tag{2}$$

The total number of authentic users accepted by the system is defined as genuine acceptance rate (GAR), i.e., GAR = 1 − FRR. Equal error rate (EER) is a point

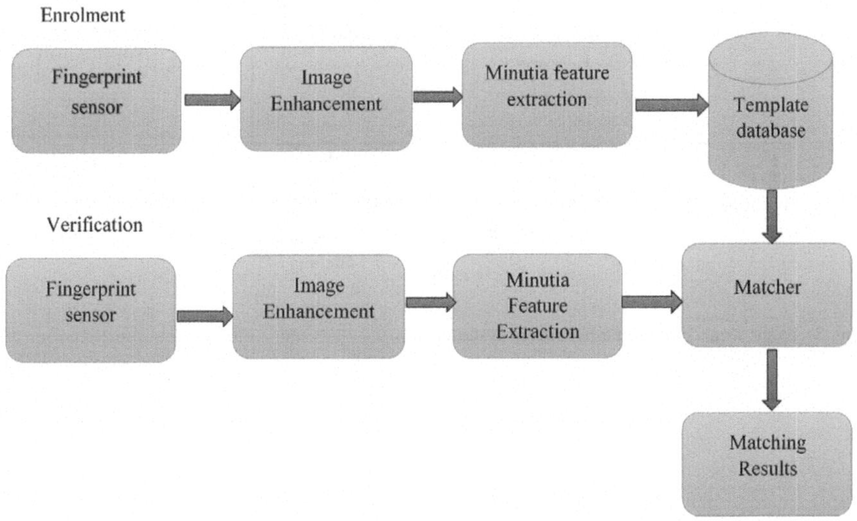


**Fig. 2** Enrollment and verification stages in a fingerprint biometric system

on the graph where FAR equals FRR [3]. Although biometrics is used to provide security to many applications, it is itself vulnerable to attacks. The vulnerabilities of a biometric system can be grouped into two classes: intrinsic vulnerabilities that are caused by the system itself due to failure of sensor or any other module in the biometric system and external vulnerabilities caused by an attacker deliberately. The effect of such vulnerabilities causes intrusion in the system [4]. In 2001, Ratha et al. proposed 8 weak points in a biometric system that can be attacked by the attacker. These eight basic sources of attack are: spoofing attack on the sensor, replay attack on the link between sensor and feature extractor, Trojan horse attack on feature extractor, the interception attack on the link between feature extractor and matcher, attack on matcher, the modification attack on the template database, attack on the channel between matcher and the database, and overriding the authentication results according to the will of hacker. Among these eight security weak sources, the security of the stored template is very critical and more prone to attacks. So the security of template stored in the database is a challenging task (Fig. 3).

Rest of the paper is organised in the various sections as given: The Sect. 2 describes various template security schemes. The Sect. 3 discusses a brief literature review on existing template security schemes. The Sect. 4 provides the analysis of different template security schemes, and finally, the conclusion is given in the Sect. 5.

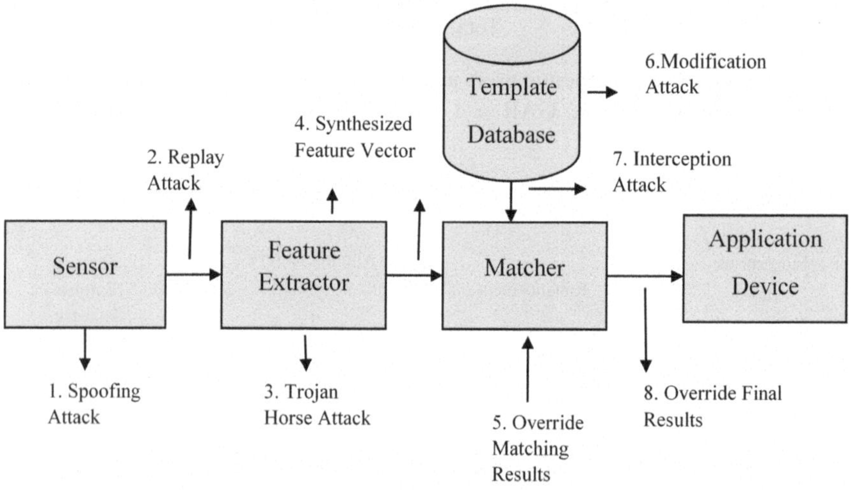


**Fig. 3** Eight weak points in a biometric system [4]

## 2 Taxonomy of Template Security Schemes

There are four major characteristics that a template security scheme must follow for being an ideal scheme. These include diversity, revocability, security, and performance. But satisfying all these requirements is a challenging task. There are two schemes that have been devised for template security, namely the feature transformation approach and biometric cryptosystem.

### *2.1 Feature Transformation Based Schemes*

In the feature transformation scheme, a function known as transformation function (f) is applied on the biometric template along with a secret key (k) and the converted template is stored in the database. Feature transformation scheme can be further classified into two approaches: invertible (bio-hashing) and non-invertible transforms (robust hashing). The main difference between invertible and non-invertible transform is that in invertible transformation scheme the function is defined by a password that needs to be securely preserved by the user, whereas in non-invertible transform a one-way function is used to transform the template. Invertible transformation is also referred as salting, and non-invertible transformation is also called cancellable biometrics. Cancellable biometrics was brought together so that the biometric template could be replaced with a newer one if it was being corrupted by the third-party attacker [1].

### *2.2 Biometric Cryptosystem Based Schemes*

Biometric cryptosystem involves a cryptographic secret key and using that key for encryption of the template. Biometric cryptosystem is categorized into key binding (fuzzy vault and fuzzy commitment) and key generation approach (secure sketch fuzzy extractor) [5]. Key binding is the technique where a secret key is bound to the template within a cryptographic framework and the key cannot be decoded without the prior knowledge of the template. In key generation, a key is derived from the biometric template to perform the encryption. Various template security schemes are shown in Fig. 4 (adapted from [5]).

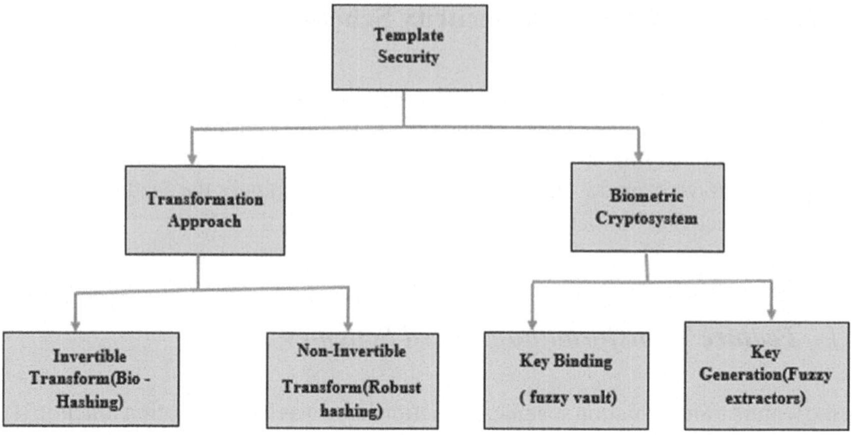


**Fig. 4** Classification of template security schemes

## 3 Literature Review

A numerous research work has been done for the security of template data to protect it from masquerade and other kind of attacks and is discussed below:

Barman et al. [6] came up with a new protocol for authentication using fingerprint-based fuzzy commitment scheme. They used three factors for their key agreement scheme which includes password, smart card, and personal biometric. The security of the scheme was verified using real-or-random model for session key security and Burrows–Abadi–Needham logic for mutual authentication. The formal security verification was carried out using AVISPA. The proposed scheme also shows resistance to various kinds of known attacks like man-in-the-middle attack, denial-of-service attack, etc. [6].

Selwal and Gupta [7] proposed a novel method called octet indexing method to secure and reduce the storage space of the template database. The novel method is based on transforming the feature vectors of fingerprint and hand geometry into binary feature vectors using the bio-hashing technique. The binary feature vectors are the fused feature vectors, and an octet indexing technique is applied to the fused binary feature vectors to secure the template and reduce its storage overhead by approximately 50%. The presented novel scheme has a recognition rate of 98.4% and an equal error rate of 0.48% [7].

Jin et al. [8] proposed a locality-sensitive hashing based on ranking called the index of max (IoM) hashing for protecting the template. In the given scheme, biometric feature vector is transformed into a discrete max ranked hash code. The proposed scheme shows two realizations of IoM hashing, namely Gaussian random projection and uniformly random permutation. The analysis proves that the scheme is secure to existing attacks and satisfies revocability and unlinkability criteria. They used FVC2002 and FVC2004 fingerprint databases for their experiment [8].

Gomez-Barrero et al. [9] proposed a scheme to compare variable length data using homographic encryption technique. In this scheme, the data is stored or exchanged only in encrypted form. The new variable length-based algorithm is combined with the fixed length technique to acquire better comparison accuracy. This scheme satisfies the unlinkability property of biometrics [9].

Kaur and Sofat [10] designed a multimodal biometric system where they fused traits of fingerprint and face using feature-level fusion to provide security to the template by using fuzzy vault scheme. The ordered feature set is made compatible with the fuzzy vault by converting it to unordered feature set. The performance of the scheme is calculated on the basis of FAR and FRR by varying the degree of the polynomial from the range of 8–14 [10].

Selwal and Gupta [2] portrayed four conceptual designs to enhance the inter-class variation of two modalities, a strong fingerprint biometric combined with the hand geometry trait. Fuzzy-based analytic hierarchy technique was employed to evaluate the proposed conceptual designs over five decision factors. Their results show that fusing fingerprint and hand geometry features using feature-level fusion are most favorable with an overall ranking of 0.73438, and out of the five decision parameters, template security is found to be of supreme importance with a rating of 0.3940 [2].

Kaur [11] designed a secure fuzzy vault scheme to protect the fingerprint vault from correlation attack by combining fuzzy vault scheme with Hadamard transformation technique. Fast Walsh–Hadamard transformation is applied to the set of minutia point and the chaff points. The experimental results show that FAR is 2% and FRR is 11% after applying Hadamard transformation technique [11].

Sarala et al. [12] proposed a system in which it was shown that the performance of a fuzzy vault system remains unchanged even after the blend substitution attack. They launched a blend substitution attack on the fuzzy vault system and designed a mechanism to detect this attack and provide corrective measures. The presented attack detection mechanism comprises vault database linked to a server containing polynomial genuine points, which checks for any change in the database continuously. The fuzzy vault is recovered back after the attack by substituting the attack points with the genuine points in the server. The obtained average recognition rates are GAR of 86.1389% and FAR of 0.4039% [12].

Lafkih et al. [13] investigated the vulnerabilities of the fuzzy vault scheme. The fuzzy vault scheme was tested practically with low-resolution fingerprint and facial images of genuine users against the masquerade attack. The attacker can gain access to a fuzzy vault with 100% probability even if the alteration level of the images is high for face images, but for the fingerprint images the alteration level has to be low for the attacker to gain access [13].

Dang et al. [14] combined two new components, chaff point generator and verifier into fuzzy vault scheme that is capable of detecting any change in the fuzzy vault and thus protecting the template from blend substitution attack possible in CRC-based fuzzy vault scheme. Continuous hashing and linear projection are used to generate chaff points during enrollment. The experimental results show a 12% increase in GAR with main polynomial degree 8 [14].

Feng et al. [15] have proposed a masquerade attack algorithm which is a combination of perceptron learning and hill climbing algorithm to create a synthetic face image from the transformed binary template and stolen token. They have implemented their algorithm on two scenarios, when the binarization algorithm is known and when it is unknown. In the first case, when binarization algorithm is known, the attack algorithm achieves 100 and 98.3% recognition rate for CMU PIE and FRGC databases, respectively. In the second case, when binarization algorithm is unknown, the algorithm achieves 20.59–85.29% and 15.14–46.57% recognition rate for same databases [15].

Prasad and Santhosh Kumar [16] proposed a system in which they generated the fingerprint template by constructing M rectangles of different orientation angles around each reference minutia, then selected the minutia that fall in the rectangle, and calculated the neighboring relation that is applied on a plane to generate bit string which is converted into complex vector and transformed using user's pin. The accuracy of this system is better as it considers every minutia in a fingerprint for template generation [16].

Moujahdi et al. [17] introduced a new technique for template protection to ensure revocability, diversity, security, and better performance. They used the information extracted from the fingerprint minutia to construct special spiral curves and stored these curves in the database rather than the actual template. The matching is performed by applying similar procedure to query template and using Hausdorff distance to find the difference between the database template and the query template [17]. Chin et al. [18] proposed a multi-biometric system using fingerprint and palm print to protect the fused template using three-stage hybrid feature transformation method. In their system, they first obtained a fused template of fingerprint and palm print using feature-level fusion. Then, using the random tiling method unique features were extracted from fused characteristics of a unique specific key. Then equal-probable $2^n$ discretization is used to construct a bit string template from unified feature vector [18].

Mihailescu [19] devised a new enrollment method to secure biometric template. The proposed scheme is based on a cryptographic method called hash chaos-based cryptography. A session key was generated using Rossler map and a pseudo-random generator. The session key is then passed to the hash function that is used in the enrollment scheme. This scheme provides almost same computational security as that provided by Davies–Meyer scheme [19].

Nguyen et al. [20] proposed and implemented a novel algorithm for generation of chaff points so as to reduce the time complexity of the fuzzy vault scheme. In this paper, a new system was employed where a fingerprint image was split into blocks and different points were generated randomly in those image cells. The point in the image cells can be chaff points only if it is unique and the distance between this point and the adjacent points is larger or equal to the threshold. The experimental results show that the EER comes out to be 2.4 and 1.9% for FVC2002-DB1A and FVC2002-DB2A [20].

Jin et al. [21] proposed a method in which a revocable fingerprint template is generated by transforming a set of minutia into a bit string using polar grid base 3 tuple quantization technique. The work is carried out in four steps which include translation and rotation of minutia, tuple-based quantization, bit string generation, and user-specific permutation and finally performs matching. The system proved to have better performance on the tested databases except FVC2004DBI with 15% EER [21]. Bhatnagar et al. [22] presented a new secure watermarking technique to secure the template from attacks. They used nonlinear chaotic map and Heisenberg decomposition in their technique for the diffusion of biometric templates. Their results have proven to the secure template against different kinds of attacks like resizing attack and rotation attack [22].

Cappelli et al. [23] introduced an approach to reconstruct the fingerprint images from the original minutia template to study the success rate of masquerade attacks against eight fingerprint matching algorithms. This approach has shown that it is possible to generate fingerprint images which have ridge patterns very close to original fingerprint patterns and thus a successful masquerade attack can be performed with 72.8% of success rate at a high level of security and 83.9% at medium level of security [23].

Jain et al. [24] in a paper named "Biometric template security: challenges and solutions" described various security issues that can be encountered after storing the template. They evaluated some of the security breaches that can possibly compromise template information like hill climbing attack and masquerade attack. Then, they highlighted the possible solutions to protect the biometric template from these attacks which include techniques like watermarking, steganography, cancellable biometrics, and cryptosystem [24].

## 4 Analysis of Template Security Schemes

The comparative analysis of various template protection schemes is shown in Table 1. The comparison is performed on the basis of the technique employed, type of fusion technique used, type of modality, performance, and datasets used. The analysis clearly shows that most of the template security schemes are used for fingerprint biometrics and the techniques used for fingerprint may not be applicable to other biometric traits.

Various attacks like correlation attack, substitution attack, etc., has been observed which affect the templates. In order to counter these attacks, different transformation and cryptosystem-based template security schemes has been employed. Fuzzy vault, a cryptosystem-based approach has been analysed closely, and it has been observed that the polynomial reconstruction is still a challenge in the case, where the polynomial is of higher degree. Hence, fuzzy vault may be a topic of research in the research community.

**Table 1** A Summary of biometric template security schemes

| S. No. | Technique | Type of biometrics and fusion type | Biometrical traits | Author | Performance |
|---|---|---|---|---|---|
| 1. | Key agreement fuzzy commitment scheme | Unimodal biometrics | Fingerprint | Barman et al. [6] | – |
| 2. | Octet indexing method | Multimodal biometrics and feature-level fusion | Fingerprint and hand geometry | Selwal and Gupta [7] | EER = 0.48% |
| 3. | Index of max hashing | Unimodal biometrics | Fingerprint | Jin et al. [8] | FVC2002 and FVC2004 databases used |
| 4. | Homographic encryption scheme | Multi-biometrics | Online signature and fingerprint | Gomez Barrero et al. [9] | EER = 0.12% |
| 5. | Fuzzy vault scheme | Multimodal biometrics | Fingerprint and face with feature-level fusion | Kaur and Sofat [10] | FRR = 8.8% FAR = 0 (when window size = 1) |
| 6. | Fuzzy-based analytic hierarchy technique | Multimodal biometrics | Fingerprint and hand geometry with feature-level fusion | Selwal and Gupta [2] | Ranking = 0.73438 |
| 7. | Fuzzy vault with Hadamard transformation | Unimodal biometrics | Fingerprint | Kaur [11] | FAR = 2% FRR = 11% |
| 8. | Fuzzy vault system | Unimodal biometrics | Fingerprint | Sarala et al. [12] | GAR = 86.1389% FAR = 0.4039% |
| 9. | Fuzzy vault | Multimodal biometrics | Fingerprint and face | Lafkih et al. [13] | – |
| 10. | Chaff point generator and verifier in fuzzy vault | Unimodal biometrics | Fingerprint | Dang et al. [14] | GAR = 12% increase |

(continued)

**Table 1** (continued)

| S. No. | Technique | Type of biometrics and fusion type | Biometrical traits | Author | Performance |
|---|---|---|---|---|---|
| 11. | Perceptron learning and hill climbing algorithm | Unimodal biometrics | Face | Feng et al. [15] | Recognition rate = 100% and 98.3% (when binarization algorithm known) for CMU PIE AND FRGC databases resp. |
| 12. | Fingerprint shell | Unimodal biometrics | Fingerprint | Prasad and Santhosh Kumar [16] | EER = 4.28 and 1.45 for FVC2002-DB1 and DB2 resp. |
| 13. | Multi-line neighboring relation | Unimodal biometrics | Fingerprint | Moujahdi et al. [17] | EER = 0.62%, 1.33%, and 2.64% for FVC 2002 DB1, DB2, and DB3 resp. |
| 14. | Three-stage hybrid transformation method | Multimodal biometrics | Fingerprint and palm print | Chin et al. [18] | EER = 10% for fusing F3 and P1 database |
| 15. | Hash chaos-based cryptography | Unimodal biometrics | General approach | Mihailescu [19] | – |
| 16. | Chaff point generation in fuzzy vault | Unimodal biometrics | Fingerprint | Nguyen et al. [20] | EER = 2.4% and 1.9% for FVC2002-DB1A and FVC2002-DB2A resp. |
| 17. | Polar grid base 3 tuple quantization technique | Unimodal biometrics | Fingerprint | Jin et al. [21] | EER = 15% |
| 18. | Watermarking technique | Unimodal biometrics | Fingerprint | Bhatnagar et al. [22] | PSNR = 42.5376 dB for FVC2002 database |

(continued)

**Table 1** (continued)

| S. No. | Technique | Type of biometrics and fusion type | Biometrical traits | Author | Performance |
|---|---|---|---|---|---|
| 19. | Masquerade attack | Unimodal biometrics | Fingerprint | Cappelli et al. [23] | Success rate = 72.8% at high level security and 83.9% at medium level security |
| 20. | Biometric template security | Unimodal biometrics | Fingerprint | Jain et al. [24] | – |

## 5 Conclusion

In this paper, various template security schemes mainly based on the fingerprint biometric trait to protect the template stored in the database have been reviewed. It has been perceived that there is no ideal template protection scheme that maintains all the four characteristics of biometrics, which include revocability, security, performance, and diversity all at once. It was also realized that uni-biometrics is mostly selected as the type of biometrics. This can be because in multi-biometric system extra overhead is to be faced to get different biometric traits into a common domain so that fusion technique could be applied easily. But it is certain to mention that multi-biometrics is more secure than uni-biometrics. Despite of all these template security schemes, template data is still vulnerable to attacks. Therefore, there is a need to design more robust schemes both for unimodal and multimodal biometric systems to provide better security to the biometric systems.

## References


1. Mwema, J., Kimwele, M., Kimani, S.: A simple review of biometric template protection schemes used in preventing adversary attacks on biometric fingerprint templates. **20**, 12–18 (2015). ISSN: 2231-2803
2. Selwal, A., Gupta, S.K.: Fuzzy analytic hierarchy process based template data analysis of multimodal biometric conceptual designs. Procedia Comput. Sci. **85**, 899–905 (2016)
3. Jain, A.K., Ross, A. A., Nandakumar, K.: Introduction to Biometrics (Google eBook), p. 311 (2011)
4. Ratha, N.K., Connell, J.H., Bolle, R.M.: An analysis of minutiae matching strength. In: International Conference on Audio- and Video-Based Biometric Person Authentication, pp. 223–228 (2001)
5. Nandakumar, K., Jain, A.K., Nagar, A.: Biometric template security. EURASIP J. Adv. Signal Process. (2008)
6. Barman, S., Das, A.K., Member, S., Samanta, D.: Provably Secure Multi-Server Authentication Protocol using Fuzzy Commitment. IEEE Access PP, 1 (2018)

7. Selwal, A., Gupta, S.K.: Low overhead octet indexed template security scheme for multi-modal biometric system. J. Intell. Fuzzy Syst. **32**, 3325–3337 (2017)
8. Jin, Z., Lai, Y., Hwang, J.Y., Kim, S., Teoh, A.J.: Ranking Based Locality Sensitive Hashing Enabled Cancelable Biometrics: Index-of-Max Hashing, 6013 (2017)
9. Gomez-Barrero, M., Galbally, J., Morales, A., Fierrez, J.: Privacy-Preserving Comparison of Variable-Length Data With Application to Biometric Template Protection, vol. 5 (2017)
10. Kaur, M., Sofat, S.: Fuzzy Vault Template Protection for Multimodal Biometric System, pp. 1131–1135 (2017)
11. Kaur, M.: Secure Fingerprint Fuzzy Vault Using Hadamard Transformation to Defy Correlation Attack (2016)
12. Sarala, S.M., Karki, M.V., Yadav, D.S.: Blended Substitution Attack Independent Fuzzy Vault for Fingerprint Template Security (2016)
13. Lafkih, M., Lacharme, P., Rosenberger, C., Mikram, M., Ghouzali, S.: Vulnerabilities of Fuzzy Vault Schemes Using Biometric Data with Traces, pp. 822–827 (2015)
14. Dang, T.K., Nguyen, M.T., Truong, Q.H.: Chaff Point Generation Mechanism for Improving Fuzzy Vault Security, pp. 147–153 (2015)
15. Feng, Y.C., Lim, M.H., Yuen, P.C.: Masquerade attack on transform-based binary-template protection based on perceptron learning. Pattern Recognit. **47**, 3019–3033 (2014)
16. Prasad, M.V.N.K., Santhosh Kumar, C.: Fingerprint template protection using multiline neighboring relation. Expert Syst. Appl. **41**, 6114–6122 (2014)
17. Moujahdi, C., Bebis, G., Ghouzali, S., Rziza, M.: Fingerprint shell: secure representation of fingerprint template. Pattern Recognit. Lett. **45**, 189–196 (2014)
18. Chin, Y.J., Ong, T.S., Teoh, A.B.J., Goh, K.O.M.: Integrated biometrics template protection technique based on fingerprint and palmprint feature-level fusion. Inf. Fusion **18**, 161–174 (2014)
19. Mihailescu, M.I.: New enrollment scheme for biometric template using hash chaos-based cryptography. Procedia Eng. **69**, 1459–1468 (2014)
20. Nguyen, T.H., Wang, Y., Nguyen, T.N., Li, R.: A Fingerprint Fuzzy Vault Scheme Using a Fast Chaff Point Generation Algorithm (2013)
21. Jin, Z., Jin Teoh, A.B., Ong, T.S., Tee, C.: Fingerprint template protection with minutiae-based bit-string for security and privacy preserving. Expert Syst. Appl. **39**, 6157–6167 (2012)
22. Bhatnagar, G., Wu, Q.M.J., Raman, B.: Biometric template security based on watermarking. Procedia Comput. Sci. **2**, 227–235 (2010)
23. Cappelli, R., Lumini, A., Maio, D., Maltoni, D.: Evaluating Minutiae Template Vulnerability, pp. 174–179 (2007)
24. Jain, A.K., Ross, A., Uludag, U.: Biometric template security: challenges and solutions. In: Secure Watermarking Multimedia, vol. 4675, pp. 629–640 (2002)

# Effect of Blurring on Identification of Aerial Images Using Convolution Neural Networks

**Palak Mahajan, Pawanesh Abrol and Parveen K. Lehana**

**Abstract** The emergence of deep learning in the field of computer vision has led to extensive deployment of convolutional neural networks (CNNs) in visual recognition systems for feature extraction. CNNs provide learning through hierarchical inferencing by providing multilayer architecture. Due to high processing capability of CNNs in multidimensional signals like images, they are considered to be predominant artificial neural networks. CNNs are extensively used in computer vision such as in image recognition where the intent is to automatically learn features followed by generalization and eventually recognizing the learned features. In this paper, we investigate the efficiency of CNNs: AlexNet and GoogLeNet under the effect of blurring which occurs frequently during image capturing process. Here, Gaussian blurring is used since it minimizes the noise embedded into the image. For experimentation, UC Merced Land Use aerial dataset is used to evaluate CNNs' performance. The focus is to train these CNNs and classifying an extensive range of classes accurately under the influence of Gaussian blurring. Accuracy and loss are the parameters of classification considered for evaluating the performance of CNNs. Experimental results validated the susceptibility of CNNs towards blurring effect with GoogLeNet being more fluctuating to varied degrees of Gaussian blurring than AlexNet.

**Keywords** Deep learning · Convolution neural networks · Gaussian blurring · Image classification

P. Mahajan (✉) · P. Abrol · P. K. Lehana
University of Jammu, Jammu and Kashmir, India
e-mail: palak.mahajan18@gmail.com

P. Abrol
e-mail: pawanesh.abrol@gmail.com

P. K. Lehana
e-mail: pklehana@gmail.com



P. K. Singh et al. (eds.), *Proceedings of ICRIC 2019*, Lecture Notes in Electrical Engineering 597, https://doi.org/10.1007/978-3-030-29407-6_34

## 1 Introduction

Exploitation of algorithms in computer vision has led to tremendous progress in machine learning [4] with the expansion of deep learning [1] techniques. Convolution neural networks (CNNs) [15] and state-of-the-art of deep learning are heavily used for semantic tasks of computer vision like image classification and object identification. CNNs are employed in numerous application areas like medical, surveillance, agriculture, flood monitoring, etc. The ability of the CNNs to extract high-level features by applying statistical learning on huge amount of dataset to obtain significant representation of input space makes it outstanding over other classical machine learning algorithms. CNNs not only enhance the accuracy but also offer to be used as feature extractor for various object detection and identification tasks. However, the increased state-of-the-art accuracy of CNNs comes with the cost of high computational complexity. AlexNet [11] and GoogLeNet [17] are remarkable CNNs being used over wide variety of datasets for classification purposes. For a given dataset, CNNs are more robust compared to typical feature extracting algorithms like HOG [5], SIFT [13], BOVW [21], and the like. CNNs' capability to learn features and to apply them on datasets of different modality makes them efficient in various application areas.

Since CNNs require greater amount of dataset for training purpose, many times the dataset collected contains significantly distorted images. Image blurring [9] is a common degradation observed in digital images that occurs during imaging process. Even a high-end camera sometimes delivers blurred image either due to manual error or due to improper focal setting of capturing device. A blurred image often contains reduced amount of information that hampers the quality of image which it intends to reciprocate. Image blurring refers to loss of sharpness in an image. The mathematical model of blurred image can be represented using image degradation model [18] as convolution of original image with the blur kernel and is shown by Eq. (1).

$$g(x, y) = f(x, y) * h(x, y) + n(x, y) \tag{1}$$

where $g(x, y)$ is blurred image, $f(x, y)$ is original image, $h(x, y)$ denotes blurring function, $*$ indicates convolution, and $n(x, y)$ is noise which is considered to be nullified since the degradation is due to blurring only.

In this paper, blurring is performed via Gaussian blurring that modifies value of pixels according to average of neighbor pixels. The values of the blurred pixels and the neighboring pixels can be computed using Eq. (2) represented as follows [9]:

$$G(x, y) = \frac{1}{2\pi\sigma^2} e^{-(x^2+y^2)/2\sigma^2} \tag{2}$$

where $\sigma$ represents blur factor, $e$ represents Euler number, and $(x, y)$ represents horizontal and vertical distance with respect to center pixel.

In Eq. (2), the initial distance values for $x$ and $y$ will be 0 for center pixel and subsequently will increase as the distance increases w.r.t center pixel. Figure 1 represents the 3-dimensional Gaussian function. Here, the bell-shape distribution is at

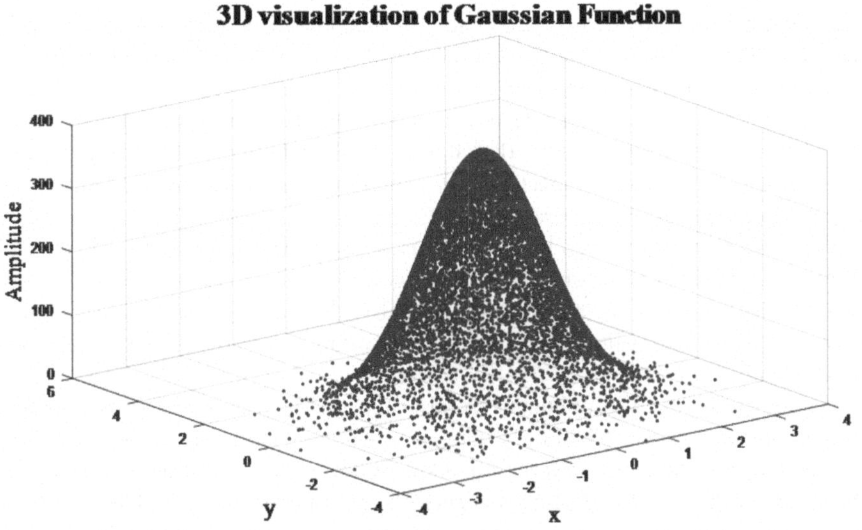


**Fig. 1** Bell-shaped distribution of 3-D Gaussian function

contour center point which can be achieved using Eq. (2). Accordingly, the convolution matrix is formed using these distribution values.

This paper focuses on examining the performance of CNNs: AlexNet and GoogLeNet on image classification under the influence of Gaussian blurring. The intension is to observe according to precept how blurring hampers the classification accuracy for image recognition purposes. The loss function has also been computed to evaluate classification error which is difference between actual and predicted results. Experiments are performed on benchmark aerial dataset with widely different characteristics, i.e., UC Merced Land Use[1] dataset containing 2100 images respectively.

This paper is organized as follows: Sect. 2 reviews the literature briefly. Section 3 presents different CNN architectures followed by Sect. 4 in which the experiments and results have been discussed. Conclusion is given in Sect. 5.

## 2 Background Work

The perception quality of images has always been prime focus of computer vision applications. With the growth of machine leaning and deep networks, vast research is going to examine the influence of distortions. LeCun et al. [12] gave an insight of CNNs in various application areas of computer vision. A comprehensive overview of

[1]Publicly available at http://vision.ucmerced.edu/datasets/landuse.html.

deep learning was provided by Sze et al. [16]. It summarized the hardware requirements of deep networks. The key points for designing the net through trade-off between various hardware architectures were highlighted.

In [19], CNN was used for image restoration. The sharp images of the face were stored using deep residual network followed by the addition of facial regularization technique. Similarly, impact of low-resolution images on face recognition in surveillance systems was expressed in [20].

A CNN-based automatic object detection network was developed by Sevo and Avramovic [14] on high-resolution aerial images. The network was trained and tested on GPU that increased the processing time. Also, the classification accuracy of the network was verified using object detection algorithm to precisely detect categories to which an object belongs.

A comparative analysis of human visual system and deep networks towards image degradation was generalized in [7]. The classification pattern was verified for various distortions like contrast variation, additive noise, etc., to test the efficiency of the system. Results showed decreased performance of deep networks like AlexNet, VGG, and GoogLeNet with decreasing signal-to-noise ratio. A similar study was comprehended by Gerhard et al. [8] to acknowledge the impact of image degradation towards local irregularities in the image. The effect of Gaussian filter on CNN for semantic segmentation has been referred in [6]. The kernels of various layers of CNN are modified for different lengths for varied convolution layers. The loss function has been estimated to compute the system performance.

In [2], AlexNet and GoogLeNet were trained to classify images on sketch dataset of ImageNet. However, the drawing lines of the sketches were often confounded with objects since CNNs were unable to abstract the learned concepts to the sketches, due to which the networks performed poorly, thereby degrading the classification accuracy of the system.

The brief literature reviewed showed the scope of exploring convolution nets with respect to various image distortions. The contribution of this paper is to provide an assessment of performance of CNNs under different levels of Gaussian blurring. The dataset considered here comprises high-quality images, and it is augmented by introducing varied levels of Gaussian blurring. The work of this paper demonstrates impact of blurring effect on classification performance of the CNNs.

## 3 CNN Architecture

A CNN architecture comprises several layers comprehending multiple neurons. The structure of each neuron estimates an activation function as shown by Eq. (3) [3]:

$$f(x) = \varphi\left(\mathbf{w}^{\mathrm{T}} x + b\right) \tag{3}$$

where $x$ represents input provided to each neuron, $w$ represents weight vector, $b$ is bias, and $\varphi$ represents a function encompassing nonlinearity in the network.

**Table 1** Summary of CNNS: AlexNet and GoogLeNet

| CNN | Developed by | No. of layers | No. of parameters |
|---|---|---|---|
| AlexNet [11] | Alex Krizhevsky, Geoffrey Hinton, Ilya Sutskever | 8 | 61 million |
| GoogLeNet [17] | Google | 22 | 7 million |

In CNN, the image itself is input to network for image identification tasks. CNN's layered architecture [12] enables output of one layer to act as inputs to succeeding layer. The foundation of CNN depends on convolution layer that performs convolutional filtering with filter described by the weights. Stride and padding are two parameters that control how convolution occurs around input. To minimize computation and memory requirements, the vector w representing weight is generally shared between neurons instead of assigning separate weight to the neurons. This vector connects towards neurons from preceding layer into a specific region. CNN contains layer for handling nonlinearity of data which is called Rectified linear unit (ReLU) that is one of its prime layer. To reduce spatial size of the image, max pooling layer is incorporated in CNN. As the architecture proceeds, fully connected layer is infused in which all the nodes are connected with the nodes of the previous layer. The latter stage is softmax layer that normalizes outputs of each unit between 0 and 1 in view that they aggregate to 1. This aids the output layer to act as probability distribution by means of each neuron to conform to probability for a well-defined class. Primarily, output is computed over a dataset which is then evaluated for well-defined class labels. The parameters in CNN are trained via large dataset of input images. A cost function is associated that specifies the likelihood of the predictions made with ground truth. The gradient [10] is estimated which is the essence of cost function that allows propagation backwards throughout and assigning optimal weights to the neurons.

Several CNN architectures have been discussed in this section, and the performance of AlexNet and GoogLeNet has been evaluated in this paper. Table 1 summarizes the architectural details of mentioned CNNs.

## *3.1 AlexNet*

The architecture of AlexNet made available in [11] triumph sensation in the deep learning for accomplishing a boosted performance for ImageNet Large Scale Visual Recognition Challenge (ILSVRC) 2012. The architecture is represented in Fig. 2.

In computer vision, AlexNet's architecture popularized CNNs for classifying ImageNet database comprises around 1.2 million high-resolution images with 1000 categories. It is implemented through 2 GPUs. The network architecture [11] consists of eight layers among which there are five convolution layers and followed by three fully connected layers. The final layer outputs the class labels. Normalization is

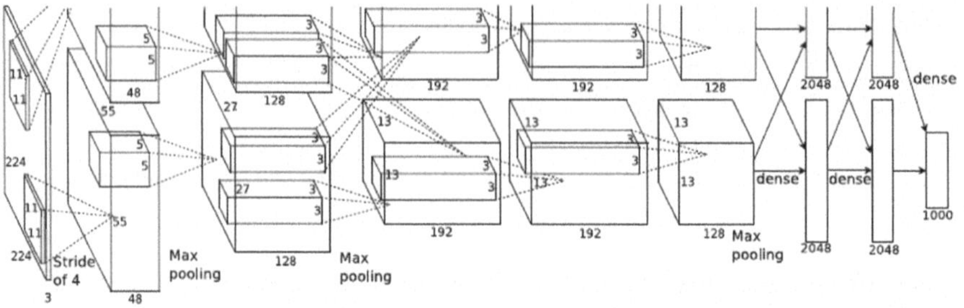


**Fig. 2** Detailed architecture of AlexNet with various layers and multiple GPU implementation [11]

applied in the form of response normalization layer to the first two convolution layers, while dropout is applied after each layer during the last two fully connected layers. Max pooling layers are succeeded by response normalization layers and fifth convolutional layer. Nonlinearity is integrated into the network through ReLU to the output of every convolutional and fully connected layer. For training the net, stochastic gradient descent is used with 128 batch size, 0.9 momentum, and 0.0005 weight decay.

## *3.2 GoogLeNet*

The GoogLeNet architecture published in [17] is the deepest and complex CNN. A 22-layer CNN works as multilevel feature extractor. It includes inception module shown in Fig. 3 that concatenates the filters with varied sizes and dimensions within a single filter.

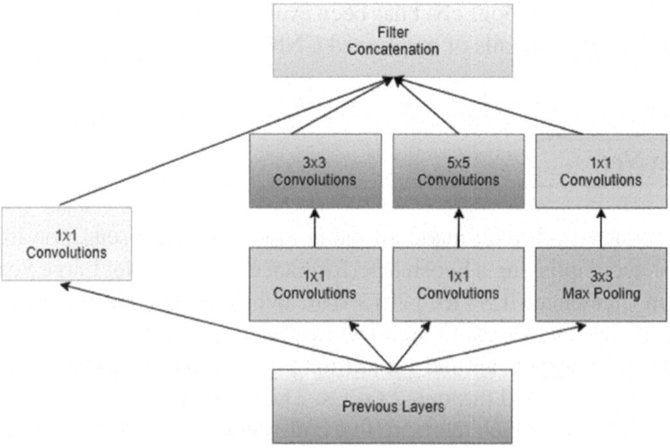


**Fig. 3** Illustration of Inception3a module of GoogLeNet [17]

The architecture comprises two convolution layers following two pooling layers and nine inception layers that indeed entails six convolution layers and one pooling layer, respectively. As per statistics for ImageNet challenge, GoogLeNet is prevailing CNN for ILSVRC challenge with top 5 classification error of 5.5% in comparison with AlexNet's top 5 classification error 15.3%. The GoogLeNet was designed specifically with computational efficiency in mind, in order to execute the net on devices with limited memory and computational resources.

## 4 Results and Discussion

This section illustrates the capability AlexNet and GoogLeNet for image classification. The implemented architecture of AlexNet and GoogLeNet are as shown in Fig. 4 representing convolution, pooling layers, kernel size, and inception module, respectively. Classification competency of CNNs is evaluated on two performance parameters: accuracy and loss. Accuracy defines percentage of classes correctly classified to entire number of classes in dataset. Loss on the other hand is not percentage, rather its summation of errors for each prediction made. The experiment is set up on Intel(R) Core (TM) i5-3210 CPU @3.40 GHz processor with 6 GB RAM. The platform used to implement is MATLAB R2018a on Windows 10.

### *4.1 Data Collection*

For validating the performance, the aerial scene classification is applied on UC Merced Land Use dataset respectively. In UC Merced, the dataset comprises manually extracted images from the USGS National Map Urban Area Imagery collection.

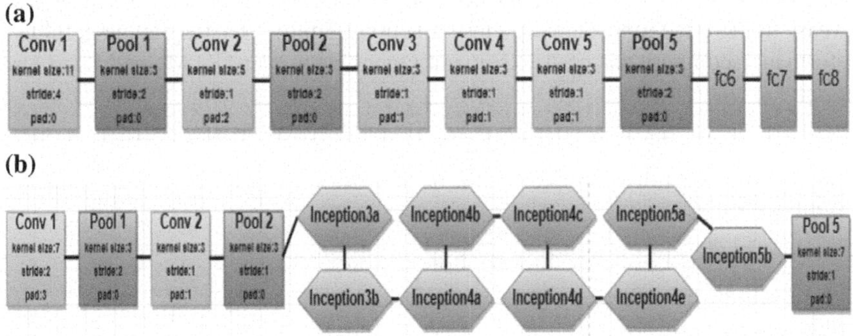


**Fig. 4** Implemented architecture of CNNS: **a** AlexNet containing 5 convolution layers, 3 pooling layers, and 3 fully connected layers. **b** GoogLeNet comprising 2 convolution layers, 3 pooling layers, and 9 inception layers, respectively

Figure 5 illustrates few ground truth images for twenty-one land use categories. It comprises twenty-one aerial scene categories with 256 × 256 color images with 1-ft/pixel resolution. The images in this dataset cover overlapping categories, and each category contains 100 images, thereby creating a dataset of 2100 images. The extensive categories of the dataset are defined as follows: agriculture, airplane, baseball diamond, beach, building, chaparral, dense residential, forest, freeway, golfcourse, harbor, intersection, medium residential, mobile homepark, overpass, parking lot, river, runway, sparse residential, storage tanks, and tennis court, respectively.

## *4.2 Observation*

On the basis of the experiment being conducted, the output has been generalized. Influence of Gaussian blurring on the classification performance of the nets is studied. The kernel of Gaussian blurring has been varied by fluctuating the standard deviation from 1 to 7, and accordingly, the size of filter window has been adjusted. Figure 6 represents a close look of Gaussian blurred images at various degrees of deviation in class category of airplane.

CNNs are trained for multiple iterations (with varied epochs) and classification parameters: Accuracy and loss have been calculated. Table 2 concludes the performance of CNNs: AlexNet and GoogLeNet in terms of accuracy and loss function.

As represented by the graph, AlexNet shows smooth downfall in accuracy as blurring increases, and GoogLeNet on the other hand shows more fluctuations in terms of accuracy. Also, loss rate increases evenly in AlexNet compared to GoogLeNet. The reason being due to limited resources, GoogLeNet achieved validation point earlier before the completion of its iterations.

The results represent CNNs are susceptible to blurring. The blurring alters the filter output of foremost convolutional layer that propagates to higher layer, thereby hindering the result at final layer. Further, as we decrease the quality of image by increasing the degree of blurring, the accuracy of the CNNs decreases and there is an increase in the error represented by loss function. Figure 7 represents the graphical performance comparisons of AlexNet and GoogLeNet over UC Merced Land Use dataset.

(a) Agriculture

(b) Airplane

(c) Baseballdiamond

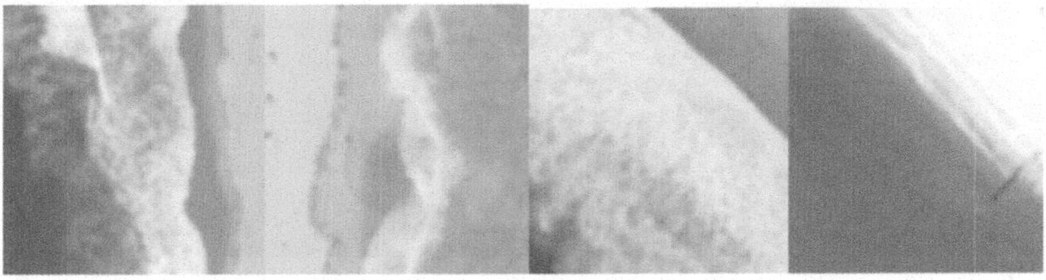

(d) Beach

(e) Building

**Fig. 5** UC Merced dataset for 21 categories, 4 samples per category has been shown

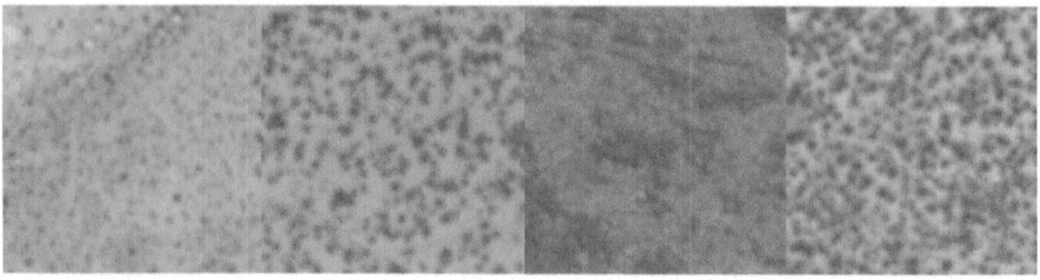

(f) Chaparral

(g) Dense residential

(h) Forest

(i) Freeway

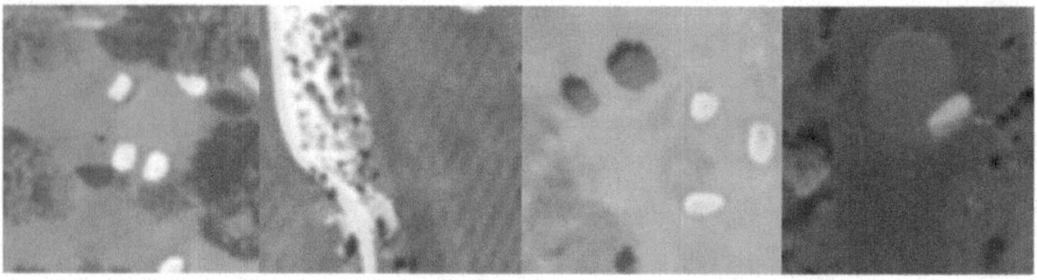

(j) Golfcourse

**Fig. 5** (continued)

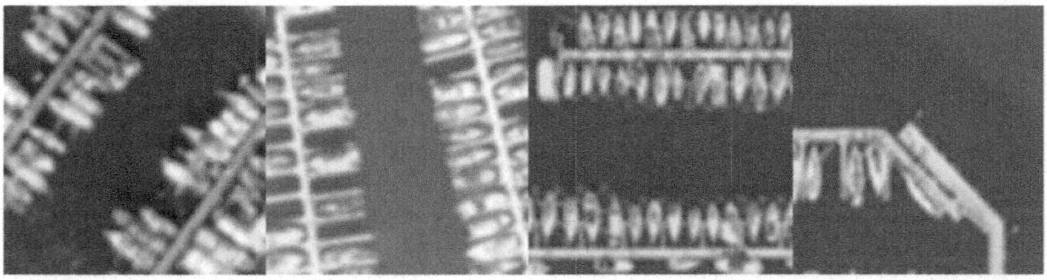

(k) Harbor

(l) Intersection

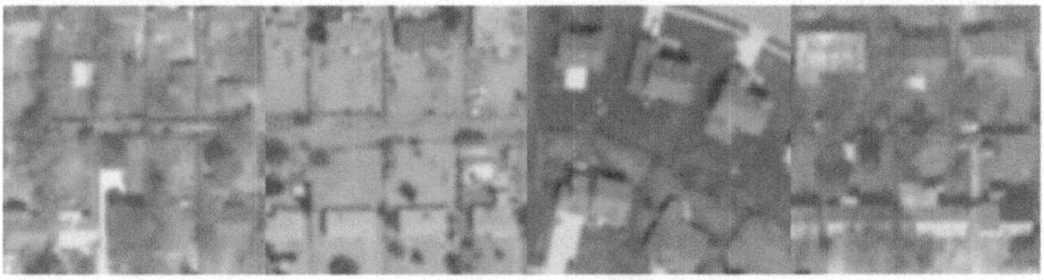

(m) Medium residential

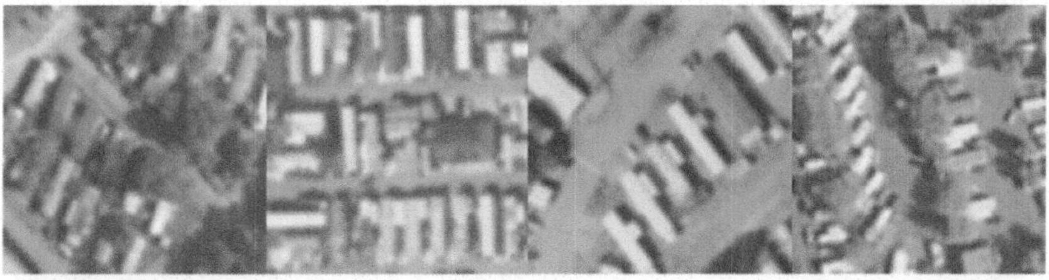

(n) Mobile homepark

(o) Overpass

**Fig. 5** (continued)

(p) Parkinglot

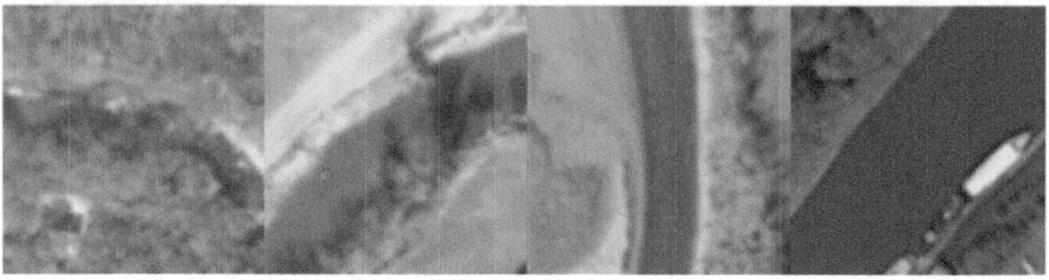

(q) River

(r) Runway

(s) Sparse residential

(t) Storagetanks

**Fig. 5** (continued)

(u) Tenniscourt

**Fig. 5** (continued)

**Fig. 6** Blurred images with different degrees of Gaussian blurring: $\sigma = 0$ (top left), $\sigma = 2$ (top right), $\sigma = 5$ (bottom left), and $\sigma = 7$ (bottom right)

**Table 2** Experimental results of AlexNet and GoogLeNet over UC Merced Land Use dataset

| Blurred image | AlexNet | | GoogLeNet | |
|---|---|---|---|---|
| | Accuracy (%) | Loss | Accuracy (%) | Loss |
| $\sigma = 0$ | 92.06 | 0.3105 | 71.43 | 0.9130 |
| $\sigma = 1$ | 83.45 | 0.5437 | 76.9 | 0.7534 |
| $\sigma = 2$ | 83.97 | 0.4430 | 82.06 | 0.5101 |
| $\sigma = 3$ | 78.05 | 0.5489 | 79.02 | 0.7312 |
| $\sigma = 4$ | 74.6 | 0.8645 | 65.22 | 0.8911 |
| $\sigma = 5$ | 76.67 | 0.6901 | 71.11 | 0.8500 |
| $\sigma = 6$ | 72.54 | 0.8098 | 62.67 | 1.0229 |
| $\sigma = 7$ | 72.38 | 0.9012 | 51.11 | 1.5050 |

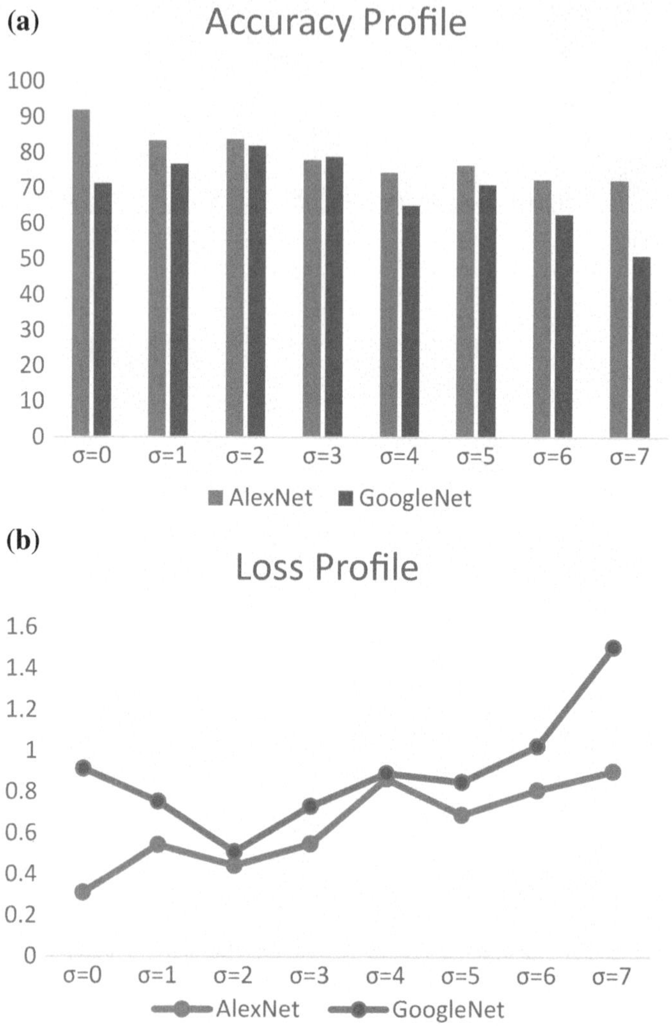


**Fig. 7** Accuracy and loss performance of AlexNet and GoogLeNet towards Gaussian blurring

## 5 Conclusion

This paper exploits the structure of deep convolution networks for various degrees of Gaussian blurring. The CNNs, AlexNet and GoogLeNet, are implemented and trained from the scratch. The performance evaluation is done on UC Merced Land Use dataset to evaluate the influence of Gaussian blurring on CNNs. The experimental results validate the utility of CNNs in the presence of blurring as image degradation. It can be inferred from results that the decreased performance of CNNs under the effect of blurring is not restricted to a specific CNN model but is universal to both

AlexNet and GoogLeNet. With AlexNet showing a smooth deviation in classification parameters with varied values of Gaussian blurring effect, GoogLeNet exhibits inconsistent change in accuracy and loss values. A solution to this can be training CNNs on low-quality images with diverse range of blurring along with increased number of samples in dataset.

Limited system resources like computing power and time have limited training of these nets in terms of how far it could have gone for training, due to which the nets are implemented for few epochs for each net before it plateaued. In future work, the experimental analysis can be extended using graphics processing units (GPU) to accelerate the feature learning process of CNNs.

## References


1. Arel, I., Rose, D.C., Karnowski, T.P.: Deep machine learning—a new frontier in artificial intelligence research. IEEE Comput. Intell. Mag. **5**(4), 13–18 (2010)
2. Ballester, P., Araujo, R.M.: On the performance of GoogLeNet and AlexNet applied to sketches. In: Proceedings o AAAI Conference on Artificial Intelligence, pp. 1124–1128, Arizona (2016)
3. Bishop, C.M.: Pattern Recognition and Machine Learning. Springer, New York (2007)
4. Ding, S., Li, H., Su, C., Yu, J., Jin, F.: Evolutionary artificial neural networks: a review. Artif. Intell. Rev. **39**(3), 251–260 (2013)
5. Dalal, N., Triggs, B.: Histograms of oriented gradients for human detection. In: Proceedings of the IEEE Computer Society Conference on Computer Vision and Pattern Recognition, pp. 886–893 (2005)
6. Gu, Y., Wu, Q., Li, J., Cheng, K.: Gaussian filter in CRF based semantic segmentation. arXiv preprint arXiv:1709.00516 (2017)
7. Geirhos, R., Janssen, D.H.J., Schuut, H.H., Rauber, J., Bethge, M., Wichmann, F.A.: Comparing deep neural networks against humans: object recognition when the signal gets weaker. arXiv preprint arXiv:1706.06969v2 (2018)
8. Gerhard, H.E., Wichmann, F.A., Bethge, M.: How sensitive is the human visual system to the local statistics of natural images? PLoS Comput. Biol. **9**(1), e1002873 (2013)
9. Gedraite, E.S., Hadad, M.: Investigation on the effect of a gaussian blur in image filtering and segmentation. In: Proceedings of ELMAR, Croatia (2011)
10. Kingma, D.P., Ba, J.L.: ADAM: a method for stochastic optimization. arXiv preprint arXiv: 1412.6980 (2017)
11. Krizhevsky, A., Sutskever, I., Hinton, G.E.: ImageNet classification with deep convolutional neural networks. In: Advances in Neural Information Processing Systems, pp. 1106–1114 (2012)
12. LeCun, Y., Kavukcuoglu, K., Farabet, C.: Convolutional networks and applications in vision. In: Proceedings of IEEE International Symposium on Circuits and Systems, pp. 253–256, France (2010)
13. Panchal, M., Panchal, S.R., Shah, S.K.: A comparison of SIFT and SURF. Int. J. Innov. Res. Comput. Commun. Eng. **1**(2), 323–327 (2013)
14. Sevo, I., Avramovic, A.: Convolutional neural network based automatic object detection on aerial images. IEEE Geosci. Remote Sens. Lett. **13**(5), 740–744 (2016)
15. Simonyan, K., Zisserman, A.: Very deep convolutional networks for large-scale image recognition. arXiv preprint arXiv:1409.1556 (2015)
16. Sze, V., Chen, Y., Yang, T., Emer, J.: Efficient processing of deep neural networks: a tutorial and survey. arXiv preprint arXiv:1703.09039v2 (2017)

17. Szegedy, C., Liu, W., Jia, Y., Sermanet, P., Reed, S., Anguelov, D., Erhan, D., Vanhoucke, V., Rabinovich, A.: Going deeper with convolutions. In: Proceedings of the IEEE Conference on Computer Vision and Pattern Recognition, pp. 1–9 (2015)
18. Vijay, R.: Image deblurring using convolutional neural network. IOSR J. Electr. Commun. Eng. **11**(5), 7–12 (2016)
19. Wang, L., Li, Y., Wang, S.: DeepDeblur: fast one-step blurry face images restoration. arXiv preprint arXiv:1711.09515v1 (2017)
20. Xu, X., Liu, W., Li, L.: Low resolution face recognition in surveillance systems. J. Comput. Commun. **2**, 70–77 (2014)
21. Yang, Y., Newsam, S.: Bag-of-visual-words and spatial extensions for land-use classification. In. Proceedings of SIGSPATIAL International Conference on Advances in Geographic Information Systems, pp. 270–279 (2010)

# PSO-Tuned ANN-Based Prediction Technique for Penetration of Wind Power in Grid

**Vijay Kumar, Yash Pal and M. M. Tripathi**

**Abstract** Today, world is paying more attention on those types of energy sources that create minimum pollution and fulfill the gap between demand and supply. Continuous increase in the power demand of consumers from various fields such as residential, industrial, and commercial, it is difficult to procure the additional supply from conventional sources with maintaining the pollution standard. So power-producing companies/agencies invest lot of fund in the development of such sources of energy which is nearly pollution-free and available from nature. In this regards, the alternative of conventional sources may be renewable energy sources. Out of different available renewable sources such as solar, wind, biomass, and small hydro, wind can be considered one of the good sources for the generation of power. Today, living standard of any country can be recognized by per capita energy consumption by its people. As power production from renewable sources may lead to minimum possible pollution, operating cost, and mostly freely and abundant availability, it will work as a major driving factor for some countries of the world to spend maximum available energy fund in the development of such mechanism/technique that will able to generate energy from renewable sources. Although power generation from wind has many advantages, major drawbacks are its intermittent nature, frequency instability, and continuous availability with certain threshold speed that is capable for power generation at all places. This paper describes the combined technique of PSO and ANN for forecasting of speed and power of wind to penetrate it in grid. The proposed method is applied on Indian wind power sector, and its results are compared with simple ANN and ANN-SVM methods.

V. Kumar (✉)
IMS Engineering College, Ghaziabad, India
e-mail: vijay5686@gmail.com

Y. Pal
Department of Electrical Engineering, NIT Kurukshetra, Kurukshetra, Haryana, India
e-mail: yashpal@nitkkr.ac.in

M. M. Tripathi
Department of Electrical Engineering, DTU, New Delhi, India
e-mail: mmtripathi@gmail.com

 
P. K. Singh et al. (eds.), *Proceedings of ICRIC 2019*, Lecture Notes in Electrical Engineering 597, https://doi.org/10.1007/978-3-030-29407-6_35

**Keywords** Prediction technique · Wind power prediction · Particle swarm optimization (PSO) · Artificial neural network (ANN) · Maximum absolute percentage error (MAPE)

## 1 Introduction

Today, power demand and its supply have become a serious concern for progress of any country. Nowadays, power generation with traditional sources has very dangerous effect on human being and environment and its availability is also limited in nature. Considering above problems and to balance future power demand, we have to move toward new kind of technology/method that is able to generate power from renewable sources [1]. According to the report of central electricity authority of January 2019, total installed power in India from all resources is 347.22 GW. Renewable power contributes 72.53 GW that is around 20% of total power generation in India [2]. With the development all around the world in the form of industrialization, nowadays to fulfill the power demand of any country becomes a critical factor. The above-mentioned problems can be reduced up to some level through implementing contribution of non-conventional sources in power generation. From renewable point of view, power generation from wind can be regarded as one of the better alternative solutions. Main drawbacks with wind power generation is its variable speed nature, availability to everywhere, and reliability. During power prediction from wind, we mostly focus on the speed forecasting that must be accurate and with minimum possible error as power is highly dependent on speed. To overcome the above problems, there will be requirement of such type of techniques that will provide prediction result with highest possible accuracy. The share of different renewable sources in India is shown in Table 1.

As there is remarkable growth in wind sector during last 5 years and result of this now, India is on 4th rank in the world for power generation from wind. The development of wind power year-wise is represented in Fig. 1.

In restructured/regulated electricity market, ISO tries to insure minimum possible per unit cost of power generation through resolving problem related to unit commitment, while in deregulated electricity markets, ISO tries to insure maximum social welfare for producer and consumer. When wind power plant operates in isolated

**Table 1** Share of renewable sources

| Source | Share (MW) |
|---|---|
| Wind | 34614.16 |
| Solar | 24031.64 |
| Small | 4507.55 |
| Biomass | 8859.12 |
| Total | 72062.14 |

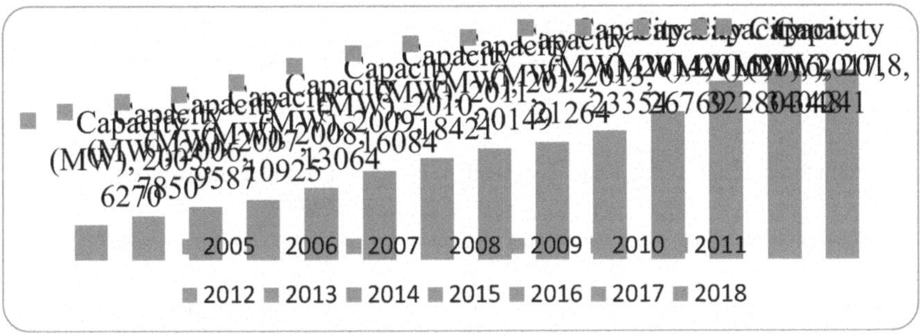


**Fig. 1** Yearly growth of wind power in India (MW)

state, it faces some problems such as variation in generation and frequency fluctuations. So, combined working of renewable sources such as wind, photovoltaic, and waste biomass may minimize above problems by taking situation of the power imbalance problem [3]. The critical issue with wind is variable output due to continuous varying speed and frequency variation [4]. Nowadays, many countries of the world have certain regulated policies to overcome the problem of pollution effects due to conventional methods of power production, insuring higher contribution of renewable resources [5]. Current work is distributed into seven sections in which the first section is introduction, the second gives the details of prediction technique, the third section is about ANN working, the fourth section explains working of particle swarm optimization, Sect. 5 discusses different error factors, Sect. 6 gives results, and the last section is conclusion.

## 2 Wind Forecasting Methods

Forecasting is a method/process of predicting the future value with the help of past as well as present data and trends analysis. Forecasting can be categorized into:.

2.1 Time-based
2.2 Historical data/weather data-based
2.3 Technique-based/model-based.

### *2.1 Timescales-Based Classification [6, 7]*

Timescale-based wind energy forecasting with duration and different application fields is shown in Table 2.

**Table 2** Classification based on timescale

| Type | Time slot | Application |
|---|---|---|
| Ultra-short-term forecasting | Below 1 h | Electricity price clearing, grid operation |
| Short-term forecasting | 1 h to day ahead | Load planning, operational security |
| Medium-term forecasting | Day ahead to 1 week | Unit commitment, reserve decision |
| Long-term forecasting | Monthly to yearly or above | Maintenance, design, feasibility analysis |

### 2.2 *Historical Data-Based Forecasting*

Historical-based classification further is categorized into two types: first is statistical, and second is learning. Statistical technique will try to convert the dependency of current calculated value in the form of preset time series value of past year's value. During short-term forecasting, statistical can be better, but main drawback with this is that as we increase time interval, there will be increment in error [8–13]. Historical forecasting is further divided as represented in Fig. 2.

At present, researchers use many methods/techniques in the field of wind power forecasting like ARIMA, persistence, statistical, ARMA model, NWP, and soft computing method (ANN, SVM, fuzzy, SVR). Different techniques are suitable for different time slots such that some technique is more favorable for long-term, while others are more suitable for short- and medium-term forecasting.

## 3 Artificial Neural Network (ANN)

Initially, neural network was mainly applied for the prediction of electricity load and price in deregulated market, but now it is used in the field of forecasting of wind parameters to improve its reliability into smart grid [14]. Although neural network has capability to minimize the variables during training process that will save computation time [15]. For time series data estimation that involves the different parameters, ANN provides satisfactory response as compared to other methods. Efficiency of ANN can be improved by involving nonlinear autoregressive exogenous model (NARX) during modeling phase, as it takes the present value of exogenous input to reflect the relation with past value; mathematically, basic NARX can be represented as follows:

$$P_n = R(y_{t-1}, y_{t-2}, y_{t-3}, \ldots, n_t, n_{t-1}, n_{t-2}, n_{t-3}, \ldots) + e_t \quad (1)$$

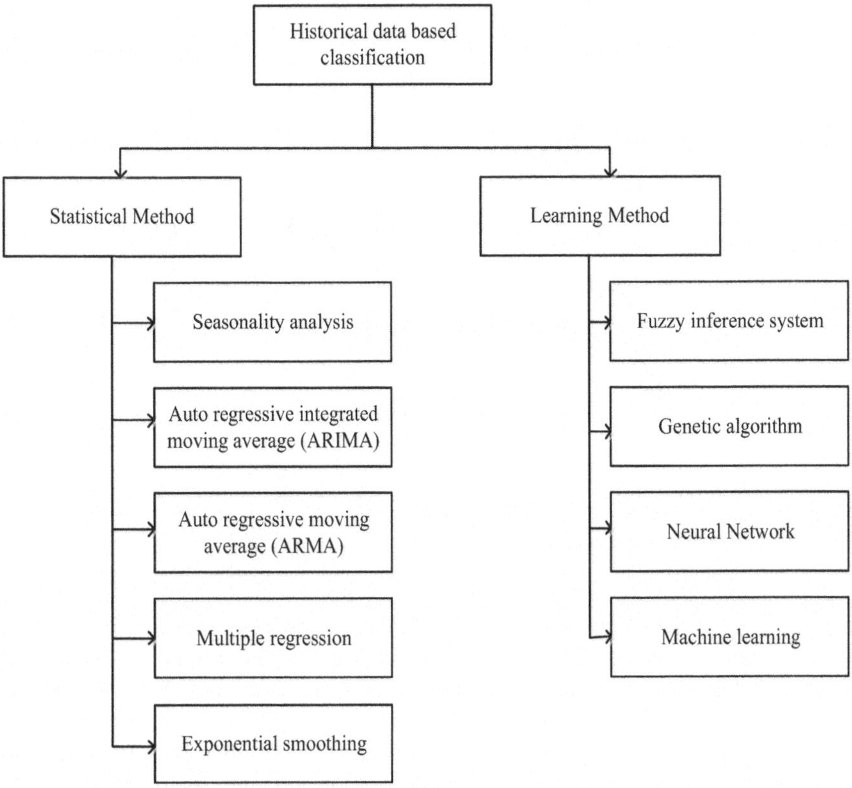


**Fig. 2** Classification based on historical data

$y$ is variable of interest, $n$ represent determination variable, and $e_t$ indicate error. Multilayer ANN structure shown in Fig. 3 involves inputs layer and hidden layer corresponding to output. Network can be represented by the following expression:

$$(K, t) = \left(K^{(1)}, t^{(1)}, K^{(2)}, t^{(2)}\right) \tag{2}$$

$K^{(1)}_{mn}$ is associated weight of unit $n$ in layer l, to the unit m in layer $m + 1$. This multilayer structure uses past value of input variables to predict the future value.

## 4 Particle Swarm Optimization (PSO)

Generally, particle swarm optimization (PSO) is used to maximize/minimize a problem that is unconstrained type to evaluate $k^*$ such that $f(k^*) \leq f(k)$ for all real vectors $k$, objective function of PSO can be represented by $f$: Rd $\rightarrow$ $R$ that is some time also known as PSO fitness function. PSO is intelligence heuristic method and mostly

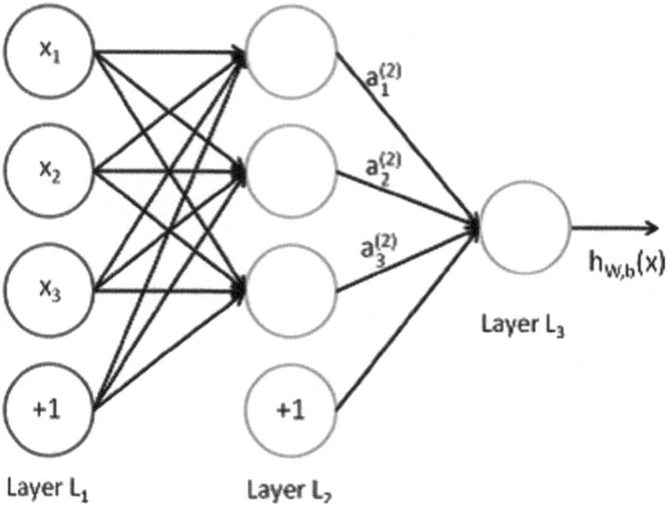


**Fig. 3** Multilayer neural network structure

inspired by behavior of bird flocks. Like GA, PSO is also population-based method; different states of its algorithm are represented via population that will try to find out suitable termination point through step-by-step modification. In PSO, population may be represented by $P = \{p_1, p_2, p_3, \ldots, p_n\}$ and most feasible result represents the swarm, where $p_1, \ldots, p_n$ are particles of feasible solutions. The PSO method observes the "space" where particles can move in group and provide best possible solutions. Figure 4 represents the working of PSO.

## 5 Error Parameters Estimation

The performance of any prediction technique/model is decided by the estimation of its error parameters such as mean absolute error, bias error, root mean square error, and MAPE. The above indicating error parameters are represented in Eqs. 3, 4, and 5.

(i) RSME can be calculated as

$$\text{RSME} = \sqrt{\frac{1}{N}\sum_{j=1}^{n}(p_{mj} - p_{cj})^2} \tag{3}$$

**Fig. 4** Working of PSO

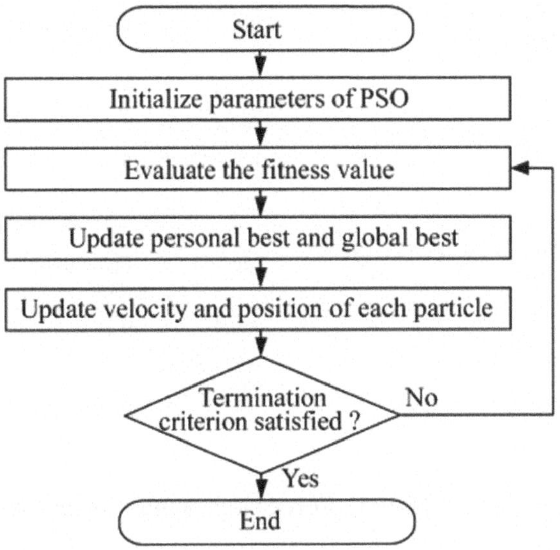


(ii) MBE can be calculated as

$$\mathrm{MBE} = \frac{1}{N}\sum_{j=1}^{n}\left(p_{mj} - p_{cj}\right) \tag{4}$$

(iii) MAE can be calculated as

$$\mathrm{MAE} = \frac{1}{N}\sum_{j=1}^{n}\left|(P_{mj} - P_{cj})\right| \tag{5}$$

where $P_{mj}$ is $j$th actual power, $P_{cj}$ is $j$th predicted power, and $n$ is the how many times observation is taken.

## 6 Results and Discussion

The corresponding MAPE for wind speed and wind power of all three methods is shown in Tables 3 and 4 (Fig. 5).

**Table 3** MAPE (%) wind speed

| Method | Prediction time (hourly) | Prediction time (day ahead) |
|---|---|---|
| ANN | 8.28 | 9.67 |
| ANN-SVM | 6.46 | 7.36 |
| PSO-ANN | 5.33 | 6.28 |

**Table 4** MAPE (%) wind power

| Method | Prediction time (hourly) | Prediction time (day ahead) |
|---|---|---|
| ANN | 9.18 | 10.17 |
| ANN-SVM | 7.89 | 8.12 |
| PSO-ANN | 6.18 | 6.81 |

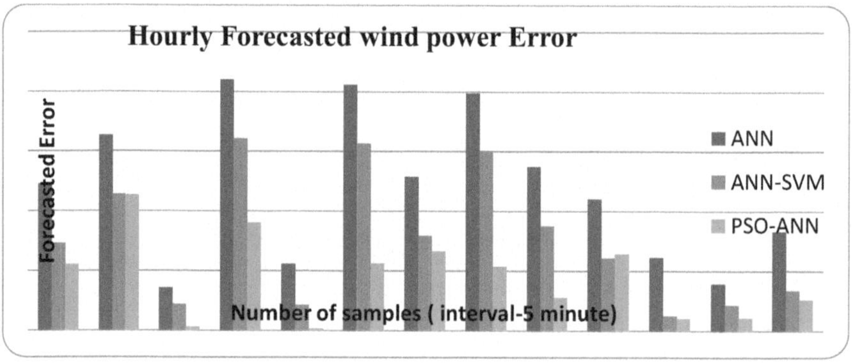


**Fig. 5** Predicted wind power error

## 7 Conclusions

The results obtained from ANN, ANN-SVM, and PSO-ANN during short-term prediction based on hourly and day ahead time slot are shown in Figs. 6, 7, and 8, and the corresponding MAPE value of the above method is represented in Tables 3 and 4. By observing the graph and table, it is clear that short-term prediction of wind speed and power with PSO-tuned ANN provides better result in terms of MAPE and other error parameters as compared to other two methods, i.e., simple ANN and SVM-tuned ANN, so PSO-tuned ANN is better than simple ANN and SVM-tuned ANN method.

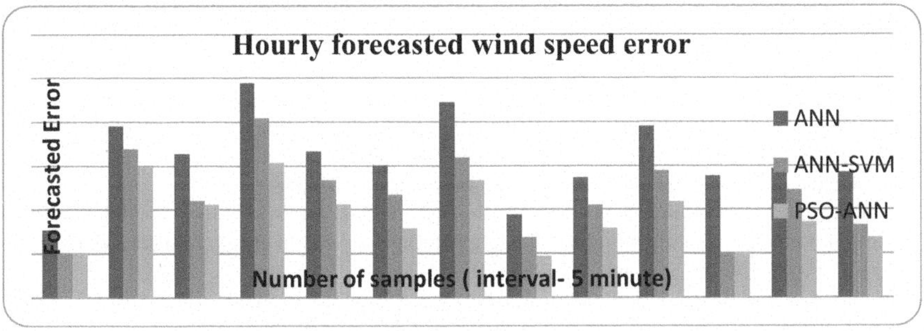


**Fig. 6** Predicted wind speed error

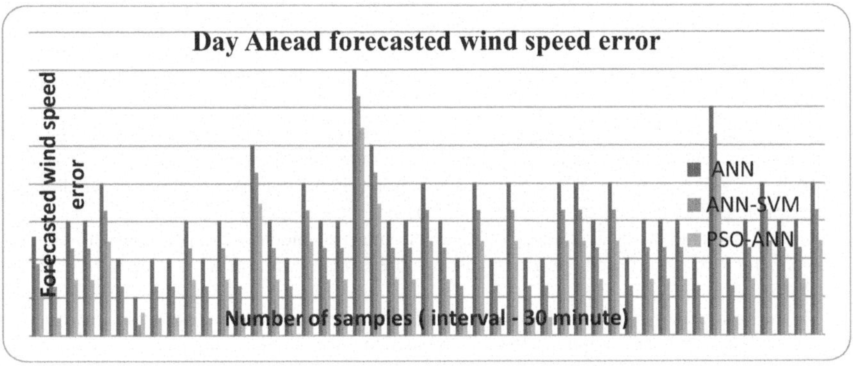


**Fig. 7** Predicted day ahead wind speed error

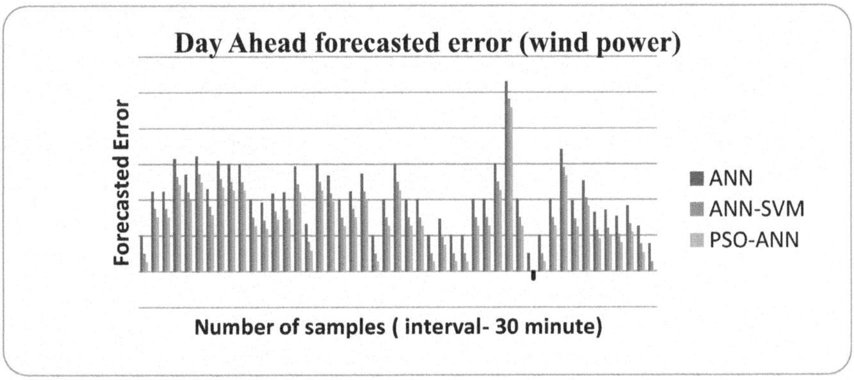


**Fig. 8** Predicted day ahead wind power error

## References


1. Majumder, R., Mukherjee, I., Tudu, B., Paul, D.: Review on feasibility of wind energy potential for India. In: Proceeding of 1st International Conference on Non-Conventional Energy (ICONCE 2014), pp. 12–18, 16–17 Jan 2014, WB, India
2. http://cea.nic.in. Executive Summary of Power Sector, Nov 2018
3. Parastegari, M., Hooshmand, R.-A., Khodabakhshian, A., Zare, A.-H.: Joint operation of wind farm, photovoltaic, pump-storage and energy storage devices in energy and reserve markets. Electr. Power Energy Syst. **64**, 275–284 (2015)
4. Suvire, G.O., Mercado, P.E.: Relationship between energy storage devices and wind farm sizes. IEEE Biennial Congress of Argentina (ARGENCON), pp. 588–593 (2014)
5. Chen, N., Qian, Z., Nabney, I.T., Meng, X.: Wind forecasts using gaussian processes and numerical weather prediction. IEEE Trans. Power Syst. **29**(2), 656–665 (2014)
6. Foley, A.M., Leahy, P.G., Marvuglia, A., McKeogh, E.J.: Current methods and advances in forecasting of wind power generation. Renew. Energy **37**, 1–8 (2012)
7. Xu, Q., He, D., Zhang, N., Kang, C.: A short-term wind power forecasting approach with adjustment of numerical weather prediction input by data mining. IEEE Trans. Sustain. Energy 1283–1291 (2015)
8. Azad, H.B., Mekhilef, S., Ganapathy, V.G.: Long-term wind speed forecasting and general pattern recognition using neural networks. IEEE Trans. Sustain. Energy 546–553 (2014)
9. Zhao, X., Wang, S.X., Li, T.: Review of evaluation criteria and main methods of wind power forecasting. Energy Procedia **12**, 761–769 (2011)
10. Ren, Y., Suganthan, P., Srikanth, N.: A novel empirical mode decomposition with support vector regression for wind speed forecasting. IEEE Trans. Neural Netw. Learn. Syst. **27**(8) (2016)
11. Yadav, H.K., Pal, Y., Tripathi, M.M.: Photovoltaic power forecasting methods in smart power grid. In: IEEE India Conference (INDICON) (2015)
12. Hoverstad, B., Tidemann, A., Langseth, H., Ozturk, P.: Short term load forecasting with seasonal decomposition using evolution for parameter tuning. IEEE Trans. Smart Grid **6**, 1904–1913 (2015)
13. Yang, L., He, M., Zhang, J., Vittal, V.: Support-vector-machine enhanced markov model for short-term wind power forecast. IEEE Trans. Sustain. Energy **6**, 791–799 (2015)
14. Catalão, J.P.S., Pousinho, H.M.I., Mendes, V.M.F.: An artificial neural network approach for short-term wind power forecasting in Portugal. In: 15th International Conference on Intelligent System Applications to Power Systems, ISAP'09 (2009)
15. Philippopoulos, K., Deligiorgi, D.: Application of artificial neural networks for the spatial estimation of wind speed in a coastal region with complex topo-graphy. Renew. Energy **38**(1), 75–82 (2012)

# A Comprehensive Review on Face Recognition Methods and Factors Affecting Facial Recognition Accuracy

**Shahina Anwarul and Susheela Dahiya**

**Abstract** As of late, the need for biometric security framework is elevated for giving safety and security against frauds, theft, and so on. Face recognition has gained a significant position among all biometric-based systems. It can be used for authentication and surveillance to prove the identity of a person and detect individuals, respectively. In this paper, a point-by-point outline of some imperative existing strategies which are accustomed to managing the issues of face recognition has been introduced along with their face recognition accuracy and the factors responsible to degrade the performance of the study. In the first section of this paper, different factors that degrade the facial recognition accuracy have been investigated like aging, pose variation, partial occlusion, illumination, facial expressions, and so on. While in the second section, different techniques have been discussed that worked to mitigate the effect of discussed factors.

**Keywords** Face detection · Face recognition · Biometric security framework · Authentication · Surveillance

## 1 Introduction

Face detection and recognition is one of the pivotal authentication systems based on biometric which can be used for the authentication process as well as surveillance. With the rapid increase in frauds day by day, face recognition is becoming a crucial system for us. Various applications of an efficient face recognition system are forensics, identification of criminals, surveillance, fraud exclusion, etc. A lot of research has been done on both national and international levels, but even after continuous research, a truly resilient and efficacious system is not available that can perform

S. Anwarul (✉) · S. Dahiya
University of Petroleum and Energy Studies, Dehradun, India
e-mail: shahinaanwarul@gmail.com

S. Dahiya
e-mail: susheela.iitr@gmail.com



P. K. Singh et al. (eds.), *Proceedings of ICRIC 2019*, Lecture Notes in Electrical Engineering 597, https://doi.org/10.1007/978-3-030-29407-6_36

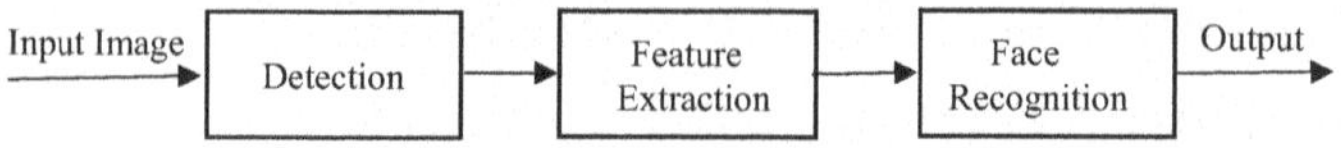


**Fig. 1** Steps of face detection and recognition

well in both normal and real-time conditions. Facial recognition has always been a very complex and challenging task. Its actual test lies in outlining an automated framework which parallels the human capacity to perceive faces. Be that as it may, there is confinement to the human capacity when it manages a lot of obscure appearances. Consequently, a programmed automatic electronic framework with relatively higher recognition accuracy and fast processing is required.

Face detection and recognition process consists of three steps in Fig. 1 [1]:

1. **Face detection**: To identify the faces in an image or video using landmark points on the face like eyes, nose, mouth, etc.
2. **Feature extraction**: Alignment, normalization of the faces for better recognition exactitude.
3. **Face recognition**: To recognize a particular person in an image or video by matching with the database.

This paper is divided into five major sections. The first section gives the brief introduction of face detection and recognition. In the second section, different factors that degrade the facial recognition accuracy have been discussed. The third section gives a brief introduction of methods involved in face detection and recognition along with its importance. The next section gives a brief review of the algorithms used till now for the appearance-based, feature-based, and hybrid method with their recognition accuracy. Lastly, it concludes with the findings from this research.

## 2 Factors Affecting Face Recognition Accuracy

Face recognition from images and videos is a formidable task. A lot of research has been done to achieve 100% precision, but still we are not getting satisfactory results in view of the various factors confronting this framework. It has been found that the factors that degrade the accuracy of the face recognition systems are: occlusion, low resolution, noise, illumination, pose variation, expressions, aging, and plastic surgery [2–4]. These factors can be classified into two categories: intrinsic and extrinsic factors [2]. Intrinsic components incorporate the physical state of the human face like aging, facial expression, plastic surgery, and so on, influencing the system, while extrinsic factors are responsible to change the appearance of the face like occlusion, low resolution, noise, illumination, and pose variation given in Fig. 2.

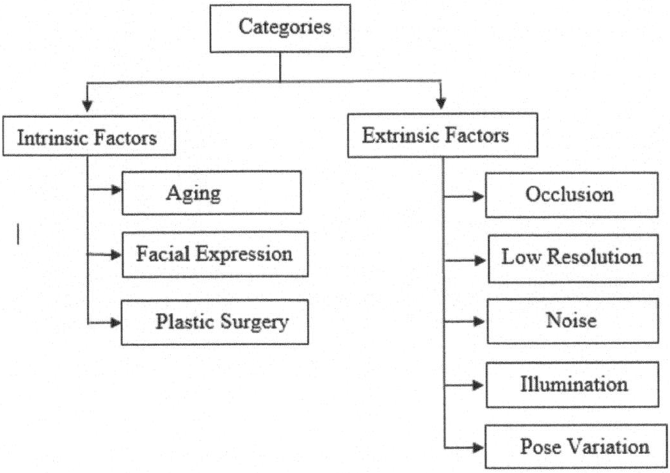


**Fig. 2** Categories of factors affecting face recognition accuracy

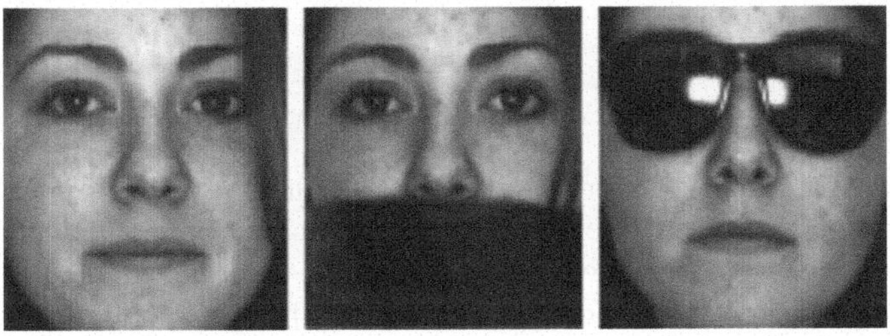

**Fig. 3** Partial occlusion in the face [6]

## 2.1 Occlusion

Partial occlusion is one of the major challenges of face recognition. It would be difficult to recognize a face if some part of the face is missing. For example, sunglasses and specs can hide eyes, earrings and hairs can hide ears, scarfs can hide half face, mustaches, and beard of boys can hide half of the face, and so on which can be shown in Fig. 3. These factors can deteriorate the performance of the system. Different approaches are being proposed by the researchers to overcome these problems [5].

## 2.2 Low Resolution

The pictures taken from surveillance video cameras comprise small faces; thus, its resolution is low as shown in Fig. 4. To compare the low-resolution query image with the high-resolution gallery image is a challenging task. Such a low-resolution picture

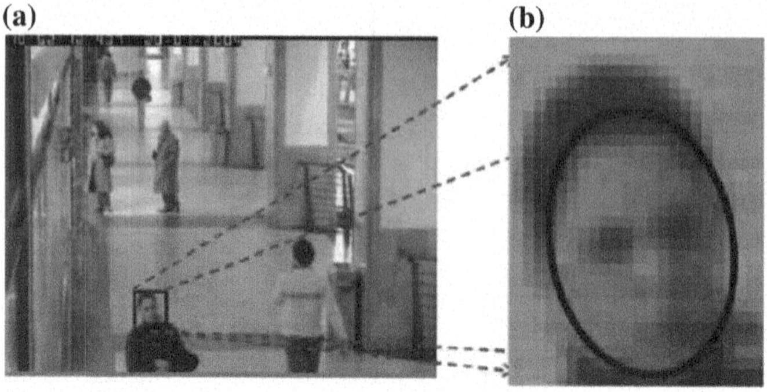


**Fig. 4** Frame from surveillance video. **a** video, **b** captured face [7]

comprises exceptionally constrained data as the greater part of the points of interest are lost. This can drastically degrade the recognition rate. Distinctive methodologies are being proposed by the researchers to tackle with this issue [5, 7].

## *2.3 Noise*

Digital images are inclined to various types of noise. This noise leads to poor detection and recognition accuracy. Noise can be introduced in images via different ways that are dependent on image creation. The pre-handling is an imperative factor in the general face detection and recognition framework [8]. Figure 5b illustrates the salt-and-pepper noise present in an image.

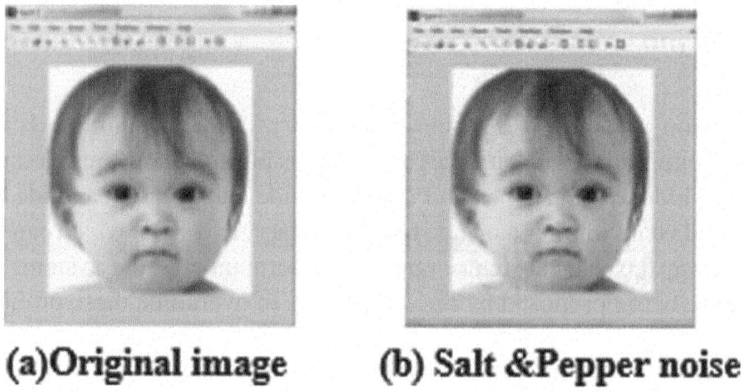


**Fig. 5** Noise in face image [8]

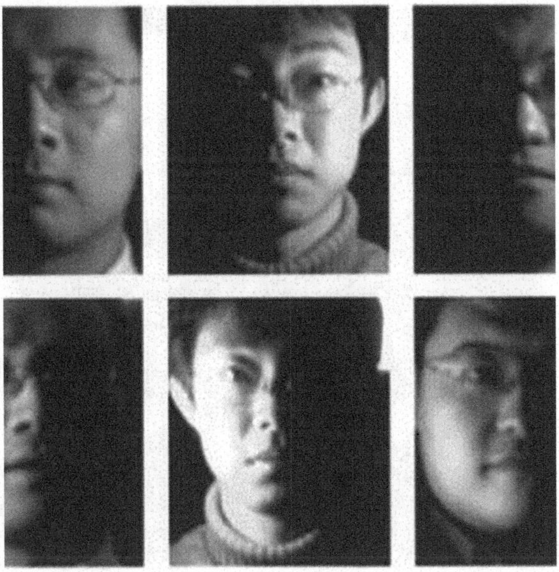

**Fig. 6** Effect of illumination in face images [14]

## 2.4 Illumination

The variations in illumination can drastically degrade the performance of the face recognition system. The reasons for these variations could be background light, shadow, brightness, contrast, etc. The images taken in different lighting conditions are shown in Fig. 6. Different approaches related to illumination are being discussed in [9–14].

## 2.5 Pose Variation

Different pose distribution is also one of the major concerns of the face recognition system. Frontal face reconstruction is required to match the profile face with the gallery face [15]. This reconstruction is required because a database image consists of frontal view and non-frontal profile face can generate faulty results. Different approaches proposed by the researchers to convert the non-frontal face to frontal face could increase the recognition accuracy [15, 16]. How pose variation degrades the performance of the algorithm dramatically is being discussed by researchers in the proposed approaches [11, 17]. Different pose distributions of an individual are shown in Fig. 7.

**Fig. 7** Pose variations [18]

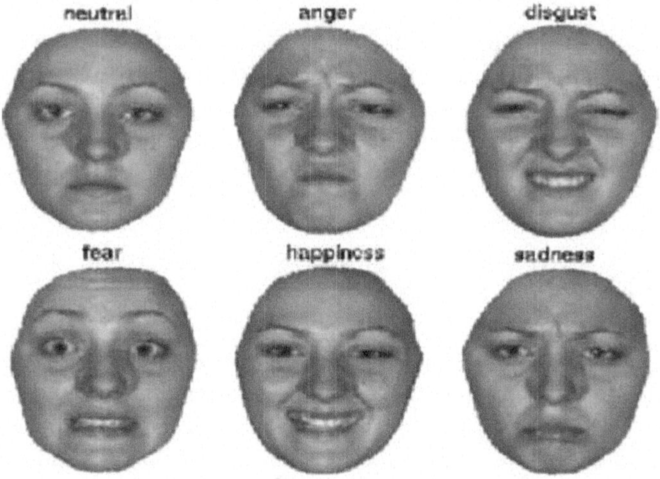


**Fig. 8** Different facial expressions [3]

## 2.6 *Expressions*

With the help of facial expressions, we can express our feelings as shown in Fig. 8. It changes the geometry of the face. A small variation in the face can create vagueness for face recognition system. Facial expression is a rapid signal that is easily affected due to the contraction of muscles and changes the facial features like eyebrows, cheeks, mouth, etc. Continuous research is being done for face recognition by taking facial expression into consideration [9, 13].

## 2.7 Aging

Aging is one of the natural components affecting face recognition systems as it swings to be a wreck for an algorithm. The face is a mix of skin tissues, facial muscles, and bones. At the point, when muscles contract, they result in the twisting of facial highlights. Be that as it may, maturing causes critical changes in facial appearances of an individual, e.g., facial surface (wrinkles and so forth) and face shape with the progression of time. The face recognition frameworks ought to be sufficiently skilled in considering this requirement [13, 19, 20]. Different texture of faces of the same person at different ages is given in Fig. 9.

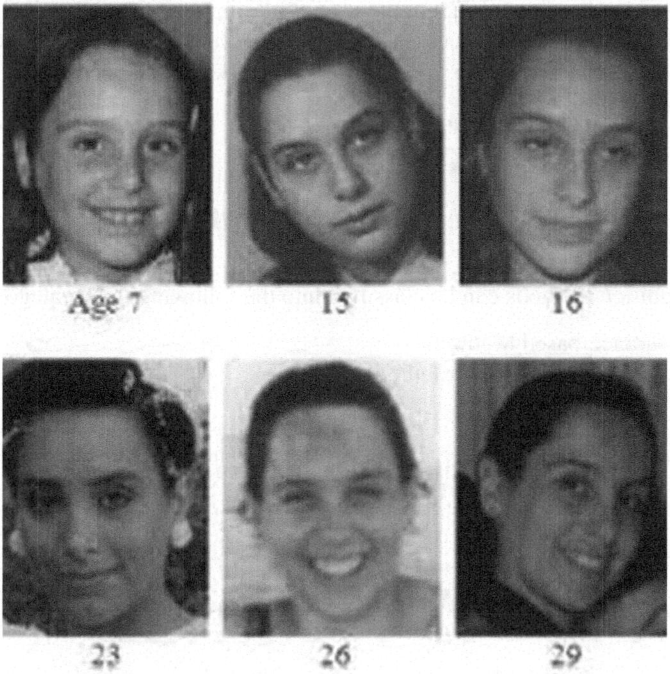


**Fig. 9** Aging variations [20]

**Fig. 10** Surgery effects [21]

## 2.8 *Plastic Surgery*

It is also a major concerned factor that affects the face recognition accuracy. Many incidents happened that because of accidents, numerous people have experienced plastic surgery and their faces will be unknown to the existing face recognition framework. Mostly criminals adopt the idea of plastic surgery to bury their identity. So, as discussed in [21], we need an identification system that is capable of recognizing faces even after reconstructive surgery. The effect of plastic surgery is shown in Fig. 10.

# 3 Classification of Face Recognition Methods

Face recognition methods can be classified into the following three categories [22]:

(i) Appearance-based Methods
(ii) Feature-based Matching Methods
(iii) Hybrid Methods

The classification of face recognition methods and the algorithms used in the above-discussed methods is shown in Fig. 11.

## 3.1 *Appearance-Based Methods*

Appearance-based methods are also known as the holistic method where a whole face is matched with the gallery face. The advantages offered by holistic approach are:

(i) They concentrate on only limited regions or points of interest without destroying any of the information in images.
(ii) No knowledge of geometry and reflectance of faces is required.

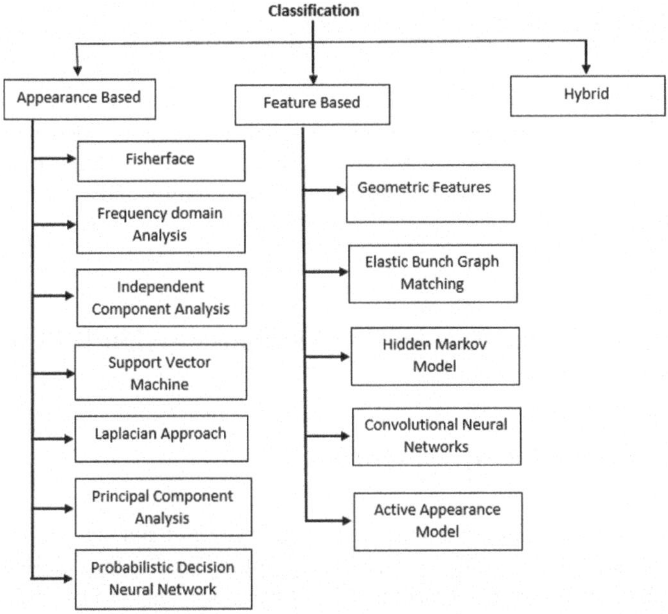


**Fig. 11** Classification of face recognition algorithms

(iii) Recognition is simple as compared to other matching approaches.
(iv) Fast and easy to implement.

But, they are quite prone to the limitations caused by facial variations such as illumination, pose, and expressions. Also, recognition accuracy is low in an unconstrained environment.

## *3.2 Feature-Based Matching Methods*

In feature-based matching methods, matching of probe face with gallery face has done using local features of the face like nose, eyes, and mouth. The location and local statistics of the extracted features has fed into the structural classifier to recognize the face.

In feature-based matching methods, recognition accuracy is better than appearance-based approaches. However, high memory usage is required in these methods and also a large amount of clean and labeled data for training.

### *3.3 Hybrid Methods*

Hybrid methods are a combination of both appearance-based methods and feature-based matching methods. Both appearance-based and feature-based methods are amalgamated to increase the recognition rate. However, this method is complex and difficult to implement as it is the combination of both feature- and appearance-based approaches.

## 4 A Review of Face Recognition Methods Used in Different Studies

### *4.1 Appearance-Based Approaches*

A lot of research has been done under this category. The main algorithms used for these methods are SIFT algorithm [17], PCA algorithm [10, 23–26], AdaBoost, LDA, elastic bunch graph matching algorithms [24], Fisherface, and SVD techniques [25].

Khan et al. [23] proposed an automatic face recognition system. In the proposed system, the PCA eigenface algorithm has been used for face recognition. Experimental results demonstrate that the proposed approach achieved 86% recognition accuracy in a constrained environment and 80% recognition accuracy in an unconstrained environment. The limitations of the proposed system are that it does not provide better results for low-resolution videos and large pose variations [23].

In Chawla and Trivedi [24], a comparison between AdaBoost, PCA, LDA, and elastic bunch graph matching algorithms for face recognition has been illustrated with their drawbacks, benefits, success rate, and other factors. After the comparative study, it has been found that PCA has the highest success rate (85–95%) but it is not suitable for video datasets. All the discussed algorithms are suitable only for smaller or simple databases.

Abdullah et al. [10] proposed an application of face recognition approach for criminal detection. In the proposed method, principal component analysis has been used to recognize the criminal faces. The results showed 80% recognition accuracy, but a lot of testing is needed for the proposed work.

Dhameja et al. [25] used a combination of PCA, Fisherface, and SVD techniques for face recognition. The proposed system was tested on AT&T face database and achieved approximately 99.5% recognition rate by the leaving-one-out method and 93.92% by the hold-out method.

Kavitha and Mirnalinee [15] proposed an algorithm to frontalize the non-frontal faces. Fitting, mirroring, and stretching operations are done to obtain the frontal face. Experiments are done on FERET, HP, LFW, and PUB-FIG datasets to prove the proposed approach reconstruction accuracy. The proposed approach can handle only up to $\pm 22.5^\circ$ pose variation.

In Gao and Jong [17], the authors proposed a face recognition algorithm which is based on multiple virtual views and alignment error. In the proposed work, Lucas Kanade algorithm and SIFT algorithm were used. FERET dataset has been used to evaluate the performance of the proposed approach. The proposed approach reported approximately 38% improvement in the performance when compared with the results obtained from other face recognition algorithms. But the proposed approach can handle only up to $\pm 60^\circ$ pose distribution. Also, recognition accuracy is poor beyond $40^\circ$. Time complexity is high in comparison with other comparing algorithms.

The work proposed by Ahonen et al. [13] used Local binary pattern to represent the face. The method was applied on FERET dataset. The results showed that the proposed approach achieved 97% accuracy for images with different facial expressions but did not perform well for other factors [13].

In [2], a comparative research has been done by Sharif et al. [2] on different approaches and applications of face recognition used till 2014. Different approaches discussed in this paper are based on eigenface, Gabor wavelet, neural network, and hidden Markov model. In this review paper, it has been concluded that (i) eigenface-based methods work well only for frontal faces, (ii) Gabor wavelet-based methods do not provide better recognition accuracy due to redundancy in Gabor filters, (iii) neural network-based methods improve the recognition accuracy, but it requires a large amount of time for the training purpose, (iv) support vector machine is slow in classification, but in combination with other approaches it generates good results.

## *4.2 Feature-Based Approaches*

The following algorithms such as face recognition through geometric features, elastic bunch graph matching, hidden Markov model, convolutional neural networks, and active appearance model are rooted in this classification [23].

Luo et al. [9] proposed unsupervised deep adaptation methods. They adopted a deep unsupervised domain adaptation neural network. They explained that the training dataset for supervised learning methods should be labeled and cleaned. The study reported approximately 17% improvement in the performance when compared with the results obtained from different face recognition algorithms. The method was applied to GBU and FERET datasets. CASIA-Webface was used to train the DCNN. But, the results presented were not sufficiently detailed for video datasets. The proposed research focused only on different illumination, expression, and age.

The key thought of the study by Zhou and Lam [19] was to provide an age-invariant face recognition model. The algorithm works on three steps: local feature extraction, identity subspace, and feature fusion using canonical correlation analysis (CCA) for face matching. Experimental results were conducted on FGNET, MORPH, and CACD datasets to present the efficacy of the proposed approach. They achieved approximately 51% improvement in recognition accuracy when applied on FGNET dataset and 13% improvement on CACD and MORPH datasets.

The paper authored by Gosavi et al. [27] dealt with the review of different feature extraction techniques. Different feature extraction approaches like geometric-based feature extraction, appearance-based feature extraction, template-based feature extraction, and artificial neural networks for face recognition are reviewed in this study. Among discussed approaches, template-based feature extraction provides more recognition accuracy. Backpropagation neural network and convolutional neural network are more efficient than other feature extraction approaches based on neural network. It has concluded that the discussed approaches were not as much as efficient [27].

In the work proposed by Fu et al. [5], a deep learning model based on guided CNN has been used for cross-resolution and occluded image recognition. The proposed system consists of two sub-models: one is used as a guide which is already trained with high-resolution images and the other is used as a learner for low-resolution input images. CASIA-Webface dataset and LFW dataset were used for training and testing purposes, respectively. This procedure accurately recognizes the cross-resolution and partially occluded images. They achieved 9% improvement in case of cross-resolution images and 4% improvement in case of partially occluded images. The proposed approach requires a lot of training as for each image in the database; the system needs to be trained for corresponding low-resolution image.

Bhavani et al. [28] used sparse representation technique for the recognition of faces in a video. Viola–Jones algorithm was used to detect the faces in an image. All the experiments have done on a real-time dataset. They achieved 75–80% recognition accuracy. However, they can work only for frontal face images.

Kakkar and Sharma [29] provided an adequate face recognition system for criminal identification by using Haar feature-based cascade classifier and local binary pattern histogram. The drawback of their system is that it can recognize only frontal faces [29].

The research done by Harsing et al. [30] proposed a new approach called entropy-based volume SIFT (EV-SIFT) for the recognition of surgery faces. The system was evaluated for the following surgeries: blepharoplasty, brow lift, liposhaving, malar augmentation, mentoplasty, otoplasty, rhinoplasty, rhytidectomy, and skin peeling. They achieved different recognition rate for the different types of plastic surgeries [30].

The contemplation in the work proposed by Sameem et al. [31] was to provide an efficient face detection and recognition system using Viola–Jones algorithm and MSAC algorithm. To perform experiments, Graz 01 dataset was used. Recognition accuracy of the proposed approach was compared with the approach proposed by Hong et al. [32] and achieved a 10% improvement in the recognition rate. It has been concluded that the proposed system is capable of handling images in an unconstrained environment.

Gaikawad and Sonawane [33] provided a different perspective on face detection and recognition from video surveillance. They have not implemented their approach to evaluate the algorithm on the dataset. However, it has been accomplished that the suggested method can cope with illumination, pose, shape, resolution, and plastic surgery.

Ding and Tao [34] proposed a framework based on convolutional neural networks (CNN) to cope with the challenges/issues related to video-based face recognition. Trunk-Branch Ensemble CNN (TBE-CNN) model has been used to confront the pose and occlusion variations. Mean distance regularized triplet loss (MDR-TL) function has been used to train TBE-CNN. The success of the proposed method was evaluated not only on single dataset but also for multiple video datasets like COX face, PaSC, and YouTube faces. They achieved approximately 95% recognition accuracy on YouTube faces database, 96% recognition accuracy on PaSC dataset and 99.33% accuracy for V2V, 98.96% for V2S, and 95.74% for S2V on COX database. In BTAS 2016 Video Person Recognition Evaluation, their approach acquired the first position. The proposed approach efficiently handles the problems like blur, partial occlusion, and poses variations.

Ding et al. [11] deduced a pose-invariant face recognition model by using multi-task feature transformation learning scheme to train the transformation dictionary for each patch. Experimental results showed that the proposed method achieved 100% recognition accuracy for $\pm 44^\circ$ yaw angle on CMU-PIE dataset, 98% accuracy for $+ 65^\circ$ on FERET dataset, approximately 87.75% recognition accuracy for $\pm 60^\circ$ pose angle with illumination on MULTI-PIE dataset, 98% recognition accuracy for $\pm 45^\circ$ pose angle with recording session on MULTI-PIE dataset, 91.78% accuracy on LFW challenging dataset. It has been concluded that the proposed approach outperforms single-task-based methods. They were the first to use multi-task learning in pose-invariant face recognition.

In [35], Hu et al. investigated the reason for the promising performance of convolutional neural network in face recognition. They proposed 3 CNN architectures of different sizes: CNN-S (Small), CNN-M (Medium), and CNN-L (Large). The experimental results showed that the proposed approach is better than some existing approaches but also provides worse result than other existing state-of-the-art approaches. In last, they concluded that network fusion can greatly enhance the performance of face recognition systems. Metric learning is also responsible to boost the face recognition accuracy.

Huang et al. [36] introduced a video database for face recognition called COX face database. They showed the effectiveness of the introduced database in comparison with other existing video datasets. They also proposed a point-to-set correlation learning (PSCL) method to prove the face recognition accuracy on proposed COX face database. Experimental results showed that the proposed approach achieves higher accuracy in comparison with other existing approaches on COX database. In last, it has been concluded that video-based face recognition is more complex when compared to image-based face recognition. More efforts can be done to improve the recognition accuracy of the proposed approach.

To authenticating the user for mobile payments, Wang et al. [12] have introduced face recognition approach based on deep reinforcement learning with convolutional neural networks. The proposed method provides 100% recognition accuracy when the gamma correction value is 1. In conclusion, the proposed approach outperforms other existing CNN-based approaches. Q-table learning was incorporated in the presented approach to fine-tune the previous model in different lighting conditions.

### *4.3 Hybrid Approach*

Hybrid approach is an amalgamation of both the appearance- and feature-based methods.

Banerjee et al. [16] contended that deep learning techniques for facial recognition. They also explained the question of whether to train the system for multiple poses is beneficial or to frontalize the profile face is propitious. They also compared the face recognition accuracy using their proposed frontalization algorithm with existing Hassner and Enbar [37] frontalization algorithm and simple 2D alignment (i.e., no frontalization) on PaSC video dataset and CW image dataset. Viola–Jones algorithm was used for face detection. Generic 3D face model was used to frontalize the profile image. Image correction and postprocessing was done to obtain the final frontalized image. They used a supervised learning approach for face recognition pipeline. CMU multi-PIE dataset with the combination of different facial landmarking algorithms was used to calculate the frontalization success rate. With the help of experimental results, they concluded that all the discussed frontalization methods along with facial landmarking algorithms experienced high failure rates beyond 40° yaw angle. It has also been concluded that the usefulness of frontalization is dependent on the facial recognition system used and may not always provide better results.

Fathima et al. [38] proposed a hybrid approach using Gabor wavelet and linear discriminant analysis for face recognition. AT&T, MIT-India, and Faces94 datasets are used to evaluate the recognition rate of the proposed approach. They achieved 88% recognition accuracy on AT&T, 88.125% on MIT-India, and 94.02% on Faces94 datasets. Results showed that the proposed approach provides better results in comparison with Gabor and efficient for the unconstrained environment.

In [39], Lei et al. [39] proposed an efficacious face detection system. The system works in two phases: detection phase and the recognition phase. In the detection phase, they used modest AdaBoost algorithm that maintains a low computational cost. Improved independent component analysis approach was used in the recognition phase. Hausdorff distance was used in the recognition phase to calculate the similarity measure between the face and other objects present in an image. An experiment has done on CMU-MIT face database to evaluate the detection rate of the algorithm. Experimental results showed that the proposed approach provides better detection rate in comparison with other existing face detection algorithms. The proposed approach can be used for other object detection and recognition tasks.

Research gaps of some of the above-discussed approaches are listed in Table 1.

**Table 1** A tabular representation of face detection and recognition methods used in different studies

| | Authors | Year | Database | Algorithm used | Recognition accuracy | Research gap |
|---|---|---|---|---|---|---|
| Appearance-based approaches | Khan et al. [23] | 2018 | NCR-IIT facial database and real-time video stream | PCA algorithm | 69–86% | (1) PCA eigenface algorithm is used for face recognition but it does not perform well for unconstrained environments<br>(2) If the face recognition algorithm is not accurate, then the whole system will fail<br>(3) Does not provide better results for low-resolution video and pose variation |
| | Abdullah et al. [10] | 2017 | Real-time video stream | PCA algorithm | 80% | (1) PCA algorithm is used for face recognition, but it is not efficient for video datasets<br>(2) PCA works well on frontal faces<br>(3) Other factors are not considered<br>(4) No dataset used to evaluate the performance of the algorithm<br>(5) Only 80% recognition accuracy |
| | Gao and Jong [17] | 2015 | FERET | Lucas Kanade, SIFT, two-phase alignment error | 99.5%, ±15°<br>26.0%, ±60° | (1) The proposed approach can handle only up to ±60° pose distribution<br>(2) Recognition accuracy is poor beyond 40°<br>(3) It is not feasible to generate multiple virtual views of all the images present in the database<br>(4) Time complexity is high in comparison with other comparing algorithms |

(continued)

**Table 1** (continued)

| | Authors | Year | Database | Algorithm used | Recognition accuracy | Research gap |
|---|---|---|---|---|---|---|
| Feature-based approaches | Luo et al. [9] | 2018 | CASIA-Webface used for training<br>GBU, FERET | Unsupervised transfer learning method, maximum mean discrepancy | 98.70–98.93% | (1) Profile image and the database image should have the same alignment (i.e., not work well if the alignment of the profile and database image is different)<br>(2) All the experiments are done on image datasets, not on video datasets<br>(3) Focused only on different illumination, expression, age but not on different poses, occlusion |
| | Fu et al. [5] | 2017 | CASIA-Webface<br>LFW | Guided convolutional neural network, the loss function | 91.9–97.1% | (1) A lot of training is needed (i.e., for each image in a database, we have to train the system for corresponding low-resolution image)<br>(2) Focus only on cross-resolution and partial occlusion |
| | Huang et al. [36] | 2015 | COX face database | Point-to-set correlation learning (PSCL) | 50.96–53.26% | (1) The proposed database does not contain images for the evaluation of the following factors such as aging, occlusion, and plastic surgery<br>(2) Does not provide effective results as required |

(continued)

**Table 1** (continued)

| | Authors | Year | Database | Algorithm used | Recognition accuracy | Research gap |
|---|---|---|---|---|---|---|
| Hybrid approaches | Banerjee et al. [16] | 2018 | PaSC videos and CW images, CMU multi-PIE dataset | Supervised learning, Viola–Jones, generic 3D model | 88.45–97.28% | (1) All the discussed frontalization methods experienced high failure rates beyond the 40° yaw angle<br>(2) Focused only on a pose and somewhere on illumination and expression<br>(3) Supervised learning approach is used, so it requires clean and labeled training data |
| | Fathima et al. [38] | 2015 | AT&T, MIT-India and Faces94 datasets | Gabor wavelet and linear discriminant analysis | 88–94.02% | (1) Does not work well for faces with different pose distribution |
| | Lei et al. [39] | 2009 | CMU-MIT face database | Modest AdaBoost, improved independent component analysis, Hausdorff distance | 95.20% | (1) Experiments are not performed to evaluate the recognition rate<br>(2) A large number of images required to train the system for the recognition stage |

## 5 Conclusion

After two decades of continuous research, it has been observed that a lot of research is going on in face detection and recognition area but still we are not getting satisfactory recognition accuracy and results. The classifications are broadly classified on appearance-based approaches, feature-based approaches, and hybrid approaches. All the approaches have their own advantages and disadvantages depending upon their area of application and datasets used.

Appearance-based methods do not provide better results for low-resolution video and pose variation. The method in which a supervised learning approach is used provides satisfactory results, but it requires a large amount of clean and labeled training data to train the system for the recognition stage.

Most of the methods do not perform well for non-frontal faces, i.e., faces with different pose distribution. In addition, the factors such as aging, occlusion, and plastic surgery also affect the recognition accuracy of the system.

After the literature review, it has been concluded that despite continuous research, an efficient face recognition system is required that can perform well in constrained as well as the unconstrained environment.

## References

1. Yu, S., Lu, Y., Zhou, J.: A survey of face detection, extraction and recognition. J. Comput. Inf. **22**, 163–195 (2003)
2. Sharif, M., Naz, F., Yasmin, M., Shahid, M.A., Rehman, A.: Face recognition: a survey. J. Eng. Sci. Technol. Rev. **10**(2), 166–177 (2017)
3. Patel, R., Rathod, N., Shah, A.: Comparative analysis of face recognition approaches: a survey. Int. J. Comput. Appl. **57**(17), 50–61 (2012)
4. Khade, B.S., Gaikwad, H.M., Aher, A.S., Patil, K.K.: Face recognition techniques: a survey. Int. J. Comput. Sci. Mob. Comput. **5**(11), 65–72 (2016)
5. Fu, T., Chiu, W., Wang, Y.F.: Learning guided convolutional neural networks for cross-resolution face recognition. In: 2017 IEEE 27th International Workshop on Machine Learning for Signal Processing (MLSP) (2017)
6. Zhang, W., Shan, S., Chen, X., Gao, W.: Local Gabor binary patterns based on Kullback–Leibler divergence for partially occluded face recognition. IEEE Sig. Process. Lett. **14**(11), 875–878 (2007)
7. Zou, W.W.W., Yuen, P.C.: Very low resolution face recognition problem. IEEE Trans. Image Process. **21**(1), 327–340 (2012)
8. Tin, H.H.K.: Removal of noise by median filtering in image processing. In: 6th Parallel and Soft Computing (PSC 2011) Removal (2011)
9. Luo, Z., Hu, J., Deng, W., Shen, H.: Deep unsupervised domain adaptation for face recognition. In: 13th IEEE International Conference on Automatic Face & Gesture Recognition, pp. 453–457 (2018)
10. Abdullah, N.A., et al.: Face recognition for criminal identification: an implementation of principal component analysis for face recognition. In: The 2nd International Conference on Applied Science and Technology 2017 (ICAST'17) (2017)
11. Ding, C., Member, S., Xu, C., Tao, D.: Multi-task pose-invariant face recognition. IEEE Trans. Image Process. **24**(3), 980–993 (2015)

12. Wang, P., Lin, W., Chao, K.-M., Lo, C.-C.: A face-recognition approach using deep reinforcement learning approach for user authentication. In: Fourteenth IEEE International Conference on e-Business Engineering, pp. 183–188 (2017)
13. Ahonen, T., Hadid, A., Pietika, M.: Face description with local binary patterns: application to face recognition. IEEE Trans. Pattern Anal. Mach. Intell. **28**(12), 2037–2041 (2006)
14. Li, S.Z., Member, S., Chu, R., Liao, S., Zhang, L.: Using near-infrared images. IEEE Trans. Pattern Anal. Mach. Intell. **29**(4), 627–639 (2007)
15. Kavitha, J., Mirnalinee, T.T.: Automatic frontal face reconstruction approach for pose invariant face recognition. In: Proceedings of the 4th International Conference on Recent Trends in Computer Science & Engineering, Elsevier, vol. 87, pp. 300–305 (2016)
16. Banerjee, S., et al.: To frontalize or not to frontalize: do we really need elaborate pre-processing to improve face recognition ? In: IEEE Winter Conference on Applications of Computer Vision, pp. 20–29 (2018)
17. Gao, Y., Jong, H.: Cross-pose face recognition based on multiple virtual views and alignment error. Pattern Recognit. Lett. **65**, 170–176 (2015)
18. Singh, R., Vatsa, M.: A mosaicing scheme for pose-invariant face recognition. IEEE Trans. Syst. Man Cybern. B Cybern. **37**(5), 1212–1225 (2007)
19. Zhou, H., Lam, K.: Age-invariant face recognition based on identity inference from appearance age. J. Pattern Recognit. **76**, 191–202 (2018)
20. Li, Z., Park, U., Jain, A.K.: A discriminative model for age invariant face recognition. IEEE Trans. Inf. Forensics Secur. **6**(3), 1028–1037 (2011)
21. Mun, M., Deorankar, P.A.: Implementation of plastic surgery face recognition using multimodal biometric features. Int. J. Comput. Sci. Inf. Technol. **5**(3), 3711–3715 (2014)
22. Dass, R., Rani, R., Kumar, D.: Face recognition techniques: a review. Int. J. Eng. Res. Dev. **4**(7), 70–78 (2012)
23. Khan, A., et al.: Forensic video analysis: passive tracking system for automated Person of Interest (POI) localization. IEEE Access **6**, 43392–43403 (2018)
24. Chawla, D., Trivedi, M.C.: A comparative study on face detection techniques for security surveillance. In: Advances in Computer and Computational Sciences, pp. 531–541 (2018)
25. Dhamija, J., Choudhary, T., Kumar, P., Rathore, Y.S.: An advancement towards efficient face recognition using live video feed. In: International Conference on Computational Intelligence and Networks (CINE), pp. 53–56 (2017)
26. Baykara, M., Daş, R.: Real time face recognition and tracking system. In: International Conference on Electronics, Computer and Computation, pp. 159–163 (2013)
27. Gosavi, V.R., Sable, G.S., Deshmane, A.K.: Evaluation of feature extraction techniques using neural network as a classifier: a comparative review for face recognition. Int. J. Sci. Res. Sci. Technol. **4**(2), 1082–1091 (2018)
28. Bhavani, K., et al.: Real time face detection and recognition in video surveillance. Int. Res. J. Eng. Technol. **4**(6), 1562–1565 (2017)
29. Kakkar, P., Sharma, V.: Criminal identification system using face detection and recognition. Int. J. Adv. Res. Comput. Commun. Eng. **7**(3), 238–243 (2018)
30. Harsing, A., Talbar, S.N., Amarsing, H.: Recognition of plastic surgery faces and the surgery types: an approach with entropy based scale invariant features. J. King Saud Univ. Comput. Inf. Sci. 1–7 (2017)
31. Sameem, M.S.I., Qasim, T., Bakhat, K.: Real time recognition of human faces. In: International Conference on Open Source Systems & Technologies, pp. 62–65 (2016)
32. Hong, Y., Li, Q., Jiang, J., Tu, Z.: Learning a mixture of sparse distance metrics for classification and dimensionality reduction. In: IEEE International Conference on Computer Vision, pp. 906–913 (2011)
33. Gaikawad, P.A.D., Sonawane, P.P.D.: An efficient video surveillance system using video based face recognition on real world data. Int. J. Sci. Eng. Technol. Res. **5**(4), 1245–1250 (2016)
34. Ding, C., Tao, D.: Trunk-branch ensemble convolutional neural networks for video-based face recognition. IEEE Trans. Pattern Anal. Mach. Intell. **40**(4), 1002–1014 (2018)

35. Hu, G., et al.: When face recognition meets with deep learning: an evaluation of convolutional neural networks for face recognition. In: IEEE International Conference on Computer Vision Workshops, pp. 384–392 (2015)
36. Huang, Z., et al.: A benchmark and comparative study of video-based face recognition on COX face database. IEEE Trans. Image Process. **24**(12), 5967–5981 (2015)
37. Hassner, T., Enbar, R.: Effective face frontalization in unconstrained images. In: IEEE Conference on Computer Vision and Pattern Recognition, pp. 4295–4304 (2015)
38. Fathima, A.A., Ajitha, S., Vaidehi, V., Hemalatha, M., Karthigaiveni, R., Kumar, R.: Hybrid approach for face recognition combining gabor wavelet and linear discriminant analysis. In: IEEE International Conference on Computer Graphics, Vision and Information Security, pp. 220–225 (2015)
39. Lei, Z., Wang, C., Wang, Q., Huang, Y.: Real-time face detection and recognition for video surveillance applications. In: 2009 World Congress on Computer Science and Information Engineering Real-time, pp. 168–172 (2009)

# Detection of Eye Ailments Using Segmentation of Blood Vessels from Eye Fundus Image

**Parul Datta, Shalli Rani and Deepika Koundal**

**Abstract** Eyes are vital part of the body which can be affected by many diseases that lead to vision loss. Glaucoma is one such eye disease that may cause vision loss. There are multiple reasons for vision loss which may be due to the appearance of unwanted blood vessels that can be caused by high level of glucose in the blood composition. This abnormal growth or change in behavior of the blood vessels represents underlying indicators of problems associated with eye diseases such as diabetic retinopathy. Hence, early detection of eye ailments can be expedited with the help of various image processing technologies. The first step after image acquisition is the processing of images to extract features that exactly match the disease under observation. This paper attempts to evaluate the blood vessels using different segmentation algorithms and introduce an improved version of the vessel algorithm. The evaluation of segmentation approaches shows that Otsu clustering algorithm is performing best as compared to other state-of-the-art techniques using eye fundus images.

**Keywords** Fundus images · Blood vessel segmentation · Diabetic retinopathy · Convolution filters

## 1 Introduction

Glaucoma is a group of eye diseases that are caused due to many reasons. When the optic nerve is damaged due to increased pressure, infection, blocked blood vessels,

P. Datta (✉) · D. Koundal
Chitkara University School of Engineering and Technology, Chitkara University, Solan, Himachal Pradesh, India
e-mail: parul.datta@chitkarauniversity.edu.in

D. Koundal
e-mail: deepika.koundal@chitkarauniversity.edu.in

S. Rani
Chitkara University Institute of Engineering and Technology, Chitkara University, Rajpura, Punjab, India
e-mail: shalli.rani@chitkara.edu.in



P. K. Singh et al. (eds.), *Proceedings of ICRIC 2019*, Lecture Notes in Electrical Engineering 597, https://doi.org/10.1007/978-3-030-29407-6_37

inflammation, or injury in the eyes, it leads to this condition. Correlation between the diabetic condition and glaucoma has been widely found around the world. Subjects with diabetic condition are twice likely to develop glaucoma as compared to the person who is not suffering from diabetic condition [1]. It is estimated that more than 3 million people living in USA are affected from glaucoma and in India 1.2 million people are suffering from this eye condition. Another medical condition due to diabetes (type 1 and type 2) which may happen is called diabetic retinopathy (DR). The symptoms include blurring of vision, having spots/floaters while watching around. In many cases, the subject can feel dark area/empty areas while observing around their surroundings [2, 3]. The DR can be classified into two groups. The first one is called the non-proliferative diabetic retinopathy and second one is proliferative diabetic retinopathy. In the first case, the blood vessels do not grow and get damaged. And in the second case, the blood vessels overgrow abnormally [4–6]. All these conditions can lead to complications that include vitreous hemorrhage, retinal detachment, glaucoma, and full blindness. Hence, paying attention to the changes in eye sighting is important. If someone is a smoker or is suffering from high glucose/high blood pressure, getting screened once a year is recommended. There are different types of modalities which are commonly used for detecting eye diseases such as optic coherence tomography (OCT) and fundus photography.

OCT is a noninvasive way to examine eye diseases [7, 8]. This method of examination uses light waves to take cross-sectional images of the retina. The images can be analyzed for detecting glaucoma, cataracts, bleeding in vitreous, retina diabetic, age-related eye complications, macular edema, and many more. The technology has undergone many changes since it was first applied to taking cross-sectional images of the retina. The resolution of the images taken by the system has improved due to the introduction of broadband light sources. It has also given birth to the development of portable probes that aid the doctors in their workflow. The use of dyes has made the process more easy, accurate, and useful for detecting foreign particles in the eyes as the images generated have more detail. Technically, the ophthalmic procedures follow a protocol to getting accurate 'region' of an eye for a specific eye ailment. For example, usually, the most common scanning protocol followed in the detection of glaucoma imaging includes the scan size of 3.4 mm around the optic nerve head and six equally spaced radial scans through the macula (6 mm) and optic nerve (4 mm) [9]. Based on the protocols, the images are generated for the detection of glaucoma. Similarly, other modalities of the eyes are examined based on the image scanning protocols [10]. OCT became obsolete with time and hence was replaced by fundus photography.

Fundus photography is a branch of photography that focuses on the images of the eyes [11, 12]. This branch of photography can help ophthalmic telemedicine [5] become more powerful. This camera helps to document the retina, the tissues in the eye, and the neurosensory veins and capillaries in the eye. It is performed with the help of many light color filters such as red, blue, or special colored dyes (fluorescein, indocyanine green) [13–16]. In a real sense, these devices are a kind of digital microscope that is attached to a digital camera that has some prefigured filters to maximize the capturing of eye retina. These digital microscopes are extremely helpful in the diabetic screening process, and now, with the advent of the many wireless technologies, these microscopes can now be remotely controlled using Internet connectivity.

The microscope provides the magnification and focuses on eyes at a specific angle (typically 30°) [17, 18], while the camera produces a chance to capture the eye retina at 2.5 times the normal size of the retina. This makes the screening process easy for evaluation. This camera, if equipped with a wide-angle lens, helps to produce high-quality images. These images are normally captured after the dilation of the eyes, and the quality of images depends on the technical mastery of the person operating the camera equipment. In this process, the images may suffer from defects such as blurred focus, white color imbalance, improper exposure, noise, and significant unwanted artifacts. The alignment of the stereo images and problems related to lens correction and aberration may be required to be rectified before the images screening can really happen. Typically, a diseased condition in the eyes can be classified into four types:

1. **Blood Vessels**: Blood vessels are tube-shaped canals that help the body to circulate the blood fluid to and from all organs of the body. These tube structures are divided into four categories which include arteries, veins, capillaries, and sinusoids. The vessels that take away blood from the heart are called arteries and their branches are called arterioles. The canals that bring blood back to the heart are called veins and its branching is called veins. The capillaries make the bridges between the arteries and veins. The working of sinusoids is similar to capillaries but they present only in liver, spleen, and bone marrow, whereas the capillaries are not. The formation of unwanted blood vessels can lead to eye complications. Images of an unhealthy eye will show the formation of unwanted blood vessels. Just like an expert, the image processing algorithms have the ability to spot, mark, and segment such healthy growth of blood vessels. In this paper, we review the segmentation method that would be best suited for blood vessel segmentation.
2. **Exudates**: Development of irregular and bright color spots in the eyes is an indicator of eye ailments. These need to be segmented and analyzed by the automated systems for detection of the specific eye diastasis.
3. **Hemorrhages**: Chunks of blood appearing as red blobs. Visually, it can be understood as a high degree of eye redness. The condition may be conjunctiva or glaucoma. For constructing a computerized system, the first step is to segment such red blobs.
4. **Microaneurysm**: Blood filled bulges in the artery walls of the eyes. It is a kind of swelling that needs to be segmented for the construction of an automated algorithm for its detection.

Rest of the paper is organized as follows: Sect. 2 presents image processing in blood vessel segmentation. Section 3 gives the literature review. Section 4 highlights the current gaps and challenges being faced. Section 5 gives the materials and methods being used in this paper. Section 6 shows the segmentation results and analysis. Finally, Sect. 7 concludes the paper with conclusion and future scope.

## 2 Image Processing in Blood Vessel Segmentation

This section gives an overview of the image processing operations, methods, and algorithms that help to build datasets, repositories to automate the process of remote telemedicine in eye care. The medical condition of the eyes may be glaucoma, cataract, diabetic retinopathy, macular degeneration, astigmatism, or any other type of eye complications. But, it is clear from the study of procedures, methods, and medical protocols that each eye ailment requires a specific type of diagnostic workflow. The block diagram shown in Fig. 1 gives generic flow of information on how by tracking the changes in blood vessels can lead to the diagnoses of eye ailments.

The normality test defines the normal shapes, thickness, size, and color texture properties of the blood vessels for a normal or the healthy subject. Any deviation from the standard deviation can lead to diagnoses of eye problems. Hence, in clinical examinations, automatic image segmentation of the blood vessels aids in measuring the damage done by the disease condition on the eye. This helps to delineate pathological regions in the eye and help in image-guided support to overcome the eye complications and medical conditions. Accurate segmentation of the blood vessels can be used to obtain quantitative information on the physiological changes in the eyes. The procedure typically followed in the image-guided support is as shown in Fig. 1. All the images are required to undergo the image preprocessing operation so that problems related to over/under focus and noise are handled properly. The next step is to get all the regions that represent the blood vessels. Last but not least, the shape, thinness, and texture of the blood vessels are analyzed for diagnosing the health of the blood vessels. If there is deviation from the normal pattern of growth in terms of thickness, shape, and texture of the vessels (arteries and veins), the normality test will fail and the subject is said to suffer from some medical condition. The health of the blood vessels in the eyes depends upon many factors. Typically, when a person is under some stress due to a disease in the eyes, the blood vessels undergo morphological, color, and textural changes. These changes (features/characteristics of diseases condition) can be captured using imagery hardware and software, and later on, subjected to machine learning and optimization algorithms to almost replace the examination process done these days manually. Significant work can read from

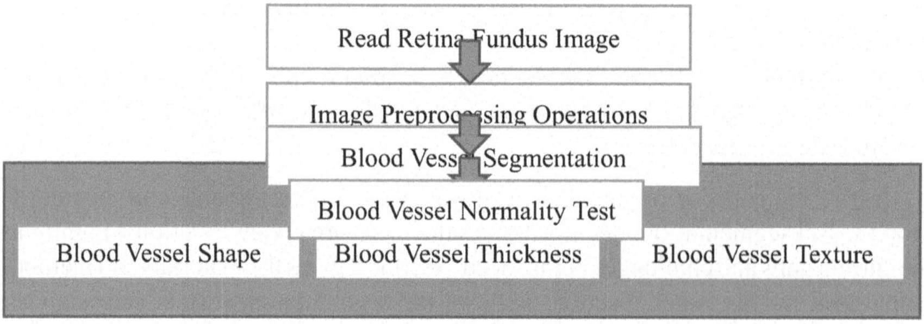


**Fig. 1** Eye normality test

the contemporary literature on the use of neural networks, support vector kernels, regression-based algorithms, and trees for supervised classification of health and unhealthy eyes. Most of the research work is focused on fine-tuning these existing algorithms. This forces to dwell on the evaluations of the previous algorithms before constructing new systems of diagnoses, detection, and facilities on medical modalities. Moreover, the experience shared by many researchers shows that the configuration and tuning of machine learning algorithms are a tiresome yet critical undertaking, as the performance of an algorithm can be dependent on the optimal configurations of hyperparameters. This can be done by initiating explorative studies on performance and optimization of the algorithms with new datasets. These explorative evaluation studies of an algorithm can be manual or automatic in nature to fine-tune the parameters of the optimization and machine learning algorithms. In the context of detecting eye complications with features of blood vessels, the prerequisite is to have high-quality image dataset of eye areas having blood vessels demarcated or segmented.

## 3 Literature Review

A medical examination of the retina images shows that fundus eye image will have arteries as well as veins. The image will have the optic nerve and a disk. Other than this, eye may have exudates and watery tissues. The person might be suffering from single or multiple ailments or he/she may be at different stages of eye diseases. As mentioned earlier, the expertise in handling the camera also matters in terms of clarity, quality, resolution, and number of artifacts in the images. Due to all these factors, it is amply clear that to segment each type of region of interest (ROI) (optic nerve, optic disk [19, 20], blood veins, blood vessels, exudates [21], noise, artificial camera artifacts, etc. [22]), there is a need to employ multiple types of segmentation approaches, algorithms, and methods to extract the blood vessels. The survey of the available dataset for blood vessel segmentation shows that most of the images, when acquired by the fundus cameras, are quite big (typically $3000 \times 2000$ pixels) in size and need to be downsampled or resized with the loss of information from the images. By doing cropping, the borders and other unwanted regions are removed. It is, in fact, a way of rescaling the images to suit the purpose. The authors in [23] subject the fundus image to color balance, contrast adjustments, and brightness adjustments [20] to improve the chance of getting accurate segmentation results. In current literature, artificial padding around the mask has also been used as preprocessing step [23].

Pieces of evidence come up of authors [24–27] constructing several filters and image processing pipelines to extract features that too manually. This is attributed to the fact that it required a lot of elimination and removal of many objects to finally extract the blood vessels. Symmetric, as well as asymmetric, filters have been used as a precursor to run the thresholding methods to get the blood vessels of the eyes.

Adaptive thresholding [22] and local entropy [28] have been frequently used as helping procedures to get the blood vessels. The author has used the concept of 'tracing lines' [29, 30] to find the vessels. This method is similar to the procedure of skeletonization.

The authors in [31] have used the concepts from differential geometry to get the segments of vessels from eye images. In such an approach, it is defined that the two 'principal curvatures' [31] at a given point of a surface are the eigenvalues. As the blood vessels branch out and form its shapes, the concepts from differential geometry become helpful in the location of the blood vessels. The authors have claimed that the curvelet-based edge detection can be used with good results for the extraction of blood vessels. Cellular Potts models have been extensively used for modeling cellular networks in biomedicine. The current literature gives knowledge of the use of this automata theory [32] and Potts modeling for understanding blood and cell vessels morphology. This approach helps to understand the shape of the cells and mathematical models to understand the 'morphogenesis' of cells, which means the behavior of cells in terms of shape and pattern as they grow or die. This is extremely important for detecting unnatural growth of blood vessels in the eyes. Edge detection methods such as Canny and Sobel have been ignored in favor of Kirsch operator [33] method for segmenting the blood vessel boundaries. It was also found that Sobel and Canny also do not perform well when compared with top-hat transformation methods of image extraction [27]. White top-hat approach defines the difference between the opening structural element (shape fits or misses the shapes in the image) and the input image. In case of black top-hat transformation approach, there is a comparison between the closing operations. It appears that authors have ignored the output given by the edge detection algorithms such as pewit, Canny, and Sobel and have work in the favor of using kirsch's template concept for building the boundaries and edges of the blood vessels. But, the authors [24, 34, 35] have used morphological mathematics to segment the vessels. Morphological operations such as open, closing, and dilating have been used along with the pipeline of filters and thresholding methods. Clustering algorithms such as ISODATA, fuzzy [19], and k-means that help in thresholding have also been used to get prominent blood vessels. Random field models have been used by authors [36, 37] for extracting blood vessels with good success ratio. The health of the vascular network of the eyes is most important for the person to remain mobile and independent in life. An unwanted growth or thickening of the blood veins is not a good sign of health. To overcome this challenge, the authors in [38] have used the convolution networks and deep learning approach to extract the blood vessel health extract. Machine learning algorithms are also used to automatically differentiate between the artery and veins.

Publically available dataset includes DRIVE [24], STARE [24, 39], DIARETDB1 [39–42] CHASE DB1 [43], Kaggle retinopathy dataset [44], and HRF [45]. All of these datasets have been made to promote collaborative research work in relation to eye-related diseases. These datasets have high-quality images that can be used to construct machine learning algorithms. These image sets also contain the images marked by the experts based on which the ground truth can be validated. Neural networks [35, 38, 46, 47] and deep learning algorithms [48–50] also find a place in the current literature related to the automation of the eye malfunction detection.

## 4 Current Gaps and Challenges

1. **Feature Separation**: Separation of medically relevant blood vessels from other parts (boundaries of the eye morphology) for eye disease detection is a challenge. The pixel intensity and color variation values are most important features that can help to distinguish the blood vessels from the boundaries and tissues of eyes, but at the same time, these features make difficult to separate the blood vessels due to tight range values.
2. **Quality of Images**: To improve the quality of eye images, the scan time of 'region of interest' needs to be optimal to acquire medical grade quality of images. Due to the small scanning time, the quality of the images sometimes is not good. Second, the expertise level of the person taking images also impacts the quality of the images.
3. **Under- or Over-Focused Images**: To compensate for the artifacts and noise for quality improvement, an application smoothing can lead to loss of details in the images and ultimately may lead to reduced fidelity of the image.

The previous algorithms have been using a pipeline of filters for identifying the intersections of the blood vessels in the retina and then applying multi-range global thresholding algorithms to collect the ROI, that is, blood vessels. Application of edge detection mask templates sometimes misses certain blood vessel boundaries due to a sensitivity of noise and artifacts in the fundus image. Then, overgrowth of blood vessels leads to different orientations of lines due to which the segmentation algorithms miss edges. This is true especially in cases where the pixel value being supported is of the small range. In such cases, a convolution matrix such as Sobel and Canny may lead to better detection of blood vessels in combination with the adaptive thresholding methods. Application of contrast stretching algorithm may help to increase the gradient between the soft and hard edges and consequently help in better blood vessel segmentation. The issue is which combination of filter and thresholding method(s) will produce medical grade segmentation of the blood vessels.

## 5 Materials and Methods

This work entails to automatic analysis of the eyes; this section gives information on the algorithm used for segmenting the blood vessels. This section gives information about the details of the image segmentation algorithm that has been developed to

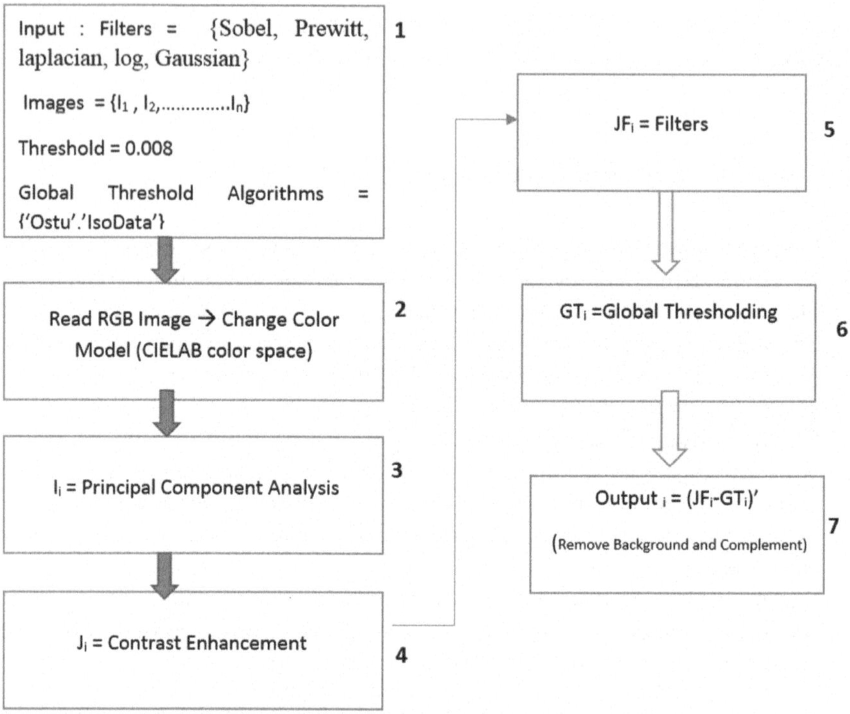


**Fig. 2** Logical flow of segmentation

improve the previous methods of extraction of the blood vessels. The implementation of the process can be understood from the block diagram in Fig. 2.

## 5.1 Dataset

For this research work, DRIVE [24] dataset has been used. The dataset contains images obtained from the retinopathy screening in the Netherlands. The 400 subjects are between the age of 25 and 90 years. The images were acquired. The camera equipment used for collecting the data used was Canon CR5 non-mydriatic 3CCD camera [2]. All the image items in the database have been used in real life for clinical diagnoses. The identity of the subject's information has been anonymized to obey the conventions and rules of privacy. Figure 2 shows segmentation flow.

**Step 1:**
The input to the segmentation process includes the eye fundus images, array of gradient filters, array of the global thresholding algorithm functions and constant

threshold values. The detection of the blood vessels depends on how accurately we can compute the gradients or the image derivative of various segments that make the complete image.

**Step 2:**
The value of a rate of pixel intensity change gives an opportunity to detect boundaries and edge of the objects in the images. By converting the image color model to CIELAB color space, the gradient becomes more prominent. This is done in Step 2. But, due to addition of unwanted artifacts and noise, the image derivative change abruptly leads to detection of false edges or segments of images. Therefore, after the preprocessing steps such as denoising, there is a need to compute the pixel intensity difference based on forward, backward, and central differences from a reference point. This is done by using image 2D convolution filters (Step 5) including Prewitt, Laplacian of Gaussian (LoG), Laplacian, Sobel, and mean filters. Table 1 provides information on the filter attributes used in this research work. All of these convolution kernels based on the concept of derivative filtering help to detect the areas of change in the images.

**Step 3:**
The third step is to construct the algorithm in such a manner that the image process does not consume too much of computational and storage resources. For this, each

**Table 1** Properties of gradient convolution filters

| S. no | Convolution gradient Kernel | Size | Filter attributes/Kernel approximation matrix | Descriptions |
|---|---|---|---|---|
| 1 | Prewitt [52] | 3 × 3 | $Gx = [+1\ 0\ -1;\ +1\ 0\ -1;\ +1\ 0\ 1]$ (*X*-direction)<br>$Gy = [+1\ +1\ +1;\ 0\ 0\ 0;\ -1\ -1\ -1]$ (*Y*-direction) | Horizontal edge emphasizing filter |
| 2 | LoG filter [51, 33] | 5 × 5 | [0 0 −1 0 0<br>0 1 −2 1 0<br>1 2 −16 2 1<br>0 1 2 1 0<br>0 0 1 0 0] | Laplacian of Gaussian [53] filter. Since the derivative of image is very sensitive to the noise and artifacts, it is preferred to smooth the image by using Gaussian filter |
| 3 | Laplacian filter | 3 × 3 | [0 1 0<br>1 −4 1<br>0 1 0] | Approximates the two-dimensional Laplacian operator [53] |
| 4 | Low-pass Gaussian filter | 3 × 3 | Sigma = 0.1 | Gaussian low-pass filter [54] |
| 5 | Sobel | 3 × 3 | $Gx = [-1\ 0\ 1;\ -2\ 0\ 2;\ -1\ 0\ 1]$<br>$Gy = [-1\ -2\ -1;\ 0\ 0\ 0;\ 1\ 2\ 1]$ | Horizontal edge emphasizing filter |
| 6 | Mean | 9 × 9 | ones(*n*(1),*n*(2))/(*n*(1)**n*(2)) | Based on the average intensity of the image |

image is subjected to the principal component analysis (PCA) mathematics to extract the coefficients. But, before this step, the color model of the image matrix is changed to get better approximation of pixel intensities for segmentations. This helps to transform each image into an invariant matrix and reduces its digital size. The steps involved in running the principal component analysis are as follows:

(1) Compute the covariance of image matrix '$I_i$'.
(2) Sort the computed eigenvectors and respective values in decreasing order.
(3) Select first '$k$' eigen to construct new matrix with '$k$' dimensions. This step helps to identify the principal directions of variations between the image matrix.
(4) Run transformation on the original image '$n$' dimension into new '$k$' dimensions.

**Step 4 and Step 5** are contrast enhancement and application of filters as briefly discussed above.

**Step 6:**
In this step, global thresholding methods are applied to collect and group the pixels that have an intensity greater than the level or threshold found by the automatic thresholding methods. This step also includes removal of the pixels that are too small to be considered as blood vessels. These two types of methods were compared viz, ISODATA and Otsu. The Otsu method of clustering is based on the principals of reducing interclass variation. The ISODATA method of pixel clustering is based on splitting and merging groups based on the standard deviation. Following are the results of these methods with various combinations of edge derivatives/kernels mentioned in Table 1.

**Step 7:**
Elimination of the background pixels (filtered image '$JF$' generated using filters mentioned in Table 1) to arrive at the intensity values to match the blood vessels intensity. This is done with the help of set theory using the subtraction method:

$$I_i = JF_i - J_i \tag{1}$$

where $JF_i$ is the filter-generated image matrix containing blood vessel pixels, $J$ is the image matrix with maximized (stretched) contrast.

Last but not least, the resultant image is subjected to mathematical operation of complement to obtain the final segmented image.

## 6 Segmentation Results and Analysis

It is apparent from the visual inspection that segmentation done by the Otsu method of global thresholding in combination with the mean filter seems to be the best out of all the combinations evaluated. The blood vessels are more distinct, and there is clarity of the shapes of the blood vessels. Clearly, the gradient Prewitt, Laplacian, and LoG convolution masks do not perform well to get the correct ROI. This is evident from the below images.

Figure 3 shows results of segmentation algorithms applied to the images. Figure 3c, f shows visually that blood vessels are prominent and have been properly segmented. It appears that the Laplacian, LoG, and Sobel filters in combination with Otsu and ISODATA do not produce appropriate markings to extract the blood vessels. The Sobel (Fig. 3h) operator is unable to smooth out the image data, and consequently, the Otsu and ISODATA methods fail to group the pixels. In case of Laplacian and LoG filters, the segmentation fails miserably. This may be attributed to the fact that second-order derivate of the special space is unable to detect region of rapid changes in gradient of the image. An independent application of Gaussian smoothing as shown in Fig. 3b produces better output in combination with Otsu. The output of the mean filter in combination with the ISODATA or Otsu is producing good quality of segmentation. It seems that mean/average filter is able to reduce the noise and smooth image better as compared to the other filters. Due to this fact, the image (Fig. 3c) is rendered with better segmentation output.

Figure 4 shows results of segmentation algorithms applied to the images that have right angle. These results are similar to the results we got in case of left angle images as shown in Fig. 3. It seems that the type of noise handled by the mean/average filter is better as compared to the other filters. It is apparent that there are rare pixel values that are unrepresentative in nature in these images. And mean filter did not require interpolating new pixels on the edge. If this would have had happened, the edges would have been blurred leading to blurry blood vessels in appearance.

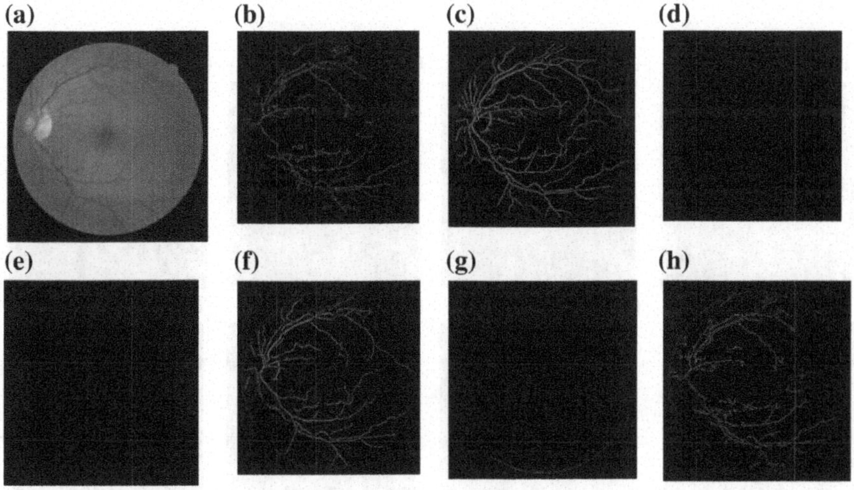


**Fig. 3** Left angle segment results. **a** Original image, **b** Gaussian, **c** ISODATA and mean, **d** Laplacian, **e** LoG, **f** Otsu and mean, **g** Prewitt, **h** Sobel

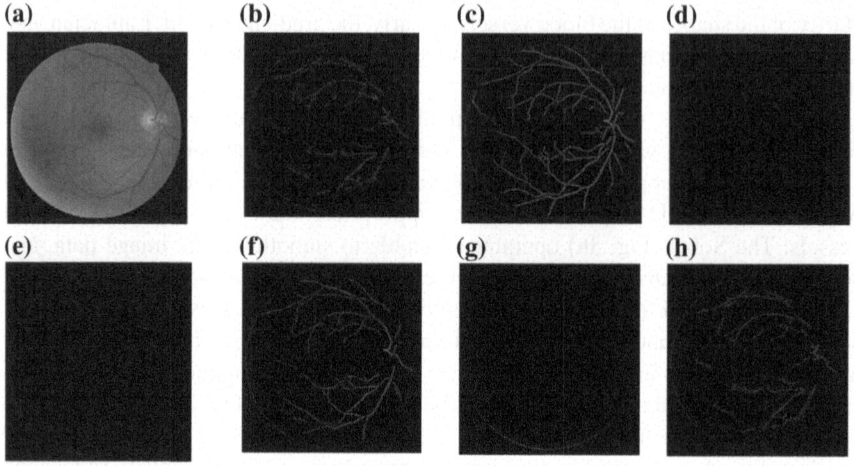


**Fig. 4** Right angle segment results. **a** Original image, **b** Gaussian, **c** mean and ISODATA, **d** Laplacian, **e** LoG, **f** Otsu and mean, **g** Prewitt, **h** Sobel

Figure 5 shows that Prewitt (Fig. 5g), LoG (Fig. 5e), and Laplacian filters (Fig. 5d) do not work in combination with any of the thresholding methods. The output of the Sobel in combination with the Otsu produces unsatisfactory results. All this can be attributed to the fact that these filters are computing gradient close to zero because of which the image loses all its bright pixels. In case of mean/average filter, which acts like a low-frequency filter, produces low-quality segmentation when it is run in combination with ISODATA. The reason is that it is overdoing interpolation to

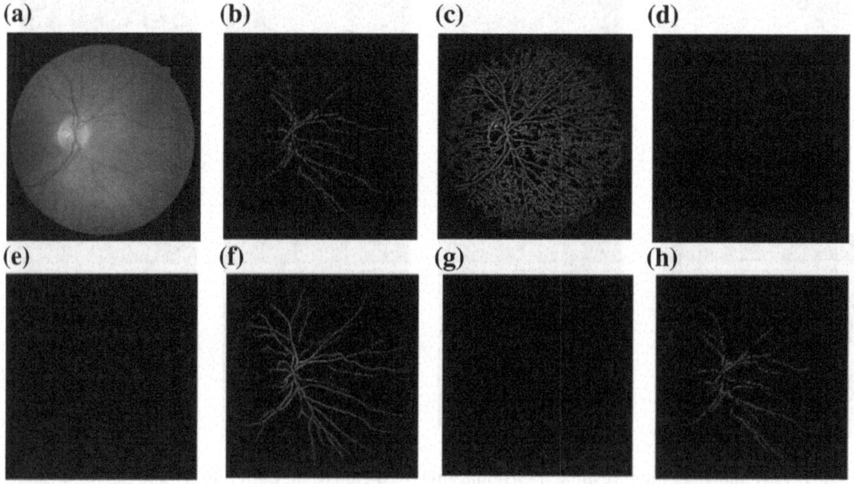


**Fig. 5** Front angle segment results. **a** Original image, **b** Gaussian, **c** mean and ISODATA, **d** Laplacian, **e** LoG, **f** Otsu and mean, **g** Prewitt, **h** Sobel

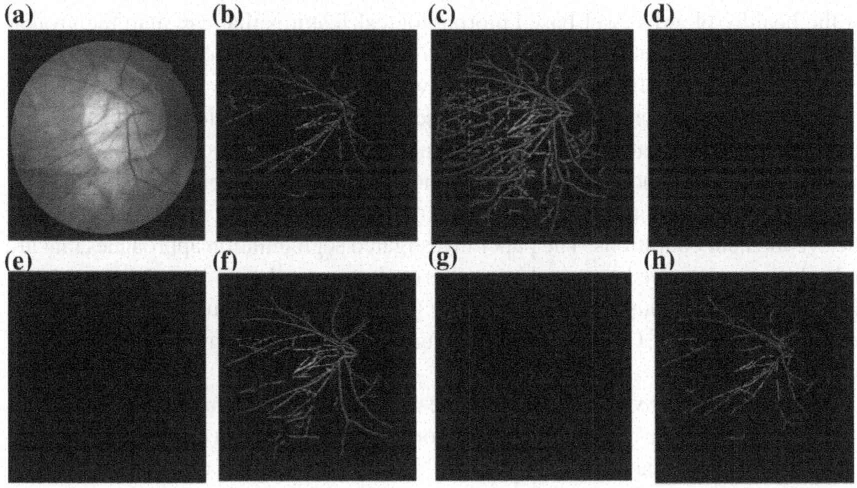


**Fig. 6** Left angle segment results. **a** Original image, **b** Gaussian and ISODATA, **c** ISODATA and mean, **d** Laplacian and ISODATA, **e** LoG and Otsu, **f** Otsu and mean, **g** Prewitt and Otsu, **h** Sobel and Otsu

compute the blood vessels and pixels get grouped that way. The combination of Gaussian filter (Fig. 5b) and ISODATA and the combination of Sobel (Fig. 5h) and Otsu give comparable output. It can be, however, noted that the best combination here is Otsu with mean filter (Fig. 5f) with size of $9 \times 9$ grid.

It can also be observed from Fig. 6c that the combination of mean filter and ISODATA oversegments the blood vessels. If we evaluate the performance of Sobel convolution, its performance in combination with ISODATA as well as with Otsu is fair in nature, but not satisfactory in nature. This can be observed in the case of the combination of Gaussian with the ISODATA and Otsu. LoG Fig. 6e, Laplacian Fig. 6d, Prewitt Fig. 6g do not work at all. It can be observed in this case that low-quality false boundaries are created leading to poor quality of segmentation in case of Gaussian plus ISODATA filters (Fig. 6b). This can be attributed to the fact that a front view of the eye has more clutter, texture, and noise. The best segmentation is produced by (Fig. 6f) Otsu global thresholding and mean filter combination.

It can be safely concluded that this evaluation shows that Otsu global thresholding algorithm is performing better as compared to the use of ISODATA when it is working along with gradient filters such as mean, Sobel, and Gaussian filters.

## 7 Conclusions and Future Scope

It is hard to differentiate between the blood vessels that are pathological in nature from the normal ones. Constant efforts and progression can be seen in this context, but a considerable amount of focus is required to find methods to validate the assumptions of the healthy blood vessel-based morphological features that can map the ground truth (normal or pathological). This work is limited to just segmentation of the blood vessels and not classification of the blood vessels.

In this research work, an explorative study has been done to identify the segmentation algorithm(s) that can accurately segment the blood vessels in the images of the eye. The purpose of the segmentation is to help in the construction of the systems that help in eye ailment diagnosis, eye disease treatment planning, and final execution of the eye medical conditions. The paper investigated segmentation approaches that are hybrid in nature as we have used image derivatives convolution kernels in combinations with the data clustering/thresholding (ISODATA and Otsu) algorithms. It was found that the use of Otsu global clustering/thresholding algorithm and mean filter works best.

The second objective was to identify the weakness of the segmentation algorithms so that fine-tuning the best-performing algorithm can be done to optimize the output. The purpose was to also find a way by which the accuracy of the best-performing algorithm can be improved to the maximum level. From this study, it is clear that a single approach to blood segmentation leads to poor results. Preprocessing is essential and combinational approach is best for developing algorithms of segmentation of blood vessels. After elaborated evaluation and observations following are the conclusions that can be drawn from this study:

1. The current literature shows a wide variety of vessel segmentation algorithms, but none of them are universally applicable to all kinds of eye medical image data. There is always a need to fine-tune, optimize, or construct new segmentation approaches for a particular set of images.
2. Advancements in medical imagery are producing a higher quality of resolution of images. This can help in developing automatic algorithms that can help to detect pathological blood vessel with the help of better segmentation and mathematical evaluation.
3. At the same time, higher computing power is also being adopted in the context of medical imaging processing. Due to these advancements, parallelization of image segmentation algorithms can be done to improve the overall efficiency of image processing.

## References


1. Pardeshi, G., Kakrani, V.: Mobile based primary health care system for rural India. Int. J. Nurs. Educ. **3**(1), 61–68 (2011)
2. Piyasena M.M.P.N., et al.: Development and validation of a diabetic retinopathy screening modality using a hand-held nonmydriatic digital retinal camera by physician graders at a tertiary-level medical clinic: protocol for a validation study. JMIR Res. Protoc. **7**(12), e10900 (2018)
3. Panwar, N., et al.: Fundus photography in the 21st century—a review of recent technological advances and their implications for worldwide healthcare. Telemed. e-Health **22**(3), 198–208 (2016)

4. Phillips, Z.F., et al.: Multi-contrast imaging and digital refocusing on a mobile microscope with a domed LED array. PLoS ONE **10**(5), e0124938 (2015)
5. Mohammadpour, M., Heidari, Z., Mirghorbani, M., Hashemi, H.: Smartphones, tele-ophthalmology, and VISION 2020. Int. J. Ophthalmol. **10**(12), 1909–1918 (2017)
6. Woodward, R., Matimba A.: Gr up SM tele-ophthalmology: opportunities for widespread access to eye care (2016)
7. Gabriele, M.L., et al.: Optical coherence tomography: history, current status, and laboratory work. Invest. Ophthalmol. Vis. Sci. **52**(5), 2425 (2011)
8. Fujimoto, J., Swanson, E.: The development, commercialization, and impact of optical coherence tomography. Invest. Ophthalmol. Vis. Sci. **57**(9), OCT1-OCT13 (2016)
9. Goodkin, M.L., Grewal, D.S., Greenfield, D.S.: Three-dimensional high-speed optical coherence tomography for diagnosis of hypotony maculopathy after glaucoma filtration surgery. J. Glaucoma **19**(6), 349–355 (2010)
10. Wang, Z., Xian, J., Man, F., Zhang, Z.: Diagnostic imaging of ophthalmology : a practical atlas. (Radiologist)
11. Litvin, T.V., Bresnick, G.H., Cuadros, J.A., Selvin, S., Kanai, K., Ozawa, G.Y.: A revised approach for the detection of sight-threatening diabetic macular Edema. JAMA Ophthalmol. **135**(1), 62 (2017)
12. Raman, R., Srinivasan, S., Virmani, S., Sivaprasad, S,. Rao, C., Rajalakshmi, R.: Fundus photograph-based deep learning algorithms in detecting diabetic retinopathy. Eye (Nov. 2018)
13. Staurenghi, G., Pellegrini, M., Invernizzi, A., Preziosa, C.: Diagnosis and detection. Dev. Ophthalmol. **58**, 39–62 (2017)
14. Saine, P.J., Tyler, M.E.: Ophthalmic Photography: Retinal Photography, Angiography, and Electronic Imaging. Butterworth-Heinemann (2002)
15. Benson, R.C., Kues, H.A.: Fluorescence properties of indocyanine green as related to angiography. Phys. Med. Biol. **23**(1), 017 (1978)
16. Ohtsubo S., Kusano, M.: Indocyanine green fluorescence properties. In: ICG Fluorescence Imaging and Navigation Surgery, pp. 9–202016. Springer Tokyo, Japan
17. Tavakoli, M., Quattrini, C., Abbott, C., Kallinikos, P., Marshall, A., Finnigan, J.: Corneal confocal microscopy a novel noninvasive test to diagnose and stratify the severity of human diabetic neuropathy. Diabetes Care **33**(8) (2010)
18. Nitoda, E., et al.: Correlation of diabetic retinopathy and corneal neuropathy using confocal microscopy. Curr. Eye Res. **37**(10), 898–906 (2012)
19. Chowdhury A.R., Banerjee, S.: Detection of cotton wool spots from retinal images using fuzzy C means. (2015)
20. Youssif, A.A.-H.A.-R., Ghalwash, A.Z., Ghoneim, A.A.S.A.-R.: Optic disc detection from normalized digital fundus images by means of a vessels' direction matched filter. IEEE Trans. Med. Imaging **27**(1), 11–18 (2008)
21. Zhou, W., Wu, C., Yi, Y., Du, W.: Automatic detection of exudates in digital color fundus images using superpixel multi-feature classification. IEEE Access **5**, 17077–17088 (2017)
22. Na'am, J., Harlan, J., Putra, I., Hardianto, R., Pratiwi, M.: An automatic ROI of the fundus photography. Int. J. Electr. Comput. Eng. **8**(6), 4545–4553 (2018)
23. Niemeijer, M., Abràmoff, M.D., van Ginneken, B.: Image structure clustering for image quality verification of color retina images in diabetic retinopathy screening. Med. Image Anal. **10**(6), 888–898 (2006)
24. Siva Sundhara Raja, D. Vasuki, S.: Automatic detection of blood vessels in retinal images for diabetic retinopathy diagnosis. Comput. Math. Methods Med. **2015**, 1–12 (2015)
25. Tufail, A., et al.: Automated Diabetic retinopathy image assessment software: diagnostic accuracy and cost-effectiveness compared with human graders. Ophthalmology **124**(3), 343–351 (2017)
26. Huang, H., Ma, H., van Triest, H.J.W., Wei, Y., Qian, W.: Automatic detection of neovascularization in retinal images using extreme learning machine. Neurocomputing **277**, 218–227 (2018)

27. Marupally, A.G., Vupparaboina, K.K., Peguda, H.K., Richhariya, A., Jana, S., Chhablani, J.: Semi-automated quantification of hard exudates in colour fundus photographs diagnosed with diabetic retinopathy. BMC Ophthalmol. **17**(1), 172 (2017)
28. Chanwimaluang T., Fan, G.: An efficient blood vessel detection algorithm for retinal images using local entropy thresholding. In: Proceedings of the International Symposium on Circuits and Systems, 2003. ISCAS'03, vol. 5, pp. V-21–V-24 (2003)
29. Sumathy, B., Poornachandra, S.: Feature extraction in retinal fundus images. In: 2013 International Conference on Information Communication and Embedded Systems, ICICES 2013 (2013)
30. Can, A.H., Shen, H., Turner, J.N., Tanenbaum, H.L., Roysam, B.: Rapid automated tracing and feature extraction from retinal fundus images using direct exploratory algorithms. IEEE Trans. Inf. Technol. Biomed. (1999)
31. Morales, S., Naranjo, V., Angulo, J., Fuertes, J.J., Alcañiz, M.: Segmentation and Analysis of Retinal Vascular Tree from Fundus Images Processing
32. Devillard, F., Heit, B.: Architectural model of a biological retina using cellular automata. J. Comput. Commun. **02**(14), 78–97 (2014)
33. Bhadauria, H.S., Bisht, S.S., Singh, A.: Vessels extraction from retinal images
34. Mookiah, M.R.K., Acharya, U.R., Chua, C.K., Lim, C.M., Ng, E.Y.K., Laude, A.: Computer-aided diagnosis of diabetic retinopathy: a review. Comput. Biol. Med. **43**(12), 2136–2155 (2013)
35. Schmidt-Erfurth, U., Sadeghipour, A., Gerendas, B.S., Waldstein, S.M., Bogunović, H.: Artificial intelligence in retina. Prog. Retin. Eye Res. **67**, 1–29 (2018)
36. Orlando, J.I., Prokofyeva, E., Blaschko, M.B.: A discriminatively trained fully connected conditional random field model for blood vessel segmentation in fundus images. IEEE Trans. Biomed. Eng. **64**(1), 16–27 (2017)
37. Fu, H., Xu, Y., Wong, D.W.K., Liu, J.: Retinal vessel segmentation via deep learning network and fully-connected conditional random fields. In: 2016 IEEE 13th International Symposium on Biomedical Imaging (ISBI), pp. 698–701 (2016)
38. Sharma, S., Maheshwari, S., Shukla, A.: An intelligible deep convolution neural network based approach for classification of diabetic retinopathy. Bio-Algorithms Med-Syst. **14**(2) (Jun. 2018)
39. Roychowdhury, S.: Classification of large-scale fundus image data sets: a cloud-computing framework
40. Kauppi T., et al.: Diaretdb1—Standard Diabetic Retinopathy Database. In: Tomi Kauppi, Valentina Kalesnykiene, Joni-Kristian Kamarainen, Lasse Lensu, Iiris Sorri, Asta Raninen, Raija Voutilainen, Juhani Pietilä, Heikki Kälviäinen, and Hannu Uusitalo. IMAGERET project, 2007
41. Fadafen, M.K., Mehrshad, N., Razavi, S.M.: Detection of diabetic retinopathy using computational model of human visual system. Biomed. Res. **29**(9) (2018)
42. Tamilarasi, M., Duraiswamy, K.: Automatic detection of microaneurysms using microstructure and wavelet methods. Sadhana—Acad. Proc. Eng. Sci. **40**(4), 1185–1203 (2015)
43. Leopold, H.A., Orchard, J., Zelek, J.S., Lakshminarayanan, V., Cheriton, D.R.: PixelBNN: augmenting the PixelCNN with batch normalization and the presentation of a fast architecture for retinal vessel segmentation
44. C. healthcare foundation EyePACS: Diabetic Retinopathy Detection | Kaggle dataset. Kaggle. Kaggle (2018)
45. Kolar, R., et al.: Retinal vessel segmentation by improved matched filtering: evaluation on a new high-resolution fundus image database. IET Image Process. **7**(4), 373–383 (2013)
46. Pratt, H., Coenen, F., Broadbent, D.M., Harding, S.P., Zheng, Y.: Convolutional neural networks for diabetic retinopathy. Procedia Comput. Sci. **90**, 200–205 (2016)
47. Oliveira, A., Pereira, S., Silva, C.A.: Augmenting data when training a CNN for retinal vessel segmentation: How to warp? In: 2017 IEEE 5th Portuguese Meeting on Bioengineering (ENBENG), pp. 1–4 (2017)
48. Liskowski, P., Krawiec, K.: Segmenting retinal blood vessels with deep neural networks. IEEE Trans. Med. Imaging **35**(11), 2369–2380 (2016)

49. Melinscak, M., Prentasic, P., Loncaric, S.: Retinal vessel segmentation using deep neural networks
50. Fu, H., Xu, Y., Lin, S., Kee Wong, D.W., Liu, J.: DeepVessel: Retinal Vessel Segmentation Via Deep Learning and Conditional Random Field, pp. 132–139. Springer, Cham (2016)
51. Mishra J., Nirmala, S.R.: Detection of cotton wool spots in retinopathy images: a review. IOSR J. VLSI Signal Process. (IOSR-JVSP) **8**(3)
52. Ganesan P., Sajiv, G.: A comprehensive study of edge detection for image processing applications. In: International Conference on Innovations in Information, Embedded and Communication Systems (ICIIECS), pp. 1–6 (2017)
53. Marr, D., Hildreth, E.: Theory of edge detection. Proc. R. Soc. London. Ser. B. Biol. Sci. **207**(1167), 187–217 (1980)
54. Hassan, S.S.A., Bong, D.B.L., Premsenthil, M.: Detection of neovascularization in diabetic retinopathy. J. Digit. Imaging **25**(3), 437–444 (2012)

# Multi-focus Image Fusion: Quantitative and Qualitative Comparative Analysis

**Shiveta Bhat and Deepika Koundal**

**Abstract** Multi-focus Image Fusion (MFIF) is a technique that combines multiple images to obtain a composite image in which all the objects are in-focus and have improved image quality. More information is stored by the focused image than that of the information stored by the source image. MFIF provides fused images which can be used for various image processing tasks like target recognition, feature extraction, and segmentation. There exists number of MFIF techniques in spatial as well as transform domain such as Stationary Wavelet Transform, Discrete Wavelet Transform, and Principal Component Analysis. In this paper, comparative analysis of various MFIF techniques which are used to fuse multi-focused images is done. Qualitative as well as quantitative evaluation has been carried out for various MFIF techniques. MFIF provides a fused image which helps for high resolution of vision. Various challenges/issues related to the existing MFIF techniques are also highlighted and will be helpful in the future.

**Keywords** Multi-focus images · SWT · DWT and PCA

## 1 Introduction

Acquiring an image within which all the objects are in-focus is difficult for sensors due to the limited depth of field (DOF) of the optical lenses [1]. The most important factor behind this is that the objects which are at a specific distance from the camera are sharp as well as focused, whereas the other objects are likely to be blurred [2]. To resolve the problem of limited DOF, a low-cost method is used that helps in obtaining an image in which all the objects are having same focal length. It is done

S. Bhat
Chitkara University School of Engineering and Technology, Chitkara University, Solan, Himachal Pradesh, India
e-mail: bhatshiveta707@gmail.com

D. Koundal (✉)
Department of Virtualization, School of Computer Science, University of Petroleum and Energy Studies, Dehradun, Uttrakhand, India
e-mail: koundal@gmail.com



P. K. Singh et al. (eds.), *Proceedings of ICRIC 2019*, Lecture Notes in Electrical Engineering 597, https://doi.org/10.1007/978-3-030-29407-6_38

by combining at least two images with dissimilar focal length [1]. The resultant image obtained is known as fused image which is having all the objects in-focus. Thus, Multi-focus Image Fusion (MFIF) is a technique which consolidates defocused input images to achieve a single-focused image having more information than the source images with enhanced image quality [3]. Image fusion (IF) extracts valuable information present in the source images and emphasizes on the helpful information without introducing any irregularity in the focused image [4]. The main objective of MFIF is to lower the redundancy in the output and provide improved geometric corrections and also enhance those features which are not visible in either of the images [5]. From the literature, it has been found that MFIF has many applications in the fields like medical image processing [6], micro-image fusion [7], visual sensor networks (VSN) [8], remote sensing [9], visual power patrol inspection [10], optical microscopy [11], and biochemical analysis [12]. Image fusion is in general executed at three different levels of information extraction namely, pixel level, feature level, and decision level [13]. Pixel level is the simplest of all which performs IF at the lowest level. It combines the values and intensities of two input images and based on its average provides a single resultant fused image [14]. In this, simple mathematical functions like mean or maximum are applied to the values of a pixel of source image so as to generate a fused image [15]. This type of fusion has certain demerits as it produces a blurring effect and smoothen the sharp edges of the fused image. Feature level is the middle level among the image fusion levels. In this type of fusion, features such as textures and edges are taken out initially from the source images and these features based fusion is performed to match certain selection criteria [16]. Decision-level image fusion techniques merge information at advanced level of abstraction. For the purpose of extracting information from source images, the source images are processed individually. In decision-level image fusion, all-in-focus image is obtained by fusing the objects of the source image employing an appropriate fusion algorithm [16].

According to literature, MFIF techniques can also be categorized as spatial domain and transform domain [15]. Spatial domain methodologies directly deal with image pixels and pixel intensity-based operations [17]. In this, the defocused source images are directly fused in the spatial domain. To achieve desirable output, manipulation of source images pixel values done in spatial domain [18]. Some examples of spatial domain include Intensity Hue Saturation (HIS), Principal Component Analysis (PCA), and Brovey Methods [3]. Spatial domain methods are accompanied with certain limitations like image blurring, spatial distortion infused image [16]. These problems can be solved using transform domain methods. In transform domain, transformation of input image into a frequency domain takes place and then Fourier transform is calculated. After obtaining the Fourier transform of the image, IF is implemented followed by performing the inverse Fourier transform in order to obtain the resultant image [19].

However, these methods are computationally very complex, and for their execution, more memory is required. Some of the transform domain methods used in MFIF algorithms are Discrete Cosine Transform (DCT) [20], Pulse-Coupled Neural

Network (PCNN) [21], and Discrete Wavelet Transform (DWT) [22]. These methods are more complex.

There exist various spatial as well as transform domain fusion techniques in literature that have been applied to multi-focused images in order to develop a single composite resultant image with all the objects having same focal plane. Some commonly used IF techniques from literature such as Wavelet Transform, DCT, Neural Network, and PCA which are applied on real multi-focused images are discussed briefly in the subsequent subsections.

Qiguang et al. developed an adaptive Pulse-Coupled Neural Network (PCNN)-based method for MFIF [23]. Texture information as well as the edges were preserved using this method. It overcame the drawback of adjusting parameters faced by traditional PCNN. Further, Li et al. presented a region-based IF method using PCNN [24]. In this method, PCNN was used to segment the source images and then the segmented parts contributed for developing the final fused image. The presented method achieved better results, i.e., less blurring even for images with misregistration or any movement in the object. It was less sensitive to noise and was computationally simple. Furthermore, Zafar et al. investigated a fusion method based on DCT for multi-exposure as well as multi-focus images [25]. This method was also applicable for images obtained from multi-exposure as well as for hardware-based in-camera execution to colored images.

Wei et al. reported a method based on Dual-Tree Complex Wavelet Transform (DT-CWT) for MFIF [26]. Fusion of low as well as high-frequency components was performed separately. Low contrast and blocking effects problem was resolved using this method. Also, ringing effect was avoided with the help of this technique. In addition, Saeedi et al. presented an approach for MFIF grounded on Double Density Dual-Tree Discrete Wavelet Transform (DD-DTDWT) with fuzzy classifier [27]. This method was shift invariant and had good frequency decomposition. This method was helpful in solving the problems like shift variance and poor directionality usually faced in DWT. This method outperformed standard wavelet transform. Afterward, Haghighat et al. reported a scheme for MFIF based on DCT [20, 28]. The variance in the DCT was calculated in order to get a real-time fusion method for MFIF. This method provided the improved quality of fused image. Also, the complexity of this method in real-time application was reduced. This method was a little bit weak on the boundaries.

Yang et al. gave a technique based on DWT for fusion of multi-focused images [29]. This method used maximum sharpness-based algorithm for coefficients of low-frequency, and for coefficients of high-frequency, maximum energy-based algorithm was used. Further, Phamila et al. reported a MFIF method based on DCT for fusing multi-focus images [30]. This method was energy efficient and extremely simple. The main drawback of DCT is that it is complex and is little bit weak on the boundaries. Furthermore, Jiang et al. developed a method for MFIF by combining SWT, fuzzy sets, and Local Spatial Frequency (LSF) so as to achieve the goal of obtaining a high-quality fused image [31]. Input images were decomposed into a sub-images sequence with the help of SWT, and the membership matrix was obtained from these sub-images using Gaussian Median Filter (GMF). This method was able to address the problem of uncertainty or fuzziness and provided better fusion results.

Zhao et al. reported an MFIF method based on Multi-Level Deeply Supervised Convolutional Neural Network as MLCNN [32]. This method was capable of capturing both low-frequency as well as high-frequency content which was helpful in fusing two or more images. This algorithm was developed with a natural enhancer which provided good fusion results even in the presence of misregistration, anisotropic blur, and non-focus areas. Immensely, misregistered source images were not properly fused. In addition, Yang et al. developed a method for MFIF by combining Robust Sparse Representation (RSR) with an adaptive Pulse-Coupled Neural Network (PCNN) to achieve better results for both registered and misregistered images [33]. This method outperformed present state-of-art techniques, but its efficiency was needed to be improved.

Farid et al. reported a method for MFIF based on Content Adaptive Blurring (CAB) [34]. This method was helpful in extending the depth of field of images. It provided better results than other state-of-art techniques. Further, Aymaz et al. developed an IF scheme based on super-resolution-based hybrid technique [35]. This method was helpful in achieving better results such as better visual perception, less distortion, and more clear edges.

This paper focuses on the comparative analysis of existing state-of-art techniques used for MFIF. This study is structured in five sections. Section 2 highlights the materials and methods used. Section 3 concludes the experimental results and discussion. Section 4 summarizes the conclusion.

## 2 Materials and Methods

### *2.1 Material*

For quantitative comparison, various evaluation metrics are used and discussed below:

**Signal to Noise Ratio (SNR) [36]**

SNR measures the ratio between noise of the fused image and the information it contains. It is a measure between the resultant and the reference image. When the fusion results are better, the value of SNR is high. The value of SNR can be calculated using Eq. (1).

$$\mathrm{SNR} = 10\log 10 \frac{\sum_{i=1}^{P}\sum_{i=1}^{Q} Rij^2}{\sum_{i=1}^{P}\sum_{j=1}^{Q}\left(R_{ij} - F_{ij}\right)^2} \tag{1}$$

where $P \times Q$ depicts image size or pixel totality of the image, ($i$)th pixel value of the reference image is denoted by $R_{ij}$, and $F_{ij}$ represents the ($j$)th pixel value of the source image.

**Root Mean Square Error (RMSE) [36]**
RMSE is defined as an evaluation metric which is helpful in measuring the changes per pixel due to image fusion techniques. RMSE is calculated using Eq. (2) as given below

$$\text{RMSE} = \sqrt{\frac{1}{P \times Q} \sum_{i=1}^{P} \sum_{j=1}^{Q} \left(R_{ij} - F_{ij}\right)^2}. \tag{2}$$

When RMSE approaches to zero, there it indicates that there is a similarity in between the resultant image and the reference image. With the decrease in similarity, the value of RMSE increases.

**Percentage Fit Error (PFE) [36]**
PFE is calculated as the norm of the variation among the corresponding pixels of the image used for reference and fused image to the norm of the reference image. PFE can be expressed by Eq. (3):

$$\text{PFE} = \frac{\left\| R_{ij} - F_{ij} \right\|}{\left\| R_{ij} \right\|} * 100 \tag{3}$$

where $\|\cdot\|$ denotes the norm operator. PFE approaches to zero when the resultant fused image and the reference image are exactly analogous to each other.

**Peak Signal to Noise Ratio (PSNR) [36]**
PSNR determines the image quality by comparing similarity among the resultant fused image and the reference image. Higher is the PSNR's value better are the fusion results. PSNR is calculated using Eq. (4) as given below:

$$\text{PSNR} = 10 \log 10 \frac{L^{\;2}}{\text{MSE}} \tag{4}$$

where the amount of gray levels is denoted using "$L$".

**Standard Deviation (SD) [36]**
SD of gray images reflects its clarity and contrast. The value of SD can be computed using Eq. (5) as given below:

$$\text{SD} = \sqrt{\frac{1}{P \times Q} \sum_{i=1}^{P} \sum_{j=1}^{Q} (F_{ij} - \text{MEAN})^2} \tag{5}$$

Higher the value of SD noisier the resultant image is. Images with smaller contrast are most likely to be affected by noise.

## 2.2 Methods

**Discrete wavelet transform (DWT)**
DWT is a signal representation which provides the frequency component of a signal at a specific instance of time [37]. It also provides the directional information in the decomposition levels of the signal. It varies for both the spatial as well as frequency range. It is implemented using lifting and filter bank approach [38]. Low Pass Filter (LPF) l[m] and High Pass Filter (HPF) h[m] are used to decompose the image and get approximation and detailed coefficients. It is decimated by a factor of two [39]. Components of low frequency in the image are preserved by the LPF l[m], and the HPF h[m] preserves the high-frequency components of the filter. LPF provides the row details, and HPF provides the column details of the image. DWT is applied in each dimension separately [38]. The single-level wavelet decomposition results in four sub-bands namely LL, LH, HL, and HH representing low-resolution approximation, horizontal, vertical, and diagonal information of the input image, respectively [40]. In DWT, downsampling takes place because of which it is shift invariant, and any movement of object in the image deteriorates the performance of the wavelet [38].

**Stationary Wavelet Transform**
SWT is translation invariant, and it solves the invariance faced in DWT [40]. It is different from DWT as it is shift invariant, and instead of down sampling, SWT took up samples process [40]. In SWT, convolution of input image is done in the same way as done in DWT, but no decimation is done to the passed signal and therefore the number of coefficients is twice than the samples in the input signal [40]. SWT is more realistic for detecting the changes in signal and pattern recognition. The size of image using SWT does not reduce and thus the information in the focused image is also preserved [41].

**Principal Component Analysis (PCA)**
PCA is a tool of mathematics that is based on linear algebra [42]. This method helps in transforming the uncorrelated variables from correlated variables, and these uncorrelated variables are termed as principal component. PCA is capable of decreasing the dimensionality, data representation, and feature extraction.

# 3 Results and Discussion

For experimental results, several frequently used benchmark/standard multi-focus image pairs are used which are publicly available online. For comparative analysis,

DWT, SWT, PCA, DWT + PCA, and SWT + PCA are implemented on a pair of multi-focused images. The subjective as well as objective evaluation is performed for these techniques. Figure 1 illustrates the visual results of felora image with different MFIF techniques.

In Fig. 1, the subfigure (a) depicts the reference image, subfigure (b) illustrates the source image 1, and (c) illustrates the source image 2. The resultant fused images obtained using DWT, SWT, DCT, PCA, DWT + PCA, and SWT + PCA fusion technique are shown in subfigure (d–i), respectively. From subfigure (d), it can be analyzed that there is a smoothening and blurring effect in the resultant images as these methods have poor directionality, whereas subfigure (f) obtained using PCA fusion technique illustrates that the resultant fused image is not sharp as the reference

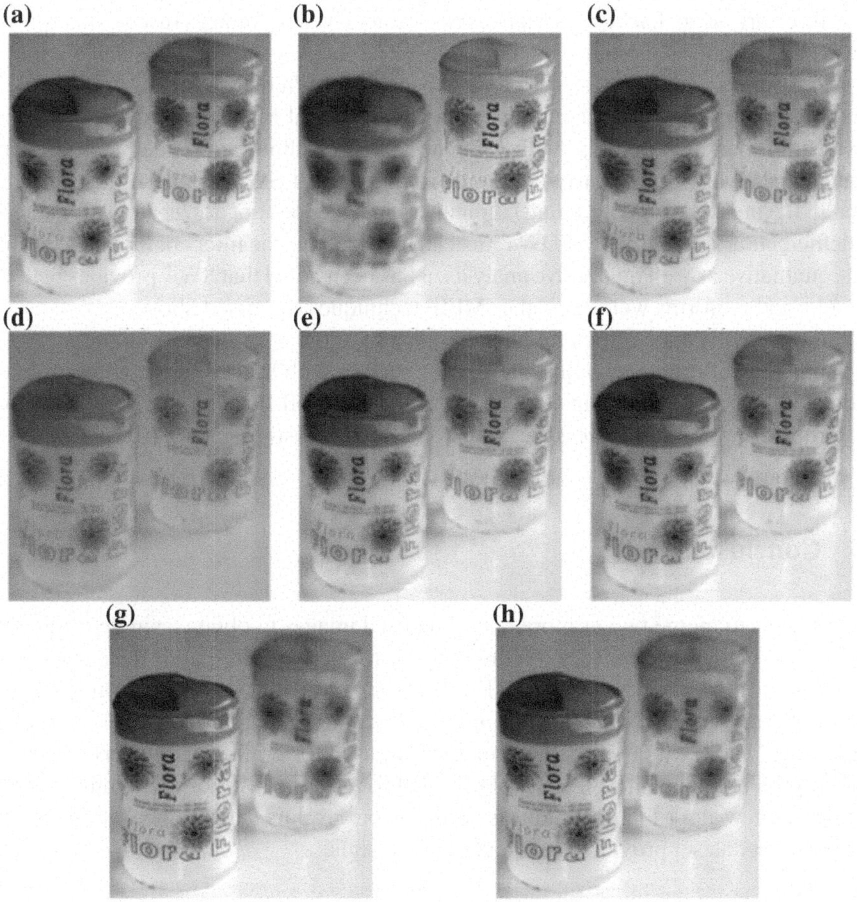

**Fig. 1** Comparison of multi-focused felora image fusion by various techniques. **a** Reference, **b** source image 1, **c** source image 2, **d** DWT, **e** SWT, **f** PCA, **g** DWT + PCA, and **h** SWT + PCA

**Table 1** Performance comparison of felora image for different fusion algorithms

| Technique | SNR | RMSE | PFE | PSNR | SD |
|---|---|---|---|---|---|
| DWT | 19.916 | 17.635 | 10.097 | 35.701 | 53.966 |
| SWT | 24.890 | 19.946 | 5.694 | 38.188 | 57.385 |
| PCA | 24.259 | 10.695 | 6.123 | 37.872 | 53.506 |
| DWT + PCA | 23.696 | 11.412 | 6.534 | 37.591 | 53.777 |
| SWT + PCA | 24.288 | 10.659 | 6.103 | 37.887 | 53.595 |

image. In subfigure (g) and (h) which are obtained using DWT + PCA and SWT + PCA, respectively, only one part of the image is clearly visible, whereas the other part is blurred because of the overshadowing of the useful data. From subfigure (e), it can be clearly seen that better visual results are achieved for images fused using SWT fusion technique as these methods are translation as well as time invariant. Various types of evaluation metrics for felora image using different MFIF techniques are listed in Table 1. For the quantitative analysis of several MFIF techniques, both the evaluation parameters, i.e., "with reference" and "without reference" metrics are calculated. It has been clearly observed from table that SWT has performed better for all the metrics as SWT is translation invariant and does not downsample the filter coefficients as done in case of DWT, instead it upsamples the filter coefficients. From the qualitative as well qualitative analysis, it has been noted that SWT provides better results and performs well than other MFIF techniques discussed above.

In the past years, various researchers have reported a lot of work in the area of MFIF using different techniques to fuse the images with limited depth of focus, but some of the issues remain unaddressed. Thus, a novel MFIF technique is required to be proposed which can address the issues faced by the existing state-of-art techniques.

## 4 Conclusion

MFIF aims to merge two or more out-of-focused images to obtain a single resultant image having better quality and more useful information than that of the defocused images. In this paper, various MFIF techniques of spatial and transform domain are compared to each other. Additionally, both the evaluation metrics, i.e., "with reference" and "without reference" metrics are used for evaluating the performance of these techniques. Thus, it is concluded that for fusion of two or more multi-focus images, a new algorithm/technique is needed to be developed that can be familiarized to different dataset providing efficient fusion results.

## References


1. Liu, Y., Liu, S., Wang, Z.: Multi-focus image fusion with dense SIFT. Inf. Fusion **23**, 139–155 (2015)
2. Tang, H., Xiao, B., Li, W., Wang, G.: Pixel convolution neural network for multi-focus image fusion. Inf. Sci. **433–434**, 125–141 (2018)
3. Kaur, G., Kaur, P.: Survey on multifocus image fusion techniques. In: International Conference on Electrical, Electronics, and Optimization Techniques (ICEEOT), 2016
4. Virk, S.R.: Review of image fusion techniques. Int. Res. J. Eng. Technol. (IRJET) 2 (2015)
5. Sharma, M.: A review: image fusion techniques and applications. Int. J. Comput. Sci. Inf. Technol. 7 (2016)
6. Wang, Z., Ma, Y.: Medical image fusion using m-PCNN. Inf. Fusion **9**, 176–185 (2008)
7. Jiang, Z.G., Han, D.B., Chen. J., Zhou, X.K., A wavelet based algorithm for multi-focus micro-image fusion. In: Proceedings of the Third International Conference on Image and Graphics (ICIG) (2004), pp. 176–179
8. Sujatha, K., Punithavathani, D.S.: Optimized ensemble decision-based multi-focus image fusion using binary genetic Grey-Wolf optimizer in camera sensor networks. Multimed. Tools Appl. **77**, 1735–1759 (2018)
9. Simone, G., Farina, A., Morabito, F.C., Serpico, S.B., Bruzzone, L.: Image fusion techniques for remote sensing applications. Inf. fusion **3**, 3–15 (2002)
10. Chen, Z, Wang, D, Gong, S., Zhao, F.: Application of multi-focus image fusion in visual power patrol inspection. In: 2nd Advanced Information Technology Electronic and Automation Control Conference (IAEAC) (2017), pp. 1688–1692
11. Song, Y., Li, M., Li, Q., Sun, L., A new wavelet based multi-focus image fusion scheme and its application on optical microscopy. In: International Conference on Robotics and Biomimetics (ROBIO) (2006) pp. 401–405
12. Plas, R.V.D., Yang, J., Spraggins, J., Caprioli, R.M.: Image fusion of mass spectrometry and microscopy: a multimodality paradigm for molecular tissue mapping. Nat. Methods **12**, 366–372 (2015)
13. Yang, Y., Zheng, W., Huang, S.: Effective multifocus image fusion based on HVS and BP neural network. The Sci. World J. **2014**, 1–10 (2014)
14. Kaur, H., Rani, E.J.: Analytical comparison of various image fusion techniques. Int. J. Adv. Res. Comput. Sci. Softw. Eng. 5 (2015)
15. Siddiqui, A.B., Rashid, M., Jaffar, M.A., Hussain, A., Mirza, A.M., Feature classification for multi-focus image fusion. Int. J. Phys. Sci. **6**, 4838–4847 (2011)
16. Garg, R., Gupta, P., Kaur, H.: Survey on multi-focus image fusion algorithms. In: Recent Advances in Engineering and Computational Sciences (RAECS) (2014), pp. 1–5
17. Nejati, M., Samavi, S., Karimi, N., Soroushmehr, S.R., Shirani, S., Roosta, I., Najarian, K.: Surface area-based focus criterion for multi-focus image fusion. Inf. Fusion **36**, 284–295 (2017)
18. Kannan, K., Perumal, A.S., Arulmozhi, K.: Optimal decomposition level of discrete, stationary and dual tree complex wavelet transform for pixel based fusion of multi-focused images. Serbian J. Electr. Eng. **7**, 81–93 (2010)
19. Sahu, D.K., Parsai, M.P.: Different image fusion techniques–a critical review. Int. J. Mod. Eng. Res. (IJMER) **2**, 4298–4301 (2012)
20. Haghighat, M.B.A., Aghagolzadeh, A., Seyedarabi, H.: Multi-focus image fusion for visual sensor networks in DCT domain. Comput. Electr. Eng. **37**, 789–797 (2011)
21. Wang, Z., Ma, Y., Gu, J.: Multi-focus image fusion using PCNN. Pattern Recogn. **43**, 2003–2016 (2010)
22. Pajares, G., Cruz, J.M.: A wavelet-based image fusion tutorial. Pattern Recogn. **37**, 1855–1872 (2004)
23. Miao, Q., Wang, B.: A novel adaptive multi-focus image fusion algorithm based on PCNN and sharpness. In: Sensors, and Command, Control, Communications, and Intelligence (C3I) Technologies for Homeland Security and Homeland Defense (2005), pp. 704–713

24. Li, M., Cai, W., Tan, Z.: A region-based multi-sensor image fusion scheme using pulse-coupled neural network. Pattern Recogn. Lett. **27**, 1948–1956 (2006)
25. Zafar, I., Edirisinghe, E.A., Bez, H.E.: Multi-exposure & multi-focus image fusion in transform domain. In: IET International Conference on Visual Information Engineering (2006), pp. 606–611
26. Wei, S., Ke, W.: A multi-focus image fusion algorithm with DT-CWT. In: International Conference on Computational Intelligence and Security (2007), pp. 147–151
27. Saeedi, J., Faez, K., Mozaffari, S.: Multi-focus image fusion based on fuzzy and wavelet transform. In: Progress in Pattern Recognition, Image Analysis, Computer Vision, and Applications, vol. 5856, pp. 970–977 (2009)
28. Haghighat, M.B.A., Aghagolzadeh, A., Seyedarabi, H.: Real-time fusion of multi-focus images for visual sensor networks. In: 6th Iranian Conference on Machine Vision and Image Processing (2010), pp. 1–6
29. Yang, Y.: A novel DWT based multi-focus image fusion method. In: International Conference on Advances in Engineering, Netherlands, pp. 177–181 (2011)
30. Phamila, Y.A., Amutha, R.: Discrete cosine transform based fusion of multi-focus images for visual sensor networks. Sig. Process. **95**, 161–170 (2014)
31. Jiang, Q., Jin, X., Lee, S.J., Yao, S.: A novel multi-focus image fusion method based on stationary wavelet transform and local features of fuzzy sets. IEEE Access **5**, 20286–20302 (2017)
32. Zhao, W., Wang, D., Lu, H.: Multi-focus image fusion with a natural enhancement via joint multi-level deeply supervised convolutional neural network. IEEE Trans. Circ. Syst. Videos Technol. (online available) (2018, in press)
33. Yang, Y., Yang, M., Huang, S., Ding, M., Sun, J.: Robust sparse representation combined with adaptive PCNN for multifocus image fusion. IEEE Access **6**, 20138–201351 (2018)
34. Farid, M.S., Mahmood, A., Al-Maadeed, S.A.: Multi-focus image fusion using content adaptive blurring. Inf. Fusion **45**, 96–112 (2019)
35. Aymaz, S., Kose, C.: A novel image decomposition-based hybrid technique with super-resolution method for multi-focus image fusion. Inf. Fusion **45**, 113–127 (2019)
36. Balasubramaniam, P., Ananthi, V.P.: Image fusion using intuitionistic fuzzy sets. Inf. fusion **20**, 21–30 (2014)
37. Swathi, N., Bindu, E., Naidu, V.P.: Pixel level image fusion using fuzzylet fusion algorithm. Int. J. Adv. Res. Electr., Electr. Instrum. Eng. pp. 261–269 (2013)
38. Nazir, A., Ashraf, R., Hamdani, T., Ali, N.: Content based image retrieval system by using HSV color histogram, discrete wavelet transform and edge histogram descriptor. In: International Conference on Computing, Mathematics and Engineering Technologies (iCoMET), pp. 1–6 (2018)
39. Qayyum, H., Majid, M., Anwar, S.M., Khan, B.: Facial expression recognition using stationary wavelet transform features. Math. Prob. Eng. (2017)
40. Singh, D., Garg, D., Singh Pannu, H.: Efficient landsat image fusion using fuzzy and stationary discrete wavelet transform. The Imaging Sci. J. **17**, 108–114 (2017)
41. Jin, X., Jiang, Q., Yao, S., Zhou, D., Nie, R., Lee, S.J., He, K.: Infrared and visual image fusion method based on discrete cosine transform and local spatial frequency in discrete stationary wavelet transform domain. Infrared Phys. Technol. **88**, 1–2 (2018)
42. Helonde, M.R., Joshi, M.R.: Image fusion based on medical images using DWT and PCA methods. Int. J. Comput. Tech. **2**, 75–79 (2015)

# Computer-Assisted Diagnosis of Thyroid Cancer Using Medical Images: A Survey

**Vatsala Anand and Deepika Koundal**

**Abstract** Thyroid cancer is the common cancer which can be found mostly in women as compared to men around the world. Thyroid gland is a butterfly-shaped gland that is located around the voice box. Earlier, doctors used to evaluate thyroid cancer manually, but now they are using computer-aided diagnosis (CAD) system for automatic detection. As incidence rate of thyroid cancer is increasing day by day, therefore, a better technology is required for its earlier detection. There are different types of imaging modalities, such as magnetic resonance imaging (MRI), ultrasound (US), and computerized tomography (CT), which are utilized for early detection of diseases. This paper presents and discusses the major trends for an exhaustive overview of thyroid nodule detection, segmentation, classification, and feature extraction techniques. The approaches used in CAD are summarized with their advantages and disadvantages.

**Keywords** Thyroid · Cancer · Computer-aided diagnosis · Classification · Segmentation · Medical images

## 1 Introduction

Cells are basic building blocks of human body that are used for building various tissues as human body is formed of various cells. Sometimes these old cells do not expire, and meanwhile, newer cells grow in the body which go out of management and become lump or cancer. These cancerous cells can be formed at any part of the body and can spread by moving into nearby normal tissues [1]. Abnormal cancerous cells can be categorized as benign or malignant. Benign cancer may grow but does

V. Anand
Chitkara University Institute of Engineering and Technology, Chitkara University, Rajpura, Punjab, India
e-mail: vatsala.anand@chitkara.edu.in

D. Koundal (✉)
Department of Virtualization, School of Computer Science, University of Petroleum and Energy Studies, Dehradun, Uttrakhand, India
e-mail: koundal@gmail.com



P. K. Singh et al. (eds.), *Proceedings of ICRIC 2019*, Lecture Notes in Electrical Engineering 597, https://doi.org/10.1007/978-3-030-29407-6_39

not spread to different tissues of body, whereas malignant cancer can spread to other tissues also. There are different types of cancers that are found in human body such as kidney, lung, cervical, skin, brain, liver, thyroid, bladder, ovarian, and breast. Among all cancers, the general type of cancer that is growing day by day is thyroid cancer [2]. Thyroid is a butterfly-shaped endocrine gland that is found within neck, close to larynx (voice box). These are commonly found in adults. When abnormal cells start growing in thyroid gland, then lumps are formed which lead to cancer which is known as thyroid cancer. In the year 2018, an estimation of 53,990 adults were diagnosed with thyroid cancer. It is a common type of cancer found in women in the range of age 20–34 years [3]. Also, about 2% of thyroid cases occur in teens and children. An increase in rate of 4% of cancer was observed from year 2005 to 2014. The nationwide-calculated new cases of thyroid cancer are 56,870, and the death estimation is 2,060 in the year 2017. The incidence of all new cancer cases is 3.4%, and the cancer death is 0.3% [4]. The rate of latest cases of thyroid cancer was 14.2%, and the number of deaths numbering was 0.5% per 100,000 men and women per year. In 2014, there were an estimated 726,646 people with thyroid cancer living in the USA. Nearly, three out of four cases are found in women. Percentage of thyroid cancer occurring in children is about 2%. Therefore, early detection of cancer is required which helps in accurate diagnosis. There are different types of modalities used for diagnosis of cancer such as magnetic resonance imaging (MRI), computed tomography (CT), and ultrasound (US) images.

US images are most widely employed because they are portable, suitable, safer, and low cost. CT scanning is trouble-free, accurate, and non-invasive as it provides details of various tissues. The focus of this survey is to summarize various methods for thyroid cancer. Usually, the computer-aided design (CAD) system for thyroid cancer diagnosis includes four stages as shown in Fig. 1:

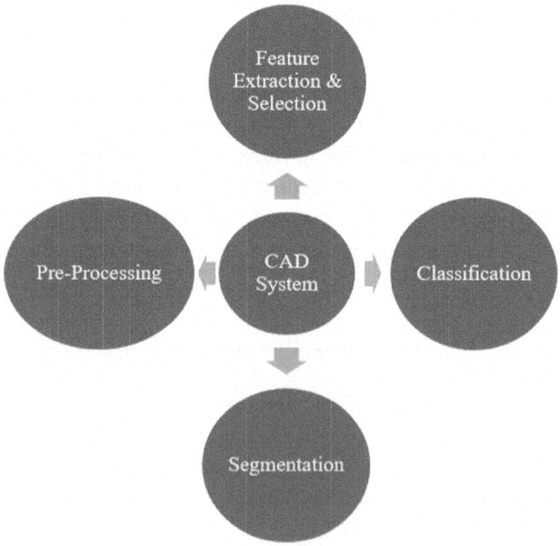


**Fig. 1** Block diagram of CAD system

1. Image pre-processing: It is used for enhancing the contrast of object (nodule) and suppressing the undesired distortions as well as to reduce the noise. After pre-processing, these images are used to segment the region of nodule.
2. Image segmentation: Segmentation is the procedure of dividing the regions of image and to extract the objects from background. In this, regions of thyroid gland that consist of nodules are segmented so that it can provide information if the nodule is benign or malignant.
3. Feature extraction and selection: In this stage, features from images are extracted and then a set of meaningful descriptors is generated. A feature set of tissue is used to categorize benign and malignant nodules.
4. Classification: Various classifiers are used to classify malignant or benign nodules using different classification strategies.

Rest of the paper is structured as follows. Section 2 describes pre-processing, and segmentation is presented in Sect. 3; classification is given in Sects. 4 and 5 shows discussion, and Sect. 6 shows conclusion.

## 2 Pre-processing

Pre-processing of medical images consists of image enhancement and noise reduction. Speckle noise is mostly found in medical images and is known as multiplicative noise which is generated by different number of scatterers. Generally, images show fuzziness that include vague tumor boundaries and also show low contrast between cautious area and neighboring tissues [5]. Therefore, surroundings of cancer in images are difficult to analyze due to speckle noise. In order to remove noise, it is important to enhance the quality of an image. Table 1 shows different pre-processing approaches.

In literature, various authors have reported several denoising and enhancement techniques for thyroid medical images.

Babu et al. introduced a method for removal of the noise and which preserved the structure of image. It was used to conquer the issue of estimation of variance of noise and signal. Furthermore, filtering algorithm based on fuzzy logic was given. It was analyzed that denoising techniques which were used previously had treated only suppression of noise. To differentiate between noise and edges, this method has to select tuning parameter that varies according to nature and amount of noise in image [6]. Kim et al. introduced speckle reduction method using the patch recurrence (SRPR) method, and comparison was done with different approaches of speckle reduction like Lee filter, Frost filter, Optimized Block-wised Non-Local Mean (OBNLM), and Non-linear Coherent Diffusion (NCD). These techniques had explored for small patches and removed speckle noise using minimum mean square error (MMSE) estimator. This method also improved the accuracy of image. They had not assisted computer-aided diagnosis, and also the accuracy of image can be improved [8]. In 2015, Koundal et al. introduced Nakagami-based noise removal method in which speckle suppression is performed by minimization problem and was solved using Lagrange and

**Table 1** Various pre-processing approaches for thyroid imaging

| References | Technique | Dataset | Results with metrics |
|---|---|---|---|
| [6] | Adaptive fuzzy logic approach, weighted averaging filter | US images | SSIM = 0.6041<br>USDAI = 1.3914 |
| [7] | Adaptive binary morphological operation | Natural, US simulated images | USDAI = 1.2967<br>SSIM = 0.5945 |
| [8] | SRPR | US images | Normal_CNR = 13.99<br>Cyst_CNR = 1.11<br>Nodule_CNR = 6.24 |
| [9] | SRR | US images | Acc = 92%<br>Sensitivity = 98.14<br>PRI > 0.83<br>Boundary error < 1 mm |
| [10] | FESR | 30 simulated images | SSNR = 23.36 ± 2.55<br>CNR = 1.70 ± 0.15<br>MSSIM = 0.57 ± 0.01 |
| [11] | Nakagami-based noise removal | 71 images (synthetic = 21, US images = 50) | UQI = 0.7440<br>FSIM = 0.8778<br>EPI = 0.8505<br>MSSIM = 0.7897<br>VIF = 0.3667<br>Time(s) = 3.10 |
| [12] | Variational model based on gamma distribution | 71 images (synthetic images = 21 and US images = 50) | UQI = 0.8606<br>FSIM = 0.8790<br>EPI = 0.8718<br>MSSIM = 0.8099<br>VIF = 0.3565 |
| [13] | Enhancement method | 15 simulated and US images | CNR = 421.80 for simulated images<br>CNR = 573.65 for US images |
| [14] | Interscale wavelet analysis method | 63 US images | Phantom image:<br>SI = 21%<br>S/mse = 17.9/18.3<br>US image:<br>SI = 23%<br>S/mse = 15.4/16.3 |
| [16] | Convex variational model | Synthetic and real US images | SNR = 17.21 dB<br>ReErr = 0.039 |
| [17] | Bayes shrink techniques | 2 synthetic and 30 US images | MSD = 69.10<br>NSD = 42.78 |
| [18] | AWMTVR | Synthetic images | FOM = 0.9599<br>SNR = 74.2363 dB<br>NMSE = 0.1141<br>MSSIM = 0.9177<br>FSIM = 0.9892<br>Q-index = 0.4903 |

(continued)

**Table 1** (continued)

| References | Technique | Dataset | Results with metrics |
|---|---|---|---|
| [19] | Bilateral filtering | US images | MSE = 53.19<br>SNR = 19.92 dB<br>PSNR = 30.87 dB<br>AD = 0.0221<br>SI = 1.32e-5 |
| [20] | PLVQNN | CT images | Acc = 98.34%<br>Sensitivity = 88.43%<br>Specificity = 99.07% |

*Acc* Accuracy
*AWMTVR* Adaptive weighted multiplicative total variation regularization
*CNR* Contrast-to-noise ratio
*DSC* Dice coefficient
*EPI* Edge preservation index
*FOM* Figure of merit
*FESR* Feature enhanced speckle reduction
*FSIM* Feature similarity
*LRSG* Locally regularized Savitzky–Golay filter
*MSD* Mean square difference
*MAE* Mean absolute error
*MSSIM* Multiscale structural similarity index matrix
*MSE* Mean squared error
*NMV* Noise mean value
*NMSE* Normalized mean squared error
*PPV* Positive predictive value
*PRI* Probabilistic rand index
*PFOM* Pratt's figure of merit
*PLVQNN* Progressive learning vector-quantization neural network
*ReErr* Relative error
*RMSE* Root mean square error
*SRR* Similar reflective region
*SNR* Signal-to-noise ratio
*SI* Speckle index
*PSNR* Peak signal-to-noise ratio
*SRPR* Speckle reduction method using the patch recurrence
*SSIM* Similarity quality measurement
*S/mse* Signal-to-mean square error ratio
*SSNR* Speckle signal-to-noise ratio
*SRR* Similarly reflective regions
*TPR* True-positive rate
*USDAI* Ultrasound despeckling assessment index
*UQI* Universal quality index
*VIF* Visual information fidelity

Chambolle's projection methods. Quantitative results indicated that the Nakagami total variation (NTV) approach achieved more suppression of speckle noise in homogeneous areas. The evaluation revealed that the process was fairly vigorous for visual improvement of real ultrasound images for additional actions like object recognition and image segmentation. However, author has not employed results on additional metrics [11]. Huang et al. have given convex variational approach for decreasing speckle noise in US images [16]. The proposed method was incorporated with Bregman iterative and variable splitting method. The optimization problem was solved by the integration of Bregman iterative method and variable splitting method for edge preservation and speckle reduction. However, staircase effect occurs in total variation method [15]. Another total variation-based model was presented by Hacini et al. that combined a multiplicative regularization function and a total variation formulation-based data misfit function. However, this method only used the conjugate gradient method for updating the approximate solution [18]. Keerthivasan et al. invented a fuzzy speckle reduction method based on directional differences and histogram in which noisy pixels were categorized on the basis of magnitude of noise in thyroid ultrasound images. However, the method did not preserve the subtle details as classification of pixels is sensitive to noise [21]. A speckle reducing anisotropic diffusion (SRAD) approach was presented by Yu et al. that involved a coefficient of variation which was dependent on noise. However, this method had used only Lee and Frost filters [22]. Zhao et al. introduced a method based on normalized cut which combined anisotropic diffusion model and fraction differential [23]. This had improved the accuracy of segmentation results and also reduced large extent of computation in designing the weight matrix of the modified normalized cut. With this method, speckle noise was reduced, and important edges were preserved in the ultrasound images. However, optimization of multiple controlling parameters was difficult which affected the versatility of method. The method was not able to retain certain features like small cysts in ultrasound images. Jaganath et al. gave despeckling algorithm based on adaptive weighted average on density current equations. The denoising filter changed the values of threshold adaptively which depend on the magnitude of noise and nature of image. The method reduced complexity of denoising procedure that was based on single-scale image denoising [24]. Guo et al. presented a method by using pulse inversion (PI). In this, interlaced iterative frequency compounding (IIFC) method is introduced. This method had given good performance of speckle suppression. It enhanced frame rate of frequency compounded speckle noise removal approach. It maintains high frame rate. However, more metrics can be used for evaluation [25]. Malutan et al. reported ultrasound image denoising and estimated the unknown mixing matrix inverse and after that shrinkage operator was applied to determine independent components. However, the results of this matrix can be enhanced by parameters adjustment of independent component analysis (ICA) algorithm [26]. In this, various techniques for denoising of images have been analyzed by using different types of modalities like CT, MRI, US, etc. Moreover, these drawbacks and strengths are also discussed.

## 3 Segmentation

Image segmentation is second important step in image processing after pre-processing. For developing a successful CAD system, many issues need to be resolved such as automatic Region of Interest (ROI) generation and the diagnosis of nodules. Accuracy of segmentation is an important factor for the extraction of various types of features. For further classification, different methods of segmentation vary widely as those depend on specialized applications and image modalities as shown in Table 2.

In literature survey, various segmentation methods are mentioned which are discussed as follows. Savelonas et al. presented a level set method named as Variable Background Active Contour (VBAC) [27]. The method utilized the background regions in order to decrease consequence of intensity inhomogeneity, speckle noise, calcifications, and tissue textures presented in the thyroid ultrasound images. The VBAC method was able to find multiple nodules, and they could be used without pre-processing. It offered greater accuracy, edge independency, and topological variations in comparison to active contour without edges (ACWE) model. The difficulty in VBAC method was that it was device dependent (images acquired from the same ultrasound machine using different settings or from different ultrasound imaging devices) and required a set of different parameters with optimum values. The drawback of this was that the work had included the object's incorporation in homogeneity and also the textural features embedment that supervises contour evolution. The further improvement on VBAC was introduced by Maroulis et al. who considered the information from sparse background regions and handled the intensity inhomogeneity in thyroid ultrasound images more effectively [28]. This model has the capability of multiple nodule delineation and was robust to noise. However, its application is limited to delineate hyper- or hypo-echoic nodules. To conquer drawbacks of VBAC, Iakovidis et al. developed genetic algorithm (GA) which was incorporated with VBAC for thyroid segmentation in ultrasound images by automatic parameter tuning [31]. However, the GA-VBAC framework required time for training phase with multiple ultrasound images. The drawback of this method was that they had not enhanced framework by training phase speed that contributed to possibility of training with numerous US images. Edge detection method that was wavelet-based was also presented by Tsantis et al. for delineation of thyroid nodules in US images [43]. The technique was positioned on the multiscale edge representation for boundary detection via thresholding modulus-maxima (M-M) chains across scales. The technique smoothens the image at various scales and detected the sharp variation points on each smoothened image via the modulus maxima. The M-M points corresponding to the same edge were grouped to form one-dimensional chains. The chains those propagated to coarser scales were considered as contours, and those did not propagate were related to noise or texture which was suppressed. The major limitations of this method are that it was not able to preserve the texture and finer details as these are suppressed along with noise. In 2006, the Hybrid Multiscale Model (HMM) [28] was presented which combined the edge detection method using wavelet and Hough transform for nodule segmentation in spite of their echogenicity.

**Table 2** Various segmentation approaches for thyroid image analysis

| References | Technique | Dataset | Results with metrics |
|---|---|---|---|
| [27] | VBAC | US images | OV = 88.8% |
| [28] | HMM | 40 US images | Roundness = 0.77<br>Nodular area = 0.89<br>Concavity = 0.75 MAD = 90.77% Acc = 91.83% |
| [29] | JET | 20 ultrasound thyroid images (benign = 10, malignant = 10) | TP = 85%, SI = 79% |
| [30] | PSO, RBF | US images | Acc = 96.76% NPV = 97.47%<br>Sensitivity = 89.73%<br>PPV = 94.88%<br>Specificity = 98.59% |
| [31] | GA-VBAC | 71 longitudinal in vivo digital images | OV = 92.5% |
| [32] | Neutrosophic clustering | 80 texture images and medical images | TPR = 98.86% OV = 96.10%<br>MAD = 75% ME = 0.010% |
| [33] | TBD, LBP, *k*-NN | 39 thyroid US images | Sensitivity = 78% Acc = 82%<br>Specificity = 81% |
| [34] | SNDRLS | US images | TP = 95.4% FP = 7.3<br>OV = 93.1% DC = 94.2%<br>MAD = 1.8 pixels HD = 0.7 pixels |
| [35] | VBAC | US images | OV = 94 ± 0.01% |
| [36] | Radon-based | US images | Acc = 89.4% |
| [37] | JET | 74 thyroid US images | OV = 92.9 ± 3.9% DC = 96.3 ± 2.1% |
| [38] | TND | US thyroid images and videos | Specificity = 62%<br>Sensitivity = 62% |
| [39] | SVM | US images | Specificity = 86.6%<br>Sensitivity = 93.8%<br>Maximum geometric mean = 90.1% |

(continued)

**Table 2** (continued)

| References | Technique | Dataset | Results with metrics |
|---|---|---|---|
| [40] | SVM, decision tree | US images | Specificity = 98.5% Acc = 97.5% PPV = 89.1% Sensitivity = 89.6% NPV = 98.3% |
| [41] | Region growing | US images | NPV = 98.04% Acc = 96.52% PPV = 89.14% Specificity = 97.61% Sensitivity = 91.58% |
| [42] | CNN | US images | FP = 0.0669 ± 0.0032% DC = 0.9224 ± 0.0027% TP = 0.915 ± 0.0077% HD = 0.6228 ± 0.14145 |

*Acc* Accuracy
*CNN* Convolutional neural network
*DC* Dice coefficient
*FP* False positive
*FN* False negative
*GA-VBAC* Genetic algorithms variable background active contour
*HMM* Hybrid multiscale model
*HD* Hausdorff distance
*JET* Joint echogenicity texture
*k-NN* *k*-nearest neighbor
*LBP* Local binary patterns
*MSE* Mean square error
*ME* Misclassification error
*MAD* Mean absolute distance
*NPV* Negative predictive value
*OV* Overlap value metric
*PSNR* Peak signal-to-noise ratio
*PPV* Positive predictive value
*PSO* Particle swarm optimization
*RBF* Radial basis function
*SNDRLS* Spatial neutrosophic distance regularized level set
*SI* Similarity index
*SVM* Support vector machine
*SNR* Signal-to-noise ratio
*TND* Thyroid nodule detector
*TBD* Thyroid boundaries detection
*TN* True negative
*TPR* True-positive ratio
*TP* True positive
*VBAC* Variable background active contour

Hence, this is not capable of segmentation of malignant and benign nodules. Also, this method is very sensitive to noise. In contrast, region-based active contour did not require previous knowledge about the shapes and boundaries of the nodule and lead to noise-tolerant image segmentation. Keramidis et al. had given another algorithm named as Thyroid Boundaries Detection (TBD) for thyroid boundary detection of thyroid boundaries. It uses linear binary patterns (LBPs) for feature extraction and $k$-NN classifier for boundaries detection [33]. The detection of nodules was done by textural feature extraction. Histograms of LBP encode the textural characteristics of the thyroid tissue. This scheme had increased the accuracy of nodule detection by providing training to the classifier. Savelonas et al. presented a joint echogenicity texture (JET) model that checked LBP distributions and intensity of image [37]. This model integrated the improvements of VBAC for the delineation of hypo-echoic and iso-echoic nodules. A drawback of this model was that it was not able to categorize certain structures like real nodules and blood vessels. It was a semi-automatic method that required the input as a rough ROI within the thyroid gland. Kollorz et al. presented the segmentation method using power watersheds for thyroid nodules segmentation in ultrasound image [44]. It was a semi-automatic method which is only able to segment benign nodules and suffered from over segmentation. In 2005, Maroulis et al. presented CAD detection approach of thyroid nodules in US images [45]. This technique achieved detailed delineation of nodules and had shown faster convergence and improved accuracy. The drawback of this method was that it does not comprise textural features embedment by detecting the non-hypo-echoic nodules to supervise contour evolution. In 2014, Gireesha et al. have reported thyroid nodule segmentation in which watershed algorithm was used. The drawback of this was that they had not used three-dimensional sonographies [46]. Moreover, the performance of CAD system can also be enhanced by adding different features. Ganesh et al. presented an automated detection of nodular boundaries of thyroid from US [47]. It included contrast enhancement, smoothing, and segmentation as sequential stages. It was not successful in detecting boundaries of US image. A speckle reduction filter was required so as to improve the diagnostic examination. This method had shown efficient delineation of thyroid nodules. The drawback of this is that identification of different nodules is not done. Koundal et al. proposed CAD system for segmentation and speckle noise reduction of thyroid US images. This method consists of three features, such as speckle removal, automatic Region of Interest (ROI) generation, and automatic segmentation of nodule. This is a general method and also can be applied to other US images [48].

## 4 Classification

After pre-processing and segmentation, features are extracted to classify images into benign/malignant. The Thyroid Image Reporting and Data System (TIRADS) is standardized US characterization data system of thyroid nodules for clinical management. Table 3 shows classification approach for thyroid image analysis.

**Table 3** Classification approach for thyroid image analysis

| References | Technique | Dataset | Results with metrics |
|---|---|---|---|
| [49] | SVM with GA feature selection PNN with GA | First dataset = 215, second dataset = 7200 | PNN with GA: Acc = 96.8%, SVM with GA: Acc = 99.05% |
| [50] | ANN | 187 instances (137 for training and 50 for testing purposes) | BPA = 92, RBF = 80 LVQ = 98 |
| [51] | ESTDD | – | Acc = 95.33% |
| [52] | AIRS | – | AIRS = 94.82% IG-AIRS = 95.90% |
| [53] | 3NN-Par, FED IC plain, Bayes | – | 3NN-Par = 94.20% FED IC plain = 96.10%, Bayes = 94.80% |
| [54] | SVM, k-NN, Bayesian | 13 thyroid images | SVM: Acc = 84.62%, TP = 8, TN = 3, FN = 2, FP = 0, Specificity = 1, Sensitivity = 80%, PPV = 1, NPV = 60%, GM = 1.34 |
| [55] | Feed-forward network | 497 images | Acc = 68%, TP = 294 TN = 44%, FN = 129 FP = 30%, Specificity = 59.5%, Sensitivity = 69.5%, PPV = 90.7%, NPV = 25.4% |
| [56] | ACWE, GLCM | 39 US images | Acc = 89.74% Specificity = 91.67% Sensitivity = 88.89% PPV = 96.0%, NPV = 78.57% |
| [57] | Decision-tree algorithm SVM | US images | NPV = 98.3%, Acc = 97.5% Specificity = 98.5% Sensitivity = 89.6% PPV = 89.1% |
| [58] | SVM, ELM | 50 thyroid images (30 = nodule, 20 = normal images) | SVM = 84.78% ELM = 93.56% |
| [59] | Decision-tree algorithm, SVM, ACM | – | Sensitivity = 93.6% Specificity = 98.8% Acc = 97.7%, NPV = 98.2%, PPV = 93.7% For ACM Sensitivity = 67.3% Specificity = 89.1% Acc = 86.5%, NPV = 94.6%, PPV = 53.4% |

(continued)

**Table 3** (continued)

| References | Technique | Dataset | Results with metrics |
|---|---|---|---|
| [60] | Feed-forward neural network | 10 US images | Sensitivity = 89.06% Specificity = 98.90% Acc = 96.51%, FP = 1.09%, FN = 10.93% |
| [46] | ANN, SVM | 40 thyroid images (30 benign nodules and 10 Malignant nodules). 28 textural features | TP = 29%, TN = 8%, FP = 2%, FN = 1, PPV = 93.54%, NPV = 88.88% Sensitivity = 96.66% Specificity = 80.00% Acc = 92.50% |
| [40] | Decision-tree algorithm | US images | Acc = 97.7% Sensitivity = 93.6% Specificity = 98.8% PPV = 93.7% NPV = 98.2% |
| [61] | LDA, SVM | – | 0.4559 ± 0.1200% |
| [62] | Fuzzy C-means Hard C-means | – | FCM = 83.7% HCM = 78.1% |

*ANN* Artificial neural network
*Acc* Accuracy
*AIRS* Artificial immune recognition system
*ACM* Active contour model
*BPA* Back propagation algorithm
*ELM* Extreme learning machine
*ESTDD* Expert system for thyroid disease diagnosis
*FN* False negative
*FCM* Fuzzy C-means
*FP* False positive
*GA* Genetic algorithm
*TN* True negative
*HCM* Hard C-means
*k-NN* *k*-nearest neighbor
*LDA* Linear discriminant analysis
*LVQ* Learning vector quantization
*NPV* Negative predictive value
*PNN* Probabilistic neural network
*PPV* Positive predictive value
*RBF* Radial basis function networks
*SVM* Support vector machine

In literature survey, various authors have reported classification techniques and also analyzed different aspects. Chang et al. presented a support vector machine (SVM) technique which was used to identify components in nodular region. Moreover, the proposed method can be applied to classify more components of nodules. However, proposed method can also persist to derive a delicate probe which will be essential to valid the process [59]. Chang et al. discussed an automatic process for component classification [57]. For identification of nodular region components, support vector machine (SVM)-based hierarchical classification method was used. The method reduced the time which was necessary for radiologists to analyze nodules and manually delineate them and achieve higher accuracy.

## 5 Discussion

In this survey, a study for CAD systems for thyroid images is analyzed with their pros and cons. These consist of pre-processing, segmentation, feature extraction, and classification. The presence of fuzziness and uncertainty in medical images generally provides varying results. Segmentation of thyroid nodules with high accuracy and their classification is still one of the major challenges in thyroid medical images due to lack of robustness of various methods applied to different image datasets. Application of different methods to different image datasets or even to the same image dataset often leads to interobserver or intraobserver variability. Nowadays, more attention is paid towards 3-D US images. This 3-D can provide all-inclusive information of thyroid nodule as compared to 2-D characteristics. The advantages of 3-D US images are understandable in CAD system as it is better for processing more amounts of data in short time that reduced the variability of observations and workload of radiologists. Moreover, researchers can also extend the work toward Doppler US images.

There are different factors on which the image analysis of thyroid depends. As researchers use different image datasets for analyzing thyroid nodules, they only examine them locally even though the datasets are not annotated appropriately. However, a global dataset is required for analyzing images.

Besides this, different performance metrics employed on same or different types of medical images leads to different results. In order to overcome these interobserver/intraobserver variations in the thyroid medical image analysis, a standard benchmark is needed for experimental results and analysis. Moreover, a robust CAD system is required to provide better accuracy and efficiency. It will improve the performance and will help to differentiate between benign and malignant nodules accurately. This survey will be helpful in providing feedback to the endocrinologists. Further, it would involve the volume estimation of thyroid from segmented thyroid region based on malignancy risk factor and classification of thyroid nodules in medical images.

## 6 Conclusion

In this survey, an attempt has been made to review various image modalities. Different techniques for denoising of images have been analyzed by using different types of modalities like CT, MRI, US, etc. Moreover, their drawbacks and strengths are also discussed. Image segmentation is second important step in image processing after pre-processing. Accuracy of segmentation is an important factor for the extraction of various types of features. For further classification, different methods of segmentation vary widely as those depend on specialized applications and image modalities as described in literature. In future, researchers are working on different images for good results.

**Conflict of Interest** There is no conflict of interest.

## References


1. Endocrineweb Homepage: https://www.endocrineweb.com/conditions/thyroid/thyroid-nodules/. Last accessed 2018/11/15
2. Cancer Treatment Centers of America Homepage: www.cancercenter.com/cancer/. Last accessed 2018/11/17
3. Cancer.Net Homepage: https://www.cancer.net/cancer-types/thyroid-cancer/statistics. Last accessed 2018/11/21
4. Cancer Stat Facts Homepage: https://seer.cancer.gov/statfacts/html/thyro.html. Last accessed 2018/11/21
5. Slideshare Image Pre-processing: https://www.slideshare.net/ASHI14march/image-pre-processing. Last accessed 2018/11/22
6. Babu, J.J.J., Sudha, G.F.: Adaptive speckle reduction in ultrasound images using fuzzy logic on Coefficient of Variation. Biomed. Signal Process. Control **23**, 93–103 (2016)
7. Babu, J.J.J., Sudha, G.F.: Non-subsampled contourlet transform based image denoising in ultrasound thyroid images using adaptive binary morphological operations. IET Comput. Vis. **8**(6), 718–728 (2014)
8. Kim, M., Song, T.: Speckle reduction of ultrasound B-mode image using patch recurrence. In: International Conference on Biomedical Engineering and Systems, pp. 1–5 (2016)
9. Narayan, N.S.: Speckle patch similarity for echogenicity-based multiorgan segmentation in ultrasound images of the thyroid gland. IEEE J. Biomed. Health Inf. **21**(1), 172–183 (2017)
10. Kang, J., Lee, J.Y., Yoo, Y.: A new feature-enhanced speckle reduction method based on multiscale analysis for ultrasound b-mode imaging. IEEE Trans. Biomed. Eng. **63**(6), 1178–1191 (2016)
11. Koundal, D., Gupta, S., Singh, S.: Nakagami-based total variation method for speckle reduction in thyroid ultrasound images. Proc. Inst. Mech. Eng., Part H: J. Eng. Med. **230**(2), 97–110 (2016)
12. Koundal, D., Gupta, S., Singh, S.: Speckle reduction method for thyroid ultrasound images in neutrosophic domain. IET Image Process. **10**(2), 167–175 (2016)
13. Morin, R.: Motion estimation-based image enhancement in ultrasound imaging. Ultrasonics **60**, 19–26 (2015)
14. Tsantis, Stavros.: Inter-scale wavelet analysis for speckle reduction in thyroid ultrasound images. Comput. Med. Imaging Graph. **31**(3), 117–127 (2007)

15. Toonkum, P., Chinrungrueng, C.: Speckle reduction of ultrasound images based on locally regularized Savitzky-Golay filters. In: 12th International Conference on Electrical Engineering/Electronics, Computer, Telecommunications and Information Technology (ECTI-CON), pp. 1–5. IEEE (2015)
16. Huang, J., Yang, X.: Fast reduction of speckle noise in real ultrasound images. Signal Process. **93**(4), 684–694 (2013)
17. Elyasi, I., Pourmina, M.A.: Reduction of speckle noise ultrasound images based on TV regularization and modified bayes shrink techniques. Optik-Int. J. Light Electr. Optics **127**(24), 11732–11744 (2016)
18. Hacini, M., Hachouf, F., Djemal, K.: A new speckle filtering method for ultrasound images based on a weighted multiplicative total variation. Signal Process. **103**, 214–229 (2014)
19. Nugroho, H.A., Nugroho, A., Choridah, L.: Thyroid nodule segmentation using active contour bilateral filtering on ultrasound images. In: International Conference on Quality in Research (QiR), pp. 43–46. IEEE (2015)
20. Chang, C.-Y., Hong, Y.-C., Tseng, C.: A neural network for thyroid segmentation and volume estimation in CT images. IEEE Comput. Intell. Mag. **6**(4), 43–55 (2011)
21. Keerthivasan, A., Jaganath Babu, J., Sudha, G.F.: Speckle noise reduction in ultrasound images using fuzzy logic based on histogram and directional differences. In: International Conference on Communications and Signal Processing (ICCSP), pp. 499–503. IEEE (2013)
22. Yu, Y., Acton, S.T.: Speckle reducing anisotropic diffusion. IEEE Trans. Image Process. **11**(11), 1260–1270 (2002)
23. Zhao, J., Zhang, L., Tian, H.: Segmentation of ultrasound images of thyroid nodule for assisting fine needle aspiration cytology. Health Inf. Sci. Syst. **1**(1) (2013)
24. Babu, J.J.J., Sudha, G.F.: A density current modeled adaptive weighted average despeckling filter for ultrasound thyroid images. Indian J. Sci. Technol. **9**(46), 1–11 (2016)
25. Guo, W., Wang, Y., Yu, J.: Ultrasound harmonic imaging with reducing speckle noise by an interlaced iterative frequency compounding approach. Biomed. Eng. Inf. 34–39 (2015)
26. Măluţan, R., Terebeş, R., Germain, C., Borda, M., Cîşlariu, M.: Speckle noise removal in ultrasound images using sparse code shrinkage. In: E-Health and Bioengineering Conference (EHB), pp. 1–4. IEEE (2015)
27. Maroulis, D.E., Savelonas, M.A., Iakovidis, D.K., Karkanis, S.A., Dimitropoulos, N.: Variable background active contour model for computer-aided delineation of nodules in thyroid ultrasound images. IEEE Trans. Inf. Technol. Biomed. **11**(5), 537–543 (2007)
28. Tsantis, S., Dimitropoulos, N., Cavouras, D., Nikiforidis, G.: A hybrid multi-scale model for thyroid nodule boundary detection on ultrasound images. Comput. Methods Prog. Biomed. **84**(3), 86–98 (2006)
29. Du, W, Sang, N.: An effective method for ultrasound thyroid nodules segmentation. International Symposium on Bioelectronics and Bioinformatics (ISBB), pp. 207–210. IEEE (2015)
30. Chang, C.-Y., Lei, Y.-F., Tseng, C.-H., Shih, S.-R.: Thyroid segmentation and volume estimation in ultrasound images. IEEE Trans. Biomed. Eng. **57**(6), pp. 1348–1357 (2010)
31. Iakovidis, D.K., Savelonas, M.A., Karkanis, S.A., Maroulis, D.E.: A genetically optimized level set approach to segmentation of thyroid ultrasound images. Appl. Intell. **27**(3), 193–203 (2007)
32. Koundal, D.: Texture-based image segmentation using neutrosophic clustering. IET Image Process. **11**(8), 640–645 (2017)
33. Keramidas, E.G., Iakovidis, D.K., Maroulis, D., Karkanis, S.: Efficient and effective ultrasound image analysis scheme for thyroid nodule detection. In: International Conference Image Analysis and Recognition, pp. 1052–1060. Springer, Berlin, Heidelberg (2007)
34. Koundal, D., Gupta, S., Singh, S.: Automated delineation of thyroid nodules in ultrasound images using spatial neutrosophic clustering and level set. Appl. Soft Comput. **40**, 86–97 (2016)
35. Savelonas, M.A., Iakovidis, D.K., Dimitropoulos, N., Maroulis, D.: Variable background active contour model for automatic detection of thyroid nodules in ultrasound images. IEEE Trans. Inf. Technol. Biomed. **11**(5), 17–20 (2007)

36. Savelonas, M.A., Iakovidis, D.K., Dimitropoulos, N., Maroulis, D.: Computational characterization of thyroid tissue in the radon domain. Comput.-Based Med. Syst. 189–192 (2007)
37. Savelonas, M.A., Iakovidis, D.K., Legakis, I., Maroulis, D.: Active contours guided by echogenicity and texture for delineation of thyroid nodules in ultrasound images. IEEE Trans. Inf. Technol. Biomed. **13**(4), 519–527 (2009)
38. Keramidas, E.G., Maroulis, D., Iakovidis, D.K.: TND: a thyroid nodule detection system for analysis of ultrasound images and videos. J. Med. Syst. **36**(3), 1271–1281 (2012)
39. Ma, J., Luo, S. Dighe, M., Lim, D.-J. Kim, Y.: Differential diagnosis of thyroid nodules with ultrasound elastography based on support vector machines. In: IEEE International Ultrasonics Symposium, pp. 1372–1375 (2010)
40. Chang, Chuan-Yu, Hsin-Cheng Huang, and Shao-Jer Chen.: Automatic thyroid nodule segmentation and component analysis in ultrasound images. Biomed. Eng.: Appl., Basis Commun. **22**(2), 81–89 (2010)
41. Chang, C.-Y., Lei, Y.-F., Tseng, C.-H., Shih, S.-R.: Thyroid segmentation and volume estimation in ultrasound images. IEEE Trans. Biomed. Eng. **57**(6), 1348–1357 (2010)
42. Ma, J., Wu, F., Zhao, Q., Kong, D.: Ultrasound image-based thyroid nodule automatic segmentation using convolutional neural networks. Int. J. Comput. Assist. Radiol. Surg. **12**(11), 1895–1910 (2017)
43. Tsantis, S., Glotsos, D., Kalatzis, G., Dimitropoulos, N., Nikiforidis, G., Cavouras, D.: Automatic contour delineation of thyroid nodules in ultrasound images employing the wavelet transform modulus-maxima chains. In: 1st International Conference from Scientific Computing to Computational Engineering, pp. 8–10 (2004)
44. Kollorz, E., Angelopoulou, E., Beck, M., Schmidt, D., Kuwert T.: Using power watersheds to segment benign thyroid nodules in ultrasound image data. In Bildverarbeitung für die Medizin 124–128 (2011)
45. Maroulis, D.E., Savelonas, M.A., Karkanis, S.A., Iakovidis, D.K., Dimitropoulos, N.: Computer-aided thyroid nodule detection in ultrasound images. Comput.-Based Med. Syst. 271–276 (2005)
46. Gireesha, H. M., S. Nanda.: Thyroid nodule segmentation and classification in ultrasound images. Int. J. Eng. Res. Technol. 2252–2256 (2014)
47. Ganesh, P., Babu, J.: Automated thyroid nodule segmentation algorithm for ultrasound images. Int. Conf. Signal Process. **3**(3), 85–90 (2014)
48. Koundal, D., Gupta, S., Singh, S.: Computer aided thyroid nodule detection system using medical ultrasound images. Biomed. Signal Process. Control **40**, 117–130 (2018)
49. Saiti, F., Naini, A.A., Shoorehdeli, M.A., Teshnehlab, M.: Thyroid disease diagnosis based on genetic algorithms using PNN and SVM. Bioinf. Biomed. Eng. 1–4 (2009)
50. Shukla, A., Tiwari, R., Kaur, P., Janghel, R.R.: Diagnosis of thyroid disorders using artificial neural networks. IEEE Adv. Comput. Conf. 1016–1020 (2009)
51. Keleş, A., Keleş, A.: ESTDD: expert system for thyroid diseases diagnosis. Expert Syst. Appl. **1**, 242–246 (2008)
52. Polat, K., Şahan, S., Güneş, S.: A novel hybrid method based on artificial immune recognition system (AIRS) with fuzzy weighted pre-processing for thyroid disease diagnosis. Expert Syst. Appl. **32**(4), 1141–1147 (2007)
53. Pechenizkiy, M., Tsymbal, A., Puuronen, S., Patterson, D.: Feature extraction for dynamic integration of classifiers. Fundamenta Informaticae **77**(3), 243–275 (2007)
54. Singh, N., Jindal, A.: A segmentation method and comparison of classification methods for thyroid ultrasound images. Int. J. Comput. Appl. **50**(11), 43–49 (2012)
55. Malathi, M., Srinivasan, S.: Classification of ultrasound thyroid nodule using feed forward neural network. World Eng. Appl. Sci. **8**(1), 12–17 (2017)
56. Nugroho, H.A., Rahmawaty, M., Triyani, Y., Ardiyanto, I.: Texture analysis for classification of thyroid ultrasound images. In: Electronics Symposium (IES), pp. 476–480 (2016)
57. Chang, C-Y, Huang H.-C., Chen, S.-J.: Thyroid nodule segmentation and component analysis in ultrasound images. Biomed. Eng.: Appl., Basis Commun. **22**(2), 910–917 (2009)

58. Selvathi, D., Sharnitha, V.S.: Thyroid classification and segmentation in ultrasound images using machine learning algorithms. In Signal Processing, Communication, Computing and Networking Technologies (ICSCCN), pp. 836–841 (2011)
59. Chang, C.-Y., Huang, H.-C., Chen, S.-J.: Automatic thyroid nodule segmentation and component analysis in ultrasound images. Biomed. Eng.: Appl., Basis Commun. **22**(2), 81–89 (2010)
60. Garg, H, Jindal, A.: Segmentation of thyroid gland in ultrasound image using neural network. In Computing, Communications and Networking Technologies (ICCCNT), pp. 1–5 (2013)
61. Kim, H.-C., Ghahramani, Z.: Bayesian Gaussian process classification with the EM-EP algorithm. IEEE Trans. Pattern Anal. Mach. Intell. **28**(12), pp. 1948–1959 (2006)
62. Amasyalı, S.A., Albayrak, F.S.: Fuzzy c-means clustering on medical diagnostic systems. In: International 12th Turkish Symposium Artificial intelligence and neural networks (2003)

# A Novel Approach of Object Detection Using Point Feature Matching Technique for Colored Images

**Manvinder Sharma, Harjinder Singh, Sohni Singh, Anuj Gupta, Sumeet Goyal and Rahul Kakkar**

**Abstract** For computer vision, image matching is an essential trait which includes scene or object recognition. Detection using point feature method is much effective technique to detect a specific target instead of other objects or within clutter scene in an image. It is done by comparing correspondence points and analyzing between cluttered scene image and a target object in image. This paper presents novel SURF algorithm that is used for extracting, describing, and matching objects in colored images. The algorithm works on finding correspondence points between a target and reference images and detecting a particular object. Speeded-up robust features (SURF) algorithm is used in this study which can detect objects for unique feature matches and which has non-repeating patterns. This approach of detection can robustly find specified objects between colored cluttered images and provide constriction to other achieving near real-time performance.

**Keywords** SURF · Object recognition · Objects capture · Matching technique

M. Sharma (✉) · S. Singh (✉)
Department of Electronics and Communication, Chandigarh Group of Colleges, Landran, Mohali, Punjab, India
e-mail: manvinder.sharma@gmail.com

S. Singh
e-mail: sohnibagga5@gmail.com

H. Singh (✉)
Department of Electronics and Communication, Punjabi University, Patiala, India
e-mail: harjinder@pbi.ac.in

A. Gupta (✉)
Department of Computer Science and Engineering, Chandigarh Group of Colleges, Landran, Mohali, Punjab, India
e-mail: anuj21@hotmail.com

S. Goyal (✉) · R. Kakkar (✉)
Department of Applied Science, Chandigarh Group of Colleges, Landran, Mohali, Punjab, India
e-mail: cgccoe.appsc.sug@gmail.com

R. Kakkar
e-mail: cgccoe.appsc.rk@gmail.com



P. K. Singh et al. (eds.), *Proceedings of ICRIC 2019*, Lecture Notes in Electrical Engineering 597, https://doi.org/10.1007/978-3-030-29407-6_40

# 1 Introduction

The detection of objects is basically an engineering domain that constitutes the laptop vision of objects present in pictures and videos that are digital in nature. The pedestrian and face detection are one of the most well-researched fields of object detection [1]. Laptop vision, recovery of the image, in police video investigation, etc., are some of the prominent applications of object detection. Object detection involves technique that is capable of discovering illustrations of objects that are real world in nature and also that are analogous to the objects present in the image or video that can be buildings, faces, etc. In order to recognize the object instances, we use the algorithm for extracting and learning the options that is typically employed in object detection [1].

Detection of objects in the sequence of pictures and videos is the main job. We humans can easily detect or acknowledge the object varying in any size in the picture without any effort, even when the image is translated, turned, or rotated in many read points [2]. The objects can be recognized by using different kinds of models which include the algorithmic rule of Viola–Jones, SURF and MSER admiring models [12], template matching, and algorithms based on boosted learning and image segmentation.

Vision is one of the strongest senses of humans and most of the knowledge that the human brain acknowledges from the environment is with the vision itself [3]. A human vision understands, perceives different objects, and recognizes distinct faces encircling in the surroundings with the least effort.

## *1.1 Computer Vision*

Computer vision is a scientific field whose main objective is to deal with the capability of computers that how they can perform the gain while understanding the high-level useful information from the images and videos. And furthermore, it is concerned with the analysis and extraction of images. All these functions are gift to human vision, and therefore, computer vision forms a perspective in engineering by performing all these tasks [11].

The applications include medical imaging, industrial control, video streaming, etc., though the applications involving the computer vision are dependent on the human sensory system [4]. Vision systems have the capability of attention in which it can closely distinguish the relevant objects through its salient features.

### 1.2 *Prerequisite of Object Recognition*

It is possible to split the human vision in two main phases: the first one being the low level and the second is high-level vision. It is very crucial to recognize the boundary that exists between these two phases of human vision. The retina collects the data being visualized in the low-level phase at the rate of bits per second which is the first step. Now, in the next step, it is very important to extract the information regarding the knowledge perceived in the previous step where it is further transmitted to the high-level phase to continue its operations. Basically, the gathering and extraction of the knowledge are done in this step.

Practically, in the high-level vision phase, the image contents are acknowledged meaning matching the illustrative scene points with the objects containing information that is very well memorized [10]. The human vision can recognize the huge number of objects in less time though there is large amount of visual data processing which is combinational in nature.

Only the human brain neurons have this extraordinary capability of recognizing such vast degree of data in such a response rate and that is why the human vision system exhibits high effectiveness due to the human brain performance. Visual attention is the very first preprocessing step that allows the access of the data [11]. After detecting the important elements in the visible object, the computer vision owing to its high-level tasks ponders on the particular locations.

## 2 Related Work for Object Detection

The following steps are included:

**Object representation**: In the beginning of object detection, the first step is to interpret or visually explain the object of interest. Object may be marked or painted with its appearance and form. Several methods are discussed for marking.

Points—Object of interest in image is marked by a degree, i.e., collection of points or center of mass. If the trailing objects occupy tiny space in the image, then this method of representation is appropriate [13].

Primitive geometric forms—In this representation method, object of interest is marked by conic, parallelogram, etc. This method is appropriate when the objects are straightforward or rigid.

Skeletal form models—Medial axis remodel is used to extract object skeleton. Model is used as form illustration for representation of objects [17, 18].

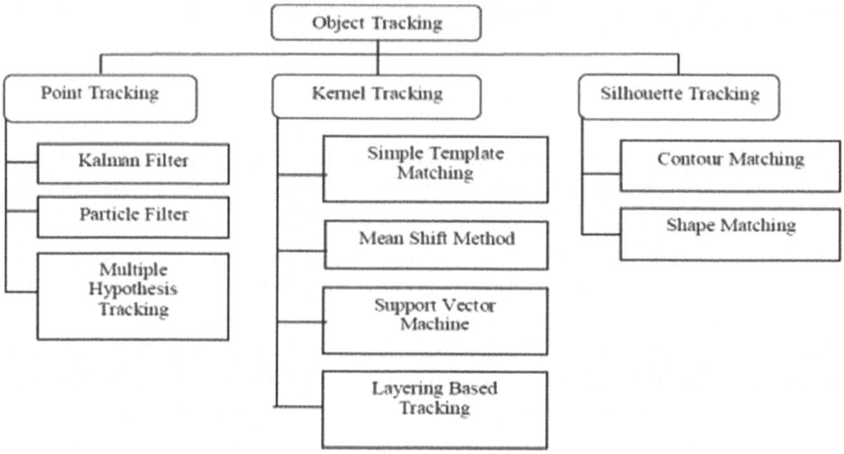


**Fig. 1** Object tracking

Articulated form—In this representation, two or more sections that are connected by some flexible joints like legs, hands, etc., that are connected by joints are marked by integral element victimization ellipse or cylinder. Kinematic motion models are used to define the relationship between the joint angles, etc. [14–16].

**Object tracking**: The objects of interest are marked using any method discussed above. For object tracking, the various techniques are shown in Fig. 1.

Point Tracking—The object of interest is tracked by feature points. Kalman filters and Multiple hypothesis tracking are used to detect and track the object of interest in frame [9].

Kernel Tracking—Kernel tracking is used to track object of interest in moving frames as well. It computes the motion of object of interest frame to border [8].

Silhouette Tracking—This technique is also used to detect object in moving frame. It generates the model of object supported to previous frame.

## 3 Speeded-Up Robust Feature (SURF)

SURF algorithm also works on same principles like scale invariant feature transform (SIFT), but details in each step are different in SURF [19]. SIFT uses parallel filters and lower approximated Gaussian Laplace with difference of Gaussian to detect a particular object, while SURF approximates laplacian of Gaussian with box filter. The advantage of using box filter in SURF is convolution with box filters can be easy calculated, and also, it can work parallely for different scales. Figure shows difference in SIFT and SURF [7]. SURF algorithm performs fast computation of operators using box filters. This enables the algorithms to be used for real-time applications like object recognition and tracking [20]. SURF algorithm works in three steps namely, feature detection, feature description, and feature matching. Figure 2 shows the difference between SIFT and SURF.

In the feature detection process, local maxima in the image which is determined by determinant of Hessian are used to select interest point candidates. If the response is above a provided threshold value, these candidates are validated, and using quadratic fitting, both location and scale of these candidates are redefined. For a megapixel image, few hundred interest points are detected [6].

In the feature description step, description is built of neighborhood of each interest point which is invariant to changes in viewpoint [21]. Haar wavelets that is local gradient orientation distribution is used to achieve rotation invariance. Based on Haar wavelet coefficients, 64-dimensional descriptors are built in this step.

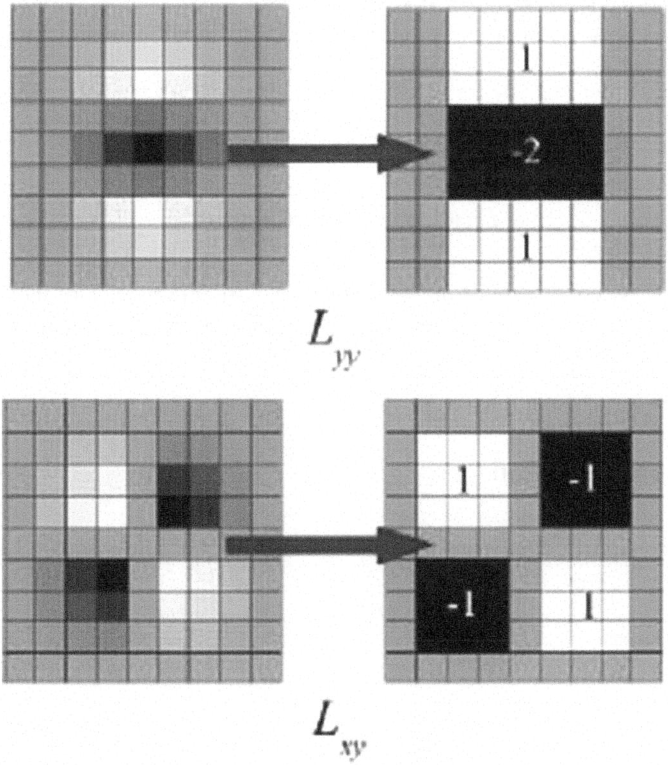


**Fig. 2** SIFT and SURF approach

For image indexation, object detection or image registration, while considering image matching process [5], local descriptors which were done in the second step are matched. By computing the Euclidean distance between all matching paris exhaustive comparison is performed. To reduce mismatching combined with optional RANSAC-based method, nearest neighbor distance ratio matching is done.

Filtering images with a square and sum of image using integral image give faster result, and the formula is given as

$$S(x, y) = \left( \sum_{i=0}^{x} \sum_{j=0}^{y} I(i, j) \right) \tag{1}$$

SURF algorithm uses Hessian matrix blob detector to detect object of interest. For a given point $p = (x, y)$ in an image $I$, the Hessian matrix $H(p, \sigma)$ can be rewritten as

$$H(p, \sigma) = \begin{pmatrix} L_x(p, \sigma) & L_{xy}(p, \sigma) \\ L_{yx}(p, \sigma) & L_{yy}(p, \sigma) \end{pmatrix} \tag{2}$$

where $p$ is point in image and $\sigma$ is scale.

$L_{xx}(p, \sigma)$ is convolution of second-order derivative of Gaussian with the image $I(x, y)$ at point $x$.

$$\begin{aligned} L_{xx}(X, \sigma) &= I(X) * \frac{\partial^2}{\partial x^2} g(\sigma) \\ L_{xy}(X, \sigma) &= I(X) * \frac{\partial^2}{\partial xy} g(\sigma) \end{aligned} \tag{3}$$

$L_{xx}(x, \sigma)$ is convolution of image with second-order derivative of the Gaussian. SURF detection works on non-maximal suppression of the determinant of Hessian matrix.

## 4 Feature Detection

Feature detection use comparing each pixel before and after sum of squared difference

$$E(u, v) = \sum_{(x,y) \in W} [I(x + u, y + v) - I(x, y)]^2 \tag{4}$$

$$\approx \sum_{(x,y)\in W} \left[ I(x, y) + [\, I_x \; I_y \,] \begin{bmatrix} u \\ v \end{bmatrix} - I(x, y) \right]^2 \tag{5}$$

This can be rewritten as

$$\approx \sum_{(x,y)\in W} \left[ [\, I_x \; I_y \,] \begin{bmatrix} u \\ v \end{bmatrix} \right]^2 \tag{6}$$

where $E(u, v)$ is error by SSD.

## 5 Algorithm

Figure 3 shows steps involved for implementation of proposed algorithm. Firstly, the object of interest, i.e., reference image, which is colored image, is read. Then, the cluttered image is read which is colored and having many objects in scene. Both images are converted into grayscale so that SURF algorithm can be implemented. Then, feature extraction is done by SURF algorithm which works in three steps detection, description, and matching. In first step detection, the unique features are automatically identified. It approximates Gaussian kernel, and its spatial derivatives by uniform kernels and the local maxima in rectangular of Hessian distribution are used to select interest points. In description, each interest point is uniquely described which does not depend on rotation and scale. In matching, the convolutions of second-order derivatives are matched. One hundred and fifty strongest feature points were taken for reference image, and 350 strongest feature points were taken in cluttered images. Then, matching of these strongest feature points is done, and firstly, putatively matched features were displayed then matching point pairing is done with removal of outlines. The calculation of geometric transform and the transformation relating the matched points was done, while eliminating outliers. Finally, locating the object in scene is done using matched points. Another reference image is analyzed using same algorithm. The approach was implemented for colored images and gave satisfied results.

## 6 Results of Proposed Method

See Figs. 4, 5, 6, 7, 8, 9, 10, 11, 12, 13, 14 and 15.

**Fig. 3** Steps for proposed SURF algorithm

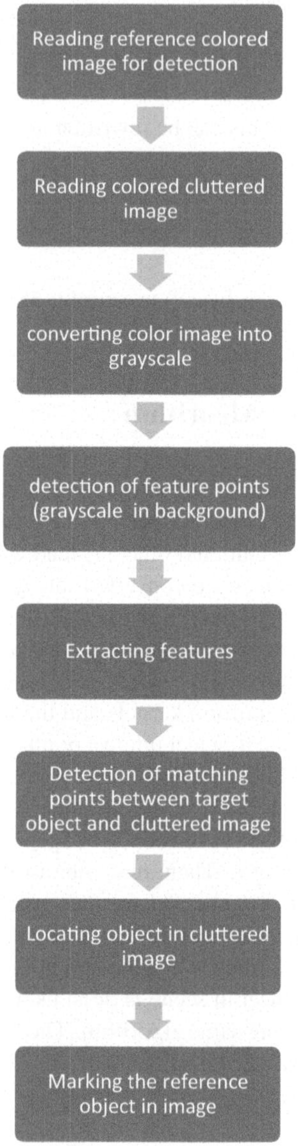

**Fig. 4** Object of interest (reference image)

**Fig. 5** Cluttered image with many objects

**Fig. 6** Feature extraction using SURF to reference image taking 150 strong features

## 7 Conclusion

In this article, we have implemented point feature matching technique for object recognition in colored image using SURF approach. SURF algorithm achieves fast and comparable to other algorithms for image matching. The article presents basic notations and mathematical concepts for extracting features and detecting object of interest. The algorithm is modeled and simulated for finding a specific object in cluttered image, for example, a particular object in many objects image. The related work has been done on grayscale images, for detection the object of interest the images taken were in grayscale. The proposed work is capable for detecting the

**Fig. 7** Feature extraction of cluttered image taking 350 strong features

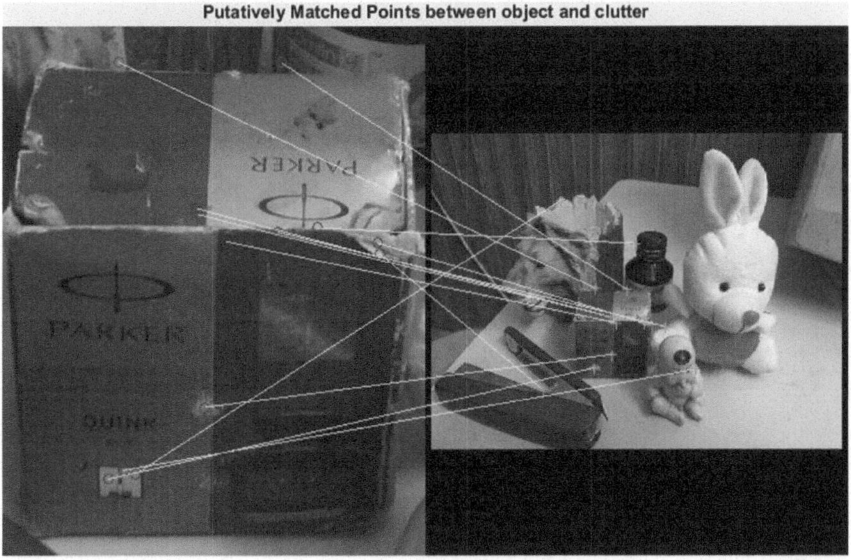


**Fig. 8** Putatively matched points between reference image and cluttered image

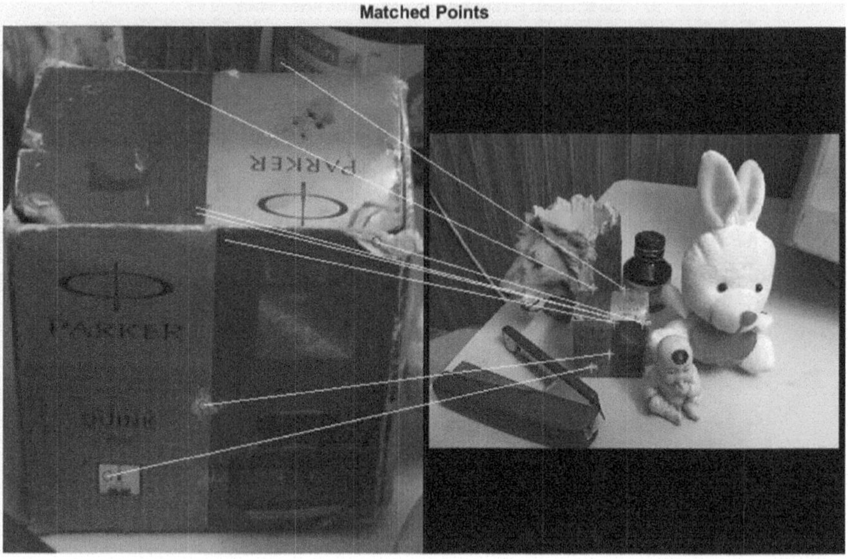


**Fig. 9** Matched points between reference image and cluttered image in-liner points

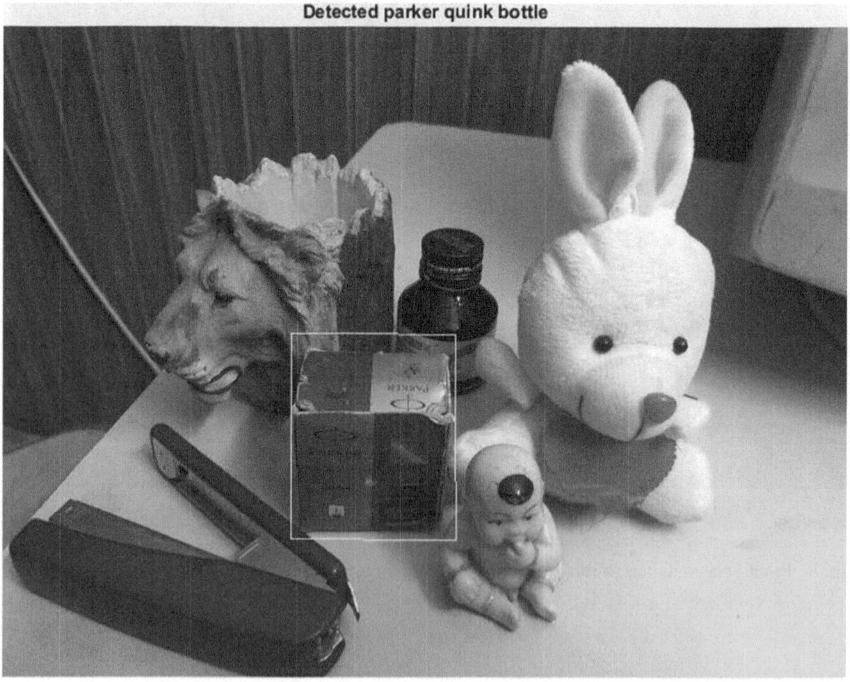


**Fig. 10** Detected object of interest in cluttered image

**Fig. 11** Another object of interest

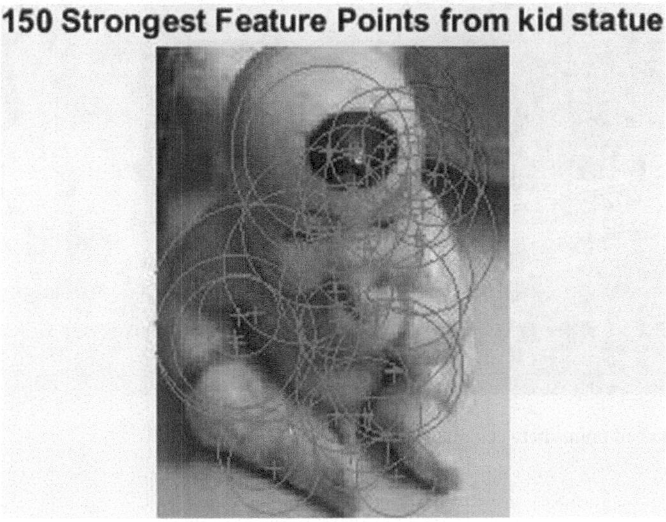


**Fig. 12** Feature extraction of reference image

object in colored image. The approach works well for uniformly colored objects and objects containing regular patterns. The fast computation enables this technique for real-time applications such as object detection and tracking. Algorithm can also be used for fingerprint matching.

**Fig. 13** Putatively matched points

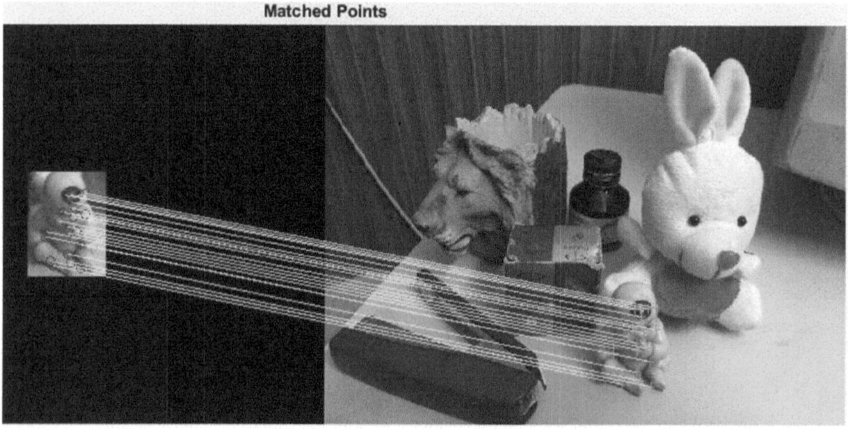


**Fig. 14** Matched points between reference and cluttered image

**Fig. 15** Detected object in cluttered image

## References


1. Yilmaz, A., Javed, O., Shah, M.: Object tracking: asurvey. Acm Comput. Surv. (CSUR) **38**(4), 13 (2006)
2. Comaniciu, D., Ramesh, V., Meer, P.: Kernel-based object tracking. IEEE Trans. Pattern Anal. Mach. Intell. **25**(5), 564–577 (2003)
3. Yu, S.X., Shi, J.: Segmentation given partial grouping constraints. IEEE Trans. Patt. Analy. Mach. Intell. **26**(2), 173–183 (2004)
4. Ballard, D., Brown, C.: Computer Vision. Prentice-Hall (1982)
5. Rakibe, R.S., Patil, B.D.: Background subtraction algorithm based human motion detection. Int. J. Sci. Res. Publ. (2013)
6. Chauhan, A.K., Krishan, P.: Moving object tracking using Gaussian mixture model and optical flow. Int. J. Adv. Res. Comput. Sci. Softw. Eng. (2013)
7. Joshan Athanesious, J., Suresh, P.: Systematic survey on object tracking methods in video. Int. J. Adv. Res. Comput. Eng. Technol. (IJARCET) October 2012, 242–247 (2012)
8. Parekh, H.S., Thakore, D.G., Jaliya, U.K.: A Survey on object detection and tracking methods. Int. J. Innov. Res. Comput. Commun. Eng. **2**(2) (2014)
9. Fernandez-Sanchez, E.J., Diaz J., Ros, E.: Background subtraction based on color and depth using active sensors. Sensors (2018)
10. Chi, Q., et al.: A reconfigurable smart sensor interface for industrial WSN in IoT environment. IEEE Trans. Ind. Inf. **10**(2), 1417–1425 (2014)
11. Zimmermann, K., Hurych, D., Svoboda, T.: NON-rigid object detection with local interleaved sequential Alignment. In: Proceedings of IEEE, vol. 36 April (2018)
12. Leordeanu, M., Sukthankar R., Sminchristia, C.: Generalized boundaries from multiple image interpretations. In: Proceedings of IEEE, vol. 36 (2014)
13. Bobick, A.F., Davis, J.W.: Object Capturing in Cluttered Scene using Point Feature Matching. In: IEEE Proceedings on Pattern Analysis and Machine Intelligence, March 2016, pp. 257–260 (2016)
14. Leordeanu, M., Sukthankar R., Sminchristia, C.: Generalized boundaries from multiple image interpretations. In: Proceedings of IEEE, vol. 36 (2017)

15. S. U. R. F. (SURF): Herbet bay, andreas ess, tinne tuytelaars, luc van gool. Elsevier preprint (2008)
16. University of Koblenz-Landau, Tech. Rep., (2009)
17. Pedersen, J.T.: Study Group SURF: Feature Detection & Description. Department of Computer Science, Aarhus University (2011)
18. Rakibe, R.S., Patil, B.D.: Background subtraction algorithm based human motion detection. Int. J. Sci. Res. Publ. May (2013)
19. Fernandez-Sanchez, E.J., Diaz J., Ros, E.: Background subtraction based on color and depth using active sensors. Sensors July 13 (2013)
20. Parekh, H.S., Thakore, D.G. Jaliya, U.K.: A survey on object detection and tracking methods||. Int. J. Innov. Res. Comput. Commun. Eng. **2**(2), (February 2014)
21. Zimmermann, K., Hurych, D., Svoboda, T.: NON-rigid object detection with local interleaved sequential Alignment. In: Proceedings of IEEE, vol. (2014)

# E-Learning Cloud and Big Data

# Behavior Study of Bike Driver and Alert System Using IoT and Cloud

**Punit Gupta and Prakash Kumar**

**Abstract** This paper presents a smart and safe bike riding system to provide a safe and an intelligent driving features with accidental, speeding and rash driving alerts using fog computing. The system is based on the Ethernet-based 2nd Generation Intel Galileo Board. This intelligent system will be embedded in the upcoming bikes and motorcycles to prevent speeding, determine driver behavior and rash driving accidents. The whole idea of the system is to generate an alert to the user and provide caution alert to the user about their driving statistics and warn them as necessary. The system is embedded with various sensors like accelerometers, gyroscope, and GPS to make this system an intelligent one. The proposed outcome of the system aims as multiple benefits of preventing accidents, maintaining the ride statistics and getting the directions for the ride. Smart bike is an IoT-based ride system. In today's world, everything is getting automated.

**Keywords** Internet of thing (IoT) · Power consumption smart devices · Home automation · Fog computing · Cloud computing

## 1 Introduction

The future of the Internet of things and fog computing increases the horizons of our vision and also enabling the public to access and contribute rich resources about probably everything ever existed in the world. The success of the convergence of the internet of things and the world is giving the capability to share experience and personalized insights with the humans. IoT (Internet of things) has shown great potential for integration with the place which needs manual labor which can be at some points time taking and inaccurate. Information distribution may be rewarded through

P. Gupta (✉)
Department of Computer and Communication Engineering, Manipal University Jaipur, Jaipur, Rajasthan, India
e-mail: punitg07@gmail.com

P. Kumar (✉)
Jaypee Institute of Information Technology, Noida, India
e-mail: kprakash91@gmail.com



P. K. Singh et al. (eds.), *Proceedings of ICRIC 2019*, Lecture Notes in Electrical Engineering 597, https://doi.org/10.1007/978-3-030-29407-6_41

inducements, thus transforming the Internet of things from mere manufacturing commercial applications to an intelligent infrastructure that would reduce human labor and could accomplish the task incomparably less time and most accurately. This will allow us in trading of enriched information and accelerate business innovation. Being open source and end-user programming will enable people to share to the Internet of things with data, valued resources and functionalities.

The Internet of things is in itself the summation of various technologies based on wireless means, system setup which store data retrieved from various hardware nodes (sensors). Internet connects the above-mentioned systems and services for useful inference and inference-based decision making for bringing the generated data to use. This integration has led to quick deduction and compilation of results along with the more rapid creation of efficient frameworks for boosting applications in the industry which include automation, higher risk tolerance and quicker delivery of services.

This interconnected device layout coupled with efficient and robust methods of data transfer has not only resulted in the accelerated computation of traditionally high resource and time-consuming tasks like service-threshold evaluation and alert generation. This has greatly reduced the risk factors associated with the increase in the user database of these services, mainly addressing the issues of high-end investment required for installation and commencement of these services, also mitigating the chances of low user satisfaction owing to more accessible and cheaper device interconnection setup and low maintenance costs involved for the subscription and troubleshooting of any of the associated services. The Internet of things has found out its application in almost all the industries ranging from healthcare to risk management. Its biggest advantage is the rate of automation provided by it, which has helped generate real-time data conclusions using commercial grade material, saving countless hours of effort and also many human lives. Based on one such application, this paper proposes an IoT-based smart bike ride system that focuses on preventing the increasing road accidents caused due to speedy and rash driving driver behavior. This intelligent system will maintain the statistics as well as generate alerts when necessary about the riding style and hence warn the user.

Paper is structured as follows: Sect. 2 presents motivation and problem statement. Section 3 presents existing work from the field of IoT and smart devices for vehicles like car, bike and trains, etc. Section 3 discourses the problem statement and proposed model to detect rash driving and alert system. Section 4 describes the simulation setup, scenarios and results as compared to normal and safe driving. Section 5 concludes the work with result outcomes and future work.

## 2 Motivation

Internet of things has solved various real-world problems over the years and has automated various fields of surveillance and alert systems. These systems consist of various interconnected device collaboratively collecting live data and all data been analyzed for alerts. A similar system has been developed for surveillance of vehicles like car and bikes, but these systems only take track of vehicle position and do not take into consideration driver rash driving behavior. Existing systems are not capable of generating alert for driver about the inconsistent behavior in driving. So to study and analyze the driving behavior, an IoT-based device for the bike is designed to study the biker behavior and also generate alert about it driving behavior. The system is also capable to detect any accident and generate alert for family members and authority.

## 3 Related Work

A lot of research and development is going in the field of Internet of things which includes from connecting to devices, collecting data from them, analyzing those large bytes of data and performing the required operation on the data and generates the desired output. Many IoT devices in all the fields including medical and healthcare, home automation, agriculture, etc., have been developed and the same goes for our smart petrol pump solution system. Some of the related IoT devices are discussed below.

## 4 eCall

eCall [1] is a service in Europe which serves as an example of an authority undertaking which aims to increase telematics of vehicles. By the time of mentioning, the grand European coalition had set its target to install the eCall model by 2015 [2]. This entire system is fixed inside a vehicle and is attached to the air-bag system to sense a collision in real time, and it is provided with a user-operated button to ask for assistance for a co-passenger with body issues or to notify the concerned authorities of some other vehicle. It has been equipped with a global positioning service incoming-acceptor for finding out the vehicle and its anticipated route before the mishappening took place. eCall is triggered by itself when the embedded sensors or trigger-attachments find out of a damaging collision. A basic overview of its working has been shown in Fig. 1.

After this service is executed for the first time, the contact sequence of the European Union (which is 112), sets-up a cellphone link to the apt quick response unit quarters (or public safety assuring centers), sending the detail of the happening (or minimum set of data → MSD) to the savior teams (also incorporating the time of occurrence), the to-the-point location of the affected vehicle and the direction in which the vehicle headed. A eCall can be initiated by thyself on pressing a button given on the vehicle, for example, by an individual who had a look at the scene.

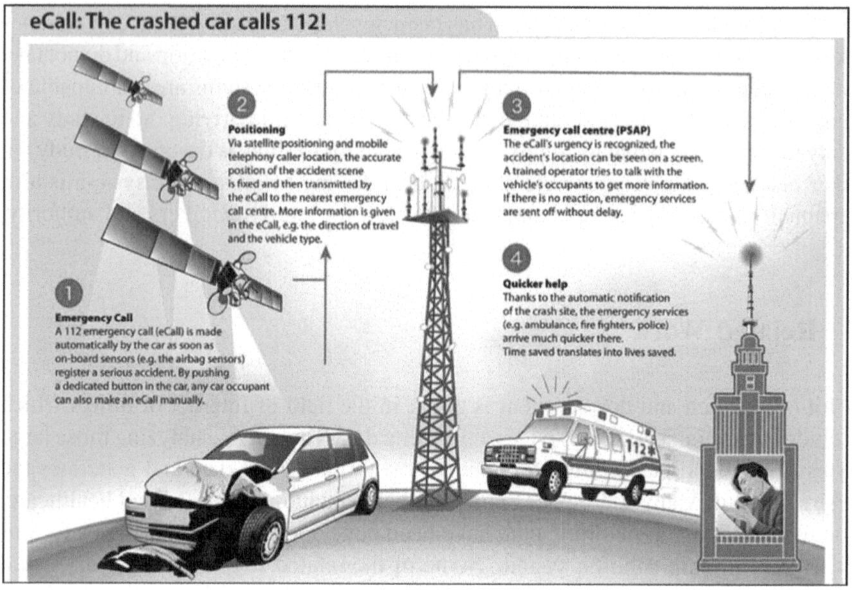


**Fig. 1** eCall system

## 5 Contran 245 [3]

In Brazil, a legislation was presented which demanded all fresh vehicles manufactured or brought to Brazil to be installed with GPRS-enabled locating systems for bringing down car thefts. This idea is on its way to fruition.

This bill or act has the ability to transform Brazil into one of the biggest telematic market in the universe, including car tracing and other services, containing simple tracing as needed by the act to features including fleet caretaking and setup and emotion response monitoring for insurance issues.

## 6 Miroad

Here, Johnson et al. [4] mentioned of which places vehicle driving styles into average, hostile and ultra-hostile. Like the above cases, sensors like accelerometers, gyroscopes, cameras and magnetometers) on an iPhone were used with the data generated into a singular classification program based on the dynamic time warping (DTW) algorithm. That system "spoke," i.e., gave an audible output for alert generation, i.e., if the threshold values were exceeded. In the USA, a significant amount of insurance companies has begun implementing service consumption-based insurance to reduce the costs for these customers and also to advocate secure driving.

In Italy, gains of service consumption-based insurance have resulted in a new law. Private companies like Octo Telematics provide recommendations for insurance and other tasks relating to the automotive industry by accumulating vehicle location and accident data along with driver behavior information. To support its pay-as-you-drive policies.

Some developments [6–10] have taken place in the utilization of mobile devices (cell-phones) for calculation of various parameters and alert provision. As an example, an application on the Android platform made by Chirgupta et al. [5] which provides quality charts of that specific driver. They had incorporated an accelerometer, a global positioning service sensor and the inbuilt camera device for capturing the feed. Ranges for accelerating securely were mentioned in the code and values breaking beyond their extremes (thresholds) marked the happening of the associated event.

## 7 Proposed Model

This section introduces a smart and safe riding intelligent system that provides the driver, along with the concerned authorities (parents, etc.) with safe and intelligent driving features including that of accident alerts, rash driving, and speeding alerts. The system is based upon on an Inter Galileo Gen 2 (Quark Processor IoT kit) based system which acts as the transmission-brain of the whole system and Amazon Web service EC2 as cloud controller for analysis. The system is connected to the network via Internet and a database is maintained consisting of the whole ride details, the speed statistics, bike tilt, GPS location and alerts if any. This system helps the one concerned person in relationship to bike driver and the concerned authority to study and stop rash drivers. The sensors used are GPS module, gyroscope and accelerometer which detail us upon the velocity as well as the tilt of the motorcycle. Gyroscope determines the tilt angle and the rotation of the motorbike, and with the help of accelerometer, the velocity of the motorbike is calculated. Figure 2 shows a very basic layout of our proposed model.

Our system responds to the speed alert when a motorbike crosses a threshold speed which can be determined by the accelerometer. It can also detect whether the person is doing the rash driving with the help of sudden rotational change readings in the gyroscope. The gyroscope can determine the rotational reading of the bike and its tilt angle, and thus can help to detect the rash driving occurrence. The proposed system focuses on the guidelines set by the local official authorities, which include maximum speed, minimum speed and sudden acceleration which are altogether combined with the measured tilt angle to generate a tested and calibrated parameter which single-handedly generates an alert for rash driving which is sent to the user and stored in the database. The authorities which can utilize the alert and stored data include parents, government officials in case of any lawsuit filing, the data can act as a conduit for showing the statistics of the moment of event occurrence, thereby helping legal undertakings proceed at an accelerated rate. Also, the intelligent system is also able

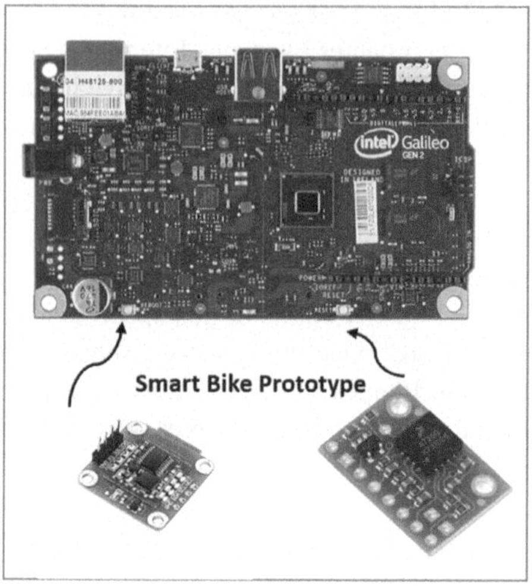


**Fig. 2** Physical layout cum component setup of the proposed model

to determine the accidents through a sudden change in an orientation of the motorbike and will let you know if you are over speeding in an area where the speed limit is restricted.

Proposed model is shown in Fig. 3 allows a user to upload the ride data to cloud Web server using cloud software as a service (SaaS) and Amazon as a service

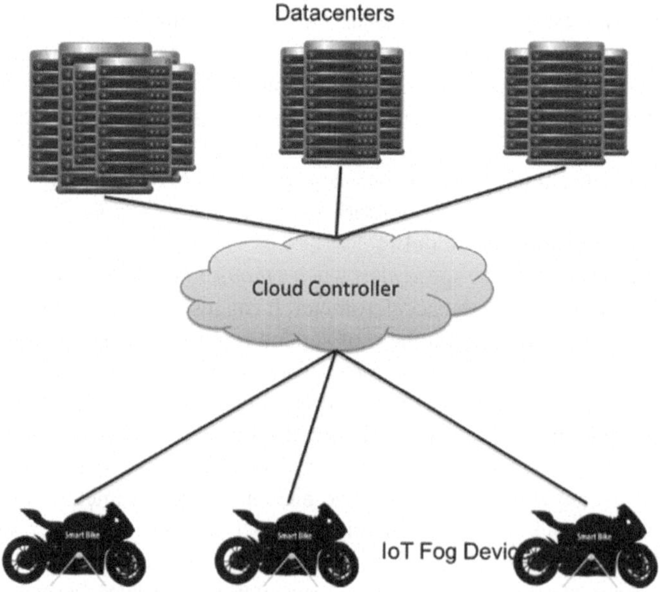


**Fig. 3** Proposed architecture

provider. Data uploaded to the Web server is bike velocity, change in acceleration with bike tilt and GPS coordinates to the Web server. The Web server is responsible to compute the data according to the threshold value and generate alert according. Alert generation includes the first phase of work. Alerts are sent to the authorized person in case of any rules violation listed below.

The second phase includes comparing the data with the data feed of a normal driver and analyze the behavior of the driver by comparing the both. Analyzing includes find the percentage of drastic difference in normal and rash driver velocity data, acceleration data and tilt degree data, if either of parameter change is too high the alert will be generated. The data will be useful to the authority to rate the driving skills of the bike driver and if found to not suitable a strict measure could be taken. Similarly, it will help the driver family to take a look at the students and other members of the family.

Cloud service helps to log and analyze a large set of logged data parallel and cloud can scale the service if a large number of devices is connected to it. Cloud service is also responsible to send the data with analyzed data to the authorities for effective actions. In analysis section, cloud service provider checks for anomalies in recorded data like access tilt in a bike with slow speed and many more rule sets discoursed in the next section. Figure 4 illustrates the plan to approach the studying of the behavior of the vehicle's movement.

The entire process can be executed in parallel, in the first case, it can be used for direct data retrieval and alert generation. In the second case, location tracking can be enabled using the GPS available on the user's end. In this case, guidelines issued by the local governing authorities shall influence the threshold values for an alert

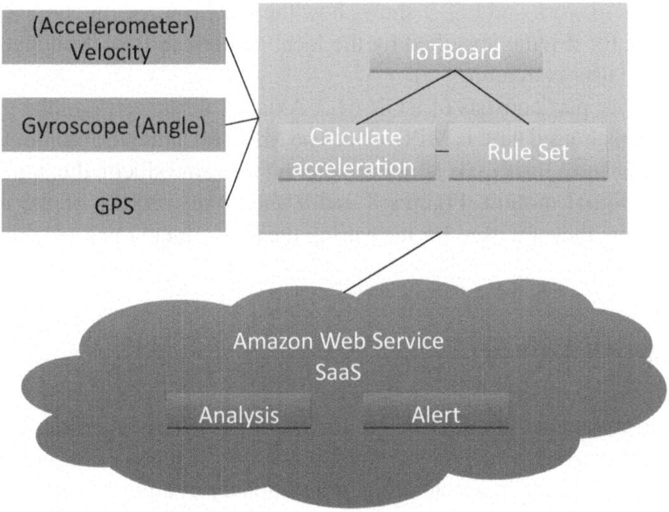


**Fig. 4** Proposed behavior study model

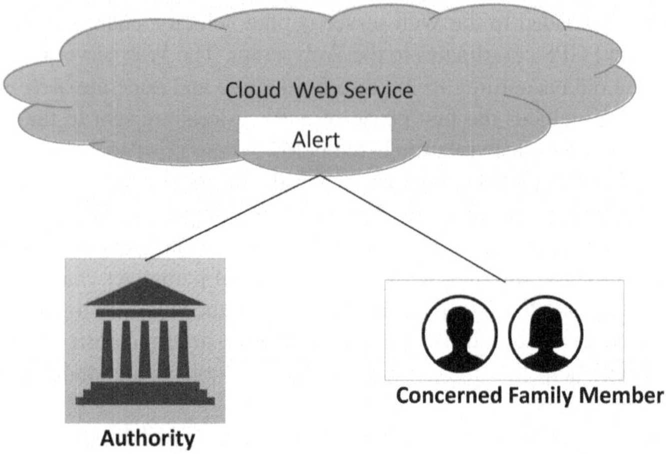


**Fig. 5** Proposed model for the alert

generation. Figure 5 shows the structure of the model which generates alerts for the user.

The software part consists of a database server which helps to maintain the statistics of the ride details. The user can also set the GPS details for the ride using his smartphone by connecting it via the Bluetooth. The system uses the Internet server to send the details of the location of the bike in case of accidents or rash driving. Real-time drive-statistics (with various alert parameters recorded) are sent via the Internet for complete data logging and future reference. Figure 6 shows how the system intends to work and Fig. 7 shows how the system works in conjunction with the guidelines for driving specified by the local authorities (which includes details of the road the user is treading on).

The hardware part consists of a GY-61 3-Axis accelerometer and the gyroscope sensor. The sensor used here is MPU6050. The MPU-6050 devices combine a three-axis gyroscope and a three-axis accelerometer on the same silicon die, together with an on-board digital motion. Figures 8 and 9 show the working of the individual sensors and how they are used for measuring the above-mentioned parameters.

## *7.1 Proposed Rule Set*

This section frames out and deduces four conditions for categorizing any form of driving as rash driving, which include crossing a defined speed limit, sudden stop of a speeding vehicle, sudden acceleration of a stationary or almost stationary vehicle, and excess tilt of the bike on that specific instantaneous velocity at that moment of time which may result in skidding and sudden stop may surprise the neighbor vehicles resulting in crash.

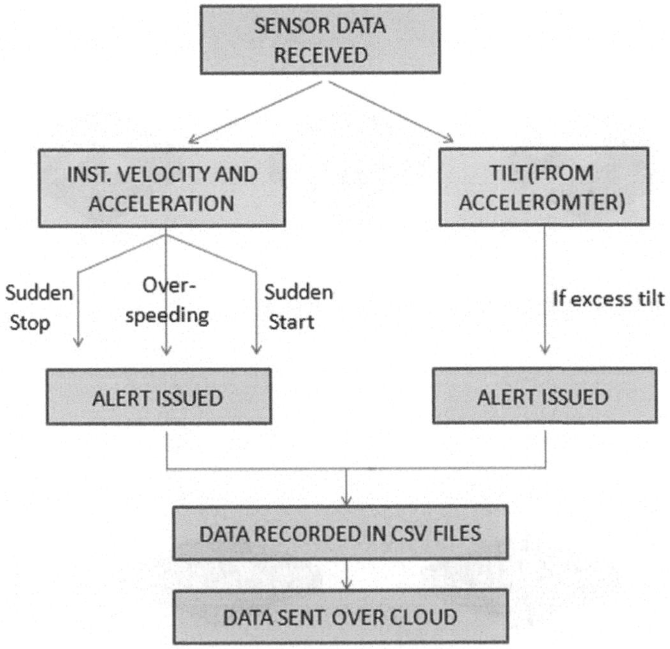


**Fig. 6** Pictorial representation of the working of the system

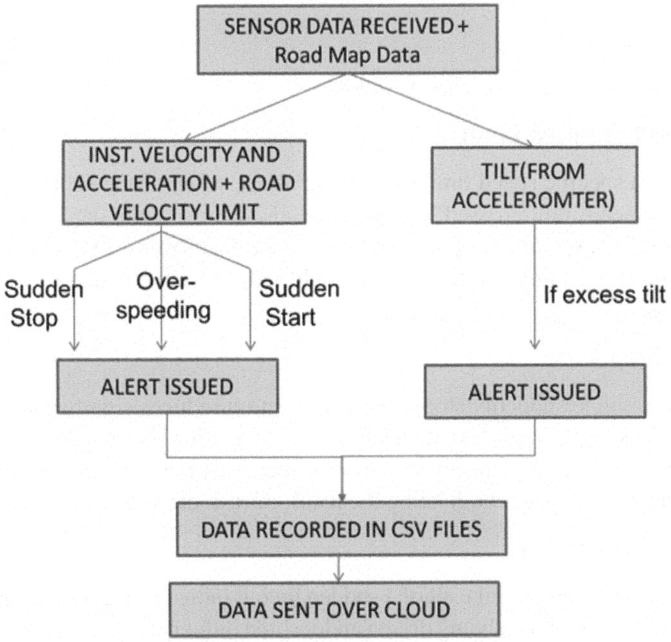


**Fig. 7** Pictorial representation of the working of the system

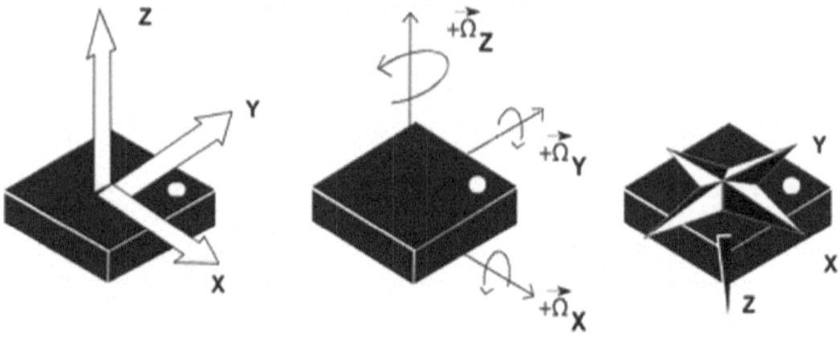


**Fig. 8** Dynamics of the gyroscope for angle measurement

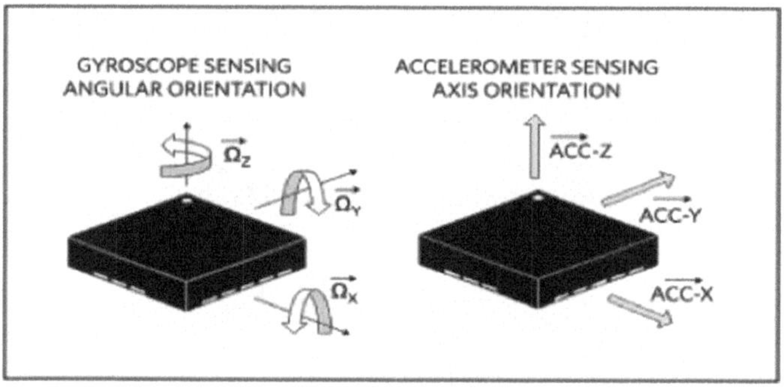


**Fig. 9** Axial positioning of the sensors

Set of rules and conditions are as follows:

- **Crossing The Speed Limit**

This defines a specific speed limit of 70 km/h above which if the vehicle goes, then an alert of "speed limit crossed" is generated and stored in the database. The data generated is for a specific time instant and if continued over a long period of time (more than 5–10 continuous iterations), would indicate intentional or self-imposed rash driving.

- **Sudden Stop**

If a moving vehicle suddenly stops (i.e., acceleration reduces significantly when its final velocity is highly reduced), then it indicates that either the vehicle was suddenly stopped via application of heavy brakes or either it as bumped into another solid, heavier object, either way, both being the result of rash driving.

- **Sudden Acceleration**

If a person accelerates the bike all of a sudden then it indicates that either the vehicle was being mishandled (i.e., being driven under some influence, etc.) or it was bumped from behind by another object both indicating toward rash driving by any of the associated parties.

**Fig. 10** Map Downtown Manhattan (Lower Manhattan, New York, NY, USA)

- **Excess Tilt Of The Bike**

If at any specific speed, if the vehicle tilts (about its center of mass) excessively for a prolonged time period then it indicates that the person is sure to get his vehicle flipped, his knees raptured, or bump into any other object or vehicle having a probability of causing heavy damage to life and property.

- **Authority Rules from Map**

The road layout taken for simulating the vehicle driving was in Downtown Manhattan (Lower Manhattan, New York, NY, USA), as shown in Fig. 10. The simulation was performed in broad daylight conditions with the presence of mild traffic. The starting point of the road trip was from the 259 Adams Street near Brooklyn Bridge (40° 41′ 45.6″N, 73° 59′ 19.3″W) to 6th Ave, Brooklyn, NY, USA (40° 38′ 14.5″N, 74° 00′ 53.7″W) with an estimated ride time of 11 min. The total length of the road covered for making the observations was 6.8 miles. The road comprised of an estimated amount of 14 turns (including both sharp and mild turns).

- **Specifications of Hardware Used**

Microcontroller: Inter Galileo Gen 2 (Quark Processor).

Sensors: GY-61 3-Axis accelerometer, Grove Rotary Angle Sensor v1.2

## 8 Experimental Results

The proposed IoT-based smart bike system has been simulated with sensors been attached on a bike with GPS connectivity. The testing is done over roadmap over Downtown Manhattan (Lower Manhattan, New York, NY, USA) with all the rules

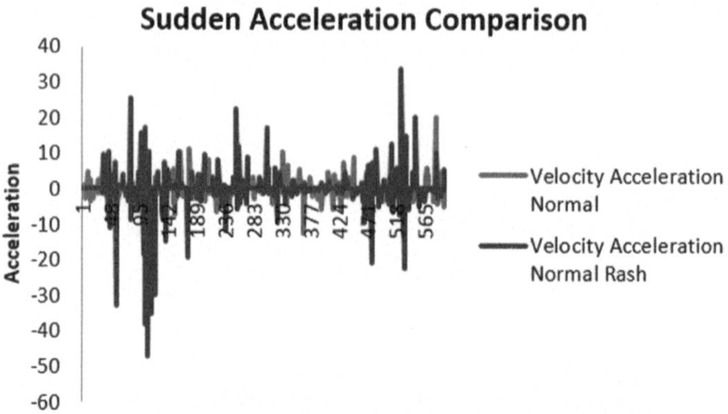


**Fig. 11** Sudden of acceleration change

of maximum speed at the specific point of the road from Google maps as shown in Fig. 10. Figure 11 shows a comparison of acceleration change in the driving of normal driver and rash driving as shown with blue and red color in the graph. The figure shows a drastic fluctuation in speed as compared to the smooth change in velocity in the normal driver.

Figure 12 shows alert generated due to a sudden change in acceleration as shown in green and red shows the acceleration change due to rash driving as compared to a dataset of normal driver. Figure 13 shows alert generated due to a sudden change in velocity resulting in over speeding and an alert is generated due to over speeding as compared to the authority speed provided by Google maps, if the speed greater than the threshold. Figure 14 shows alert generated due to sudden stop resulting in a high probability of bike disbalancing and an alert is generated due to this as compared normal driver shown in blue color. Figure 15 shows alert generated due to excess tilt in bike resulting in a high probability of bike disbalancing and an alert is generated due to this as compared normal driver shown in blue color and high

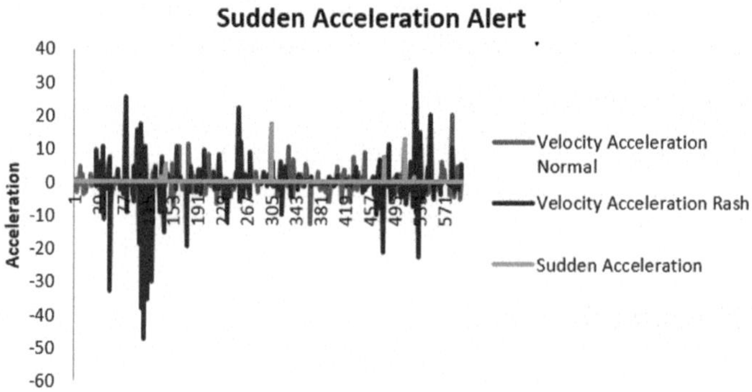


**Fig. 12** Sudden acceleration alert

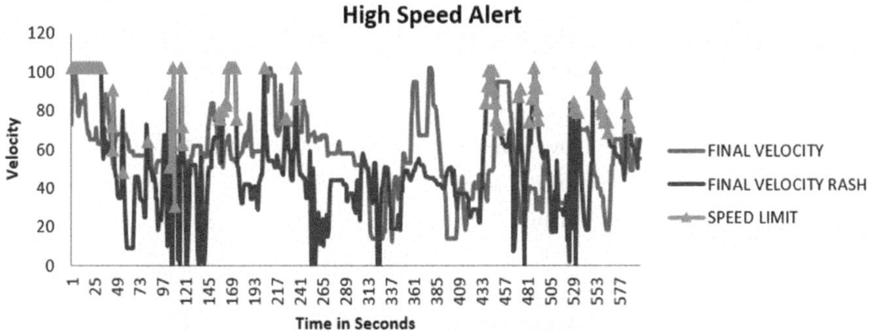


**Fig. 13** Alert generated due to a sudden change in velocity

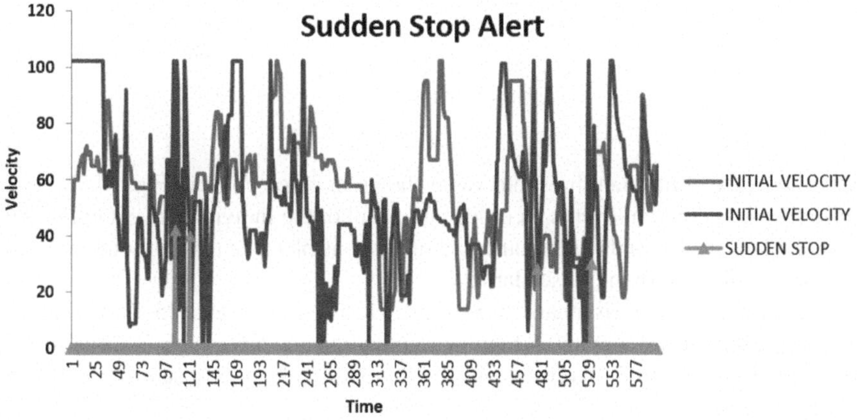


**Fig. 14** Sudden stop alert

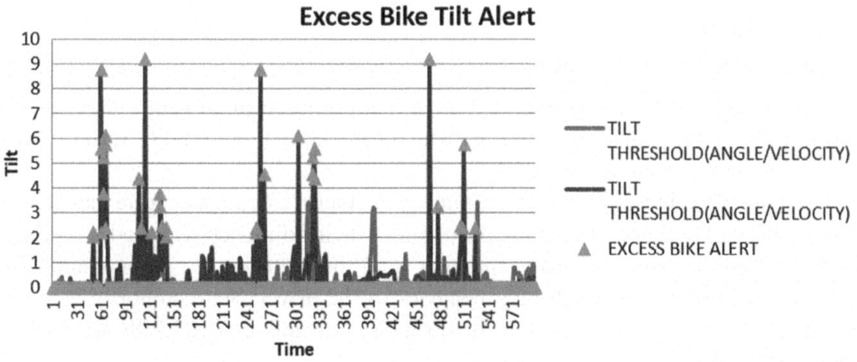


**Fig. 15** Sudden excess tilt in bike

**Table 1** Event and violation count recorded

| Violation type | Total | Count of violation |
|---|---|---|
| Tilt | 599 | 34 |
| En stop | 599 | 4 |
| Over speeding | 599 | 115 |
| Sudden acceleration | 599 | 6 |
| High acceleration | 599 | 125 |

**Table 2** Analysis output

| Violation type | % violation |
|---|---|
| Tilt | 5.68 |
| Sudden stop | 0.67 |
| Over speeding | 19.20 |
| Sudden acceleration | 1.00 |
| High acceleration | 20.87 |

peaks show excel tilt than threshold value shown in green color in Fig. 14. Table 1 illustrates total count of events recorded and count of the individual event out of the total. Table 2 corresponds to the outcome of analysis phase in term of total violated percentage after removing false alarms.

The output of analysis phase from Figs. 11, 12, 13, 14 and 15 show that 19 percentage high in velocity, 20% high in acceleration and 5.6% high in tilt ration as compared to the normal driver, which is sufficient to take action against the driver and send an alert.

## 9 Conclusion

From experimental result section, it is clear that proposed intelligent system is the best in its kind for providing better alerts and can detect more type of rash driving as compared to normal driving skills.

The main idea of this system is to minimize the road accidents which are increasing day-by-day by alerting and warning the driver of their riding behavior and providing them the best security necessary and also send the alerts to the parents and one concerned person about the driving behavior of the driver. State authority can use this data for gripping and finding the driver with these unfit driving skills and can be used to study the driving behavior or a specific area.

## References

1. Pinart, C., Calvo, J.C., Nicholson, L., Villaverde, J.A.: ECall-compliant early crash notification service for portable and nomadic devices. In: Vehicular Technology Conference. VTC Spring 2009. IEEE 69th 2009 Apr 26, pp 1–5. IEEE, 2009
2. Zhang, Y., Lin, W.C., Chin, Y.K.: A pattern-recognition approach for driving skill characterization. IEEE Trans. Intell. Transp. Syst. **11**(4), 905–916 (2010)
3. Chen, K., Lu, M., Fan, X., Wei, M., Wu, J.: Road condition monitoring using on-board three-axis accelerometer and GPS sensor. In: Communications and Networking in China (CHINACOM), 2011 6th International ICST Conference on 2011 Aug 17, pp. 1032–1037. IEEE, 2011
4. Johnson, D.A., Trivedi, M.M.: Driving style recognition using a smartphone as a sensor platform. In: Intelligent Transportation Systems (ITSC), 2011 14th International IEEE Conference on 2011 Oct 5, pp. 1609–1615. IEEE, 2011
5. Jalali, S.: M2M solutions—design challenges and considerations. In: Intelligent Computational Systems (RAICS), 2013 IEEE Recent Advances in 2013 Dec 19, pp. 210–214. IEEE, 2013
6. Kalra, N., Chugh, G., Bansal, D.: Analyzing driving and road events via smartphone. Int. J. Comput. Appl. **98**(12), 5–9 (2014)
7. Lea, R., Blackstock, M.: City hub: a cloud-based iot platform for smart cities. In: Cloud Computing Technology and Science (CloudCom), 2014 IEEE 6th International Conference on 2014 Dec 15, pp. 799–804. IEEE, 2014
8. Schietekat, J.M., Booysen, M.J.: Detection of reckless driving in the Sub-Saharan informal public transportation system using acceleration-sensing telematics. In: EUROCON, 2013 IEEE 2013 Jul 1, pp. 597–601. IEEE, 2013
9. Poojary, S.V., Rashmi, M., Shetty, S.: Humps and pothole detection and alerting system for safe journey. Int. Res. J. Eng. Technol. (IRJET) **03**(05) May 2016
10. Chu, H.L., Raman, V., Shen, J., Choudhury, R., Kansal, A., Bahl, V.: In-vehicle driver detection using mobile phone sensors. In: ACM MobiSys 2 Apr 2011

# E-Learning Web Accessibility Framework for Deaf/Blind Kannada-Speaking Disabled People

**A. B. Rajendra, N. Rajkumar, Sharath N. Bhat, T. R. Suhas and Shree Poorna N. Joshi**

**Abstract** E-learning is one of the best tools to support an individual's education system worldwide. There is a demand to create an e-learning site in Kannada for disabled people to remove accessibility barriers. In this paper, we have introduced a new framework for Web accessibility, providing easy access and lifelong learning to the Web site for Kannada's deaf/blind speakers. The main purpose of this framework is to support Kannada's deaf/blind by using the language of Kannada sign, speech text, and Kannada Moon code. Kannada-speaking disabilities can learn effectively in this framework.

**Keywords** E-learning · Accessibility · Kannada · Deaf/blind

## 1 Introduction

The Web provides new access for people with disabilities to information and communication technology. Web accessibility involves primarily the use of the Web by people with disabilities. Web accessibility in particular means that people with different skills understand, perceive, interact and navigate the Web [1]. However, for people with different skills, most Web software tools are not adequately accessible, making it difficult or inaccessible for them to contribute to the Web. What does it mean that there is a need for attempts to build a system that facilitates Web access for people, especially those who can do it differently? Worldwide, there is widespread recognition that users with different abilities have the same right to approach information technology [2]. This paper proposes a Web accessibility framework with most types of disabilities (deaf/blind/deaf-blind) to make access to the Web differently for people in Karnataka. The proposed framework enables them to use the different Web

A. B. Rajendra (✉) · N. Rajkumar · S. N. Bhat · T. R. Suhas · S. P. N. Joshi
Department of Information Science and Engineering, Vidyavardhaka College of Engineering, Mysuru, Karnataka, India
e-mail: abrajendra@vvce.ac.in

A. B. Rajendra · N. Rajkumar
Accendere Knowledge Pvt. Ltd., New Delhi, India



P. K. Singh et al. (eds.), *Proceedings of ICRIC 2019*, Lecture Notes in Electrical Engineering 597, https://doi.org/10.1007/978-3-030-29407-6_42

sites at significantly lower cost, time and effort. The proposed framework extracts the Web site's content in an accurate format suitable for various users.

This paper is organized accordingly. Section 2 focuses on related work literature. Section 3 presents the Kannada Moon code for persons with disabilities, Sect. 4 focuses on the Web accessibility framework for deaf/blind persons with disabilities, Sect. 5 conclusion and future works.

## 2 Literature Review

The study was conducted to calculate people with certain disorders and to "access" the Web [3]. What researchers mean by "access," though, is uncertain. It could simply mean that a computer can be connected to cyberspace at home or at work or that the individual in question can make use of active cyberspace. Around 43.3% of the world's population has some kind of disorders. In the same study, 56.7% of people without disorders have access to cyberspace. There is a significant difference [4].

The study is classified according to:

- Deafness: This study contains accessible systems for the deaf to use.
- Blindness: includes accessible systems which blind people can use.
- Deaf-Blindness: contains accessible systems which deaf-blind people can use.

Ramesh and Nagaraj [5] proposed a new interpreting methodology for the Kannada language by means of a sign symbol to easily identify words for the alphabet. The Kannada Sign Language (KaSL) helps people to communicate with them by means of hand gestures in a deaf-blind manner. The KaSL must be captured and saved in the database and used the feature extraction and classification method to recognize and interpret words. The KaSL has tested various data sets captured. A paper on Sign Language Recognition for Indian as English and Hindi has also been presented by Umang and Aarti [6]. They captured their hand and transformed them, using MATLAB, into speech and text. Vikram et al. [7] proposed the use of the British Sign Language (BSL) as a new way of developing a wearable manual to detect hand gesture. This method improves the accuracy of hand gesture detection.

In Rajapandian et al. [8], a new system has been proposed to help the deaf, blind and deaf-blind interact easily with ordinary people. Joao et al. [9] spoke about the support for education architecture for communication blind and deaf, which integrates real-time sign language and language conversion.

Moustafa et al. [10] a bidirectional SVBiComm system, for Sign Voice, runs bidirectionally for the deaf, blind, dumb and normal people through the use of master learning, in a speech to text. In order to enable various education applications to be accessible, Efthimiou and Fotinea [11] have worked on the Greek sign language for deaf users in a platform environment. The multimedia dictionary construction tool for Indian Sign Language (ISL) was proposed by Dasgupta et al. [12]. This system does not provide KaSL.

Chung-hsien et al. [13] proposed an innovative Chinese Taiwanese Sign Language (TSL) sentence generator for deaf people. Mateen et al. [14] worked on Microsoft Kinect V2 to translate deaf people's sign language. He then dubbed the sign language and suggested that 3D animation SL be used as a deaf talk. The result of accuracy is 87% for speech in SL and 84% for speech in SL for deaf speech. Salian et al. [15] proposed the capture and processing of the SL movement in real time through MATLAB. In this machine, techniques for learning were used to improve response time and precision. Fuat et al. [16] worked on the capsule network to help the deaf and the dump to communicate with the face, hand and lipreading through SL. About 97% of the accuracy results were validated for capsule network SL recognition.

American Sign Language (ASL) supporting deaf children's learning, Shirali-Shahreza et al. [17] suggested a system to recognize the SL using the PHP scripting language and vice versa. Zulfiqar Ali et al. [18] described their goal of interpreting the real-time SL by using mobile cameras as a conversion medium between deaf-blind, deaf, blind, dumped and normal. Muttaki et al. [19] used the machine learning method to detect and identify the Bangla Sign Language (BaSL) using Oriented Gradients Histogram and support vector machine.

## 3 Kannada Moon Code for Disabled People

For Kannada-speaking deaf-blind people, the researcher had some difficulty in creating the Web site for Kannada. They have studied all the communication techniques possible to help the survivors who speak Kannada. For Kannada alphabets, the researcher used a moon code font as shown in Fig. 1. They built Kannada Moon topic on Kannada font development. Kannada Moon is the new font's original version

| | 0 | 1 | 2 | 3 | 4 | 5 | 6 | 7 | 8 | 9 | A | B | C | D | E | F |
|---|---|---|---|---|---|---|---|---|---|---|---|---|---|---|---|---|
| U+0C8x | □ | ◌ಁ | ◌ಂ | ◌ಃ | □ | ಅ | ಆ | ಇ | ಈ | ಉ | ಊ | ಋ | ಌ | | ಎ | ಏ |
| U+0C9x | ಐ | | ಒ | ಓ | ಔ | ಕ | ಖ | ಗ | ಘ | ಙ | ಚ | ಛ | ಜ | ಝ | ಞ | ಟ |
| U+0CAx | ಠ | ಡ | ಢ | ಣ | ತ | ಥ | ದ | ಧ | ನ | | ಪ | ಫ | ಬ | ಭ | ಮ | ಯ |
| U+0CBx | ರ | ಱ | ಲ | ಳ | | ವ | ಶ | ಷ | ಸ | ಹ | | | ◌಼ | ಽ | ◌ಾ | ◌ಿ |
| U+0CCx | ◌ೀ | ◌ು | ◌ೂ | ◌ೃ | ◌ೄ | | ◌ೆ | ◌ೇ | ◌ೈ | | ◌ೊ | ◌ೋ | ◌ೌ | ◌್ | | |
| U+0CDx | | | | | | ◌ೕ | ◌ೖ | | | | | | | | ೞ | |
| U+0CEx | ೠ | ೡ | ◌ೢ | ◌ೣ | | | ೦ | ೧ | ೨ | ೩ | ೪ | ೫ | ೬ | ೭ | ೮ | ೯ |
| U+0CFx | | ೱ | ೲ | | | | | | | | | | | | | |

**Notes**

1.^ As of Unicode version 11.0

2.^ Grey areas indicate non-assigned code points

**Fig. 1** Official Unicode Consortium code for Kannada [20]

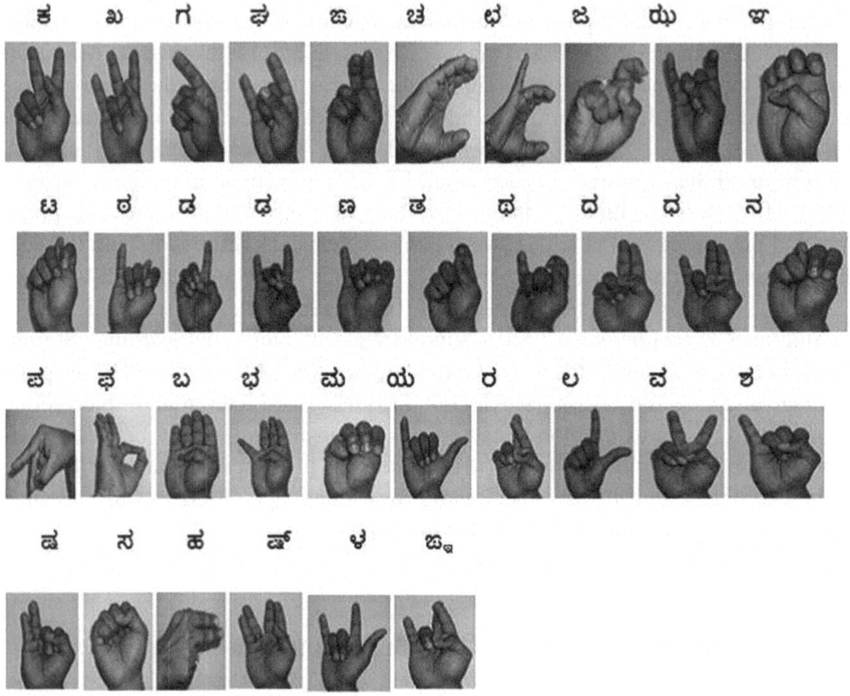


**Fig. 2** KaSL symbol samples [5]

and wishes to have several Kannada updates. The Kannada language read system is from left to right. The people with the surrounding blind have no difficulty with their mother tongue as Kannada. So in the proposed Kannada-speaking framework, the researcher used this Kannada Moon code to help Kannada's Web access. The researcher uses the Fontlab application to implement the look of the Kannada Moon code. The benefits of hand gesture and moon code help the disabled to understand as shown in Figs. 2 and 3.

## 4 Proposed Web Accessibility Framework for Deaf/blind Kannada-Speaking Disabled People

The proposed system and its mechanisms work together to create the information material for the deaf-blind in combined techniques. In Fig. 4, the system architecture proposed at three levels consists of retrieval of databases, application logic and presentation of GUI.

In Fig. 5, explains the working of the Web accessibility framework proposed for the deaf and blind Kannada speakers with disabilities on the client and the server side.

**0C80** **Kannada** **0CEA**

**Various signs**

- 0C80 ಀ KANNADA SIGN SPACING CANDRABINDU
- 0C81 ◌ಁ KANNADA SIGN CANDRABINDU
- 0C82 ◌ಂ KANNADA SIGN ANUSVARA
- 0C83 ◌ಃ KANNADA SIGN VISARGA
- 0C84 ಄ KANNADA SIGN SIDDHAM
  - = siddhirastu
  - • used at the beginning of texts as an invocation
  - → A8FC ꣼ devanagari sign siddham
  - → 111DB 𑇛 sharada sign siddham

**Independent vowels**

- 0C85 ಅ KANNADA LETTER A
- 0C86 ಆ KANNADA LETTER AA
- 0C87 ಇ KANNADA LETTER I
- 0C88 ಈ KANNADA LETTER II
- 0C89 ಉ KANNADA LETTER U
- 0C8A ಊ KANNADA LETTER UU
- 0C8B ಋ KANNADA LETTER VOCALIC R
- 0C8C ಌ KANNADA LETTER VOCALIC L
- 0C8D <reserved>
- 0C8E ಎ KANNADA LETTER E
- 0C8F ಏ KANNADA LETTER EE
- 0C90 ಐ KANNADA LETTER AI
- 0C91 <reserved>
- 0C92 ಒ KANNADA LETTER O
- 0C93 ಓ KANNADA LETTER OO
- 0C94 ಔ KANNADA LETTER AU

**Consonants**

- 0C95 ಕ KANNADA LETTER KA
- 0C96 ಖ KANNADA LETTER KHA
- 0C97 ಗ KANNADA LETTER GA
- 0C98 ಘ KANNADA LETTER GHA
- 0C99 ಙ KANNADA LETTER NGA
- 0C9A ಚ KANNADA LETTER CA
- 0C9B ಛ KANNADA LETTER CHA
- 0C9C ಜ KANNADA LETTER JA
- 0C9D ಝ KANNADA LETTER JHA
- 0C9E ಞ KANNADA LETTER NYA
- 0C9F ಟ KANNADA LETTER TTA
- 0CA0 ಠ KANNADA LETTER TTHA
- 0CA1 ಡ KANNADA LETTER DDA
- 0CA2 ಢ KANNADA LETTER DDHA
- 0CA3 ಣ KANNADA LETTER NNA
- 0CA4 ತ KANNADA LETTER TA
- 0CA5 ಥ KANNADA LETTER THA
- 0CA6 ದ KANNADA LETTER DA
- 0CA7 ಧ KANNADA LETTER DHA
- 0CA8 ನ KANNADA LETTER NA
- 0CA9 <reserved>
- 0CAA ಪ KANNADA LETTER PA
- 0CAB ಫ KANNADA LETTER PHA
- 0CAC ಬ KANNADA LETTER BA
- 0CAD ಭ KANNADA LETTER BHA
- 0CAE ಮ KANNADA LETTER MA
- 0CAF ಯ KANNADA LETTER YA
- 0CB0 ರ KANNADA LETTER RA
- 0CB1 ಱ KANNADA LETTER RRA
- 0CB2 ಲ KANNADA LETTER LA
- 0CB3 ಳ KANNADA LETTER LLA
- 0CB4 <reserved>
- 0CB5 ವ KANNADA LETTER VA
- 0CB6 ಶ KANNADA LETTER SHA
- 0CB7 ಷ KANNADA LETTER SSA
- 0CB8 ಸ KANNADA LETTER SA
- 0CB9 ಹ KANNADA LETTER HA

**Various signs**

- 0CBC ◌಼ KANNADA SIGN NUKTA
- 0CBD ಽ KANNADA SIGN AVAGRAHA

**Dependent vowel signs**

- 0CBE ◌ಾ KANNADA VOWEL SIGN AA
- 0CBF ◌ಿ KANNADA VOWEL SIGN I
- 0CC0 ◌ೀ KANNADA VOWEL SIGN II
  - ≡ 0CBF ◌ಿ 0CD5 ◌ೕ
- 0CC1 ◌ು KANNADA VOWEL SIGN U
- 0CC2 ◌ೂ KANNADA VOWEL SIGN UU
- 0CC3 ◌ೃ KANNADA VOWEL SIGN VOCALIC R
- 0CC4 ◌ೄ KANNADA VOWEL SIGN VOCALIC RR
- 0CC5 <reserved>
- 0CC6 ◌ೆ KANNADA VOWEL SIGN E
- 0CC7 ◌ೇ KANNADA VOWEL SIGN EE
  - ≡ 0CC6 ◌ೆ 0CD5 ◌ೕ
- 0CC8 ◌ೈ KANNADA VOWEL SIGN AI
  - ≡ 0CC6 ◌ೆ 0CD6 ◌ೖ
- 0CC9 <reserved>
- 0CCA ◌ೊ KANNADA VOWEL SIGN O
  - ≡ 0CC6 ◌ೆ 0CC2 ◌ೂ
- 0CCB ◌ೋ KANNADA VOWEL SIGN OO
  - ≡ 0CCA ◌ೊ 0CD5 ◌ೕ
- 0CCC ◌ೌ KANNADA VOWEL SIGN AU

**Virama**

- 0CCD ◌್ KANNADA SIGN VIRAMA
  - • preferred name is halant

**Various signs**

- 0CD5 ◌ೕ KANNADA LENGTH MARK
- 0CD6 ◌ೖ KANNADA AI LENGTH MARK

**Additional consonants**

- 0CDE ೞ KANNADA LETTER FA
  - ≈ KANNADA LETTER LLLA
  - • obsolete historic letter
  - • name is a mistake for LLLA

**Additional vowels for Sanskrit**

- 0CE0 ೠ KANNADA LETTER VOCALIC RR
- 0CE1 ೡ KANNADA LETTER VOCALIC LL

**Dependent vowels**

- 0CE2 ◌ೢ KANNADA VOWEL SIGN VOCALIC L
- 0CE3 ◌ೣ KANNADA VOWEL SIGN VOCALIC LL

**Reserved**

*For viram punctuation, use the generic Indic 0964 and 0965.*

- 0CE4 <reserved>
  - → 0964 । devanagari danda
- 0CE5 <reserved>
  - → 0965 ॥ devanagari double danda

**Digits**

- 0CE6 ೦ KANNADA DIGIT ZERO
- 0CE7 ೧ KANNADA DIGIT ONE
- 0CE8 ೨ KANNADA DIGIT TWO
- 0CE9 ೩ KANNADA DIGIT THREE
- 0CEA ೪ KANNADA DIGIT FOUR



**Fig. 3** Enumerated versions of the Unicode Standard of Kannada [21]

**Fig. 3** (continued)

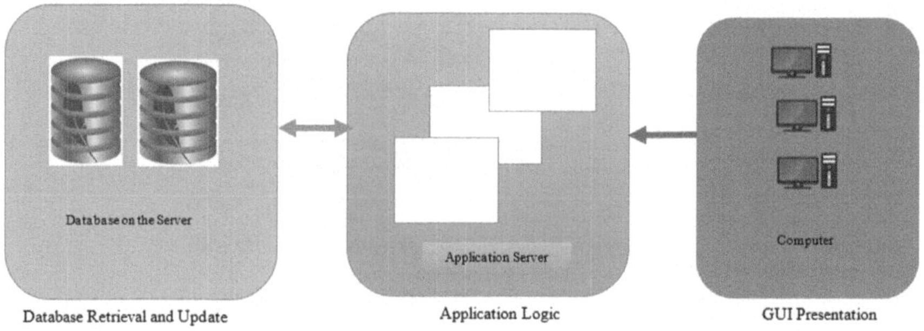


**Fig. 4** Three-tier proposed system architecture

*Step 1: Client-side module User Interface*: When the client enters the required URL first, he/she decides on the type of disability. The server-side modules then process the request and return the answer to the user.

*Step 2: Module for Server Side*

- Validator User ID: Check the login data of the user and determine the user type.
- Data Library of users: includes all the information of the user; Name, password, e-mail address and type of user disabled.
- Scanner Web site: this is the initial module on the device's server side. The scanned content is then passed on to the acceptable module and opens the specified web page and scans the complete contents.
- Meta-Language Management: This module summarizes the steps that researchers followed to be prepared to use every web page's meta-language.
- Tags Library: Scan and insert all web pages in the database for further processing within the contents generator module.
- Page Content Generator: Two submodules: 1-Web Page Content Extractor and 2-Web Tags Filter.
  - Web content extractor extracts the content of the Kannada web page.

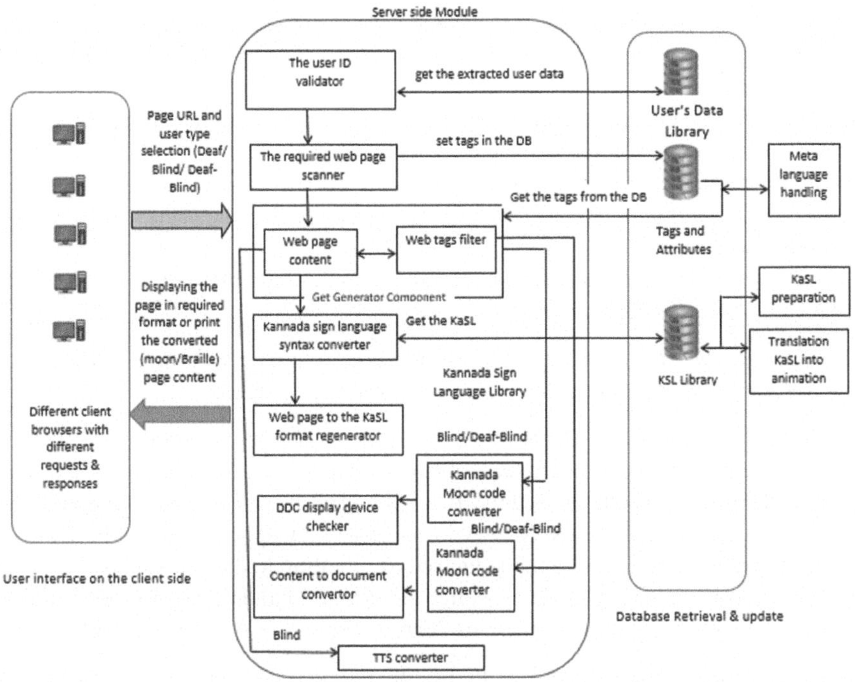


**Fig. 5** Proposed web accessibility framework for deaf/blind Kannada-speaking disabled people

  – Web tags filter module eliminates the extracted content of each tag and returns only the pure content of the page. Then the extracted content is transferred to the next modules.

- KaSL syntax Converter: Interpret KaSL synonymous SL words from the KaSL library on the specified web page.
- KaSL Library Preparation Module: The KaSL in its vocabulary and descriptive linguistics is no different from the Kannada language. The primary requirement is to create a dictionary system for the exchange of data with KaSL specialists, experts and reading various KaSL references. The second step is to obtain a piece of writing in an extremely specific field and to apply the knowledge of KaSL. All words must be converted to their roots in order to induce words that are translated into KaSL from any page. The synonymous words should then be declared. Then check the synonymous existence of KaSL.
- Interpretation of KaSL Words in the Animations module: Once the words in the KaSL module are assembled. It is necessary to declare the gestures of each word and keep the corresponding word in the database. Then the animation of each word is meant by gestures of the word. In the information, the animations are finally reconstructed.

- KaSL library: It contains almost 6,300 Kannada gestures close to the most common Kannada words. The KaSL syntax device output is then transferred to the module in question.
- KaSL format Web Page Regenerator: This module converts signs to an appropriate Web format by recreating web page content with tags.
- Text to Speech Converter (TTS): TTS is an external service used in the framework. If the user is a (blind person), he/she will choose whether the page will be loudly spoken or not.
- Converter code takes the user input and gets an acceptable output in the blind language selected (Braille or Moon).
  - Kannada Moon Code Converter: Moon Code Converter web page content.
  - Braille Converter: Moon code converter web page content.
- Display Device Checker (DDC): display of the output on the appropriate device; DDC checks whether or not an acceptable display device (Moon Display Device) is blocked by a personal computer. If it is blocked, it will transfer the output. If no device is plugged in; a special error message may appear to be a blind user's spoken error message and may be viewed as vibrating to the deaf-blind user.
- Content to Document Converter: Enter the device module (Moon Code/Braille) and create a flying computer file with highly descriptive Web site content.

The Kannada-speaking people with disabilities can access and remove barriers to effective learning with the KaSL animation system.

## 5 Conclusion

The proposed framework addresses three disability varieties: deaf, blind and deaf-blind. The library consists of all Kannada words and corresponding hand gesture KaSL animations. The system uses Kannada Moon code for the deaf-blind people with KaSL animation library and is considered as reliable; as they are very small animated gif images. The KaSL library has more benefits than another sign language. By this system, deaf-blind Kannada-speaking disabled people can access for their lifelong learning with less cost, effort and time. In the future, the researcher can take this method for implementation in another sign language for their learning.

**Acknowledgements** The authors express gratitude toward the assistance provided by Accendere Knowledge Management Services Pvt. Ltd. in preparing the manuscripts. We also thank our mentors and faculty members who guided us throughout the research and helped us in achieving the desired results.

## References


1. Thatcher, J., Burks, M.R., Heilmann, C., Kirkpatrick, A., Lauke, P.H. Lawson, B., Henry, S.L., Regan, B., Rutter, R., Urban, M., Waddell, C.: Web accessibility: web standards and regulatory compliance, Part 1 Chapter 1, ISBN-10: 1-59059-638-2, ISBN-13: 978-1-59059-638-8 (2006)
2. Miró-Borrás, J., Bernabeu-Soler, P.: Text entry in the E-commerce age: two proposals for the severely handicapped. J. Theor. Appl. Electr. Commer. Res., **4**(1) (2009). ISSN 0718-1876 Electronic Version
3. Survey of Income and Program Participation (SIPP), Available: http://www.census.gov/sipp/, Retrieved: 18/5/2018
4. Association for Volunteer Administration: Understanding web accessibility. J. Volunt. Adm. **21**, 36–39 (2003)
5. Ramesh, M.K., Nagaraj, H.: New methodology for translation of static sign symbol to words in Kannada Language. Int. J. Comput. Appl. **121**, 25–30 (2015)
6. Umang, P., Aarti, G.A.: Moment based sign language recognition for indian languages. In: 2017 International Conference on Computing, Communication, Control and Automation (ICCUBEA), pp. 1–6. IEEE (2017)
7. Vikram, M.S., Vinay, N.K., Shruti C., M., Suma, M., Ambika, D.R.: Virtual talk for deaf, mute, blind and normal humans. In: Proceedings—2013 Texas Instruments India Educators' Conference, TIIEC 2013, pp. 316–320 (2013)
8. Rajapandian, B., Harini, V., Raksha, D., Sangeetha, V.: A Novel approach as an aid for blind, deaf and dumb people. In: 2017 IEEE 3rd International Conference on Sensing, Signal Processing and Security (ICSSS), pp. 403–408 (2017)
9. Joao, U., Tiago, O., Paula Maria, E., Nuno, E., Fernando Maciel, B.: ACE assisted communication for education: Architecture to support blind & deaf communication. In: IEEE Global Engineering Education Conference, EDUCON, pp. 1015–1023 (2018)
10. Mariam Moustafa, R., Nada Gamal, M., Rania, A., Abdel, A., Abul, S.: SVBiComm: sign-voice bidirectional communication system for normal, "deaf/ dumb" and blind people based on machine learning. In: 2018 1st International Conference on Computer Applications & Information Security (ICCAIS), pp. 1–8. IEEE (2018)
11. Efthimiou, E., Fotinea, S.-E.: An environment for deaf accessibility to educational content. In: New Trends in ICT and Accessibility—Proceedings of the 1st International Conference in Information and Communication Technology and Accessibility, ICTA 2007, pp. 125–130 (2007)
12. Dasgupta, T., Shukla, S., SandeepKumar, Diwakar, S., Basu, A.: A Multilingual Multimedia Indian Sign Language Dictionary Tool. In: Proceedings of the 6th Workshop on Asian Language Resources. pp. 57–64 (2008)
13. Chung-hsien, W., Yu-hsien, C., Chi-shiang, G.: Text generation from Taiwanese sign language using a PST-based language model for augmentative communication. IEEE Trans. Neural Syst. Rehabil. Eng. **12**, 441–454 (2004)
14. Mateen, A., Mujtaba, I., Rafia, M., Sana, K.: Deaf talk using 3D animated sign language. In: SAI Computing Conference, pp. 330–335 (2016)
15. Salian, S., Dokare, I., Serai, D., Suresh, A., Ganorkar, P.: Proposed system for sign language recognition. In: 2017 International Conference on Computation of Power, Energy, Information and Communication (ICCPEIC), pp. 58–62 (2017)
16. Fuat B E Ş, E., Merve Ayyüce, K., Bolat, B.: Recognition of sign language using capsule networks. In: 2018 26th Signal Processing and Communications Applications Conference (SIU), pp. 5–8 (2018)
17. Shirali-Shahreza, M., Shirali-Shahreza, S.: Encouraging persons with hearing problem to learn sign language by internet websites. In: Eighth IEEE International Conference on Advanced Learning Technologies. pp. 1036–1037 (2008)
18. Zulfiqar Ali, M., Muhammad Uzair, A., Talha Hussain, S., Zahid Abbas, B., Umer, A.: Real time translator for sign languages. In: Proceedings—2017 International Conference on Frontiers of Information Technology, FIT 2017, pp. 144–148 (2018)

19. Muttaki, H., Tanvir Hossain, S., Mrinmoy, D.: A machine learning based approach for the detection and recognition of Bangla sign language. In: 2016 International Conference on Medical Engineering, Health Informatics and Technology (MediTec), pp. 1–5 (2016)
20. "Unicode character database," The Unicode Standard, Retrieved: 10/3/2018
21. "Enumerated Versions of the Unicode Standard," The Unicode Standard, Retrieved: 10/3/2018

# Real-Time Prediction of Development and Availability of ICT and Mobile Technology in Indian and Hungarian University

**Chaman Verma, Zoltán Illés and Veronika Stoffová**

**Abstract** An experimental study was conducted to predict the development and availability (DA) of the latest information and communication technology (ICT) and mobile technology (MT) in Indian and Hungarian University. A primary dataset with 328 instances and 16 features was analyzed using four supervised machine learning algorithms such as support vector machine (SVM), artificial neural network (ANN), random forest (RF), and logistic regression (LR). The dataset was trained and tested using hold out and *K*-fold cross-validation methods with classifiers. Further, to compare the performance of classifiers, *T*-test at 0.5 significant level was also applied. Feature mapping was also achieved by applying principal component analysis (PCA) to enhance the prediction accuracy of classifiers. The findings of the study conclude the feature extraction using PCA enhanced the prediction accuracy of each classifier except SVM with tenfolds at 0.5 thresholds of variance. Also, it is revealed that within the real time of 1.4 s ANN attained stable and highest accuracy in the prediction of DA of ICT and MT in the University of both countries. *T*-test implies the significant difference between RF and others in prediction accuracy. Also, a significant difference is found in ANN and others considering the processor time to train model for real-time prediction.

**Keywords** Artificial neural network · Feature extraction · Machine learning · Real time · Principal component analysis

---

C. Verma (✉) · Z. Illés
Eötvös Loránd University, Budapest, Hungary
e-mail: chaman@inf.elte.hu

Z. Illés
e-mail: illes@inf.elte.hu

V. Stoffová
Trnava University, Trnava, Slovakia
e-mail: NikaStoffova@seznam.cz



P. K. Singh et al. (eds.), *Proceedings of ICRIC 2019*, Lecture Notes in Electrical Engineering 597, https://doi.org/10.1007/978-3-030-29407-6_43

## 1 Introduction and Related Work

Machine learning (ML) is the process of estimating unknown dependencies or structures in a system using a limited number of observations and it is used in data mining applications to retrieve hidden information and used in decision-making [1]. In ML, for classification, and regression problem various classifiers can be used for learning decision trees, rules, Bayes networks, artificial neural networks, and support vector machines, and different knowledge representation models can be used to support decision-making methods [2]. In addition to statistical methods [3, 4], the use of ML in trending in the educational domain in the prediction of various targets in educational datasets [5–8]. The support vector machine (SVM) is a supervised learning model presented for binary classification in both linear and nonlinear versions [9] and it performs classification by constructing an *N*-dimensional hyperplane that optimally separates the data into two categories [10]. Random forest (RF) is a combination of tree predictors such that each tree depends on the values of a random vector sampled independently and with the same distribution for all trees in the forest [11]. Artificial neural network (ANN) is simple mathematical models defining a function $f: X \rightarrow Y$ or distribution over $X$ or both $X$ and $Y$, but sometimes models are also intimately associated with a learning algorithm or learning rule [12]. The binary logistic regression (LR) is much suitable for a binary classification problem [13]. Recently, students' demographic features such as residence state and gender were predicted using supervised machine learning classifiers [14–16]. Dimension reduction (DR) is a way to a reduced number of features in the given dataset. Feature extraction is the type of DR that can be implemented using principal component analysis (PCA) which transforms or projects a space composing of many dimensions into a space of fewer dimensions. PCA is a mapping method which transforms several (possible) correlated features into a (smaller) number of uncorrelated features called principal component.

Use of ML to provide real-time prediction is also popular and trending. The real time is the time level of computer responsiveness that a user senses as sufficiently immediate or that enables the computer to keep up with some external process. An experimental study was conducted to predict the age group of the University's' students and an idea of real-time prediction was added [17]. According to [18], real-time tasks are produced due to the occurrence of either internal or external events. In real-time systems, the absolute deadline for task begins with time zero and the relative deadline is with respect to the task released time. Real-time predictive models of European school students' nationality to monitoring online ICT access and ICT-based activities were also suggested by [19]. The prediction of Indian and Hungarian University student's attitude for real time was also conducted by [20]. The present study is conducted to predict future development and availability (DA) of ICT and MT in Indian and Hungarian University in real time. To achieve this goal, we compared predictive models using *T*-test and CPU time is also calculated. By implementing the presented predictive models, we can also query across the entire dataset online or query a subset of the dataset for a suitable match in the real-time prediction of

DA of ICT and MT. By deploying DA predictive models into the real-time module of the institution's Web site followings benefits may be occurred.

1. To identify the latest ICT and MT techniques are developing or not at the institute.
2. To detect the latest ICT and MT resource facilities are available or not to the stakeholder.
3. To institute administration to aware about the status of development and availability reporting by the student in real time.

The present study is categorized into five major sections. Section 1 explains the basic introduction of presented work with little literature about machine learning and real-time, and Sect. 2 discusses research design and methodology. Section 3 explores the experimental results, and Sect. 4 discusses predictive model evaluation and debates on the results of experiments. Section 5 concludes the essence of the present study with a significant recommendation for future work.

## 2 Research Design and Methodology

### *2.1 Dataset and Preprocessing*

A primary dataset has been gathered using stratified random sampling from Indian and Hungarian University with Google form and direct discussions at a different scale. The questionnaire is divided into five parts: The first one is demographic and the other four belong to ICT and MT parameters represented by attitude (6 features), development-availability (16 features), educational benefits (9 features) and usability (6 features). Out of 331 students, 169 students belong to the Eötvös Loránd University of Hungary and 162 students belong to the Chandigarh University of India. The overall dataset consists of 331 instances and 46 features which are related to the 4 major ICT parameters and demographic characteristics.

### *2.2 Feature Extraction*

Using self-reduction, 9 features related to the demographic characteristics such as age, sex, locality, country are removed, and 21 features that belong to other 3 ICT parameters are also eliminated because our focused parameter is DA. The missing values are handled with Replace Missing Value filter in the Weka tool. Based on the mean score of student's responses, we framed three distinct classes named Yes equals to 1, No equals 2, and Don't know equals 3 under attribute DA status. Only three instances were removed belong to the class Don't know due to a minority. Finally, the target attribute DA status has two values named development and availability (DA) and non-development and availability (NDA). Afterward, a total of 328 instances with 16 features are analyzed using the unsupervised PCA filter. PCA performed transformation and provided 15 features which are combinations 5 features. The maximum and minimum standard deviation (SD) of 1st ranked feature and 7th ranked feature is calculated 1.75 and 0.69. Based on 0.5 thresholds ($1.751 * 0.5 = 0.875$), we considered features having SD is greater than 0.875. Hence, we found only 7 features to be trained and tested using classifiers and named as PCA-7.

### 2.3 Training, Testing, and Validation

Firstly, training ratio or holdout method is applied on the unextracted dataset in which splitting task performed randomly in two distinct subsets whereas the first subset from where classifier tries to extract knowledge and second set is used to the tested extracted information. Hence, the ratio to train and test dataset is used as 66–44%. Secondly, *K*-fold cross-validation is also applied at various folds such as 2, 4, 6, 8, and 10 to validate the test dataset to train the maximum. To build predictive models, the unextracted dataset is trained, tested and validated using holdout and *K*-fold methods. The accuracy of trained models is compared at various folds using *T*-test at 0.5 level of confidence. Afterward, the extracted dataset PCA-7 is also tested and validated with tenfold cross-validation at 0.5 variance thresholds. Again, the accuracies of extracted trained models are compared using *T*-test at 0.5 level of confidence.

### 2.4 Classifiers

To predict the development and availability of ICT and MT at University, the unextracted dataset is trained with four supervised machine learning classifiers such as support vector machine (SVM), multilayer perceptron(ANN), random forest (RF), and logistic regression (LR) in Weka 3.9.3 tool.

### 2.5 Performance Metrics

To evaluate the performance of the predictive model in binary classification, we used the following metrics:

1. Accuracy: The number of correct predictions of DA status from overall predictions.
2. Error: The number of incorrect predictions of DA status from overall predictions.

3. Receiver operating characteristics curve (ROC): It displays a graphical curve which shows the true positive rate (Sensitivity) on the $y$-axis and false positive rate (1-Specificity) at $x$-axis with various thresholds.
4. Area Under ROC: To show the accuracy of models' area under the curve of ROC is also suitable.
5. $F$-score: It is harmonic mean of precision and recall which also states the significance of the predictive model is calculated by the formula

$$F = 2 * (\text{Recall} * \text{Precision})/(\text{Recall} + \text{Precision}) \tag{1}$$

6. Kappa Static: The Cohen's Kappa is statistical which determines the agreement among instances in the dataset. The formula to calculate Kappa is given below

$$\text{Kappa} = (\text{Calculated accuracy} - \text{Expected accuracy})/(1 - \text{Expected accuracy}) \tag{2}$$

### 2.6 Real Time

According to the goal of the study, there is a requirement of prediction of the development and availability of ICT and MT at University; we need to calculate the process (CPU) time to training the model. For this experiment application of Weka is appropriate which helps to calculate and compare the CPU time to train a predictive model by classifiers. At a 0.5 significant level, $T$-test is also applied to compare this real time to predict the DA of ICT and MT at the University.

## 3 Experiments and Result Discussions

### 3.1 Experiment-I

In this experiment, we used the holdout method to train and test the unextracted dataset using four classifiers. The training ratio 66–44% was considered to test and train dataset. Usually, a random split occurs on data set under the mining task of test data. The accuracy of supervised machine learning classifiers was also compared using $T$-test at a 0.5 significant level.

Data from Fig. 1 reflects the accurate comparison of four classifiers compared with $T$-test for DA prediction of ICT and MT. It is found that two classifiers named LR and SVM gained the highest and same accuracy of 96.4% in prediction. The minimum accuracy is achieved by RF which is 75%. According to $T$-test, we found a significant difference between the accuracy of RF and the rest of three classifiers in DA prediction.

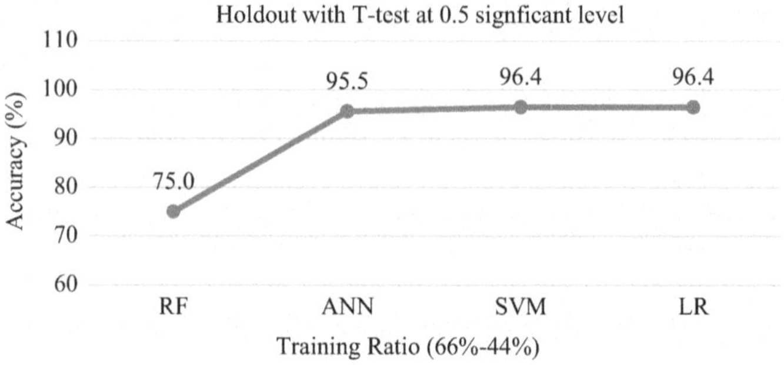


**Fig. 1** Train and test using holdout with *T*-test at a 0.5 significant level

## *3.2 Experiment-II*

This experiment is conducted with the help of the *K*-fold cross-validation method to validate the test data. For this, the unextracted dataset is validated using four classifiers. In this, k considered as train set and *k*-1 considered as a test set. In this experiment, we selected *k* value such as 2, 4, 6, 8, and 10 for the dataset and trained models using classifiers to examine DA prediction accuracy. To compare the accuracy of classifiers, we also applied *T*-test at a 0.5 significant level. *T*-test found a significant difference in prediction accuracy of RF and ANN; SVM and ANN; LR and RF; LR and SVM. Data from Fig. 2 evident that RF and SVM's accuracies are directly proportional to the *k* value and the accuracy (96.2%) of ANN is stabilized at 6 and 8-fold. Later, at tenfold, it is increased by 0.4. The maximum prediction accuracy (97.1%) is attained by SVM and LR at tenfold. The minimum prediction accuracy (89.8%) is gained by RF at fourfold.

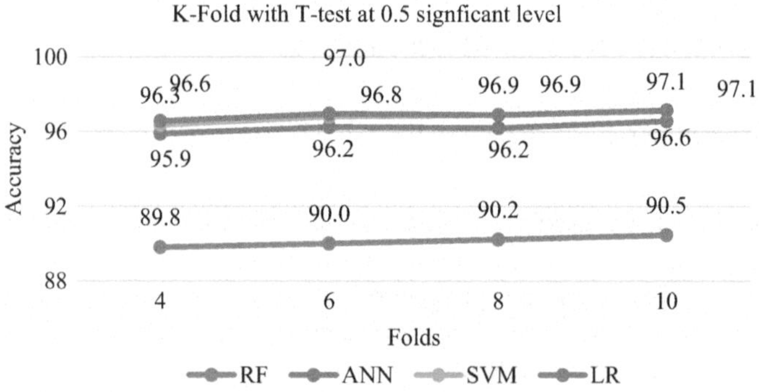


**Fig. 2** Validation using *K*-fold with *T*-test at a 0.5 significant level

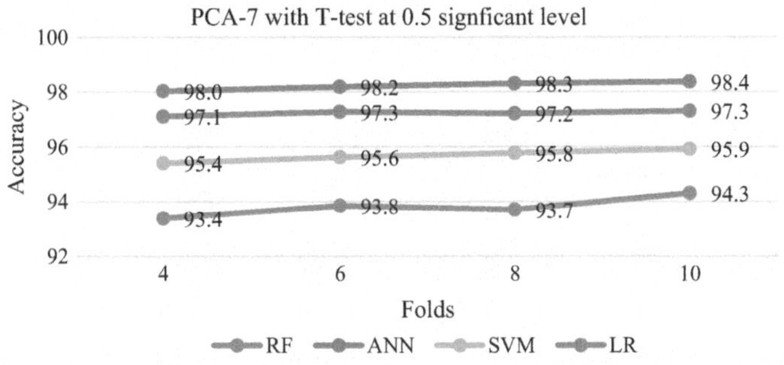


**Fig. 3** Extracted dataset train and test using tenfold with *T*-test at 0.5 significant level

## 3.3 Experiment-III

In this experiment, the dataset is extracted using the PCA method. Later, extracted dataset are tested and validated using *K*-fold with varying k value. Also, DA prediction accuracy is significantly tested using *T*-test at 0.5 level of significance. Out of 15 features, PCA provided 07 features which are combinations 5 features. At 0.5 thresholds, we considered only those features whose variance values are greater than 0.875.

Data from Fig. 3 shows the accurate comparison of the extracted dataset of four classifiers compared with *T*-test at a 0.5 significant level. It is revealed that PCA-7 significant enhances the DA prediction accuracy of all classifiers except SVM only. One hand, the accuracy of SVM decreases with 1.2%, and another hand, ANN accuracy enhanced by 1.8% at tenfold which was found the maximum in the unextracted dataset. Hence, it is concluded that feature extraction enhanced DA prediction accuracy (98.4%) of ANN. Also, *T*-test with feature extraction did not find a meaningful difference between RF and SVM; ANN and LR in prediction accuracy.

## 3.4 Experiment-IV

To present a real-time significant model, this experiment compared the induced CPU time to predict the DA of ICT and MT at the University. For this, we tested and validated extracted dataset using *T*-test at 0.5 significant level to keep in view CPU training time in seconds. In Fig. 4, primary y-axis denotes accurate prediction count of DA of ICT and MT, and the secondary y-axis shows CPU time in seconds. The x-axis shows the comparison of CPU time induced to train model by each classifier on the extracted dataset. For both of classes DA and NDA, the identical prediction is found by ANN. Out of 328, a total of 322 instances are predicted correctly, and induced CPU time is calculated 1.45 which is maximum time.

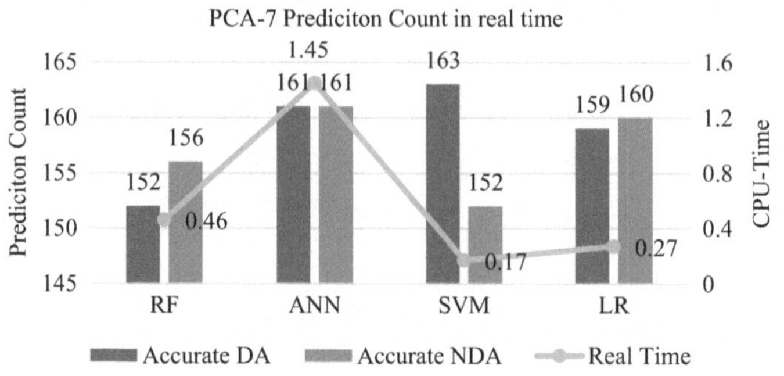


**Fig. 4** Real-time prediction count at tenfold with *T*-test at 0.5 significant level

The minimum CPU time 0.17 s is induced by SVM to train PCA-7 with the prediction of 315 instances. The minimum prediction count for DA is 152 and for NDA is 156 provided by RF. Further, *T*-test found a significant difference between ANN's CPU time and the other three classifiers.

## 4 Model Evaluation

To evaluate the DA predictive model, four performance metrics are compared by running an experiment with *T*-test at a 0.5 significant level. The results are shown in Table 1.

We displayed the various performance measures to evaluate the significance of models. The strongest association among instances is proven by excellent kappa static 0.97 of ANN which signifies the strength of prediction. Highest *F*-score 0.98 of ANN also proved strong balanced between the precision and the recall. The RF has maximum prediction error and ANN has minimum prediction error which are calculated as 5.7 and 1.7, respectively. The area under the curve is 1 which also proved the significance of the ROC curve of ANN and SVM classifier.

Data from Fig. 5 shows the ROC curve which compared the sensitivity of real-time models at various thresholds for the DA class. The significant TP rate starts from 0.75 and ends to 0.99 with updating thresholds. Also, can be seen at thresholds

**Table 1** Performance metrics at the tenfold of PCA-7 at 0.5 thresholds

| Classifier | Kappa static | *F*-score | Area under ROC | Error |
|---|---|---|---|---|
| RF | 0.89 | 0.94 | 0.99 | 5.7 |
| ANN | 0.97 | 0.98 | 1 | 1.7 |
| SVM | 0.92 | 0.96 | 0.96 | 4.1 |
| LR | 0.95 | 0.97 | 1 | 2.7 |

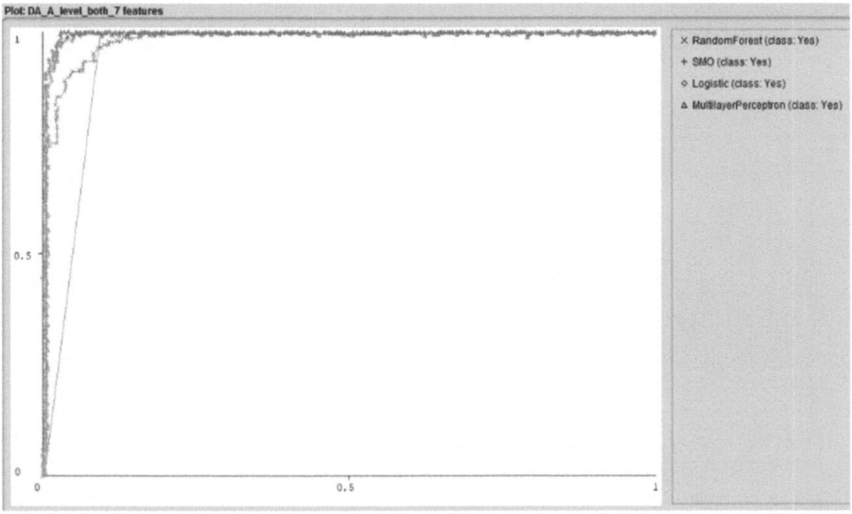


**Fig. 5** ROC of DA prediction at tenfold

0.2 the sensitivity is high 0.95 and the FP rate is 0.05 which reveals the significance of the real-time predictive models.

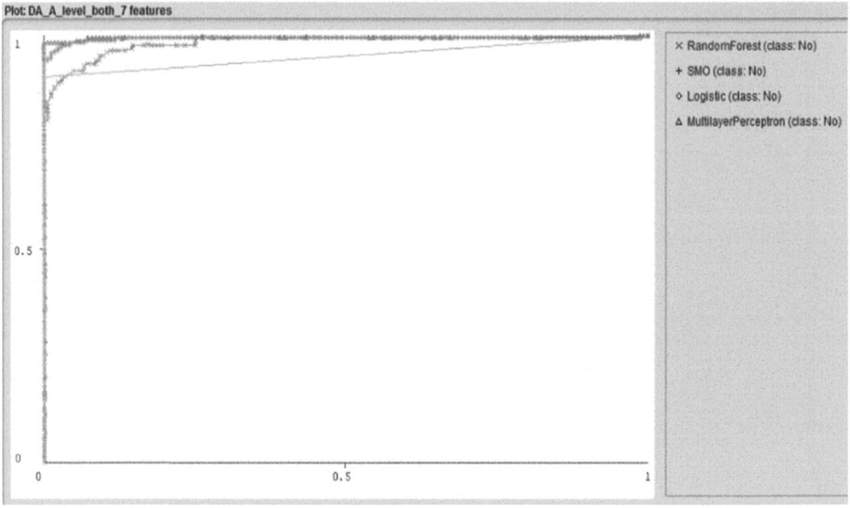


**Fig. 6** ROC of NDA prediction at tenfold

Figure 6 displays the ROC curve which compared the sensitivity of real-time models at various thresholds for NDA class. Here, we have found that the ANN classifier's TP rate higher than others and it starts from 0.90 and ends to 0.99 with updating thresholds. The real-time model started sensing at thresholds 0.01 with the highest TP rate 0.96 and the FP rate is 0.04 which exposes the significance of the real-time predictive model. Accordingly, ANN with PCA-7 outperformed the others to predict the DA and NDA of ICT and MT in both countries.

## 5 Conclusion

To predict the DA, four experiments are conducted on the primary dataset using four supervised machine learning classifiers with *T*-test at a 0.5 significant level. The dataset is trained, tested, and validated using holdout and *K*-fold techniques. To extract features from datasets PCA is also applied to enhance the prediction accuracy of classifiers. In the holdout method, LR gained the highest accuracy (96.4%) as compared to others. Further, *T*-test found a significant difference between the accuracies of RF and the rest of three classifiers at a 0.5 significant level. In the *k*-fold method, highest prediction accuracies (97.1%) are given by SVM and LR at tenfold. Here, *T*-test also found a significant difference in prediction accuracy of RF and ANN; SVM and ANN; LR and RF; LR and SVM. Additional, after applied PCA-7 the accuracies of all classifiers are enhanced except SVM classifier. It is also concluded that the highest prediction of DA in 1.45 s is gained by ANN classifier with PCA-7. It is also revealed that the lowest CPU time 0.17 s is induced by SVM to train PCA-7 with 95.9% prediction accuracy. Further, *T*-test at 0.5 significant level also found a significant difference between real-time induced by ANN and CPU time of the remaining three classifiers. Therefore, due to the highest accuracy of 98.4%, it is proved that ANN is best suited for the real-time prediction of DA of ICT and MT in the University of both countries. The future research is recommended to implement deep neural network models to enhance the prediction accuracy with minimizing the training time.

**Acknowledgements** The first author thanks Tempus Public foundation of Hungary to sponsoring his Ph.D. study related to this research work. Also, his project is sponsored by the Hungarian Government and Co-financed by the European Social Fund under the project "Talent Management in Autonomous Vehicle Control Technologies (EFOP-3.6.3-VEKOP-16-2017-00001)".

## References

1. Bonnaccorsi, A.: On the relationship between firm size and export intensity. J. Int. Bus. Stud. **23**(4), 605–635 (1992)
2. Clerc, M.: The swarm and the queen: towards a deterministic and adaptive particle swarm optimization. In: Proceedings of the IEEE Congress on Evolutionary Computation (CEC), pp. 1951–1957, IEEE (1999)
3. Verma, C., Dahiya, S.: Gender difference towards information and communication technology awareness in indian universities. SpringerPlus **5**, 1–7 (2016)
4. Verma, C., Dahiya, S., Mehta, D.: An analytical approach to investigate state diversity towards ict: a study of six universities of punjab and Haryana. Indian J. Sci. Technol. **9**, 1–5 (2016)
5. Kabakchieva, D.: Student performance prediction by using data mining classification algorithms. Int. J. Comput. Sci. Manag. Res. **1**(4), 686–690 (2012)
6. Maria Teresa, C., Maria, Noel R.: Prediction of university students academic achievement by linear and logistic models. The Spanish J. Psychol. **2**(1), 275–288 (2015)
7. Kolo, D.: A decision tree approach for predicting students' academic performance. Int. J. Educ. Manag. Eng. **5**, 12–19 (2015)
8. Verma, C.: Educational data mining to examine mindset of educators towards ICT knowledge. Int. J. Data Min. Emerg. Technol. **7**, 53–60 (2017)
9. Deshmukh, B., Patil, A., Pawar, B.: Comparison of classification algorithms using weka on various datasets. Int. J. Comput. Sci. Inf. Technol. **4**(2), 85–90 (2011)
10. Cheu, R.L., Srinivasan, D., Tian E.: Support vector machine models for freeway incident detection. In: Intelligent Transportation Systems, Proceedings. vol. 1, pp. 238–243. IEEE (2003)
11. Kumar, S., Ramesh.: Comparative analysis of weka data mining algorithm random forest, random tree and lad tree for classification of indigenous news data. Int. J. Emerg. Technol. Adv. Eng. **5**(1), 507–517 (2015)
12. Minsky, M., Papert, S.: Perceptrons: An Introduction to Computational Geometry. MIT Press (2017)
13. Verma, C., Ahmad, S., Stoffová, V., Illés, Z., Dahiya, S.: Gender prediction of the european school's teachers using machine learning: preliminary results. In: International Advance Computing Conference, pp. 213–220. IEEE, India (2018)
14. Verma, C., Ahmad, S., Stoffová, V., Illés, Z.: Forecasting residence state of indian student based on responses towards information and communication technology awareness: a primarily outcomes using machine learning. In: International Conference on Innovations in Engineering, Technology and Sciences, IEEE, India (2018, in press)
15. Verma, C., Stoffová, V., Illés, Z., Dahiya, S.: Binary logistic regression classifying the gender of student towards computer learning in European schools. In: THE 11th Conference of Ph.D. students in computer science, pp. 45. Szeged University, Hungary (2018)

16. Verma, C., Stoffová, V., Illés, Z.: An ensemble approach to identifying the student gender towards information and communication technology awareness in european schools using machine learning. Int. J. Eng. Technol. **7**, 3392–3396 (2018)
17. Verma, C., Stoffová, V., Illés, Z.: Age group predictive models for the real time prediction of the university students using machine learning: Preliminary results. In: International Conference on Electrical, Computer and Communication, In. Press., IEEE, India (2019)
18. Verma, C., Stoffová, V., Illés, Z.: Rate-Monotonic Vs Early Deadline First Scheduling: A Review. In: International Conference on education technology and computer Science in building better future, pp. 188–193. University of Technology and Humanities, Poland (2018)
19. Verma, C., Ahmad, S., Stoffová, V., Illés, Z. Singh, M.: National identity predictive models for the real time prediction of european schools students: preliminary results. In: International Conference on Automation, Computational and Technology Management, IEEE, London (2019, in press)
20. Verma, C., Illés, Z., Stoffová, V.: Attitude prediction towards Ict and mobile technology for the real-time: an experimental study using machine learning. In: The 15th International Scientific Conference eLearning and Software for Education, university Politehnica of Bucharest, Romania (2019, in press)

# A Web Extraction Browsing Scheme for Time-Critical Specific URLs Fetching

**Sunita and Vijay Rana**

**Abstract** Web browsing is the need of the hour present state of the art problem of presenting the specific result to the users. This paper works towards creating an efficient search engine that removes stop words, extract meaningful words and form clusters of highest frequency words and final stage present the result in terms of URLs. The proposed system is divided into three phases: In the first phase, pre-processing is performed by eliminating the stop words. The outcome of this phase is a reduced query. In the second phase, the extraction of meaningful words with a frequent word or similar word replacement is applied. In the last phase, meaningful URLs are fetched through location-sensitive searching and then presented within the same interface. The result is presented in term of URLs fetched and execution time it takes to fetch the results.

**Keywords** Web browsing · Pre-processing · Clustering

## 1 Introduction

Today, the Internet has become the heart and soul of the modern era. Almost everything required by the users can be located over the Web. In the early stage, [1] Web browsers present relevant as well as irrelevant material to the user based on the query presented. Rank-based search engines that are an invention of the modern era presents a user with the result of allocating rank to contents. Higher contents matched within the query with the URL contents will yield the output at first place. Earlier such search engines are paid.

Sunita (✉)
Department of Computer Science, Arni University, Kangra, India
e-mail: sunitamahajan2603@gmail.com

V. Rana
Department of Computer Science, Sant Baba Bhag Singh University, Jalandhar, India
e-mail: vijay.rana93@gmail.com

 
P. K. Singh et al. (eds.), *Proceedings of ICRIC 2019*, Lecture Notes in Electrical Engineering 597, https://doi.org/10.1007/978-3-030-29407-6_44

Nowadays, competition is high and due to that reason, search engines are also evolving. The most popular search engine like Google, Alta-Vista, etc. are free to use and provide a user with the [2, 3] rank-based results. The most common problem with these search engines is a specific result, which is not generated in response to the user query. Overall execution time in such a situation is generally high. This work is focused primarily on the specific result in response to the user query.

The rest of the paper is organized as under: Sect. 2 gives the literature survey, Sect. 3 gives the problem definition, Sect. 4 gives the proposed system, Sect. 5 gives the result and performance analysis and Sect. 6 gives the conclusion and future scope and last section gives the references.

## 2 Literature Survey

This section puts a light on the existing mechanism used within the browsing process to provide efficient results to the user.

In this discussed various mechanisms of extracting meaning from the presented query strings, the sense embedding [4] using retrofitting mechanisms seems to be need of the hour and discussed through this literature effectively. Learning vector is maintained in order to determine the problems if any within the query string. The pre-processing to words is performed through the learning vector. Similarity graph calculation is performed to reduce execution time while browsing Websites. Fetched URLs are maintained within the buffer and finally, these results are displayed to the user using the Web browser.

In proposing a cluster of fake account detection [5] in social media, supervised machine-learning pipeline approach is used . Key features used in this case depend upon name, email, address and company.

The rank-based algorithm [6], [7] is proposed for semantic Web search. In this algorithm, the criteria-based information derived from the huge semantic environment and then user query is analyzed. It mainly utilizes page relevance and then provides a relevance score for a Web page. The page relevance measure involved graph-based representations along with the probability-aware approach. The results show that cost reduction and accuracy are better. But it does not base on Web repositories and multiple ontologies.

In spam-based Web search [8, 9] technologies are used and it is used to detect Web spam. It firstly detects the content features then non-spam pages are identified. After that spam pages that are made by spammers are detected. The results show that it has helped in spam detection. But it does not focus on the semantics which is utilized in searching and also there are no appropriate methods used for detecting content features.

A weighted page rank algorithm [10] in a mobile system that is used to link the structure of various Web pages and calculate the rank of the pages. If a page has more outgoing links, then it has the highest rank. This rank is used to give probability about the particular page when a user query is given. In this algorithm, the current rank of the page is utilized for estimating the probability.

An algorithm [11] that analyzes the structure of Web page links and the authority is provided in that link. Then, according to the user query, information is stored in authority pages. It works in two steps: firstly, sampling is done, and then iterative calculation is performed to solve the user query. It calculates the rank of pages.

Described ontology-based techniques [12] are based on index and relationships. This provides a better search and also provide pages that are based on user interest. It also utilizes semantics for searching the Web pages according to a user query. It only displays those pages that achieved the relationship to the user query.

In describing a system that utilizes an OWL technique for semantic illustration [13] and utilized for monitoring use. In this data, recuperation method is used for highlighting the user interest and then semantic comparability is tested. It gives the group estimation that highlights data, which is according to a user query.

Intelligent Web service [14, 15] uses ontology and retrieves the information in a précised manner. It provides an intelligent agent that analysis user query and gives data related to it according to most searches. It utilizes mining and shortlists the Web pages that are semantically related to it. It decreases irrelevant search result and précised knowledge discovery is made.

## 3 Problem Definition

The existing literature focuses on rank-based searching mechanisms by looking at the user query. The stop word elimination and meaningful word extraction mechanism are missing. This causes high noise and generally results fetching through the Web. To solve the problem, stop word fetching and elimination along with meaningful word extraction must be used. This work is proposed through this literature using stop work elimination and meaningful word extraction phase. Also clustering is done for improving execution time.

## 4 Proposed Work

From the literature survey, it is extracted that work has been done towards pre-processing of user query, but least amount of work is done towards location-based clustering mechanism that emphasizes on pre-processing phase for keyword extraction, and word processing and classification of user query includes location-sensitive site extraction procedure. The entire work of the proposed system is categorized into phases.

**Phase 1: Identifying Tokens from a User Query**

This phase consists of extracting the meaningful work, which exists in the dictionary representing meaningful words. All these words are known as tokens. In addition to identifying tokens, it also identifies the misspelled words and suggests corrections. The corrected words are replaced with existing URL words.

**Phase 2: Removing Stop Words**

Stop words removal becomes a need of the hour to reduce the time required to perform searches. These words are bound to ignore by the search engine. These words will be removed in the proposed mechanism. To accomplish this, stop words dictionary is maintained. Once the stop words are removed, a length of query string is subsequently reduced.

**Phase 3: Extracting Keywords**

Keywords in Web searching represent the most profound search words. The frequency of occurrence of keywords in URL is high as compared to a normal word. Thus, keywords identification in the proposed system is achieved using a statistical measure known as a model. The highest frequency of the word will be directly proportional to the probability of keywords.

**Phase 4: Forming Clusters**

This phase presents determine from other browsers. In this phase, the most probable clustering mechanism is proposed. This is a simple mechanism, in which keywords extracted from the query string are stored within the dataset. These keywords are accompanied by the count variable. This variable increases as the same keywords appear again within the URL string. In the case, Euclidean distance is evaluated corresponding to each keyword. Threshold distance is also maintained. In case, Euclidean distance is less than the threshold distance, then keywords are collected within the cluster.

**Phase 5: Location-sensitive URL Result**

The algorithm for the proposed system is given as under

---

**Algorithm MPV_Cluster**

---

- URL represents the user input which is stored within ‘U’ variable
- ‘db’ indicates the dictionary of meaningful words.
- Extracted_Tokens indicates the meaningful tokens.
- Extracted_stop is the variable for maintaining Query without stop words
- Stop is the database for stop words.
- MPV is the historical information of keywords searched.
- Cluster is the group of keywords.

---

**Start**

```
(a)  Read the URL from the query string
       U=URL(Query_String)
   Phase to extract tokens from the query string
(b)  For i=1: length(U)
          For j=1: length(db)
          if (U(i)==db(j))
   Extracted_Tokens=Extracted_Token+" "+U(i)
               Break
               End of if
               End of for
   End of for
   Phase to remove stop words and extracting keywords
(c) Extracted_stop= Extracted_Tokens
(d)  For i=1: length(Extracted_stop)
          For j=1: length(stop)
          if (Extracted_stop(i)!=stop(j))
   Without_stop= Without_stop+" "+Extracted_stop(i)
                 End of if
                 End of for
    End of for
(e)  Phase to find most probable clustering
        For i=1:length(Without_stop)
        For j=1: length(MPV)
           if Without_stop(i)==MPV(j)
    Counti=Counti+1
              End of if
              End of for
  End of for
```

**(f)** Finding distance in terms of count and storing the result with corresponding cluster index

```
        For i=1: length(Without_stop)
           if (count(Without_stopi)<=Threshold)
 Clusteri= Without_stopi
               End of if
End of for
```

**(g)** Using Google GPS key retrieving location sensitive URLs
**(h)** Retrieve website URLs corresponds to Without_stopi and print result in terms of execution time.
**Stop**

---

Next section gives the performance analysis and result corresponding to algorithm given above.

## 5 Performance Analysis and Discussion

The performance of this system is predicted in terms of execution time and several URLs fetched. The execution time is predicted in units of sec's and the number of URLs is presented as a collection of Websites which are efficient and specific enough for user satisfaction. The simulation is conducted in the Open MVC with asp.net platforms. The result of existing and proposed kinds of literature is compared and better result is obtained in the case of the proposed system.

This phase, although additional time in dissimilar word handling, yields a better result in terms of meaningful URL fetching. User query initially is passed through tokenization where the entire query is parsed and compared against the token dictionary. The process yields meaningful and dissimilar words. The query is then formed again from the tokens. The tokens, then compared against the stop word dictionary to eliminate them from the user query. Once stop words eliminated from the string, the spell checking phase checks the words and proposes corrections. In case the user accepts the corrections, words in strings are replaced. This will conclude the pre-processing phase.

Keyword extraction is critical since meaningful information processing is achieved only if keyword extraction is successful. Keyword extraction phase compares the extracted words after the correction phase. Correction phase gives the optimal result and the keyword extraction phase consumes less time since the correction is not a part of an extraction. This determines keyword extraction and correction enhances performance in terms of time consumed in browsing. The parametric results of pre-processing and keyword extraction phase without clustering when browsing is performed by the user is listed in Table 3. Execution time parameter indicates the total time it takes to produce the result in the form of maximum possible URLs. Although limited Websites are used for the purpose of simulation, but still result of time consumption is less than 2 s for each user query (Table 1).

The result produced by the proposed mechanism performs order by determining the most frequent keywords searched.

The plots in terms of number of URLs fetched of existing literature without clustering are given as under (Fig. 1).

This will set-up locality of reference to enhance the speed with which searching operation is being performed. The rank-based mechanism is termed as most probable clusters and during searching, the only relevant cluster is required to be searched.

**Table 1** Parametric result of a number of keywords fetched without clustering from user query and execution time

| No. of keywords | No. of URLs | Execution time (s) |
|---|---|---|
| Apple | 1 | 1.03 |
| | 8 | |
| | 0 | |
| Bank | 3 | 1.85 |
| | 0 | |
| | 0 | |
| Interest | 1 | 1.08 |
| | 6 | |
| | 0 | |
| Java | 1 | 0.85 |
| | 2 | |
| | 0 | |
| Orange | 1 | 1.78 |
| | 0 | |
| | 0 | |
| Giant | 2 | 1.94 |
| | 0 | |
| | 0 | |

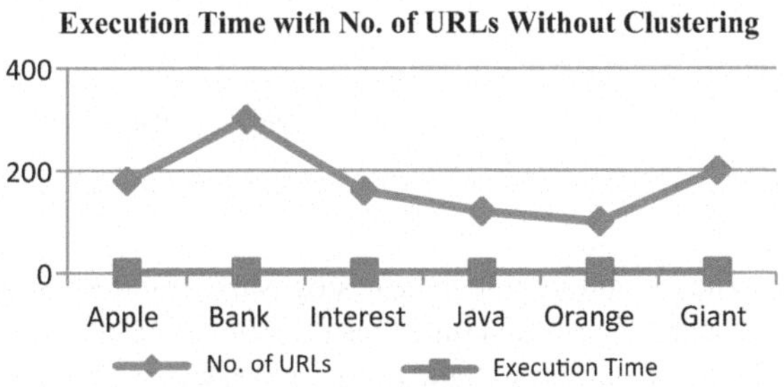


**Fig. 1** Number of URL fetched without clustering

The result corresponding to the keywords searched and rank assigned is given in the Table 2.

As the rank is allotted and a cluster is formed, the execution time required is subsequently reduced. The number of keywords although increased, execution time is reduced. Table 3 predicts the execution time with the clustering mechanism employed (Fig. 2).

Clustering set-up locality of reference and allows a searching process to be simplified with less complexity. Using the said mechanism, only those clusters which are likely to contain the specified keyword is searched and the rest of the cluster are ignored causing least time consuming during the browsing of information.

**Table 2** Most probable clustering (MPC) with rank allocation

| Queries | Keyword | Rank |
|---|---|---|
| Q1 | Apple | 4 |
| Q2 | Bank | 3 |
| Q3 | Interest | 5 |
| Q4 | Java | 4 |
| Q5 | Orange | 3 |
| Q6 | Giant | 5 |

**Table 3** Execution time with clustering

| No. of keywords | No. of URLs | Execution time (s) |
|---|---|---|
| Apple | 100 | 0.85 |
| Bank | 200 | 1.03 |
| Interest | 120 | 0.65 |
| Java | 200 | 0.30 |
| Orange | 500 | 0.98 |
| Giant | 100 | 1.04 |

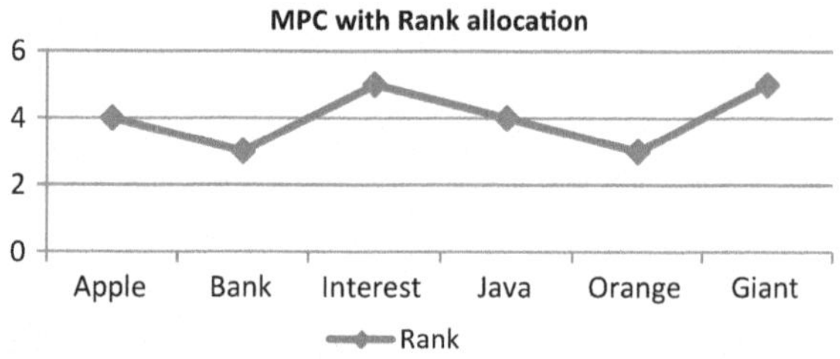


**Fig. 2** MPC with rank allocation

The execution time in browsing can cause mass users to attract the search engine or reject it. Execution time thus plays a critical role during browsing. The number of URLs fetched through the proposed system is given as under (Fig. 3).

The execution time of the fetched URLs through the proposed and existing system is given as under.

The objective of the proposed browsing scheme is to reduce the complexity during searching for URL over the Web. Google API's plays a critical role in our simulation work. Locations sensitive API's are employed to give the result specific to the location that also limits the search space causing reduced execution time.

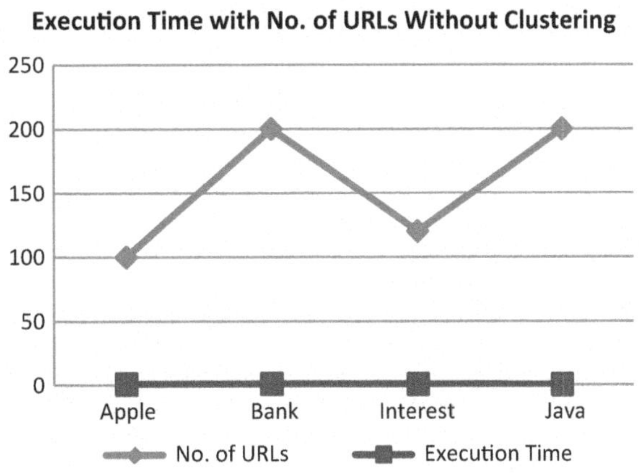


**Fig. 3** Number of URL fetched with clustering

## 6 Conclusion and Future Scope

The volume of search space usage during Web browsing results in high execution time during a URL retrieval operation. Execution time reduction can cause mass users to interact with the browser. The proposed work aims to reduce execution time by the use of most probable clustering mechanisms along with the user query correction mechanism makes it useful to look for meaningful and specific URLs. The result in terms of execution time with clustering shows improvement. Direct interaction of a user during word correction allows better communication and specific URL results.

An additional advantage of the proposed mechanism is location-sensitive Web URL is fetching that is obtained using the location API's provided by Google. In future, the proposed work implication in a real-time environment can be tested and execution time can be further improved using a high degree of specificity through redundancy check and elimination procedure.

## References

1. Yang, D., Song, J.: Web content information extraction approach based on removing noise and content-features. In: Proceedings—2010 International Conference on Web Information Systems and Mining, WISM 2010, vol. 1, pp. 246–249 (2010)
2. Wanaskar, U.H., Vij, S.R., Mukhopadhyay, D.: A hybrid web recommendation system based on the improved association rule mining algorithm. J. Softw. Eng. Appl. **2013**(August), 396–404 (2013)
3. Singh Chouhan, J., Gadwal, A.: Improving web search user query relevance using content based page rank. In: IEEE International Conference on Computer, Communication and Control, IC4 2015 (2016)
4. Panchenko, A.: Senses for WSD and Enrichment of lexical Resources. http://compling.hss.ntu.edu.sg/ (2018)
5. Freeman, D.M., Hwa, T.: Detecting Clusters of Fake Accounts in Online Social Networks Categories and Subject Descriptors. IEEE Access (2015)

6. Witten, I.H., Moffat, A., Bell, T.C.: Managing Gigabytes : compressing and Indexing Documents and Images (1999)
7. Aguilar, J., Valdiviezo-dı, P., Riofrio, G.: Original article A general framework for intelligent recommender systems, pp. 147–160. ACM (2017)
8. Sharma, A.: Spam filtering using k mean clustering with local feature selection classifier. IJCA **108**(10), 35–39 (2014)
9. Zaman, Z.: Spam Detection In Social Media Employing Machine Learning Tool for Text Mining. IEEE Access (2017)
10. Pavalam, S.M., Raja, S.V.K., Jawahar, M., Akorli, F.K.: Web crawler in mobile systems. IJMLC **2**(4), 531–534 (2012)
11. Naaz, S.: Analysis of web pages through link structure. IJCA **122**(11), 22–26 (2015)
12. Lee, T., Chun, J., Shim, J., Lee, S.: An ontology-based product recommender system for B2B marketplaces. Int. J. Electron. Commer. **11**(2), 125–155 (2006)
13. Chen, Y.S., Chang, W.H., Fang, H.M., Yeh, Y.M., Cheng, R.S.: A context-aware reasoning framework with OWL for mobile web information acquisition. J. Internet Technol. **11**(2), 203–214 (2010)
14. Makvana, K., Shah, P., Shah, P.: A novel approach to personalize web search through user profiling and query reformulation. In: 2014 International Conference on Data Mining and Intelligent Computing ICDMIC (2014)
15. Rana, V.: Optimizing performance of user web browsing search. In: International Conference on Advanced Informatics for Computing Research, pp. 230–239. Springer, Singapore (2018)

# A Comparative Study of Famous Classification Techniques and Data Mining Tools

**Yash Paul and Neerendra Kumar**

**Abstract** Data mining is the procedure or technique of drawing out the facts and patterns hidden in huge sum of data and converts it into a readable and understandable form. Data mining has four main modules like classification, association rule analysis, and clustering and sequence analysis. The classification is the major module and is used in many different areas for classification problems. Classification process gives a summary of data investigation which may be utilized to develop models or structures, telling different classes or predict future data trends for improved understanding of the data at maximum. In this survey, various data mining classification techniques and some important data mining tools along with their advantages and disadvantages are presented. Data classification techniques are classified into three categories namely, Eager learners, Lazy learners, and other Classification techniques. Decision tree, Bayesian classification, Rule based classification, Support Vector Machines (SVM), Association rule mining and backpropagation (Neural Networks) are eager learners. The *K*-Nearest Neighbor (KNN) classification and Case Based Reasoning (CRT) are lazy learners. Other classification techniques include genetic algorithms, fuzzy logic and Rough Set Approach. Here six important data mining tools, basic Eager learner, Lazy learner and other classification techniques for data classification are discussed. The aim of this article is to provide a survey of six famous data mining tools and famous different data mining classification techniques.

**Keywords** Fuzzy logic · Decision tree · Bayesian tree · Frequent pattern · Association rule · Support vector machine

---

Y. Paul (✉)
Ph.D. School of Informatics, Eötvös Loránd University, Budapest, Hungary
e-mail: yashpaul1234567@gmail.com

N. Kumar
John von Neumann Faculty of Informatics, Óbuda University, Budapest, Hungary
e-mail: neerendra.kumar@phd.uni-obuda.hu

Department Computer Science & IT, Central University of Jammu, Jammu, India



P. K. Singh et al. (eds.), *Proceedings of ICRIC 2019*, Lecture Notes in Electrical Engineering 597, https://doi.org/10.1007/978-3-030-29407-6_45

# 1 Introduction

Data mining utilizes investigation packages to determine unidentified, concealed and legitimate patterns along with their associations from very large data sets. The analyzing tools comprised of mathematics, statistics, machine learning (ML) methods, etc. Therefore, data mining is not only dealing with gathering and organization the data but, also includes investigation or examination and forecasting of the data. There are quite a lot of applications of ML and the important one of which is data mining. Classification comprises of two steps [1]. Step one is, (Training phase/Learning step), where a classifier or model is constructed from the training examples set, which consists of the database records along with class labels. Step two comprises of application of model built from step 1 on unseen data with no class label to predict the class label. Because the class label is given along with the training data, the step one is also called supervised learning or classification. Whereas, when class label attribute is not given to classify the tuples, this step is called unsupervised learning or clustering and can be formulated as $y = f(x)$, where $x$ is tuple to be tested and $y$ is its predicted class label. People are usually committing mistakes during analysis or while establishing the relationship between number of instances, Therefore ML algorithms practically can be applied to these above-mentioned problems (mainly classification and prediction) successfully, which improves the effectiveness, accuracy, robustness, etc. of the system and the designs of the machines. In this paper, the focus of machine learning techniques and their applications is to perform supervised tasks. Under three above-mentioned categories of classifiers such as Eager, Lazy and other classification techniques, eight different, important and popular classification techniques or classifiers along with their advantages and disadvantages are discussed. Under lazy and eager category and three algorithms are discussed as other classifiers.

## *1.1 Eager Learners*

Eager classifiers are those classifiers or models which are constructed from a known set of training examples before classifying unseen test examples. Therefore, we can say summarize about these type of classifiers that learned or construct models or classifiers are already trained and are waiting eagerly to classify unobserved tuples without class label i.e. test data. In this section, six basic eager learners along with their properties are discussed.

**Table 1** Training set for decision tree

| Name | Age | Credit-rating | Student | Buys_CD |
|---|---|---|---|---|
| Johan | Young | Fair | Yes | Yes |
| Rohan | Young | Fair | No | No |
| Merry | Mid_age | Excellent | No | Yes |
| Tina | Senior | Fair | No | No |
| ___ | ___ | ___ | ___ | ___ |

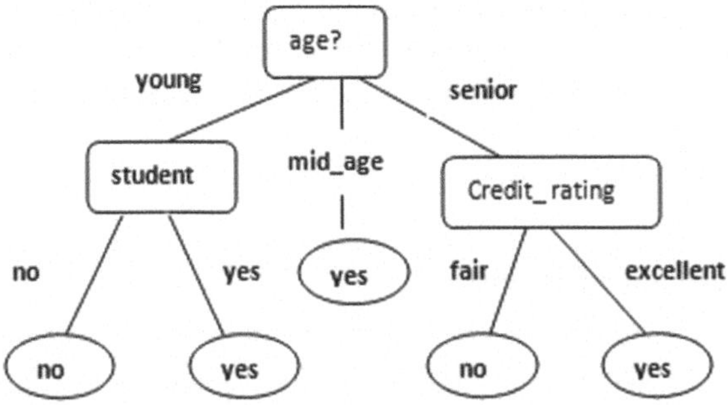


**Fig. 1** An example of decision tree

## *1.2 Decision Tree (DT) Classification*

This model looks like a well-known structure called *flow chart* where interior nodes of the structure signify various tests on attributes and branches denotes the outcomes of the tests performed on interior nodes. Finally leaf nodes or terminal nodes of the tree stores a class label [1, 2].

Decision tree (DT) structure for training set given in Table 1 is shown in Fig. 1. The basic *D*-tree algorithm is designed by using top-down, divide-conquer mode and is based on greedy approach.

## *1.3 SLIQ and SPRINT*

In 1980 J. Ross Quinlan gave a basic well-known *D*-tree algorithm called as ID3 (Iterative Dichotomiser) [3–6]. Later, he provided C4.5, which was the descendant of ID3. ID3 and C4.5 use greedy (non-backtracking) technique. The trees are designed by using top-down, recursive, and divide and conquer way. C4.5 produces accurate results and takes less model building time, has less searching time and uses less memory to execute large program. But it suffers from Empty branches, insignificant branches and over fitting. On the other hand, ID3 gives more accurate results as compared to C4.5, usually work with nominal attributes without missing values, with reduced false alarm and omission rates, high discovery or detection rate with less the memory utilization [7]. But it is time-consuming while searching an item requires more memory as compared to C4.5 to execute large program. CART (Classification and Regression Tree) was discovered by L. Breiman, J. Friedman, R. Olshen and C. Stone and is employed to produce binary DT. The above-mentioned algorithms, e.g. ID3 and C4.5 work efficiently when the given data sets for the classifications are small because there is a limitation that all training examples must be located in the primary memory. But this is not feasible when the training data set has millions of training records or samples. Therefore, SLIQ [8] (*Supervised Learning In Quest*, where Quest is the *Data Mining*.) and SPRINT [3] (Scalable PaRallelizable INndution of Decision Tree) are two algorithms which are capable of handling training data sets that are excessively huge to fit into primary memory. The strong point of these algorithms is that they will knob both categorical and continuous data.

### *1.4 Rain Forest and BOAT*

The efficiency and scalability of SLIQ depend on its memory resident data structure, where in SPRINT memory issue is resolved, but still there is a requirement of a hash tree relative in amount to the training data set. This becomes more costly when the size of training data set increases. Rain Forest algorithm [1, 9] further increase the scalability and efficiency of decision tree induction by maintaining AVC ("Attribute-value, class label") for all attribute, at every tree node, which describes the training records at the node. It can use any selection measure like Gini index, gain ratio [1], etc. and is more effective than SLIQ and SPRINT. The group of each and every AVC-sets at node $N$ is designated as AVC group of $N$ nodes and Rain Forests has number of techniques to handle the AVC group that could not store into main memory.

**BOAT** [10] (Bootstrapped Optimistic Algorithm for Tree Construction) has totally dissimilar approach to scalability and has no special data structures required at all. It uses "bootstrapping" a statistical method to create number of small subsets of the training set in such a way that each subset get store into primary memory. Each sample or subset is then utilized to build the tree structure which results in many small trees structures. These subtrees are processed and a new tree $T$ will be created, which is very similar to the tree that could have been produced from the whole training data if it accommodates into the memory. BOAT may work with all attribute selection measures like Information Gain, Gini Index and Gain Ratio, etc. and follows the principle of purity of partitions. Advantage of BOAT as compared to other above-mentioned technique is that it requires only two scans of $D$ (dataset).

## 1.5 Bayesian Classifiers

Bayesian models are also called as Statistical model and are based on Baye's theorem. These classifiers predict class membership probabilities which mean what is the probability of a given example associated to a specific class [1, 2, 11, 12]. There are two types of Bayesian classifiers:

i. Naïve Bayesian classifiers
ii. Bayesian belief network

**Naïve Bayesian classifiers**: "It works on the assumption that the cause of a feature value in a known class is not dependent on the features values of the other features and this hypothesis is branded as class *conditional independence*". Because of this assumption the computation concerned is very simple. Bayesian belief networks are graphical structures, which permit the demonstration of dependencies among various subsets of features or attributes. Bayesian classifiers predict classes by using Baye's theorem. Baye's theorem, gives a method to compute the posterior probability, $P(H/Y)$, from $P(H)$, $P(Y/H)$, and $P(Y)$

Therefore,

(1)

Description of Baye's theorem.

i. $Y$ is a tuple or "evidence".
ii. $H$ is hypothesis, e.g. the data tuple $Y$ pertains to a particular class $C$.
iii. $P(H)$ is known as priori probability, of $H$, it is the probability that the hypothesis $H$ holds given the explanation of tuple $Y$.
iv. $P(H/Y)$ is known as posteriori probability of $H$ trained on $Y$.
v. $P(Y)$ is known as priori probability of $Y$.
vi. $P(Y/H)$ is a posteriori probability of $Y$ trained on $H$.

Because of their ease in allowing all attributes to contribute in the final decision equally and independently Naive Bayesian [1] classifiers are well-liked in machine learning field.

*Advantages*:

i. Theoretically, Naive Bayesian classifiers are simple and have the least error rate in contrast to every other classifiers. But, in practice it is not mandatory that this is always true.
ii. Bayesian classifiers give a theoretical explanation for other models that clearly not utilizing Baye's theorem.

*Disadvantages*:

i. The main disadvantage of Naive Baye's classifier is its very strong statement and imagination about the shape or distribution of the data, Due to this assumption sometimes it may give very bad results.

ii. Another problem happens due to data scarcity.

**Bayesian Belief Networks**: If "class conditional independence" conditions are satisfied, then the naïve Bayesian classifier is the best classifier as compared to all other classifiers in terms of accuracy. But in practice, dependencies will continue living among variables. Bayesian belief networks indicate joint conditional probability distributions [13, 14]. They permit class conditional independencies condition among variables or attributes. Finally, the learning can be done on these structures of causal relationships.

*Components*:

i. A directed acyclic graph (DAG)
ii. A group of conditional probability tables

In the Directed Acyclic Graph (DAG) Nodes are signify as random variables and these variables may be discrete or continuous. Each one arc shows a probabilistic dependence. An arc starting a node *X* to node *Y*, demonstrate that *X* is parent or direct predecessor of node *Y*, and *Y* is called a successor of *Y*. Figure 2 shows a straightforward belief network, for six Boolean variables and associated conditional probability table (CPT).

*Advantages and Disadvantages*:

(a)

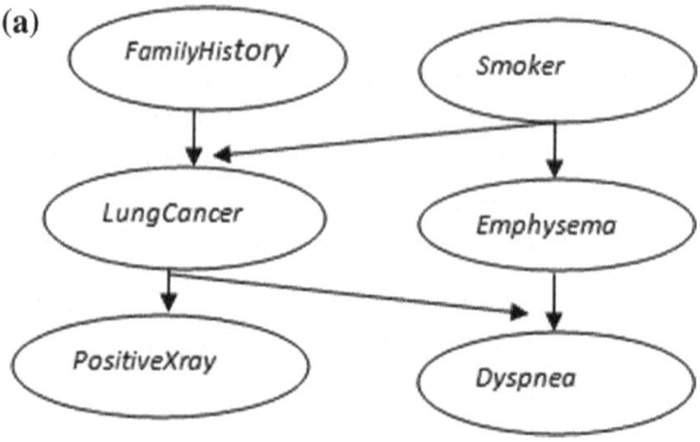


(b)

| | FH.S | FH.~S | ~FH.S | ~FH.~S |
|---|---|---|---|---|
| LC | 0.8 | 0.5 | 0.7 | 0.1 |
| ~LC | 0.2 | 0.5 | 0.3 | 0.9 |

**Fig. 2** **a** An anticipated causal model, characterized by a DAG. **b** Table contains values of probabilities for Lung Cancer (LC) show feasible grouping of the values of its parent nodes, FamilyHistory (FH) and Smoker (S) [1]

i. It can handle incomplete data set
ii. It allows one to learn about casual relationship
iii. It readily facilitate use of prior knowledge
iv. Construction of graph is complex

In *summary*, unlike Naive Bayesian classifiers, Decision Tree classifiers can combine with other provisions and can yield extraordinary outcomes for several domains. However, training process of a Decision Tree model is very intricate and not simple and some times the number of trees generated may cross the limit and become hard to control the situation [1]. Bayesian networks give a natural representation for conditional independence and get better the classification performance by taking out the inappropriate features, resulting in short computational time. Unfortunately, Naive Baye's structures require extremely huge number of examples to acquire excellent results. In Bayesian classifier there is no correct way to choose a prior information, and if we do not proceed with care, disingenuous results can be generated.

## 1.6 Backpropagation (Neural Networks)

Back propagation is a learning algorithm of famous model called neural network and it performs learning scheduled on multilayer Feed Forward Neural Network [11, 15–17]. A neural network comprises of coupled input/output components, where weights are assigned to each connection, connecting these units. A multilayer feed-forward neural network [1] along with weights on few connections is shown in Fig. 3.

Each layer comprises of units and these units are known as input units, Middle units output units. The inputs to the network match to the feature considered or calculated for each training tuples. The above network has one layer of output units and is called one-layer network, if it has two layers of output units then this is called as two-layer neural network and so on. Similarly, when a network contains two hidden layers such network is considered as three-layer neural network, and same nomenclature is used for further increase in layers accordingly. The gist of feed forward is that there are no weight cycles back in direction to an input component

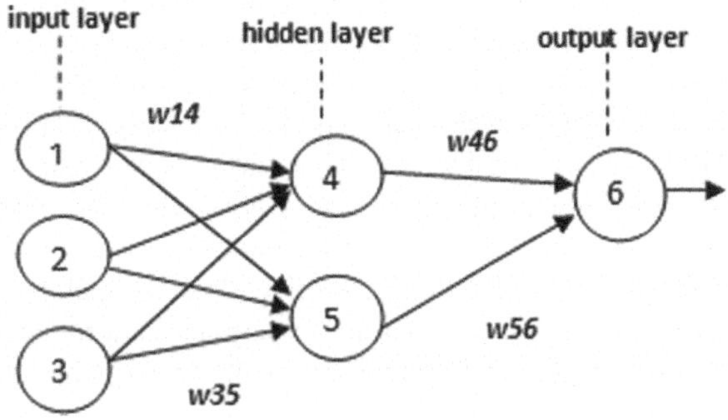


**Fig. 3** Simple multilayer feed-forward neural network [17]

or to an output component of prior layer. Neural network is taught by changing the weights, as a result accurate class label can be predicted. Due to these connections between the units such category of learning is called connectionist learning. Each one output component takes a weighted summation of the outputs from components in previous layers and applies a nonlinear function to the weighted sum.

**Advantages**

i. The powerful advantage of the neural networks is that they high acceptance of noisy data and they are capable of classification of examples over which they are not educated.
ii. They can be applied when we have tiny familiarity about relations among features and classes.
iii. Unlike decision tree algorithms, they are well-matched to constant inputs and outputs values and support Parallelism among various processes.
iv. They may be utilized in handwritten character recognition, laboratory medicine, pathology, and to train a computer to pronounce English text.
v. They able to model the class forecasting as a nonlinear permutation of the inputs, i.e. they perform nonlinear regression.

**Working of Backpropagation**. Backpropagation can be trained by analyzing training examples recursively, matching up to the network's forecast or prediction for all data examples with the real already identified class label of the training example. For each one training example the weights are altered in order to limit the mean squared error between the system's expectation and the real objective value. These adjustments are prepared in the backward way (out layers to first hidden layer) and thus the backpropagation.

### *1.7 Support Vector Machines (SVM)*

Support vector machines [1, 18] is a very fast and new classification technique for very large (millions of support vectors) linear and nonlinear data sets.

Working Mechanism of SVM: It uses a nonlinear mapping to convert the actual training data into a higher proportions and dimensions. A hyperplane a "decision boundary", separating the tuples of one class from another, is searched in these new higher dimensions. These hyperplanes are searched by the SVM by using support vectors i.e. "Essential training tuples" and margins [19–22]. SVM for separable linear data sets: The simplest case where the classes (two) are linearly separable, i.e. there is a line that can be drawn to take apart the entire tuples of class $+1$ (Buys_CD $=$ yes) from every single one of the tuples of class $-1$ (Buys_CD $=$ no). For example the Fig. 4 shows B1 and B2 are linearly separable data.

In case of 3-D, we look for hyperplane as a replacement for a line. Through the learning stage, the SVM seeks for the hyperplane having biggest margin because

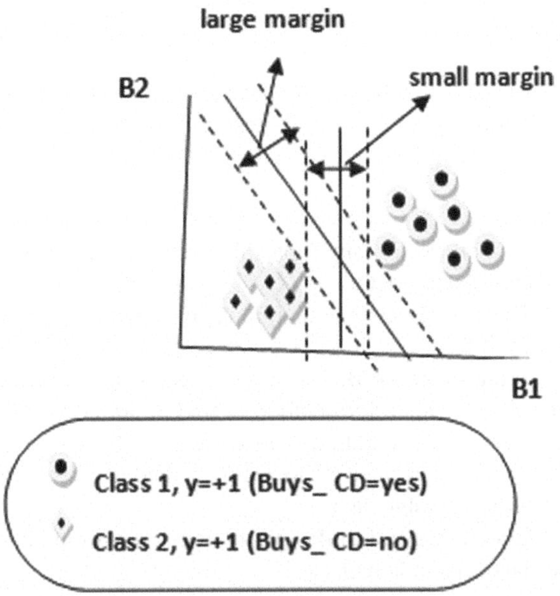


**Fig. 4** 2-D linearly separable training data

larger margin is more precise at classifying test examples than hyperplane having smaller margin and the large margin is called Maximum Marginal, Hyperplane (MMH). The mathematical equation for MMH is:

(2)

The weight vector, , here, $n$ = overall features, b is just a scalar. Let us we have two attributes B1 and B2 having values x1 and x2 for tuple $X$ ($X$ = (x1, x2)), Training tuples are 2-D. If b is considered as a weight, then Eq. (2) can be written
As:

(3)

The point positioned on top of the separating plane accepts

(4)

And the point positioned beneath the separating plane accepts

(5)

The hyperplane describe the surfaces of the margin will be summarized as

(6)

(7)

Example that lies on or above $H_1$ belongs to $+1$class and an example that lies on or below $H_2$ belongs to $-1$ class. On combining (4) and (5) we get

(8)

Support vector, any tuples that fall on hyperplanes or and satisfy Eq. (8) are called support vectors.

The size of maximal margin is given by the formula , Where is called Euclidean norm of $W$ i.e. .

**Training of SVM**. The support vectors are generated by rewriting the Eq. (6) by using Lagrangian formulation to find finding support vectors using Karush–Kuhn–Tucker (KKT) conditions. Once MMH and support vectors are found the SVM can be trained. A trained SVM can be called as linear SVM if it is used to categorize linearly separable data.

**Classifying test tuples**. With Lagrangian formulation the MMH will be modified as the decision boundary:

(9)

Here , are class label and support vector, is test tuple, and are numeric parameters calculated automatically by classifier. To test a tuple , feed to (9) and the test out the sign of the end result. If sign of the result is plus then is on or above the MMH and example belongs to class $+1$ (buys_CD = yes). But if the sign is minus, then is on or under the MMH which suggests the example belongs to $-1$ class (buys_CD = no).

There is dot product between the

In Lagrangian formulation

**SVM for linearly not separable data**. This is the case when no straight line like in linear SVM, can exist that would separate the classes. The linear SVM can be extended in two steps to find the solution of such situation and is capable to find nonlinear decision boundary i.e. nonlinear hypersurface. In first step, we change the original data into high dimensions by applying nonlinear mapping. In second step, look for hyperplane which is linear separating in new dimensional space, and then solve the problem by using linear SVM.

*Kernel Function*: Without knowing the clever trick to map our space to a higher dimension, this may lead a very computationally expensive transformation and there can large number of new dimensions too. Repeating this for all vectors in the database leads to a lot of efforts. Therefore this motivates us for cheaper solution and kernel function trick help us to reduce this computation and gives cheaper solution.

SVM needn't bother with the real vectors to work its electrification, it is really can be done by just with the dot products. It implies that it avoids the cost computations of the new measurements or dimensions. Let the new space we required is:

The dot product in that space will be:

Then we tell the SVM to perform its tasks, but by means of the new dot product and we name this a kernel function. That is the kernel trick, which bypass a lot of pricey calculations. In general, when we have linear kernel, the classifier is linear. However, when kernel is nonlinear then a nonlinear classifier with no transformation of the data.

**Advantages**:

i. SVM has a regularization parameter, which helps to avoid over-fitting.
ii. When *C* and *r* parameters are chosen approximately good then SVMs provide a good out-of-sample generalization.
iii. SVM uses the kernel trick, by which machine acts smartly to solve the problems.
iv. SVMs give a single solution because the optimality problem here is convex and this is an advantage contrast to Neural Networks, which may give number of solutions related with local minima.
v. The kernel implicitly contains a non-linear transformation.

**Disadvantages**:

i. One common drawback of non-parametric techniques like SVMs is, not have transparency of results.
ii. The parameters optimization and kernel selection is challenging, if parameters are not chosen carefully unaccepted results will be produced. Practically it may be the most severe difficulty with is the elevated algorithmic complexity and wide memory requirements.

In summary, the kernel trick, in fact, is not a part of support vector machines, and it may be utilized with other linear classifiers e.g. logistic regression. A SVM only concerns of finding the decision boundary. In many applications, it is found that in large number of applications SVM is very perfect classifier in terms of accuracy as compared to other classifiers. It has less overfitting, robust to noise, more famous in text classification problems especially in high-dimensional spaces. On the other hand, SVM is a binary classifier and to achieve multi-class classification, pair-wise classifications can be utilized, i.e. one class against all others, for all classes which is computationally costly and slow down the speed of the device.

## 1.8 Rule-Based Classification

To classify a test tuples, rule-based classifier [1, 23–25] is constructed from a set of IF-THEN rules, and these rules may be constructed from a decision tree or directly from the raining data set.

**IF-THEN Rules-based classification**. An IF-THEN rule can be written as:

IF condition THEN conclusion. The IF-part or left-hand side of a rule is called rule antecedent or precondition. The THEN-part or right-hand side is called rule consequent. Rule *R* can be assessed by its coverage and accuracy. Given a tuple, *X*, from data set *D*, let n covers—the number of tuples covered by *R*; *n* correct—the number of tuples correctly classified by *R*, and |*D*|—the number of tuples in *D*. Coverage and Accuracy of a rule *R* are:

## 1.9 Associative Classification

Before discussing the various methods of associative classification, we must be familiar with the necessary ideas of association rule mining and such terms and rules are mentioned below.

Frequent patterns or itemsets [1], itemsets, substructures, or subsequences that repeat frequently in a data set are called frequent patterns. Each item in a market basket can be represented by presence (1) or absence (0) of items, and such Boolean vectors (0,1) can be analyzed to know the buying patterns of items that frequently purchased or associated by the customers. The association or relationship between the frequently purchased itemsets can be represented in the form of a rule called as association rule [26]. For example, if we want to know the information about the customers who purchase CD also have a tendency to purchase head phone at the similar time is signified by the following association rule,

Support and confidence are to measures which based on some threshold value, determine the interestingness of an association rule [27]. Support depicts is usefulness and confidence depicts the confidence or certainty of a revealed rule. The confidence 70% means that 70% of the total clients who bought a CD also purchased the head phone, and support of 20% means that 20% customers are those customers who purchased CD and head phone together.

Methods of associative classification.

Association rule mining includes two steps:

i. Frequent itemsets mining, where patterns are searched to find frequent patterns.

ii. Rule generation, here Association rules are produced from the frequent patterns.

There are three important associative methods are given below:

i. Classification based association (CBA)
ii. Classification based on multiple association Rules (CMAR)
iii. Classification based on positive association rules (CPAR)

**CBA** [28]. Like Apriority algorithm, it also uses the iterative approach to generate frequent patterns the total number of passes needed are the same as span of the longest rule found. Classifiers are trained only from the strong association rules i.e. rules having support and confidence values are above or equal to the minimum support and confidence threshold values. The classifier stores and organized the rules according to decreasing precedence based on their confidence and support. To classify a tuple *Y*, It chooses the rule which satisfies and matches tuple *Y*. But when more than two rules are matched tuple *Y*, it selects rule having highest confidence. When no rule matches, default rule, i.e. rule having lowest precedence, is used to classify a test tuple *Y*. Rues having identical antecedent (precedence), the rule with maximum confidence is used to characterize that group of rules. The performance of CBA is good than C4.5 on good number of data sets.

**CMAR** [29]. Its approach to finding frequent patterns and generate association rules are different than CBA. It uses *F-P* (frequent pattern) tree to generate frequent patterns and complete set of strong rules association rules. It also uses another tree for efficient storage and retrieval of rues, by applying some pruning techniques on the rules based on some criteria. It requires only two database scans to generate frequent patterns. Association rules are discussed in details in [28–31].To classify a tuple *Y*, It chooses the rule which satisfies and matches tuple *Y*. But when more than two rules are matched to *Y*, instead of selecting one rule for classification, it selects multiple rules. It segregates the matched rules into grouping as per their class labels, all rules within a group allocate a common class label and each group has dissimilar class labels as well. The strongest group among groups can be identified by applying weighted $\chi 2$ measure on each group. Finally the test tuple Y is assigned to strongest group for classification, and because of this, CMAR has higher Accuracy than CBA.

**CPAR** [26]. Both CBA and CMAR generate candidate association rules, which needs to further rectify, to generate strong association rules for classifiers. This process is time-consuming. To generate rules it uses FOIL (first-order inductive learner) algorithm, which generates rules to differentiate *yes* examples (CD_buys = yes) from *no* examples (CD_buys = no). When there are more than two classes in the set then FOIL is applied to each class. In this multiclass problem the entire examples of class *C* may be treated as positive examples and the remaining examples are treated as negative examples. To classify a tuple *Y*, it works similar to the CMAR, but instead of selecting a complete strong group for classification it selects best *k*-rules from each group for classifying test tuple *Y*. Such type of selection keeps away from the pressure of lower ranked rules. It generates extreme few rules than CMAR and has much more effective for huge data sets.

# 2 Lazy Learners

Lazy learners learn or gain from its neighbors. Lazy learner holds up until last moment before performing any model improvement or growth for the classification of a specified test example or query in hand. The famous lazy learners that are discussed below are *K-Nearest-Neighbor (KNN) and Case Based Reasoning (CBR) classifiers*.

## *2.1 K-Nearest-Neighbor Classifiers (KNN)*

KNN has high computing power and is greatly used in pattern recognition and classify a tuple by comparing it with the training tuples (instances) in database, whose nature or type is similar to it [1, 32, 33]. The "closeness" between two instances may be measured by using different distance measures, e.g. Manhattan Distance (MD), Euclidean Distance (ED) etc. Euclidean distance between two points' $x_i$, and $x_j$ is:

For categorical attribute, the distance is measured by comparing the corresponding values of the feature in example $X_1$ with the example $X_2$. The problem with these classifiers is that they endure from meager accuracy when noisy or extraneous attributes are given. They are also slow when the size of the classifying tuples or data set is large.

## *2.2 Case Based Reasoning (CBR)*

These sorts of classifiers utilize a data set of solutions of problem for taking care of the problem Instead of storing training tuples as instances like in KNN, they accumulate the examples or cases for statement solving as difficult figurative descriptions [34, 35]. To classify a new case or tuple, the classifiers will first verify whether exactly the same training case exits. In the event that yes then answer for that case is returned and if not it looks through the training tuples or cases having segments like those of the new case. Problem with this classifier is that it is difficult to find good similarity metric and appropriate methods for merging the solutions. On the other hand, other issues consist of the selection of most important features for indexing training cases and the progress of effective indexing techniques. The advantage of this technique is that it effortless to know and implement. Training process is exceptionally fast and well suited for multilevel classes and also Robust to noisy training data [35].

# 3 Other Classification Algorithms

These techniques are less preferable but in few applications, these are very efficient than the techniques until we have studied. The various such techniques are given below.

## 3.1 *Genetic Algorithms (GA)*

The inspiration of genetic algorithm [26, 36] is originated from natural evolution. In GA initially population is to be made and the population comprises of rules which are generated randomly which can be encoded in bit string.

Some important points about Genetic Algorithms:

i. Based on the idea of the continued existence of the fittest, another population is framed that comprises of the fittest guidelines in the existing population and children values of these principles too.
ii. The classification accuracy is used to find the fitness of the rule.
iii. Mutation and crossover are two genetic operators and mutation is utilized to construct children or offspring.
iv. In crossover, the substring from pair of guidelines is exchanged to create a fresh set of rules/guideline.
v. In mutation, arbitrarily pointed bits in a string of a rule are reversed.
vi. Parallelization is very easy to achieve in such algorithms and can be used for classification as well as for optimization problems.

## 3.2 *Rough Set Approach*

The rough set approach [37–39] can be utilized to find structural relationship in noisy or information. The Rough Set Theory depends on the foundation of equivalence classes inside the specified training examples that structure the equivalence classes and are unclear or noisy. There are a few classes in the given real-world information, which can't be recognized by available set of attributes. We can utilize the rough sets generally to characterize such type of classes. For a specified class C, the rough set definition is estimated by two sets as:

**Lower Approximation of C**: The lower estimation of *C* comprises of the considerable number of data examples, which are stand on the knowledge of the feature, are sure to have a place with class *C*.
**Upper Approximation of C**: The upper estimation of *C* comprises of every example that depend on knowledge of attributes and can't be portrayed as not having a place with *C*.

### 3.3 Fuzzy Set Approaches

**Fuzzy** means ambiguity and this phenomenon occurs when the boundaries of classes in the data are not clear.

Fuzzy Set idea was proposed by Lotfi Zadeh as an alternative of theory known as two-value logic (CRISP, where boundaries are clear) and probability and this theory is also known as Possibility theory where the membership lies in the interval [0,1]. Because of this theory, we can work with high level of abstraction and can deal with imprecise dimension of data. The word "YOUNG" an example of fuzzy system because an individual 25 is young and other who is 27 years old is also young [1].

## 4 Conclusion

Various classification techniques and data mining tools along with their advantages and disadvantages are discussed. When data set is small and noise-free KNN classifier is good choice for classification problems. SVM are frequently and widely used classifiers in all classifications tasks. Classification methods, in general, are strong in modeling communications. We have seen Decision Trees and Bayesian Network and by and large have different working style, when one is very accurate the other is not for the same problem and vice versa. But on the hand, rule-based and decision tree classifiers have a similar working profile. Therefore it motivates the searchers for integration of different classifiers into one, where the best features of two or more classifiers are integrated into a single algorithm which gives more precise, certain and accurate results. Many combined methods have been given and anticipated, but it is still not clear which one is the best method which meets all requirements. Study also makes the reader become familiar with six popular data mining and big tools along with their freely available resources.

Please note that the first paragraph of a section or subsection is not indented. The first paragraphs that follow a table, figure, equation, etc. do not have an indent, either.

Subsequent paragraphs, however, are indented.

## References

Han, J., Kamber, M.: Data Mining: concepts and Techniques, 2nd edn, Morgan Kaufmann Publishers (2006)

Weiss, S.M., Kulikowski, C.A.: Computer Systems that Learn: Classification and Prediction Methods from Statistics, Neural Nets, Machine Learning, and Expert Systems. Morgan Kaufmann, Burlington (1991)

Murthy, S.K.: Automatic construction of decision trees from data: a multi-disciplinary survey. Data. Min. Knowl. Discov. **2**, 345–389 (1998)

Quinlan, J.R.: C4.5: Programs for Machine Learning. Morgan Kaufmann (1993)

Breiman, L., Friedman, J., Olshen, R., Stone. C.: Classification and Regression Trees. Wadsworth International Group (1984)
Kamber, M., Winstone, L., Gong, W., Cheng, S., Han, J.: Generalization and decision tree induction: efficient. Classification in data mining. In: Proceedings of 1997 International Workshop Research Issues on Data Engineering (RIDE'97), pp. 111–120. Birmingham, England (1997)
Kalpana, R., Bansal, K.L.: Comparative study of data mining tools. Int. J. Adv. Res. Comput. Sci. Softw. Eng. **4**(6), 216–223 (2014)
Shafer, J., Agrawal, R., Mehta, M.: SPRINT: a scalable parallel classifier for data mining. In: Proceedings of 1996 International Conference on Very Large Data Base (VLDB'96), pp. 544–555. Bombay, India (1996)
Gehrke, J., Ramakrishnan, R., Ganti. V.: Rainforest: a framework for fast decision tree construction of large datasets. In: Proceedings of 1998 International Conference Very Large Data Bases (VLDB'98), pp. 416–427. New York, NY (1998)
Gehrke, J., Ganti, V., Ramakrishnan, R., Loh, W.-Y.: BOAT—optimistic decision tree construction. In: Proceedings of 1999 ACM-SIGMOD International Conference on Management of Data (SIGMOD'99), pp. 169–180. Philadelphia, PA (1999)
Mitchell, T.M.: Version spaces: a candidate elimination approach to rule learning. In: Proceedings of 5th International Joint Conference on Artificial Intelligence, pp. 305–310. Cambridge, MA (1977)
Duda, R.O., Hart, P.E., Stork, D.G.: Pattern Classification, 2nd. edn. Wiley (2001)
Heckerman, D.: Bayesian networks for knowledge discovery. In: Fayyad U.M., Piatetsky-Shapiro G., Smyth P., Uthurusamy R. (eds.) Advances in Knowledge Discovery and Data Mining, pp. 273–305. MIT Press (1996)
Pearl, J.: Probabilistic Reasoning in Intelligent Systems. Morgan Kauffman (1988)
Rumelhart, D.E., Hinton, G.E., Williams, R,J,: Learning internal representations by error propagation. In: Rumelhart D.E., McClelland J.L. (eds.) Parallel Distributed Processing. MIT Press (1986)
Rosenblatt, F.: The perceptron: a probabilistic model for information storage and organization in the brain. Psychol. Rev. **65**, 386–498 (1958)
Russell, S., Norvig, P.: Artificial Intelligence: a Modern Approach. Prentice Hall (1995)
Minsky, M.L., Papert, S.: Perceptrons: an Introduction to Computational Geometry. MIT Press (1969)
Mezard, M., Nadal, J.P.: Learning in feedforward layered networks: the tiling algorithm. J. Phys. **22**(12), 2191 (1989)
Boser, B., Guyon, I., Vapnik, V.N.: A training algorithm for optimal margin classifiers. In: Proceedings of Fifth Annual Workshop on Computational Learning Theory, pp. 144–152. ACM Press: San Mateo, CA (1992)
Burges, C.J.C.: A tutorial on support vector machines for pattern recognition. Data Min. Knowl. Disc. **2**, 121–168 (1998)
Vapnik, V.N., Chervonenkis, A.Y.: On the uniform convergence of relative frequencies of events to their probabilities. Theory Probability Appl. **16**, 264–280 (1971)
Vapnik, V.N.: The Nature of Statistical Learning Theory. Springer-Verlag (1995)
Vapnik, V.N.: Statistical Learning Theory. Wiley (1998)
Clark, P., Niblett, T.: The CN2 induction algorithm. Mach. Learning **3**, 261–283 (1989)
Chen, M.S., Han, J., Yu, P.S.: Data mining: an overview from a database perspective. IEEE Trans. Knowledge Data Eng. **8**, 866–883 (1996)
Li L., Dong, G., Ramamohanrarao. K.: Making use of the most expressive jumping emerging patterns for classification. In: Proceedings of 2000 Pacific-Asia Conference on Knowledge Discovery and Data Mining (PAKDD'00), pp. 220232. Kyoto, Japan (2000)
Quinlan, J.R.: Learning logic definitions from relations. Mach. Learn. **5**, 139–166 (1990)
Major, J., Mangano, J.: Selecting among rules induced from a hurricane data base. J. Intell. Info. Syst. 39–52 (1995)

Liu, B., Hsu, W., Ma. Y., Integrating classification and association rule mining. In: Proceedings of 1998 International Conference on Knowledge Discovery and Data Mining (KDD'98), pp. 80–86. New York, NY (1998)

Li, W., Han, J., Pei, J.: CMAR: accurate and efficient classification based on multiple classification rules. In: Proceedings of 2001 International Conference on Data Mining (ICDM'01), pp. 369–376. San Jose, CA (2001)

Ziarko, W.: The discovery, analysis, and representation of data dependencies in databases. In Piatetsky-Shapiro G., Frawley W.J. (eds.) Knowledge Discovery in Databases, pp. 195–209. AAAI Press (1991)

Cios, K., Pedrycz, W., Swiniarski, R.: Data Mining Methods for Knowledge Discovery. Kluwer Academic Publishers (1998)

Fix, E., Hodges, J.R.: Discriminatory analysis non-parametric discrimination: consistency properties. In: Technical Report 21–49-004(4), USAF School of Aviation Medicine, Randolph Field, Texas (1951)

Riesbeck, C., Schank, R.: Inside Case-Based Reasoning. Lawrence Erlbaum (1989)

Agrawal, R., Imielinski, T., Swami, A.: Mining association rules between sets of items in large databases. In: Proceedings of 1993 ACM-SIGMOD International Conference on Management of Data (SIGMOD'93), pp. 207–216. Washington, DC (1993)

Yin, X., Han, J.: CPAR: classification based on predictive association rules. In: Proceedings of 2003 SIAM International Conference on Data Mining (SDM'03), pp. 331–335, San Francisco, CA (2003)

Mitchell, M.: An Introduction to Genetic Algorithms. MIT Press (1996)

Goldberg, D.: Genetic Algorithms in Search, Optimization, and Machine Learning. Addison-Wesley (1989)

Mehta, M., Agrawal, R., Rissanen, J.: SLIQ: a fast scalable classifier for data mining. In: Proceedings of 1996 International Conference on Extending Database Technology (EDBT'96), pp. 18–32. Avignon, France (1996)

Dong, G., Li, J.: Efficient mining of emerging patterns: discovering trends and differences. In: Proceedings of 1999 International Conference on Knowledge Discovery and Data Mining (KDD'99), pp. 43–52. San Diego, CA, (1999)

Pawlak, Z.: Rough Sets, Theoretical Aspects of Reasoning about Data. Kluwer Academic Publishers, Dordrecht, Netherlands (1991)

# Necessary Information to Know to Solve Class Imbalance Problem: From a User's Perspective

**G. Rekha and Amit Kumar Tyagi**

In many real-world applications, class imbalance problem is the most attentive (also a major challenging) problem for machine learning (ML). The traditional classification algorithms assume evenly distributed in the underlying training set. In class imbalanced classification, the training set for one class called (majority class) far exceed the training set of the other class called (minority class), in which, the more often interesting class is minority class. We need to increase minority class samples than majority class samples in analysing of a datasets (related to an application). This is a hot problem in the past several decades. This article tries to provide as much as information to know or work about class imbalance problem with a detail description (from a user's perspective). For this, we include several articles from a reputed publication like IEEE, ACM, Elsevier, Wiley, etc. Hence, this work will help a lot to all the future researchers to find out or a summary (about their interest) with respect to this class imbalance problem (raising in several applications).

Class imbalance · Majority class · Minority class · Training set

## 1 Introduction

Recently, in machine-learning and data-mining domain [1], the class imbalance problem has been recognized a serious problem, which required solution with a higher accuracy, i.e. sampling of datasets of respective classes should be highly accurate.

G. Rekha (✉)
Department of Computer Science and Engineering, Koneru Lakshmaiah Education Foundation, Vaddeswaram, Guntur Andhra Pradesh, 522502, India
e-mail: gillala.rekha@klh.edu.in

A. K. Tyagi
Department of Computer Science and Engineering, Lingaya's Vidyapeeth, Faridabad, Haryana 121002, India
e-mail: amitkrtyagi025@gmail.com

 
P. K. Singh et al. (eds.), *Proceedings of ICRIC 2019*, Lecture Notes in Electrical Engineering 597, https://doi.org/10.1007/978-3-030-29407-6_46

This problem occurs when the training data is not equally distributed among different classes. In several real-world's applications, such as like fault diagnosis, anomaly detection, medical diagnosis, detection of oil spillage in satellite images, face recognition, text classification, protein sequence detection, and many others [2], the existence of this problem is crucial. The standard classifiers generally perform poorly in these cases. Classifiers usually ignore the minority class by overwhelming the majority class. But the minority class is often of more interest and more importance, therefore, it called for a strong urgency to be recognized. In the past decade, several works have been done with respect to class imbalance problem [3]. These methods can be categorized into three groups:

i. External approach: In the external approach, the data is pre-processed in order to balance the dataset to overcome the effect caused by imbalanced set.
ii. Internal approach: proposing new algorithms or change existing ones to handle the class imbalance problem at the time of classification.
iii. Hybrid approach: Combination of both external and internal approaches.

The drawback of internal approaches is of being algorithm explicit, whereas external approaches are independent of the classifier used. In general, class imbalance problem has been identified in several application areas/domains varying from education, biomedical, management, and many more. The solutions proposed in past are helpful but not enough to address this problem. The techniques proposed are highly depending on the nature of the data in a skewed distribution.

Hence, the remaining paper is organized as: Section. 2 summarizes existing analyzing techniques to solve the class imbalance problem. Further, several domains (applications) have been noticed (included) with respect to imbalance data (having class imbalance problem) in Sect. 3. Then, Sect. 4 discusses several evaluation models for addressing the class imbalance problem in brief. Further, Sect. 5 describes several possible research directions. Later, we discussed this problem (class imbalance problem) in Sect. 6 with one (at least) real-world example with respect to user's perspective and get several interesting facts or things (like accuracy paradox, overfitting, etc.). In last, this work is concluded with some future remarks in Sect. 6.

## 2 Existing Analysis Techniques for Solving the Class Imbalance Problem

In the past decade, many algorithms have been proposed by several researchers to address (or solve) the class imbalance problem in classification. In this section, we present an overview of all the techniques for imbalanced learning techniques. Broadly, the three basic strategies for dealing with class imbalance learning are preprocessing, algorithm-centric and ensemble methods. In the pre-processing approach, resampling of sample space is done to balance the data. Apart from resampling, the feature selection method is also adopted to optimize the feature space.

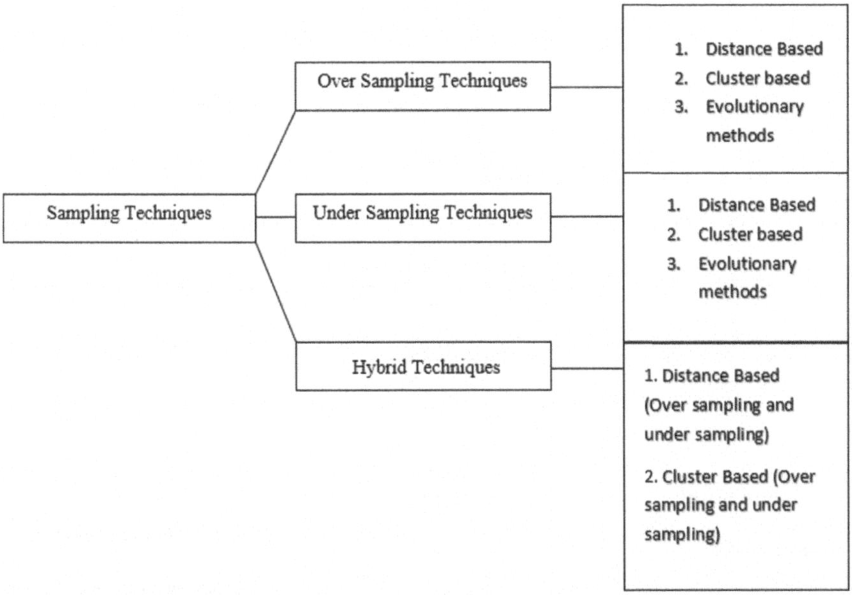


**Fig. 1** Different sampling methods to solve the class imbalance problem

**Pre-processing approach**: Before building the classification model, the data is preprocessed to balance the imbalanced data. Under preprocessing, resampling of data is done to obtain the balance in the sample space [4].

In general, the resampling techniques classified into three categories depending on the balancing techniques used to handle skew distribution. Figure 1 discusses respective three categories in brief.

- Under-sampling Technique: The imbalanced dataset is balanced by removing the irrelevant samples in the majority class. Random under-sampling (RUS) is the most effective (and mostly used in under-sampling) and simplest sampling technique [5]. Generally, RUS randomly eliminates the samples, a collection of datasets (i.e., which belongs to majority class), to balance the skew distribution in the respective datasets or with minority class's samples.
- Over-sampling Technique: An over-sampling technique generates synthetic data samples for the minority class to balance the skew distribution. Random over-sampling (ROS) and synthetic minority over-sampling technique (SMOTE) are simple and widely used techniques [6].
- Hybrid Technique: The combination of both over-sampling and under-sampling approaches is used to balance the skew distribution.

**Feature Selection**: **Compared to Pre-processing techniques**, significantly fewer work was done in past considering feature selection for class imbalance problem. In general, the aim of feature selection is to select a subset of $k$ features from the entire

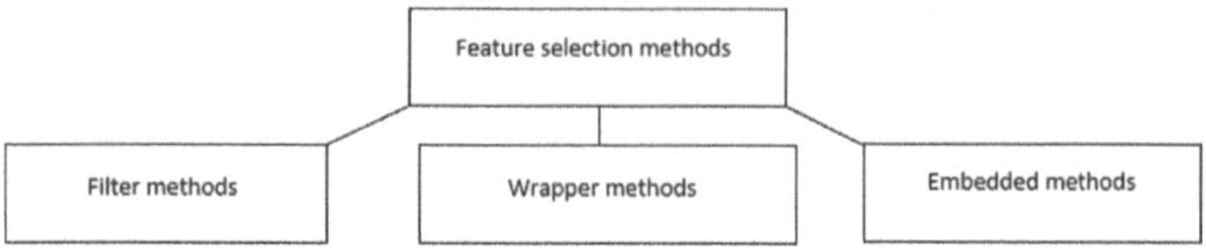


**Fig. 2** Different feature selection techniques

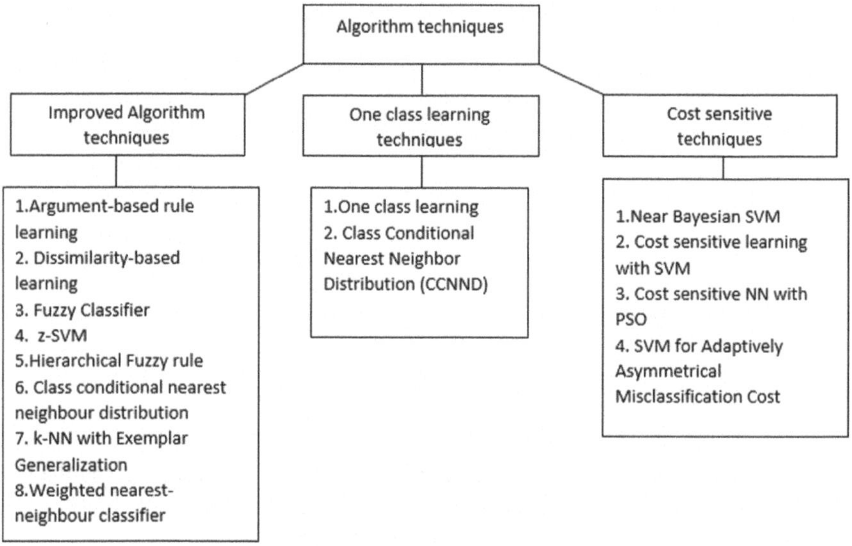


**Fig. 3** Different algorithm-centric approaches

feature space. The selection of important features will result in the optimal performance of the classifier [7]. Here, feature selection is divided into three processes, i.e., filters, wrappers. and embedded methods (see Fig. 2).

**Algorithm-Centric Approach**: To improve the classification performance for imbalanced data, an attempt by the research community was to improve the learning ability of the existing classification algorithm. In the past decade, a plenty of improved classifiers such as support vector machine (SVM) [8], neural network (NN), *k*-nearest neighbor (KNN), decision tree (DT), rule-based classifiers, Naïve Bayes (NB), and many more have been proposed for handling imbalanced datasets in classification. Hence, all techniques/algorithms (as an interrelation) are discussed or showed in Fig. 3.

**Ensemble Methods**: Ensemble methods or classifiers, combine several base classifiers to improve the performance than using a single classifier. Nowadays, among all existing methods, one popular method is used mostly to solve this class imbalance problem, i.e., which is called as ensemble classifiers. The different ensemble techniques are bagging, boosting, and hybrid methods. While implementing ensemble methods [7], the different base classifiers used in the literature are support vector

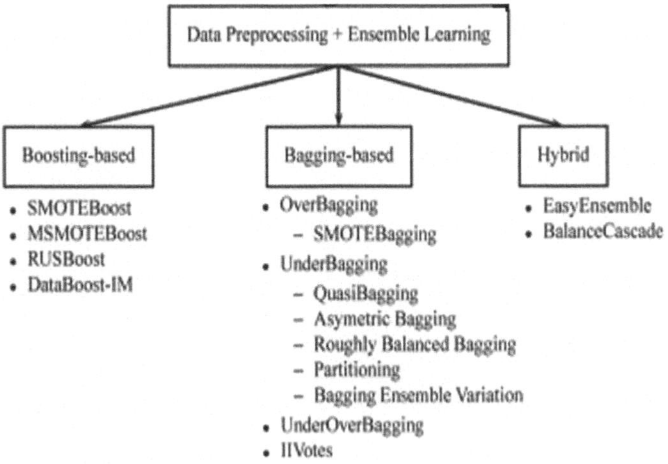


**Fig. 4** Different ensemble methods

machine (SVM), neural network (NN), NB, decision trees including C4.5, CART, random forest, and rule-based classifiers.

Further from Fig. 4, we can say that boosting-based method includes techniques like SMOTEBoost, MSMOTEBoost, RUSBoost, and DataBoost-IM, whereas, bagging-based method includes several techniques like OverBagging, Underbagging, and UnderOverBagging, etc. In last, the hybrid method includes only two methods (till now) which are EasyEnsemple and BalanceCascade.

Hence, this section discusses several existing analysis techniques to solve the class imbalance problem. Now, in the next section, we will discuss about several domains (applications) with respect to class imbalance data.

## 3 The Existence of Imbalanced Data Classification in the Different Application Domains

The data-mining and machine-learning methods have been widely used in automated systems for analyses as a massive amount of data are generated currently. The prediction of the future actions based on past actions is an important problem. Prediction often consists of rare actions. Such rare actions data are usually an imbalance in nature. More efforts are required by the learning algorithms to learn such rare elements in the data. Class imbalance problem is existing for different application domain ranging from engineering to management (see Table 1).

Hence, this section discusses about several interesting applications (related to real world) which have imbalance problem of dataset (in a large dataset). Now, next section will discuss about several evaluation models for addressing class imbalance problem.

**Table 1** Imbalanced data problem with respect to several applications

| Domain | Application | Approach | Use cases |
|---|---|---|---|
| Biomedical engineering | Medicine and Biology for healthcare | Detection and prediction of abnormal structure in chemical and biomedical activities | Disease diagnoses, early warning of diseases, and detection of proteins [9] |
| Chemical engineering | Physical and life sciences | Conversion of cells into useful forms | Prediction of gene expression and drug resistance [10] |
| Financial management | Financial problems | Planning, procuring and control of resources of financial firms | Fraud detection in Credit card application [11], stock market prediction, and credit loan application |
| Information technology | Information devices | Identifying interesting events generated from information devices | Software defect prediction [12], evaluating the quality of software, and network intrusion detection |
| Web technology | Web application | Analyzing the Web data | Sentiment analysis, QoS prediction of Web services [13], detection of fake Websites |
| Energy management | Energy consumption and production | Planning and operation of energy consumption and production units | Availability of power distribution system [14], fitness levels of power transformers, fraud detection in energy field |
| Security | Security management system | Effective planning, organizing and controlling of security measures | Risk detection, analysis of risk in identifying crimes, and detecting unusual events from video surveillance [15] |
| Education | Education systems | Effective understanding of educational issues | Detection of student failure rate [16] |

**Table 1** (continued)

| Domain | Application | Approach | Use cases |
|---|---|---|---|
| Business | Business management system | Planning, organizing, and controlling an organization as while as staffing, leading | Customer relationship management (CRM) [17] data to analyse the customer behavior, preferences and need, detection/prediction of customer churn |
| Industry | Industrial manufacturing system | Identifying and solving the manufacturing problems pertaining to industries | Machinery fault diagnosis and defect detection in an automated inspection system [18] |

**Table 2** Confusion matrix

| | Predicted | |
|---|---|---|
| Actual | TP | FN |
| | FP | TP |

# 4 Evaluation Model for Addressing Class Imbalance Problem

As disused above, class imbalance problem consists two classes majority and minority, whereas, we need to focus to increase balance among minority class (positive), i.e., among an imbalance dataset, we need strongly to balance minority class of data as an urgency to get accurate results. In general terms, it is a popular problem in artificial intelligence/machine learning, where the total number of samples related to one class (i.e. positive or minority class) is far less than the total number of another class (i.e. negative or majority class).

**Samples of Minority Class < Samples of Majority Class**

So, we used several techniques to solve this (class) imbalance problem using changing class distribution (using under-sampling techniques, over-sampling and advanced sampling), features selection, classifiers level (via manipulating classifiers internally, cost-sensitive learning, one-class learning), and ensemble method (refer Sect. 2). The standard metric to measure the classifier performance is by using confusion matrix (cost matrix) as shown in below table.

The acronym for each cell in the confusion matrix (represented in Table 2) is as following:

**Table 3** Metrics used in confusion metrics

| Metrics | Formula |
|---|---|
| Accuracy | (TP + TN)/(TP + FN + TN + FP) |
| Error rate = 1-Accuracy | (FP + FN)/(TP + TN + FP + FN) |
| True positive rate (TPrate)/Recall/sensitivity | TP/(TP + FN) |
| True negative rate (TN rate)/specificity | TN/(TN + FP) |
| False positive rate (FP rate) | FP/(TN + FP) |
| False negative rate (FN rate) | FN/(TP + FN) |
| Precision/positive predictive value (PPvalue) | TP/(TP + FP) |

- TP (True Positive): The number of positive samples correctly classified as positive.
- TN (True Negative): The number of negative samples correctly classified as negative.
- FP (False Positive): The number of negative samples incorrectly classified as positive.
- FN (False Negative): The number of positive samples incorrectly classified as negative.

These terms (i.e. metrics TP, TN, FP, and FN) have been discussed in Table 3. These metrics are the mostly used metrics for evaluating the performance of the classifiers. Among them, accuracy and error rate are mostly used one-performance metric among all metrics (i.e. to solve class imbalance problem). Accuracy assesses the overall efficiency of the classifier for binary class problems.

- Precision/positive predictive value is used to measure the correctness of the positive samples (out of total positive samples how many are correctly classified as positive).
- True Positive Rate (TP rate)/Recall/Sensitivity is a measure of completeness or it specifies the accuracy of positive samples.
- True Negative Rate (TN rate)/Specificity is a measure of actual negative samples correctly identified as negative.
- False Positive Rate (FP rate) is measured as number of incorrect positive predicted samples divided by the total number of negative samples. It is calculated as 1-specificity.
- False Negative Rate (FN rate) is the measure of proportions of positive samples, which are incorrectly classified as negative samples.

Considering the class-imbalanced datasets, accuracy gives more importance to the majority/negative classes than the minority classes. This leads to a difficulty for classifier to learn from minority class. For example, if we consider a problem where only 1% of the samples belong to the positive class (or minority class), high accuracy of 99% is achievable by predicting the negative class (or majority class) for all samples. Yet, all minority class samples, the rare, and more interesting samples, are misclassified. Moreover, , in the past decade, additional metrics were proposed for imbalanced datasets.

**F-measure**: It is also known as *F* Score/*F*1 Score, and also it can be defined as "a weighted harmonic mean of precision and recall." It can be defined as:

$$F\text{−measure} = \frac{2 * \text{Recall} * \text{Precision}}{\text{Recall} * \text{Precision}}$$

The value of *F*-measure increases proportionally with an increase of precision and recall. The higher the value of *F*-measure, the better the classifier working on positive samples.

**Geometric Mean (G-Mean)**: *G*-Mean is an interesting measure because it computes the prediction of the accuracies of the two classes, attempting to maximize them with good balance. It takes into consideration the sensitivity and specificity and provides a balance between classification performance on both majority and minority classes.

$$G\text{−mean} = \sqrt{\text{Sensitivity} * \text{Specificity}}$$

The prediction of poor performance on the positive samples may lead to low value of *G*-mean in spite of negative samples are correctly classified by the classifier. Two popular graphical tools used in imbalanced domains are the receiver operating characteristics (ROC) curve and the corresponding area under the ROC curve (AUC). Note that Provost et al. [19] proposed ROC and AUC as alternatives to accuracy.

**Receiver Operating Characteristics (ROC) curve**: The ROC curve gives graphical trade-off between TPrate (benefit) and FPrate (the cost). The curve gives true positive rate as a function of false positive rate. The performance of a classifier is represented by a single point in the ROC space. The more inclined the curve is toward the upper left corner, the better is the classifier's ability in discriminating positive (minority) and negative (majority) classes. The ROC curves are good ways to compare models or set of models.

**Area Under Curve (AUC)**: The area under the ROC curve is a summary indicator of ROC curve. It provides the evaluation of the best model by sorting the performance of all models, on average. Higher the value of AUC, better the prediction model. The below Table 4 shows the scale for the interpretation of AUC.

Hence, this section discusses several evaluation models in Table 4 and metrics like ROC, *G*-Mean, etc., to measure performance of mechanism to solve class imbalance problem. Now, next section will investigate several future directions with respect to class imbalance problem.

**Table 4** Scale for the interpretation of AUC

| AUC value | Model performance |
|---|---|
| 0.9–1.0 | Excellent |
| 0.8–0.9 | Very good |
| 0.7–0.8 | Good |
| 0.6–0.7 | Fair |
| 0.5–0.6 | Poor |

## 5 Possible Research Directions

Today, solving class imbalance problem in a feasible and valid way is a challenging task. A reach problem is needed to solve in an affordable and accurate manner. Also, a research problem is needed to help someone (society) with its solution. Then we call that particular problem is a feasible and correct problem. Keeping this thing in our mind, in this section, we explain several possible research directions in class imbalance problem.

- **Ensemble Learning**: Ensemble-based algorithms have been employed to improve the class imbalance problem. The performance of the ensemble learning algorithms is based on the individual classifiers accuracies, and also the diversity in handling the single problem between all classifiers. The diversity is measured as a degree to which the classifiers make different decisions on one problem. In [20], authors considered diversity while building ensemble classifiers. But careful study still needed to address the diversity problem in the future.
- **Variation in Imbalanced datasets**: in the past decade, hundreds of algorithms proposed to handle the class imbalance problems and they demonstrated outstanding performance. All the algorithms in general treated imbalance datasets constantly and handled it using versatile algorithms. But, in real-world scenario, the imbalanced datasets have variation in terms of imbalanced ratio, a number of dimension/features, and the number of classes. In [7], the author suggested that using a specific learning algorithm is inefficient to handle such variations in the data. So, adaptive learning is the new research direction in class imbalance problem.
- **Imbalance Data Streams**: The availability of massive amount of data leads to big data analytics. The most important challenge is to handle and answer to streaming and fast-moving data. Online learning has received higher attention from community/researchers belonging to machine-learning and data-mining domain/area. So, when learning from an online stream, the main difficulty arises is online class imbalance learning. In [21] the author discussed the problems arise with online class imbalance problems like lack of prior knowledge in identifying minority data classes and majority classes. Since the requirements of quick and accurate responses needed for data that arrive at a different time interval in big data era, imbalanced data stream becomes a new and popular research direction.

- **Semi-Supervised Learning**: In most of the data analysis, collection of massive data is cheaper but it is expensive to label the samples to train the classifier. In big data, it is quite common to find massive data with few labeled samples and plentiful unlabeled samples. For class-imbalanced datasets, it is difficult and expensive for semi-supervised learning to extract class label for minority samples. Active learning is used in general to predict the label of unlabeled samples. For handling imbalanced datasets, few active learning algorithms have been proposed in past. More research work needed in investigation of different active learning for selecting and utilization of informative samples in imbalanced datasets.
- **Application Areas**: The application in which the imbalanced datasets exist are management science, decision-making, emergency management, natural disasters (public health incidents/social security incidents), IoT (Internet of things), Big Data, and security management.

Hence, this section discusses about several research directions related to class imbalance problem. Now, next section will discuss a real-world problem and will share several interesting facts (regarding to class imbalance problem) form a user's perspective.

## 6 Open Discussion

The class imbalance problem is typically a classification problem with the skewed distribution of class labels. It can be a binary classification problem (with two class labels) or multi-class classification problem (with more than two class labels). As discussed in Sect. 4, smaller class (minority or positive) is highly interested to balance to get accurate results. But the question is how we will identify which class is majority or minority. So, for that, this section provides a real-world example to under this problem in clear-cut manner. For example, in a medical diagnosis of a rare disease where there is a critical need (or huge requirement) to identify such a rare medical condition among the normal populations. Any errors in diagnostic will bring stress and further complications to the patients. The physicians could not afford any incorrect diagnosis since this could severely affect the patients' well-being and even change the course of available treatments and medications. Thus, it is crucial that a classification model should be able to achieve higher identification (i.e. accuracy) rate on the rare probabilities of occurring diseases/occurrences (minority class) in datasets. In another interesting example, suppose a patient visits to a doctor and he takes several features in consideration and based on that doctor told that you will die tomorrow. Then this patient visits to a different doctor and this doctor uses similar features (which first doctor used) then he concluded that you are ok. Now patient is confused. Now, if this patient visits to a third doctor and this doctor also used similar features then he may get a different result and will tell to patient that you may or may not die tomorrow (i.e., 50–50 chances). So, for such false positive, true positive, false negative, true negative, etc., we require confusion metric (see Sect. 4) to represent such decisions/data.

In another example, suppose a new disease is raised somewhere, where about 0.01% of people have contracted the disease. Then, we collect data of 60,000 people, which includes six people with the disease. Now here, if we want to make a model that, given data of a new person, predicts whether that person has contracted the disease, we can easily build a model that scores really high for accuracy on the training set. Our model easily can show/predict that no person has the disease or no person who is not exactly equal to one of our known disease carriers. But based on just predicting that everyone is healthy has an accuracy of 99.99%. Then such models will become useless (due to less sampling of data). Apart from that, if we re-sample and analyze this data again then this time we will get different accuracy results. Hence, accuracy is a paradox for users, i.e. getting high accuracy does not mean that classes are balanced now. So, we need to make our models like that they will be able to train and find patterns in our imbalanced data. So, in the future/whenever we have imbalanced data/classes, our model can learn to choose the majority class instead of reacting to the data. Also, we can try several other processes (can avoid accuracy paradox) like collecting more data for sampling, changing several metrics for respective data, or resampling our data again and again (several times, we did not get two consecutive similar results), generating synthetic samples, using different–different algorithms to balance our data, etc. Moreover this, there are several approaches (to solve class imbalance problems) like balancing class imbalance (via training a set with equal numbers of classes by sampling out surplus class data points), anomaly detection (taking smaller class as anomalies and using such algorithms for classifications), and generating minority class samples (using the distributions of features in minority class, sample out more datapoints). Note that some algorithms are less prone to the problems typically associated with unbalanced classes, as long as there are some samples of each class in the training data. As discussed above, accuracy is a paradox in machine-learning or data-mining domain, here it means "when our model's accuracy measures and give a response to us that we have received excellent accuracy (such as 90%), but in actual, this received accuracy is only reflecting the underlying class distribution." With a large number of dataset and analysing this datasets with more number of features, we may get good accuracy than the less number of datasets with consisting similar (or less) features. In this, classification accuracy is the first metrics, we use to solve our classification *problems*. Note that a false sense of performance can create overfitting in a trained data.

Hence, this section shows a clear-cut vision to class imbalance problem and effect of good accuracy (with less resampling of data) and overfitting in datasets. Now, next section with conclude this work in brief and will show a future path to the future researchers (interested one) in this respective (hot) problem.

## 7 Conclusions

As discussed above, we reached to a conclusion that class imbalance problem has identified as biggest problem (especially in machine learning) to solve in many practical applications/domains. Today's machine learning is used to learn several facts/extracting knowledge from imbalanced data sets. In Sect. 2, we have discussed several techniques to balance/solve this (class) imbalance problem using changing class distribution (using under-sampling techniques, over-sampling and advanced sampling), features selection, classifiers level (via manipulating classifiers internally, cost-sensitive learning, one-class learning), and ensemble method (refer Sect. 2). Hence, the purpose of this paper is to show a clear understanding to a reader/researcher to find related facts regarding/required to solve a class imbalance problem. This work also explains several domains where class imbalance problem arises and metric used to measure performance of techniques (which is used to solve a respective problem). Also, we end this work by showing several future research directions to the future researchers (with respect to respective problem). Remember that, sometime we may not get much improvement or much accurate result (performance of the induced classifier) through balancing an imbalanced dataset. Also, many learning systems are not much interested to find differences in class distributions. Hence, we need a clearer and deeper understanding of how class distribution affects each phase of the learning process for more learners. A deeper understanding of the basics will help us to design better methods for dealing with the problem of learning with skewed class distributions.

**Acknowledgements** This research is funded by the Koneru Lakshmaiah Education Foundation, Lingaya's Vidyapeeth and Anumit Academy's Research and Innovation Network (AARIN), India. The authors would like to thank Koneru Lakshmaiah Education Foundation, Lingaya's Vidyapeeth and AARIN, India, an education foundation body and a research network for supporting the project through its financial assistance.

## References

. Kotsiantis, S., Kanellopoulos, D., Pintelas, P.: Handling imbalanced datasets: a review. GESTS Int. Trans. Comput. Sci. Eng. **30**(No 1), 25–36 (2006)

. Longadge, R., Dongre, S.: Class imbalance problem in data mining review. arXiv preprint arXiv: 1305.1707 (2013)

. Ali, A., Shamsuddin, S.M., Ralescu, A.L.: Classification with class imbalance problem: a review. Int J Adv Soft Comput Appl **7**(3), 176–204 (2015)

. López, V., Fernández, A., García, S., Palade, V., Herrera, F.: An insight into classification with imbalanced data: Empirical results and current trends on using data intrinsic characteristics. Inf. Sci. **250**, 113–141 (2013)

. Tahir, M.A., Kittler, J., Mikolajczyk, K., Yan, F.: A multiple expert approach to the class imbalance problem using inverse random under sampling. In: International Workshop on Multiple Classifier Systems, pp. 82–91. Springer, Berlin, Heidelberg (2009)

. Chawla, N.V., Bowyer, K.W., Hall, L.O., Kegelmeyer, W.P.: SMOTE: synthetic minority over-sampling technique. J. Artif. Intell. Res. **16**, 321–357 (2002)
. Yijing, L., Haixiang, G., Xiao, L., Yanan, L., Jinling, L.: Adapted ensemble classification algorithm based on multiple classifier system and feature selection for classifying multi-class imbalanced data. Knowl.-Based Syst. **94**, 88–104 (2016)
. Tang, Y., Zhang, Y.Q., Chawla, N.V., Krasser, S.: SVMs modeling for highly imbalanced classification. IEEE Trans. Syst. Man Cybern. Part B (Cybern.) **39**(1), 281–288 (2009)
. Herndon, N., Caragea, D.: A study of domain adaptation classifiers derived from logistic regression for the task of splice site prediction. IEEE Trans. Nanobiosci. **15**(2), 75–83 (2016)
. Raposo, L.M., Arruda, M.B., de Brindeiro, R.M., Nobre, F.F.: Lopinavir resistance classification with imbalanced data using probabilistic neural networks. J. Med. Syst. **40**(3), 69 (2016)
. Zakaryazad, A., Duman, E.: A profit-driven Artificial Neural Network (ANN) with applications to fraud detection and direct marketing. Neurocomputing **175**, 121–131 (2016)
. Tan, M., Tan, L., Dara, S., Mayeux, C.: Online defect prediction for imbalanced data. In: 37th IEEE International Conference on Software Engineering (ICSE), 2015 IEEE/ACM, vol. 2, pp. 99–108. IEEE (2015)
. Li, Z., Bin, Z., Ying, L., Yan, G., Zhi-Liang, Z.: A Web service QoS prediction approach based on collaborative filtering. In: Services Computing Conference (APSCC), IEEE Asia-Pacific, pp. 725–731. IEEE (2010)
. Xu, L., Chow, M.Y., Taylor, L.S.: Power distribution fault cause identification with imbalanced data using the data mining-based fuzzy classification $ E $-algorithm. IEEE Trans. Power Syst. **22**(1), 164–171 (2007)
. Wang, Y., Li, X., Ding, X.: Probabilistic framework of visual anomaly detection for unbalanced data. Neurocomputing **201**, 12–18 (2016)
. Márquez-Vera, C., Cano, A., Romero, C., Ventura, S.: Predicting student failure at school using genetic programming and different data mining approaches with high dimensional and imbalanced data. Appl. Intell. **38**(3), 315–330 (2013)
. Lessmann, S., Voß, S.: A reference model for customer-centric data mining with support vector machines. Eur. J. Oper. Res. **199**(2), 520–530 (2009)
. Cateni, S., Colla, V., Vannucci, M.: A method for resampling imbalanced datasets in binary classification tasks for real-world problems. Neurocomputing **135**, 32–41 (2014)
. Fawcett, T.: An introduction to ROC analysis. Pattern Recogn. Lett. **27**(8), 861–874 (2006)
. Wang, S., Yao, X.: Diversity analysis on imbalanced data sets by using ensemble models. In: IEEE Symposium on Computational Intelligence and Data Mining, CIDM'09, pp. 324–331. IEEE (2009)
. Ghazikhani, A., Monsefi, R., Yazdi, H.S.: Online neural network model for non-stationary and imbalanced data stream classification. Int. J. Mach. Learn. Cybernet. **5**(1), 51–62 (2014)

# Suicidal Ideation from the Perspective of Social and Opinion Mining

**Akshma Chadha and Baijnath Kaushik**

**Abstract** Social media is a way of communicating with others and its popularity is growing worldwide. It has a lot of influence on its users. People read various posts and get affected by it. Suicide is one of the major health issues on social media which influence others to do the same. The number of suicides is increasing day by day. Thus, a need arises to find or develop a way to control suicides through social media. Machine learning is being widely used by many researchers for this purpose, with the help of psychiatrists. A lot of studies have been done in this field. In this paper, we have reviewed the existing work in this field inferring their limitations so that further work can be carried out.

**Keywords** Suicide · Suicidal ideation · Depression · Anxiety

## 1 Introduction

Social media is a huge platform to express one's feelings. Social networking sites have become a part of almost every adult. It has a positive as well as negative impacts. People can communicate on various topics, which can be helpful in business activities. But people can also spread rumors or false information [1]. So we need to have a check on this, because of that social media analysis is a complex problem. The usage of social networking is increasing day by day. The individuals can express themselves using social networking. The people may have self-injurious thoughts, which may lead to suicide, as shown in Fig. 1. Nowadays the number of suicide cases with respect to social media is increasing. Suicide is very common in military personnel, social media and medical records help to prevent them [2]. The number of suicides is increasing every year. Anxiety leads to suicidal attempts and is considered as one of the risk factors of suicide evidence by many organizations [3]. Suicide is to escape from oneself and all the negative thoughts associated. To avoid suicide one should avoid the feeling of anxiety. The rate can be decreased by preventing suicide. The people having suicidal ideation share things like that with their family and friends.

A. Chadha (✉) · B. Kaushik
Shri Mata Vaishno Devi University Katra, Katra, India
e-mail: akshma2013@gmail.com



P. K. Singh et al. (eds.), *Proceedings of ICRIC 2019*, Lecture Notes in Electrical Engineering 597, https://doi.org/10.1007/978-3-030-29407-6_47

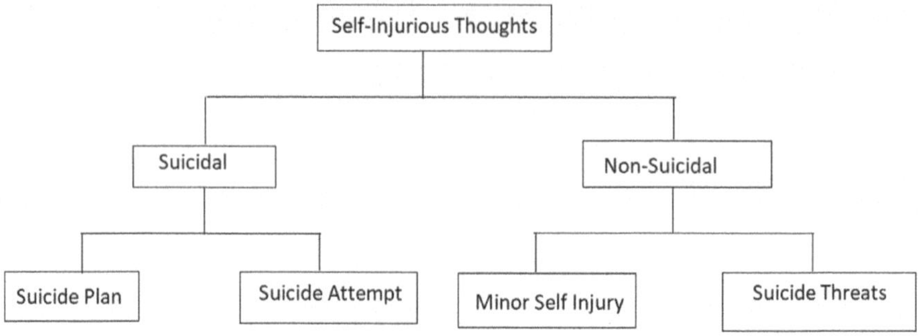


**Fig. 1** Self-Injury thoughts

Nowadays people are more likely conveying their feelings on social media sites, by keeping a check on the behavior of an individual having suicidal thoughts. This can be done by keeping an eye on their social media life. Suicide is a major health issue. The number of suicidal behavior is increasing day by day by getting the way, how individuals are having suicidal ideation, we can prevent the suicide from occurring. E-health intervention can help people having suicidal ideation by sending videos or commenting on their posts. E-health intervention includes identifying people with suicidal ideation, helping them through a guided and unguided medium [4]. Mental sickness is a massive cause of disorder globally. There are no particular laboratory tests for detecting depression. We detect it based on the reviews and feelings of the patients and their family and friends. Social media can be used as a tool for diagnosing depression. Usually, young people have the ideation to harm themselves. These are the people, who use the Internet much more than any other peer group.

This can be prevented by examining responses to Nonsuicidal Self-Injury (NSSI) videos. Suicide does not just depend on an individual, it is affected by the environment and social life. The mood of the public depends merely on the posts on social media sites [5]. Depression is one of the causes of burden in the countries. Snapchat is also used for the prevention of suicide. As the number of suicides is increasing day by day, the reason behind it must be found [6]. This can be done by using machine learning. The steps used in performing it is shown in Fig. 2.

The remainder of the paper is organized in the following sections. Section 2 describes the literature survey on this topic. Section 3 describes data collection in existing work, which describes how the data is collected previously for this work. Section 4 describes findings, which tell about the limitations in the previous work. Section 5 describes the conclusion from the study and Sect. 6 gives ideas for the future work.

We have studied about suicidal ideation on social media data, as very little work has been done in this field. By examining the individuals with suicidal ideation, we can protect them from doing so and treat them with the help of psychologists, friends, and family. This field needs a lot of work to be done. Some of the work that has already been done is studied.

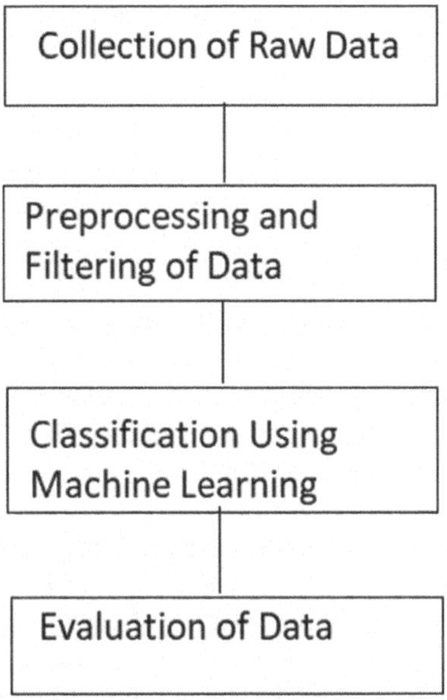


**Fig. 2** Steps for classifying suicidal and nonsuicidal data using machine learning

## 2 Literature Review

Stephen P. Lewis et al. has taken YouTube videos to ascertain risk and benefits of Nonsuicidal Self-Injury (NSSI). Two coding rubrics were formed, first, it focuses on the essence of remarks that are posted as a reply to YouTube videos, second rubric analyzes the most persistent comments from the first analysis. The result includes an intention to recover, presently looking for help or therapy to recover, and linking the video to recover. The comments are much more powerful if they come from a similar source, i.e. the young adults will be affected by the messages of youth [7].

Hong-Hee Won et al. has done work to anticipate national suicide numbers with the help of social media data. The data had been figured out from the Weblog posts. The economic data includes consumer price index, unemployment rate, and stock index valuations were drawn out from Korea Composite Stock price index.(KNSO) The data has been split into a 12-year training set for identifying the important predictor variables and creating a prediction model, and a one-year validation set for figuring out the model. The number of suicides is evaluated with respect to observed, predicted intervals, which is the celebrity suicide periods. The author made use of univariate linear regression analysis using training set. The dependent variable was logarithm transformed to satisfy a normal distribution assumption in the regression analysis. The multivariate regression model was formed using these selected variables determined in the training set. The author made use of *R* software [8].

Helen Christensen et al. use E-health intervention as a means to identify individuals at risk of suicide and then offer help through the Web. E-health intervention for suicide prevention is classified as screening whether people might be at risk for suicide or a mental health problem, Web applications both guided and unguided have been developed to provide psychological interventions to assist in reducing suicidal behavior, and thirdly, a person is considered to be at risk of suicide because of the nature of their social media use [4].

Amayas Abboute et al. have done the research in three parts: (1) To retrieve messages from Twitter related to suicide, (2) A classification model is built to trigger alert for messages at high risk, (3) Present those messages to psychiatrists or professionals. First, vocabulary is built for the topics related to suicide, depression, sadness, etc. Tweets with these keywords are collected through Twitter API and the tweets of people already committed suicide were also collected to be taken as proved cases. The messages are classified as risky and non-risky manually by some professional. After that automatic classification is done using six classifiers: JRIP, IBK, IB1, J48, Naive Bayes, and SMO. The results are tested via WEKA. Two validation was applied leave-oneout validation (LOO) and also ten-fold cross-validation. The classifier that proved to be the best was Naïve Bayes in both the validations with an accuracy of 63.15% in LOO and 63.27% in 10-CV. Then, it is discovered that the tweets related to depression were mainly in the non-risky category, so these were not taken into consideration. This tool can effectively help in suicide prevention [9].

Gualtiero B. Colombo et al. made a list of friends and followers of the authors of suicidal tweets. Their mutual friends are also found and a directed graph is made between them to see if there is any relation between all of them. It is found that a community of people exists who have suicidal ideations. A retweeted graph is also made, which tells about the amount of risk. At last both the graphs are combined to calculate the level of propagation between users with suicidal ideation. The relation is found between authors of the tweet and the people who retweeted, i.e. it is found that the retweet is made by a friend/follower or it is someone other than the community, which will be a higher degree of risk. In particular, retweets to be considered are those which are not belonging to the suicidal set (authors with suicidal content). Graph characteristics of Twitter users with suicidal ideation were found [10] (Table 1).

Bridianne O'Dea et al. selected a team of human coders who finds out the tweets with suicidal ideations and classify them in three categories, i.e. strongly concerning,

**Table 1** Keywords/search terms

| | | |
|---|---|---|
| Want to die | End it all | Kill myself |
| I hate myself | Have nothing to | Wish could just fall asleep |
| Want to end it all | I'm sorry that I am leaving | I am worthless |
| Asleep and never wake | Die in my sleep | Don't want to live |
| Hate myself | Want to sleep forever | Life is too hard |
| My life is pointless | Ready to die | Need to die |
| Suicide | Take my own life | Want to be dead |

possibly concerning, and safe to ignore. In case of doubt, put the tweet in a default class. A machine classifier is built that can classify tweets in the three categories. The scikit-learn toolkit is used. term frequency weighted by inverse document frequency (TDIDF) was used in spite of simple frequency. Two machine-learning classifiers were used for text classification support vector machine (SVM) and logistic regression. The precision scores were greater than 75% for all the classes [11].

Jessica R. Barrett et al. collected the data on non-fatal self-harm under the SHIELD service improvement project for self-harm. The Clinical Record Interactive Search (CRIS) tool is used to search de-identified electronic health records. The full notes of health records are searched for those which have 'hits' on selected keywords. Each record is coded into various categories based on 'clinicians' notes: false positive, goodbye note, help-seeking note, etc. The data is then merged with the self-harm data generated by SHIELD to get presentation information. This information is used to identify which presentation has left a paper note. Those having partners are more likely to leave a note. Those with a family history of suicide are less likely to leave a note. Higher Beck Suicide Intent Scale (BSIS) score of those who leave a paper note shows that paper notes are riskier. Data loss at various stages of data extraction is a limitation because the final sample contains a very less number of records [12].

Kelly Soberay et al. studied about suicide as a very common problem in military personnel nowadays. Machine learning is useful in detecting individuals at risk of suicide in social media posts. The ability to predict suicide requires to look into various risks which are quite complex, machine learning is a better option over traditional techniques. Machine learning is explored longitudinally and with a large number of suicides, attempters to see the accuracy of suicide prediction. Another approach to predicting suicide posts on social media is(Autoregressive Integrated Moving Average (ARIMA) to identify trends in communication about depression and suicide. Twitter is used extensively to be used for analyzing social media content, as the data on Twitter is available publically. A comparison is made between nonsuicidal twitter posts and suicidal twitter posts, a strong linguistic post was represented by higher word counts, use of self-referencing, greater anger. Intense, violent, and growing anger are significantly related to suicide. Users who are more related to suicide-related content are more connected to the ones having suicidal behavior, this is like a communicable disease. This can majorly affect the youth [2].

Kate H. Bentley et al. has done a meta-analytical review to ascertain that anxiety disorder is a significant risk factor for suicidal ideation and behavior. According to various national agencies, anxiety disorder is a risk factor. According to many studies, anxiety disorders are theoretically and empirically proved to be a risk factor for suicidal ideations. The literature survey is done by using the following terms longitudinal, longitudinally, prediction, follow-up, self-injury, suicide, etc., for the meta-analysis random-effects model was used, to keep the heterogeneity within and between the studies. The effect sizes for any anxiety construct predicting suicide attempts and deaths are estimated. Orwin's fail-safe $N$ was used to estimate the number of missing studies with ORs of 1.00 to lower the magnitude of effect sizes to negligible values. Egger's test of intercept was also employed to predict the standardized effect. A funnel plot is used to represent the standard error. For diagnosing accuracy 39 suicide attempt cases with anxiety constructs and 26 death cases with anxiety constructs were included. I2 statistic is used to check heterogeneity within meta-analysis. Various analysis indicated that anxiety disorders are a predictor of suicidal ideation and attempt but not death. Anxiety disorders are not likely to serve as a risk factor for suicidal ideation and behavior, they act as risk factors for a shorter period of time [3].

M. Johnson Vioules et al. states that change in behavior is a major risk factor for suicide especially when that change is related to some pain or loss. Nowadays people express their feelings abruptly on social media, so there must be some way to detect this negative change in behavior which is known as change point detection problem in data mining. To solve this problem, martingale framework is used for change point detection especially in unlabeled data. Sudden increase or decrease in friends and followers can be a risk factor, an individual at risk may spend a lot of time on social media or some do not like to talk to anyone. Replying or retweeting content related to mental health is also a contributing risk factor. The behavior can also be identified by the timing of the posts and the text of the post which tell the current mood and mental health, this type of behavior is known as post-centric as it is related to the post. To classify the text, two classifiers are used: (1) natural language processing method and (2) distress classifier. The main advantage of this study is that it considers the individual's behavior with respect to his own history and not with some other individual [13].

Patricia A. Cavazos-Rehg et al. deals with Tumblr as a social media site. Young people have an interest in posts related to depression and mental health problems. Individuals having suicidal ideation try to be friends on social media with those who post self-harm videos. The comments on these kinds of videos are encouraging and there is no recovery mentioned in it. Tumblr is a microblogging site where people having common interests forms a community. In contrast to Facebook and Twitter where people make friends with their family, co-workers, friends, etc., in Tumblr people get connected to those whom they will not even know but share common interests. It may be possible that a community in Tumblr talks about depression as their keen interest. This study examines the depression-related posts on Tumblr, the most popular posts related to suicide, self-injury, depression is examined at a higher priority. Patricia A. Cavazos-Rehg searched Tumblr accounts related to suicide and self-harm. After the accounts have been searched, the characteristics of these accounts are viewed and the engagement of these accounts with other users is checked. SAS Proc SURVEYSELECT procedure was used to select posts randomly. Member of research team having expertise in mental health studied these posts to prepare a list of themes that are relevant based on the topics observed. The research assistants viewed the posts and coded them [14].

Pete Burnap et al. deals with different classes related to suicidal ideation. Twitter data is classified into different classes using support vector machine (SVM), decision tree (rule-based), and Naïve Bayes. The baseline experiments needed refining. Rotation forest ensemble approach was used which enhances the performance. SVM was best in terms of performance. The maximum *F*-measure using rotation forest was 69% [15].

Bart Desmet et al. deals with finding emotions using suicide notes. They make use of both natural language processing (NLP) and machine-learning techniques. SVM is used as a classifier to predict the presence of emotions. Memory-based shallow parser (MBSP) was used for pre-processing of data. Spelling correction was done using TICCL, a corpus-based spelling correction system. Bootstrap resampling was also used to determine the *F*-scores. The difference between the original and spell-checked datasets are very small [16].

## 3 Data Collection in Existing Work

Crowdsourcing is used to collect labels (a mechanism to collect behavioral data from the population) [17].

The data is collected from Twitter API. First, a vocabulary is made of the type of text related to suicide. The post was human annotated by crowdflower. These were used to find content having suicidal thoughts. Term frequency inverse document frequency (TFIDF) was applied to the datasets to find the terms associated with suicide in each dataset that was not present in the other [10].

Data collection is done using twitter public API and then filtered according to the given criteria. API is used with a tool developed by CSIRO to find out tweets having keywords related to suicide with the name and picture of the profile [11].

Data collection is done using the public API offered by Twitter. After that, the data is filtered based on the given keywords. As there is a huge volume of data, Amazon Web Service is used to process the data [18]. Crimson Hexagon's ForSight software is used to access Twitter data for terms related to depression and suicide. It has partnered with Twitter to provide a full census of public tweets. The Website http://hashtagify.me/ was used to identify hashtags related to depression and suicide. Some keywords must be used as an exclusion criterion like suicide bombers [19].

## 4 Findings

Finding individuals with suicidal ideation on social media is a trending topic, a lot of research is going on this. A literature review has been done and various limitations have been found. The focus is to be done on these so that various limitations can be solved as research problems. The limitations are written in a tabulated form in Table 2.

**Table 2** Limitations found

| S no. | Paper title | Author name | Year of publication | Limitations |
|---|---|---|---|---|
| 1 | Helpful or Harmful? an Examination of Viewers' Responses to Nonsuicidal Self-Injury Videos on YouTube | Stephen P. Lewis | 2012 | (1) Demographic data from people who responded to NSSI videos were unavailable<br>(2) The people considered are usually young adults which may not be true |
| 2 | E-Health Interventions for Suicide Prevention | Helen Christensen | 2013 | (1) Most of the work is done on the publically available network like Twitter and work is not done on private network like Facebook |
| 3 | We Feel: Mapping Emotion on Twitter | Mark E. Larsen | 2015 | (1) The period of data collection is quite small<br>(2) Large z-scores were observed so the additional analysis is required<br>(3) Suicide rates showed a positive relation with cheerfulness and negative relation with affection, rage, and neglect |

(continued)

**Table 2** (continued)

| S no. | Paper title | Author name | Year of publication | Limitations |
|---|---|---|---|---|
| 4 | Using Social Media to Monitor Mental Health Discussions-Evidence from Twitter | Chandler McClellan | 2016 | (1) Better filtering to the data must be applied so that the data can be refined<br>(2) There is no way to differentiate between expected and unexpected shocks<br>(3) The finding cannot be applied to a wider population |
| 5 | Anxiety and its Disorders as Risk Factors for Suicidal Thoughts and Behaviors: A Meta-Analytic Review | Kate H. Bentley | 2016 | (1) Anxiety is a very important risk factor of suicidal ideation and behavior but it is not studied under those conditions to establish the relationship<br>(2) A large number of individuals must be included in the study of a larger follow-up to detect suicidal behavior |
| 6 | Detection of Suicide-Related Posts in Twitter Data Streams | M. Johnson Vioules | 2017 | (1) The author presents the result of only 2 twitter users, testing should be performed on more users timelines<br>(2) The parameter setting of the martingale framework can be improved |

## 5 Conclusion

As the use of social media is increasing day by day, the data associated with it is also increasing and the need to study topics related to social media communication is also increasing [1]. Suicidal thought is a very big problem related to public health. The number of suicides is increasing globally. To prevent suicide, we must find out the risk factors related to it. Finding the risk factors will help to find the individuals at risk so that it can be treated and prevented [3]. Social media is used to monitor the mental health of various personalities. It is helpful in spreading public health messages globally [20]. The people having mental illness are the ones who live in poverty and are not much educated. Because of the things they have to suffer due to mental illness, as the treatment of it is very costly. As the tweets in twitter are public, so anyone can read it and can get affected by that. People use hashtags in twitter which helps people to view posts related to any of the topics. Social media is a wide platform to get information related to mental health. To identify communication patterns related to depression and suicide on Twitter, the data is collected by using hashtags related to depression and suicide [19]. Mental illness is a major disorder worldwide. There are not proper laboratory tests to detect mental illness, it can be diagnosed based on the symptoms asked from family and friend. Social media helps diagnose mental illness in regard to these challenges. People frequently post on social media. These posts express their mood, behavior, etc., which helps to diagnose the mental health of the person [17]. A platform is to be created for psychiatrists to find the tweets based on suicidal topics. Suicide is a very big economic issue [21]. The main focus was to find people with suicidal ideation as early as possible. This will have a very good social impact. The life of many people can be saved by an early detection of their suicidal thoughts. This can be done by contacting the person directly or talking to their relatives [9]. Twitter is a major source for detecting suicidality in people. Suicides can be prevented by understanding the way people with suicidal ideations talk. Not all individuals with suicidal ideation attempt to die but they are at high risk of killing themselves. Some people who have suicidal ideation talk about it with their friends and family on Twitter and can be stopped from doing that [22]. Online methods should be there to prevent suicide by extracting the tweets with suicidal ideation and people having such ideas [11]. Many studies have been taken to see the mental health of a person especially depression. In most developing countries, depression is caused due to the burden [18]. The rate of suicide is increasing worldwide. There must be some way to prevent suicides. Suicide can have any of the three risk factors (1) environmental factors, (2) health factors, and (3) historical factors. Suicide behavior can be warned by talks, behavior, and mood. The first step toward suicide is to find its risk factors. The individuals at risk do not like to take professional help, so informal means like social media can be used to help them as they are more comfortable in writing their thoughts on social media compared to talking to someone on face [13].

## 6 Future Scope

Finding suicidal ideation on social media is a tricky topic and a lot of work needs to be done in this field. There is a large scope in this. We should reach the youth who talk about Nonsuicidal Self-Injury videos through the Internet. By using some more words with the keywords, the results can be improved. The set of annotated posts were limited. We must be dealing with more than one hop away neighbors [10]. Expanding the range of search terms related to suicide. Some offline measures should also be taken into consideration: family and friends, questionnaires and consultation from clinics. An analysis should be done of the twitter accounts of those who have already committed suicides. The work should be done on more social media sites like Facebook, Snapchat, Instagram, etc.

## References

1. Stieglitz, S., Mirbabaie, M., Ross, B., Neuberger, C.: Social media analytics–challenges in topic discovery, data collection, and data preparation. Int. J. Inf. Manage. **39**, 156–168 (2018)
2. Social media and machine learning in suicide prevention (2017)
3. Bentley, K.H., Franklin, J.C., Ribeiro, J.D., Kleiman, E.M., Fox, K.R., Nock, M.K.: Anxiety and its disorders as risk factors for suicidal thoughts and behaviors: A meta-analytic review. Clin. Psychol. Rev. **43**, 30–46 (2016)
4. Christensen, H., Batterham, P., O'Dea, B.: E-health interventions for suicide prevention. Int. J. Environ. Res. Public Health **11**(8), 8193–8212 (2014)
5. De Choudhury, M., Kiciman, E., Dredze, M., Coppersmith, G., Kumar, M.: Discovering shifts to suicidal ideation from mental health content in social media. In: Proceedings of the 2016 CHI Conference on Human Factors in Computing Systems, pp. 2098–2110. ACM (2016)
6. Chhabra, N., Bryant, S.M.: Snapchat toxicology: social media and suicide. Ann. Emerg. Medicine **68**(4), 527 (2016)
7. Lewis, S.P., Heath, N.L., Sornberger, M.J., Arbuthnott, A.E.: Helpful or harmful? An examination of viewers' responses to nonsuicidal self-injury videos on YouTube. J. Adolesc. Health **51**(4), 380–385 (2012)
8. Won, H.H., Myung, W., Song, G.Y., Lee, W.H., Kim, J.W., Carroll, B.J., Kim, D.K.: Predicting national suicide numbers with social media data. PloS one **8**(4), e61809 (2013)
9. Abboute, A., Boudjeriou, Y., Entringer, G., Azé, J., Bringay, S., Poncelet, P.: Mining twitter for suicide prevention. In: International Conference on Applications of Natural Language to Data Bases/Information Systems, pp. 250–253. Springer, Cham (2014)
10. Colombo, G.B., Burnap, P., Hodorog, A., Scourfield, J.: Analysing the connectivity and communication of suicidal users on Twitter. Comput. Commun. **73**, 291–300 (2016)
11. O'Dea, B., Wan, S., Batterham, P.J., Calear, A.L., Paris, C., Christensen, H.: Detecting suicidality on Twitter. Internet Interventions **2**(2), 183–188 (2015)
12. Barrett, J.R., Lee, W., Shetty, H., Broadbent, M., Cross, S., Hotopf, M., Stewart, R.: 'He left me a message on Facebook': comparing the risk profiles of self-harming patients who leave paper suicide notes with those who leave messages on new media. BJPsych Open **2**(3), 217–220 (2016)
13. Vioulès, M.J., Moulahi, B., Azé, J., Bringay, S.: Detection of suicide-related posts in Twitter data streams. IBM J. Res. Dev. **62**(1), 7–1 (2018)
14. Cavazos-Rehg, P.A., Krauss, M.J., Sowles, S.J., Connolly, S., Rosas, C., Bharadwaj, M., Grucza, R. and Bierut, L.J.: An analysis of depression, self-harm, and suicidal ideation content on Tumblr. Crisis (2016)

15. Burnap, P., Colombo, G., Amery, R., Hodorog, A., Scourfield, J.: Multi-class machine classification of suicide-related communication on Twitter. Online Soc. Netw. Media **2**, 32–44 (2017)
16. Desmet, B., Hoste, V.: Emotion detection in suicide notes. Expert Syst. Appl. **40**(16), 6351–6358 (2013)
17. De Choudhury, M., Gamon, M., Counts, S., Horvitz, E.: Predicting depression via social media. ICWSM **13**, 1–10 (2013)
18. Larsen, M.E., Boonstra, T.W., Batterham, P.J., O'Dea, B., Paris, C., Christensen, H.: We feel: mapping emotion on Twitter. IEEE J. Biomedical Health Informatics **19**(4), 1246–1252 (2015)
19. McClellan, C., Ali, M.M., Mutter, R., Kroutil, L., Landwehr, J.: Using social media to monitor mental health discussions–evidence from Twitter. J. Am. Med. Inform. Assoc. **24**(3), 496–502 (2017)
20. Luxton, D.D., June, J.D., Fairall, J.M.: Social media and suicide: a public health perspective. Am. J. Public Health **102**(S2), S195–S200 (2012)
21. Fu, K.W., Cheng, Q., Wong, P. W., Yip, P.S.: Responses to a self-presented suicide attempt in social media. Crisis (2013)
22. Pak, A., Paroubek, P.: Twitter as a corpus for sentiment analysis and opinion mining. LREc **10**(2010), 1320–1326 (2010)

# Clustering of Tweets: A Novel Approach to Label the Unlabelled Tweets

**Tabassum Gull Jan**

Twitter is one of the fastest growing microblogging and online social networking site that enables users to send and receive messages in the form of tweets. Twitter is the trend of today for news analysis and discussions. That is why Twitter has become the main target of attackers and cybercriminals. These attackers not only hamper the security of Twitter but also destroy the whole trust people have on it. Hence, making Twitter platform impure by misusing it. Misuse can be in the form of hurtful gossips, cyberbullying, cyber harassment, spams, pornographic content, identity theft, common Web attacks like phishing and malware downloading, etc. Twitter world is growing fast and hence prone to spams. So, there is a need for spam detection on Twitter. Spam detection using supervised algorithms is wholly and solely based on the labelled dataset of Twitter. To label the datasets manually is costly, time-consuming and a challenging task. Also, these old labelled datasets are nowadays not available because of Twitter data publishing policies. So, there is a need to design an approach to label the tweets as spam and non-spam in order to overcome the effect of spam drift. In this paper, we downloaded the recent dataset of Twitter and prepared an unlabelled dataset of tweets from it. Later on, we applied the cluster-then-label approach to label the tweets as spam and non-spam. This labelled dataset can then be used for spam detection in Twitter and categorization of different types of spams.

Spam labelling · Clustering · Tweets

## 1 Introduction

Twitter is one of the most popular social media platforms with almost 313 million monthly active users and 500 million tweets per day are posted. On the basis of the

T. G. Jan (✉)
Department of Computer Science & Technology Central University of Punjab, Bathinda, India
e-mail: tabassumgull2012@gmail.com

 
P. K. Singh et al. (eds.), *Proceedings of ICRIC 2019*, Lecture Notes in Electrical Engineering 597, https://doi.org/10.1007/978-3-030-29407-6_48

user traffic, the global ranking of Twitter is 12 among all the websites available on Twitter. Twitter spams have long been a critical issue that Twitter spam detection is the trending research area nowadays. Most of the work has been carried out in the field using different machine learning techniques; however, the main focus of the recent works carried is based on the application of machine learning techniques into Twitter spam detection, categorization of spammers in Twitter, types of spams in Twitter, etc. The problem with the existing techniques is that they are based on supervised machine learning algorithms, and for that very purpose, there is a need for labelled dataset. To label the data manually is a tedious task and it needs a lot of manpower and time. Manual labelling of tweets also leads to inter-observer variability because for a person to label a tweet as spam and non-spam depends on his or her nature of interest. For one person, it can be spam, and for others, it won't be. Also, the datasets available are very old and small in size, and for machine learning algorithms to achieve a higher range of accuracy, the size of the training dataset should be large enough so that the learning algorithm learns the patterns very well and later on predicts the results accurately. Hence, there exists a need to design a recent dataset of Twitter that is based on unsupervised labelling or semi-supervised labelling. A lot of work has been carried out in the field of analysing the tweets but to design a recent dataset of Twitter as large as one can and then assign labels to these tweets using semi-supervised machine learning techniques has not been yet done.

In the recent past, there are various methods suggested to detect the spam and non-spam tweets in the Twitter. Some of them focused on content-based features of tweets only, some on account-based features, some did the semantic analysis of the tweets and some did combination of them. However, given the noisy contents of tweets along with the possible heterogeneity in the content of tweets, it is difficult to detect the spam tweets from non-spam tweets. With an increase in security credentials needed for using Twitter, spammers have also become smarter to evade the detection during the credential validation process. So to detect such spammers in order to avoid violation of privacy (by posting spam posts) is the need of the hour. First of all, the need is to design and prepare a recent dataset of Twitter. In this paper, we have used the Twitter Streaming APIs to download the most recent 3240 tweets corresponding to a particular tweet ID. We have collected the tweet dataset corresponding to the legitimate users like cricketers, Bollywood actors, actresses, musicians, singers, politicians, etc. The advantage of designing the recent dataset of Twitter is that it can be better used to analyse the patterns to trace the spammers in Twitter. Further, the problem of spam drift will be almost solved because here we will be able to better study the trends in the tweets of spam posts and how attackers are becoming smarter to evade the detection. Also, older datasets were limited in terms of features, but here we have analysed the maximum possible features regarding Twitter spam detection.

As cited in the aforementioned discussions, we will discuss first the collection of the recent dataset of tweets using streaming APIs. The process and technique are used to extract the features of the tweets. And finally, the novel semi-supervised approach is used to label the tweets as spam and non-spam. Our method is quite a new approach in this area, also quite different from the existing techniques. First, we seek to understand the information available in the tweet, the meaning of each feature and how its value aids in spam labelling. We have performed extensive pre-processing of the tweets in order to eliminate the unnecessary constructs or fields related to spam and non-spam labelling so that we can focus on features that have higher weight with regard to spam labelling observed representation of tweets ensures high intra-class similarity between the tweets and low inter-class similarity. Further, in order to group similar tweets belonging to a particular spam category, we propose the cluster-then-label approach for tweet labelling. We have graphically shown the tweets in different clusters and with different spam labels.

The noteworthy contributions of this novel approach used for labelling the unlabelled tweets are summarized as:

- We seek to represent the pre-processed tweets using as a novel approach of cluster-then-labelling.
- We introduce a semi-supervised approach for labelling the large dataset of tweets.
- We have processed the dataset from Twitter using Streaming APIs by extracting maximum possible features from the tweets that can be used for different purposes in future.

So the main aim of this paper is to label the unlabelled tweets using semi-supervised approaches. In the next section, we will discuss the machine learning approach used to label the tweets as spam.

## 2 Spam Detection in Twitter

Spam can be defined briefly as sending unsolicited messages (in the form of texts, videos, pictures) to other users. Those messages could have malware, advertisement, malicious link, etc. In the context of OSN, spam can take any of the forms, for example, fake profiles, fraudulent reviews, clickjacking, advertisement, malicious links and personally identifiable information. Hence, receiving unsolicited messages is a privacy violation. Since dependence on online social networks like Twitter, Facebook, LinkedIn, etc. has spread their Web in each and every walk of life whether it is communication, entertainment, marketing, advertisement, news, academics, business, etc. Twitter Web is growing very fast, so Twitter spam is a huge problem that needs to be addressed as the popularity of Twitter greatly depends on the quality and integrity of contents contributed by users. Unfortunately, this popular online social networking site, that is, Twitter has attracted numerous spammers to post spam content which pollutes the community and a big loss to an organization at the other end. In today's digital and social networking era, social spamming is more successful than traditional methods such as email spamming by using a normally hidden social relationship between users. Detecting spam is the first foremost and very critical step in the battle of fighting spam. Twitter spam has long been a very critical and difficult problem that needs to be resolved. This spam brings huge losses to site operators and has an adverse impact on subsequent processing of data such as user behaviour analysis, data mining and resource recommendation. So far, researchers have done a lot of work in spam detection in Twitter still, it is not accurate and completed yet. Different researchers developed a series of machine learning based methods and blacklisting techniques to detect spamming activities on Twitter. In this section, we will discuss the works done in the field of spam detection by different researchers using different techniques.

Twitter has become a popular aspect of social networking spams due to its susceptibility and vulnerability to attacks. These social network platforms face threats imposed by spam accounts that propagate advertisements, phishing sites, fraud, etc. Such spam activities negatively affect normal users' experience and adverse to subsequent processing of user data. These spam behaviours bring huge losses to site operators and adverse to subsequent processing of data such as user behaviour analysis, data mining and resource recommendation. Till now, various researchers have developed a series of machine learning based methods and blacklisting techniques to detect the spamming activities on Twitter. According to research works carried in this field so far, different methods and techniques have achieved an accuracy of almost 80% and more. Since the Web of Twitter is increasing day by day and due to its enormous growth, many users begin to trust this social platform *f* information interchange. And that is the reason, users share the information in tweets regarding the happy moments, tragedies, accidents, celebrations, etc. and thus communicate with each other. Owing to such big popularity and fame, attracts malicious users to attack it. Attacks in the form of spam posts, scam, phishing sites, malware downloading and posting pornographic content have been reported on Twitter. This is all because tweets are small in length and attackers usually use shortened URLs in tweets so as to force a common user to click on it and thus achieve their goals (such as redirect the Twitter user to external attack servers, etc.).

Also because of the problems of spam drift and data fabrication techniques, to detect spam activities in real-life scenarios using machine learning based methods have become difficult. Spam in Twitter is the huge problem of information security. Spam accounts threaten user's privacy as spammers use it for their malicious aims. Spams can cause annoyance to users as well wastage of time as well networking resources. So spam detection is the mandatory and emerging concept in the field of information security. Spam detection in Twitter is based on some statistical properties of tweets like number of followers, number of followings, words per tweet, retweet count, links per tweet, repeated words in tweet, user mentions, etc. After feature engineering is done, results are analysed by simply applying the supervised machine learning algorithms like naive Bayes, support vector machine, decision tree, etc. All the detection models use labelled dataset and then analyse these statistical properties to perform Twitter analysis. In this paper, we will design an approach that uses unlabelled dataset and statistical properties of tweets for spam detection and categorization. In the next section, we will concisely discuss the existing methods used to label the tweets and the techniques on the basis of which tweet spam labelling is done.

## 2.1 Related Works

Twitter has become a popular aspect of social networking spams due to its susceptibility and vulnerability to attacks. These social network platforms face threats imposed by spam accounts that propagate advertisements, phishing sites, fraud, etc. Such spam activities negatively affect normal users' experience and adverse to subsequent processing of users data. These spam behaviours bring huge losses to the site operators and adverse to subsequent processing of data such as user behaviour analysis, data mining and resource recommendation.

Several machine learning approaches and algorithms were applied in past research works carried in the field of Twitter spam detection in which they extracted statistical features of tweets and formed a training dataset. After designing the dataset, the use of account and content-based features like length of tweet, digit count in tweet, no. of followers, no. of characters in tweets, account age, no of followings, account reputation, friends count, favourites count, etc. were made to detect spam and spammers [2]. They analysed the results using supervised machine learning algorithms like support vector machine. Further, some researchers trained RF classifier [12] and then used this classifier to detect spam on social networking sites like Twitter, Facebook and Myspace as discussed in [2, 12]. Features discussed in these research works can be manipulated easily by mixing spam tweets with normal tweets, purchasing more followers, texting users to please accept the friend request so as to make a balance between different ratios used for spam detection in Twitter. Some researchers also proposed robust features which were based on a social graph so that feature modification can be avoided and fabrication of tweet features is least possible. A sender and receiver concept was used by [11] in which they extracted the feature's distance and connectivity between tweet sender and receiver in order to find out whether it is spam or non-spam. Each and every terminology in Twitter is represented as a graph, for example, users are represented as nodes of the directed graph, and relation between two users (followers, following, etc.) is represented as an edge pointing towards the node. Due to this work, performances of various classifiers were greatly improved. A more robust feature such as local clustering coefficient, betweenness centrality and bidirectional links ratio was calculated for detection of spam tweets [14].

The researchers of [5] proposed a set of techniques of machine learning with semantic analysis for classifying the sentence and product reviews based on Twitter data. It contributes to the sentiment analysis for customers' review classification which is helpful to analyse the information in the form of the number of tweets, the content of the tweets, where opinions are highly unstructured and are either positive or negative, or neutral. After pre-processing the dataset, they extracted and selected the feature vector list and thereafter applied machine learning based classification algorithms (naive Bayes, maximum entropy and SVM) along with the semantic orientation based WordNet which extracts synonyms and similarity or duplicity for the content feature. The naive Bayes technique which gives us a better result than the maximum entropy and SVM is being subjected to the unigram model which gives a better result than using it alone. An improvement in accuracy from 88.2 to 89.9% was noticed when the semantic analysis WordNet was applied.

The researchers [7] proposed an inductive-learning method for the detection of Twitter spammers and apply a random forest approach to a limited set of features that were extracted directly from Twitter traffic. The most important achievement was the reduction in the feature set. That is why experimental results revealed that the proposed method outperforms existing approaches at that time. The researchers of [3] proposed a method for detecting spam tweets in Twitter using a data stream clustering algorithm. They analysed various features for tweet spam detection like follow rate, features based on content (such as repetitious tweets, HTTP links, replies and mentions, trending topics) and generally considered features like followers number, following number, follow rate, tweet number, retweet ratio, available link number on tweet. It labels data from spam accounts as spam and legitimate accounts as non-spam. For pre-processing, they used software called RapidMiner. The output tweets are then given input to MOA software (where they implemented DenStream algorithm for clustering the spam tweets). Experimental evaluation shows that when the algorithm is set properly, accuracy and precision will improve in comparison with previous works done using classification algorithms.

Researchers of [6] presented a new method using extreme learning machine (ELM), a supervised machine, for detecting spam accounts in social networks through their behavioural characteristics. They also verified the detectability of spam accounts through experimental evaluation. Following the behavioural analysis, they also highlighted the results that show the importance of social interaction features when distinguishing between normal users and spam accounts.

The researchers of [9] proposed a spam detection approach for Twitter based on sentimental features. They performed experiments on a collection of the dataset of 29K tweets with 1K tweets for 29 most trending topics of 2012 on Twitter. They evaluated the usefulness of said approach by using five classifiers, i.e. Bayes network, naive Bayes, random forest, support vector machine (SVM) and J48. Naive Bayes, random forest, J48 and SVM spam detection performance improved with the proposed feature combination. The results also demonstrated that proposed features provide better classification accuracy when combined with content and user-oriented features.

Every tweet comprises of different statistical properties like number of followers, number of words per tweet, number of hashtag included in the tweet, number of URLs in tweet, whether shortened URL or not. Different machine learning algorithm uses these characteristics of tweet to detect whether a particular profile is spam or no spam. First of all, these statistical properties of tweets are extracted. These properties then help us to differentiate between spam and non-spam users. Then with spam samples, a training data is formed. This training data trains the classifier which in turn detects the spam users. Since the properties of tweets vary over time, additional properties of the user profile based on content, user behaviour, the sentiment of the text, account-based features are incorporated.

The researchers [1] developed spam profiles detection models based on a set of simple, easily calculated and publicly available features in Twitter. They extracted the set of features from the available profile information regardless of the language used in user tweets. They collected and analysed a dataset of 82 Twitters profiles. With feature engineering, they investigate ten binary and simple features that can be used to classify spam profiles. Moreover, a feature selection process was utilized to identify the most influencing features in the process of detecting spam profiles. Two methods used for feature selection were Relief and Information Gain. Four machine learning algorithms were used to develop the detection models and two feature selection methods are applied to identify the most influencing features in the detection process. The work showed that promising results can be obtained using the naive Bayes and decision trees classifiers. The results reveal that suspicious words and the repeated words have a high influence on the accuracy of the detection process regardless of the language used in the tweets of the user.

Further, the research works of [13] explored the issues related to the current issues in Twitter, causes and drawbacks of it. Further, they also studied Twitter spam detection techniques and proposed a new classification method that addresses these issues and came up with their suggested solutions based on deep learning algorithms. For the purpose of performance evaluation, they firstly collected a labelled dataset of 376,206 spam and 73,836 non-spam tweets from a 10-day ground-truth dataset with more than 600 million real-world tweets. From this collected dataset, for pre-processing word vector technique was used and after that, the tweets were converted into high-dimensional vectors. Experimental results reveal that all the algorithms applied at that time performed well. And almost all performance evaluation metrics have a value higher than 81%, and most of them are more than 90%.

With recent and new advancements in knowledge and technologies approaches used to detect spammers in social networks also become mature and got advanced. But the reality behind the scenes is spammers who are also becoming more overs-mart day by day by developing and using more sophisticated mechanisms to evade detection. With this advent, new approaches have been proposed by researchers to address these problems.

The researchers [4] presented a hybrid approach for detecting automated spammers in Twitter by amalgamating community-based features with other statistical feature categories, namely metadata, content and interaction-based features and so on. They used 19 different features, including six newly defined features and two redefined features for training three machine learning classifiers, namely random forest, decision tree and Bayesian network, on a real dataset that comprises non-spam users and spammers. Further, they also analysed the discrimination power of different feature categories. They concluded that interaction- and community-based features are the most effective and result yielding for spam detection, whereas metadata-based features are proven to be the least effective.

The researchers of [8] proposed a cluster-then-label semi-supervised learning approach for pathology image classification. This method was proposed basically to identify high-density regions in the data space. The idea behind this technique was to first cluster data space into clusters and then assigns labels to these clusters using semi-supervised technique by using small amounts of labelled dataset. The Euclidian distance between the labelled data point pi and unlabelled data point qi was compared with core radii ei of the different identified clusters. The weakness of this approach was inter-observer variability when labelling the ground truth of 1500 patches between two pathologists. The Kappa agreement coefficient between the two was $k = 0.77$. Also, an insufficient number of labelled data points (1 patient) leads to failure of clustering method because of scare data available. The researchers of [10] proposed a semi-supervised spam detection technique for Twitter stream by taking into account the features of the tweet level. The proposed model has two modules: spam detection module operating in real time and model update that operates in batch mode. Further, spam detection module has four detectors, namely blacklisted URLs, near-duplicate detector, reliable non-spam (ham) detector to label trusted users posts, multiclassifier-based detector which labels the remaining tweets. Experimental observations show that confidently labelled clusters and tweets make the system effective in capturing spamming patterns.

## 3 System Overview

This section briefly explains the proposed system architecture and its shown diagram as shown in Fig. 1. In particular, the whole framework of labelling the unlabelled tweets consists of following broad components: (I) Repository of tweets is constructed initially by collecting the dataset of tweets using streaming APIs. The dataset is collected using the tweet IDs that are publicly available on the Internet. (II) The tweets are subsequently processed (in order to reduce the noise in tweets in the form of irrelevant tweets) with the aim of maximum possible feature extraction. (III) A small labelled dataset is constructed from these tweets using the strategy of manual inspection and statistical analysis of tweet features. (IV) Finally, a novel approach of clustering is applied on the whole dataset in order to group the tweets into different categories based on tweet similarity and statistical features of tweets (like ratio of followers to followings, reputation, number of user mentions, no. of hashtags and so on) and assign the respective label to the tweets on the basis of concentration of which labelled point is greater.

### *3.1 Tweet Pre-processing*

Considering the length constraint of the tweet into consideration, there is a little information in most cases available to judge a tweet as spam or non-spam because of the presence of short noisy words, hashtags and abbreviations used by the user. It has been noticed that tweet more apparently contains meaningless data most of the times. So, to cope up with this noisy data in tweet and extract more information from the tweets we need to do pre-processing. Kindly note that for experiments in this paper, we have taken or selected the tweets in US English in terms of JSON objects

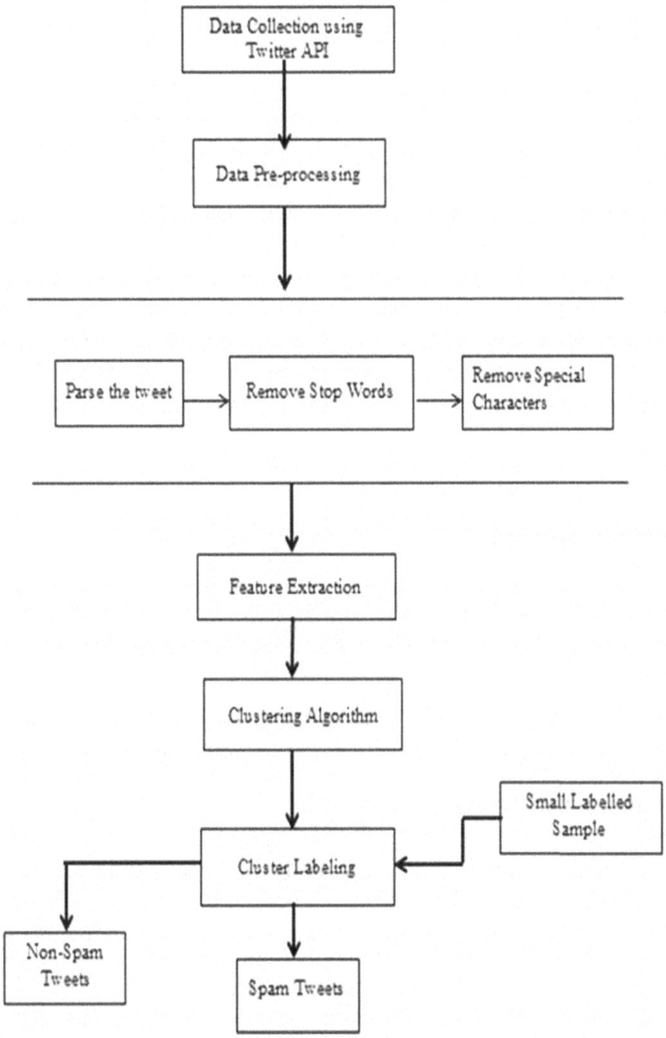


**Fig. 1** Overall design

retrieved from Twitter using the Streaming APIs. Let us suppose we have collected English raw tweets from Twitter using Streaming APIs initially.

The pre-processing stages carried out on these tweets are discussed in the following subsection:

**Remove URLs from Tweet content:** URLs in tweet content usually refer to the linked web page. Since we are interested to find the spam related information within the tweet content. So, we choose to remove URL links from tweet content as we have a separate feature extracted in our dataset for links present in the tweet so links here in the content seems no beneficial.

**Removing Retweeting and @ Tags in Tweet content:** Retweeting is the action of the Twitter user to repost an already posted tweet by other user and on the other hand @ tag in tweets refer to usernames of users to which this post is related. Since these also add noise and their irrelevance in tweet content regarding spam, we choose to remove such things also because we have extracted separate feature is tweet retweeted or not and the number of users mentioned in the tweet.

**Remove Hashtags:** Hashtags are used by used prior to some interesting facts or phrases and can be placed at anywhere in the tweet whether beginning, middle or end and any number of times. Phrases preceded or succeeded with hashtags can help us in finding the user's interest. Hashtagged phrases are already extracted as a feature, so here hashtags need to be removed from tweet content.

### *3.2 Tweet Clustering*

In tweet clustering phase, the processed tweets were clustered into clusters using the unsupervised K-means algorithm. The whole process of tweet clustering is summarized in the following steps:

1. Combine the processed tweets of all users into a single data frame.
2. Extract features from each tweet and the process of feature extraction include:
   - Tokenization: This breaks the individual tweet content into words called as tokens.
   - TFIDF: This applies the TFIDF algorithm to create feature vectors from the tokenized tweet texts.
   - Hash TFIDF: This applies a hashing function to the tokenized vectors.
3. Apply the K-means clustering algorithm on hashed vectors.
4. Evaluate and analyse the results of the K-means clustering by identifying tweet membership to clusters, perform dimensionality reduction and plot the clusters.

### *3.3 Features Extracted*

On the basis of extracted feature set, we have different types of spam detection features of tweets like account-based, content-based, graph-based, tweet-based, the user profile-based, etc.

**Tweet-Based Features include**:

1. No. of likes
2. No. of retweets
3. No. of replies
4. Created at
5. Location (within the tweet)
6. URL
7. Text
8. Unique mentions
9. Hashtags.

**User-Based Features include**:

1. Username
2. Profile photograph
3. Location
4. Creation date
5. No. of tweets
6. Favourites count
7. Followers count
8. Friends count
9. No. of retweets
10. No. of lists
11. Verified
12. Protected.

**Calculated Features include**:

1. Word count in the profile description
2. User reputation
3. Tweet count
4. Duplicate URLs
5. Tweet frequency
6. Follower ratio
7. URL ratio
8. Hashtag ratio
9. Mention ratio
10. Unique mention ratio
11. Retweet ratio
12. Suspicious words
13. Text to links ratio.

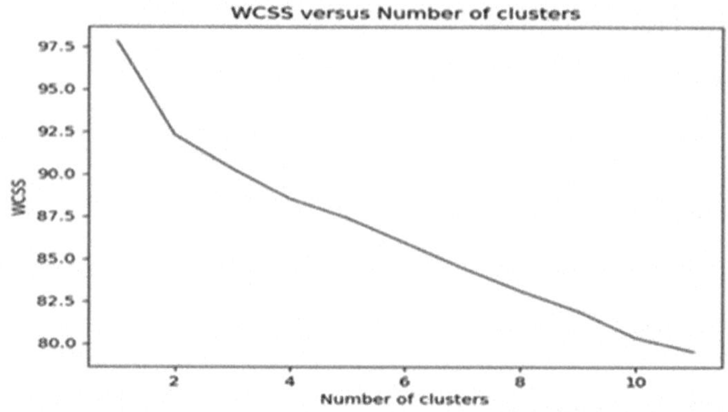


**Fig. 2** Plot of WCSS versus number of clusters

## *3.4 Cluster Labelling*

In cluster labelling, we are assigning label to each cluster that we obtain as an output of the *K*-means clustering algorithm. We assign the labels to cluster by following a semi-supervised approach. First of all, we assume that in our dataset, we have a small number of labelled tweets. These tweets are labelled following the process of manual labelling and statistical feature analysis of tweets. Then in each cluster, we calculate the concentration of each labelled entity. Assign that label to the whole cluster whose label concentration is greater than that of the others. Thus, following this semi-supervised cluster-then-label approach, we are able to assign labels to a huge dataset. Further, we can continue this procedure to assign labels to each spam category.

# 4 Experiments and Results

After clustering tweets into different clusters, we analysed the results by plotting the clusters. We plotted the cluster points and calculate the within-cluster sum of squares parameter which will tell us how well are the points in the cluster classified or grouped as shown in Fig. 2. We have plotted a graph between WCSS and number of clusters and it showed that on increasing the number of clusters in the clustering algorithm, the variation of tweets grouped in a cluster is less and vice versa.

Further, we also plotted the tweets after clustering so as to show the label assigned to each tweet on the basis of tweet similarity and statistical properties of tweets. Figure 5 below illustrates the different ways of spam categories that can be obtained used the proposed cluster and label approach of spam labelling.

On plotting the 50 most recently used words in case of both spam as well as non-spam tweets, we observe that spammers use the same word number of times. Just

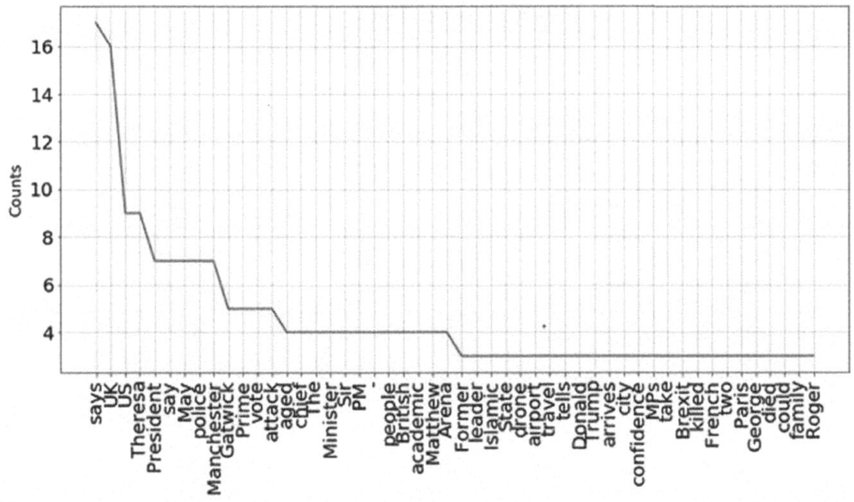


**Fig. 3** Fifty most recently used words in non-spam tweets

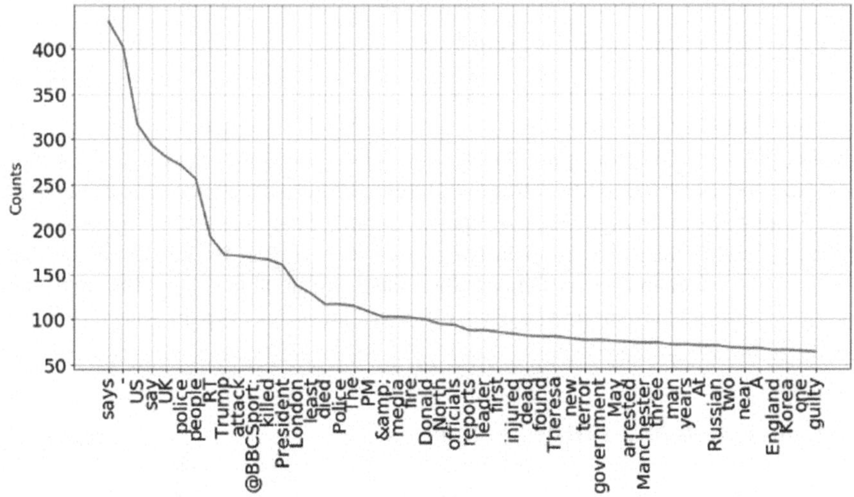


**Fig. 4** Fifty most recently used words in spam tweets

like, the word "says" is used in both spam as well as non-spam tweets, but its count is greater in spam tweets than in non-spam tweets. Also, we have analysed that the concentration of suspicious words is more in spam tweets than in non-spam tweets. The plot of 50 most recently words used in spam and non-spam posts is shown in Figs. 3 and 4, respectively.

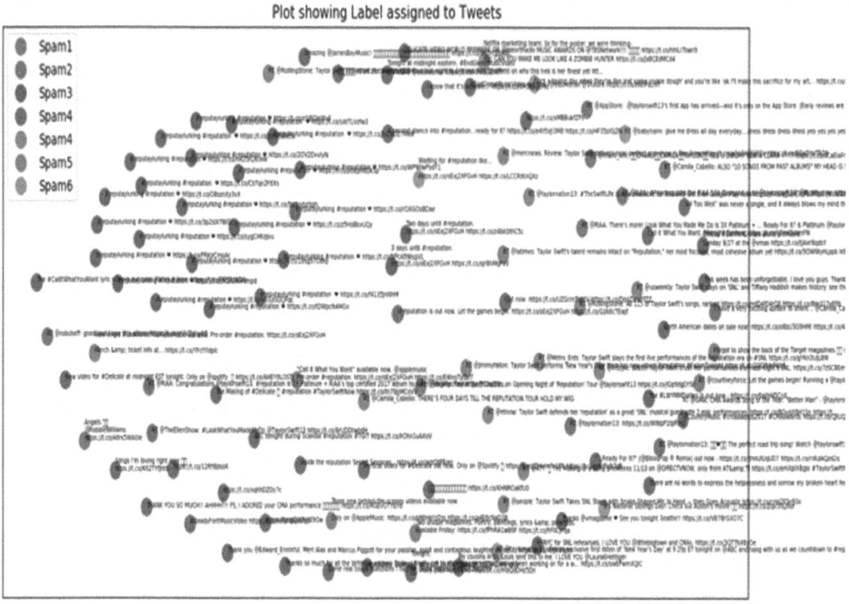


**Fig. 5** Plot showing label assigned to tweets

After clustering the tweets into clusters, the labels assigned to different spam labelled posts we design a spam dataset and follow the same approach of cluster-then-label to label the spam tweets into different categories just as we have shown in Figs. 3 and 4 above. In Figs. 3 and 4, Spam1, Spam2, Spam3 and so on up to Spam6 are the different categories of spam tweets detected. Hence, we can conclude that following the novel approach of cluster-then-label, we are able to label the spam and non-spam posts and also assign labels to different categories of spam as well (Fig. 5).

## 5 Conclusion and Future Scope

In this paper, we have explained a novel cluster-then-label approach to label the large unlabelled dataset of tweets. This method comprises of four stages. The first stage is based on the collection of recent dataset of tweets from Twitter using Streaming APIs. Subsequently, we performed pre-processing on the tweets extracted from Twitters JSON les. Then, we created a small labelled dataset of tweets on the basis of which we can apply the semi-supervised cluster-then-label approach to the dataset to label the unlabelled tweets based on the concentration of labelled points in the cluster. Our experiments showed the effectiveness of this proposed approach. Hence, we conclude that using this novel approach, we can perform spam labelling and hence spam detection as well.

Further, this research dataset can later be used for different purposes as we have extracted all the maximum possible features of the tweets during preparation of the dataset. Further, we can perform Twitter spam account detection based on the different features using this dataset. We can in future analyse the patterns that spam posts usually follow in order to trace the root of spam posts. We can in future check the accuracy of different machine learning algorithms on this dataset regarding spam detection and categorization. This labelled dataset can then be used in future for different purposes like sentiment analysis of tweets to check the behaviour of the user posting tweets, mood analysis of tweets, spam URL analysis, etc.

## References


Ala'M, A.Z., Faris, H., et al.: Spam profile detection in social networks based on public features. In: 2017 8th International Conference on information and Communication Systems (ICICS). pp. 130–135. IEEE (2017)

Benevenuto, F., Magno, G., Rodrigues, T., Almeida, V.: Detecting spammers on Twitter. In: Collaboration, Electronic Messaging, Anti-abuse and Spam Conference (CEAS). vol. 6, p. 12 (2010)

Eshraqi, N., Jalali, M., Moattar, M.H.: Detecting spam tweets in twitter using a data stream clustering algorithm. In: 2015 International Congress on Technology, Communication and Knowledge (ICTCK). pp. 347–351. IEEE (2015)

Fazil, M., Abulaish, M.: A hybrid approach for detecting automated spammers in twitter. IEEE Trans. Inf. Forensics Secur. **13**(11), 2707–2719 (2018)

Gautam, G., Yadav, D.: Sentiment analysis of twitter data using machine learning approaches and semantic analysis. In: 2014 Seventh International Conference on Contemporary Computing (IC3). pp. 437–442. IEEE (2014)

Liu, C., Wang, G.: Analysis and detection of spam accounts in social networks. In: 2016 2nd IEEE International Conference on Computer and Communications (ICCC). pp. 2526–2530. IEEE (2016)

Meda, C., Bisio, F., Gastaldo, P., Zunino, R.: A machine learning approach for twitter spammers detection. In: 2014 International Carnahan Conference on Security Technology (ICCST). pp. 1–6. IEEE (2014)

Peikari, M., Salama, S., Nofech-Mozes, S., Martel, A.L.: A cluster-then-label semi-supervised learning approach for pathology image classification. Sci. Rep. **8**(1), 7193 (2018)

Perveen, N., Missen, M.M.S., Rasool, Q., Akhtar, N.: Sentiment based twitter spam detection. Int. J. Adv. Comput. Sci. Appl. (IJACSA) **7**(7), 568–573 (2016)

Sedhai, S., Sun, A.: Semi-supervised spam detection in twitter stream. IEEE Trans. Computational Soc. Syst. **5**(1), 169–175 (2018)

Song, J., Lee, S., Kim, J.: Spam filtering in twitter using sender-receiver relationship. In: International workshop on recent advances in intrusion detection. pp. 301–317. Springer (2011)

Stringhini, G., Kruegel, C., Vigna, G.: Detecting spammers on social networks. In: Proceedings of the 26th annual computer security applications conference. pp. 1–9. ACM (2010)

Wu, T., Liu, S., Zhang, J., Xiang, Y.: Twitter spam detection based on deep learning. In: Proceedings of the Australasian Computer Science Week Multiconference. p. 3. ACM (2017)

Yang, C., Harkreader, R., Gu, G.: Empirical evaluation and new design for fighting evolving twitter spammers. IEEE Trans. Info. Forensics Sec. **8**(8), 1280–1293 (2013)

# Performance Analysis of Queries with Hive Optimized Data Models

**Meghna Sharma and Jagdeep Kaur**

**Abstract** The processing of structured data in Hadoop is achieved by Hive, a data warehouse tool. It is present on top of Hadoop and helps to analyze, query, and review the Big Data. The execution time of the queries has drastically reduced by using Hadoop MapReduce. This paper presents the detailed comparison of various optimizing techniques for data models like partitioning and bucket methods to improve the processing time for Hive queries. The implementation is done on data from New York Police Portal using AWS services for storage. Hive tool in Hadoop ecosystem is used for querying data. Use of partitioning has shown remarkable improvement in terms of execution time.

**Keywords** Big Data · Hadoop · Hive · Partitioning · Bucket methods

## 1 Introduction

The Big Data is the latest buzzword in today's IT world. It refers to the collection of data that is expanding at tremendous speed. Some instances of Big Data are the data generated by Facebook, YouTube, and Twitter in the form of text, images, videos, audios, etc. The different e-commerce-related websites, healthcare-related websites and apps, and data generated by IoT are contributing to this data explosion [1]. In the last few years, Big Data was generated from different domains, and meaningful information extracted from these sources has contributed to real-time decision making for the organizations.

M. Sharma (✉) · J. Kaur
The NorthCap University, Gurugram, Haryana, India
e-mail: meghnasharma@ncuindia.edu

J. Kaur
e-mail: jagdeep_kaur82@rediffmail.com



P. K. Singh et al. (eds.), *Proceedings of ICRIC 2019*, Lecture Notes in Electrical Engineering 597, https://doi.org/10.1007/978-3-030-29407-6_49

The traditional database techniques cannot handle the Big Data due to many reasons. Firstly, the traditional techniques cannot maintain confidentiality and accuracy of the high-quality data. Secondly, with the traditional techniques, it is difficult to understand the relationships between data items. Thirdly, the data storage is a big challenge with traditional technologies. Fourthly, the traditional techniques cannot provide the analyzing and querying capability to extract exact information from the data [2]. The Big Data technologies help to overcome all these challenges. This paper deals with the analyzing and querying part of the Big Data. The major Big Data technologies are Pig, Jaql, and Hive. The focus of this paper is on analyzing Hive queries. Hive is a data warehouse system that helps in querying data using SQL-like language. It was designed by Facebook to make the usage of Apache Hadoop easy. In Hive, the data is represented in the form of tables; whereas in Apache Hadoop, data is managed in the form of files in the HDFS. The tables in Hive are divided into partitions and buckets [3]. The paper is organized as follows. Section 2 covers the features of Hive. Section 3 covers various optimization methods. Section 4 presents the related work. The methodology and experimental analysis are covered in Sects. 5 and 6, respectively. Section 7 shows the results. Section 8 consists of conclusion.

## 2 Features of Hive

Hive gives higher level of abstraction. It works with a SQL-like language known as HiveQL which helps the users to access the data store in Hadoop's HDFS or Hbase [4]. According to authors in [5], Hive is designed for large-scale processing only. It works by transforming queries into MapReduce jobs. The features can be summarized as:

- The Apache Hive works on the distributed storage.
- It facilitates easy data extraction, transformation, and loading (ETL) operations.
- It enables the processing of different formats like structured data, unstructured data, and semi-structured data.
- It uses HiveQL which is a declarative language and has SQL-like structure.
- Hive has an in-built feature of schema-on-read and multiple schemas which postpones the schema application till the data is read [6].

Apart from these useful features, there are some limitations of using this technology. For example, it is recommended for online analytical processing (OAP) and not for online transaction processing (OTP). It does not support update and delete operations. Instead of these limitations, Hive is very popular tool for querying the large datasets using Hadoop. Although Pig is also available to perform the Map and Reduce tasks due to similarity with SQL, Hive has become more popular. The operation and execution of Hive queries result in high speed and efficiency in processing. This is achieved with two optimization methods in Hive: Partitioning and Bucketing. These methods are discussed in the next section.

## 3 Hive Optimized Data Models

The different data applications like social networking sites, banking sites, search engines, applications on clouds, IoT-based applications, etc. are producing large amount of data [7]. The data model of Hive consists of tables, partitioning, and buckets. The quick response of the large amount of data is achieved by partitioning and bucket concepts. Hive has two types of tables: Internal/Managed table and External table. When the data is loaded in the internal table, the data is moved into Hive warehouse directory. As the data is temporary on deleting a table, all the data and its metadata are lost. The external tables are recommended when data is to be used outside Hive also. Even on dropping the external table, the data remains intact and only the metadata is deleted. The two important optimized data models are partitioning and bucketing. Partitioning divides tables into partitioned columns like department, year, etc. It becomes easy to query data in part of partitioned data. The partitions are further divided into buckets. It works on the basis of hash functions of some columns. The optimization is achieved by partitioning, bucketing, and using file formats like OCR and Parquet. These are further explained as follows.

### *3.1 Partitioning*

Partitioning makes the execution of queries faster. It is achieved by dividing the table into smaller chunks with the help of partition key columns. For instance, in an employee table having attributes emp_id, emp_name, emp_sal, emp_dept, emp_designation. Partition is created for each department, i.e., emp_dept. Hive stores each partition separately and scans the desired partition hence making the query response fast. Partitioning is performed on columns or collection of columns so that a less number of partitions are produced. If partition on emp_id is generated, it will result in large number of partitions and will increase the response time of the queries. So, buckets are used for the emp_ids.

### *3.2 Bucketing*

Bucketing is also used to divide the data into smaller and more manageable parts. The main difference between partitioning and bucketing is that bucketing uses the hash function of the column to create various 'buckets'. The buckets use some hashing algorithms for searching records in the bucket.

### *3.3 File Formats ORC and Parquet*

The optimal data storage is also a prime requirement for optimal execution of Hive queries. There are two file formats used for optimal storage.

#### 3.3.1 ORC (Optimized Row Columnar) File Format

Here, the data is stored in a columnar fashion. That means, in a column-major format, the data is stored in contiguous form. The advantage of columnar format is that when the query deals with only small numbers of columns, it will scan the concerned columns only instead of scanning the whole data set. This in turn will decrease the seek time and hence performance is enhanced.

#### 3.3.2 Parquet File Format

This file format is used for processing complex and nested data structures. Compression on per column basis is done using this format. The metadata is written in the end of the file and can be used for processing large files.

## 4 Related Work

Big Data is generated by different applications. For analyzing and querying, Big Data different technologies are used. Many researchers and communities have used the Big Data technologies and published their findings. Researchers and practitioners are actively involved in the usage of different technologies like Pig, Hive, Jaql, etc. for querying the files on HDFS of Hadoop's MapReduce. The authors show in [8] that how Hadoop has reduced the load from the networks and the servers. Hadoop's file system is presented in [9]. The Apache HBase and its architecture are presented in [10]. The researchers have discussed in [11] about the MapReduce framework and how the Map and Reduce classes work to support parallelism. Apart from this, the different collection systems are also discussed like Apache Sqoop [12], Flume [13], and Chukwa [14]. Storm, an open-source system handles real-time data as compared to Hadoop, is presented in [15].

# 5 Methodology

A set of queries are executed with real-time dataset taken from New York Police Portal using AWS service, on a Hive instance executing on an Intel Core i5 machine with 2.4 GHz processor and 4 GB of RAM.

## *5.1 Datasets*

This dataset describes the various parking violation codes in New York and lists the fines imposed on the defaulters. The data is transferred from the above link to s3 bucket on Amazon AWS to bring it to HDFS and then accesses through Hive queries. The dataset consists of forty-three attributes as shown in Table 1.

## *5.2 Problem Statements*

The table is created for the dataset and the following queries are executed:

1. Display all the distinct registration states where parking tickets are issued for the year 2017.
2. Display all total number of parking violation tickets during the year 2017 for each state ordered by registration state in descending order.
3. Display the plate type for first three registration states with maximum number of violation tickets in the year 2017.

**Table 1** Dataset

| Attribute | Type |
|---|---|
| Summons Number | Number |
| Plate ID | Plain Text |
| Registration State | Plain Text |
| Plate Type | Plain Text |
| Issue Date | Date and Time |
| Violation Code | Number |
| Vehicle Body Type | Plain Text |
| Vehicle Make | Plain Text |
| Issuing Agency | Plain Text |
| Street Code1 | Number |
| Street Code2 | Number |
| Street Code3 | Number |
| Vehicle Expiration Date | Number |
| Violation Location | Plain Text |
| Violation Precinct | Number |
| Issuer Precinct | Number |
| Issuer Code | Number |
| Issuer Command | Plain Text |
| Issuer Squad | Plain Text |
| Violation Time | Plain Text |
| Time First Observed | Plain Text |
| Violation County | Plain Text |
| Violation in Front of or Opposite | Plain Text |
| House Number | Plain Text |
| Street Name | Plain Text |
| Intersecting Street | Plain Text |
| Date First Observed | Number |
| Law Section | Number |
| Subdivision | Plain Texts |
| Violation Legal Code | Plain Text |
| Days Parking In Effect | Plain Text |
| From Hours In Effect | Plain Text |
| To Hours In Effect | Plain Text |
| Vehicle Color | Plain Text |
| Unregistered Vehicle? | Plain Text |
| Vehicle Year | Number |
| Meter Number | Plain Text |

(continued)

**Table 1** (continued)

| Attribute | Type |
|---|---|
| Feet From Curb | Number |
| Violation Post-Code | Plain Text |
| Violation Description | Plain Text |
| No Standing or Stopping Violation | Plain Text |

## 6 Experimental Analysis

The concept of partitioning is demonstrated on the dataset. The table is created without partition and then with partition.

Now, data is organized as directories/folders partitioned according to attribute value, and the searching of data becomes easy when the attribute is mentioned in the WHERE clause.

This has resulted in increase in execution time at the time of creation. After creating the tables, the data is fetched into it from the dataset. Now, the three different queries corresponding to three different operations are executed with partition and without partition. To perform the first operation to display all the distinct registration states where parking tickets are issued for the year 2017, the query and its execution without partitioning are shown in Fig. 1. The same operation is performed with partitioning and it is shown in Fig. 2. The number of mappers used without partitioning is eight

```
hive> select distinct(registration_state) from
nyc_parking_violations where instr(issue_date,'2017')>0;
Query ID = ec2-user_20181129161717_747bc7d4-8977-
4531-adb5-805dcdb6a00d
Total jobs = 1
Launching Job 1 out of 1
Number of reduce tasks not specified. Estimated from
input data size: 92
In order to change the average load for a reducer (in
bytes):
Stage-Stage-1: Map: 8   Reduce: 92   Cumulative CPU:
443.01 sec   HDFS Read: 668502 HDFS Write: 195 SUCCESS

Time taken: 349.182 seconds, Fetched: 65 row(s)
```

**Fig. 1** Query 1 without partitioning

```
hive> select distinct(registration_state) from
nyc_parking_violations_partitione
d where issue_year=='2017';

Query ID = ec2-user_20181129161010_8e42845e-
d3a5-4b0d-a637-384178e13809
Total jobs = 1
Launching Job 1 out of 1
Hadoop job information for Stage-1: number of
mappers: 4; number of reducers: 16
Time taken: 98.116 seconds, Fetched: 65 row(s)
```

**Fig. 2** Query 1 with partitioning

and the number of reducers is ninety-two. But with partition, the number of mappers and reducers is reduced to four and sixteen, respectively.

So, 65 rows are fetched to complete this operation as evident from Figs. 1 and 2. Now, the second operation is to display all total number of parking violation tickets during the year 2017 for each state ordered by registration state in descending order.

In second query, the operation is performed through the GROUP BY and ORDER BY clause. The snapshot without partitioning and with partitioning is shown in Figs. 3 and 4.

As shown in Fig. 3, the execution time is 363.745 s when no partitioning is used. The same query, when executed with partition, shows drastic reduction in the execution time as shown in Fig. 4.

The third operation to display the plate type for first three registration states with maximum number of violation tickets in the year 2017 is executed in the form of query three as shown in Fig. 5

Now, the same query when executed with partitioning it results in reduction of the execution time as shown in Fig. 6.

```
hive> select count(distinct
summons_number),registration_state as
no_of_tickets from nyc_parking_violations
where instr(issue_date,'2017')>0 and
(street_code1 is NULL or street_code2 is NULL
or street_code3 is NULL) group by
registration_state order by  registration_state
desc;
Hadoop job information for Stage-1: number of
mappers: 8; number of reducers: 92
Stage-Stage-2: Map: 1  Reduce: 1   Cumulative
CPU: 5.39 sec   HDFS Read: 39096 HDFS
Write: 10 SUCCESS
Total MapReduce CPU Time Spent: 7 minutes
21 seconds 100 msec
Time taken: 363.745 seconds, Fetched: 2 row(s)
```

**Fig. 3** Query 2 without partitioning

```
hive> select count(distinct summons_number) as
no_of_tickets from nyc_parking_violations_partitioned
where issue_year=='2017' and (street_code1 is NULL or
street_code1==0 or street_code2 is NULL or
street_code2==0 or street_code3 is NULL ) group by
registration_date order by registration_date;
Query ID = ec2-user_20181114181111_64a46add-244e-4467-
8b33-1ccea8b0ab49
Total jobs = 1
Launching Job 1 out of 1
Number of reduce tasks determined at compile time: 1
Ended Job = job_1542218942943_0001
MapReduce Jobs Launched:
Stage-Stage-1: Map: 3  Reduce: 1   Cumulative CPU: 70.09
sec   HDFS Read: 35708 HDFS Write: 8 SUCCESS
Total MapReduce CPU Time Spent: 1 minutes 10 seconds 90
msec
OK
==========================================================
         Time taken: 58.77 seconds, Fetched: 1 row(s)
```

**Fig. 4** Query 2 with partitioning

```
hive> select a.registration_state,plate_type from
nyc_parking_violations a join (select
registration_state,count(summons_number) as cnt from
nyc_parking_violations where issue_year == '2017' group
by registration_state order by cnt desc limit 3) b on
a.registration_state=b.registration_state limit 5;
MapReduce Jobs Launched:
Stage-Stage-1: Map: 4  Reduce: 16
SUCCESS
Stage-Stage-2: Map: 1  Reduce: 1   Cumulative CPU: 5.31
sec   HDFS Read: 11686 HDFS Write: 156 SUCCESS
Stage-Stage-5: Map: 8   Cumulative CPU: 57.32 sec
HDFS Read: 215272 HDFS Write: 280 SUCCESS
Time taken: 680 seconds, Fetched: 5 row(s)
```

**Fig. 5** Query 3 without partitioning

```
hive> select a.registration_state,plate_type from
nyc_parking_violations_partitioned a join (select
registration_state,count(summons_number) as cnt from
nvc_parking_violations_partitioned where issue_year ==
'2017' group by registration_state order by cnt desc limit
3) b on a.registration_state=b.registration_state  limit 5;
MapReduce Jobs Launched:
Stage-Stage-1: Map: 4  Reduce: 16   Cumulative CPU:
101.39 sec   HDFS Read: 125312 HDFS Write: 3017
SUCCESS
Stage-Stage-2: Map: 1  Reduce: 1   Cumulative CPU: 5.31
sec   HDFS Read: 11686 HDFS Write: 156 SUCCESS
Stage-Stage-5: Map: 8   Cumulative CPU: 57.32 sec
HDFS Read: 215272 HDFS Write: 280 SUCCESS
Time taken: 180.815 seconds, Fetched: 5 row(s)
```

**Fig. 6** Query 3 with partitioning

## 7 Results

The queries are executed on a dataset having 11.7 million rows. The execution is presented in the tabular form in Table 2. The performance of Hive queries with partitioning shows remarkable decline in execution time.

This can be represented in the form of a graph as in Fig. 7 and the difference in times is quite remarkable.

**Table 2** Execution time

| | Without partitioning | With partitioning |
|---|---|---|
| Query 1 | 349.182 s | 98.116 s |
| Query 2 | 363.745 s | 58.77 s |
| Query 3 | 680 s | 180.815 s |

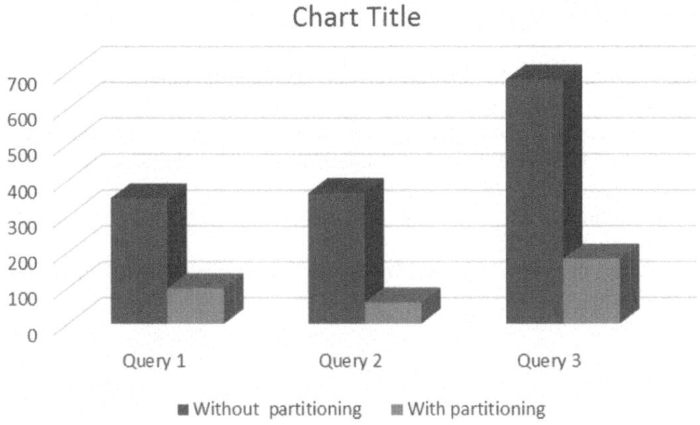


**Fig. 7** Performance analysis

## 8 Conclusion

Hive is found suitable for efficient processing of data. With the use of partitioning techniques on the dataset, the queries were executed faster as compared to without partitioning. The authors have taken queries using GROUP BY and ORDER BY clauses. Also, queries involving joins with nesting are also tested. The future work consists of using the concept of bucketing on the partitioned dataset and then to analyze the execution time as compared to usage of partitioning. These set of queries can also be used for larger dataset with different types of data.

## References

1. Pen, H.D., Dsilva, P., Mascarnes, S.: Comparing HiveQL and MapReduce methods to process fact data in a data warehouse. In: 2017 2nd International Conference on Communication Systems, Computing and IT Applications (CSCITA), pp. 201–206. IEEE (2017, April)
2. Oussous, A., Benjelloun, F.Z., Lahcen, A.A., Belfkih, S.: Big data technologies: a survey. J. King Saud Univ. Comput. Inf. Sci. (2017)
3. Shaw, S., Vermeulen, A.F., Gupta, A., Kjerrumgaard, D.: Hive architecture. In: Practical Hive, pp. 37–48. Apress, Berkeley, CA (2016)
4. Sakr, S.: Big data 2.0 processing systems: a survey. Springer International Publishing (2016)
5. Bansal, H., Chauhan, S., Mehrotra, S.: Apache Hive Cookbook. Packt Publishing Ltd (2016)

6. Loganathan, A., Sinha, A., Muthuramakrishnan, V., Natarajan, S.: A systematic approach to big data. Int. J. Inf. Comput. Technol. **4**(9), 869–878 (2014)
7. Zikopoulos, P., Eaton, C.: Understanding big data: analytics for enterprise class Hadoop and streaming data. McGraw-Hill Osborne Media (2010)
8. Usha, D., Jenil, A.: A survey of big data processing in perspective of Hadoop and MapReduce. Int. J. Curr. Eng. Technol. **4**(2), 602–606 (2014)
9. Elgazzar, K., Martin, P., Hassanein, H.S.: Cloud-assisted computation offloading to support mobile services. IEEE Trans. Cloud Comput. **4**(3), 279–292 (2016)
10. Coronel, C., Morris, S.: Database systems: design, implementation, & management. Cengage Learning (2016)
11. Lydia, E.L., Swarup, M.B.: Big data analysis using Hadoop components like Flume, MapReduce, Pig and Hive. Int. J. Sci. Eng. Comput. Technol. **5**(11), 390 (2015)
12. Vohra, D.: Using Apache Sqoop. In: Pro Docker, pp. 151–183. Apress, Berkeley, CA (2016)
13. Hoffman, S.: Apache Flume: Distributed Log Collection for Hadoop. Packt Publishing Ltd (2015)
14. Shireesha, R., Bhutada, S.: A study of tools, techniques, and trends for big data analytics. IJACTA **4**(1), 152–158 (2015)
15. Mazumder, S.: Big data tools and platforms. In Big Data Concepts, Theories, and Applications, pp. 29–128. Springer, Cham (2016)

# A Review on Scalable Learning Approches on Intrusion Detection Dataset

**Santosh Kumar Sahu and Durga Prasad Mohapatra**

**Abstract** There has been much excitement recently about Big Data and the dire need for data scientists who possess the ability to extract meaning from it. Data scientists, meanwhile, have been doing science with voluminous data for years, without needing to brag about how big it is. But, now those large, complex datasets should process smartly. As a result, it improves productivity by reducing the computational process. As a result, Big Data analytics takes a vital role in intrusion detection. It provides tools to support structured, unstructured, and semi-structured data for analytics. Also, it offers scalable machine learning algorithms for fast processing of data using machine learning approach. It also provides tools to visualize a large amount of data in a practical way that motivates us to implement our model using scalable machine learning approach. In this work, we describe a scalable machine learning algorithm for threat classification. The algorithm has been designed to work even with a relatively small training set and support to classify a large volume of testing data. Different machine learning approaches implemented and evaluated using intrusion dataset. The data is normalized using the min–max normalization technique, and for SVM classification, data transforms into sparse representation for reducing computational time. Then using Apache Hive, we store the processed data into HDFS format. All the methods except the neural network are implemented using Apache Spark. Out of all the approaches, the fine KNN approach outperforms in terms of accuracy in a reasonable computational time, whereas the Bagged Tree approach achieves slightly less accuracy but takes less computational time for classifying the data.

**Keywords** Intrusion detection · Apache Spark · Big Data · Machine learning · SVM · KNN · Neural network · Ensemble approach

---

S. K. Sahu (✉) · D. P. Mohapatra
Department of CSE, NIT Rourkela, Rourkela, Odisha, India
e-mail: santoshsahu@hotmail.co.in

D. P. Mohapatra
e-mail: durga@nitrkl.ac.in



P. K. Singh et al. (eds.), *Proceedings of ICRIC 2019*, Lecture Notes in Electrical Engineering 597, https://doi.org/10.1007/978-3-030-29407-6_50

# 1 Introduction

We live in a digital world where data is increasing rapidly due to advance in technology, mostly the use of sensors (IoT), and ease of Internet Technology. The IoT in embedded devices used in all sectors to accurately gather information, real-time control, and operation of the different task without human interventions. As a result, a vast amount of data is generated, and it is a big challenge to store, process, and analyze the data for decision making. The term "Big Data signifies the sheer volume, variety, velocity, and veracity of such data." Big Data [1] is structured, unstructured, and semi-structured or heterogeneous. It becomes difficult for computing systems to manage "Big Data" because of the immense speed and volume at which it is generated. Conventional data management, warehousing, and data analysis system fizzle to analyze the heterogeneous data for not only processing but also storing and retrieving the data. The need to sort, organize, analyze, and systematically offer this critical data leads to the rise of the much-discussed term, Big Data.

According to Gartner, data is growing at a rate of 59% every year. The growth of the data can be depicted regarding the following four Vs:

**Volume**: It refers to the amount of data generated by the organization or individual. The size of data in most of the organization is approaching exabyte and may increase to zettabyte in the coming years.
**Velocity**: It describes the rate at which data is generated, captured, and shared. An enterprise can capitalize on data only if it is captured and shared in real time. Most of the traditional approach face problems associated with data, which keeps adding up but can not be processed quickly. They generally take batch processing or manual processing that takes several hours or days for analysis.
**Variety**: The data is being generated at a breakneck pace from different sources and forms such as structured, unstructured, and semi-structured. To deal with such type of data in the form of flat files, images, videos, audios, and other sources for decision making is a complex task.
**Veracity**: It refers to the uncertainty of data, i.e., whether they obtained data is correct or consistent. Among the complete information which is generated, we require to select the accurate and consistent data for data processing and decision-making process.

Traditional approaches process the data specific to a region or point of interest. They may not process and interpret the data in real time. But to improve the data analytics, it is required to integrate every data from every source that is correct and consistent should be used in predicting analysis which is a challenging task. Big Data technologies are vital in supplying correct interpretation, which may additionally cause extra concrete decision making resulting in greater operational efficiencies, cost discounts, and decreased risks for the commercial enterprise. To harness the energy of huge facts, we require an infrastructure that could manage and vast manner volumes of structured and unstructured information in real time and can maintain security and protection.

Big Data analytics reforms the ways to conduct business in many ways, such as it improves the decision making and business process management. In this paper, it is used to classify the intrusions using various machine learning approaches that are available in Apache Spark and Mahout distributions. As per the study, the bottleneck during processing does not occur in distributed processing. As per the MapReduce functions, the data calls the program for processing. Due to the size of the program is negligible as compared to the size of data, the probability of bottleneck can be avoided in the large-scale analytics process.

As per the literature study, the methods related to Apache Spark/Mahout used in our research. Due Curse of Dimensionality, the predictive models overfitted, biased, and not able to provide high accuracy. To avoid the curse of dimensionality, Big Data used in this experiment that deals with high dimension, variety, veracity as well as velocity data as mentioned in Fig. 1. Besides, the Big Data analytics also parallelizes the execution efficiently with maintain fault tolerant and the bottleneck of data during implementation.

The big advantage is that the program comes to the data for processing. In conventional data processing, the program calls the data as argument that leads towards bottleneck and requires high computational resources during execution. The Big Data Processing is quite different. The size of the program is very less as compared to the Data. Therefore, this type of processing avoid bottleneck and not suffers curse of dimensionality. The MapReduce and HDFS (Hadoop File System) provide reliability, flexibility, high performance, and efficiency in storing, managing, monitoring,

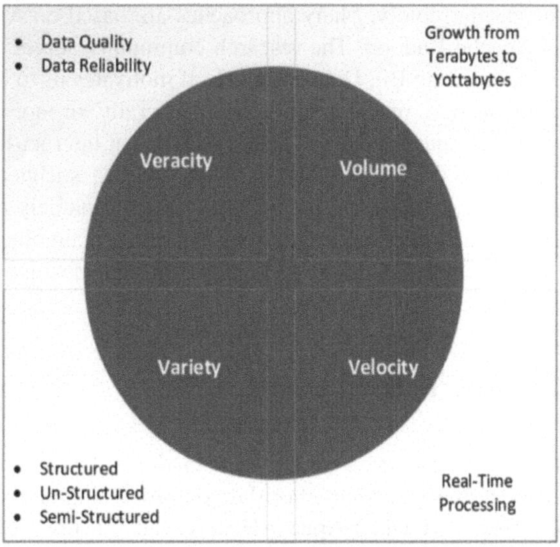


**Fig. 1** Elements of Big Data (four Vs)

**Table 1** Review of different scalable approaches used on intrusion detection

| Authors | Year | Big data approaches for intrusion detection |
|---|---|---|
| Gupta and Kulariya [2] | 2016 | A framework for fast and efficient intrusion detection using Apache Spark |
| Keegan et al. [3] | 2016 | Review on different intrusion detection techniques using cloud |
| Kulariya et al. [4] | 2016 | Performance analysis of intrusion detection schemes using Apache Spark |
| Mavridis and Karatza [5] | 2017 | Performance evaluation of cloud-based logfile analysis with Apache Spark |
| Kumari et al. [6] | 2016 | Anomaly detection in network traffic using K-mean clustering |
| Hsieh and Chan [7] | 2016 | Detection DDoS attacks based on neural network using Apache Spark |
| Mavridis and Karatza [8] | 2015 | Logfile analysis in the cloud with Apache Spark |
| Rathore et al. [9] | 2016 | Hadoop based real-time intrusion detection for high-speed networks |

processing, and visualizing the data [1]. It motivates us to implement the network log analysis process using Big Data analytics. To detect the intrusion, scalable machine learning approaches used to experiment on different supervised learning approach and find the best method among all. The detail algorithms and data flow described in subsequent sections

Table 1 shows the recent works in intrusion detection using Big Data technology. The study shows that researchers working in this area using new technology are increasing rapidly. Many approaches are based on Apache Spark and MapReduce of Apache Hadoop. The research community solves various problems of different domains using Big Data approach. It motivates us to implement intrusion detection using the state-of-the-art approach. Generally, an enormous amount of network packets is generated due to the large-scale use of internet-based applications day by day. To process the high volume and velocity data, such techniques are utmost essential. Therefore, in this experiment, Big Data tools namely Apache Spark and Mahout are used in the intrusion detection process. The main objective is to find a better model that provides more accurate results, low false positive and less computational time.

## 2 Experimental Details

In this experiment, a single-node Hadoop cluster is setup virtually along with its ecosystem applications. For data preprocessing and store into Hadoop Distributed File System (HDFS), Apache Hive is used. The raw well-logs are parsed, and we have designed a schema to save them into HDFS. There are 41 attributes in each instance present in the dataset out of which three are categorical data. The categorical data is assigned by constant for analysis. Sahu et al. [10] gave details about the preprocessing of the dataset for further analysis. Similarly, the output class is broadly represented by four attack classes. The normal class is also considered for analysis. The second class is a combination of the four threat types (one (R2L), two (U2R), three (Probe), and four (DoS)). The detail statistics of the data used in this experiment is depicted in Table 3.

KDD_Corrected [11] and NSLKDD [12] datasets are used in this experiment. The dataset is preprocessed as per [10]. Figure 2 described the data and process flow of our architecture. First, the raw logs processed and give input via hive script to store into HDFS. For data processing task, Apache Spark is used. Apache Spark is an open-source distributed general-purpose cluster-computing framework. It achieves high performance for both batch and streaming data, using a state-of-the-art DAG scheduler, a query optimizer, and a physical execution engine. It offers over 80 high-level operators that make it easy to build parallel apps. And, you can use it

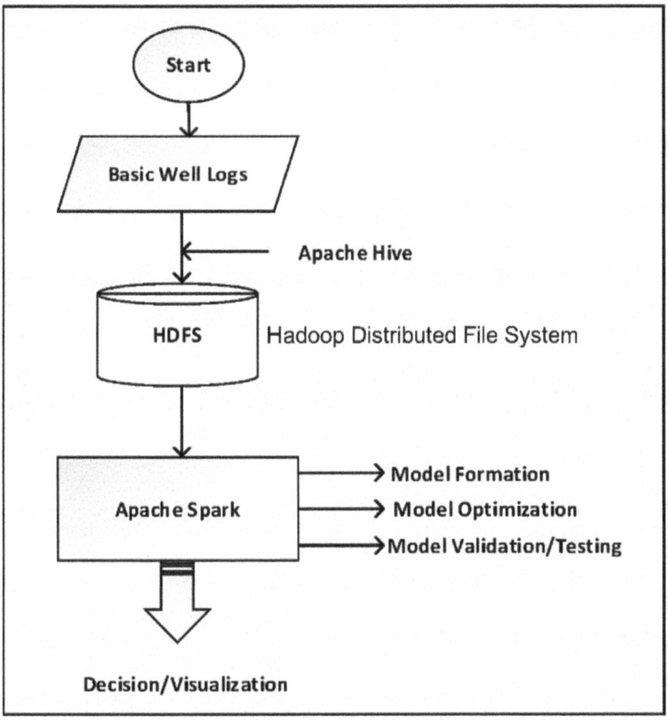


**Fig. 2** Data and process flow diagram

**Table 2** Hardware/software specifications

| | |
|---|---|
| Operating system | Windows 10 and CentOS for Hadoop cluster |
| Data Handling | Hadoop and Flat Files |
| Processing | Apache Spark, Matlab, Excel |
| Visualization | Matlab and Excel |
| Processor | Intel Xeon |
| RAM | 64 GB |

interactively from the Scala, Python, R, and SQL shells. In this experiment, Scala programming is used for classification of intrusion classes. Hadoop cluster offers data storing and MapReduce operation. The remaining processing is carried using Spark. The support vector machine, neural network, KNN, and ensemble approaches used to classify the rock types. The model parameters optimized by the searching approach. It was validated and tested using the existing labeled instances. Out of sixteen well-log data, ten are used for training the model, two are used for validation/cross-validation, and the remaining four are used in testing the model. The detailed hardware and software specifications used in this experiment are given in Table 2.

## 3 Results and Discussion

Supervised learning approaches used in this experiment for attack classification. The result of each method is discussed below:

**Support Vector Machine (SVM)**: Different SVM approaches are applied and tested with the input data. The data is transformed into a sparse format before being fed to the SVM models to avoid overfitting.

The model parameters also optimized and validated using validation data. The output of SVM approaches is listed in Table 3. The cubic SVM method gives 99.7%

**Table 3** Prediction results of different SVM approaches

| Sl. No. | SVM approach | Accuracy in % | Time in sec | Prediction average rate (obs/sec) |
|---|---|---|---|---|
| 1 | Linear SVM | 97.9 | 106.61 | 17,000 |
| 2 | Quadratic SVM | 99.6 | 78.213 | 31,000 |
| 3 | Cubic | 99.7 | 1888.4 | 50,000 |
| 4 | Fine Gaussian | 99.5 | 105.93 | 8600 |
| 5 | Medium Gaussian | 99.5 | 61.55 | 16,000 |
| 6 | Coarse Gaussian | 97.9 | 97 | 8200 |

accuracy in the 1888 s. The confusion matrix and various performance evaluation of the model are shown in Fig. 3. The first cell of the confusion matrix shows the number of instances which are correctly classified as negative and the third cell represents the number of positive instances that are correctly classified. The remaining cells are teaching the class that is wrongly classified. The ratio between the true positive rate and false negative rate is depicted in Fig. 4. Similarly, the ratio between positive predictive value (PPV) and false discovery rate (FDR) shown in Fig. 5.

**K-Nearest Neighbor (KNN)**: The K-nearest neighbor algorithm also provides a stable and useful result. Sahu et al. [13] discussed how KNN is an efficient learning approach for threat classification on different intrusion datasets. The KNN approach is implemented as per conventional way of data processing. But in this experiment, the KNN approach is implemented in apache spark using Scalable approach. Table 4 depicts the accuracy and time taken by the variant models using KNN approach. As compared to other methods, it takes less time in prediction and also achieves a satisfactory accuracy rate. The fine KNN approach achieved a 99.86% accuracy rate in 465 s as compared to other KNN approaches. The KNN classification technique achieves high classification accuracy using NSLKDD dataset. The Fig. 6 shows the confusion matrix, Fig. 7 visualize the several performance assessment of the model, Fig. 8 depicts the ratio between TPR vs FNR, and Fig. 9 shows the Area Under Curve(AUC) of the Fine KNN Model. The Fine KNN model provides a better classification result as compared to other KNN Approaches with less computational time.

**Ensemble Approach**: As per literature study, this approach provides a stable and better result as compared to a single learning approach. In this approach, multiple weak learners are combined to solve the task with higher accuracy and a low false positive rate. Hence, we applied different ensemble method in this experiment. The

| True Class | | | |
|---|---|---|---|
| | -1 | 11707 | 34 |
| | 1 | 51 | 13398 |
| | | -1 | 1 |
| | Predicted Class | | |

| Measure | Value (%) | Derivations |
|---|---|---|
| Sensitivity | 99.566253 | TPR = TP / (TP + FN) |
| Specificity | 99.746873 | SPC = TN / (FP + TN) |
| Precision | 99.710416 | PPV = TP / (TP + FP) |
| Negative Predictive Value | 99.62079 | NPV = TN / (TN + FN) |
| False Positive Rate | 0.2531269 | FPR = FP / (FP + TN) |
| False Discovery Rate | 0.2895835 | FDR = FP / (FP + TP) |
| False Negative Rate | 0.0043375 | FNR = FN / (FN + TP) |
| Accuracy | 99.662565 | ACC = (TP + TN) / (P + N) |
| F1 Score | 99.638282 | F1 = 2TP / (2TP + FP + FN) |
| Matthews Correlation Coefficient | 99.32321 | TP*TN - FP*FN / sqrt((TP+FP)*(TP+FN)*(TN+FP)*(TN+FN)) |

**Fig. 3** Confusion matrix and performance evaluation parameters of cubic SVM model

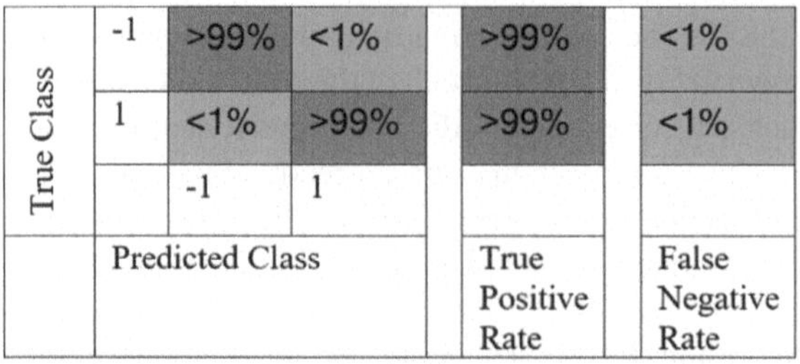


**Fig. 4** Ratio between TPR and FNR matrix cubic SVM model

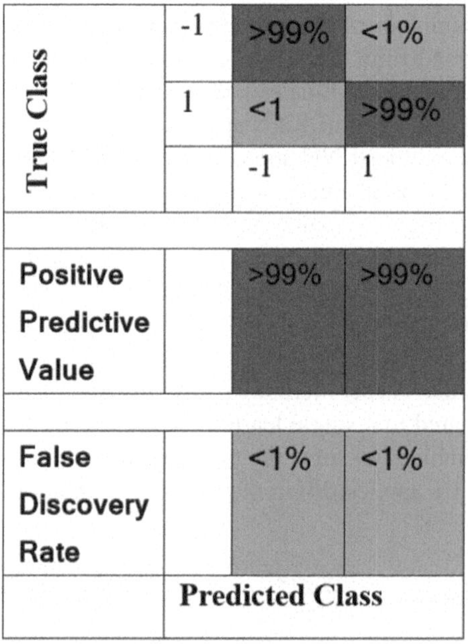


**Fig. 5** PPV versus FDR cubic SVM model

result of these approaches is given in Table 5. Figure 10 shows the confusion matrix using Bagged Tree ensemble approach. If time considers as performance evaluation criteria, then all the ensemble approaches achieved a stable accuracy near to 100% with a minimum average time and the number of observations per second also very high as compared to other approaches. Figure 11 shows the ratio between true positive rate and false negative rate. Similarly, Fig. 12 shows the ratio between positive predictive value and false discovery rate. As per Fig. 12, the positive and negative prediction rate is more than 99% that indicated a better classification result. Figure 13 describes the area under curve of the two classes and it touches the top-left corner that indicated an excellent classification accuracy with minimal error rate.

**Table 4** Prediction results of different KNN approaches

| Sl. No. | KNN approach | Accuracy in % | Time in sec | Prediction Avg. Rate (obs/sec) |
|---|---|---|---|---|
| 1 | Fine KNN | 99.9 | 465 | 200 |
| 2 | Medium KNN | 99.5 | 459 | 200 |
| 3 | Coarse KNN | 98.3 | 460.43 | 200 |
| 4 | Cosine KNN | 99.5 | 452.67 | 200 |
| 5 | Cubic KNN | 99.4 | 2358.7 | 33 |
| 6 | Weighted KNN | 99.7 | 454.49 | 200 |

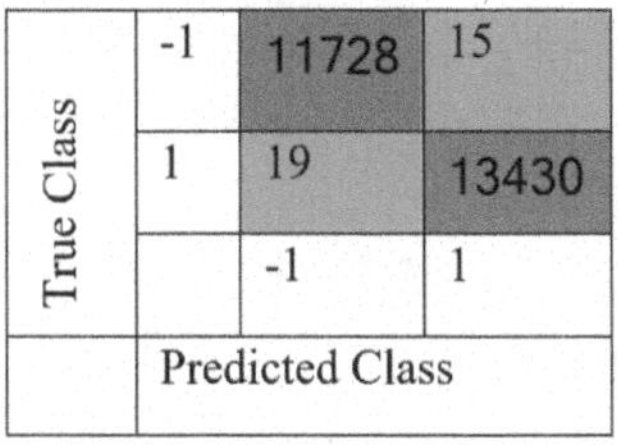


| Measure | Value(%) | Derivations |
|---|---|---|
| Sensitivity | 99.838257 | TPR = TP / (TP + FN) |
| Specificity | 99.888434 | SPC = TN / (FP + TN) |
| Precision | 99.872264 | PPV = TP / (TP + FP) |
| Negative Predictive Value | 99.858726 | NPV = TN / (TN + FN) |
| False Positive Rate | 0.1115656 | FPR = FP / (FP + TN) |
| False Discovery Rate | 0.1277357 | FDR = FP / (FP + TP) |
| False Negative Rate | 0.0016174 | FNR = FN / (FN + TP) |
| Accuracy | 99.865037 | ACC = (TP + TN) / (P + N) |
| F1 Score | 99.855258 | F1 = 2TP / (2TP + FP + FN) |
| Matthews Correlation Coefficient | 99.728999 | TP*TN - FP*FN / sqrt((TP+FP)*(TP+FN)*(TN+FP)*(TN+FN)) |

**Fig. 6** Confusion matrix and performance evaluation parameters of Fine-KNN model

**Fig. 7** Ratio between True Positive Rate and False Negative Rate matrix Fine-KNN Model

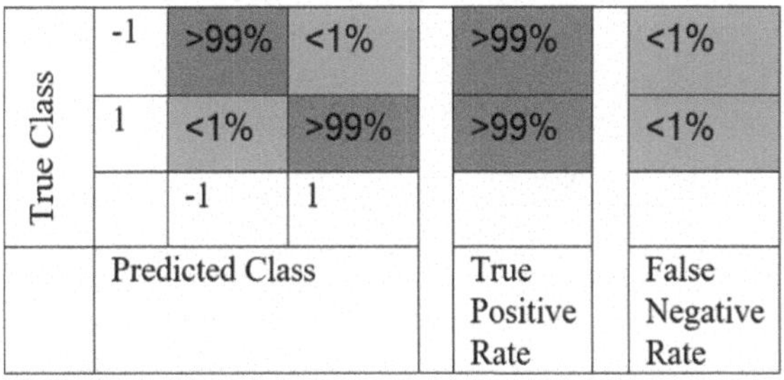


The Bagged Tree approach outperforms among these ensemble methods. The main advantage of this approach is more stable, robust, and adaptive and can apply in a variety of data with minimum time as given in Table 5.

**Neural Network (NN)**: The advantages of NN include adaptive learning, Self-organization, real-time operation, and fault tolerant. As a result, the neural network with twenty hidden layers considered in this experiment to classify the threat classes.

Figure 14 shows the four-confusion matrix that describes the training, validation,

**Fig. 8** PPV versus FDR of Fine-KNN Model

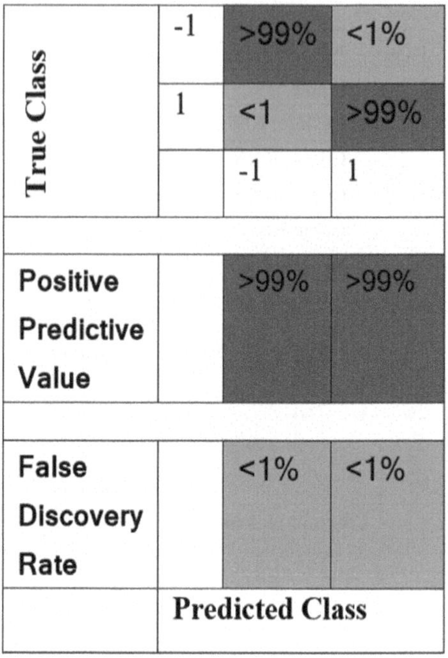


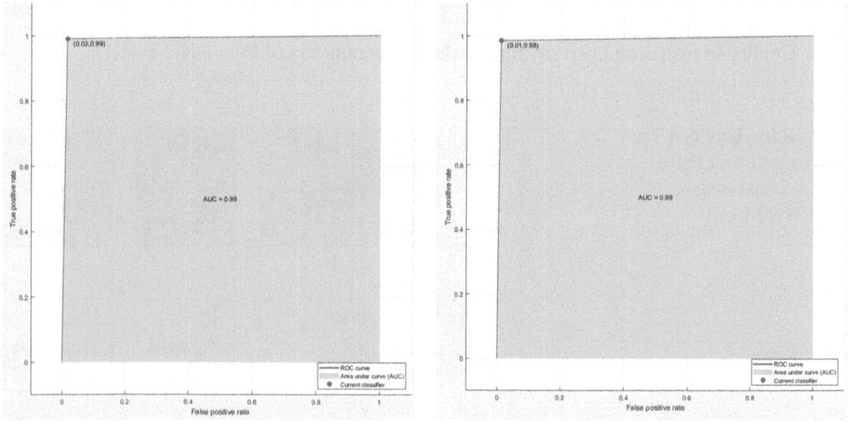


**Fig. 9** Area under curve (AUC) when +ve (normal) and −ve (attack) class of Fine-KNN Model

**Table 5** Prediction results of different ensemble approaches

| Sl. No. | Ensemble approach | Accuracy in % | Time in sec | Prediction Avg. Rate (obs/sec) |
|---|---|---|---|---|
| 1 | Boosted Tree | 99.7 | 66.623 | 19,000 |
| 2 | Bagged Tree | 99.8 | 42.849 | 15,000 |
| 3 | Subspace Discriminant | 95.7 | 53.389 | 9400 |
| 4 | Subspace KNN | 81.2 | 14.396 | 8600 |
| 5 | RU Boosted Trees | 99 | 68.922 | 21,000 |

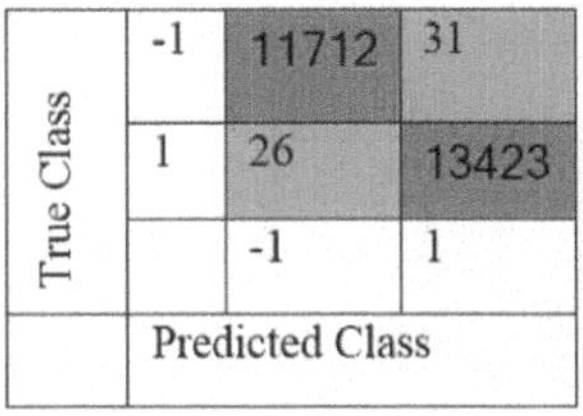


| Measure | Value (%) | Derivations |
|---|---|---|
| Sensitivity | 99.778497 | TPR = TP / (TP + FN) |
| Specificity | 99.769585 | SPC = TN / (FP + TN) |
| Precision | 99.736013 | PPV = TP / (TP + FP) |
| Negative Predictive Value | 99.806677 | NPV = TN / (TN + FN) |
| False Positive Rate | 0.2304147 | FPR = FP / (FP + TN) |
| False Discovery Rate | 0.2639871 | FDR = FP / (FP + TP) |
| False Negative Rate | 0.002215 | FNR = FN / (FN + TP) |
| Accuracy | 99.773738 | ACC = (TP + TN) / (P + N) |
| F1 Score | 99.757251 | F1 = 2TP / (2TP + FP + FN) |
| Matthews Correlation Coefficient | 99.54586 | TP*TN - FP*FN / sqrt((TP+FP)*(TP+FN)*(TN+FP)*(TN+FN)) |

**Fig. 10** Confusion matrix and performance evaluation parameters of Bagged Tree model

testing, and overall accuracy of the NN module. In the learning process, the output is consistent and robust. It provides a stable output in all cases. Similarly, the receiver operating characteristics (ROC) curve is given in Fig. 15 which is plotted the true positive rate (TPR) against the false positive rate (FPR) at various threshold settings during training, testing, and validation process.

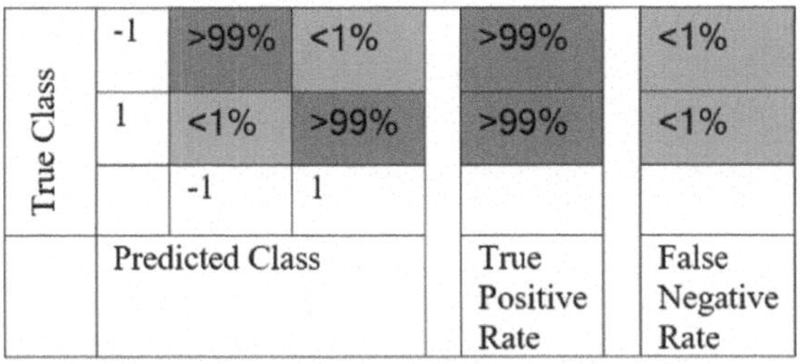


**Fig. 11** Ratio between TPR and FNR matrix of Bagged Tree model

The +ve and −ve classes are more accurately classified as it is showing in Fig. 15 that it is very close to the top-left corner of the graph. The overall training, testing, and validation process are satisfactory as per the result obtained using different supervised learning approach.

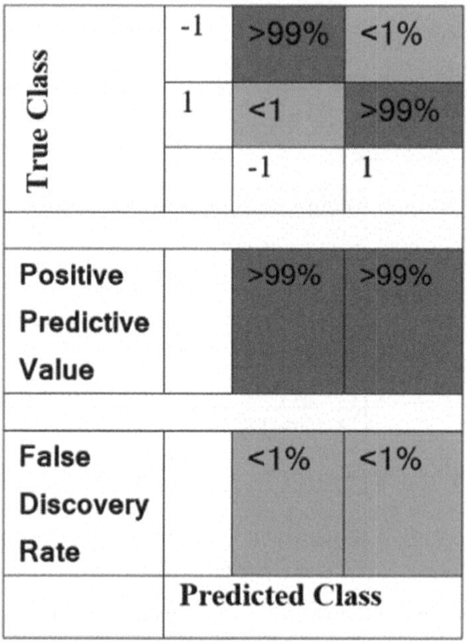


**Fig. 12** PPV versus FDR of Bagged Tree model

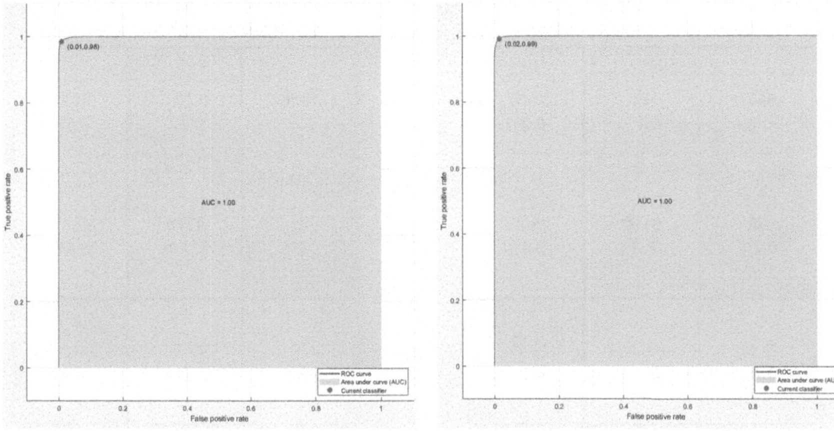


**Fig. 13** Area under curve (AUC) when +ve (normal) and −ve (attack) class of Bagged Tree model

## 4 Conclusion

The businesses, governments, and other agencies around the world adapted data-intensive, decision-making processes increasing day by day. Nowadays, it is a big challenge to acquire, process, and interpret the extensively large and variety of data. Big Data analytic techniques are used to deal with high volume, variety, and velocity of data. It effectively stores, process, and visualizes the data in an effective manner. Drawing upon implementation experiences of early adopters of Big Data technologies across multiple industries; this paper focuses on the issues and challenges faced during data analytics by network packet analysis. It also described the data flow from raw to the processed data format used for predictive analysis. The most advanced scalable machine learning approaches are applied and visualized using Apache Spark and Mahout. In our future work, we will apply the machine learning approaches using Big Data analytics to solve other feasible problems related to intrusion detection system. The input data is divided as 60% training data, 20% validation, and 20% testing data. We found the training, validation, and testing error in term of various performance evaluation parameters and confusion matrix. The predictive models provide a stable output both on the low and high volume of data. Due to we are using Big Data analytics, the models are free from the curse of dimensionality. The final classification accuracy by using all techniques is various from 99.7 to 99.9%. Therefore, we conclude that the scalable machine learning algorithms are outperforming in threat classification using intrusion dataset. The fine KNN approach provides highest accuracy, whereas the Bagged Tree ensemble approach classifies the date with a higher rate. The error rate is very less in all the approaches.

In our future work, more datasets are considered for analysis with supervised as well as unsupervised approaches to achieve better result.

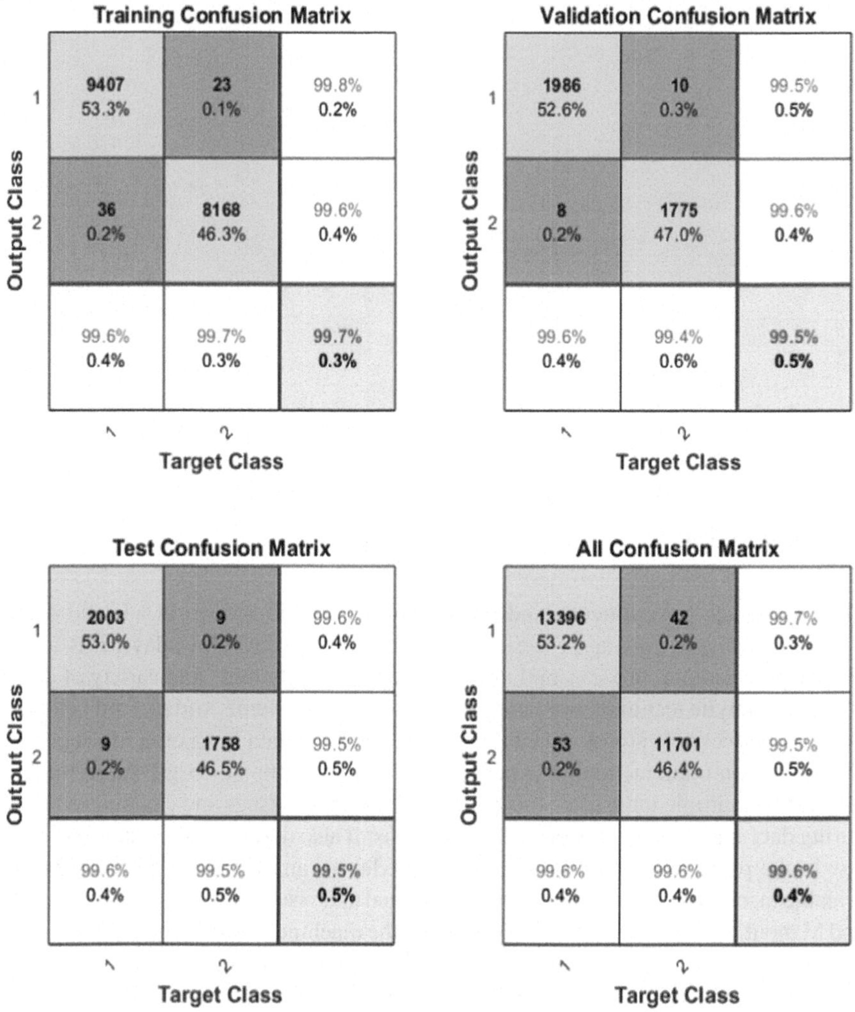


**Fig. 14** Confusion matrix during training, validation, testing and overall model evaluation using Neural Network

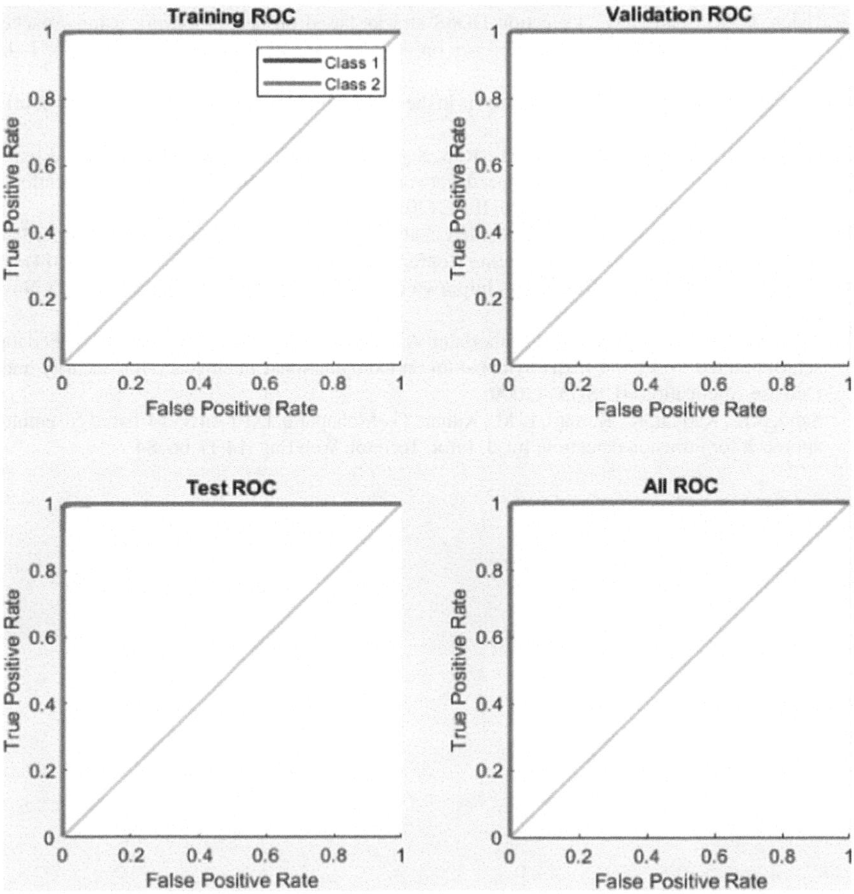


**Fig. 15** ROC during training, validation, testing and overall model evaluation using Neural Network

# References


1. NSLKDD.: Dataset Homepage, https://www.unb.ca/cic/datasets/nsl.html
2. Gupta, G.P., Kulariya, M.: A framework for fast and efficient cyber security network intrusion detection using apache spark. Procedia Comput Sci **93**, 824–831 (2016)
3. Keegan, N., Ji, S.-Y., Chaudhary, A., Concolato, C., Yu, B., Jeong, D.H.: A survey of cloud-based network intrusion detection analysis. Human-centric Comput. Inf. Sci. **6**(1), 19 (2016)
4. Kulariya, M., Saraf, P., Ranjan, R., Gupta, G.P.: Performance analysis of network intrusion detection schemes using Apache Spark. In: 2016 International Conference on Communication and Signal Processing (ICCSP), pp. 1973–1977. IEEE (2016)
5. Mavridis, I., Karatza, H.: Performance evaluation of cloud-based log file analysis with Apache Hadoop and Apache Spark. J. Syst. Softw. **125**, 133–151 (2017)
6. Kumari, R., Singh, M.K., Jha, R., Singh, N.K.: Anomaly detection in network traffic using K-mean clustering. In: 2016 3rd International Conference on Recent Advances in Information Technology (RAIT), pp. 387–393. IEEE (2016)

7. Hsieh, C.-J., Chan, T.-Y.: Detection DDoS attacks based on neural-network using Apache Spark. In: 2016 International Conference on Applied System Innovation (ICASI), pp. 1–4. IEEE (2016)
8. Mavridis, I., Karatza, E.: Log file analysis in the cloud with Apache Hadoop and Apache Spark (2015)
9. Rathore, M.M., Paul, A., Ahmad, A., Rho, S., Imran, M., Guizani, M.: Hadoop based real-time intrusion detection for high-speed networks. In: 2016 IEEE Global Communications Conference (GLOBECOM), pp. 1–6. IEEE (2016)
10. Sahu, S., Sarangi, S., Jena, S.: A detailed analysis on intrusion detection datasets. In: 2014 IEEE International Advance Computing Conference (IACC), pp. 1348–1353. IEEE (2014)
11. KDD_Corrected Dataset Homepage, http://www.springer.com/lncs (2016). Accessed 21 Nov 2016
12. Tavallaee, M., Bagheri, E., Lu, W., Ghorbani, A.: A detailed analysis of the KDD CUP 99 data set. Submitted to Second IEEE Symposium on Computational Intelligence for Security and Defense Applications (CISDA) (2009)
13. Sahu, S.K., Katiyar, K., Kumari, K.M., Kumar, G., Mohapatra, D.P.: An SVM-based ensemble approach for intrusion detection. Int. J. Infor. Technol. Web Eng. **14**(1), 66–84

# Assessing Drivers for Telecom Service Experience—Insights from Social Media

**Arpan Kumar Kar and Kanupriya Goyal**

**Abstract** In current times, the telecommunications industries are one of the tremendous factors to increase the economy, more so on the developing country like India. The telecom industry is one of the interesting industries to study, not only due to its technological development and its policies but also due to the high rate of development of this industry over the past few years and a significant factor to increase economy of the nation. Customer loyalty has become an important factor for both manufacturers and service providers in increasing competition for customers in today's customer-centred era. The present study aims to understand the factors affecting customer loyalty. The findings indicated that information security, customer support, and responsiveness have a positive relationship between customer loyalty. Data are collected from 4 lakh tweets from Twitter by using popular hashtags and @ mention of telecommunication firms in Twitter. Topic modelling and sentiment mining were done on these tweets. The statistical analysis indicated that responsiveness, information security, and customer support play a significant factor for customer loyalty in the telecommunication industry.

**Keywords** Telecommunication · Twitter analytics · Social media analytics · Customer loyalty · User-generated content

## 1 Introduction

For the success of telecommunication industries, it is significant for the service provider to maintain a positive relationship with the customer in a challenging and competitive market. Aydin [1] highlighted that in the telecommunication industry, the main condition for protecting the service provider is through winning customer loyalty in this competitive market. Therefore, the objective of this research is to determine drivers for customer loyalty towards the telecommunication service providers

A. K. Kar (✉) · K. Goyal (✉)
DMS, Indian Institute of Technology Delhi, New Delhi, India
e-mail: arpan.kumar.kar@gmail.com

K. Goyal
e-mail: kanupriyag1410@gmail.com

 
P. K. Singh et al. (eds.), *Proceedings of ICRIC 2019*, Lecture Notes in Electrical Engineering 597, https://doi.org/10.1007/978-3-030-29407-6_51

in India. Mobile telecommunication market has become more competitive with the increase of a number of mobile phone services to the customer as well as the number of mobile phone service provider are expected to increase. The growth rate of the mobile telecommunication service market is significant in India for the past few years. This study attempts to understand the drivers for loyalty in telecom, based on mining discussions in social media platforms.

Sentiment analysis using Twitter data has been used in this study, which classified the positive and negative views from the tweets, and some deep mining about the positive and negative words. Subsequently, topic modelling and content analysis have been performed to map topics to factors driving service quality. Further, this mapping has been analysed statistically to validate the important factors affecting the consumption of telecommunication services in India.

The main objective of this study is to evaluate the impact of reliability, network quality, responsiveness, customer support, information security, secure communication, and call quality on customer loyalty in the telecom industry of India.

## 2 Literature Review

Here, the detailed literature review carried out with the independent variables reliability, network quality, responsiveness, customer support, information security, and on the predictor variable customer loyalty. According to Krishnamurthi [2], customer loyalty can be defined as essential components for a brand long-term viability. According to Oliver [3], loyalty of customers can be defined as a customer's confinement of making repeat purchasing and providing positive publicity through word-of-mouth. According to Oliver [4], customer loyalty can be defined as the repurchase of a preferred telecommunications service provider. Further according to Uncles [5], loyalty of customer's can be classified as:

*Attitudinal-Loyalty: where the customers owing to their pre-determined impact of a certain brand.*
*Behavioural-Loyalty: where loyalty to a brand is defined mainly with past purchasing experiences related to the use of that brand or services.*
*Co-determinants Loyalty: where loyalty is measured by probability variables such as individual's impact and their characteristics.*

According to Hur [6], customer loyalty is defined as continuous repurchase and consumption of the same services like telecom. According to Rauyruen [7], creating a loyal customer does not depend only to maintain various customers over time, but also to maintain a relationship with customers to encourage them for future purchase and level of recommendation (Table 1).

**Table 1** Factors affecting social media usage in service consumption

| S. No | Dimensions | Definitions | Literature evidence |
|---|---|---|---|
| 1. | Reliability (in) | The service provider shows a sincere interest in solving the customer's problem. Service providers deliver their services at the time it promises to do so | Parasuraman et al. [8] |
| 2. | Responsiveness (in) | Employees are never too busy to respond to requests. Employees tell exactly when service will be performed | Parasuraman et al. [8, 9] |
| 3. | Network quality (in) | Call the quality of a specifically chosen network is always good | Asaju et al. [10] |
| 5. | Customer support (in) | Easy of reporting a complaint. The speed of complaint processing. Friendliness when reporting a complaint | Kim et al. [11] |
| 6. | Information security (in) | Security of customer information can act as one of the key differentiating factors and can help increase customer loyalty | Ribbink et al. [12] |
| 7. | Call quality (in) | Call quality coverage | Kim et al. [11] |
| 8. | Secure communication (in) | Secure communication is required to reduce the risk of data loss, theft and sabotage | Castiglione et al. [13] |
| 9. | Customer loyalty (dep) | Combination of customer's favourable attitude towards the service and intention to repurchase the service | Gerpott [14], Roy et al. [15] |

# 3 Theoretical Framework

The theoretical framework of this research was taken from previous studies by Parasuraman et al. [8], Ribbink et al. [12], Gerpott et al. [14], Grover et al. [16–19], Chhonker et al. [20], and Castiglione et al. [13]. Following hypothesis has been developed as illustrated below:

*H1: Reliability has a positive impact on customer loyalty.*
*H2: Network quality has a positive impact on customer loyalty.*
*H3: Responsiveness has a positive impact on customer loyalty.*

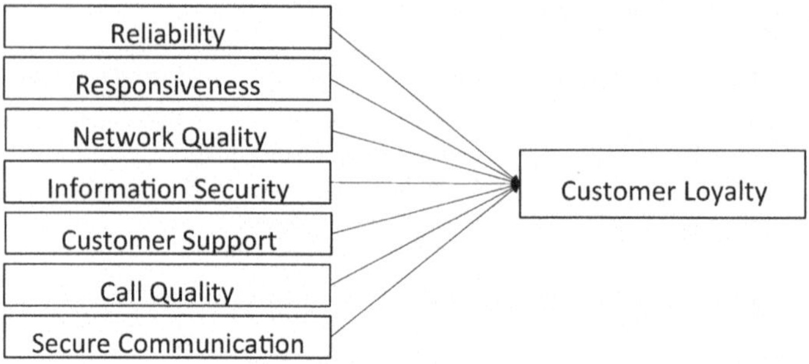


**Fig. 1** Model for assessing service quality of telecom services

*H4: Customer support has a positive impact on customer loyalty.*
*H5: Information security has a positive impact on customer loyalty.*
*H6: Secure communication has a positive impact on customer loyalty.*
*H7: Call quality has a positive impact on customer loyalty* (Fig. 1).

## 4 Methodology

In this study, primary data collected from Twitter for over a period of 12 months from January 2018 to December 2018 based on mentions of the dominant telecom service providers in India like Vodafone, Airtel, Jio, BSNL, MTNL, Idea Cellular, and Aircel. First, these collected tweets are cleaned. For analysis cleaned tweets were used to identify the sentiment of the users. These cleaned tweets are classified into two categories based on negative and positive sentiments. Then topic modelling is done to identify the topics of each negative and positive views of users. Then we apply the content analysis methodology in which seven parameters are taken an independent variable and customer loyalty taken as dependent variables. And then ANOVA and t-test are done to identify the perception of customers towards telecommunication service in India. Then multiple regression is carried out to identify the impact of the perception of customers towards telecommunication services. In this study, we used a python language to mine the Twitter data and to carry out the sentiment analysis for Telecommunication industries in India.

## 5 Overview of Findings

The collected data were transferred to SPSS for statistical tests. Multiple regression analysis was applied to test the hypothesis.

**Fig. 2** Word cloud to identify the polarity of positive and negative sentiments, respectively

## *5.1 Word Cloud*

Word cloud basically used to identify the result of polarity and emotion analysis [21]. Figure 2 illustrates the word clouds used to identify the polarity and emotion from the words which are positive and negative sentiments.

## *5.2 Multiple Regression Model Summary*

Regression analysis was used to identify the relationship between service quality and customer loyalty [22]. The model summary discloses the following result as shown in Table 2.

**Predictors**: (constant), reliability, network quality, responsiveness, customer support, information security.

Here, R2 = 0.395 shows that all seven service quality dimensions explained approximately 39.5% of the variance in overall loyalty of customers. And 0.166 is standard error of the estimate which shows the square—the root of the mean square for the residuals in the ANOVA and standard deviation of the error term.

The ANOVA results depict that the overall result is statistically significant because the value is less than 0.05 and F-value is equal to 0.108/0.028 = 3.923 as shown in Table 3.

**Table 2** Model summary

| *R* square | Adjusted *R* square | Std. the error of the estimate |
|---|---|---|
| 0.395 | 0.295 | 0.166 |

**Table 3** ANOVA

| | Model | Sum of squares | Df | Mean square | F. | Sig. |
|---|---|---|---|---|---|---|
| 1 | Regression | 0.759 | 7 | 0.108 | 3.923 | 0.002 |
| | Residual | 1.161 | 42 | 0.028 | | |
| | Total | 1.920 | 49 | | | |

**Table 4** Coefficients of regression analysis for model

| | Model | Unstandardized coefficient | | Standard coefficient | $t$ | Sig. |
|---|---|---|---|---|---|---|
| | | $B$ | Std.error | Beta | | |
| 1 | (Constant) | 0.004 | 0.712 | | 0.006 | 0.995 |
| | Reliability | −0.019 | 0.081 | −0.034 | −0.235 | 0.815 |
| | Responsiveness | −0.152 | 0.069 | −0.285 | −2.224 | 0.032 |
| | N/w quality | −0.014 | 0.082 | −0.021 | −0.166 | 0.869 |
| | Information security | 0.253 | 0.093 | 0.388 | 2.714 | 0.010 |
| | Customer support | 0.567 | 0.129 | 0.567 | 4.392 | 0.000 |
| | Secure communication | −0.008 | 0.104 | −0.010 | −0.078 | 0.938 |
| | Call quality | 0.032 | 0.101 | 0.039 | 0.314 | 0.755 |

**Dependent Variable**: customer loyalty; **Predictors**: (constant), reliability, network quality, responsiveness, customer support, information security, call quality, secure communication.

Service quality dimension information security ($\beta = 0.253, p < 0.05$) has the positive and noteworthy influence on customer loyalty. Furthermore, customer support ($\beta = 0.567, p < 0.05$) and responsiveness ($\beta = -0.285, p < 0.05$) have the positive influence on customer loyalty. While reliability, secure communication, call quality and network quality have values ($\beta = -0.034, p > 0.05$), ($\beta = -0.010, p > 0.05$), ($\beta = 0.039, p > 0.05$), ($\beta = -0.021, p > 0.05$), respectively. These dimensions have insignificant influence on customer loyalty as shown in Table 4.

## 6 Discussion

The main reason to conduct this study was to identify the impact of service quality on customer loyalty in the telecommunication sector in India. To identify the importance of service quality, researchers applied statistical tests on collected data. The outcomes show that three out of seven dimensions of service quality (responsiveness, information security and customer support) have an influence on customer loyalty. While the other four dimensions (reliability, secure communication, network quality, call quality) have not much impact on loyalty of customer's in the telecommunication sector in India. The result of hypotheses 3, 4, and 5, responsiveness customer loyalty, customer support customer loyalty, and information security customer loyalty, respectively, are supporting the outcomes between service quality and customer loyalty in the telecom industry.

The result clarifies that service quality has important in customer loyalty of the telecom industry of India. Furthermore, the result of this research will help in the telecommunication sector to improve its service quality and also to improve the reputation of their company.

## 7 Conclusion

The current research was carried out to find the constructs winch impact on customer's loyalty in the telecom industry of India. Dimensions of service quality included in the research were reliability, responsiveness, network quality, information security, customer support, secure communication, and call quality while customer loyalty was taken as dependent variables. Multiple regression analysis confirmed that three independent variables such as responsiveness, information security, and customer support had a high-positive impact on customer loyalty while reliability, secure communication, call quality, and network quality did not have a positive impact on customer loyalty. This result will help in telecom industries to improve their service quality which will increase customer loyalty and for the government to frame a policy to address this need.

**Source of Funding** This research is funded by the SEED Division of DST, Ministry of Science and Technology of India, Government of India, [SP/YO/048/2017(C)].

## References

1. Aydin, S., Ozer, G.: The analysis of antecedents of customer loyalty in the Turkish mobile telecommunication market. Eur. J. Mark **39.7**/8, 910–925 (2005)
2. Krishnamurthi, L., Raj, S.P.: An empirical analysis of the relationship between brand loyalty and consumer price elasticity. Mark. Sci. **10**(2), 172–183 (1991)
3. Oliver, R.L., Rust, R.T., Varki, S.: Customer delight: foundations, findings, and managerial insight. J. Retail. **73**(3), 311–336 (1997)
4. Oliver, R.L.: Whence consumer loyalty? J. Mark., 33–44 (1999)
5. Uncles, R.J., Frickers, P.E., Harris, C.: Dissolved nutrients in the Tweed Estuary, UK: inputs, distributions and effects of residence time. Sci. Total Environ. **314**, 727–736 (2003)
6. Hur, W.M., Park, J., Kim, M.: The role of commitment on the customer benefits—loyalty relationship in the mobile service industry. Serv. Ind. J. **30.14**, 2293–2309 (2010)
7. Rauyruen, P., Miller, K.E.: Relationship quality as a predictor of B2B customer loyalty. J. Bus. Res. **60**(1), 21–31 (2007)
8. Ribbink, D., Van Riel, A.C., Liljander, V., Streukens, S.: Comfort your online customer: quality, trust and loyalty on the internet. Manag. Serv. Qual. Int. J. **14.6**, 446–456 (2004)
9. Parasuraman, A., Zeithaml, V.A., Berry, L.L.: SERVQUAL: a multiple-item scale for measuring consumer perc. J. Retail. **64**(1), 12–40 (1988)
10. Asaju, Y., Ayisat, W., Dahalin, Z.B., Ta'a, A.: Mobile network quality of experience using big data analytics approach. In: 8th International Conference on Information Technology (ICIT), pp. 658–664. IEEE (2017)
11. Kim, M.K., Park, M.C., Jeong, D.H.: The effects of customer satisfaction and switching barrier on customer loyalty in Korean mobile telecommunication services. Telecommun. Policy **28.2**, 145–159 (2004)
12. Zeithaml, V.A., Parasuraman, A., Berry, L.L.: Delivering quality service: balancing customer perceptions and expectations. The Free Press, New York (1990)
13. Castiglione, A., Cattaneo, G., De Maio, G., Petagna, F.: Secr3t: secure end-to-end communication over 3G telecommunication networks. In: 2011 Fifth International Conference on Innovative Mobile and Internet Services in Ubiquitous Computing, pp. 520–526. IEEE (2011, June)

14. Gerpott, T.J., Rams, W., Schindler, A.: Customer retention, loyalty, and satisfaction in the German mobile cellular telecommunications market. Telecommun. Policy **25**(4), 249–269 (2001)
15. Roy, S.K., Butaney, G., Bhutan, B.: Examining the effects of customer loyalty states on the word of mouth. In: PACIS 2009 Proceedings, pp. 19–85
16. Grover, P., Kar, A.K., Ilavarasan, P.V.: Understanding nature of social media usage by mobile wallets service providers–an exploration through SPIN framework. Procedia Comput. Sci. **122**, 292–299 (2017)
17. Grover, P., Kar, A.K.: Big data analytics: a review on theoretical contributions and tools used in literature. Global J. Flex. Syst. Manage. **18**(3), 203–229 (2017)
18. Grover, P., Kar, A.K., Dwivedi, Y.K., Janssen, M.: The untold story of USA presidential elections in 2016-insights from Twitter analytics. In: Conference on e-Business, eServices and e-Society, pp. 339–350. Springer, Cham (2017)
19. Grover, P., Kar, A.K., Davies, G.: "Technology-enabled Health"–Insights from Twitter analytics with a socio-technical perspective. Int. J. Inf. Manag. **43**, 85–97 (2018)
20. Chhonker, M.S., Verma, D., Kar, A.K., Grover, P.: m-commerce technology adoption: thematic and citation analysis of scholarly research during (2008–2017). Bottom Line **31**(3/4), 208–233 (2018)
21. Aswani, R., Kar, A.K., Ilavarasan, P.V.: Search engine marketing is not all gold: insights from Twitter and SEOClerks. Int. J. Inf. Manage. **38**(1), 107–116 (2018)
22. Chatterjee, S., Kar, A.K., Gupta, M.P.: Success of IoT in Smart Cities of India: An empirical analysis. Gov. Inf. Quart. (2018) (An article in press)

# Collaborative Topic Regression-Based Recommendation Systems: A Comparative Study

**Balraj Kumar, Neeraj Sharma and Shubhangi Sharma**

**Abstract** The collaborative filtering is a very popular and powerful approach prominently being used in many research areas of computer science like recommendation systems, information retrieval, data mining, etc. When used for making recommendations, the traditional collaborative filtering methods suffer from certain problems, where the data sparsity is the significant one that causes the deterioration of the recommendation quality. In order to alleviate this issue, the research fraternity has started proposing the use of some additional domain information in formulating recommendations. In literature, different models have been proposed that make use of such kind of add-on information extensively and have also shown the promising performance than the other state-of-the-art approaches. Hence, the increasing use of add-on information is creating an overwhelming impact on the recommendation field. The piety of this article is to present a meticulous comparative study of various such recommendation models especially those which belong to the family of collaborative topic regression recommendation models in the light of several parameters and this study further leads to propose a novel recommendation prototype based on the fusion of different kinds of auxiliary domain knowledge.

**Keywords** Auxiliary domain knowledge · Collaborative filtering · Item network · Matrix factorization · Recommendation system · Social network

---

B. Kumar (✉)
School of Computer Applications, Lovely Professional University, Phagwara 144411, Punjab, India
e-mail: balraj_kr@yahoo.co.in

N. Sharma · S. Sharma
Department of Computer Science, Punjabi University, Patiala 147002, Punjab, India
e-mail: sharma_neeraj@hotmail.com

S. Sharma
e-mail: shubhangisharma1197@gmail.com



P. K. Singh et al. (eds.), *Proceedings of ICRIC 2019*, Lecture Notes in Electrical Engineering 597, https://doi.org/10.1007/978-3-030-29407-6_52

# 1 Introduction

The recommendation technology is one of the powerful mechanisms today that attempts to tackle the *information overload* problem. This problem has emerged due to the availability of massive data and information over the Internet where the searching of relevant and personalized items poses a significant challenge to the users. Thus, the recommendation system is an evolutionary step in this direction that provides efficient and most promising techniques to filter out the overload information [1, 2]. Such systems produce recommendations based on pre-recorded data [3] and generally find their deployment in commercial applications such as Netflix, Amazon, Last.fm.

In broader sense, a software system that makes a suggestion about an item to buy, subscribe, or invest may be regarded as a recommender system [4]. The recommendation approaches employed in various recommender systems can broadly be classified into three categories [5] collaborative filtering (CF), content-based (CB) filtering, and hybrid filtering. CF methods [6–8] are based on past preferences that include user ratings on items for prediction purpose. Content-based methods [9] work on user profiles or item profiles for generating recommendations. The privacy issues may cause hindrance in collecting user profiles than past records. Thus, CF methods get more preference and are considered more reliable than CB methods. Hence, the recommendation models participating here in comparative study strongly support the use of CF methods. Hybrid filtering [10, 11] is the blended approach of CF- and CB-based methods. Such type of filtering approach attempts to fuse two techniques in such a way that the benefits of one can be used to fix the drawbacks of the other.

The traditional CF methods make use of user-rating matrix to generate recommendations and generally this matrix is highly sparse. This sparsity indicates that there are very few user ratings available against huge number of items. Due to this sparsity problem, CF approaches cannot achieve quality recommendations. To alleviate this issue, the research community has started proposing the use of some additional domain information in recommendations generation. In literature, different models have been proposed that make use of this auxiliary information extensively and outperform the state-of-the-art approaches [12]. The additional information that has been incorporated in recommender systems involves item content [13], social relationships [14–16], and context [17]. Besides this, a fusion of different types of information is also in practice such as item content and social relationships [18], item content and item relations [12], item content and context [19], and context and social relationships [20, 21]. The present article too highly emphasizes the use of auxiliary domain information apart from the user-item ratings to produce quality recommendations. Thus, three recommendation models namely *Collaborative Topic Regression* (CTR) [13], *Relational Collaborative Topic Regression* (RCTR) [12], and *Collaborative Topic Regression–Social Matrix Factorization* (CTR-SMF) [18] have been identified for the present comparative study.

The purpose of this article is threefold:

- First, to introduce a family of collaborative topic regression-based recommendation models and to highlight their mutual relationships.
- Second, to present a comparative analysis of CTR, CTR-SMF, and RCTR recommendation models from CTR family.
- Third, to propose a novel recommendation prototype by integrating the properties of CTR-SMF and RCTR models for better prediction accuracy.

## 2 Background

This section provides with the background knowledge essential to get the basic insights of the recommendation field.

*Collaborative filtering*. Collaborative filtering (CF) is the oldest but the most widely adopted and successful recommendation approach in recommendation field. It recommends items to the target user on the basis of past preferences of other users with similar tastes [22]. Collaborative filtering methods are classified into two groups: memory-based and model-based methods [23]. *Memory-based methods* are those where the item recommendations are generated by exploiting the complete user-rating database, whereas the *model-based methods* are those where a user-rating database is used to fit a model first and then that model is used to make predictions [24]. In comparison with memory-based methods, model-based methods have shown promising performance in literature and this is the major reason for their high popularity.

*Matrix factorization*. Among latent-factor models, the most common approaches include matrix factorization (MF) and its extension such as probabilistic matrix factorization (PMF) [25] have demonstrated a significant performance in literature and subsequently got popularity in recent times. In MF technique, both users and items are characterized by vectors of factors deduced from item rating patterns and a recommendation is caused by high correspondence between user and item factors. Besides this, they are flexible enough for modeling the different real-life situations [26]. Social matrix factorization (SMF) [14–16] integrates social connections with matrix factorization for better recommendation performance. [16] assume that a user's taste is similar to the average taste of his friends. Thus, the purpose of SMF is to analyze the social network graph to develop $l$-dimensional feature representation of users.

*Topic modeling*. In machine learning, the topic modeling is a kind of statistical model used to discover sets of "topics" in a collection of documents based on a hierarchical Bayesian analysis of the original texts [27]. It is generally used in text-mining tool to discover hidden semantic structures in text. Latent Dirichlet allocation (LDA) is one of the simplest topic models [28]. LDA provides a way of automatically discovering topics that elucidate about the similarity of some parts of the data. It assumes that the topics are generated prior to the documents [29].

The probabilistic topic modeling [29] is a set of algorithms aiming to ascertain and annotate large collections of documents based on different themes such as culture, indoor games, international affairs, domestic industries, education. The probabilistic topic models help identify a collection of *topics* from a huge pool of documents. Here, the *topic* spans across the terms inclined toward a single subject or theme. These models are able to provide an interpretable low-dimensional representation of the documents [30]. The different tasks where they are profoundly in use include corpus exploration, classification of documents, and information retrieval.

## 3 Recommendation Models in CTR Family

This section acquaints with the family of *Collaborative Topic Regression* (CTR)-based recommendation models that recommend items to users in different domains. These models include CTR, CTR-SMF, and RCTR. Apart from these, there are two more models namely *Limited Attention Collaborative Topic Regression* (LA-CTR) [31] and *Context-Aware Collaborative Topic Regression–Social Matrix Factorization* (C-CTR-SMF2) [32]. These systems were evolved one after the other with the aim of achieving better recommendation performance. They attempted to add auxiliary information for this purpose. Figure 1 presents the progression of these models as CTR family:

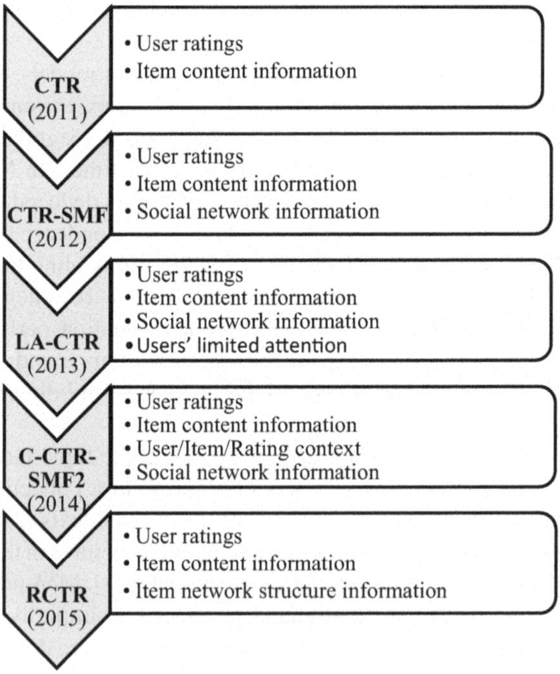


**Fig. 1** Evolution of CTR family

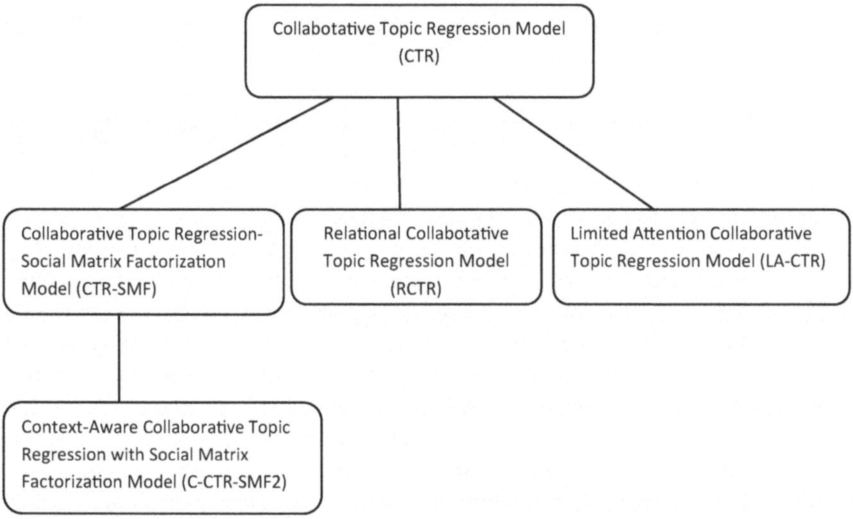


**Fig. 2** Hierarchy of CTR family

This figure highlights the auxiliary information employed by each of the recommendation models in addition to its induction year. These recommendation models hold a mutual relationship among them based on auxiliary domain knowledge where one model is derived from the other. Thus, their relationship can be best described using a hierarchy as shown in Fig. 2.

Here, CTR acts as the base model, while CTR-SMF, RCTR, and LA-CTR are the derived models. In addition to user rating and item content information used in CTR model, RCTR and CTR-SMF also integrate item relations and user relations, respectively, to deal with sparsity problem as well as to improve the prediction accuracy, while LA-CTR adds users' limited attention along with social network information. Likewise, C-CTR-SMF2 was introduced as an extension of CTR-SMF with an addition of contextual information.

To acquire more familiarity of each of these models, they are explicated as under:

(1) *Collaborative Topic Regression* (CTR) [13] implements a recommender system by integrating both the users' rating records and the item content information to recommend scientific articles to researchers. It attempts to combine the traditional CF with probabilistic topic modeling to fit a model that expends the latent topic space to elucidate the observed ratings as well as the observed words. [33] proposed a novel online Bayesian inference algorithm for CTR model which is scalable and efficient for learning from data streams. It carries out joint optimization by combining both PMF and LDA models to achieve a much tighter coupling.

(2) *Collaborative Topic Regression–Social Matrix Factorization* (CTR-SMF) [18] is a hierarchical Bayesian model, which is also an extension of CTR, and integrates social network structure into CTR using social matrix factorization methods [15] to recommend music items. This model mainly contributes in the direction to demonstrate the effectiveness of social network information to achieve better prediction accuracy and in turn better recommendation quality. It can make predictions for new/unseen items and new/inactive users of a social network. Moreover, it also touches the issue of *social information leak* (caused by final static social network) existing in many RSs employing social information.
(3) *Limited Attention Collaborative Topic Regression* (LA-CTR) [31], an extension to the basic CTR model that incorporates limited and non-uniformly divided attention and ensures to learn more accurate user preferences by taking human cognitive factors into account. LA-CTR recommends items to users in a social media by integrating users' limited attention with their ratings, their social network, and item content. In addition to interpretable user profiles, this model also tells about how much attention the user pays to others, which may further investigate why people follow others and how the information spreads over the online social network.
(4) *Context-Aware Collaborative Topic Regression–Social Matrix Factorization* (C-CTR-SMF2) [32] is also a hierarchical Bayesian model that further extends CTR-SMF to the next level and attempts to improve recommendation performance by creating a fusion of four types of information involving rating records, item content, social relationships among users, and most importantly the contextual information. The contextual information can be categorized into three classes: user context, item context, and rating context. This method uses spectral clustering to create subgroups of users and items in similar contexts. It combines LDA with SMF approach—LDA to mine item content while SMF to handle ratings and social relationships.
(5) *Relational Collaborative Topic Regression* (RCTR) [12], another extension to CTR, is a hierarchical Bayesian model which attempts the fusion of different types of auxiliary information into this model such as user ratings, the content of items, and item relations information. RCTR handles the cold-start issues (like a new user or a new item) very intelligently by making effective use of item network information. This information not only helps overcome the data sparsity problem, but also improves the recommendation accuracy. Here, the item relations are represented with a family of link probability functions, which is based on relational topic model presented in [34].

## 4 Models in Comparison

There are a large number of studies on recommendation systems which are designed keeping in view the fusion of several kinds of auxiliary domain knowledge. Auxiliary domain knowledge has a significant role to play in the design and development of recommendation systems. It has been observed that the increasing use of such information has created an overwhelming impact on the recommendation field. Beside the user-item interactions (ratings) data, this auxiliary information may also involve item content features, relational network among items, and users' social networks information. The integration of such add-on information in the recommendation model helps boost the accuracy of predictions [12].

Likewise, to enhance the prediction accuracy further and to address the prime issues of RSs like data sparsity and cold-start, etc., this paper proposes a novel recommendation prototype based on similar auxiliary information in *Section-6*. To build a strong basis for this, the paper provides an extensive comparative analysis (as given in Section-5) of recommendation models taken from CTR family excluding LA-CTR and C-CTR-SMF2 models. The reason for their exclusion is that CTR, CTR-SMF, and RCTR models provide the essential and strong background for proposing the new recommendation prototype and act as the base models for the new prototype. In other words, the proposed prototype is going to be a joint derivative of CTR-SMF and RCTR models without any direct concern of LA-CTR and C-CTR-SMF2. Thus, LA-CTR and C-CTR-SMF2 are not the part of this comparative study.

## 5 Comparison of Recommendation Models

This article aims to present a comprehensive comparative analysis among CTR, RCTR, and CTR-SMF recommendation models in the light of several parameters. These parameters are selected very carefully with an intent to cover different aspects of these models. To the best of our knowledge gathered so far, there is no such comparative analysis exists in the literature. This comparative study is of prime importance for the scientific and research fraternity as it attempts to yield some innovative and exciting research directions for future. This comparison can be viewed from two perspectives: similarities and dissimilarities in models.

### *5.1 Similarities in Models*

These recommendation models are having many features in common. These common features are highlighted and briefly described in Table 1.

**Table 1** Similarities in recommendation models

| Similarity features | Feature description |
|---|---|
| Collaborative filtering | Collaborative filtering, the most commonly used approach, is the backbone of all the three models. In CF methods hierarchy, model-based approaches are more popular and widely preferred over memory-based methods [23] due to their better performance in literature. Memory-based methods never guarantee about good prediction results |
| Auxiliary domain knowledge | All the models make use of some auxiliary domain knowledge. CTR uses user ratings with item content information. Besides this, RCTR adds item relations using a relation graph, while CTR-SMF employs social information represented by a user network. Hence, auxiliary information is used in each model |
| Sparsity and cold-start issues | The objective of each model is to improve the prediction accuracy, which in turn enhances the recommendations quality. But the sparsity and cold-start issues degrade the performance of recommendation algorithms. For this reason, each of the recommendation models attempts to alleviate these issues with the fusion of some add-on information |
| In-matrix and out-of-matrix predictions | All the three models address two prediction cases: (i) *In-matrix prediction* produces recommendation for items rated by at least one user. (ii) *Out-of-matrix prediction* makes recommendation for items without any ratings. The traditional CF techniques can address *in-matrix prediction* only, but not *out-of-matrix prediction*, because they only use rating information of other users |
| Gaussian distribution | For feedback of each user-item pair, all models use the Gaussian feedback model which characterizes the feedback in binary form |
| Evaluation metrics | *Precision* and *recall* are the two possible evaluation metrics. But in all three models, only recall is used as it is difficult to evaluate the *precision* due to zero rating for the items which indicates either the user is unaware of the item or he/she does not like it |

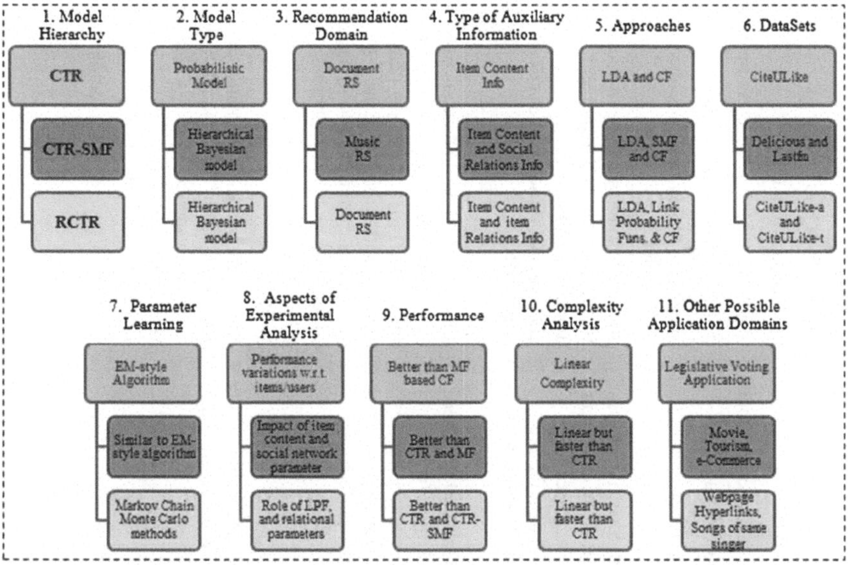


**Fig. 3** Compact view of dissimilarities among models

## *5.2 Dissimilarities in Models*

Besides the similarities, there are many dissimilarities as well that exists among these models. Figure 3 presents the compact view of dissimilarities, while the detailed view is given in Table 2.

In Fig. 3, the dissimilarities are shown parameter-wise, numbered 1 through 11, from left to right and top to bottom. Parameter 1 is the model hierarchy that corresponds to the model-wise differences presented against each model in the form of hierarchies. The blue, green, and orange color nodes are aligned with CTR, CTR-SMF, and RCTR models, respectively, to highlight their differences.

This study not only presents the approaches, datasets, strengths and weaknesses, etc., of CTR-based models, but also attempts to make aware the research community with the knowledge from basic to advance level in terms of latest theories, concepts, and practices being used in modern RSs.

# 6 Discussion and Future Directions

This study attempts to provide the adequate ground knowledge essential to investigate and explore different research perspectives based on auxiliary domain knowledge to build modern recommender systems with better prediction accuracy. For instance, one may contemplate to explore the possibilities of RCTR model for CTR-SMF

**Table 2** Dissimilarities among recommendation models

| Parameters | CTR model [13] | CTR-SMF model [18] | RCTR model [12] |
|---|---|---|---|
| Model hierarchy | It is the base model | It is derived from CTR | It is derived from CTR |
| Model type | It is the probabilistic model | It is the hierarchical Bayesian model | It is the hierarchical Bayesian model |
| Recommendation Domain | It implements the document RS | But it implements the music RS | It also implements the document RS |
| Type of auxiliary information | User ratings with item content information | Integration of user relations into CTR model | Integration of item relations into CTR model |
| Approaches | Logically incorporates item content information (via LDA—Latent Dirichlet Allocation) into CF based on latent-factor model (via PMF approach) | It fuses LDA with SMF, i.e., it uses LDA to accumulate item content in latent topic space and uses SMF methods to derive latent feature space of user from its social network graph | It attempts to integrate the relations among items (with link probability functions) into item content (via LDA) and CF-based framework (with PMF) |
| Datasets | It uses only a single dataset, namely *CiteULike* (http://www.citeulike.org/)—A bibliography sharing service | It also uses two real-world datasets, namely *Delicious* (http://www.delicious)—for bookmarking preferences & *Lastfm* (http://www.last.fm)—for music preferences | It uses two real-world datasets from CiteULike: *CiteULike-a*: the original one used in [13] without any item relations and *CiteULike-t*: the derived one from *CiteULike-a* based on seed tags and article citations from Google Scholar (http://scholar.google.com) |
| Parameter learning | To learn the maximum a posteriori (MAP) parameters, it uses an EM-style algorithm | It also develops an algorithm similar to one used in CTR for learning the MAP estimates | Apart from the MAP estimates, it also makes use of Bayesian methods for learning such as Markov Chain Monte Carlo methods, and Variational methods [35] |

(continued)

**Table 2** (continued)

| Parameters | CTR model [13] | CTR-SMF model [18] | RCTR model [12] |
|---|---|---|---|
| Aspects of experimental analysis | CTR's capabilities are compared with MF-based CF and other contemporary methods. Authors also observe its performance variations w.r.t. number of items a user likes or w.r.t. number of users that like an item | This model is compared with contemporary CF methods and also explored to analyze the impact of content parameter and user relations information on the prediction correctness | Here, the authors compare its performance with other contemporary methods in extreme sparsity cases and also try to investigate the role of link probability functions (LPF), relational parameter and other parameters in improving the prediction performance |
| Performance | By merging MF and LDA, CTR attains enhanced prediction accuracy than MF-based CF | It outperforms CTR and matrix factorization [26] since it uses user relations to better model the user latent space | Its performance is better than CTR, CTR-SMF, and other up-to-date methods in terms of lower-training time, higher-prediction accuracy and ability to learn good interpretable latent structures beneficial for recommendation |
| Complexity analysis | The complexity in CTR is linear in terms of the number of items (i.e., papers in the users' libraries) | It is faster than CTR when content parameter is small, but gets slower when the value increases. For small number of latent space dimensions, its prediction accuracy is similar to CTR having large dimensions | It requires very few learning iterations than CTR to acquire agreeable prediction performance. But in each learning iteration, it adds very nominal extra linear time to CTR |
| Other possible application domains | This model can further be examined to implement legislative voting application | Other possible domains where social network can have an effective role may include movie, tourism, e-commerce, etc. | This model can be explored for other domains as well where item networks can be found in webpage hyperlinks, restaurants run by the same owner, songs sung by the same singer |

settings with users' network [12]. After an exhaustive literature review, this section proposes a novel recommendation prototype that could possibly be derived from the prime models involved in the present comparative study.

**The Proposed Prototype**—This section proposes a novel recommendation prototype namely *Relational Collaborative Topic Regression with Social Matrix Factorization* (RCTR-SMF) by further extending the hierarchy of CTR family. This prototype attempts to integrate the features of CTR-SMF and RCTR models together to reflect a hybrid approach; hence, it gets its name as RCTR-SMF recommendation model.

Figure 4 presents the proposed recommendation prototype (in dotted lines) derived from RCTR and CTR-SMF models. The fundamental objective of this proposed prototype is to seamlessly integrate a lot of auxiliary information such as the users' rating records, item content, items' network, and social networks among users to address the key issues of RSs like data sparsity and cold-start, etc. The basic idea is to achieve better prediction accuracy which in turn would contribute to enhance the quality of recommendations.

In the figure, all the existing recommendation models primarily emphasize on improving the prediction accuracy by incorporating different types of supplementary information into the recommendation algorithm. CTR attempts to add item content to the user ratings to deal with sparsity and cold-start problems. To tackle these issues further, on the other side, both RCTR and CTR-SMF models extend the CTR by integrating network structure among items and social network information,

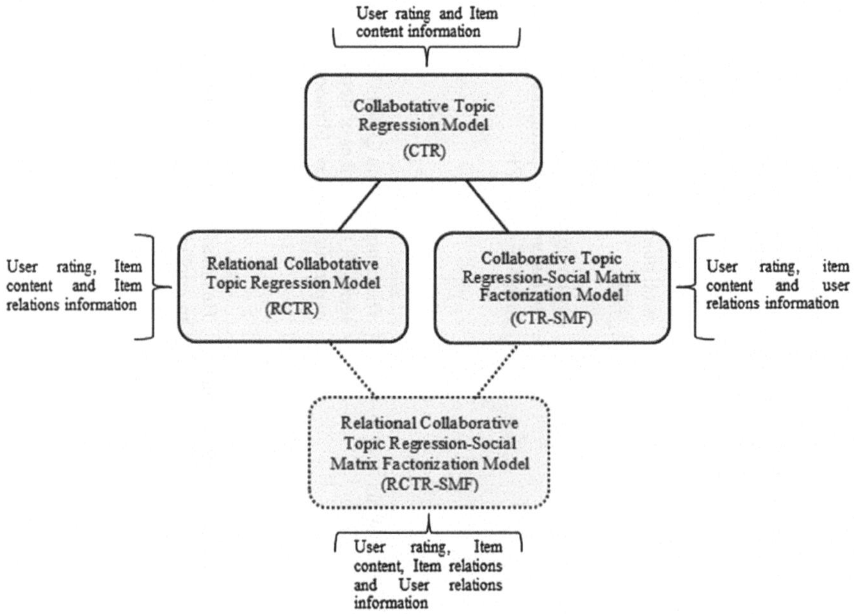


**Fig. 4** Proposed recommendation prototype

respectively, into CTR to achieve better prediction accuracy with lower training time. On the same lines, the proposed RCTR-SMF model too is supposed to expend the seamless fusion of different kinds of auxiliary domain knowledge such as user-item feedback, item content, item relations, and user relations to alleviate the sparsity and cold-start issues which would lead to better prediction accuracy and hence better quality of recommendations.

CF and LDA are the prominent approaches used in all models. Additionally, RCTR utilizes link probability functions to represent item relations and CTR-SMF employs SMF methods to represent user relations. RCTR/CTR-SMF models exploit relational information of items/users and are able to build better quality recommendations than CTR. CTR delivers better performance over MF methods, but CTR-SMF outperforms both CTR and MF. RCTR claims to be the best one as it outperforms all of them and other contemporary methods. With the fusion of variety of auxiliary information and model-based CF approaches, the proposed model is also supposed to take the recommendation performance to the next level and would surely outperform one and all present methods.

Although there are many challenges in the development and deployment of RSs, but the proposed study is expected to address only data sparsity and cold-start issues. These are the challenges that directly affect the quality of predictions and the overall performance of a system. The recall will be the only evaluation metric for the proposed model just like other models. The following are some of the possible extensions that may be explored while pursuing this problem and investigating it further for research:

- RCTR model [12] combines the tag graph and citation graph into a single graph. But it is possible to separately model them by introducing other latent variables. Thus, the behavior of the proposed model can be examined by separately modeling more than one item networks by introducing other latent variables.
- To deal with huge data of users and items, it is essential to design some distributed learning algorithm for the proposed model, which would make it scalable for large scale datasets [24].
- For social information leak problem, time stamped or evolving societal relations information [18] would be more useful to obtain quality predictions.

Lastly, the proposed RCTR-SMF would be an intelligent system in the recommendation field expected to be of great significance that may contribute toward the possible solution of information overload problem and may prove to be an exciting and motivating research area.

## 7 Conclusion

The key findings of this comparative study show that the recommendation models, CTR, RCTR, and CTR-SMF from CTR family claim to improve the recommendation quality in terms of both coverage and accuracy, and outperform the present recommendation methods. Each of these models expends some add-on information to alleviate the sparsity and cold-start problems which in turn help enhance the prediction accuracy. Thus, the increasing use of auxiliary domain knowledge is creating an overwhelming dominance on the recommendation field. This paper also proposes a novel recommendation prototype based on the fusion of different kinds of add-on information which may prove to be an exciting, innovative, and motivating research area. Moreover, it also throws some light on the future directions which are thought-provoking and may be considered worth investigating further.

## References

1. Liu, J., Jiang, Y., Li, Z., Zhang, X., Lu, H.: Domain-sensitive recommendation with user-item subgroup analysis. IEEE Trans. Knowl. Data Eng. **28**(4), 939–950 (2016)
2. Zhang, Z.K., Zhou, T., Zhang, Y.C.: Personalized recommendation via integrated diffusion on user–item–tag tripartite graphs. Physica A **389**, 179–186 (2010)
3. Sarwar, B., Karypis, G., Konstan, J., Riedl, J.: Analysis of recommendation algorithms for e-commerce. In: Proceedings of 2nd ACM Conference on Electronic Commerce, pp. 158–167 (2000)
4. Lee, J., Sun, M., Lebanon, G.: A comparative study of collaborative filtering algorithms. arXiv preprint arXiv, 1205.3193, pp. 1–27 (2012)
5. Adomavicius, G., Tuzhilin, A.: Toward the next generation of recommender systems: a survey of the state-of-the-art and possible extensions. IEEE Trans. Knowl. Data Eng. **17**, 734–749 (2005)
6. Bartolini, I., Zhang, Z., Papadias, D.: Collaborative filtering with personalized skylines. IEEE Trans. Knowl. Data Eng. **23**, 190–203 (2011)
7. Cai, Y., Leung, H.F., Li, Q., Min, H., Tang, J., Li, J.: Typicality-based collaborative filtering recommendation. IEEE Trans. Knowl. Data Eng. **26**, 766–779 (2014)
8. Shi, Y., Larson, M., Hanjalic, A.: List-wise learning to rank with matrix factorization for collaborative filtering. In: Proceedings of 4th ACM Conference on Recommender Systems, ACM, pp. 269–272 (2010)
9. Balabanović, M., Shoham, Y.: Fab: content-based, collaborative recommendation. Commun. ACM **40**, 66–72 (1997)
10. Barragáns-Martínez, A.B., Costa-Montenegro, E., Burguillo-Rial, J.C., Rey-Lopez, M., Mikic-Fonte, F.A., Peleteiro-Ramallo, A.: A hybrid content-based and item-based collaborative filtering approach to recommend TV programs enhanced with singular value decomposition. Inf. Sci. **180**(22), 4290–4311 (2010)
11. Agarwal, D., Chen, B.C.: Regression-based latent factor models. In: Proceedings of the 15th ACM SIGKDD International Conference on Knowledge Discovery and Data Mining, ACM, pp. 19–28 (2009)
12. Wang, H., Li, W.J.: Relational collaborative topic regression for recommender systems. IEEE Trans. Knowl. Data Eng. **27**, 1343–1355 (2015)
13. Wang, C., Blei, D. M.: Collaborative topic modeling for recommending scientific articles. In: Proceedings of 17th ACM SIGKDD International Conference on Knowledge Discovery and Data Mining, ACM, pp. 448–456 (2011)
14. Chen, C., Zeng, J., Zheng, X., Chen, D.: Recommender system based on social trust relationships. In: Proceedings of IEEE Ninth International Conference on e-Business Engineering (ICEBE), pp. 32–37 (2013)
15. Ma, H., Yang, H., Lyu, M.R., King, I.: Sorec: Social recommendation using probabilistic matrix factorization. In: Proceedings of 17th ACM Conference on Information and Knowledge Management, ACM, pp. 931–940 (2008)

16. Jamali, M., Ester, M.: A matrix factorization technique with trust propagation for recommendation in social networks. In: Proceedings of fourth ACM conference on Recommender systems, ACM, pp. 135–142 (2010)
17. Baltrunas, L., Ludwig, B., Ricci, F.: Matrix factorization techniques for context aware recommendation. In: Proceedings of fifth ACM conference on Recommender systems, pp. 301–304 (2011)
18. Purushotham, S., Liu, Y., Kuo, C.C.J.: Collaborative topic regression with social matrix factorization for recommendation systems. In: Proceedings of 29th International Conference on Machine Learning, Edinburgh, Scotland, UK (2012)
19. Agarwal, D., Chen, B.C.: fLDA: matrix factorization through Latent Dirichlet allocation. In: Proceedings of third ACM international conference on Web search and data mining, ACM, pp. 91–100 (2010)
20. Liu, X.: Towards context-aware social recommendation via trust networks. Web Information Systems Engineering—(WISE 2013), pp. 121–134. Springer (2013)
21. Liu, X., Aberer, K.: Soco: a social network aided context-aware recommender system. In: Proceedings of 22nd international conference on World Wide Web, pp. 781–802 (2013)
22. Schafer, J.B., Frankowski, D., Herlocker, J., Sen, S.: Collaborative filtering recommender systems. The Adaptive Web, pp. 291–324. Springer Berlin Heidelberg (2007)
23. Koren, Y.: Factorization meets the neighborhood: a multifaceted collaborative filtering model. In: Proceedings of 14th ACM SIGKDD International Conference on Knowledge Discovery and Data Mining, New York, USA, ACM, pp. 426–434 (2008)
24. Kumar, B., Sharma, N.: Approaches, issues and challenges in recommender systems: a systematic review. Indian J. Sci. Technol. **9**(47), 1–12 (2016)
25. Salakhutdinov, R., Mnih, A.: Probabilistic matrix factorization. Adv. Neural. Inf. Process. Syst. **20**, 1257–1264 (2008)
26. Koren, Y., Bell, R., Volinsky, C.: Matrix factorization techniques for recommender systems. Computer **42**, 30–37 (2009)
27. Blei, D.M., Lafferty, J.: Topic models. Text mining: classification, clustering, and applications. **10**(71) (2009)
28. Blei, D.M., Ng, A.Y., Jordan, M.I.: Latent Dirichlet allocation. J. Mach. Learn. Res. **3**, 993–1022 (2003)
29. Blei, D.M.: Probabilistic topic models. Commun. ACM **55**(4), 77–84 (2012)
30. Chang, J., Gerrish, S., Wang, C., Boyd-Graber, J.L., Blei, D.M.: Reading tea leaves: how humans interpret topic models. Adv. Neural Inf. Process. Syst. 288–296 (2009)
31. Kang, J.H., Lerman, K.: LA-CTR: A limited attention collaborative topic regression for social media. In: Proceedings of Twenty-Seventh AAAI Conference on Artificial Intelligence, pp. 1128–1134 (2013)
32. Chen, C., Zheng, X., Wang, Y., Hong, F., Lin, Z.: Context-aware collaborative topic regression with social matrix factorization for recommender systems. In: Proceedings of the Twenty-Eighth AAAI Conference on Artificial Intelligence, pp. 9–15 (2014)
33. Liu, C., Jin, T., Hoi, S.C.H., Zhao, P., Sun, J.: Collaborative topic regression for online recommender systems: an online and Bayesian approach. Mach. Learn. **106**(5), 651–670 (2017). (Springer)
34. Chang, J., Blei, D.M.: Relational topic models for document networks. In: Proceedings of International conference on artificial intelligence and statistics, vol. 9, pp. 81–88 (2009)
35. Jordan, M.I., Ghahramani, Z., Jaakkola, T.S., Saul, L.K.: An introduction to variational methods for graphical models. Mach. Learn. **37**, 183–233 (1999)

# Automatic Extraction of Product Information from Multiple e-Commerce Web Sites

**Samiah Jan Nasti, M. Asger and Muheet Ahmad Butt**

**Abstract** With the growth of e-commerce, shopping online has now become a part and parcel of every one's life. The advantage of e-commerce Web sites is that they can reach to a very large number of customers despite of distance and time limitations. The main aim of this paper is to extract the product information from various e-commerce sites. Extraction of such information can help the business organizations to fetch and attract the large number of customers to their Web site and increase profit. So, in this paper, we propose a fully automatic method which will extract and integrate information from multiple e-commerce Web sites in order to improve business decision making. The proposed method is also comparatively better at precision and recall than other methods.

**Keywords** Document Object Model (DOM) tree · Crawling · Clustering · Wrapper generation

## 1 Introduction

The booming of e-commerce in recent years began with the introduction of Automatic Teller Machines (ATM) and Electronic Data Interchange (EDI) [2]. E-commerce, a new and smart way of doing business, is a massively crowded online market, which has partially or fully affected every one's life. Online shopping has revolutionized the shopping behavior of consumers. In India, Amazon, Snap Deal and Flipkart

S. J. Nasti (✉)
Department of Computer Sciences, BGSB University, Rajouri, Jammu and Kasmir, India
e-mail: samiah.mushtaq14@gmail.com

M. Asger
School of Mathematical Sciences and Engineering, BGSB University, Rajouri, Jammu and Kasmir, India
e-mail: m_asger@yahoo.com

M. A. Butt
Department of Computer Sciences, University of Kashmir Hazratbal Srinagar, Srinagar, Jammu and Kasmir, India
e-mail: ermuheet@gmail.com



P. K. Singh et al. (eds.), *Proceedings of ICRIC 2019*, Lecture Notes in Electrical Engineering 597, https://doi.org/10.1007/978-3-030-29407-6_53

are now treated as new showrooms. These Web sites advertise and sell different kinds of products and reach to a large number of users despite of time and distance limitations. With the growing competition in the e-commerce market, customer's information is important to merchant. So, need is to understand the customer's need in this era. The extraction of product information from multiple e-commerce sites is one of the practical applications in the Web mining realm. As we know there are billions of the Web sites on the Web, which differ in their underlying structure, and visual style, but the beauty of the e-commerce Web sites is that, upon a Web page request the information about the products is embedded into the Web page. The information about the products is actually stored in the underlying database and is automatically embedded into the Web pages using scripts. In e-commerce, we mine online shopping stores as much attention is being paid to price comparison, customer attraction, customer retention, etc. Extracting such information from the Web sites can help business organizations in answering questions like identifying profitable customers and identifying those Web sites from where the customers buy frequently. In this paper, we propose an automatic method for extracting product information from various e-commerce sites. The rest of the paper is organized as follows. In Sect. 2, we present related work, in Sect. 3, we present our proposed approach, in Sect. 4 experimental results are shown, and finally, we draw conclusion in Sect. 5.

## 2 Related Work

Ansari [1] proposed an integrated architecture for an e-commerce system with Data Mining. Their system can dramatically reduce the pre-processing, cleaning and data understanding, in knowledge discovery projects.

Perner et al. [5] proposed an architecture according to users' needs and preferences by extending an e-shop into an intelligent e-marketing and selling platform. Their method uses two types of Data Mining techniques, namely classification and clustering.

In e-commerce sites, to predict the user's behavior, a new approach was proposed by [8]. In order to predict the purchase and traversal behaviour of future users, the proposed approach involves extracting of information from integrated data of past users.

Lee et al. [4] propose a model for e-commerce known as integrated path traversal patterns (IPA) and association rules for Web usage mining. This IPA model takes into consideration both the traversing and purchasing behavior of the customers at the same time. This model not only take the traversal forward information of the user, but also takes into account the users backward traversal information, which makes the model accurate and correct for capturing users purchasing and traversal behaviors.

Satokar and Gawali [6] present a personalization system, which depends on features extracted from hyperlinks, for Web search. Their personalization system which uses a weighted URL rank algorithm can help users not only to get relevant Web pages, but also domains the user is interested in it.

Kiruthika et al. [3] discusses the use of association rules in discovering patterns in Web usage mining. They propose a system in which they preprocess the Web server logs, cluster the closely related transactions, and discover the association rules. Their proposed system can help the Web site designers to improve their Web site design.

Todi et al. [7] developed an application which extracts information from e-commerce Web sites for classification of data in order to benefit both customers as well as companies. They use Naïve Bayes and Decision Trees, the two most popular supervised algorithms for classification and compared them. The results show that Decision Trees perform better than Naïve Bayes. Using such kind of applications, customers can understand the qualities of the available products and a competitor can understand how their competitors are priced.

## 3 The Proposed Architecture

We propose an automatic method for extraction of structured Web data records from multiple e-commerce Web sites. Our proposed method introduces a new extraction technique based on the similarity between the DOM tree tag node patterns. The extracted data records can be used further for mining purposes and other manipulations. It consists of five phases viz. Crawling, Pre-Processing, Structural Similarity Calculation, Clustering and Wrapper generation. These five phases are called sequentially by the main method. The overall algorithm of our proposed method and its high level architecture are shown in Figs. 1 and 2, respectively.

According to our need we create a crawler in JavaScript containing two modules only, finder and downloader. The finder receives the URL's of the Web pages and gives them to the Downloader module which in turn downloads the Web pages and

***Algorithm**: Extraction of structured data from multiple websites*

---

*Input: Set of HTML files (WebPageFile) of web documents.*
*Output: Data records*

1. ***Begin***
2. ***for** each WebPageFile*
3. *Crawl all webpages from WWW and store them into local directory.*
4. *Cleanup HTML code and generate DOM trees of webpages.*
5. *Generate Column Similarity Matrix.*
6. *Cluster the structurally similar webpages.*
7. *Extract and store information.*
8. ***end for***
9. ***End***

---

**Fig. 1** Algorithm for extraction of structured data from multiple websites

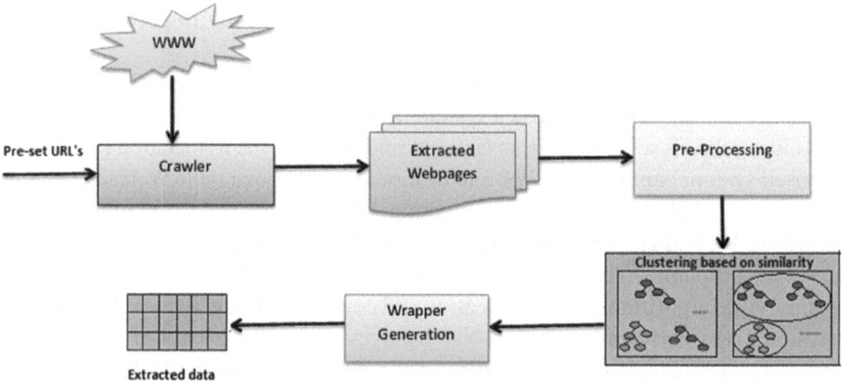


**Fig. 2** Block diagram of proposed method

stores them. The crawler maintains two queues, visited queue (VQ), which stores the visited URL's and unvisited queue, (UQ) which stores the unvisited URL's. Initially the crawler is given a seed URL where from the processing starts. The finder module extracts all the links from the current URL, pushes the current URL to the VQ, passes it to the downloader and also pushes all other links to the UQ. The finder repeats the process recursively and the downloader, downloads the Web pages and stores them in local machine. To clean the Web pages, we use HTMLCleaner-2.2 software. It is an open source HTML parser written in Java. The cleansed Web page is given to dom4jmodule which then constructs the DOM tree for that Web page.

In order to calculate structural similarity we use the Levenshtein edit-distance algorithm to compute similarity between two Web pages ($P_i$) and ($P_j$). We do not take into consideration the text of the Web pages while calculating similarity as we are only interested in their structural similarity. Initially we convert each DOM tree of a Web page into a string $S$. Accordingly, we represent two Web pages as $S_i$ and $S_j$, respectively. While transforming a DOM tree of a Web page into a string, DOM tree is traversed in depth first order and each text node is replaced by a text tag.

In the structural similarity phase the set of Web pages $\{P_1, P_2, \ldots. P_n\}$ can be represented as a string $\{S_1, S_2, \ldots, S_n\}$. We then compute the similarity square matrix $\mathcal{M}_{n*n}$ where $n$ is the number of Web pages and $\mathcal{M}_{ij}$ represents the Normalized edit distance similarity (Ned) between $P_i$ and $P_{j.}$ For calculating the similarity between two pages we would use column similarity which seems more robust. We can calculate the column similarity by using the following Eq. 1. Finally a column similarity matrix is generated where each cell denoted by $\text{pos}_{ij}$ is obtained as Colsim ($P_i$, $P_j$).

$$\text{Colsim}\left(P_i, P_j\right) = 1 - \frac{\sum_{k=1}^{n} \left|\text{pos}_{ik}\right| - \left|\text{pos}_{jk}\right|}{n} \tag{1}$$

We would use Hierarchical Agglomerative Clustering algorithm to group the Web pages. In this method we assign each element to its own cluster, then by computing the similarity between each of the clusters, we join the most similar clusters resulting into as many clusters as preferred. That means we cluster those groups where inter-element similarity is high. This inter-element of a set Ø, is measured by the auto—similarity formula. Two thresholds namely auto-wise similarity threshold denoted by ŧ1 and pairwise similarity threshold ŧ2 are measured before forming a new group. The algorithm starts with a collection $P$ of $n$ singleton clusters where each cluster represents a single Web page. We repeat the algorithm till only one cluster is left. Initially we find a pair of clusters ($c_i$, $c_j$) within which inter-element similarity is high. The inter-element similarity of a set, $S$ (φ) is calculated by following Eq. 2. Then merge the clusters ($c_i$, $c_j$) into one new cluster ($c_{i+j}$) and remove ($c_i$, $c_j$) from collection $P$ and add cluster ($c_{i+j}$) to it. Repeat the process till a Hierarchical tree known as Dendrogram is formed. The pseudo code for bottom-up clustering is given in Fig. 3

$$S(\varphi) = \frac{2}{|\varphi||\varphi - 1|} \sum_{P_i, P_j € \varphi} \text{Colsim}\left(P_i, P_j\right) \tag{2}$$

After clustering of Web pages of the same type, we use a structure based approach for automatic generation of wrappers. The Algorithm for Wrapper Generation is given in Fig. 4.

***Algorithm:** Clustering*

```
Input: pageset, ŧ1 , ŧ2
Begin
  Suppose D_ij be the distance between P_i and P_j in the pageset
  Let C be set of groups
  Initialize each page to a group
   do while (C >=1)
    Choose two webpages w1,w2 ∈ C having high auto-similarity
    Compute S(w1 U w2)
     if(S(w1 u w2) >ŧ1&&Colsim(P_i,P_j)>ŧ2 )then
       Remove w1 and w2 from C
         Let ϑ = w1 U w2
         Insert ϑ into C
        else
         Break
        endif
      endwhile
    return C
  End
```

**Fig. 3** Clustering algorithm

***Algorithm:*** *Wrapper Generation*

*Input:pageset, ŧ3*

```
Begin
Set temp ← Page having high auto-similarity and
  maximum potential nodes
Remove the selected template from the pageset
Apply sorting technique on pageset in descending order
Setcount=1
  do while(each page in pageset)
    Ned(template, page)
    S = getMatchNodes(template, page)
      for each pair (n1, n2) in S
          n1.Setcount = n2. Setcount+1
          Add nodes to template when not mapped (n1,n2)
        endfor
    endwhile
  C = ceil ((pageset.count+1)* ŧ3)
  Ignore the nodes where Setcount<C
  Return temp
End
```

**Fig. 4** Wrapper generation algorithm

The algorithm takes as input a set of Web pages and compares the Web pages to find the similarities and differences between them and generating a wrapper (union free regular expression) in this process. Initially from a cluster a Webpage is chosen as a wrapper say *W*, which satisfies the Eq. 3.

$$\max_{i=0}^{n} \frac{\sum_{j=0}^{n}\left(1 - \mathrm{Ned}\left(P_i, P_j\right)\right)}{n} \tag{3}$$

## 4 Experimental Setup and Evaluation

The approach has been implemented in Java Programming language and the experiments were conducted on laptop with 64 bit Windows 7 operating system, i5 CPU @ 3.1 GHZ, 8 GB RAM and 7200 RPM hard drive. We use precision, recall and F-score standard measures to evaluate our proposed approach. These evaluation metrics are most popular measures for evaluating Information Retrieval systems. The equations for calculating precision, recall and F-score is given by following equations:

$$\text{Precision} = \frac{\text{number of correctly extracted}}{\text{number of extracted}} \quad (4)$$

$$\text{Recall} = \frac{\text{number of extracted}}{\text{number of accessed}} \quad (5)$$

$$\text{F-score} = \frac{2 \times \text{precision} \times \text{recall}}{\text{precision} + \text{recall}} \quad (6)$$

Table 1 shows the experimental results of our proposed method. For each Web site, we start from the base URL and we set our system to 150 as target Web pages to extract. As can be seen from the table below, our proposed method achieves satisfactory precision and recall on Web data under experiment.

We use these datasets to evaluate structured Web data extraction systems. These datasets contain template generated Web pages (web pages filled with data from underlying databases). Figure 5 shows the precision and recall of these datasets and clearly reveals that our approach achieves perfect precision and nearly perfect recall.

**Table 1** Experimental results of our proposed approach

| Website | Web pages accessed | Web pages extracted | Web pages correctly extracted | Recall (%) | Precision (%) |
|---|---|---|---|---|---|
| Snapdeal.com | 150 | 139 | 136 | 92.60 | 97.84 |
| Flipkart.com | 150 | 110 | 104 | 73.33 | 94.54 |
| Homeshop18. com | 150 | 135 | 129 | 90.00 | 95.55 |
| Myntra.com | 150 | 124 | 106 | 82.66 | 85.48 |
| Bigbuy.com | 150 | 139 | 127 | 92.66 | 91.36 |
| eBay.com | 150 | 132 | 125 | 88.00 | 94.69 |
| Currys.co.uk | 150 | 127 | 123 | 84.66 | 96.85 |
| Argos.co.uk | 150 | 130 | 115 | 86.66 | 88.46 |
| Craftsvilla.com | 150 | 130 | 126 | 86.66 | 96.92 |
| Kashmirbox.com | 150 | 125 | 109 | 83.33 | 87.20 |

**Fig. 5** Precision and recall results of our approach with other datasets

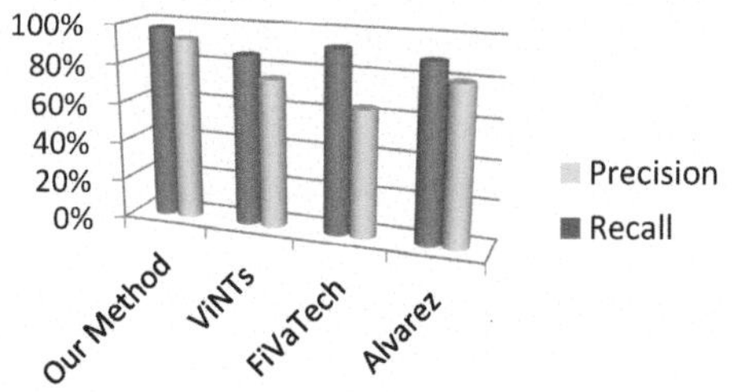

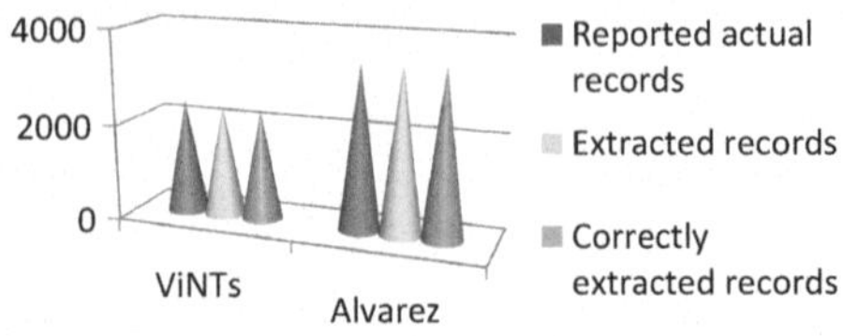


**Fig. 6** Data records reported with ViNTs-2 and Alvarez datasets

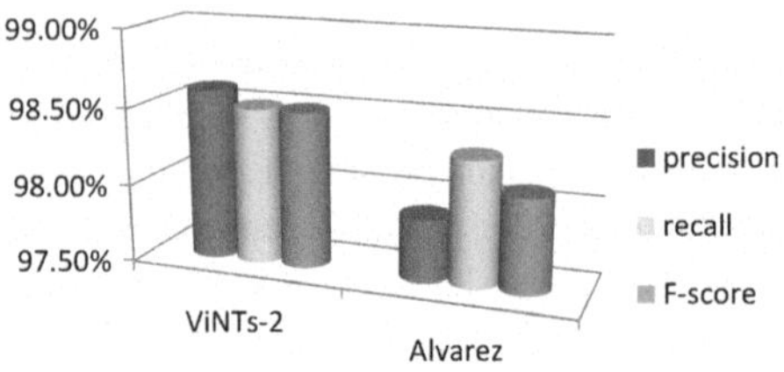


**Fig. 7** Precision, recall and F-score with ViNTs-2 and Alvarez datasets

Also extraction results of ViNTs-2 and Alvarez Datasets are given in Figs. 6 and 7, respectively.

## 5 Conclusion

We have programmed our method in Java language. Our method can be very useful and effective to our online shopkeepers. As the market of e-commerce is changing very fast. So, the shopkeepers are paying more attention on price analysis. In this chapter, we introduce a new method for extracting data records from structured e-commerce Web pages. Our method works in five phases DOM tree creation, similarity measure, clustering, wrapper generation and extraction of data. Initially the Web pages are downloaded using a Web crawler and stored in a local machine. Then the HTML Web pages are converted to DOM trees. By analyzing the DOM trees, the Web pages of the same type are clustered into groups by using structural similarity of Web pages and clustering of the Web pages is done using hierarchical agglomerative clustering. We then analyze the Web pages from each cluster to generate the extraction rules and finally the data is extracted and integrated. The proposed method makes full use of the structural similarity of Web pages. Experimental results show that our proposed method effectively extracts data records from famous e-commerce Web sites.

## References


1. Ansari, S.L.: An integrated approach to control system design. Accouting Organ. Soc. Sci. Direct **2**(2), 101–112 (1977)
2. Kalakota, R., Whinston, A.B.: Electronic Commerce: A Mangers Guide. Addison-Wesley Professional (1977)
3. Kiruthika, M., Jadhav, R., Dixit, D., Rashmi, J., Nehete, A., Khodkar, T.: Pattern Discovery Using Association Rules. Navi Mumbai, India (2011)
4. Lee, Y.-S., Yen, S.-J., Tn, G.-H., Hsieh, M.-C.: Mining travelling and purchasing behaviors of customers in electronic commerce environment. In: IEEE International Conference on e-Technology, e-Commerce and e-Service (2004)
5. Perner, P., Gollmann, D., Zhou, J.: Information and communication society. In: 5th International Conference, ICICS (2003)
6. Satokar, K., Gawali, S.: Web search result personalization using web mining. Int. J. Comput. Appl. (2010)
7. Todi, A., Agrawal, A., Taparia, A., Lakhmani, N., Shettar, R.: Classification of E-commerce data using data mining. Int. J. Eng. Sci. Adv. Technol. **2**(3), 550–554 (2012)
8. Vallamkondu, S, Grunewald, L.: Integrating purchase patterns and traversal patterns to predict HTTP requests in E-Commerce sites. In: IEEE International Conference on E-Commerce, pp. 256–263 (2003)

# Security and Privacy

# Performance Evaluation and Modelling of the Linux Firewall Under Stress Test

**Nikita Gandotra and Lalit Sen Sharma**

**Abstract** *Iptables* is a stateful packet filtering firewall in Linux that monitors ingress and outgress traffic. The filtering is performed based on rules which are conditions predetermined by the network administrators. This paper investigates the performance of *Iptables* with different rule sizes (200, 500, 1000, 5000 and 10,000) and high traffic rates for different time durations. An experimental set-up is established for evaluating the performance of *Iptables* under stress by varying the packet rates from, viz., 1000 to 8000 PPS and different time durations (30–120 s). The performance is recorded on key parameters: CPU utilisation, response rate, packet dropped, packet processing time, throughput and bandwidth. These parameters reflect the sensitivity of the firewall for managing high rates of network traffic. ClassBench is used to generate rule sets of different sizes that imitate the real-life rule sets, and the network traffic is generated by DITG, a traffic-generating tool. Finally, a mathematical model is developed that can estimate the performance of the firewall in different traffic scenarios. Also, the proposed model is tested by performing validation tests on real test bed and shows less than 10% relative error.

**Keywords** *Iptables* · Netfilter · DITG · Rule set · ClassBench · Packet rate

## 1 Introduction

*Iptables* manages Netfilter module [1] that performs stateful packet inspection (SPI) by monitoring the contents of the packet headers. *Iptables* is the only stateful packet filtering firewall inside the Linux 2.4.x and later versions. It is the reformed successor of the ipchains (Linux 2.2.x) and ipfwadm (Linux 2.0.x) systems. *Iptables* has become extremely popular among researchers due to its reliability, robustness, flexibility and huge scope for customisation. Since *Iptables* is an open-source firewall and provides

N. Gandotra (✉) · L. S. Sharma
Department of Computer Science & IT, University of Jammu, Jammu, J&K, India
e-mail: nkt_2201@yahoo.co.in

L. S. Sharma
e-mail: lalitsen@yahoo.com



P. K. Singh et al. (eds.), *Proceedings of ICRIC 2019*, Lecture Notes in Electrical Engineering 597, https://doi.org/10.1007/978-3-030-29407-6_54

high dependability for handling burst traffic [2], thus it was chosen upon the other software firewalls. *Iptables* provides facilities like network address translation and port translation (NAT/NAPT), system logging of network activities, rate limiting the traffic and packet manipulation facility like altering the TOS byte of the IP header. It works by interacting with the Netfilter hooks in the kernel's network stack. As every packet progresses through the network stack, it triggers these hooks allowing the kernel modules associated with *Iptables* to interact with the traffic at certain key points. *Iptables* controls the behaviour of traffic on the basis of firewall rules laid by the network administrator [3]. If a packet matches all the conditions specified in the rule, then the rule is applied; otherwise, default policy is applied. The default policy represents the normal behaviour of the firewall when there are no matching rules in the rule space. Thus, *Iptables* ensures that the ingress and outgress traffic conforms to the desired behaviour of the firewall.

The *Iptables* rules are stored in the form of tables, and each table is implemented as a separate module. There are three primary tables used in *Iptables*: FILTER table, NAT table and a MANGLE table [4]. The packet filtering is performed by FILTER table, and it decides whether to ACCEPT or DROP the packets; the network address translations and demilitarised zone declaration are implemented by NAT table, and the TOS, TTL and security are implemented by MANGLE table. Most commonly, FILTER table is used for implementing a security policy in the network.

This paper focuses precisely on investigating the performance of *Iptables* under stress and modelling it. Stress tests are performed by varying the packet rates (1000–8000 PPS) for different rule-set sizes (200, 500, 1000, 5000 and 10,000) and for different time durations (30–120 s). An experimental set-up was established for conducting real experiments to determine the impact of rule sets and packet rates on the performance of Linux firewall. The rule sets are generated by using ClassBench [5], and a benchmark tool for filter set generation and the network traffic is generated by using distributed Internet traffic generator (DITG) [6]. We found that the packet rates and the size of rule sets had a significant impact on the performance of firewall and the performance is traced out on the basis of key parameters: CPU utilisation, response rate, packet loss ratio, packet processing time, throughput and bandwidth. These are important indicators of firewall's performance and are useful in development of meaningful technique for characterising the performance of firewalls. Based on the observations from the experiments, we developed a statistical model to estimate the performance of the firewall so that the behaviour of the firewall could be predicted when deployed across different rule sets and traffic profiles. The model has been validated with new test cases, and it was found that the estimate of the performance of the firewall is within a relative error of 9% of the measured values.

The main contributions of this paper are: (1) None of the previous works presented a comprehensive investigation on the impact of packet rate and time duration on the performance of *Iptables*; (2) the present study examines the performance of *Iptables* for five wide-ranging rule-set sizes in contrast to single rule-set size; (3) the key parameters such as CPU utilisation, packet processing time and response time were not considered before for performance evaluation; (4) the prediction model proposed in this paper is relatively a new approach for quantifying the performance of firewall. Also, it aptly covers all the aspects of network and system performance of the firewall.

The paper is divided into the following sections: Sect. 2 discusses the background and related work that has motivated our study, while Sect. 3 presents the experiment plan and set-up in detail. In Sect. 4, the obtained performance results from the real experiments are documented and discussed. Section 5 presents our statistical model, and Sect. 6 validates our model. Finally, in Sect. 7, conclusions are drawn.

## 2 Related Work

Only a little work was published on the firewall performance modelling and analysis. Most of the existing works laid much emphasis on rule conflict detection [7, 8], optimisation of the firewall rule sets [9, 10] and developing of new architectures for improving the performance of the firewall [11, 12].

In [13], correctness and performance tests on *Iptables* have been performed. For checking the correctness, various rules were set up and frames were sent to examine the accept/reject behaviour of the firewall while for performance testing, throughput and delay were measured as a function of rule base size. The tests were limited to maximum of 500 rules in the rule base.

Niemann et al., in [14], studied the impact of ACL and QoS rules on the throughput rate per client in distributed client–server applications. The number of clients ranged from 5 to 320 and frame size from 64 to 1024 bytes. The maximum 1280 rules were considered for the experiments.

Wang et al. [15] examined the performance of two firewalls in an experimental study. Their experimental study is based on two different types of firewalls: ModSecurity and *Iptables* with the rule-set size of 6700 rules. The experiments evaluate the capacity of the firewall, impact of CPU and memory usage on performance of firewall measured by varying packet length from 512 to 7680 bytes and request rate from 500 to 3500 requests/s. They inferred that the *Iptables* acts better than ModSecurity in most cases.

Salah et al. [16] discussed the performance of network firewall based on rule positions for which an analytical model was proposed based on Markov chain. The performance of Linux Netfilter firewall was analysed for 10,000 dummy rules in the proposed model when subjected to normal traffic and DoS traffic. The model was verified using simulations and real experiments. The dummy rules were only based on MAC addresses that may not follow the real-life filter sets.

Lyu and Lau in [17] studied Linux TIS firewall packages for security and performance in terms of transaction time and latency. They designed seven security levels for security evaluation and for the performance, and HTTP and FTP services were used for measuring the performance degradation.

In [18], Cisco ASA, Packet Filter and Checkpoint SPLAT firewalls have been evaluated for their performance and security. For evaluating the performance, throughput and the maximum number of concurrent connections were considered while for security simple tests were performed to check the firewall's resistance. It was inferred that Cisco ASA shows better results than other firewalls.

Hayajneh et al., in [19], evaluated the performance of network and personal firewalls for security and performance. The performance evaluation was done on the basis of throughput, jitter, delay and packet loss, while the security evaluation was done for different types of attacks. In particular, Cisco ASA 5510, Packet Filter, Comodo and ZoneAlarm were considered, and the performance tests were done for only two packet sizes: 512 and 1460 bytes. It was inferred that Cisco ASA achieved better performance in comparison with other firewalls.

Cheminod et al. [20] studied the performance of a general-purpose firewall Fortinet 2016 and two commercial firewalls: Belden—Hirschmann 2016 and Moxa 2016 for industrial uses. In particular, the performance was evaluated for TCP/IP, Modbus/TCP protocols. They concluded that Fortinet 2016 performs better in terms of throughput and latency but does not support Modbus protocol, while the other two firewalls support Modbus but exhibit twice the latency of Fortinet 2016.

In [21], Beyene et al. assessed the performance of two firewalls SonicWall E5500 and Fortinet FortiGate-ONE on the basis of the throughput provided by the firewall. For experiments, the packet size was varied from 64 to 1518 bytes on TCP and UDP connections. Also, they developed a model SyFi to compute the throughput of the particular stateful firewall based on CPU utilisation and validated their model with another firewall HP TMSzl.

## 3 Experiment Plan

### *3.1 Experimental Set-up*

The experimental set-up consisted of four PCs connected via D-Link Web Smart DGS-1210-16 16-port switch forming a LAN. Figure 1 shows the test bed along with their corresponding IP addresses. For generating the network traffic, distributed Internet traffic generator (DITG) was used on three machines. It is an open-source traffic generator [22] which generates IPv4 and IPv6 packets and provides facility of measuring, monitoring and analysing the network statistics. It also generates different types of traffic (TCP, UDP and ICMP) with varied packet sizes and packet rates [23]. It uses NTP for monitoring the link statistics, and this does not affect the experiments. Here, the traffic of type TCP was generated with packet size of 1024 bytes since it represents majority of the traffic online. The system specifications of the PCs are discussed in Table 1.

The Linux firewall is located at 192.170.1.119 and is configured to varied filter sets under different test scenarios. ClassBench [5] is used to generate the filter sets. It is a benchmark tool for filter set generation that produces rule sets of varying sizes which represent a replica of real-life rule sets and has no duplicate rules. The

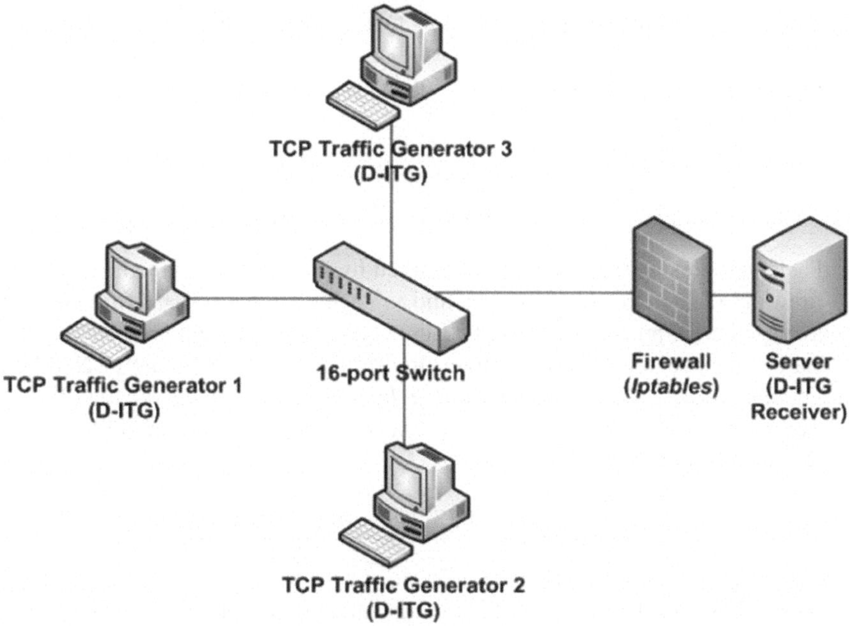


**Fig. 1** Experimental set-up

**Table 1** System specifications

| Machine | Description | Specification |
|---|---|---|
| TCP traffic generator 1 | Intel® Core™ i3-530 CPU @ 2.93 GHz CPU; 8 GB RAM | Ubuntu 16.04 LTS with DITG |
| TCP traffic generator 2 | Intel® Core™ i3-3110 CPU @ 2.40 GHz; 2 GB RAM | Ubuntu 16.04 LTS with DITG |
| TCP traffic generator 3 | Intel® Core™ i7-4770 CPU @ 3.40 GHz; 4 GB RAM | Ubuntu 16.04 LTS with DITG |
| Firewall/server | Intel® Core™ i3-3110 CPU @ 2.40 GHz; 8 GB RAM | Ubuntu 16.04 LTS |

following is a rule set with 200 rules:

$$.db\_generator - bc < rulefile\_name > 200\,2 - 0.5\,0.1 < filterset\_name >$$

The performance of the Linux firewall is evaluated on the following key parameters:

1. CPU utilisation (in %): the amount of CPU utilised by the Linux firewall during the tests. It is computed by using SAR Linux utility.

2. Response rate (times/s) is measured as rate at which replies are sent from the server.
3. Packet loss ratio (in %) furnishes the total packets dropped by the firewall and is computed as:

$$\text{PLR} = \frac{\text{Packets Dropped by Iptables} \times 100}{\text{Total Packets Received by NIC}}$$

4. Packet processing time (in μs) is measured by the amount of time taken by the firewall to search the rule space and find a matching rule. In order to measure the average kernel processing time, the Linux code for timestamp is inserted at the start of packet reception by Netfilter processing and the end point of Netfilter, i.e. in *ip_local_deliver()* and *ip_local_deliver_finish()* in *ip_input.c* file [24].
5. Throughput is measured in terms of packets processed per second (PPS).
6. Bandwidth (in Mbps)

$$\text{Bandwidth} = \frac{\text{PPS} \times 8 \times \text{size of (Packet)}}{1024 \times 1024}$$

where PPS is packets processed per second and the size of packets is in bytes.

### 3.2 *Experiment Plan*

The experiments evaluate the impact of rule-set size, packet rate and time duration on the performance of the Linux firewall. The rules are used on the INPUT chain of FILTER table to restrict the incoming traffic. The default policy is set as ACCEPT. For every incoming packet, the firewall compares it with the rules in the rule set. Finding no matching rule, the firewall applies default policy of ACCEPT and forwards it to server. Thus, firewall is set to work in the worst case when the matching rule is placed on the end of the rule space. The experiments are organised into four test scenarios and are listed in Table 2. Experiments are organised in four testing scenarios, namely Test 1, Test 2, Test 3 and Test 4. Each testing scenario indicates different time durations for which the test is executed. The experiments in a test scenario are executed for five rule-set sizes that in turn consist of eight test cases. These test cases represent the packet rate at which the traffic is sent to the server, starting from 1000 packets per second to 8000 packets per second (PPS), and are increased by 1000 PPS.

## 4 Observations and Results

The recorded key parameters for the experiments are documented in this section. There are totally 160 test cases for which the real experiments were performed.

**Table 2** Test scenario

| Test | Time duration (sec) | Rule-set size | Packet rate (packets/sec) | Test cases |
|---|---|---|---|---|
| 1 | 30 | 200, 500, 1000, 5000 & 10,000 | (1000–8000) | 40 |
| 2 | 60 | 200, 500, 1000, 5000 & 10,000 | (1000–8000) | 40 |
| 3 | 90 | 200, 500, 1000, 5000 & 10,000 | (1000–8000) | 40 |
| 4 | 120 | 200, 500, 1000, 5000 & 10,000 | (1000–8000) | 40 |

For simplicity, the results and observations are organised into the following three categories to study the effect of packet rate, time duration and rule-set size on the performance of *Iptables*.

## 4.1 *Impact of Packet Rate and Rule-Set Size on the Performance of* Iptables

In order to study the impact of packet rate and rule-set size on the performance of *Iptables*, the observations taken on a same time duration are considered. The recorded observations of all the test cases for time duration of 90 s, i.e. Test 3, are discussed in Figs. 2, 3, 4, 5, 6 and 7. Figure 2 compares the connections established per second for different packet rates on different rule-set sizes. It can be clearly seen that as the rule-set size increases, the response rate decreases steadily. Also, with the increase in packet rate, the response rate has also increased. In Fig. 3, the CPU utilisation for different rule-set sizes is depicted. It shows that the CPU utilisation increases sharply with increase in rule-set size and packet rate. Figure 4 compares the packet

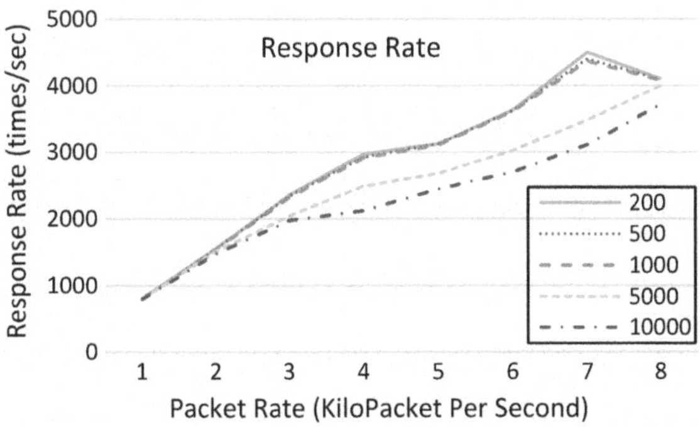


**Fig. 2** Response rate for different packet rates for Test 3

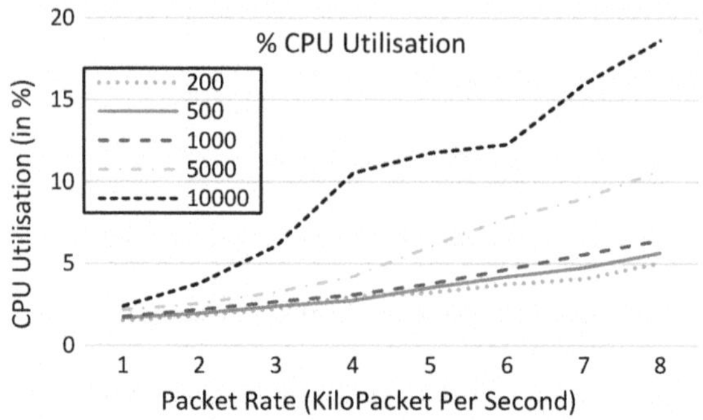


**Fig. 3** CPU utilisation (in %) for different packet rates for Test 3

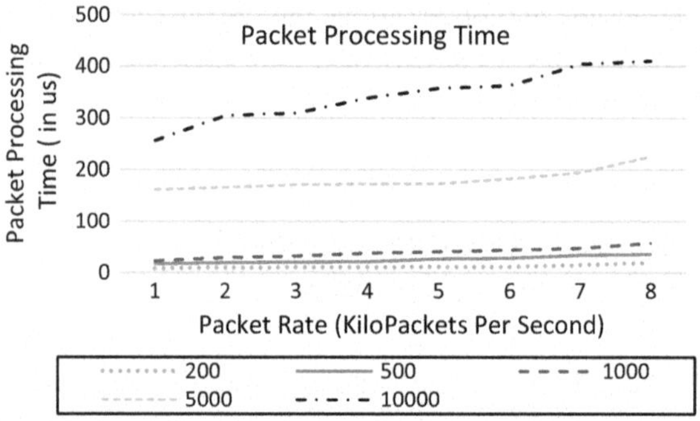


**Fig. 4** Packet processing time for different packet rates for Test 3

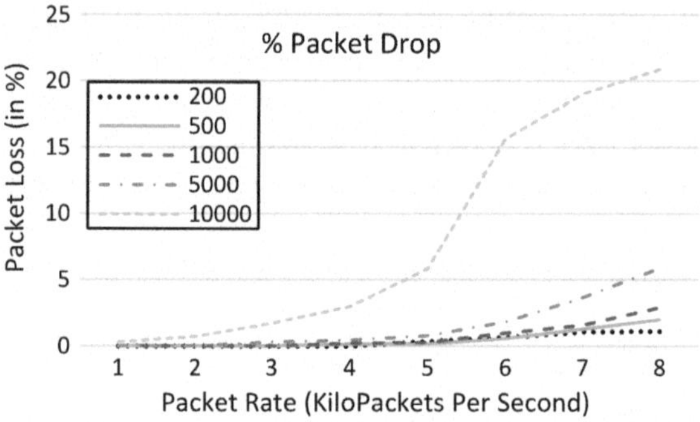


**Fig. 5** Packet loss ratio for different packet rates for Test 3

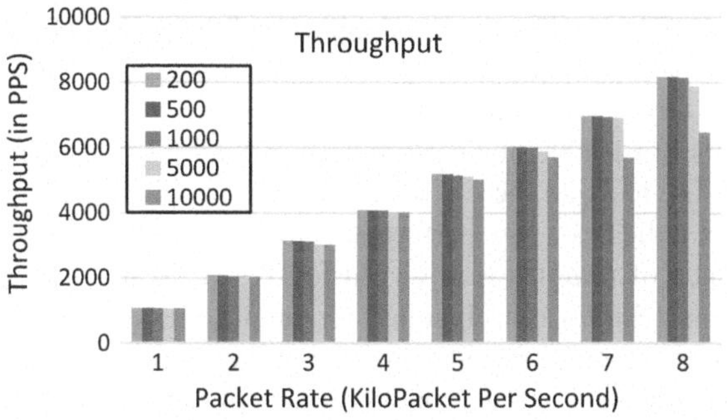


**Fig. 6** Throughput (in Mbps) for different packet rates for Test 3

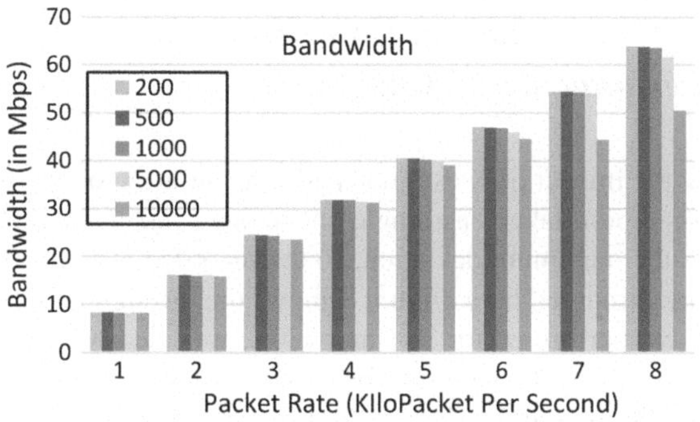


**Fig. 7** Bandwidth for different packet rates for Test 3

processing time taken for different rule sets and packet rates. It clearly indicates that the packet processing time is increased with the increase in rule-set size while for different packet rates, the values lie in close vicinity and therefore show only a little variance that can be ignored. The graph in Fig. 5 depicts the packet drop % with different rule-set sizes. It shows that when the size of rule set increases, the percentage of the packets dropped by the firewall sharply increases. Also, there is a marked increase in packet drop % with increase in packet rate. Similarly, Figs. 6 and 7 show the comparison of throughput and bandwidth with different rule-set sizes. It is seen that the throughput and bandwidth of the firewall increase with increase in packet rates while it shows a constant degradation with increase in rule-set size. Thus, from these graphs, we can infer that there exists a steady relationship between the performance of the firewall and the rule-set size and packet rate.

Thus, we can conclude that *Iptables* shows an excellent performance for rule-set size 200, 500 and 1000 as only small amount of CPU is utilised even at higher traffic rates and the packet loss of 0–2% is indicated. However, the throughput and bandwidth of the firewall has decreased slightly when the rule-set size is increased to 1000 rules. The performance of *Iptables* further gets decreased on the rule set of size 5000 as the CPU utilisation has doubled and the packet processing time has significantly increased from previous test cases. Also, there is a constant decrease in the response rate, throughput and bandwidth of the firewall. Similarly, increasing the rule-set size further to 10,000, the performance of *Iptables* gets degraded as 20% of the total packets are lost and there is a steady decrease in all the key parameters of the firewall. This indicates a low performance by the firewall. A similar behaviour in the performance of *Iptables* is observed for Test 1, Test 2 and Test 4. Thus, it is inferred that the *Iptables* is well suited for small to moderate rule sets. Also, in smaller rule sets, the *Iptables* starts to drop the packets when the packet rate reaches to 4000 packets per second or more, but as the size of rule set increases, the *Iptables* begins to drop packets early.

## *4.2 Impact of Packet Rate and Time Duration on the Performance of* Iptables

In order to study the impact of packet rate and time duration on the performance of *Iptables*, the observations taken on a single rule set are considered. Since rule-set size of 10,000 shows the maximum deviation, so the recorded tests for rule-set size for 30 s and 120 s are discussed here. Figure 8 shows the throughput loss that occurred when the tests were performed for 30 s and 120 s. It is clearly seen the throughput has decreased when the tests were run for longer duration. Also, with the increase in packet rate, the throughput has increased. Likewise, Fig. 9 compares the number of packets dropped by the firewall when tests were run for 30 s and 120 s. With the increase in time, the number of packets dropped by the firewall has significantly increased and the rise in packet rate has clearly affected the performance of *Iptables* as it begins to drop greater number of packets with the increase in packet rates.

Figures 10 and 11 compare the CPU utilisation and packet processing time taken by the firewall, respectively. The CPU utilisation has increased marginally for longer duration of test, while the packet processing time has also increased. Packet processing time shows little or no variance with different packet rates as obtained values are

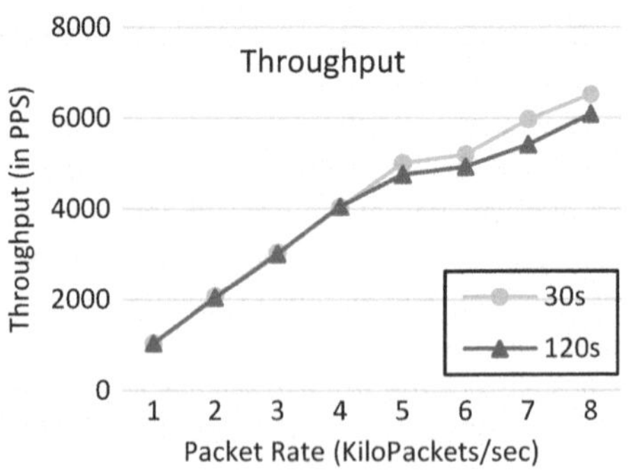


**Fig. 8** Throughput for different packet rates at 30 s and 120 s

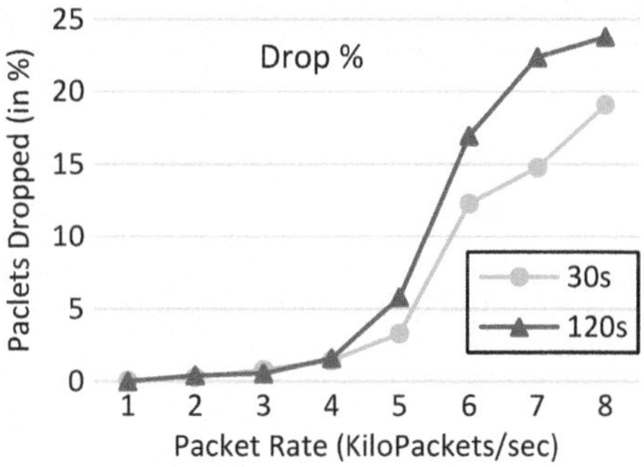


**Fig. 9** Packet loss ratio (in %) for different packet rates at 30 s and 120 s

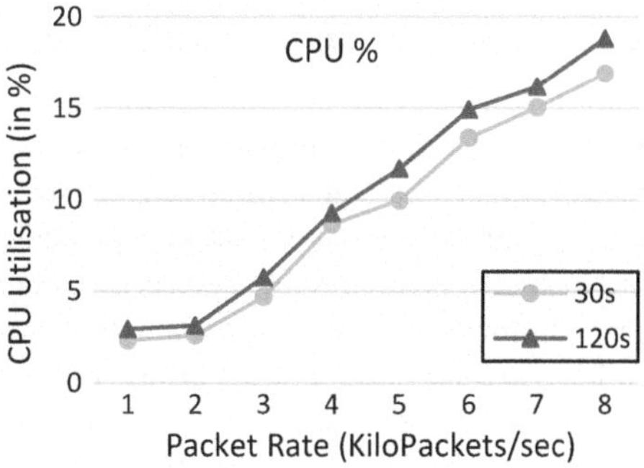


**Fig. 10** CPU utilised for different packet rates at 30 s and 120 s

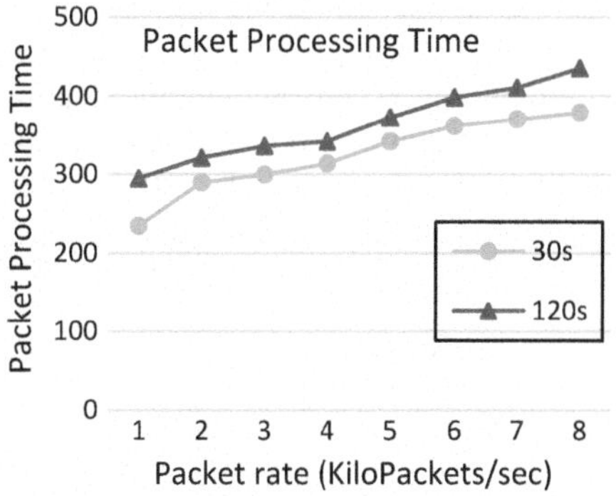


**Fig. 11** Packet processing time for different packet rates at 30 s and 120 s

closely related and shows a linear behaviour. Thus, from Figs. 8, 9, 10 and 11, it can be inferred that there is a substantial degradation in the performance of *Iptables* with respect to time duration and this is probed further in next case.

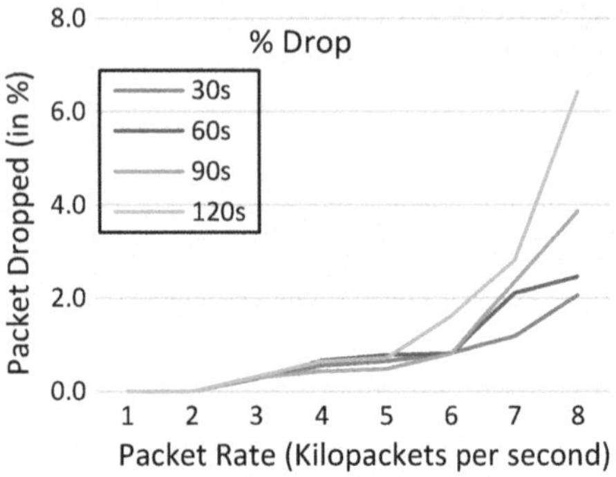


**Fig. 12** Packet loss ratio for different packet rates with 5000 rules

## *4.3 Impact of Time Duration on the Performance of* Iptables

In order to study the impact of time duration on the performance of *Iptables*, the observations taken on a single rule-set size of 5000 over different time durations are considered. Figure 12 compares the percentage of the packets dropped by the firewall. Up to packet rates of 6000 packets sent per second, there is only a marginal increase in the number of packets lost when the time duration of the tests was increased for 30 to 120 s in a step size of 30 s. Further increasing the packet rate, the number of packets lost by the firewall increases. This shows more dependence on the performance of *Iptables* on packet rates than time duration. In Fig. 13, the CPU utilisation by the firewall taken over different time durations is compared. There is only a slight increase in the CPU utilisation, and it shows no significant deviation on time duration.

Similarly, in Figs. 14 and 15, the performance of *Iptables* taken over different time durations is compared for the response rate and packet processing time. Though there exists some marginal difference for different time durations, it is of very little significance. Thus, it can be inferred that the firewall exhibits a higher relationship to packet rate than the time duration.

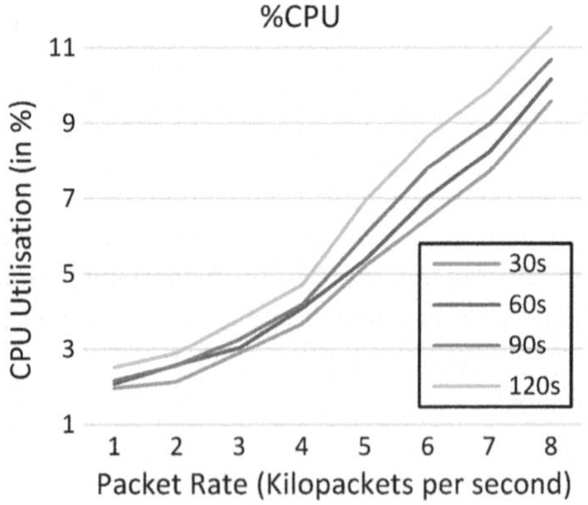


**Fig. 13** CPU utilised for different packet rates with 5000 rules

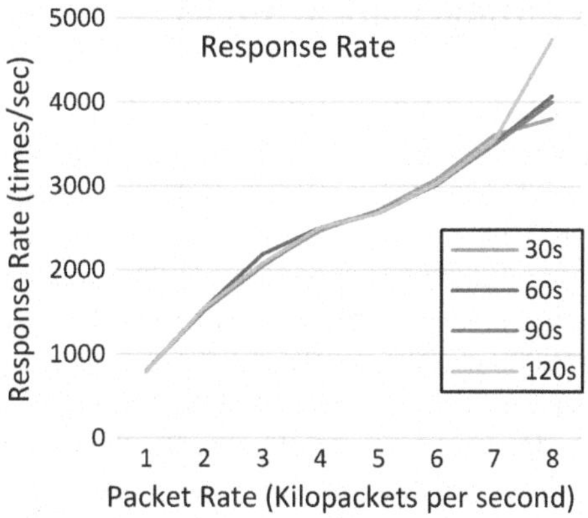


**Fig. 14** Response rate for different packet rates with 5000 rules

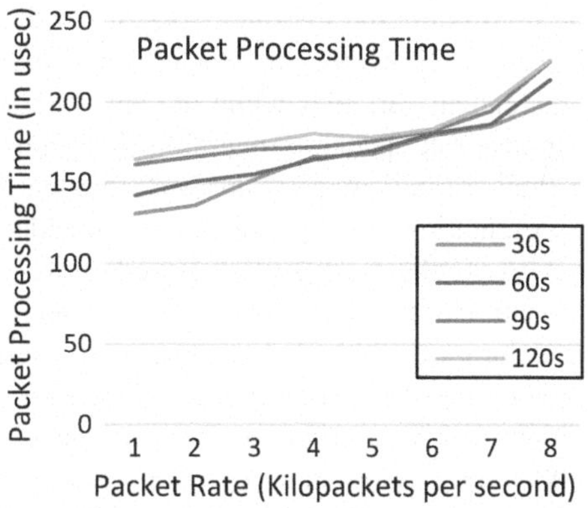


**Fig. 15** Packet processing time of different packet rates with 5000 rules

From all the cases discussed above, we conclude that there exists a significant relationship between the rule-set size, packet rate and the performance of the firewall but no such relationship exists for time duration; i.e. the performance of *Iptables* is independent of the time duration for which it is executed. The key parameters CPU utilisation, response rate, throughput and bandwidth exhibit a direct relationship with packet rate, while an inverse relationship exists with rule-set size. Likewise, the packet processing time depends only upon the size of rule set as it is the amount of time taken by the firewall to search the rule space to find a matching rule and thus exhibits a direct relationship with it. Correspondingly, when the CPU utilisation increases, the firewall takes longer to process the packets, so the number of packets lost by the firewall also increases. Therefore, the packet loss ratio shows a significant relationship with CPU utilisation instead of packet rate and rule-set size.

## 5 Mathematical Model

On the basis of observations from the real experiments, we develop a statistical model so that the behaviour of the firewall when deployed across different types of networks could be predicted. For example, in large networked enterprises, the firewall usually has 1000s of rules defined for security or as access control list with packet rate up to 5000 packets per second. For commercial application of firewall, it could be configured to a rule-set size of 10,000 rules or more with traffic flowing at packet rate of 8000 packets per second or more. Correspondingly, in the medium-to-small-scale networks, rule-set size of up to 1000 rules at slower packet rates is used. Thus, our model can prove useful to predict the performance of the firewall.

The statistical model is based on regression analysis of the observed data and applied using the Statistical Package for Social Sciences software (SPSS) version 17.0 (IBM SPSS software) [25]. The key parameters were analysed using multiple regression and various curve-fitting functions. As the performance of *Iptables* depends only on the rule-set size and packet rate, so they are considered as the independent variables while the key parameters CPU utilisation, packet loss ratio, response rate, packet processing time and throughput are the dependent variables as they are indicators of the performance of *Iptables*. For simplicity, the symbolic representation of the parameters is considered and these are defined in Table 3.

**CPU Utilisation (in %)**: Based on the observations, it was found that the CPU utilisation of the firewall depends only on the rule-set size and packet rate and not on time duration. Table 4 shows the model summary of the relationship between CPU utilisation and the rule-set size and packet rate. The correlation coefficient (R) shows a high positive correlation. Also, the level of significance for our observed data is 0.000 which is less than 0.05. Thus, there is a significant relationship between CPU utilisation and rule-set size and packet rate. So, the CPU utilisation can be estimated by the following equation

$$\mathrm{CPU} = -1.41 + 0.001 * \mathrm{RS} + 0.001 * \mathrm{PR} \tag{1}$$

**Packet Processing Time (in μs)**: Packet processing time (PPT) furnishes the amount of time taken by the firewall to search the rule space to find a matching

**Table 3** List of parameters used in the model

| Symbol | Parameter |
|---|---|
| RS | Rule-set size |
| PR | Packet rate |
| CPU | CPU utilisation |
| PPT | Packet processing time |
| PLR | Packet loss ratio |
| CPS | Response rate |
| T | Throughput |

**Table 4** Model summary for different key parameters

| | Predictors | *R* | *R* square | Adjusted *R* square | Std. error of estimate |
|---|---|---|---|---|---|
| CPU utilisation | (Constant), pckt_rate, rule_set_size | 0.887 | 0.786 | 0.784 | 1.68346 |
| Packet processing Time | (Constant), rule_set_size | 0.980 | 0.961 | 0.961 | 25.5188328 |
| Packet loss | CPU | 0.927 | 0.859 | 0.857 | 1.792 |
| Response rate | (Constant), pckt_rate, rule_set_size | 0.970 | 0.940 | 0.940 | 274.49622 |
| Throughput | (Constant), pckt_rate, rule_set_size | 0.992 | 0.983 | 0.983 | 287.57327 |

rule. It directly depends upon the number of rules, i.e. the size of rule set. Thus, the packet processing time can be computed on rule-set size and Table 4 shows its model summary. The correlation coefficient (*R* value) shows a high positive correlation, and the level of significance is $0.000 < 0.05$. Thus, the packet processing time can be computed as

$$\mathrm{PPT} = 6.873 + 0.034 * \mathrm{RS} \quad (2)$$

**Packet Loss Ratio (in %)**: The packet loss is highly influenced by % CPU utilised by the firewall. When the CPU utilisation increases, the firewall takes longer to process the packets and thus the number of packets lost by the firewall also increases. Thus, packet processing time aptly describes the packet loss ratio. The packet loss ratio fits the cubic model, and the model summary is shown in Table 4. The *R* value shows a high positive correlation between the packet processing time and the packet loss ratio. Also, the level of significance is 0.000, i.e. $< 0.05$; i.e. the model is a good fit. Thus, the packet loss ratio (PLR) can be computed as follows:

$$\mathrm{PLR} = 0.524 - 0.422 * \mathrm{CPU} + 0.096 * \mathrm{CPU}^2 \quad (3)$$

**Response Rate (times/s)**: Fig. 2 clearly depicts that as the rule-set size increases, the response rate decreases steadily. Also, with the increase in packet rate, the number of responses (connections) is increased. So, the response rate is estimated based on rule-set size and packet rate, and the model summary is shown in Table 4. The *R* value is 0.97 which shows a highly positive correlation. The level of significance is 0.000, i.e. $< 0.05$. So, response rate can be estimated by the following:

$$\mathrm{CPS} = 835.398 - (0.064 * \mathrm{RS}) + 0.46 * \mathrm{PR} \quad (4)$$

**Throughput (in PPS)**: The case 1 discussed above clearly shows that with the increase in rule-set size, the throughput of the firewall is decreased. Also, the throughput is directly dependent on the packet rate as throughput is defined as packets processed per second. So, it can be estimated based on rule-set size and packet rate, and the model summary is shown in Table 4. The value of correlation coefficient is 0.992; i.e. there exists a highly positive correlation. Also, the level of significance is 0.000, i.e. < 0.05. So, the model is a good fit and throughput can be estimated by the following:

$$T = 374.381 - (0.053 * \text{RS}) + 0.95 * \text{PR} \quad (5)$$

## 6 Validation of the Model

In this section, we validate the accuracy of our model with new test cases by comparing the predicted observations with the actual observations taken from new experiments. The statistical model is evaluated on the basis of the following new test cases:

(i) Test V1: 5000 packets are sent per second on rule-set size of 1000.
(ii) Test V2: 7000 packets are sent per second on rule-set size of 10,000.
(iii) Test V3: 9000 packets are sent per second on rule-set size of 9000.
(iv) Test V4: 2000 packets are sent per second on rule-set size of 500.

These new test cases are performed, and their recorded observations are listed in Table 5 for the five key parameters: CPU utilisation, packet loss ratio, packet processing time, response rate and throughput. The values predicted by Eqs. (1) to (5) are also listed in the table. The corresponding error in observations is computed as follows:

$$\text{Relative Error (in \%)} = \frac{(\text{Observedvalue} - \text{Predictedvalue}) \times 100}{\text{Observedvalue}}$$

From Table 5, it is clearly seen that the estimates from our model are accurate with less than 10% error. Thus, these test cases (V1, V2, V3 and V4) validate the accuracy of our model.

## 7 Conclusion and Future Scope

In this paper, we attempted to evaluate and model the performance of *Iptables* which is the only stateful packet filtering firewall inside the *Linux* 2.4.x and later kernel versions. The performance evaluation is an important aspect that determines the

**Table 5** Relative error in validation tests

| Parameter | Test case | Observed value | Predicted value | Error (in %) |
|---|---|---|---|---|
| CPU utilisation | V1 | 4.6 | 4.59 | 0.21 |
| | V2 | 15.88 | 15.59 | 1.82 |
| | V3 | 16.81 | 16.59 | 1.30 |
| | V4 | 1.23 | 1.09 | 8.38 |
| Packet loss | V1 | 0.69 | 0.61 | 9.19 |
| | V2 | 18.46 | 18.03 | 2.32 |
| | V3 | 20.67 | 20.56 | 0.54 |
| | V4 | 0 | 0.15 | 0.00 |
| Packet processing time | V1 | 39.40 | 40.873 | 3.74 |
| | V2 | 325.223 | 346.873 | 6.66 |
| | V3 | 296.332 | 312.873 | 5.58 |
| | V4 | 25.693 | 23.873 | 7.08 |
| Response rate | V1 | 3100 | 3071.348 | 0.92 |
| | V2 | 3150 | 3415.398 | 8.43 |
| | V3 | 3998.3 | 4399.39 | 9.99 |
| | V4 | 1595.98 | 1723.398 | 7.98 |
| Throughput | V1 | 5116.67 | 5071.381 | 0.89 |
| | V2 | 5966.67 | 6494.381 | 8.84 |
| | V3 | 7883.33 | 8447.381 | 7.15 |
| | V4 | 2066.67 | 2247.881 | 8.77 |

dependability of the firewall. Characterising the performance is useful in predicting the behaviour of the firewall in varied traffic profiles and rule-set sizes. The performance of the firewall was comprehensively investigated for its dependence on the filter set size and packet rate. In this paper, an experimental set-up was established and real experiments were performed by varying the rule-set size ranging from smaller rule sets (<1000) to large rule sets (≥5000) and varying traffic at packet rate starting from 1000 packets per second to 8000 packets per second.

Based on the observations from real experiments, it is established that the performance of *Iptables* is dependent on rule-set size and packet rate while it is independent of the time duration. CPU utilisation, response rate, throughput and bandwidth exhibit a direct relationship with packet rate, while an inverse relationship exists with rule-set size. Likewise, the packet processing time depends only upon the size of rule set and is independent of the packet rate. Similarly, the packet loss ratio indirectly depends upon the rule-set size and packet rate as it shows a significant relationship with % CPU utilised instead of showing a relation with packet rate and rule-set size.

Also, on the basis of the experimental observations, a statistical model was established using the regression analysis. The model was validated by evaluating new test cases, and we found that it shows conformity with 90% accuracy.

In future, the work can be expanded to study the behaviour of the *Iptables* firewall with packet classification algorithms.

## References

1. Netfilter Project, [Online]. Available: www.Netfilter.org. Accessed 1 Oct 2017
2. Su, W., Xu, J.: Performance evaluations of Cisco ASA and Linux *Iptables* firewall solutions (May 2013)
3. *Iptables*, [Online]. Available: http://www.Iptables.info/en/structure-of-Iptables.html. Accessed 7 Sept 2017
4. Andreasson, O.: [Online]. Available: http://onz.es/Iptables%20Tutorial.pdf (2001)
5. Taylor, D.E., Turner, J.S.: Classbench: a packet classification benchmark. IEEE/ACM Trans. Networking **15**(3), 499–511 (2007)
6. Avallone, S., Guadagno, S., Emma, D., Pescapè, A., Ventre, G.: D-ITG distributed internet traffic generator. In: First International Conference on the Quantitative Evaluation of Systems (QEST '04), Enschede, The Netherlands, 27–30 September 2004
7. Diekmann, C., Hupel, L., Michaelis, J., Haslbeck, M., Carle, G.: Verified *Iptables* firewall analysis and verification. J. Autom. Reasoning (2018). https://doi.org/10.1007/s10817-017-9445-1
8. Gouda, M.G., Liu, A.X., Structured firewall design. Comput. Netw. **51**(4), 1106–1120 (2007)
9. Acharya, S., Wang, J., Ge, Z., Znati, T.F., Greenberg, A.: Traffic-aware firewall optimization strategies. In: Proceedings of IEEE ICC (2006)
10. El-Atawy, A., Samak, T., Al-Shaer, E., Li, H.: Using online traffic statistical matching for optimizing packet filtering performance. In: Proceedings of IEEE INFOCOM (2007)
11. Trabelsi, Z., Zeidan, S., Masud, M.M., Ghoudi, K.: Statistical dynamic splay tree filters towards multilevel firewall packet filtering enhancement. J. Comput. Secur. Elsevier **53**, 109–131 (2015)
12. Saadaoui, A., Souayeh, N.B.Y.B., Bouhoula, A.: Automatic detection and correction of firewall misconfigurations—a formal approach. In: The 8th International Symposium on Symbolic Computation in Software Science 2017 (SCSS 2017) (2017)
13. Hoffman, D., Prabhakar, D., Strooper, P.: Testing *Iptables*. In: Proceedings of the 2003 Conference of the Centre for Advanced Studies on Collaborative research (CASCON '03), Toronto, Ontario, Canada, 6–9 Oct 2003
14. Niemann, R., Pfingst, U., Gobel, R.: Performance evaluation of Netfilter: a study on the performance loss when using Netfilter as a firewall. In arXiv preprint arXiv:1502.05487, 19 Feb 2015
15. Wang, C., Zhang, D., Lu, H., Zhao, J., Zhang, Z., Zheng, Z.: An experimental study on firewall performance: dive into the bottleneck for firewall effectiveness. In: Proceeding of IEEE 10th International Conference on Information Assurance and Security (IAS), 978-1-4799-8099-4114 (2014)
16. Salah, K., Elbadawi, K., Boutaba, R.: Performance modelling and analysis of network firewalls. IEEE Trans. Netw. Serv. Manage. **9**(1), 12–20 (2012)
17. Lyu, M.R., Lau, L.K.Y.: Firewall security: policies, testing and performance evaluation. In: Proceedings of 2000 IEEE International Computer Software and Applications Conference
18. Sheth, C., Thakker, R.: Performance evaluation and comparative analysis of network. In: Proceedings of IEEE ICDeCom (2011)
19. Hayajneh, T., Mohd, B.J., Itradat, A., Quttoum, A.N.: Performance and information security evaluation with firewalls. Int. J. Secur. Appl. **7**(6), 355–372 (2013)
20. Cheminod, M., Durante, L., Maggiora, M., Valenzano, A., Zunino, C.: Performance of firewalls for industrial applications. In: Proceedings of the 4th International Symposium for ICS & SCADA Cyber Security Research (2016)

21. Beyene, Y., Faloutsos, M., Madhyastha, H.V.: SyFi: a systematic approach for estimating stateful firewall performance. PAM 2012, LNCS 7192, pp. 74–84 (2012)
22. Botta, A., Donato, W., Dainotti, A., Avallone, S., Pescapé, A.: [Online]. Available: http://traffic.comics.unina.it/software/ITG/manual/. Accessed 16 Nov 2017
23. Mishra, S., Sonavane, S., Gupta, A.: Study of traffic generation tools. Int. J. Adv. Res. Comput. Commun. Eng. (IJARCCE) **4**(6) (2015)
24. Melara, A.J.: Performance analysis of the Linux firewall in a host. San Luis Obispo, June 2002
25. IBM SPSS Software, IBM, [Online]. Available: https://www.ibm.com/analytics/data-science/predictive-analytics/spss-statistical-software. Accessed 25 Feb 2018

# Template Security in Iris Recognition Systems: Research Challenges and Opportunities

**Shehla Rafiq and Arvind Selwal**

**Abstract** An iris recognition framework is a standout, stable, secured, and widely accepted biometric in the secured authentication infrastructure. It has been broadly perceived as one of the grounded biometrics credited to its high precision execution. In this paper, a comprehensive analysis of several iris template security techniques has been carried out together with open opportunities for further research challenges. The analysis mainly pointed out that majority of the techniques results in a trade-off among various performance and security parameters, e.g., FAR, FRR, EER, revocability, diversity, and security. The well-known open iris databases such as CASIA, IITD were accessible to complete the experiments. It also has been noticed that the template security scheme for uni-biometric system may not work for multi-biometric system. Among the two broadly available template protection techniques, feature transformation schemes outperform its counterpart cryptosystems mainly due its unlinkability and irreversibility properties.

**Keywords** Iris recognition · Cancellable biometrics · Iris template security

## 1 Introduction

Traditional identity recognition mechanisms rely on "what you possess" or "what you have" such as secret codes, ID card. But, they are effectively being lost, unable to remember, or shared. Biometrics is the branch of science that is used to create an identity of an individual depending on "who you are." Such a system exploits the behavioral and biological characteristics like fingerprint, iris, face, voice, and gait of an individual for recognition [1]. The biometric frameworks can be arranged into two types dependent on the number of attributes utilized, namely unimodal biometric system and multimodal biometric system. Further details may be found in

S. Rafiq (✉) · A. Selwal
Department of Computer Science and Information Technology, Central University of Jammu, Jammu, J&K 181143, India
e-mail: shehlarafiq5@gmail.com

A. Selwal
e-mail: arvind.cuj@gmail.com



P. K. Singh et al. (eds.), *Proceedings of ICRIC 2019*, Lecture Notes in Electrical Engineering 597, https://doi.org/10.1007/978-3-030-29407-6_55

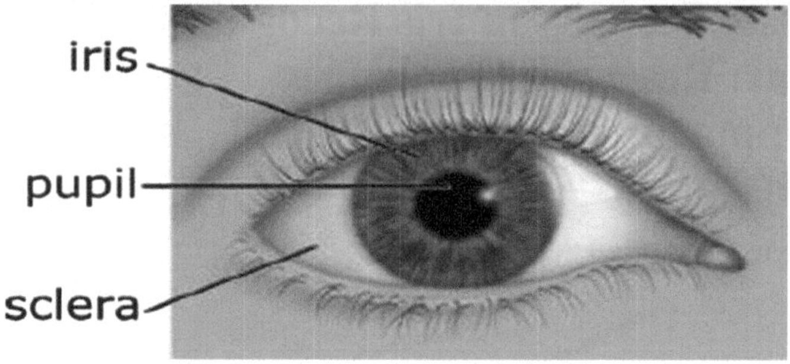


**Fig. 1** Human eye (adapted from [8])

[2, 3]. These traits are permanently associated with an individual; therefore, they need to be securely stored. As characterized by the European Association "European Union" (EU) information security order, biometric information is viewed delicate information, which implies its utilization is exposed directly to the protection conservation [4]. By providing the strong connection between an identity and the owner of an identity, the applications of biometrics are vast [5]. Generally, the applications are categorized into subsequent classes: **commercial applications** such as physical access control, ATM, and PDA, **government** uses for example airport security, nationwide identification card, and border control, **forensic** uses like criminal investigation, corpse identification, and parenthood determination [6]. Among all the biometric traits which are available today, iris is measured as one of the remarkable consistent biological trait [1]. The shape of the iris is flat and is geometrically configured by the two interdependent muscles controlling the broadness of a pupil. As appeared in Fig. 1, this trait is attributed by its discriminability and stability. The iris pattern has higher entropy than other biometric traits. The identification process and the speed of comparison is very high [7].

The algorithm which is the base of current practical automated iris recognition system has been developed by Daugman [9]. The iris images are converted to the corresponding templates as depicted in Fig. 2 [6].

The main steps in algorithm are (1) image capturing, where a raw image of individual's eye is acquired; (2) iris localization and normalization, where the iris is spotted and unwrapping of iris return to normal texture, and the operator used for localization is expressed in Eq. (1):

$$\max_{(r,a_0,b_0)} \left| G_\sigma(r) * \frac{\partial}{\partial r} \oint_{r,a_0,b_0} \frac{I(a,b)}{2\pi r} \mathrm{d}s \right| \tag{1}$$

where

$I(a, b)$ an image of an eye
$G_\sigma(r)$ the Gaussian smoothing function with scale $(\sigma)$

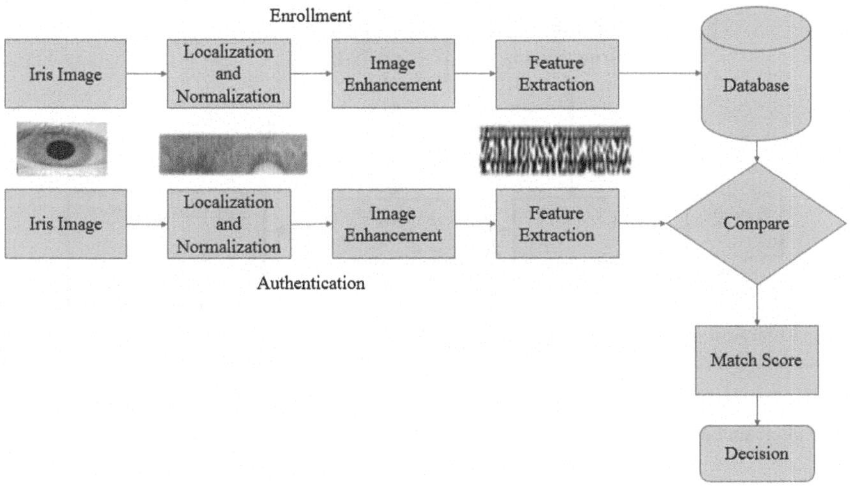


**Fig. 2** Schematic representation of typical iris identification method

| | |
|---|---|
| $R$ | radius of the circle |
| d$s$ | rounded curve of radius r and center coordinates $(a_0, b_0)$. |

(3) feature extraction, in which the features of iris are generated, i.e., binary feature vectors which are commonly called IrisCodes. In polar coordinate system, given the normalized iris image $I(\rho, \varphi)$, the demodulation and the phase quantization method can be expressed in Eq. (2) as follows:

$$k_{\{\mathrm{Re,Im}\}} = \mathrm{sgn}_{\{\mathrm{Re,Im}\}} \int_{\rho} \int_{\varphi} I(\rho, \varphi) \mathrm{e}^{-i\omega(\theta-\varphi)} \mathrm{e}^{-\frac{(r-\rho)^2}{\alpha^2}} \mathrm{e}^{-\frac{(\theta-\varphi)^2}{\beta^2}} \rho \mathrm{d}\rho \mathrm{d}\varphi \tag{2}$$

in which $k_{\{\mathrm{Re,\ Im}\}}$ could be considered just as a complex-valued bit where real and imaginary components are both 0 or 1 based on the symbol of the 2-D integral; $I(\rho, \varphi)$ is the normalized iris image; $r_0$ and $\theta_0$ are the polar coordinates of iris image; $\alpha$, $\beta$ are the multiscale 2-D wavelet size constraints; $\omega$ is the wavelet frequency; (4) matching (feature comparison), in which the Hamming distance has been used as a system of measurement for the matching process and is expressed in Eq. (3) as follows:

$$\mathrm{HD} = \frac{\|(\mathrm{IrisCodeA} \otimes \mathrm{IrisCodeQ}) \cap \mathrm{MaskP} \cap \mathrm{MaskQ}\|}{\|\mathrm{MaskP} \cap \mathrm{MaskQ}\|} \tag{3}$$

where IrisCodeP and IricCodeQ are two-phase code bit vectors each of 2048 bits; MaskP and MaskQ are mask bit vectors; ⊗ represents the Ex-OR operator that identifies the disagreeing bits across the two IrisCodes; ∩ represents AND operator that covers the noisy region. In order to cope with rotational consistencies, circular

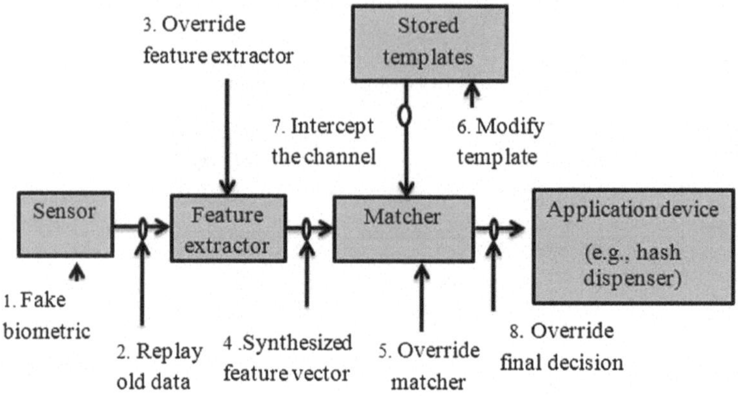


**Fig. 3** Potential attack points in a biometric system (adapted from [12])

bit shifts are used. The smallest iris distance relates the best match across the two templates [10].

## *1.1 Attacks on Iris Biometric System*

Any system including a biometric system is liable to various types of attacks. These attacks are performed at various points. Ratha et al. [11] recognized the eight points of assaults in a biometric system. These assaults include an attack at the sensor, assault on the channel across the sensor and the feature extractor module, overrule feature extraction process, tampering features extracted from feature extractor, overrule matcher, assault on a channel among the saved template and matcher, modifying the stored templates, and overriding the final decision. These are described in Fig. 3 [12].

Biometric systems are more compatible and easy to use. The irreversible link between the biometric traits of an individual and their personal records is the fundamental issue about biometrics. Biometric features cannot be revoked and replaced once they get compromised, unlike passwords and tokens [11]. Since iris is enduringly related to everyone and iris codes have extremely different information of the individual, the disclosure of iris code may lead to various types of security breaks, for example, replay assault and masquerade assault. This infers losing individuality permanently. The security and privacy issues of iris codes have been a major issue as biometric feature reveals non-changeable data of individuals in a database. Once an attacker gets access on them, their purpose of utilization gets lost, and henceforth, its security has become a part of a biometric application known as biometric template protection (BTP) [1].

## 2 Iris Template Security

A template in biometrics represents information of inherent capabilities that summarize the biometric features of an individual. The most likely deleterious or harmful attack in a biometric system is on the biometric features that are kept in the database. The templates stored in the database can be used to take unapproved access once they get replaced by an imposter user. These stolen templates can be used to create spoof attacks to gain unofficial or unauthorized admittance to the system. To design an excellent biometric template protection system, the subsequent four principles need to be fulfilled [13, 14].

(1) Diversity: To confirm the privacy of a user, it should be able to generate distinctive secure templates from the similar source to prevent the cross-matching over the database.
(2) Revocability: One should be able to cancel the compromised template and generate the new template from the original one.
(3) Security: It should be computationally difficult to acquire the actual template as of the protected/transformed template.
(4) Performance: The template security scheme must be capable of preserving the biometric estimation over the accuracy obtained in the original counterparts.

The template protection techniques are broadly classified into two types; namely cancellable biometrics and biometric cryptosystem (Fig. 4).

**Cancellable Biometrics/Feature Transformation**: Ratha et al. [11] introduced the notion of cancellable biometrics. Cancellable biometrics deforms the biometric features with a purpose to compare features of a trait in transformed domain during authentication and enrollment as represented in Fig. 5 [12].

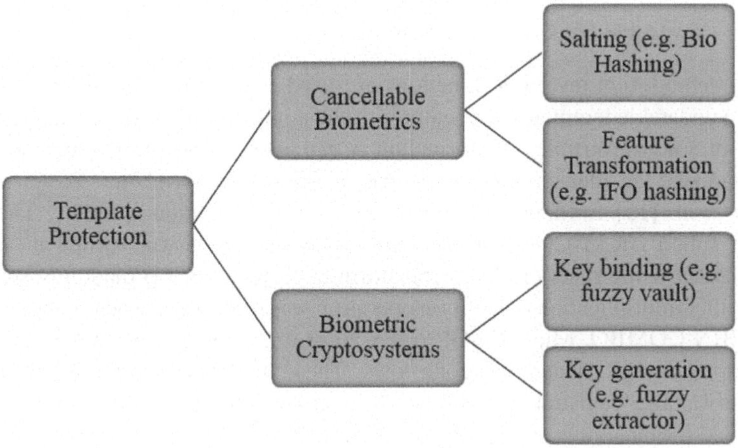


**Fig. 4** Classification of iris template security (adapted from [12])

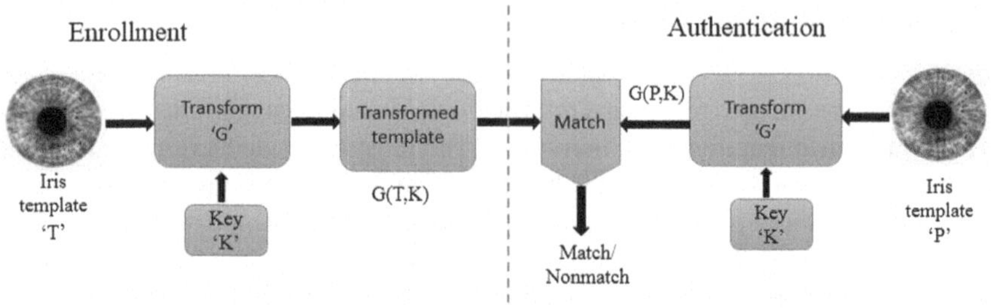


**Fig. 5** Iris biometric template after feature transformation technique (as adapted from [12])

Cancellable biometrics is of two types based on the approach of transform used as follows:

(i) Biometric salting method for iris template

In biometric salting, any invertible transform of biometric template is taken as salting, and the transform is invertible so the protection of transform platform is very important. If the transform pattern is compromised, an opponent could recuperate the actual template, and the performance of the system is lost. As the key is specific to the user, multiple templates are available for the same biometric data. The revoking and the replacing of compromised template became easy by using an alternative user-specific key [14].

Zuo et al. [15] forthput a salting technique for grayscale unwrapped iris. In this method, a totally synthetic pattern was either added or multiplied to iris pattern and is called GRAY-SALT. In BIN-SALT method, IrisCode is mixed with the key pattern by an Ex-OR operation. The problem with these methods was they suffered from an alignment problem.

(ii) Non-invertible transformation methods for Iris template:

In this method, non-invertible functions are used to transform biometric data. To produce updatable templates, the features of the applied transform are changed. The benefit of a non-invertible transformation is that even if transformed templates are compromised; the attackers cannot recreate all biometric data [14].

Zuo et al. [15] forthput two non-invertible transformation methods. The first method called GRAY-COMBO, the unwrapped iris image was shifted in a row-wise manner using random offset (transform key), then on two randomly selected rows, either multiplication or addition operation was performed. The second method called BIN-COMBO, where Ex-OR or Ex-NOR operation was performed between any two randomly selected iris code rows. These methods have the advantage of being advantage "registration free" which means that the IrisCodes do not need to be aligned for matching.

**Biometric Cryptosystems**: Biometric cryptosystems were initially designed with an aim of either securing a cryptographic key to a biometric or directly produce a cryptographic key from the biometric trait. The biometric cryptosystem requires some public information (which is also called as helper data) about the biometric

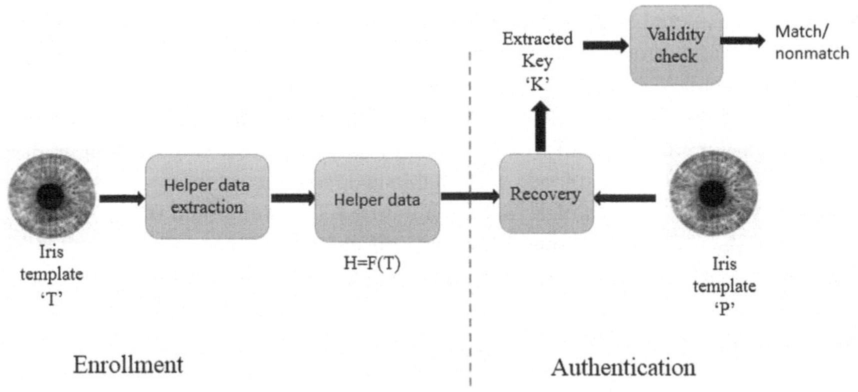


**Fig. 6** Enrollment and authentication mechanism in biometric cryptosystems (as adapted from [12])

template to be stored. Biometric cryptosystems are, therefore, also known as data-based helper methods. Helper data do not reveal any worthy information about the original templates. Biometric comparisons are accomplished by authenticating key validities. The authentication process output is either a key or a message of failure as depicted in Fig. 6 [12].

Depending in which manner helper data is achieved, biometric cryptosystems are classified as key binding and key generation systems. In key binding biometric cryptosystems, helper data is obtained by binding a chosen key (that is not dependent on any biometric features) to a biometric template. By using only helper data, it is computationally hard to get the biometric template or the key [14]. This method avoids the intraclass variations in biometric information [12]. Fuzzy commitment scheme as for iris biometric cryptosystem was illustrated by Hao et al. [16] Biometric cryptosystems are designed in such a way that they are not to provide revocability and diversity. Biometric cryptosystems stored data or helper data in the key generation is derived only from the biometric template, and the helper data reveals no information about the original template. The cryptographic key is directly obtained from the helper data and the biometric query [12, 16].

## 3 Review of Literature

This section exhibits a point by point study of different template protection schemes, and furthermore extensive examination of technologies, open issues and challenges is presented.

In [17], tokenless-based approach was used for protecting IrisCode, referred to as BioEncoding. To be precise, the samples of iris images were captured, and the corresponding IrisCodes were generated and are collected in b binary vectors. Then, they extracted the "consistent bits" from b vectors and stored in a consistent bit vector

K (a bit is considered consistent if it remains constant across b binary vectors) [18]. The positions of the consistent bit vectors in K were collected in another vector called as position vector $V \in \{1, 0\}^m$. The bits in the consistent bit vector K were grouped into address words having fixed length $n$, and a randomly generated sequence s was generated of length $2^n$. Finally, each word in K is plotted to a single bit value in $R \in \{1, 0\}^p$. The protected BioCode was established from the set of $p = \frac{m}{n}$ addressed a bit where m is the length of K, and then the BioCode was stored on the database rather than the original IrisCode. Moreover, different BioCodes were generated by changing the random sequence $R$ for all users; therefore, revocability property was satisfied. Their works also revealed that there was no degradation in performance pertaining to its initial parts.

Jenisch and Uhl [19] use block permutation and remapping of iris texture as a scheme for template protection. The iris texture was partitioned into fixed-sized blocks. These blocks were regrouped according to a permutation key. After that, they duplicate some of the blocks by remapping operation to provide the protection against a stolen key. They also showed that 60% of initial iris image can be rebuilt from the compromised template.

Alvarez Marino et al. [20] forthput a crypto-biometric scheme based on fuzzy extractor on iris templates. In their work, they associated a key K to a user for authentication using her own biometric template. If a user was correctly verified, the key K would be returned to her. During the enrollment phase, the key was concealed in the coefficients of a polynomial having degree d, and during the verification phase, the polynomial must be reconstructed in order to recover the secret or key. Lagrange interpolation process was used for reconstruction. They had analyzed the characteristics and efficiency for choosing utmost applicable parameters for offering high level of security in the scheme. Gayathri and Ramamoorthy [21] proposed a multi-biometric system that fuses the information of palm print and iris biometric trait to check whether the feature level fusion can attain better recognition performance than mono-biometric modality. To be specific, the input images of palm print and iris were fed as input to Gabor texture feature extraction to extract selected features. As the feature vectors obtained from different methods are in different sizes, he features from identical images may be correlated. These features are then merged by using wavelet fusion technique. Ultimately, the feature vectors were matched with the stored template using K-NN classifier. The proposed method achieves recognition accuracy of 99.2% and with FRR equal to 1.6%.

Rathgeb et al. [10, 22] proposed a template protection scheme for IrisCodes. In their work, adaptive bloom filters were instigated to attain cancellable iris biometrics. The binary feature vectors were extracted by generic iris recognition system based on a row-wise structure of normalized iris textures, i.e., the iris codes are represented as a two-dimensional binary feature vectors. The matrix of iris codes was separated into m blocks of equal size where each column consists of $m \leq h$. Later on, the entire sequence of columns of each block was transformed into corresponding locations within bloom filters [22] "A bloom filter 'b' is represented as a bit array of length $2^{m - \text{bits}}$, where at the outset all bits are set to zero," i.e., a sum of **f** distinct bloom filters. In order to map columns within two-dimensional iris codes to their decimal indexes, a transform function h is implemented. The mapping function for each column is expressed as follows:

$$b\left[h(x) \oplus p\right] = 1, \text{ as } h(x) = \sum_{i=0}^{m-1} x_i 2^i \tag{4}$$

where $p$ denotes an application-specific secret. One utmost advantage of the proposed transform is that it generates alignment-free templates, i.e., at the time comparison, the templates need not be aligned. Moreover, the bloom filter-based mixing transform was delineated to achieve the two most important requirements of irreversibility and unlinkability. The biometric performance was improved by applying multi-biometric fusion yielding equal error rates below 0.5%.

In [21], steganography-based approach was proposed to secure the iris template. Least significant bit (LSB) steganography technique was used for enhancing security. Random number-based embedding was used in LSB to provide more security. The bits are embedded into LSB's of the blue pixel only so that there will be less distortion of pixels if only one color component of RGB is used. The IrisCode bits were hidden in three least significant bits of cover image (the medium used to hide data). The 24-bit color image was used to locate the position for hiding each bit. Finally, the stego image (which contains secret information) is used as a template in the iris database. The performance of the proposed approach was calculated, and it was found that the resulting MSE and PSNR values, histogram plot, and ROC curve plot were satisfactory. However, IrisCode concealed in the cover image could be disclosed to low-level signal processing.

Dwivedi and Dey [13] proposed a cancellable iris biometric generation scheme centered on the randomized lookup table mapping. The technique first generates the rotation-free templates, and then these templates were transformed into a row vector. The row vector $X \in \{0, 1\}^{1 \times M}$ was divided into a number of blocks of size N and histograms of iris code were used to generate the value of N. The decimal vector was created by dividing the row vector where the word length = N. The size of decimal vector was given by $v = \{v_i \in [0, 2^n - 1] \mid \underline{i} = 1,\ldots,l\}$. Finally, the decimal vector was mapped to the randomly generated lookup table $T \in \{0, 1\}^{P \times Q}$ and $P \geq 2^n - 1$, and the p bits taken from every row of the lookup table was stored. The different cancellable templates could be generated by altering the value of N thus satisfying the revocability criterion.

Thul et al. [23] recommended a multimodal biometric identification system that fuses the scores of fingerprint and iris using "sum rule-based matching score level fusion" for improving the performance of system and security level. They used the phase-based image matching algorithm for poor-quality fingerprint images and Log-Gabor filter for extracting the features from iris image. After feature extraction and matching, similarity scores or distance scores are produced by the biometric system, and there is a need to transform these scores into the same nature. Finally, the scores were fused using "sum rule-based score level fusion" and were matched to a pre-decided threshold to declare whether a user is authentic or an imposter.

Khalaf et al. [24] proposed an iris template protection scheme based on enhanced Hill Cipher. Hill Cipher is a type of symmetric key algorithm and a block cipher that can be used to protect biometric template. In their work, the Hill Cipher algorithm was improved by the addition of pseudorandom generator, and iterating the encryption processes for K time and also during encryption, the multiplication operation was changed to summation. The normalized correlation coefficient (NCC) and running time had been used for checking the efficiency of encryption.

Chai et al. [1] proposed a non-invertible transformation-based cancellable iris technique, named as "Indexing-First-One" (IFO) hashing. IFO is mainly based on min-hashing that was originally used in the search engine to identify duplicate web pages and eliminate them from search results as well as in the clustering problems. To be precise, IFO hashing makes use of "n" independent hash functions $h_1$, $h_2$,..., $h_m$ where each of the independent hash function was obtained from K number of tokenized permuted IrisCode, $Y$ in a column-wise manner. IFO hashing function was derived by the procedure represented as $H(Y) = \{h_i(Y)i = 1,\ldots,n\}$.IFO hashing had been further enhanced by two different mechanisms, viz. K-order Hadamard product which was generated by conjunction all the $Y$'s, i.e., $Y^K = \Pi_{m=1}^{k}\left(Y_m^{'}\right)$ and modulo thresholding function that was inflicted to reduce the leakage of $Y$ through a threshold value $\tau$, $1 \le \tau < P$ with first selected $P$ elements from each row in the product code $\mathrm{Y}^{\mathrm{K}}$. Then ,from the selected first P elements, the index value was recorded where first one bit "1" occurred and denoted as $Q_Y$, i.e., for every $Q_Y \geq \mathrm{P} - \tau$, calculate $Q_Y^{'} = Q_Y \bmod(P - \tau)$. The threshold value resulted in a many-to-one mapping; hence, non-invertibility criterion is satisfied. IFO hashing scheme fulfilled both diversity and revocability properties. Moreover, the scheme offered very high resilience against various utmost security and privacy attacks.

Zhao et al. [18] put forward a local ranking method for iris template protection. The original iris data y was XORed with an n-bit application-specific string x, and the result was stored in a new string $z = x \oplus y$; then the result was divided into m blocks: $v = v_1, v_{2,}\ldots,v_m$, where each block had k bits. Next, the blocks were partitioned into b groups: $V = V_1, V_2,\ldots,V_{b,}$ where $V_i = \{u_{(i-1) \times t+1},\ldots,u_{i \times t} | i = 1\ldots.b\}$, and where t is group size. The blocks in each group were sorted according to their decimal values. Ultimately, the original blocks were transformed into their rank values for storage. The method also supported the two important strategies "shifting and masking" to improve recognition performance. Furthermore, the proposed method also satisfies the irreversibility, revocability, and unlinkability properties.

## 4 Interpretations Appertaining to Various Iris Recognition Techniques

The relative analysis of various template protection schemes is shown in Table 1. The most techniques that are summarized in the table are cancellable biometrics because the two important criteria, i.e., diversity and revocability that need to satisfy for an

**Table 1** A summary of different iris template security schemes

| Method | Year | Type of biometric and fusion type | Author | Performance measures and dataset |
|---|---|---|---|---|
| Tokenless cancellable biometrics | 2010 | Unimodal biometric | Ouda et al. [27] | ERR = 2.31,CASIA-IrisV3 interval dataset |
| Cancellable iris biometrics based on block remapping | 2011 | Unimodal biometric | Jenish and Uhl [19] | ERR = 1.2, CASIA V3 interval iris database. |
| Crypto-biometric scheme–Fuzzy extractor | 2012 | Unimodal biometric | Alvarez Marino et al. [20] | FAR = 4.42%, FRR = 9.67%, CASIA iris database. |
| – | 2012 | Multimodal biometric and feature level fusion | Gayathri and Ramamoorthy [21] | FRR = 1.6%, accuracy 99.2% and PolyU palmprint with IITK iris database |
| Multi-biometrics using adaptive bloom filters | 2013 | Multi-instance and feature level fusion | Rathgeb et al. [10] | EER = 0.5% and IITD iris database |
| Stenography using LSB | 2015 | Unimodal biometric | Chaudhary and Nath [28] | FAR = 0.1%,GAR = 95% and CASIA-database |
| Lookup table mapping | 2015 | Unimodal | Dwivedi and Dey [13] | Accuracy 94.26% and CASIA V3 database |
| – | 2016 | Multimodal and score level fusion | Thul et al. [23] | – |
| Enhanced Hill Cipher | 2016 | Unimodal | Khalaf et al. [24] | – |
| Cancellable biometrics using IFO | 2017 | Unimodal | Chai et al. [1] | EER = 0.16% and CASIA-V3 database |
| Local ranking | 2018 | Unimodal | Zhao et al. [18] | FAR = 0.01%,EER = 1.36% and CASIA-irisV4-Lamp,CASIA-irisV3 Interval |

excellent biometric template are not satisfied by biometric cryptosystems. Moreover, the helper data in biometric cryptosystems (fuzzy extractor) needs to be carefully designed. In biometric cryptosystems, producing a key having high stability and entropy is also very difficult [12, 14]. The cancellable biometrics is not specific to single trait meanwhile it has been effectively applied to face [25] and fingerprint [26]. Furthermore, for multi-biometric template protection scheme, the concept can be applied at feature level where mixing of templates is done [21].

## 5 Research Challenges

The techniques which has been summarized in the Table 1 are subjected to various research challenges, viz. in tokenless cancellable technique work has to be carried out for analyzing the security of the technique by considering the hill climbing attack; In the Crypto-biometric scheme(Fuzzy extractor), there is a need to create strategies aimed at extracting iris templates of more bit length, and more profound investigation of concern between the parameters of security and biometric effectiveness; Stenography using LSB—Need to increase the efficiency of algorithm and maintaining the integrity of iris template; Lookup table mapping—Higher accuracy may be achieved by up the segmentation method; Cancellable biometrics using IFO—Work could be carried to features which are in binary form and for identification purposes; Local ranking-—Algorithm should be stretched out to help different methods in iris recognition, and it additionally needs enhancement in order to help template protection for different biometrics. There are other challenges in recognition process as well like recognition of iris in less controlled backgrounds, the off angle impoverished contrast iris image recognition. To capture and recognize the iris at greater distance along, the motion of a person is as well a challenge. To improve the recognition accuracy, a small number of combined transforms for feature extraction and fusion of two or more neural networks for classification can also be put into practice.

## 6 Conclusion and Future Scope

With a regularly developing accentuation on security systems, automatic individual ID structures dependent on biometrics have been acquiring more concentrated in viable and investigated perspectives. The iris biometric is one of the immensely created individual recognizable proof ways. Since an iris is unique for an individual, the security is a noteworthy worry for iris-based frameworks. Various types of assaults are moved against an iris acknowledgment framework. This paper examined the diverse techniques proposed by researchers to protect the iris biometric template in a database. Unlike cancellable biometrics, biometric cryptosystems are not intended to offer diversity and revocability, and furthermore the security of biometric features is mandatory to restrict data spillage of helper data. There is no perfect template protection method. The application situation and requirements assume a remarkable part in the decision of a protection method. The other primary factors that affects the determination of a template protection scheme are the selected biometric, illustration of its features, and the intra-user varieties degree. A solitary template security scheme possibly not enough to fulfill the intra-user differences extent. In future, hybrid schemes that make utilization of the benefits of the diverse template security methods must be created. At last, with the emerging attention in multi-biometric and multi-factor verification techniques, methods which all the while protect multi-biometric templates as well as multi-validation aspects should be created.

## References


1. Chai, T., et al.: Cancellable iris template generation based on Indexing-First-One hashing. Pattern Recognit. **64**, 105–117 (2017)
2. Selwal, A., Gupta, S.K., Surender: Low overhead octet indexed template security scheme for multi-modal biometric system. J. Intell. Fuzzy Syst. **32**, 3325–3337 (2017)
3. Oloyede, M.O., Member, S., Hancke, G.P.: Unimodal and multimodal biometric sensing systems : a review. **3536** (2016)
4. Díaz Díaz, E.: The new European Union General Regulation on data protection and the legal consequences for institutions. Church, Commun. Cult. **1**, 206–239 (2016)
5. Gomez-barrero, M., Rathgeb, C., Galbally, J., Busch, C., Fierrez, J.: Unlinkable and irreversible biometric template protection based on bloom filters. Inf. Sci. (Ny) **370–371**, 18–32 (2016)
6. Jain, A.K., Ross, A., Prabhakar, S.: An introduction to biometric recognition. IEEE trans. Circuits Syst. Video Technol., London (2004)
7. Misztal, K., Saeed, E., Tabor, J., Saeed, K.: Iris pattern recognition with a new mathematical model to Its Rotation Detection. In: Saeed, K., Nagashmi, T. (eds.) Biometrics and Kansei Engineering 2012, pp. 43–65. Springer, New York (2012)
8. Harakannanavar, S.S.: Comparative survey of Iris recognition. In: 2017 International Conference on Electrical, Electronics, Communication, Computer, and Optimization Techniques (ICEECCOT), pp. 280–283 (2017)
9. Daugman, J.: How Iris recognition works. IEEE Trans. Circuits Syst. Video Technol. **14**, 21–30 (2004)
10. Rathgeb, C., Busch, C.: Cancelable Multi-biometrics: mixing Iris-Codes based on adaptive bloom filters. Comput. Secur. **42**, 1–12 (2014)
11. Ratha, N.K., Connell, J.H., & Bolle, R.M.: An analysis of minutiae matching strength. In: International Conference on Audio-and Video-Based Biometric Person Authentication, pp. 223–228 (2001)
12. Nandakumar, K., Jain, A. K., Nagar, A.: Biometric template security. EURASIP J. Adv. Signal Process. **2008** (2008)
13. Dwivedi, R., Dey, S.: Cancelable Iris template generation using look-up table mapping. In: 2nd International Conference on Signal Processing and Integrated Networks (SPIN), pp. 785–790 (2015)
14. Anuja, J.S., Praveen, K., Amritha, P.P.: A survey of Iris template security aspects. **119**(15), 1471–1481 (2018)
15. Zuo, J., Ratha, N.K., Connell, J.H.: Cancelable Iris biometric. In: 2008 19th International Conference on Pattern Recognition, pp. 1–4 (2008)
16. Hao, F., Anderson, R., Daugman, J.: Combining crypto with biometrics effectively. IEEE Trans. Comput. **55**, 1081–1088 (2006)
17. Rathgeb, C., Uhl, A.: A survey on biometric cryptosystems and cancelable biometrics. EURASIP J. Inf. Secur. **2011**, 1–25 (2011)

18. Zhao, D., Fang, S., Xiang, J.: Iris template protection based on local ranking. Secur. Commun. Netw. **2018**, 1–9 (2018)
19. Jenisch, S.,Uhl, A. Security analysis of a cancelable iris recognition system based on block remapping. In: 2011 18th IEEE International Conference on Image Processing, pp. 3213–3216 (2011)
20. Mariño, R.Á., Álvarez, F.H., Encinas, L.H.: A crypto-biometric scheme based on iris-templates with fuzzy extractors. Inf. Sci. **195**, 91–102 (2012)
21. Gayathri, R., Ramamoorthy, P.: Feature Level Fusion of Palmprint and Iris **9**, 194–203 (2012)
22. Rathgeb, C., Breitinger, F., Busch, C., Baier, H.: On application of bloom filters to iris biometrics. IET Biom. **3**, 207–218 (2014)
23. Thul, S.V,, Rishishwar, A., Raghuwanshi, N.: Sum rule based matching score level fusion of fingerprint and Iris images for multimodal biometrics identification, pp. 1370–1376 (2016)
24. Khalaf, E.T., Mohammed, M.N., Sulaiman, N.: Iris template protection based on enhanced hill cipher. In: Proceedings of the 2016 International Conference on Communication and Information Systems (ICCIS'16), pp. 53–57 (2016)
25. Gomez-Barrero, M., Rathgeb, C., Galbally, J., Fierrez, J., Busch, C.: Protected facial biometric templates based on local gabor patterns and adaptive bloom filters. In: 2014 22nd International Conference on Pattern Recognition, pp. 4483–4488 (2014)
26. Abe, N., Yamada, S., Shinzaki, T.: Irreversible fingerprint template using minutiae relation code with bloom filter. In: 2015 IEEE 7th International Conference on Biometrics Theory, Applications and Systems (BTAS) (2015)
27. Ouda, O., Tsumura, N. & Nakaguchi, T. Tokenless Cancelable Biometrics Scheme for Protecting IrisCodes. *2010 20th Int. Conf. Pattern Recognit.* 2–5
28. Chaudhary, S., Nath, R.: A new template protection approach for Iris recognition. In: 2015 4th International Conference on Reliability, Infocom Technologies and Optimization (ICRITO) (Trends and Future Directions), pp. 2–7 (2015)
29. Selwal, A., Gupta, S.K., Surender, Anubhuti.: Performance analysis of template data security and protection in biometric systems. In: 2015 2nd International Conference on Recent Advances in Engineering & Computational Sciences (RAECS) (2016)

# Comprehending Code Fragment in Code Clones: A Literature-Based Perspective

**Sarveshwar Bharti and Hardeep Singh**

**Abstract** As Code Clones are defined on the notion of similarity in code fragments, it is necessary to first know what a code is meant by in accordance with Code Clones. A Source Code Fragment, which is a sequence of source code lines, is the basic entity that is used to analyze similarity/relation between Code Clones. For analysis, removal, avoidance, and management of Code Clones we have to first detect clones in software systems. There are more than 40 clone detection tools that implement some clone detection techniques to detect clones, but it is not well-defined what could be the appropriate minimum threshold for Clone length and with which unit of estimation. This paper, on the basis of Code Clone literature, presents different Units of Measurement of Clone Size and a comprehensive review of minimum Clone Size based on a particular technique used in Clone Detection and also argues that a unique Unit of Measurement and Minimum Clone Size should be presented.

**Keywords** Clone · Clone coverage · Clone granularity · Code fragment · Comparison granularity · CSIR rule · Minimum clone size · Unit of measurement

## 1 Introduction

Baxter et al. [1] defined Clone as:

> A clone is a code fragment that [is] identical to another fragment.

And Koschke [2] presented the definition by Baxter, 2002 as:

> Clones are segments of code that are similar according to some definition of similarity

The term 'Clone' is utilized by the software community in two different ways [3]: 'Clone' as a noun—refers to a code fragment that is similar to one or more code

S. Bharti (✉) · H. Singh
Guru Nanak Dev University, Amritsar, Punjab 143005, India
e-mail: sarveshwar.dcsrsh@gndu.ac.in

H. Singh
e-mail: hardeep.dcse@gndu.ac.in

 
P. K. Singh et al. (eds.), *Proceedings of ICRIC 2019*, Lecture Notes in Electrical Engineering 597, https://doi.org/10.1007/978-3-030-29407-6_56

fragments. 'Clone' as a verb—indicates the act of producing a code segment (e.g. by Copy-pasting). To study the relation between code clones, and, as code clones are defined on the notion of similarity between code fragments, it is necessary to know what a code fragment is.

Walenstein et al. [4] wrote in a paper on similarity in programs:

> In order to clarify the overview [what constituted a different similarity 'type'] we shall take cues from established notions of what a "program" [or code fragment] is. Having a clean definition is critical since it is logical to assume that only when the definition of "program" is nailed down can one hope to properly pin the notion of similarity in programs.

For analysis, removal, avoidance, and management of code clones, we have to first detect clones in software systems. There are more than 40 clone detection tools [5] that implement different clone detection methods to detect clones, but it is not clear what could be the appropriate minimum threshold for clone length. So, to understand the concept of code fragment and thus the relationship between them, the first question that arises is

> How much of code can be regarded as a code segment. [3]

There are various studies in the literature that answers the issue of minimum clone size but each study provides minimum clone size based on the technique used. So, it is as yet not clear what should be the least clone size and with which unit of estimation. When the unit of estimation is picked, it ought to be chosen what could be the suitable least limit for clone length dependent on this unit.

This paper presents a comprehensive review of various units of measuring clone length and minimum clone length for respective clone detection technique used in the literature.

The rest of this paper is organized as follows: In the next segment i.e. Sect. 2, we discuss the basic definition of Code Fragment and various types of Clone Granularities as found in the literature. Section 3, presents various units of measuring clone size as found in the literature. This discussion is supported by presenting various Comparison Granularities. In Sect. 4, we have provided a review of Minimum Clone Lengths used in the literature. We presented the literature review in a tabular form depicting Clone detection techniques, unit of measurement, minimum clone size and finally references in support of our findings. In Sect. 5, we discuss the impact of minimum clone size using a study from literature. Then finally this paper ends with a conclusion and future work, and acknowledgments along with references.

## 2 Understanding Code Fragment

Bellon et al. [6] defined code fragment as:

> A Code Fragment is a tuple (f, s, e) which consists of a name of the source file f, the start line s, and the end line e, of the fragment. Both line numbers are inclusive

From the above definition, the length of the Code Fragment is determined as
*Code Fragment Length* = $e - s + 1$
(where $e$ is End Line and $s$ is Start Line).

In clone detection process, to use comparison algorithms more efficiently, the transformation of source text into an internal format is carried out. When representing source in this internal format we can identify a code fragment accordingly, e.g. in token representation, tuple ($f$, $s$, $e$) would be, $f$ as file name, $s$ as start token number, $e$ as end token number.

Now, the question is how much of contiguous source code can be contemplated as code fragment so that it can fulfill the criteria to be a clone candidate for Clone Detection. In literature, contiguous portions of source code at different levels of granularities have been used like at the level of statements, entire file, class definitions, method body and code block [3]. A granularity of Clones can be "Fixed" with predefined syntactic boundary or "Free" with no syntactic boundary, i.e. "clones are similar code fragments without considering any limit or boundary on their structure or size" [7]. Figure 1 shows different types of clone granularities found in the literature, which are broadly classified as Fixed and Free. While free granularity has no syntactic boundary, Fixed has a predefined syntactic boundary and thus can be seen at the different level of granularity. Figure 1 is created using the concepts of Clone Granularity as found in [7] and [3]. The first part of figure, i.e. free and fixed granularity types are taken from [7] and fixed granularity type is then extended using [3].

To detect clones that are useful from the maintenance perspective Zibran and Roy [3] proposed the following characteristics of chosen granularity of the code segment as motivated by their experiences and the criteria suggested by Giesecke [8]:

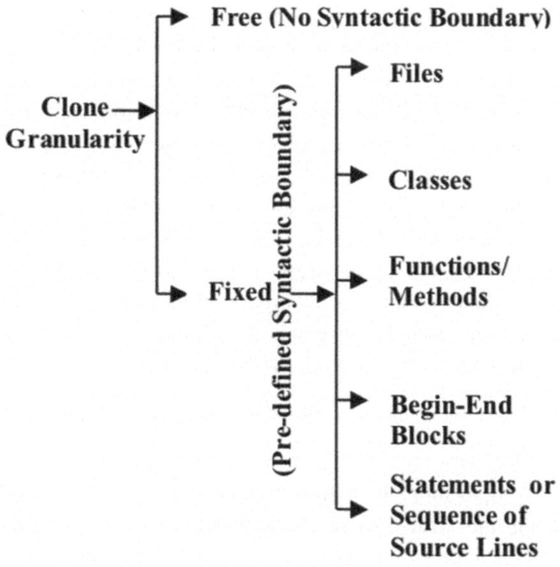


**Fig. 1** Types of clone granularities

***Coverage***: "The set of all code segments should cover maximal behavioral aspects of the software".
***Significance***: "*Each code segment should possess implementation of a significant functionality*".
***Intelligibility***: "Each code segment should constitute a sufficient amount of code such that a developer can understand its purpose with little effort".
***Reusability***: "The code segments should feature a high probability for informal reuse".

And finally suggested that code segment at the level of blocks or functions can be the utmost appropriate granularity for dealing with clones, especially for maintenance. For convenience, authors call these four characteristics as 'CSIR Rule' in rest of this paper, where *C* stands for *Coverage*, *S* for *Significance*, *I* for *Intelligibility* and *R* stands for *Reusability*.

## 3 Units of Measuring Clone Size

There are more than 40 Clone Detection Tools [5] available. Each tool implements an algorithm that works on a particular code representation. Each clone detection tool is developed using a particular clone detection technique. There are different Clone detection Techniques as found in literature viz. String-based, Token-based, Tree-based, PDG-based, Metrics-based and Hybrid Approach and each clone detection technique use a particular level of Clone Granularity for comparison. In literature, there is a number of different comparison granularities used by clone detection techniques. Figure 2 lists all such comparison granularities. For easy understanding, authors listed comparison granularities in the form of tree representation as shown in Fig. 2. This figure is created from [5]. Roy et al. [5] while comparing Clone Detection Tools, described comparison granularity as one of the technical facets and then described its various attributes viz. Line, Substring/Fingerprint, Identifiers and Comments, Tokens, Statements, Subtree, Subgraph, Begin-End Blocks, Methods, Files, and, Others. In order to stipulate a comparison of both general methods and distinct tools, they gathered citations of the identical group jointly using a category annotation, *L* for Lexical i.e. Token-based, *T* for Text-based, *S* for Syntactic i.e. Tree-based, *G* for Graph-based and *M* for Metrics-based, for each attribute. In Fig. 2 each attribute of Comparison Granularity is further categorized based on the above-mentioned category annotations.

Now, as Juergens et al. [9] described:

> Code is interpreted as a sequence of units, which for example could be characters, normalized statements, or lines.

And, as there are different units for representing code and thus different comparison granularities, so the particular type of comparison granularity used, corresponds to a particular "Unit of Measurement" for measuring Clone Length. Figure 3 shows

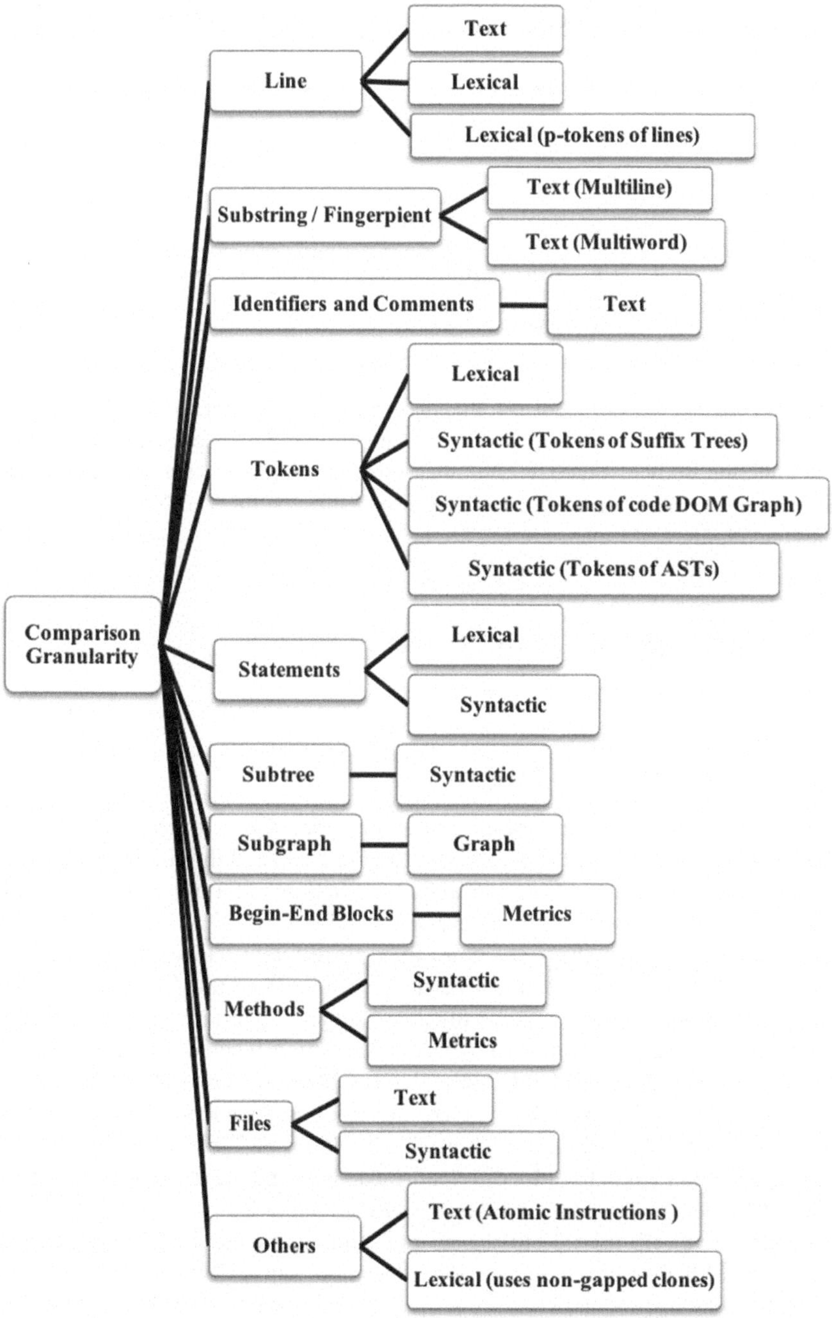


**Fig. 2** Comparison granularity (based on [5])

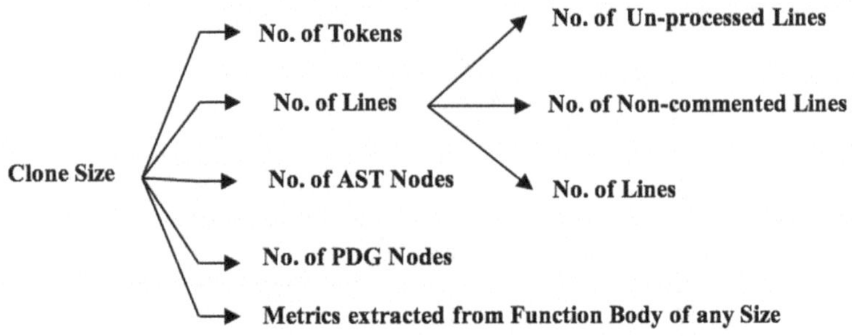


**Fig. 3** Different units of measurement of clone size

different Units of Measurement as found in literature viz. Number of Tokens, Number of Source Lines, Number of AST Nodes, Number of PDG Nodes and Metrics value from Function Body of any Size. "Number of Lines" Unit of Measurement is seen as different Units viz. A number of Un-processed Lines, Number of Non-Commented Lines and Number of Lines.

## 4 Minimum Clone Length

Now to answer the question, as pointed in Sect. 1:

> How much of code can be regarded as a code segment [3],

Authors present a summary of the Minimum Clone Size with respective Unit of Measurement, as found in the literature. Authors evaluated the findings related to minimum clone length in [7] and extended the findings in a more elaborated and systematic way in a tabular form as revealed in Table 1.

As discussed earlier, there are more than 40 different Clone Detection Tools [5] available, these tools implement particular Clone Detection Technique. In the 'Clone detection Technique', column of Table 1 authors list different Clone Detection Techniques found in the literature. Each Clone Detection Technique uses a particular unit of measuring Clone Size as listed in 'Unit of Measurement' column of Table 1. To optimize the results, every Clone Detection Tool used some minimum threshold for Clone Length. In 'Minimum Clone Size' column of Table 1, authors list the Minimum Clone Length used by different researchers and the references in support of their findings are listed in 'Citations' column of Table 1.

Kamiya et al. [10] used 30 tokens as their minimum clone length while detecting clones with their token-based clone detection tool CCFinder. Various other researchers like Kapser and Godfrey [11], Kim and Murphy [12], Jiang et al. [21], Higo et al. [22], etc. also used the same minimum clone length.

In line-based techniques, different researchers have different opinions. Bellon et al. [6] used 6 unprocessed lines as their minimum clone length, Baker [14] used

**Table 1** Summary of minimum clone size and units of measurement

| Clone detection technique | Unit of measurement | Minimum clone size | Citations |
|---|---|---|---|
| Token-based | Tokens | 30 tokens | Kamiya et al. [10]<br>Kapser and Godfrey [11]<br>Kim and Murphy [12]<br>Li et al. [13] |
| Line based | Un-processed lines | 6 lines | Bellon et al. [6] |
| | Non-commented Lines | 15 lines | Baker [14] |
| | Lines | 50 lines | Johnson [15] |
| Abstract syntax tree based | AST node | Subtree | Baxter et al. [1] |
| Program dependency graph-based | PDG node | Sub graph | Komondoor and Horwitz [16]<br>Krinke [17]<br>Komondoor and Horwitz [18] |
| Metrics-based (Function clone detection) | Function metrics | Function body of any size | Mayrand et al. [19]<br>Lague et al. [20] |

15 non-commented lines as minimum clone length and Johnson [15] used 50 lines. There are other opinions too, but the authors presented these three to give a general idea of what is the range of value for minimum clone length when considering line-based techniques.

Baxter et al. [1] introduced clone detection by means of an Abstract Syntax Tree. AST (Abstract Syntax Tree) was produced by parsing the source code and then algorithm was applied on AST to detect the Sub-Tree Clones, thus, using sub-tree as the Minimum Clone Size and AST nodes as Unit of Measurement.

In literature, a number of researchers used PDG (Program Dependence Graph) based clone detection technique for detecting clones. Komondoor and Horwitz [16] used PDG (Program Dependence Graph) to find isomorphic Sub-Graphs. Krinke [17] presented an approach to identify alike code in programs established on locating similar Sub-Graphs in an attributed directed graph. This approach was used on the Program Dependence Graph. Komondoor and Horwitz [18] defined an algorithm for extricating “difficult” set of statements. Control-Flow Graph of a method and set of nodes in that CFG that have been selected for removal are the inputs to the algorithm. At the point when the algorithm finishes, the marked nodes will form a hammock (they described it as a sub-graph of CFG that has a single entry node, and from this entry point control moves to a specific outside-exit node), and thus extricating them into a distinct method and swapping them with a method call. Thus PDG based clone detection technique used PDG Nodes as a Unit of Measurement and Sub-Graph as a Minimum Clone size.

Another Clone Detection Technique found in the literature is Metrics-Based Technique. Function Clone Detection Technique uses Function Metrics values calculated from functions to detect clones at the function level of granularity. Mayrand et al. [19] presented an identification method to automatically recognize duplicate as well as near-duplicate functions in huge size systems based on metrics extricated from the source code exploiting tool DatrixTM. This Clone discovery method utilizes 21 function metrics grouped into four points of evaluation. Each point of correlation is utilized to compare functions and determine their level of cloning. Eight cloning levels are then defined as an ordinal scale ranging from exact copy clones to distinct functions. Lague et al. [20] also used the same above mentioned Clone Detection Technique. They compared two subsequent versions of the functions on the basis of the above mentioned 21 DatrixTM metrics used by the clone detection methodology. Thus function clone detection technique, which is a metrics-based technique uses metrics as the Unit of Measurement of clone size and function body of any size for extracting metrics, so, using function body as a Minimum Clone Size.

Table 1 provides a general understanding of the minimum threshold for Clone Size used in the literature by different researchers, while the citations in Table 1 may not be complete.

## 5 Impact of Clone Length

In the previous sections, this paper discussed what a code fragment is and what is the minimum clone length used by different researchers. Now, the question arises:

Does Code Clone Length matter in Code Cloning?

To answer this question, the authors present a study by Gode et al. [23]. Gode et al. illustrated their findings, from identifying clones within an industrial C/C++software system with 1400 KLOC, as revealed in Fig. 4.

In Fig. 4, horizontal axis symbolizes Minimum Clone Length (granularity used was statements) and a vertical axis represents Clone Coverage in percent. Gode et al. defined Clone Coverage as "the percentage of source code being part of at least one clone."

To show that parameters utilized for clone identification impact the clone coverage, they chose three noticeable parameters, a minimum size of clone, the omission of generated code (*G*) and whether identifiers and literals are normalized (*N*) or not. Now from Fig. 4, it is clear that different value of minimum clone length and different combinations of the parameters have a strong effect on clone coverage that ranges from 19 to 92%. They also compared the clone coverage for different systems with minimum clone length varying but other parameters were fixed, as shown in Fig. 5.

Gode et al. observed that clone coverage of all software systems declines with growing minimum clone length. From this study, they established that exploiting

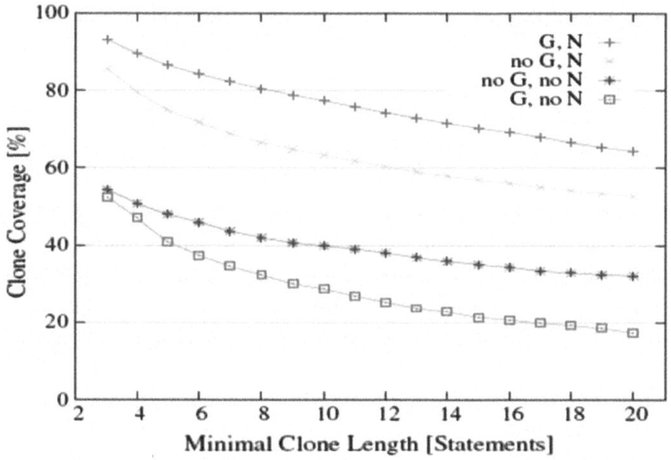


**Fig. 4** Clone coverage using different parameters [23] (figure reproduced with permission)

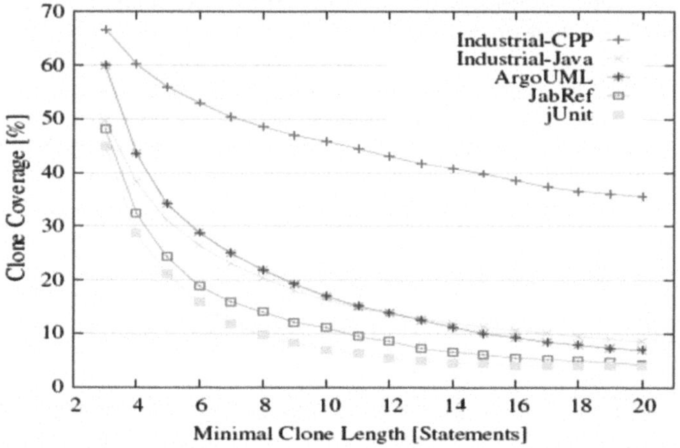


**Fig. 5** Clone coverage of different systems [23] (figure reproduced with permission)

clone coverage for comparing software systems should be taken carefully, as the result depends on the parameters used.

## 6 Conclusion and Future Work

As discussed earlier, each Clone detection Technique uses some level or type of clone granularity i.e. comparison granularity for comparing Code Fragments to detect Clones, but as discussed, there are number of different types of clone granularities and thus number of different Units of measurement for measuring clone length, so, we don't have a unique Unit of Measurement, and thus standard Minimum Clone Size. Different researchers use different 'Units of Measurement' and 'Minimum Clone Size', so, our research community should work toward this direction and make a standard for 'Unit of Measurement' and 'Minimum Clone Size' and obviously this standard must obey the 'CSIR Rule' discussed in Sect. 2, only then we can have better results from the clone detection process and thus better maintenance. To ascertain the exact extent of clone length required and thus analyze Code Clones, more comprehensive survey is required which is beyond the scope of this short paper.

Present Clone Detection Tools needs Minimum Clone Size specified by the user, so, as discussed in Sect. 5, selection of Minimum Clone Length should be done with care.

Like Code Clone Detection, calculating Minimum Clone Size can also be a Tool supported, where Minimum Clone Size will be calculated with the help of a tool, so our community should also work towards this direction.

From the literature survey we did, it is found that this matter of Unit of measurement and Clone Size found a little space in related publications, and thus authors think that this paper will be a keynote paper towards the study of Code Fragment and thus the Code Clones.

**Acknowledgements** We would like to thank University Grants Commission, a statutory body of Govt. of India for offering the financial support for this research in the form of Rajiv Gandhi National Senior Research Fellowship to the first author and also would like to thank Guru Nanak Dev University (Department of Computer Science), Amritsar for the scholastic and infrastructural support towards the current research work.

## References

1. Baxter, I.D., Yahin, A., Moura, L., Anna, M.S., Bier, L.: Clone detection using abstract syntax tree. In: Proceedings of 14th International Conference on Software Maintenance (ICSM'98), Bethesda, Mayland (1998)
2. Koschke, R.: Survey of research on software clones. In: Dagstuhl Seminar Proceedings 06301: duplication, Redundancy, and Similarity in Software (Dagstuhl 2007) (2007)
3. Zibran, M.F., Roy, C.K.: The road to software clone management: a survey (2012)
4. Walenstein, A.: Similarity in programs. In: Dagstuhl Seminar Proceedings 06301: Duplication, Redundancy, and Similarity in Software (Dagstuhl 2007) (2007)
5. Roy, C.K., Cordy, J., Koschke, R.: Comparison and evaluation of code clone detection techniques and tools: a quantitative approach. Sci. Comput. Progr. **74**(7), 470–495 (2009)
6. Bellon, S., Koschke, R., Antoniol, G., Krinke, J., Merlo, E.: Comparision and evaluation of clone detection tools. IEEE Trans. Softw. Eng. **33**(9), 577–591 (2007)
7. Roy, C.K., Cordy, J.R.: A survey on software clone detection research. Qween's University, Kingston (2007)
8. Giesecke, S.: Generic modelling of code clones. In: Dagstuhl Seminar Proceedings 06301: Duplication, Redundancy, and Similarity in Software (Dagstuhl 2007) (2007)
9. Juergens, E., Deissenboeck, F., Hummel, B., Wagner, S.: Do code clone matter? In: Proceedings of 31st International Conference on Software Engineering ICSE 2009, Vancouver, BC (2009)

10. Kamiya, T., Kusumoto, S., Inoue, K.: CCFinder: a multilinguistic token-based code clone detection system for large scale source code. IEEE Trans. Softwa. Eng. **28**(7), 654–670 (2002)
11. Kapser, C.J., Godfrey, M.W.: Supporting the analysis of clones in software systems: a case study. In: IEEE International Conference on Software Maintenance ICSM 2005 (2005)
12. Kim, M., Murphy, G.: An empirical study of code clone genealogies. In: Proceedings of the 10th European Software Engineering Conference (ESEC) held jointly with 13th ACM SIGSOFT Symposium on the Foundations of Software Engineering (FSE-13) ESEC/SIGSOFT FSE 2005 (Lisbon 2005) (2005)
13. Li, Z., Lu, S., Myagmar, S., Zohu, Y.: CP-Miner: finding copy paste and related bugs in large scale Software Code. IEEE Trans. Softw. Eng. **32**(3), 176–192 (2006)
14. Baker, B.: On finding duplication and near duplication in large software systems. In: Proceedings of the 2nd Working Conference on Reverse Engineering (WCRE'95) (1995)
15. Jonson, J.: Substring matching for clone detection and change tracking. In: Proceedings of International Conference on Software Maintenance ICSM'94, Victoria, BC (1994)
16. Komondoor, R., Horwitz, S.: Using slicing to identify duplication in source code. In: Proceedings of 8th International Symposium on Static Analysis SAS 2001, Paris (2001)
17. Krinke, J.: Identifying similar code with program dependence graphs. In: Proceedings of the 8th Working Conference on Reverse Engineering (WCRE'01), Stuttgart (2001)
18. Komondoor, R., Horwitz, S.: Effective, automatic procedure extraction. In: Proceedings of the 11th IEEE International Workshop on Program Comprehension IWPC 2003, Portland (2003)
19. Mayrand, J., Leblane, C., Merlo, E.: Experiment on the automatic detection of function clones in a software systems using metrics. In: Proceedings of International Conference on Software Maintenance (IWSM'96), Monterey (1996)
20. Lague, B., Proulx, D., Mayrand, J., Merlo, E.M., Hudepohl, J.: Assessing the benefits of incorporating function clone detection in a development process
21. Jiang, Z.M., Hassan, A.E., Holt, R.C.: Visualizing clone cohesion and coupling. In: XIII Asia Pacific Software Engineering Conference APSEC 2006, Bangalore (2006)
22. Higo, Y., Kamiya, T., kusumoto, S., Inoue, K.: Method and implementation for investigating code clones in a software system. Inf. Softw. Technol. **49**(5), 985–998 (2006)
23. Gode, N., Hummel, B., Juergens, E.: What clone coverage can tell. In: Proceedings of 6th International Workshop on Software Clones IWSC 2012, Zurich (2012)

# Mobile Edge Computing-Enabled Blockchain Framework—A Survey

**Pronaya Bhattacharya, Sudeep Tanwar, Rushabh Shah and Akhilesh Ladha**

**Abstract** Mobile edge computing (MEC) enables cloud-based services to extend to edge networks consisting of mobile base systems. MEC provides software and hardware platforms to incorporate seamless and decentralized data management schemes adjacent to base systems, thus reducing the end-to-end latency of the user. It is an integral component of the fifth-generation (5G) architecture and operates by providing innovative IT-based services. MEC spans across multiple authoritative domains where trust and interoperability among nodes is a prime concern between low power-enabled sensor nodes, as in the case of Internet of things (IoT)-based environments. The requirements of trust and interoperability make a blockchain framework applicable to MEC platform. In such platforms, miners can solve computationally expensive proof-of-work (PoW) puzzles containing mobile transactions as blocks added to immutable ledger so that a substantial amount of CPU computations and energy constraints are consumed. This article presents a systematic survey of MEC architecture and introduces a mobile blockchain framework that can be incorporated with the MEC architecture to facilitate the mining scheme. Then, the article analyzes the effects of integration of blockchain with MEC platform. Finally, concluding remarks and future work are provided.

**Keywords** Mobile edge computing · Mobile blockchain · Mining · 5G · IoT nodes

P. Bhattacharya (✉) · S. Tanwar · R. Shah · A. Ladha
Department of Computer Science and Engineering, Institute of Technology, Nirma University, Ahmedabad, Gujarat 382481, India
e-mail: pronoya.bhattacharya@nirmauni.ac.in

S. Tanwar
e-mail: sudeep.tanwar@nirmauni.ac.in

R. Shah
e-mail: rushabh.shah@nirmauni.ac.in

A. Ladha
e-mail: akhilesh.ladha@nirmauni.ac.in

P. Bhattacharya
Department of Computer Science and Engineering, Dr. A.P.J Abdul Kalam Technical University, Lucknow, Uttar Pradesh 226031, India



P. K. Singh et al. (eds.), *Proceedings of ICRIC 2019*, Lecture Notes in Electrical Engineering 597, https://doi.org/10.1007/978-3-030-29407-6_57

# 1 Introduction

The cellular infrastructures of today are facing a demand-driven explosion to provide quality of service (QoS) to various data-hungry mobile applications [1]. Earlier, mobile cloud computing (MCC) was proposed as a solution because it integrates cloud and mobile platforms to increase capabilities of mobile nodes in terms of storage and energy requirements as a centralized cloud service [2, 3]. MCC suffers from many security vulnerabilities and latency in data transmission, thus making it unsuitable for real-time services [4]. MEC addresses the challenges of MCC by designating cloud resources to edge systems within a radio network (RN). Thus, end-user accesses data through RN, and hence, user experience is enhanced as powerful computing is now possible with services like location and context awareness closer to the user [5, 6], within normally 1–2 hops. This drastically reduces end-to-end latency and solves issues related to network congestion. Figure 1 shows the MEC architecture which includes modular routing [7], network scalability [8], and platform services [9].

Thus, IoT-enabled mobile devices can access the edge servers to enhance computing capability and meet the low latency requirements as imposed by 5G [10]. The above promising architecture, however, has some serious drawbacks. In a distributed computing environment, the edge network analytics support serves as centralized support to various mobile users within an RN. Moreover, the edge network engine undergoes a highly power-intensive CPU computation due to ever-increasing data. This leads to more battery drain of mobile devices even in the presence of low powered protocols in mobile-based IoT applications like Message Queuing Telemetry Transport (MQTT) and Constrained Application Protocol (COAP). Also, as multiple authoritative domains have built their infrastructure over the cloud, we need a trust-based mechanism between the various communicating systems. Blockchain comes to the rescue by providing a trust-enabled smart edge network system [11], where a service provider can facilitate IoT-enabled mobile devices to operate via edge computing service node to support various blockchain applications.

As shown in Fig. 2, a blockchain is a distributed ledger over a public or private network that records transactions between peer nodes that do not trust each other. The information or data as transactions are hashed, verified, and mined into blocks which are added to the chain by miners based on a consensus mechanism. The addition

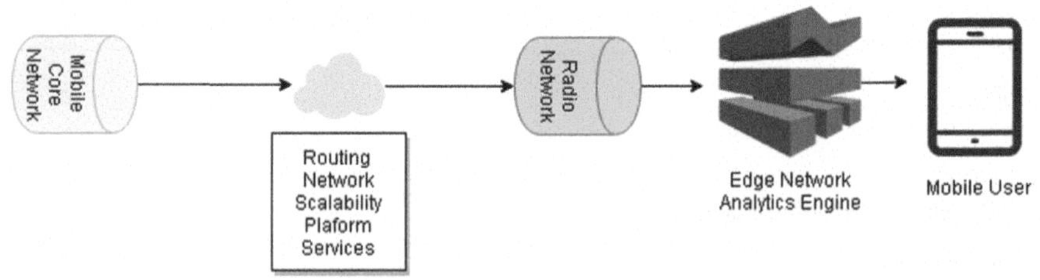


**Fig. 1** MEC architecture

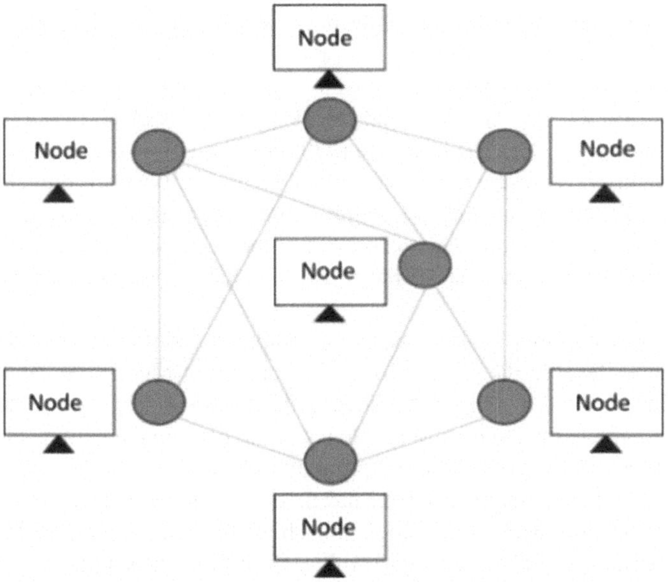


**Fig. 2** A simplified look of blockchain architecture

of blocks is always done to the longest valid chain. This rule of longest chaining allows recorded transactions in blocks to be immutable as any change to the block will change hash value leading to the invalidation of blocks. Thus, the valid chain provides a history of transactions as logs which can be verified and created at any moment in the network.

The rapid growth of business processes has led to an inevitable requirement of shifting security processes over blockchain networks as they ensure trust and transparency. Today, blockchain is gaining more agility as it is integrating with many domains like finance [12, 13] in the form of digital assets, remittance, and online payments. Also, blockchain is widely used as emerging technology in IoT [14, 15], smart contracts [16], healthcare industry [17, 18], voting [19], and verification of educational documents [20]. Further, blockchain can be used in a transactional manner in tracking tangible luxury items, intellectual property rights, and many other uses.

Blockchain achieves consistency in transactions by accounting for auditability, atomicity, and integrity of data over distributed autonomous platforms, where peer nodes do not trust each other. They are similar to distributed systems where nodes continuously check other nodes integrity using a consensus protocol to agree on a common state of the chain. The chains are cryptographically auditable as they rely on Merkle root value and order-execute architecture in which blockchain network orders the transactions first using a consensus protocol and then executes them in the listed order in all peer nodes in a sequential manner. The entities involved in the transaction performs an update to their local copy of the document which is then added by computing the hash value of the document which could be digitally signed using users' private/public key pairs and added to the chain. The validation of the transactions is done by miners which add a block to a chain. This logical chaining is done by the process of hashing of data blocks, where any block $B_i$ stores the hash of its previous block $B_{i-1}$. The hash in any $i$th block is computed as $H_i = f(\text{input}_i, \text{ID}_i, \text{Timestamp}, H_{i-1})$ where $\text{input}_i$ is the input document, $\text{ID}_i$ is the digital identifier associated with the document, Timestamp is the current timestamp value, and $H_i$ and $H_{i-1}$ are the hashes of current and previous blocks, respectively. The blocks link to form a trace back to the genesis block, thus allowing consensus in a blockchain network.

Also as shown in Fig. 3, all the hash are computed and used to form the hash at the next higher level in the chain. This is the concept of Merkle tree, and the final Merkle value is stored in a block; hence even there is a tamper in one of the blocks, it leads to complete invalidation of all blocks in the path, to the genesis block. This makes the blockchain system "tamper-proof and secure."

To add a block, a miner must solve a puzzle in a challenge-response environment by guessing starting bytes of the block in such a way that the hash of the block is smaller than the acceptable target hash value. Each block acts a puzzle for a miner

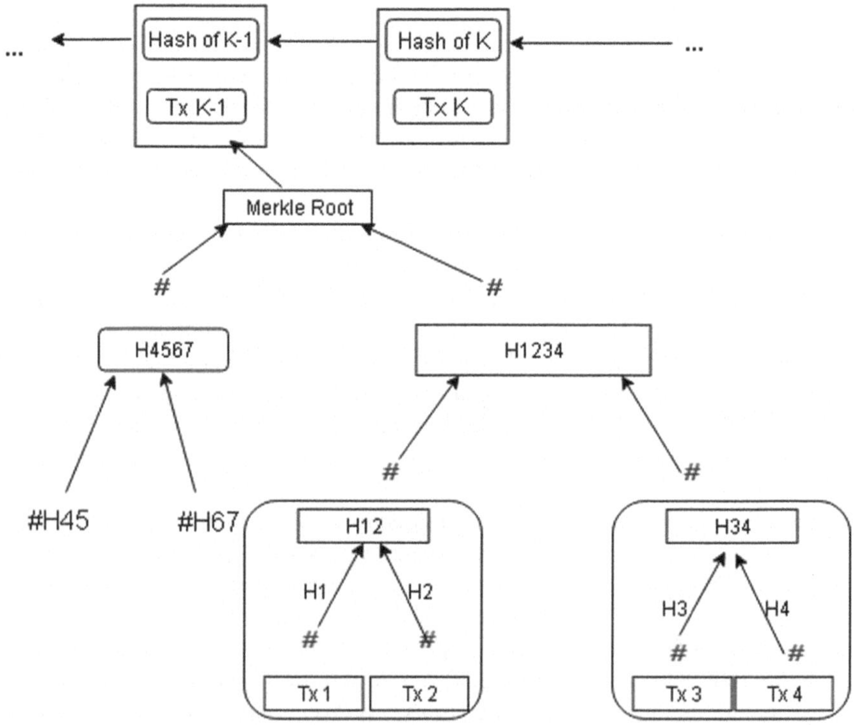


**Fig. 3** Overview of blockchain transactions and Merkle root

which is termed as nonce or difficulty value. Once the nonce is solved by a miner, the block gets appended to the existing validated chain by appending the hash value of the chain to the block. The above concept is also known as "proof-of-work (PoW)" in a blockchain network. The copies of the new block are added to all nodes in the network maintaining consensus.

The remainder of the article is organized as follows. Section 2 provides an overview of MEC architecture. Section 3 provides the technological aspects of integration of blockchain in MEC and designing of mining as a service (MaaS) in MEC architecture. Section 4 discusses the proposed framework for mobile blockchain in MEC with possible rewards schemes for miners. Finally, Sect. 5 discusses future directions and concluding remarks.

## 2 Overview of MEC Architecture

### *2.1 Modulars in MEC*

MEC refers to service environment close to the user within an RN. Thus, deploying MEC as base station improves bottlenecks and increases system robustness [4, 21]. According to the technical white paper by the European Telecommunications Standards Institute (ETSI), MEC can be categorized as on-premises, proximity, lower latency, location awareness, and network context information [22]. MEC can be implemented as a software entity such as Open vSwitch (OVS) [23]. MEC platforms include three functionalities, namely routing modular, network capability exposure modular and management modular. Routing modular is responsible for forwarding packets between RN and user. We can define a software-defined flow to smoothly conduct the offered load. Network capability exposure modular securely provides network services like location, video/voice calling through the invocation of suitable application programming interfaces (API), thus providing platform as a service (PaaS) [24]. Management modular deals with the management of local IT infrastructure in third-party applications forming infrastructure as a service (IaaS), such as OpenStack [9]. The interactions among the modulars are shown in Fig. 4.

### *2.2 The MEC Architecture*

A radio access network (RAN) is used at the lowest level of communication which facilitates the connections between the mobile devices and the edge network [25]. The RAN networks normally employ a 4G long-term evolution (LTE) and distribute a wide geographical area in smaller clusters, which are then controlled by radio network controller (RNC). The RNC is responsible to control the base station nodes and to carry out network management functions.

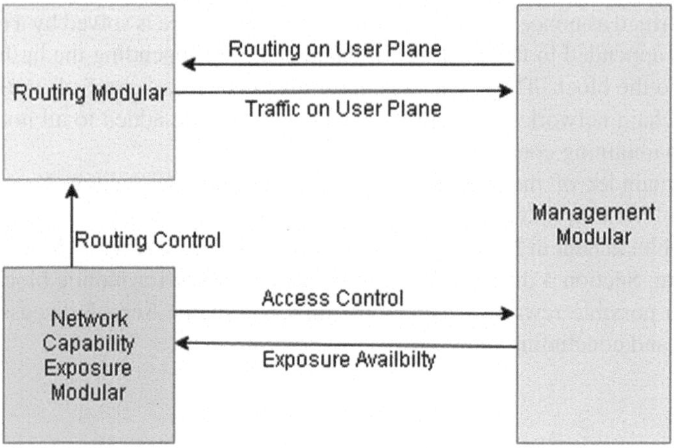


**Fig. 4** Interaction among various modulars in MEC architecture

The three-layer MEC architecture is proposed as follows:

1. **User Interface Layer**—The user layer is normally the devices which gather data like mobile, IoT sensors, social networks, and big-data applications which normally communicate with the RAN network. The applications need to transfer huge data for computation to the MEC edge servers.
2. **MEC Servers**—It is the most important part of the architecture, and it mainly consists of geo-distributed user interface layer. The user layer is normally the devices which gather data like mobile, IoT sensors, social networks, and big-data applications which normally communicate with the RAN network. The applications need to transfer huge data for computation to the MEC edge servers or virtual servers that have built-in IT capability. These MEC servers provide content offloading services where the useful content of the applications could be kept at servers and downloaded whenever required. This ensures resource optimizations and saves useful time.
3. **Cloud Servers**—The content which is only requiring heavy computations is forwarded to the cloud platform, and the results are shared back to the MEC server.

As shown in Fig. 5, at the edge of the architecture we have mobile devices which install application and process data. The applications communicate with a middle layer, which are MEC servers which is a virtualization of the cloud services and incorporates a local infrastructure, thus provide infrastructure as a service (IaaS). All the extensive computations are now performed at the MEC nodes which make a quick response to a user application. At the top level, we have the cloud-based services for computations not possible at edge level, and one deployment is performed at cloud nodes, content replicas are again maintained at MEC nodes to facilitate faster processing.

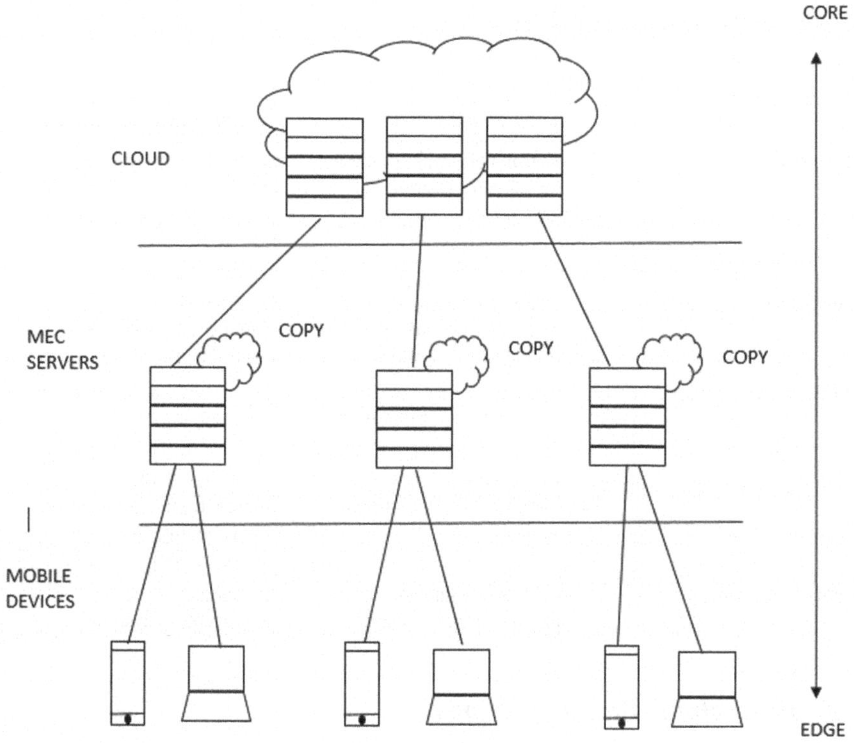


**Fig. 5** MEC architecture

# 3 Blockchain Consensus and Mining in MEC Architecture

## *3.1 Security Issues in MEC Architecture*

A joint collaboration between the European Telecommunications Standards Institute (ETSI) and Industry Specification Group (ISG) standardized the MEC architecture. MEC also operates upon 5G infrastructure-based public–private partnership (PPP) [14]. MEC can be characterized into various forms such as on-premises, proximity, lower latency, location awareness, and network context information [26]. MEC architecture suffers from security and privacy concerns. Some of them are listed in Table 1.

**Table 1** MEC security architecture issues

| Security parameters | Services violation | Possible attacks |
|---|---|---|
| Confidentiality | Location aware services to the end-user | Interception, packet sniffing of MEC and cloud channels |
| Integrity | Multi-management domains, sharing identifications in cloud servers | Authentication from an attacker on cloud platforms, masquerading sensitive information from cloud servers |
| Availability | Compromised IoT sensors operating on cloud storing user data | Distributed denial-of-service(DDoS) attacks, ripple effects |
| MEC server security | Physical security breaches, design flaws, configuration errors | DDoS Attacks, hijacking of cloud servers |
| Cloud virtualization security | Bot virtual machines created to drain out computational resources | Agent-based attacks, malfunctioning application programming interfaces (API), byzantine attacks |
| End-device security | Inject false values or information to systems | Injection attacks, compromised systems |

## *3.2 Blockchain-Based Solutions*

A blockchain-based edge computing system works in the following manner. Firstly, a blockchain user creates a transaction which can be transferred to a neighboring node. Each neighboring peer now collects the transferred transactions over a certain time period discarding the fake transactions. After the time period, the neighboring peers pack the transactions in a block mining is done by solving a difficulty based nonce called PoW. The mined blocks are now validated by a majority of peers and appended to the longest-running chain achieving consensus.

As shown in Fig. 6, the data stored in cloud servers are passed through IP routers normally employing type of service handshake parameters with client applications and passed to the MEC service providers, normally within an RN. The RN can use communication technologies like 5G and provide smaller MECs to serve smaller cells. These cells, normally called microcells, or picocells or femtocells as in the case of 5G architecture, try to provide dedicated service to smaller user groups. This acts as an edge layer to end-user devices. The transactions performed by end-users are then mined as blocks and stored in either public or private blockchain networks. Several approaches are applied to achieve trust and low powered computations as discussed below.

**Designing Complex Proof-of-Work Systems**—An attacker can create a piece of false blockchain information and bot users in the network. Then, using these bots he can create false transactions and fake blockchain information. These attacks are

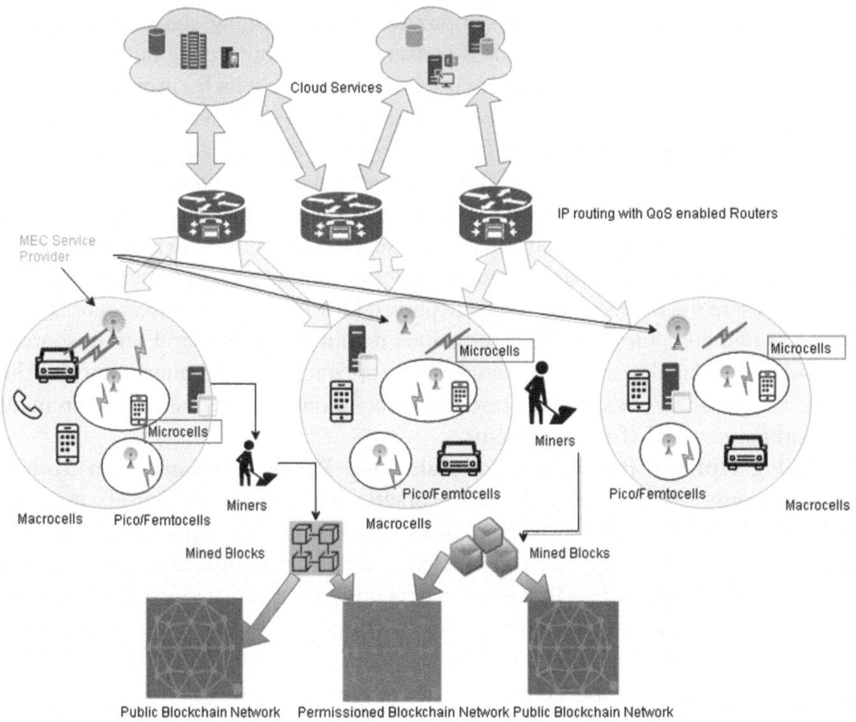


**Fig. 6** MEC-enabled blockchain network

known as Sybil attacks. One way to curb these attacks is to increase the difficulty of mining algorithm so that an attacker does not have enough computing resources in resource-constrained environments; hence, the environment cannot support the attacker. The attackers provide a nonce value that minimizes the hash value of the packet header below the required threshold of the difficulty level [27]. To manipulate the network, the attacker has to gain a majority in consensus; i.e., it has to achieve a computing power of 51% of total network power. Since the attacker has to hold 51% of the computing power in the network to manipulate the network, this becomes computationally infeasible in a resource-constrained environment.

**Designing Mining as a Service (MaaS) for Mobile Blockchain**—Considering the requirements for IoT-based environments, authors in [14] suggested the incorporation of many blockchain-based solutions that operate at low energy and lower communication overheads. Since IoT devices combine many low-powered sensor and actuator devices, exchange of information over geographically distributed environment poses a major challenge. Further, the complexity of the mining algorithm in limited energy levels of the network becomes a challenging issue. The solution to the above problem is allowing small data servers in an RN to accept offloaded jobs to execute from adjacent mobile and IoT devices [26] in a MEC-enabled blockchain environment. Allowing this local computation solves the problem of blockchain deployment in

IoT by allowing PoW difficulty based puzzles, strong hashing algorithms, encryption of data, and achieving distributed consensus. The above integration facilitates cloud hashing and achieving mining as a service(MaaS) in which a user can buy software services in the cloud, to mine blocks and generate incentives, without actually investing in installing hardware platforms. This allows miner nodes to be sufficiently closer to the edge devices, which further reduces the overall complexity and propagation latency for the end user which is suitable in resource-constrained environments, normally found in delay-sensitive IoT-based services. A suitable example would be providing authentication-enabled data privacy for smart homes and smart grid-based systems where the MaaS nodes can be deployed near to grid meters to execute smart contracts and compute resource reservations required for the user. Extra resources allocated to grid nodes are not executed as they are not part of the smart contract. The above scheme allows flexible user resource reservation, which is the key requirement in smart automated IoT-based systems.
**Designing Optimal Parameters for Balancing Resource Demands in Mobile Environments**—Due to limited energy, practically demand of all users may not be fulfilled. This leads to a resource allocation problem. Also, a particular user may define a software-defined networking (SDN) flow and have a different set of value for a service than another user. The valuation depends on certain factors like the number of transactions in a block and mining rewards. The edge computing provider can thus maximize profits by adjusting the price levels based on the demand of competing users. Thus, a direct proportionality can be achieved giving an optimal economic model for resource allocation in an edge computing environment.

## 4 Proposed Mobile Blockchain-Enabled Edge Framework

The data recorded by the sensor nodes are first sent to the edge servers that will now run the client blockchain application, thus allowing the mining of nodes on edge servers, instead of sending data to cloud platforms. The basic steps can be now computed as follows:

- Each user $U_i$, where $i = \{1, 2, 3, \ldots N\}$, will run the client-based blockchain API for recording the data on their systems. These data or readings are recorded as transactional data for these systems.
- The proposed framework will consist of $N$ users, or systems, which will act as a miner node. The role of miner nodes is to send a request for seeking computational power to the edge computing server, normally running the server version of blockchain API synchronized with the client-based API interface. The server application might be running Ethereum or Ripple for the execution of smart contracts to provide smart reservation as well as guaranteeing only the desired resource requirement. Thus, smart execution leads to the exact resource reservation to the miner nodes.

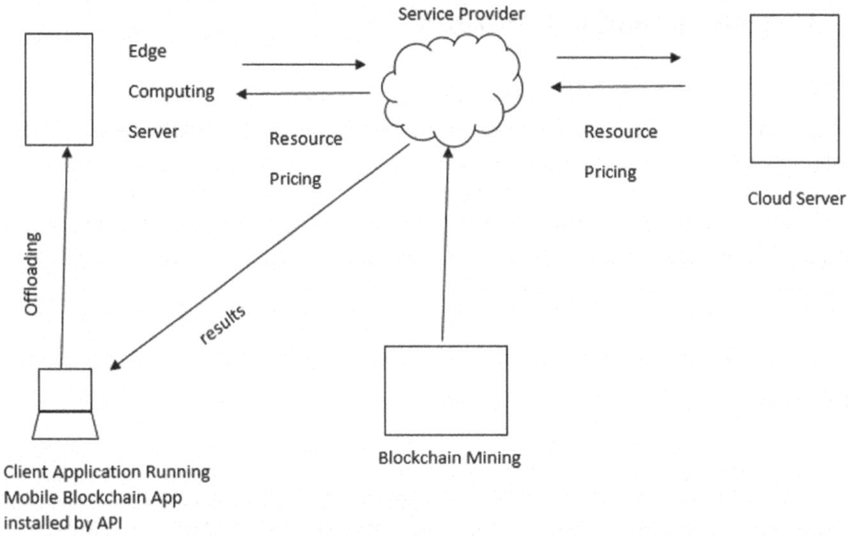


**Fig. 7** Proposed mobile blockchain-enabled edge computing framework

- The above scenario leads to offloading of computational power to the edge systems which provides the desired MaaS to the miner nodes. The miner nodes now can solve the PoW puzzles on edge nodes, and their pricing schemes are now governed by the edge nodes rather than the cloud nodes leading to the efficient design of pricing mechanisms for the miner nodes. In addition to this, MaaS also provides user infrastructure to build efficient cloud-based applications.

As shown in Fig. 7, the edge computing server or the MEC deploys the infrastructure on a service provider on which the mining as a service is performed and the reward scheme for miners is decided at the middle level. The details are then transferred to the cloud server where the resource pricing is decided for using infrastructure services and informed back to the client. The client can also occasionally offload jobs to be executed at the edge server and blocks are formed by miners once the transactions are verified and added to the chain. The resource-intensive PoW puzzles are solved by miners by taking resources from service provider; hence, the proposed architecture does not drain the limited energy or battery power of the mobile devices; thus, trust management is now added to the edge platform using blockchain network; and dually, the limited energy sources of the client node are also saved. This framework will be beneficial to operate in low-powered energy environments, namely monitoring services in IoT platforms where sensors can be installed in client nodes and monitoring can be done at MEC servers.

## 5 Conclusions and Future Work

In this article, we have proposed a blockchain mining framework that can solve complex proof-of-work (PoW) puzzles for mobile blockchain applications, especially for IoT-based mining tasks where resource optimization is a major concern. In the future, we would like to explore the results and the impact of the mining scheme and rewards and pricing of miners. We would also like to develop an efficient reward-based scheme for miners and also a consensus scheme which will simulate the block addition by solving PoW puzzles in such low-powered environments.

## References


1. Borgia, E., Bruno, R., Conti, M., Mascitti, D., Passarella, A.: Mobile edge clouds for information-centric IoT services. In: Proceedings of IEEE Symposium on Computers and Communications (ISCC), Messina, Italy, June 2016, pp. 422–428. Author, F.: Article title. Journal **2**(5), 99–110 (2016)
2. Marotta, M.A., et al.: Managing mobile cloud computing considering objective and subjective perspectives. Comput. Netw. **93**, 531–542, Oct. [Online]. Available: http://www.sciencedirect.com/science/article/pii/S1389128615003667 (2015)
3. Dinh, H.T., Lee, C., Niyato, D., Wang, P.: A survey of mobile cloud computing: architecture, applications, and approaches. Wirel. Commun. Mob. Comput. **13**(18), 1587–1611 (2013)
4. Jararweh, Y., et al.: The future of mobile cloud computing: integrating cloudlets and mobile edge computing. In: Proceedings of 23rd International Conference on Telecommunications (ICT), pp. 1–5. Thessaloniki, Greece (2016)
5. Yi, S., Li, C., Li, Q.: A survey of fog computing: concepts, applications, and issues. In: Proceedings of Workshop Mobile Big Data (Mobidata), pp. 37–42. Hangzhou, China (2015)
6. Jararweh, Y., et al.: SDMEC: software defined system for mobile edge computing. In: Proceedings of IEEE International Conference on Cloud Engineering Workshop (IC2EW), pp. 88–93 Berlin, Germany (2016)
7. European Telecommunication Standards Institute. Mobile Edge Computing Introductory Technical. Whitepaper (2019)
8. Suikkola, V.: Open exposure of telco capabilities—identification of critical success factors for location-based services in open telco. In: 6th International Conference on Wireless and Mobile Communications, pp. 202–208. IEEE Press: Valencia, Spain (2010)
9. Moreno-Vozmediano, R., Montero, R.S., Llorente, I.M.: IaaS cloud architecture: from virtualized datacenters to federated cloud infrastructures. Computer **45**(12), 6572 (2012)
10. Wong, V.W., et al.: Key technologies for 5G wireless systems. Cambridge University Press (2017)
11. Zhang, Y., et al.: Offloading in software defined network at edge with information asymmetry: a contract theoretical approach. J. Signal Process. Syst. **83**(2), 241–253 (2016)
12. Foroglou, G., Tsilidou, A.L.: Further applications of the blockchain (2015)
13. Peters, G.W., Panayi, E., Chapelle, A.: Trends in crypto-currencies and blockchain technologies: a monetary theory and regulation perspective (2015)
14. Christidis, K., Devetsikiotis, M.: Blockchains and smart contracts for the Internet of Things. IEEE Access **4**, 2292–2303 (2016)
15. Zhang, Y., Wen, J.: An IoT electric business model based on the protocol of bitcoin. In: Proceedings of 18th International Conference on Intelligence in Next Generation Networks (ICIN). pp. 184–191. Paris, France (2015)

16. Kosba, A., Miller, A., Shi, E., Wen, Z., Papamanthou, C.: Hawk: the blockchain model of cryptography and privacy-preserving smart contracts. In: Proceedings of IEEE Symposium on Security and Privacy (SP). pp. 839–858. San Jose, CA, USA (2016)
17. Peterson, K., Deeduvanu, R., Kanjamala, P., Mayo, K.B.: A blockchain-based approach to health information exchange networks (2016)
18. Vora, J., et al.: BHEEM: a blockchain-based framework for securing electronic health records, 2018 IEEE Globecom Workshops (GC Wkshps), Abu Dhabi, United Arab Emirates, pp. 1–6 (2018)
19. Wang, L., Liu, W., Han, X.: Blockchain-based government information resource sharing. In: 2017 IEEE 23rd International Conference on Parallel and Distributed Systems (ICPADS), pp. 804–809. Shenzhen (2017)
20. Srivastava, A., Bhattacharya, P., Singh, A., Mathur, A., Prakash, O., Pradhan, R.: A distributed credit transfer educational framework based on blockchain. In: IEEE 2018 2nd International Conference on Advances in Computing, Control and Communication Technology (IA3CT 2018), Allahabad, Uttar Pradesh, India, pp. 54–59 (2018)
21. Satria, D., Park, D., Jo, M.: Recovery for overloaded mobile edge computing. Futur. Gener. Comput. Syst. **70**, 138–147 (2017)
22. Patel, M., et al.: Mobile-edge computing—introductory technical white paper. In: White Paper, Mobile-Edge Computing (MEC) Industry Initiative (2014)
23. Pfaff, B., Pettit, J., Koponen, T., et al.: The design and implementation of open vSwitch. In: Networked Systems Design and Implementation (2015)
24. Beimborn, D., Miletzki, T., Wenzel, S., et al.: Platform as a Service (PaaS). Bus. Inf. Syst. Eng. **3**(6), 381–384 (2011)
25. Commun. (ICFCC), Kuala Lumpar, Malaysia, pp. 334–338. CommVerge. (2016). Radio Access Network (RAN) Optimization. Last Accessed on 19 Feb 2002. [Online]. Available: http://www.commverge.com/Solutions/SubscribersServicesManagement/RANOptimization/tabid/174/Default.aspx (2009)
26. Wu, Y., et al.: Joint traffic scheduling and resource allocations for traffic offloading with secrecy-provisioning. IEEE Trans. Vehic. Tech. **66**(9), 8315–8332 (2017)
27. Pass, R., Shi, E.: FruitChains: a fair blockchain. In: PODC'17 Proceedings of ACM Symposium, Principles of Distributed Computing, pp. 315–24 Washington, DC (2017)

# Performance Evaluation of *Snort* and *Suricata* Intrusion Detection Systems on Ubuntu Server

**Alka Gupta and Lalit Sen Sharma**

**Abstract** Network intrusion detection systems (*NIDS*) are emerging as a reliable solution in providing protection against threats to integrity and confidentiality of the information on the Internet. Two widely used open-source intrusion detection systems are *Snort* and *Suricata*. In this paper, *Snort* and *Suricata* are compared experimentally through a series of tests to identify more scalable and reliable *IDS* by putting the systems under high traffic. Results indicated that *Snort* had a lower system overhead than *Suricata* and utilized only one processor on a multi-core environment. However, *Suricata* evenly utilized all the processing elements of the multi-core environment and provided higher packet analysis rate. For malicious traffic, both *Snort* and *Suricata* dropped packets with *Snort* on the higher side for low traffic rate and size. But with large packet size and high rate of malicious input traffic, *Suricata* dropped more packets as compared to *Snort*. It was also observed that the memory utilization of *Suricata* depended on both the size of traffic and the amount of malicious traffic; whereas, memory utilization of *Snort* was independent of the input traffic.

**Keywords** *Snort* · *Suricata* · Performance · *NIDS* · Multi-threaded · Multi-core · Experiment

## 1 Introduction

Attempts to breach into information systems and networks are on a rise [1], and for combating them we need network intrusion detection and prevention system (*NIPS*) which capture and inspect network packets for signs of any malicious activity. In real-time networks, traffic may vary over a large range and keeping up with all the traffic while inspecting it is a processor-intensive as well as memory-intensive activity which may add to network latency. If *NIDPS* is unable to keep pace with the

A. Gupta (✉) · L. S. Sharma
Department of Computer Science and IT, University of Jammu, Jammu, Jammu and Kashmir, India
e-mail: alkagupta48@gmail.com

L. S. Sharma
e-mail: lalitsen@yahoo.com

 
P. K. Singh et al. (eds.), *Proceedings of ICRIC 2019*, Lecture Notes in Electrical Engineering 597, https://doi.org/10.1007/978-3-030-29407-6_58

traffic in real time, then uninspected packets are either dropped, causing problems for genuine traffic, or forwarded without checking, posing a threat to network security. Another issue with *IDS* is the amount of memory the *NIDPS* consumes. Many *NIDPS* like *Snort* buffers the network data in memory while analyzing it, and so running out of RAM to store network data can lead to dropping of packets. So, a reliable, accurate, scalable and memory-efficient *NIDPS* is required to ensure the security and interminable performance of the network at all times.

Network intrusion detection systems monitor the network and identify and report the entities that are attempting to compromise the confidentiality, integrity or availability of a resource. Intrusion prevention system, however, performs the same process, with an added ability to even block the intrusion attempts. *Snort* and *Suricata* are two open-source *NIDS/NIPS* which are extensively used for monitoring, detecting and preventing of information security attacks on networks. The proposed research work will provide accurate, detailed, current and technical information about the performance measurements of *Snort* and *Suricata*. This would help in finding the performance bottlenecks in order to improve the detection rate of the *IDS*.

## *1.1 Snort*

*Snort*, developed by Martin Roesch, was initially launched as a lightweight cross-platform packet sniffing device [2] and was later released with IDS feature in 2003. It is an open-source development methodology where users can contribute to the development of the system by suggesting modifications in source code, report bugs and suggest bug fixes. It has now evolved into a powerful intrusion detection and prevention system, and more than 5 million [3] users have downloaded it till date.

*Snort* can be configured in three main modes: sniffer, packet logger and network intrusion detection. In sniffer mode, the system reads network packets and displays them on the console. In packet logger mode, it logs the packets to the disk; whereas, in intrusion detection mode, the system monitors network traffic and analyzes it against a rule-set defined by the security analyst. The program will then perform a specific action based on what has been identified. *Snort* uses a rule-driven language which combines the benefits of signature, protocol and anomaly-based inspection methods. *SNORT* uses deep packet inspection (DPI) for examining packets over stateful packet inspection (SPI) as SPI inspects only packet header; whereas, DPI first examines the packet header, but in cases where this is not sufficient it goes on to examine the packet contents as well. In addition to this, DPI is also capable of examining the contents across multiple packets.

## *1.2 Suricata*

*Suricata* is developed by Open Information Security Foundation (*OISF*) under a project funded by the US Department of Homeland Security in order to build an alternative to *Snort*. *Suricata* is an open-source, free, fast and robust detection engine which is capable of real-time intrusion detection (*IDS*), inline intrusion prevention (*IPS*), network security monitoring (*NSM*) and offline pcap processing. It is a multi-threaded application unlike *Snort* and has an extensive and powerful rule-set and signature database for detecting a wide range of network threats. It works on *Linux, FreeBSD, Open BSD, Mac* and *Windows*.

## *1.3* Snort *Versus* Suricata

With *Snort* being the most popular and widely used IDS, the need for *Suricata* arises from the performance limitations of *Snort*'s single-threaded architecture. As *Snort* compared all incoming packets with its rule-set to identify threats and so when it is exposed to a high rate of malicious traffic, it tends to drop packets. This task of comparing packets with multiple rules can be parallelized, and so a multi-threaded comparison engine can prove out to provide higher performance and scalability with less or no packet drops. Table 1 lists some points of difference between the two.

**Table 1** Comparison of *Snort* and *Suricata IDS*

| Parameter | Intrusion detection system | |
|---|---|---|
| | *Snort* | *Suricata* |
| Multi-threaded | No | Yes |
| Operating systems | All | All |
| Developer | Sourcefire | Open Information Security Foundation |
| Rules | VRT *Snort* rules, SO rules, pre-processor rules, emerging threats rules | Emerging threats rules, VRT *Snort* rules |
| Installation | Manual or using packages | Manual or using packages |
| User-friendly | More | Less |
| Documentation | Well-documented and provides solutions to common issues | Not well-documented |
| Cost | Commercial version has a price | Free |
| GUI | Large number of compatible GUIs | Very few |
| High-speed network support | Not present | Present |

*Suricata* has an added functionality of application-aware detection rules which help in detecting protocol-specific traffic on non-standard ports and can apply protocol-specific log settings to these detections as well. This ability is also provided in *Snort* but to a small extent using OpenAppID. *Snort* also contains features unavailable in *Suricata* such as the option of hiding certain rules that are not required for inspecting particular network traffic.

## 2 Previous Work

Whitea et al. in [4] compared the performance of *Suricata* with *Snort* (single and multi-instance) on the basis of scalability and performance. They performed a total of 8600 tests by varying the number of cores used (1–24 cores), the rule-sets used for signature comparison, the workload used to obtain results and the configuration of both the IDSs. The metrics used for comparison were packets per second (PPS) as processed by each IDS, the amount of memory used by each IDS process and the CPU utilization. Results showed that both *Snort* and *Suricata* were scalable but *Suricata* outperformed *Snort* in almost all the test scenarios. *Suricata* also exhibited lower average memory usage and lower average CPU utilization.

Thongkanchorn et al. in [5] compared and analyzed the detection accuracy of three popular open-source intrusion detection systems—*Snort*, *Suricata* and Bro IDS. They studied the effect of number of active rules, different traffic rates and eight types of attacks on the evaluation efficiency of the intrusion detection systems and concluded that the use of different set of rules (active rules) for different attack types resulted in increased accuracy of the IDS. Also, Bro IDS showed better performance among other IDS systems when evaluated under different attack types and using a specific set of active rules.

In [6], *Snort*, Bro and *Suricata* are compared to know their advantages and disadvantages by following two stages of tests, scanning and penetration. They concluded that *Snort* and *Suricata* were easy to install and update rules, Bro requires the least amount of resources, and with *Suricata* the computer hung a number of times.

In [7], the performance of *Snort* and *Suricata* is analyzed packet loss at different traffic rates ranging up to 2 Gbps. The results detected a significant increase in the packet drop with the increase in traffic speed. Also with increase in the packet size, the packet loss decreased. However, the results seemed inconsistent with regard to packet size, and the results are not inferred properly.

*Snort* and *Suricata* are compared on network traffic of Naval Postgraduate School in a series of three tests [8]. Both IDSs generated false positives and false negatives which were attributed to weaknesses of the rule-sets used for the tests. *Suricata* required more memory and CPU resources than *Snort* but is scalable and can accommodate an increase in network traffic without requiring multiple instances. As the experiments are conducted on a virtual environment with a huge amount of data, the statistics were not accurate and IDS may behave different for a real network.

## 3 Experimental SetUp

The performance of *Snort* and *Suricata* is compared in network intrusion detection mode by analyzing their performance under high-speed and heavy load conditions. *Snort v2.9.12* is installed in its default configuration with 8453 rules provided by *Snort* Vulnerability Research Team (VRT). *D-ITG* is used to generate malicious traffic. The test bench consists of six computers in a star topology connected via a D-link web smart DGS-1210-16 16-port switch using 1.0 Gigabit Ethernet cable as shown in Fig. 1. Four traffic generators are used to generate high rate of traffic using D-ITG. Same traffic is sent to both the servers with *Snort* and *Suricata* to get accurate results. The hardware specifications of the systems are shown in Table 2.

Tests were conducted to evaluate and compare the performance of *Snort* and *Suricata* for different packet sizes (1400, 1024, 512 bytes) at different traffic rates

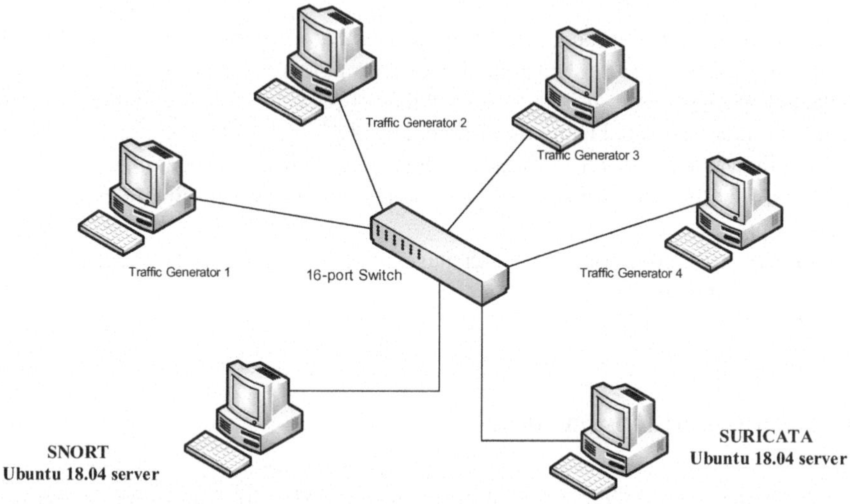


**Fig. 1** Experimental setup

**Table 2** System specifications

| Machine | Description | Specifications |
|---|---|---|
| Traffic generators 1–4 | Dell Intel(R) core(TM) i3-3110M CPU @ 2.40 GHz, 8 GB RAM | Linux Ubuntu with *D-ITG* traffic generator |
| *Snort* v2.9.12 | Hp Intel(R) core(TM) i7-4770M CPU @ 2.40 GHz, 8 GB RAM (8 cores) | Ubuntu 18.04 server |
| *Suricata* v4.1.2 | Hp Intel(R) core(TM) i7-4770M CPU @ 2.40 GHz, 8 GB RAM (8 cores) | Ubuntu 18.04 server |

for TCP. The tests were performed at different packet rates of 10,000, 20,000, 30,000, 40,000 and 50,000 packets per second. The traffic generators were sending traffic to both the servers with the two IDSs for about 60 s. The sizes of packets used for test are intentionally chosen to be below one maximum transmission unit (MTU) as Internet traffic normally consists of packets pertaining to these sizes.

*Snort* version 2.9.12 and *Suricata* version 4.1.2 have been installed in their default configurations, and they write outputs to their respective files located in the default log directory. In order to get accurate results, both the IDSs are installed on dedicated machines with no background activities. Same rules are used for both the signature-based detectors, viz. *SNORT* VRT rule-set and emerging threats rule-set. The study further evaluates the effect of two types of traffic on the performance of IDS: (1) normal traffic (with no or less than 100 alerts) (2) malicious traffic.

Four performance metrics are used to calculate and compare the performance of the two IDS in different test scenarios. These metrics are based on the parameters that impact the performance of the IDS. The evaluation parameters are as follows:

- **Packet drop**: It is the total number of packets that were dropped by the *IDS* and were not checked. More the value of packet drop, less is the performance of *IDS*.
- **Packet analysis rate**: It is the total number of packets analyzed by the *IDS* per second. It is calculated using the formula: $\frac{\text{Total packets analysed by } IDS}{60}$
- More the value of packet analysis rate, higher is the performance of *IDS*.
- **CPU utilization (in %)**: Percentage of total amount of processing resource used by the process.
- **Memory utilization (in %)**: Percentage of total amount of physical memory utilized by the process.

## 4 Results and Observations

For this experiment, TCP packets of different sizes (512, 1024 and 1400 bytes) are sent at different rates of 10,000, 20,000, 30,000, 40,000 and 50,000 for 60 s to both *Snort* and *Suricata*, and number of packets analyzed and dropped by both the IDSs are recorded. Figure 2 illustrates the performance of both IDS systems for the packet size 512 bytes for both normal and malicious traffic.

With the increase in traffic rate, the packet analysis rate of both the IDSs also increased for packet size of 512 bytes. The packet analysis rate of *Snort* was always less than *Suricata* for all traffic rates of normal and malicious traffic except for normal traffic at 30,000 packets per second. Also, packet analysis rate of normal traffic is more than that for malicious traffic for both the intrusion detection systems.

At the packet size of 1024 bytes, *Suricata* always showed better packet analysis rate at all traffic rates for both normal and malicious TCP traffic. Also, packet analysis rate of normal traffic is more than that for malicious traffic for both *Snort* and *Suricata*.

Figure 3 shows the performance for a larger packet size of 1400 bytes. Both the IDSs showed similar performance to that of packet size (1024 bytes). The number

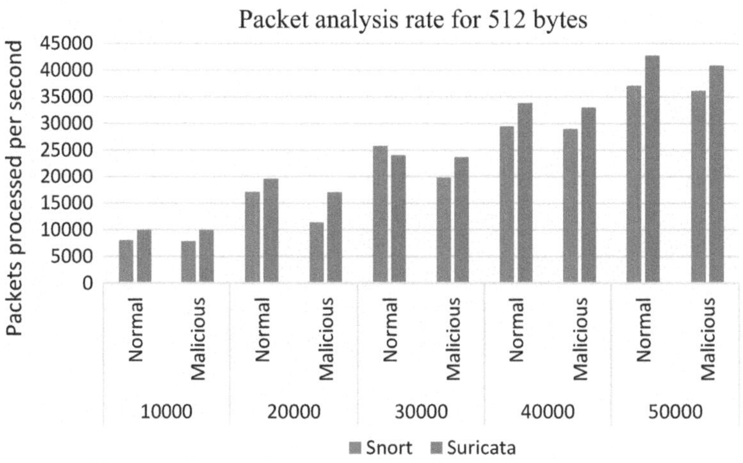


**Fig. 2** Packet analysis rate of TCP traffic of packet size of 512 bytes

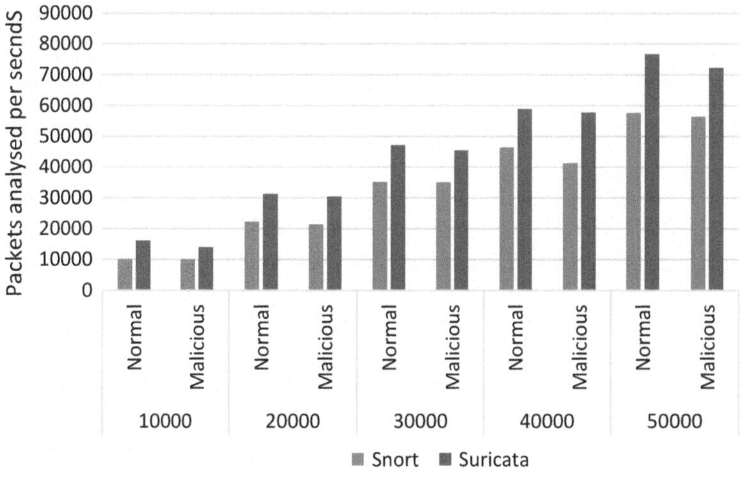


**Fig. 3** Packet analysis rate of TCP traffic of packet size of 1024 bytes

of packets analyzed by *Snort* is always less than that for *Suricata* for both normal and malicious traffic (Fig. 4).

During the tests, the CPU utilization and memory utilization have also been evaluated for both *Snort* and *Suricata* for all the test cases. In the following tables (Tables 3 and 4), CPU utilization of both the IDS systems for packet sizes of 1024 bytes and 1400 bytes is given for different traffic speeds.

For packet size of 1024 bytes, with the increase in traffic rate, the CPU utilization also increases linearly for both the IDSs. *Snort*, which is a single-threaded application, has a higher CPU utilization for malicious traffic than normal traffic as more

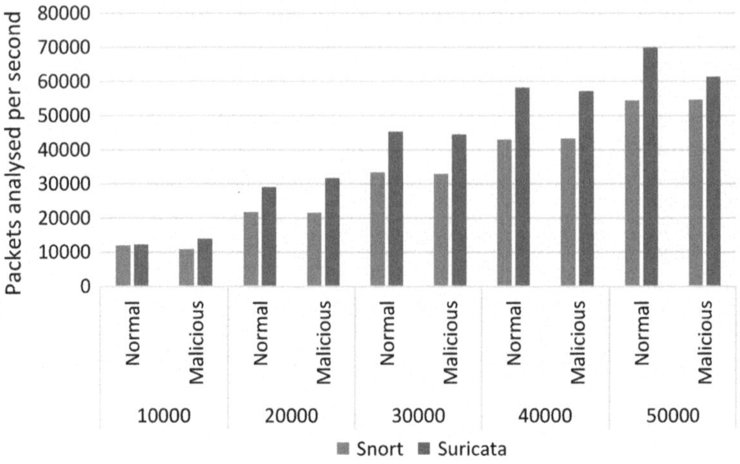


**Fig. 4** Packet analysis rate of TCP traffic of packet size of 1400 bytes

**Table 3** IDSs CPU utilization for TCP traffic—packet size of 1024

| Traffic rate (in packets per second) | *Snort* Normal | *Snort* Malicious | *Suricata* Normal | *Suricata* Malicious |
|---|---|---|---|---|
| 10,000 | 10.59 | 19.43 | 52.1 | 71.8 |
| 20,000 | 21.6 | 35.72 | 62.6 | 81.2 |
| 30,000 | 25.5 | 44.06 | 83.04 | 99.06 |
| 40,000 | 31.23 | 48.4 | 104.7 | 105.89 |
| 50,000 | 33.5 | 61.73 | 106.13 | 132.2 |

**Table 4** CPU utilization for TCP traffic—packet size of 1400

| Traffic rate (in packets per second) | *Snort* Normal | *Snort* Malicious | *Suricata* Normal | *Suricata* Malicious |
|---|---|---|---|---|
| 10,000 | 11.9 | 19.49 | 53.63 | 66.14 |
| 20,000 | 17.11 | 35.95 | 67.61 | 79.17 |
| 30,000 | 23 | 44.64 | 87.39 | 97.38 |
| 40,000 | 27.46 | 51.61 | 103.59 | 108.1 |
| 50,000 | 35.73 | 63.71 | 113.86 | 140.92 |

processing is required for identifying and logging a malicious packet on the disk. Similar behavior is shown by *Suricata*; however, as it is a multi-threaded application and the tests are carried out on eight-core system, the CPU utilization is more than 100%. In this case also, the CPU usage is more for malicious traffic than for normal traffic.

For larger packet size of 1400 bytes, CPU utilization is generally more than that for packet size of 1024 for all the four cases. Here, also for both the IDSs, CPU usage for malicious traffic is more than that for normal traffic due to more processing required in logging the alerts.

The amounts of packets that go unchecked by the *IDS* are summed up as dropped packets. More the number of dropped packets, lesser is the efficiency of the *IDS*. In Fig. 5, the malicious packets dropped by both the *IDS*s for two different packet sizes (1024 and 1400 bytes) and five different packet rates are represented. When normal traffic was sent, no or less than 1000 packets were dropped. When malicious packets were sent, the number of packets dropped was very large. For small packet size of 1024 bytes, both the *IDS*s show low packet drops but as rate increases from 30,000 packets per second, dropped packets increase from 5000 to 37,000 in 60 s.

For packet size of 1400 bytes and at higher traffic rates of 40,000 and 50,000, dropped number of packets is very high. The number of packets dropped by *Suricata* is low than *Snort* for lower traffic rates but as traffic rate reaches 50,000 packets per second, the number of packets dropped by *Suricata* showed a high jump. This shows the incapacity of both the *IDS*s to handle traffic of larger size at high rates.

Memory utilization percentage which gives the total amount of physical memory utilized by the process is given in Tables 5 and 6 for both the IDSs. The memory utilized by *Snort* is independent of traffic rate and type of input traffic; whereas, for *Suricata*, memory utilization is proportional to the amount of malicious traffic in the input.

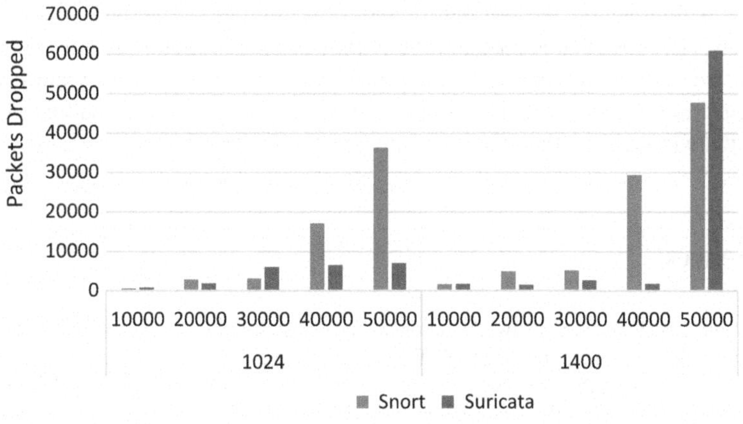


**Fig. 5** Packet drop for malicious traffic of packet size of 1024 and 1400 bytes

**Table 5** Memory utilization (in %) for packet size of 512 bytes

| Traffic rate (in packets per second) | *Snort* Normal | *Snort* Malicious | *Suricata* Normal | *Suricata* Malicious |
|---|---|---|---|---|
| 40,000 | 14.7 | 14.7 | 11 | 11.4 |
| 50,000 | 14.7 | 14.7 | 11 | 11.7 |

**Table 6** Memory utilization for packet size of 1400 bytes

| Traffic rate (in packets per second) | *Snort* Normal | *Snort* Malicious | *Suricata* Normal | *Suricata* Malicious |
|---|---|---|---|---|
| 40,000 | 14.7 | 14.7 | 11 | 11.8 |
| 50,000 | 14.7 | 14.7 | 11 | 12.2 |

## 5 Conclusion

The work presented in this paper focused on evaluating and comparing the performance of two well-known IDSs: *Snort* and *Suricata* in high-speed network. Both *Suricata* and *Snort* were evaluated on high-performance systems under various traffic speeds with different packet sizes and different types of traffic. The results revealed that due to *Suricata*'s ability of utilizing multiple cores uniformly, it showed better packet analysis rate than *Snort* in almost all the test cases and is scalable to accommodate increased network traffic. However, at high packet rate of 50,000 pps and large packet size of 1400 bytes, it showed an abrupt increase in the amount of dropped packets. Also, *Suricata*'s multi-threaded architecture requires more CPU resources than *Snort*, and it was found that the average CPU use of *Suricata* on multi-core system was nearly double of that for *Snort*. This could be attributed to the overhead required to manage multiple detection threads in *Suricata*. The memory utilization percentage of *Suricata* depends on the rate and type of input traffic unlike *Snort*.

*Snort* is lightweight and fast but limited to its ability to scale, but its processing overhead is less than that of *Suricata*. *Suricata* when deployed on single-core system is expected to show low performance. So, if resources and scalability are limited, then *Snort* still remains the first choice. But if network is scalable and processing cost is not an overhead, then *Suricata* emerges as an ideal choice.

## References

1. Agbogun, J.B., Ejiga, F.A.: Network security management: solutions to network intrusion related problems. Int. J. Comput. Inf. Technol. **02**(04), 617–625 (2013)
2. Roesch, M.: *Snort*-lightweight intrusion detection for networks. In: 13th Systems Administration Conference (LISA), Seattle, Washington, USA (1999)
3. "*Snort*," Sourcefire, [Online]. Available https://www.Snort.org/
4. Whitea, J.S., Fitzsimmonsb, T.T., Matthews, J.N.: Quantitative analysis of intrusion detection systems: *Snort* and *Suricata*. Int. Soc. Optics Photonics (2013)
5. Thongkanchorn, K., Ngamsuriyaroj, S., Visoottiviseth, V.: Evaluation studies of three intrusion detection systems under various attacks and rule sets. IEEE (2013)
6. Ridho, M.F., Yasin, F., Sulistyo, Y.: Analysis and evaluation *Snort*, bro, and *Suricata* as intrusion detection system based on Linux server. Naskah_Publikasi (2014)

7. Naidu, R.A., Avadhani, P.S.: A comparison of two intrusion detection systems. Int. J. Comput. Sci. Technol. **4**(1), 316–319 (2013)
8. Albin, E., Rowe, N.C.: A realistic experimental comparison of the *Suricata* and *Snort* intrusion-detection systems. In: 26th International Conference on Advanced Information Networking and Applications Workshops (2012)

# Global Smart Card ID Using RFID: Realization of Worldwide Human Mobility for Universal Validation

**Praveen Kumar Singh, Karan Dhawan, Neeraj Kumar and Bineet Kumar Gupta**

**Abstract** Today, human still carry number of cards to authenticate their identities across the globe. Smart cards possess potential to substitute all other existing IDs by smart card IDs at national and global level. Contemporary issues of terrorism and illegal migrants across many international borders too support the cause of seeking a viable solution. This paper has examined the evolution and necessity of human identification along with current worldwide scenario of existing national IDs. Integration of biometrics with smart card technology presents a strong authentication tool to the identity card holder. This paper offers an insight into the feasibility and technological aspects of this potential application. A recent trend of using radio-frequency identification (RFID) and biometric technologies for personal identification in e-passports and other applications too paves the way to explore a global identity solution [1]. However, apart from the technological challenges, there are policy and legal constraints imposed by various governments across the globe which acts as a barrier and has been briefly touched upon in this paper. In order to accommodate the massive global IDs, an IPv6-based numbering scheme has been proposed for identity registration and data access of every human across the globe in this research paper. The state-of-art technologies and IoT with its widespread usage facilitates this proposal of integrated global ID solution incorporating smart card with RFID and biometric technologies against a multipurpose universal ID framework [2]. This study has

P. K. Singh (✉) · B. K. Gupta
Department of Computer Application, Shri Ramswaroop Memorial University, Lucknow-Deva Road, Barabanki, India
e-mail: praveen.197505@yahoo.com

B. K. Gupta
e-mail: bkguptacs@gmail.com

K. Dhawan
Faculty of Electronics Engineering, Indian Institute of Technology, New Delhi, India
e-mail: karandhawan@yahoo.com

N. Kumar
Department of CS & IT, Babasaheb Bhimrao Ambedkar University (a Central University) Satellite Campus, Teekarmafil, Amethi, India
e-mail: neerajmtech@gmail.com



P. K. Singh et al. (eds.), *Proceedings of ICRIC 2019*, Lecture Notes in Electrical Engineering 597, https://doi.org/10.1007/978-3-030-29407-6_59

also highlighted the future promises and the major research impact of the proposed application.

**Keywords** Smart card · RFID · Biometrics · IPv6 · IoT · E-ID · Radio frequency · LF · HF · UHF · Database · Cloud computing · DNA · ISO · Standardization

## 1 Background

Identity verification has always been the endeavor of mankind. Initially, the most common method was to rely on one's memory to identify another person. Then, almost 100,000 years ago new identification methods like physical display of jewelry or other decorative goods were found in South Africa, Algeria, and Israel. Another alternative was tattooing dated back to approximately 2000 BC in Ancient Egypt [3]. Thereafter, identification evolved to the written word in 3800 BC during the Babylonian Empire. The Roman Empire further developed data collection techniques and a variety of identity documents like birth certificates and citizenship records were introduced. King Henry Vs of England introduces passports in 1414. In 1829, the British Parliament enacted the reforms of Prime Minister Robert Peel to place more emphasis on printed records which became the precursor to modern government databases that link to ID cards [4]. In the late 1870s, Sir William Herschel made a biometric breakthrough. This evolved into fingerprint classification and was later automated by the Japanese in the 1880s. By 1936, USA had also begun rolling out their Social Security number cards and some other countries too followed the suit. It was not until 1977 that the USA computerized its paper records and paved the way for "smart cards." RFID, an auto identification technology in World War II was first time used to detect aircrafts of their friends or foe by the Britishers [5]. It took almost more than five decades to see it as commercial viability mainly due to cost propositions. This technology has evolved since then tremendously [6–8]. Since the World Trade Center tragedy of September 11, 2001, there has been sea change in outlook for IDs in relation to policies by worldwide governments.

## 2 Introduction

Establishing an identity today for self is a painful process due to requirement of having separate IDs like official ID card, PAN card, library cards, canteen cards, Voter ID, Passport, Driving Licenses, Ration card, etc., for a variety of purposes [9]. A multipurpose national ID is one of the obligatory prerequisites for all the countries to offer all their basic necessities ranging from education, housing, transportation, medical services, financial transactions, etc. The major challenges which are needed to be tackled with strenuous endeavors include registration for each and every citizen, finding out technical solutions, putting down government policies as well as identifying all the legal barriers. Various governments including India have conceptualized the need of having a single identity at national level. There is a need to have an integrated ID solution at national level against the data management necessity, accountability, availability, security, cost coverage, privacy, and so on [10]. Further, a mechanism for worldwide mobility of all human beings too needs to be explored. In this study, it is proposed to incorporate existing technological platforms like smart cards, biometrics, and RFID to accomplish desired dividends.

Depending upon the degree of authentication needed to access, all institutions or services have been categorized in three different classes in this research study. Emergence of IoT is quite significant in this perspective. Cloud computing offers tremendous potential to accommodate such colossal database, and it has been proposed as a viable solution in this research study to address the requirement of database management. Ability of RFID technology to discern all the RFID tags in the vicinity within the equipment range without any requirement of line of sight distinguish it from other such applications [11, 12]. Use of RFID with smart card ID can facilitate to track a mobile user in real time. Thus, integration of these combined technical platforms of smart card, RFID, and biometrics offers a viable ID solution for any country at national level. However, to expand such solution to global level will be comprised with enormous challenges mainly due to want of frequency harmonization, standardization, unison data structure, management of the colossal database, etc.

## 3 A Brief Technological Overview

Smart card ID holds a potential for substituting all existing ID cards through a sole smart ID card to provide the craving solution. Real-time tracking with precise positioning and dynamic corrections of all types of mobile objects are now achievable through RFID. Various biometric technologies incorporate the peculiarities of voice, signature, Irish, fingerprint, face, or hand in verification of users. Several biometric algorithms have been proposed and tested to enhance the prospects of authentication [13, 14]. There is a possibility exist to combine biometrics with smart card and RFID technologies and strengthen the authentication against a fool proof solution. A very brief overview has been discussed on smart card, RFID, and biometrics in succeeding paragraphs.

### *3.1 Smart Card Technology*

A smart card is as a portable card which can compute, store, and holds the data in an embedded processor chip in a secure storage for authentication of user's identity. It utilizes an integrated circuit (IC) chip embedded with internal memory [15]. A smart card connects with a reader either through a physical contact or by a radio frequency

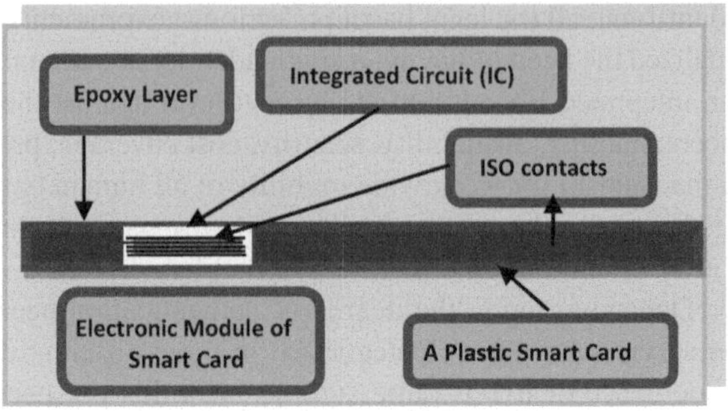


**Fig. 1** A plastic smart card's structural sideway view

(RF) through contactless interface. Embedded microcontroller allows smart cards to establish mutual authentication, biometric matching, or encryption to communicate with a reader. In contact smart cards, card reader is physically connected through electrical contacts of the smart card to read the data stored in its microcontroller chip [16]. The data stored in a contactless smart card can be read with even no physical contacts which in turn offer an excellent comfort for the user and have an added advantage in relation to contact smart cards against durability, costs, and reliability.

A structural sideway view is illustrated in Fig. 1 for a plastic smart card. When we look at its internal structure, an epoxy layer is also visible on its magnetic stripe. The emergence of IoT has transformed the entire understanding of security for Smart card technology. Identification for a user is now relates to primarily with secure authentication instead of secure identification [14]. It signifies that in endeavor of securing user's credentials for the identity authentication, a smart card offers an excellent portable platform.

## *3.2 RFID Technology*

It establishes a wireless link for objects identification. It is also called as a dedicated short-range communication. A RFID system consists of three main components explicitly a transponder, a transceiver along with an antenna. The transponder is electronically programmed and also known as RF tag. When it combines with an antenna, it forms a RFID tag whereas when transceiver and an antenna are combined together, i.e., they represent a RFID reader [17]. A RFID tag consists of microchip with an antenna which are combined together and mounted over a substrate. A RFID tag gets activated when the associated antenna emits radio signal with an aim to read/write the data over it. A simplified block diagram illustrates the processing of RFID data in Fig. 2.

A RFID reader comprises of a control module with a microcontroller chip which processes the data and one radio-frequency module. It connects to a computer which posses a data base. The RFID reader connects to an antenna emitting radio frequency

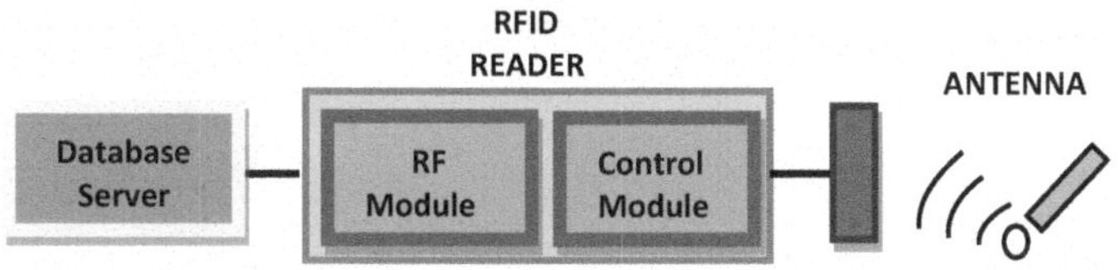


**Fig. 2** RFID reader simplified block diagram

through a RF module. The data transmission provides information with respect to the location and certain product specific information as price, date of manufacturing, size, etc.

**Use of Frequency Bands in RFID**: There are essentially three frequency bands as low frequency (LF), high frequency (HF), and ultra high frequency (UHF) which are being utilized in all RFID applications. These frequency bands vary in ranges from 30 kHz to 5.8 GHz. LF ranges are between 30–500 kHz and 10–15 MHz frequency ranges represent HF and UHF applications occupy 850–950 MHz, 2.4–2.5 GHz, and 5.8 GHz frequency ranges. LF application tags are cheaper, faster, and used in smaller applications [10, 18]. These tags have relative advantage over other types of tags as they are least affected in metal or fluids in their operation. HF tags offer better ranges; however, they are costlier. UHF tags have higher-transmission rates and provide best ranges of even more than 30 m but these are the most expensive tags. They also differ in their frequency ranges in different countries.

### 3.3 *Biometric Technology*

The biometric idiom is a combination of *bio* and *metric* with two Greek words. When they combined, it implies "measurement of life." Biometric technology pertains to employing physical as well as behavioral characteristics that can be gauged to differentiate on user from another [7]. The most prevalent biometric behavioral traits consist of signature, voice recordings, and keystroke rhythms while biometric physiological traits comprise of fingerprints, retina, facial images, hand geometry, and iris. Figure 3 depicts a fundamental biometric system. Initial process in this system is known as registration in which all individuals are needed to register in an existing database. The entire biometric trait's information of a user is fed by an algorithm which can convert this data into a template to be stored in that database. Choosing an appropriate biometric encryption (BE) technology becomes a paramount significance for all the existing biometric modalities [16]. It facilitates the security of all available biometric templates; however, they are also associated with a variety of challenges like interoperability, user acceptance, accuracy, reliability, etc.

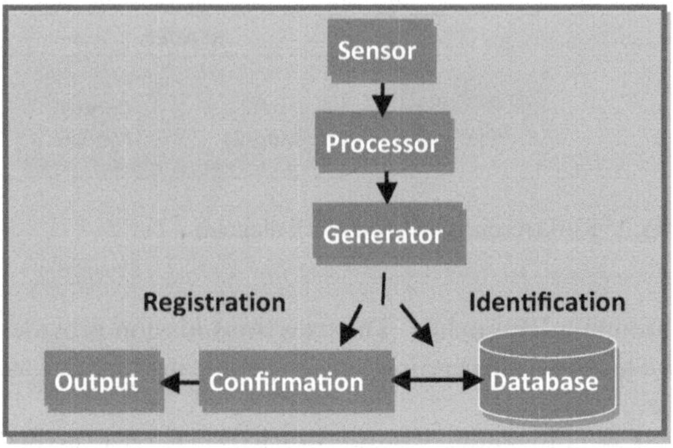


**Fig. 3** A fundamental biometric system

# 4 Necessity of Global Identification and Current Scenario

Traditional IDs like PAN cards, Passports, Voter IDs, Driving Licenses, Ration cards, and so on are more often than not offline cards. In India, the initiative against a national ID of unique identification project (UID) or "Aadhaar" card is also a passive.

ID that employs barcode-based technology having a limited security as well as data storage capability such kind of IDs prevalent in many parts of globe also pose several queries on numerous other issues similar to non-citizen residents of that particular country, specially refugees, stateless persons, asylum seekers, and many other "illegal" migrants [7]. They do make up a little yet noteworthy cross section of the population in that country. Biometric security in such kind of IDs has some limitations, for example, fingerprint biometrics can provide errors and may not be accurate mainly due to users could have inferior quality of fingerprints or cut marks or burnt, and at the same time, iris testing too have problems with blinds who could wear colored lenses and prompting false results.

Even though, there have been incredible technological developments, realization of a solo worldwide multipurpose ID is yet to be accomplished. Smart card and RFID technologies have potential to integrate the necessary considerations of a UID system to locate and track the ID users instantaneously which is yet to be explored [4, 12, 19]. Although, extensive amount of research has already been done to absorb the current realities, there is hardly any empirical study available today to qualify against the feasible suggested framework that can ensemble to substitute all existing widespread IDs into a sole multipurpose ID which has online authentication for a particular nation, and there is probably no such study exist which can claim it at global level.

Currently, there are a number of paradigms of UID's manifestations. In Estonia, their national ID offers an extensive array of e-government services, though without biometrics. Likewise, Belgium supports online government applications through their e-ID card which has been doled out with smart cards to all their citizens. In November 2010, a contactless ID was initiated in Germany. Figure 4 above illustrates an Image of New National e-ID of Spain which provides a fair idea on how the

**Fig. 4** An image of new national e-ID of Spain

new generation of smart IDs looks like. Among the entire developing nations, one of the first multipurpose ID systems was introduced by Malaysia naming MyKad [14]. It acts as an ATM card, ID card, health card, driver license travel document, public transport payment card, and electronic wallet. Today, digital identity technologies for instance biometrics and smart cards have proliferated with an approximation to 120 countries deploying electronic passports integrating these extremely secure features and almost 60 countries are implementing e-ID cards.

Currently, there are many countries with compulsory ID cards like Algeria, Argentina, China, Germany, Hungary, Netherlands, North Korea, Singapore, Turkey, Ukraine, etc. Similarly, there are countries where national identity cards are not compulsory like Austria, France, UK, Mexico, Switzerland, USA, etc. At the same time, there are several nations in the world which have no national ID cards like Australia, Canada, Japan, New Zealand, etc. With their extensive deployments across huge parts of Gulf, Europe, Latin America, and other parts part of globe, it presents interesting paradigms for the potential of e-IDs to influence millions of common lives all through emerging and developed economies [19]. Currently, diverse approaches are being followed right from a state controlled structural service and issuance of digital identities like in UAE to more decentralized arrangement with the Germany or through a partnership between private and public sectors in Sweden to develop the desired identity ecosystem.

The rapid espousal of e-governance in many countries today focuses on mobility and demonstrates the strong appeal of trusted identification technologies like smart card, RFID, and biometrics. Today, there is a turnaround in what people and worldwide governments expect to deliver from their National ID cards mainly due to enhanced technological developments which can serve groundwork of smarter IDs for their citizens [20]. The growth of these IDs means a single ID card can enable citizens of a particular nation to access their state benefits. This swiftly evolving domination of e-IDs reflects the influence of smart card and biometric technologies toward e-commerce and e-governance and also presents a strong case to integrate it with RFID and explore the feasibility to take individual identity solution at global level.

# 5 Major Research Approach and Validation

In this research study, the aim has been to build a framework consisting of some interrelated practices in respect of a single multipurpose ID and then propose a solution for a global universal ID (UID) card. Several attributes which can facilitate to build up a base in the formulation of the multipurpose UID framework like its necessity, availability, applications, apprehensions on security, privacy, authentication, need of a database, and approaches of different governments and lead private agencies have been deliberated [21].

Various available options have been considered through all the feasible resources to characterize distinctiveness of an individual. Figure 5 above illustrates the front and back end prototype model of global UID card. Realization of human mobility at global level needs an appropriate global UID card with a suitable numbering scheme which should adequately accommodate the entire existing human on the globe. At the same time, it must have feasibility to absorb all future needs including progressively growing population across the world. Keeping in view of this significant factor, an IPV6-based numbering scheme has been proposed in this research study [8]. IPv6 supposedly permits $2^{128}$ or approximate $3.4 \times 10^{38}$ addresses which have been considered sufficient to accommodate the data requirement for the proposed system.

A 32 digit hexadecimal numbering scheme permits 128 bits to define required addresses. Initial four digits have been chosen to represent country code whereas first digit could be for a continent, second digit for the type of country like United Nations member, any other global association, disputed territory, etc., and the last two digits represent the country codes. Next three octets likewise symbolize the states/provinces, district and city/town codes in a particular country [22]. Next octet represent the human object which apart from having some mandatory and common parameters for all the countries like gender, blood group, type of citizen, i.e., senior citizen/adult/child, etc., it is essentially suggestive in nature which is open to incorporate the relative requirements for a particular country.

Figure 6 depicts the proposed numbering scheme for the global UID card in which last three octets represent the human object counter. Last two digits of sixth octet has been put as a reserve to accommodate the all future requirements whereas

**Fig. 5** Proposed front and back end prototype model of global UID smart card

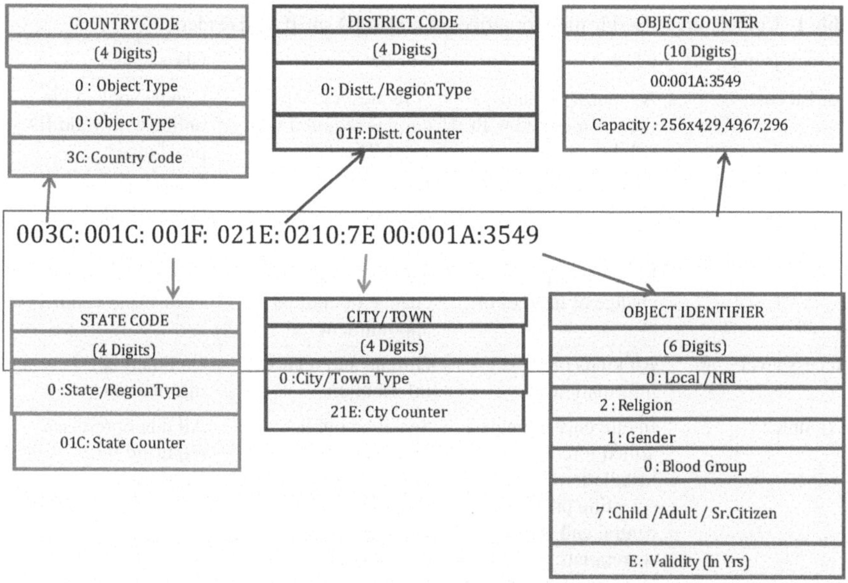


**Fig. 6** Proposed IPv6-based global UID smart card numbering scheme

last two octet belong to running counter having a capacity of 256 × 4, 29, 49, 67, 296 counters [23]. Keeping in view of current global population of approximate 7–8 billions, this counter with two more reserve digits can meet all the future needs to address the human objects.

While everyone will carry the same global UID card, therefore distinction of individual human object is proposed with three different types of smart card reader namely Class A, Class B, and Class C are illustrated in Table 1. In Class A card reader, the highest degree of security shall be ensured and it will be used to access only prominent premises like national intelligence agencies, armed forces/police installations, sensitive scientific premises and airports, etc. Class B types of card readers will be utilized to provide access to places where degree of authentication of human may not be very serious but a certain degree of checks and balances will be established for a reasonable level of security [20]. These premises will be like railway station, banks, educational institutions, etc. Class C smart card readers in the proposed system will be applicable for all other types of places where an individual will need to authenticate his/her identification to access a particular premise/service.

The most important aspect of the proposal lies in successful management of a complex global network which will be comprised of requisite database system, associated servers, and terminals in an integrated access network. Figure 7 illustrates a cloud-based network of the proposed global UID card which will consist of a cloud-based server capable of handling the multiple requests simultaneously. An appropriate activity and violation log management system with load balancing user

**Table 1** Comparative classification for proposed global ID smart card readers

| Comparison points | Class A | Class B | Class C |
|---|---|---|---|
| For License | A request on hard copy is required with valid ID proof | A request on hard copy is required with valid ID proof | Can be applied via online with valid ID proof attach |
| Authorized to | Head of company/org | Head of department/office for official use only | Everyone for commercial use |
| Validity | Renewal required on change of head of org | Renewal required on change of entitled appointment | Depend on purpose of license and validity |
| Access level | All kinds of information | Official and related information | Commercial information |
| Examples | Intelligence agencies, armed forces/police installations, sensitive scientific premises, central and state secretariats, airports, all types of financial transactions, etc. | Sensitive public premises access like railway station, banks, shopping malls, passport office, educational institutions, etc. | All other premises requiring ID authentication |

request handler will also be associated in the network [24]. A registered mobile number of each and every user will be used in this network for virtual mobile one time password (OTP) alert to authenticate the login request made by the users. Different classification of card readers as discussed above will be utilized for biometric authentication over the proposed global UID network.

In order to manage the worldwide human objects, there is a need of appropriate multiple registration system in each country which will be a connected to a centralized data center. The data access flow chart for the proposed system has been illustrated in Fig. 8. When a new registration request arrives in the system, the validity of the request is authenticated from the existing database of the system. In case of no match, a response in this respect is initiated to data network administrator whereas an OTP to the registered user is generated if the request gets authenticated and the request is logged [16, 25]. If there is any mismatch in the OTP, request is again sent back to data network administrator to inform the user. However, when the OTP is authenticated, the requisite access to the user is granted, data is processed and the resultant log is maintained by the network manager.

A significant aspect involved in this proposed global ID is to have an ability to locate and track the human object with a RFID chip attached with this card to communicate through radio waves and the use of DNA in combination with another biometric technology to offer desired level of security assurances in addition to the suitable encryption algorithm incorporated in the proposed system. For example, all Class A-based premises may need to use of DNA as one of the biometric authentications whereas in Class B and Class C premises/services can allow the user to use any

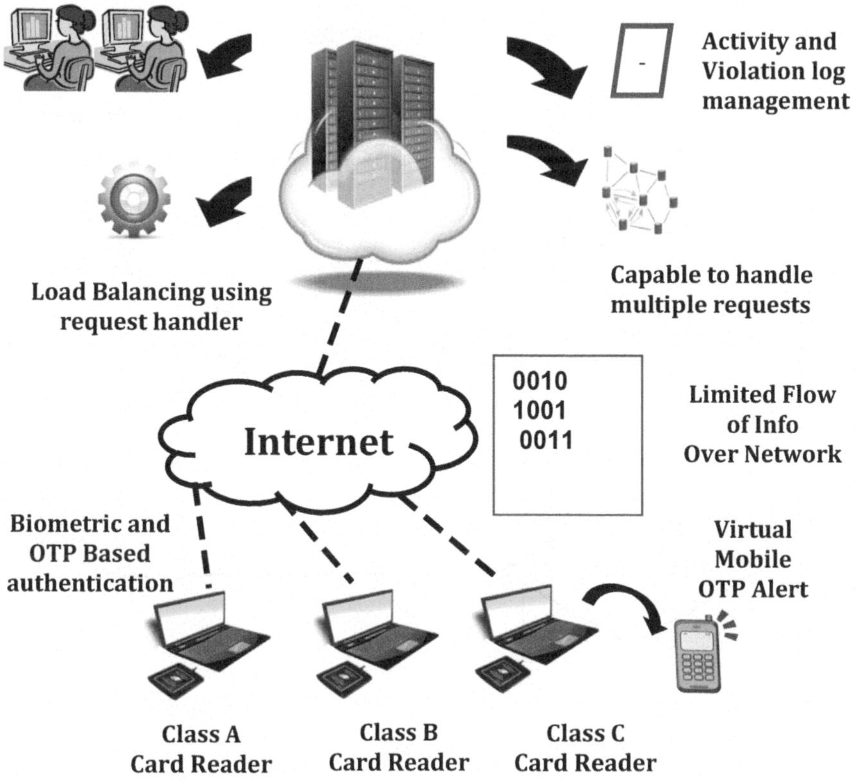


**Fig. 7** A cloud-based network proposed global UID smart card

appropriate biometric technology other than the DNA [26]. A naive and simplistic endeavor has been put in this study to use smart cards and radio-frequency identification (RFID) technologies with combination of biometric technologies to untangle convoluted social and economic quandary of worldwide mobility of a human object.

## 6 Challenges and Major Issues with Global Smart Card ID Using RFID

A concerted effort in this study has been laid on all probable allied factors to propose a viable UID solution at global level. However, there are certain impediments in realization of the global UID card like security, duplicity of efforts, frequency harmonization and standardization, etc., which needs to be overcome and to be given the due importance at various levels including the worldwide governments [9]. Some of the challenges and major issues with global smart card ID using RFID are discussed in succeeding paragraphs.

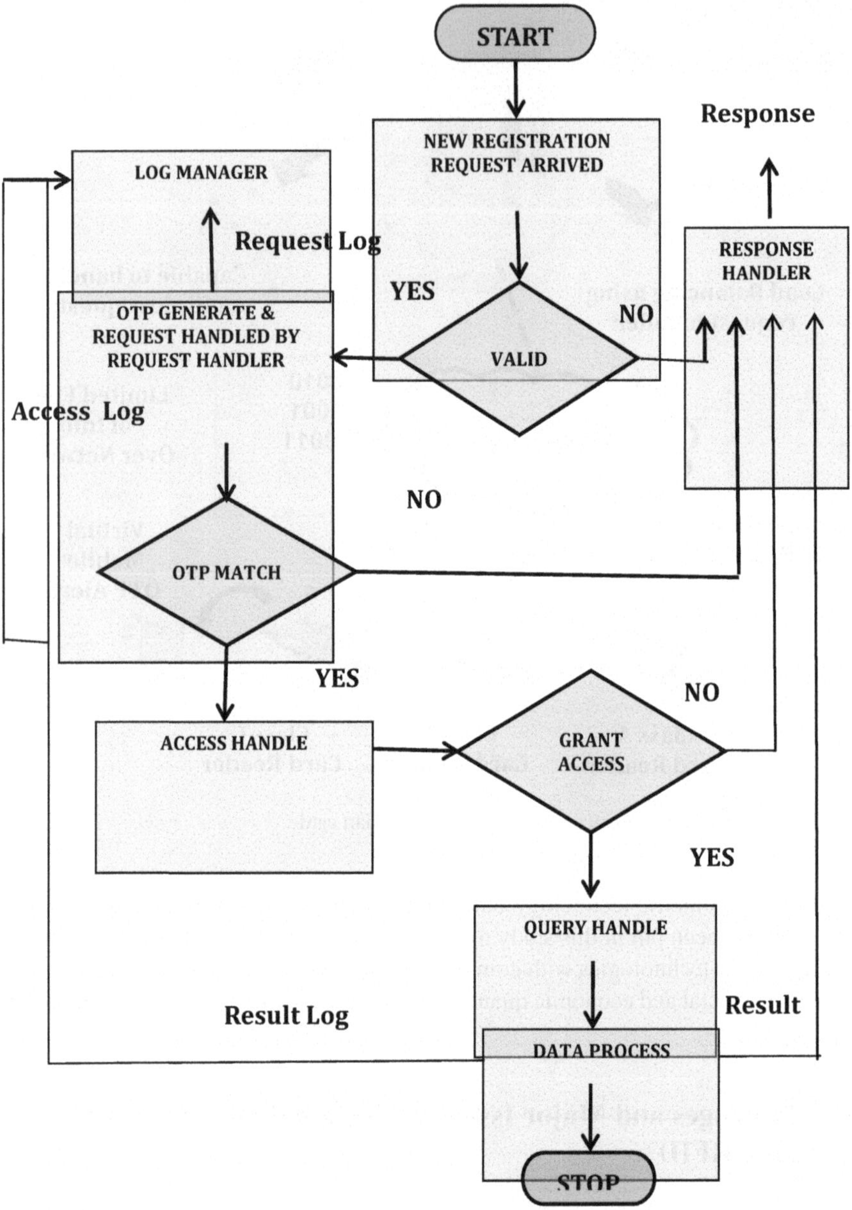


**Fig. 8** Data access flow chart for the proposed global UID smart card

## 6.1 Registrations of Users with Multiple Identities

While visualizing a global ID, the first issue we need to confront with is the registration of each and every individual on this globe. The fact that still in most of the third world and also many developing countries have been unable to register their all citizens due to socio-economic limitations, it will be a huge challenge for our application. There will be another challenge to have a single registration agency for this purpose, which may not even feasible due to differences in many states across the world [17]. To simplify the issue, we propose to establish one registration agency in each country which will be connected to a cloud-based server at centralized data center to manage the required data processing associated with this application.

## 6.2 Interoperability Requirements

In all business and state run institutions, dissemination of information and sharing the data is a key to success for any service or application. The compilation and sharing of this data by all such institutes is dictated by technological challenges and beleaguered with privacy apprehensions which certainly needs to be resolved being an important factor [26]. However, apart from drawing consensus from several segments, different worldwide RFID standards, data structure, interfaces, network protocols, etc., interoperability requirement for the proposed global ID network will be gigantic challenge to confront with by the network administrators.

## 6.3 Issues of Identity Thefts and Security

The foremost apprehension of users in any identity system is losing out personal information to unauthorized and inimical elements to misuse it and put them in trouble. Therefore, while envisaging a global ID network with different technologies like smart card, RFID, and biometrics, there must be a foolproof identity system with adequate layers of security. They need to anticipate and analyze all possible causes of forging the identity system and provision the countermeasures accordingly. Encryption technology and encryption algorithm should be carefully chosen to strengthen the smart card security [19]. Biometric templates of all the users must be adequately secured to prevent it from any unauthorized hacking. In addition, smart card and RFID data security, security of access and terminal networks, RFID security in respect of its potential deployment, front end security hazards in RF communication in RFID and the various associated network protocols will be required to thoroughly examined their implementation.

## 6.4 Depleting Trust on Biometric IDs

In many countries, a gigantic exertion is in progress to accumulate biometric data for their entire population. In India too, the collection has been rousing under the unique ID program. The foremost fear among these citizens is that their biometric database must not be part of any violations against the individual privacy. Apart from cloning, hacking, spoofing, etc., there are several other types of security and privacy issues which need to be dealt with biometric system used in the global ID card [10, 27]. The uneasiness of biometric users relates to their privacy concern against their biometric data held with the government, and some private agencies have potential to be abused for the reason other than the assent provided by the user.

## 6.5 Management of Associated Database System

In order to accommodate such colossal data of worldwide users, a centralized database system with all the requisite security must be established. If the entire user IDs will be connected to this central database on which worldwide government agencies shall be dependent, it may lead to an authoritative domain by somebody who may plant forged information. Unless the database system is wangled very cautiously, the security risk of the implied and planted data will become a very severe issue [12]. Another risk with the associated database system of global smart card ID is that when someone tries to attain a fake access who's ID can not be trusted, it will warrant a very robust security and authentication structure integrated with the database system.

## 6.6 Issue of Global Frequency Harmonization

Radio waves respond differently with diverse range of frequencies; thus, it is vital to choose the right frequency for a particular worldwide application. Different nations assign different bands of frequencies in the whole RFID radio spectrum. Different countries have different standardization, power, bandwidth usage, and data structure for their respective RFID devices in the same frequency bands. Lack of compatibility in the frequency bands for similar application in different nations becomes a major impediment to proliferate the subject application at global level [18]. No single frequency spectrum band therefore optimally fulfills all the requirements of potential and existing needs of any worldwide application dependent on a particular frequency band till the consensus at global level arrives against the use of same frequency band.

Therefore, realization of the global smart card ID with RFID needs a serious issue of frequency harmonization to be encountered with through International Telecom Union (ITU) at global level where the same set of frequencies are being used for different military and commercial RFID applications. In addition, global standardization of RFID application through ISO and EPC too demands a huge challenge in implementation of this worldwide identity application [22, 28]. The global smart card ID with RFID is a mandatory requirement in realization of worldwide human mobility in pursuit of an international or a global identity which presents a single multipurpose paperless ID.

## 7 Future Research Impact

This study proposes a prospective application of global smartcard ID with RFID and biometrics. This is a framework suggested to serve as a universal identification on multiple platforms for a worldwide user. The intended identity framework in this study is an attempt to advocate a global identity system to overcome the possibilities of forgery and theft in existing identification methods limited to national boundaries and to offer a solution to do away with requirements of multiple identity documents having manifold registrations and standards [15, 29]. Despite the fact that, there are more than a few such applications at micro level are available like e-passports, to administer a warehouse or logistics, etc., but these applications are restricted to limited confined geographical ranges.

The role of biometrics with smart card is proposed to influence in one-to-one authentication system for a user in this study. The global ID advocated in this framework study intends to minimizing fraud and corruption, augment the delivery of government services, facilitate the voting process to strengthen the democratic, and republic political system where the citizens are pivotal and to increase the overall identity security. If implemented, it shall turn out to be the leading biometric application of identification technology as on date and also will encompass far-reaching implications for all those developing countries that look to implement their national ID programs to further boost their economical and social development exertions [30]. This global identification system has potential to offer a unique ID in hexadecimal code to every human on this earth and support each and every country that look forward to extend its benefits to do away with requirement of multiple identity documents.

This study has intended to institute the thought process by combining different prevalent identity applications collectively in a single smart card by its integration with RFID and biometric technology with the inclusive support of IoT and cloud computing [31]. The users irrespective of their locations on globe will be able to carry a single identity card which can be utilized for all their authentication and financial transactions when their smart card will be linked to their global ID [4, 32]. The pertinent point in this prominent application relates to most of the developed nations who continue to be oppose even their national ID programs for diverse unusual reasons while among the developing nations, the national ID is increasingly turns out the basis for a reliable and secure environment in which the eventual goal is to have a universal ID for multipurpose application to act as the sole distinctive identifier to support manifold applications in their provinces. These may incorporate ID authentication, pension payments, land title registration, banking transactions, tax filings, voter verification, border crossing, etc. Consequently, it will facilitate a perfect platform to realize an integrated global ID solution for human mobility across the world.

## 8 Conclusion

Today, certain nations have mandatory ID cards system in their countries for their citizens and quite a few of them enforce to carry these IDs at all times. The technological proposition in this study of a universal ID system is a significant application in ongoing information technology evolutions. It is an evolution which compels worldwide governments to contemplate about distinctive identities for their citizens as an integrated entity [9, 16, 33]. A single worldwide ID card for a global human mobility is yet to be accomplished and the proposed integrated identity solution with smart card-based ID card along with RFID and a robust biometrics system does posses a potential to realize these expectations. However, it will still be subject of comprehensive analysis and to uncover contours of infallible system, it opens new possibilities for future research and technical evolutions. Nuisances of the proposed cloud-based network in this study will still require a further and thorough analysis to define threads of its overall functionality [34]. In addition, there will be numerous policy and technological concerns in worldwide implementation of this proposed global ID system which have been highlighted in this study and will need a very high order of coordination and cordial consensus from various governments across the globe.

**Acknowledgements** We would like to acknowledge the priceless suggestions offered by Prof. (Dr.) Mukul Misra, Director (Research & Consultancy) of Shri Ramswaroop Memorial University, Lucknow, India. We are also grateful to Col A. K. Singh, Ph.D., Vice Chancellor, Shri Ramswaroop Memorial University, Lucknow, India for his precious assistance in finalizing this paper. We record our earnest gratitude to Col V. K. Tiwari, Ph.D., Dr. Yashwant Singh, HoD, Faculty of Computer Science, Jammu University, India and Dr. Pradeep Kumar Singh, JP University of IT, India for their valuable propositions in this research study.

## References

1. Vowe, S., Waldmann, U., Poller, A., Türpe, S.: Electronic identity cards for user authentication promise and practice. IEEE Secur. Priv. **10**(1), 48–53 (2012 Jan/Feb)
2. Lingeswaran, K., Arul Das, S.V.: GPS based automated public transport fare collection systems based on distance travelled by passenger using smart card. Int. J. Sci. Eng. Res. (IJSER) **2**(3) (2014)
3. Singh, P.K., Kumar, N., Gupta, B.: Evolution of a framework for multipurpose ID: opportunities and challenges. Int. J. Adv. Stud. Comput. Sci. Eng. (IJASCSE) **6**(10), 1–7 (2017)

4. Weinstein, R.: RFID: a technical overview and its application to the enterprise. IT Prof. **7**(3), 25–30 (2005)
5. Markantonakis, K., Mayes, K.: An introduction to smart cards and RFIDs in secure smart embedded devices platforms and applications, pp. 4–22, Springer Science (2014)
6. Messerges, T.S., Dabbish, E.A., Sloan, R.H.: Examining smart-card security under the threat of power analysis attacks. IEEE Trans. Comput. (51), 541–552 (2002)
7. Turjman, F., Alturjman, S.: Confidential smart-sensing framework in the IoT era. J. Super Comput. https://doi.org/10.1007/s11227-018-2524-1 (2018) (Springer)
8. Zhang, S., Liu, X., Wang, J., Cao, J.: Tag size profiling in multiple reader RFID systems. In: Proceedings IEEE Conference of Computers and Communication, pp. 1–9 (May 2017)
9. Chu, J.: Applications of RFID technology [Book/Software Reviews]. IEEE Microw. Mag. **16**(6), 63–64 (2015)
10. Imani, M., Qiyasi Moghadam, A., Zarif, N., Arabnia, H.R.: A comprehensive survey on addressing methods in the internet of things (2018) (Unpublished)
11. Das, A.K.: Analysis and improvement on an efficient biometric-based remote user authentication scheme using smart cards. Inf. Secur. IET **5**(3), 145–151 (2013) (IET)
12. Alabady, S.A., Al-Turjman, F., Din, S.: A novel security model for cooperative virtual networks in the IoT era. Int. J. Parallel Prog. https://doi.org/10.1007/s10766-018-0580-z (2018) (Springer)
13. Li, K., Qi, G.J., Ye, J., Hua, K.A.: Linear subspace ranking hashing for cross-modal retrieval. IEEE Trans. Pattern Anal. Mach. Intell. **39**(9), 1825–1838 (2017)
14. Singh, P.K., Kumar, N, Gupta, B.: Smart card ID: an evolving and viable technology. Int. J. Adv. Comput. Sci. Appl. (IJACSA) **9**(3), 115–124 (2018)
15. Taherdoost, H.: Appraising the smart card technology adoption; case of application in university environment. Science Direct, Research and Development Department, pp. 104–105, Ahora Ltd Management Consultation Group, Kuala Lumpur, Malaysia (May 2017)
16. Beaver, M.: RFID technology in university ID cards, and The implications of RFID technology in university ID cards, S&T's Peer to Peer: Undergraduate Res. J. **1**(1) (2016) (Published by Scholars' Mine)
17. Reynolds, F., Nath, B., Want, R.: RFID technology and applications. IEEE CS and IEEE Comput. Soc. **5**(1), 20–22 (2006)
18. Sandhu, P., Kaur, M., Mohan, N., Sandhu, M.: RFID technology principles advantages limitations & its applications. Int. J. Comput. Electr. Eng. **3**(1), 1771–1836 (2011)
19. Liu, X.: Multi-category RFID estimation. IEEE/ACM Trans. **25**(1), 264–277 (2017)
20. Liu, X.: Multi-category RFID estimation. IEEE/ACM Trans. Netw. **25**(1), 264–277 (2017)
21. Landaluce, H., Perallos, A., Onieva, E., Arjona, L., Bengtsson, L.: An energy and identification time decreasing procedure for memory less RFID tag anti collision protocols. IEEE Trans. Wirel. Commun. **15**(6), 4234–4247 (2016)
22. Al-Kassab, J., Rumsch, W.-C.: Challenges for RFID cross-industry standardization in the light of diverging industry requirements. IEEE Syst. J. **2**(2) (2008)
23. Zhang, S., Liu, X., Wang, J., Cao, J.: Tag size profiling in multiple reader RFID systems. In: IEEE Conference Computer Communication, pp. 1–9 (May 2017)
24. Jain, A.K., Ross, A.: Bridging the gap: from biometrics to forensics, Philos. Trans. R. Soc. B **370**(1674), 1–10 (2015)
25. Bella, G., Bistarelli, S., Martinelli, F.: Biometrics to enhance smartcard security. Lecture Notes in Computer Science, vol. 3364 (2005)
26. Das, A. K.: Analysis and improvement on an efficient biometric-based remote user authentication scheme using smart cards. Inf. Secur. **5**(3), 147–150 (2013) (IET)
27. Labati, D.R., Genovese, A., Munoz, E.: Biometric recognition in automated border control: a survey. ACM Comput. Surv. **49**(2), A1–A39 (2016)
28. Lai, Y.L.: Cancellable iris template generation based on indexing-first-one hashing. Pattern Recogn. (64), 105–117 (2017)
29. Ramanathan, U.: Considering social implications of biometric registration: a database intended for every citizen in India. IEEE Technol. Soc. Mag. **34**(1), 10–16 (2015)

30. Lo, N.W., Tsai, J.L., Wu, T.C.: Novel anonymous authentication scheme using smart cards. IEEE Trans. Ind. Inf. **9**(4), 2006–2011 (2013)
31. Arjona, L., Landaluce, H., Perallos, A., Onieva, E.: Scalable RFID tag estimator with enhanced accuracy and low estimation time. IEEE Signal Process **24**(7), 982–986 (2017)
32. Daisy Premila Bai, T., Albert Rabara, S., Vimal Jerald, A.: An adaptable secure smart card architecture for internet of things and cloud computing. IJRET (5), 163–169 (2016)
33. Duan, L., Zhang, X., Wang, Z.J., Duan, F.: A feasible segment by ALOHA algorithm for RFID systems. Wirel. Pers. Commun. **96**(2), 2633–2649 (2017)
34. Ross, A.A., Nanda Kumar, K., Jain, A.K.: Handbook of Multi Biometrics. Springer (2006)

# Design of Low-Power Dual Edge-Triggered Retention Flip-Flop for IoT Devices

**Ajay Mall, Shaweta Khanna and Arti Noor**

**Abstract** With the advancement in the VLSI technology, the demand for low power consumption and high performance increased gradually. When the applications of data retention are considered, then the need for advanced memory units is taken into account. This requirement of enhanced memory units is incorporated with the concept of conservation of energy which is achieved by using low-power techniques. In digital circuits, flip-flops are the essential memory and timing elements. New methods and techniques needed to be developed for implementing energy-efficient low-power flip flops. This paper proposes dual edge-triggered flip-flop (DETFF) along with gating technique, one of the most reliable low-power techniques which provide one-time solution to low-power applications. The DETFF circuit based on gating technique is simulated using MENTOR GRAPHICS tool in 180 nm technology. This design is efficient in reducing power dissipation leading to the reduction in area and delay and subsequently leads to the high speed of the device.

**Keywords** Low power · VLSI · Data retention · Memory units · Energy efficient · Dual edge-triggered Flip-Flop

## 1 Introduction

Internet of things (IoT) is a promising technology in future. This platform has revolutionised the top industries in the world like health care, retail, safety and security, offices and factories, education, travel and banks and even financial institutions. IoT is believed to change the entire way of people communication and working. IoT results in more improved and efficient businesses. Use of IoT platform will make enterprises adopt better technology in their businesses. This helps in dropping the manufacturing cost, having a close eye on the productivity of employees, providing a great experience to the customers and ensuring safety of the stores against theft.

A. Mall · A. Noor
School of Electronics, Centre for Development of Advanced Computing, Noida, India

S. Khanna (✉)
JSS Academy of Technical Education, Noida, India
e-mail: shweta.khanna04@gmail.com



P. K. Singh et al. (eds.), *Proceedings of ICRIC 2019*, Lecture Notes in Electrical Engineering 597, https://doi.org/10.1007/978-3-030-29407-6_60

It is estimated that in the next decade IoT will produce around $4.6 trillion turnover for public sector and $14.4 trillion turnover for private sector. IoT technology will prove as a boon to the society. Till 2020 around a quarter billion vehicles will be connected to IoT making the world of transportation, a blessing to the society. The architecture of IoT consists of perception layer, network layer, middleware layer, application layer and business layer. Each of these layers plays their own part. The primary part of an IoT [1, 2] is a sensor end node connected to an IoT-based device. Sensor collects the information from the device, and then, network layer keeps this confidential information and passes it through RFID, satellite, Wi-fi, 3G, 4G or 5G depending upon the sensor type. Middleware layer processes the information provided by the lower layer and also saves that particular information as well as retrieves it whenever needed. Application layer manages the application part, i.e. smart wearable [3], smart vehicle or any other smart device. Business layer provides the data analysis capability to an IoT system whether it is any business model, any graph or flow chart.

The area of interest for this paper is the sensor end node which collects the required information from the device. The basic characteristics for sensors are its wireless connectivity, security, reliability, flexibility and self-powering operation. But the key concern for this node is its power requirements. One option to fulfil the requirements for sensors is battery. But battery-based systems suffer unconventional problems caused by it. Battery comes with limited lifespan, heavyweight and large size. Another alternative to battery-powered systems is providing power through solar cells, RF power and piezoelectricity. After providing power to the system, the next step is to conserve that power. The need for conservation of power leads to the concept of low-power techniques. With the growing demands of VLSI technology, the upcoming circuits are leading to a great level of circuit integration, high-power requirements and high clock speed. But the future concern for the advance devices is to have low power consumption [4]. This concern for low power opens the back doors for many recent technologies which are meant for low-power systems. Low-power techniques are those techniques which provide the high-end solution for energy consumption by giving the low-power option with still maintaining the system's performance [5]. The consumption of power can be explained as the power of clock is directly proportional to the square of the supply voltage.

$$P_{\mathrm{clk}} = V_{\mathrm{dd}}^2 f_{\mathrm{clk}}(C_{\mathrm{clk}} + C_{\mathrm{ff,clk}}) + f_{\mathrm{data}} C_{\mathrm{ff,data}}$$

where

| | |
|---|---|
| $P_{\mathrm{clk}}$ | Power of clock |
| $V_{\mathrm{dd}}$ | Supply voltage |
| $f_{\mathrm{clk}}$ | Clock frequency |
| $f_{\mathrm{data}}$ | Average data rate |
| $C_{\mathrm{clk}}$ | Total capacitance as per clock network |
| $C_{\mathrm{ff,clk}}$ | Capacitance of the clock path as per flip flop |
| $C_{\mathrm{ff,data}}$ | Capacitance of the data path as per flip flop. |

There are many ways in which the power dissipation can be reduced. The voltage, i.e. $V_{dd}$, can be reduced but this may lead to poor performance of the device. The total capacitance of the device will automatically reduce when the number of transistors increases thereby increasing the complexity of the circuit. In a circuit, there is linear dependence between dissipated power and frequency, because dissipated power increases linearly with respect to frequency. Switching of system clock usually occurs at very high frequency across the chip and hence has a significant impact on the power. Clock power can be two times higher than the logic power for static logic and thrice the logic power of dynamic logic. To minimise the clock power, the operating frequency of the system should be less for attaining desirable performance. The other and the most effective method is to reduce the clock frequency, i.e. $f_{clk}$ which is achieved by the use of dual edge-triggered flip-flop (DETFF) [6] and also does not degrade the system performance. DETFF reduces the clock frequency to the half of its original value, and hence, this in turn reduces the power requirement of the device. This implies that DETFF is an effective solution for low-power applications. Nowadays, data retention is one of the greatest requirements for the modern technical world. The demand for high capacity, security, reliability and low power consumption is of topmost importance for data retention applications. The use of transiently powered systems (TPS) [7] fulfils the demand for high storage capacity devices efficiently and effectively. TPS works on the concept that when the energy or power requirements are met, then it works otherwise the current state is stored and retrieved again when the energy level is met. It not only generates the required power level but also maintains it by regulation and rectification. For the purpose of data retention, earlier non-volatile random access memories (NVRAMs) are used but these drastically increase the area and delay which is not suitable for low-power applications. This paper proposes the study of DETFF [8] as its stores the state in sleep mode or provides temporary storage before passing it to NVRAM. The implementation of DETFF using pass transistors and transmission gate is not highly recommended because these circuits suffer voltage degradation and are not able to restore logic levels, hence are not suitable for low-power applications. The single edge-triggered flip-flop suffers by glitches at the output end because of excess transition time. It works for the opposite polarity, and then, the output is multiplexed at the end. But the design of DETFF reduces this problem as it provides efficient operation even at high clock frequencies [9].

## 2 Proposed Device

*A* dual edge-triggered flip-flop is simply a design which works on both positive and negative edge hence named dual edge-triggered [10] which also means that it reduces the clock frequency half to that of single edge-triggered flip-flop which works on single edge either positive or negative. Several papers have presented designs that provide improvement in various parameters, and it has been also observed that most of the power dissipation takes place in the clock network so it is required to reduce

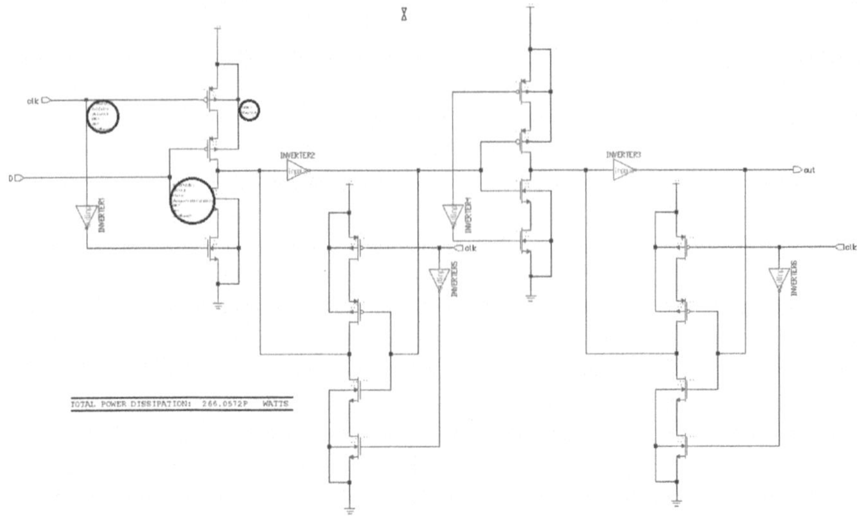


**Fig. 1** Single edge flip-flop

the clock frequency and supply voltage to reduce the overall power dissipation in the circuit. In this paper, four circuits have been analysed in which the first circuit is simple DETFF consisting of two single edge-triggered flip-flops. The individual SETFF operates at opposite clock polarity, and the output is multiplexed in both the SETFF. But this design offers some drawback; i.e., there are glitches at the output end and also the power dissipation and delay of the circuit are higher than expected. The schematic and output of single edge flip-flop are shown in Figs. 1 and 2.

The second circuit is based on dynamic design in which the circuit operation is dependent on the clock pulse being applied to the circuit if the clock pulse is low the top buffer is enabled and the bottom buffer is disabled and the output is driven to output stage. Also, the enabled and disabled circuits are isolated; therefore, power dissipation is less but the delay in the circuit is more. The schematic and output of dynamic flip-flop are shown in Figs. 3 and 4.

The third circuit avoids stacked PMOS design to provide less delay in the circuit but it dissipates more power when compared to the earlier design. When the clock is low and data is high, the output is also high. When the clock is high the bottom-most transistor is ON which makes the data bit low and then it is driven through inverter which gives the output. The schematic and output of low-voltage flip-flop are shown in Figs. 5 and 6.

The proposed circuit is the combination of two circuits, i.e. dynamic DETFF and low-voltage DETFF; both these circuits have some drawback and some advantages so in proposed circuit the advantages are been taken and the drawback is avoided which means dynamic circuit was having less power dissipation and low-voltage circuit was having less delay [11]. Also, there is retention circuit in the beginning in which there is a sleep transistor consisting of NMOS and PMOS (for clock gating)

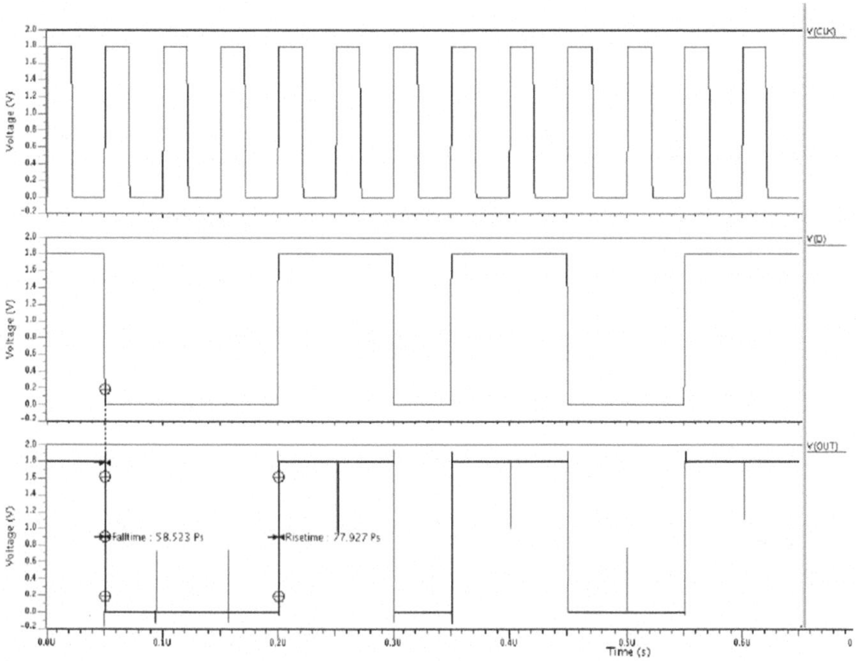


**Fig. 2** Output waveform of DETFF using single edge

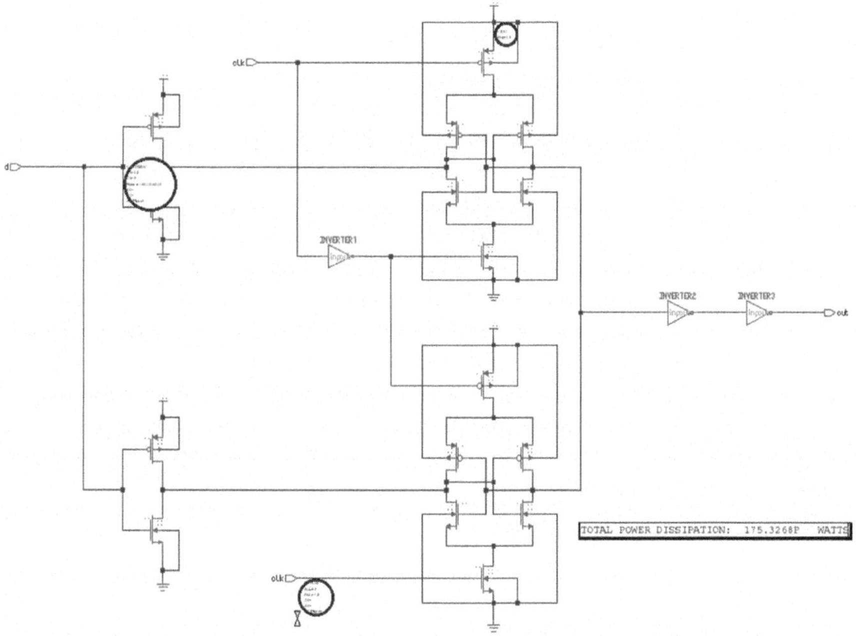


**Fig. 3** Dynamic flip-flop

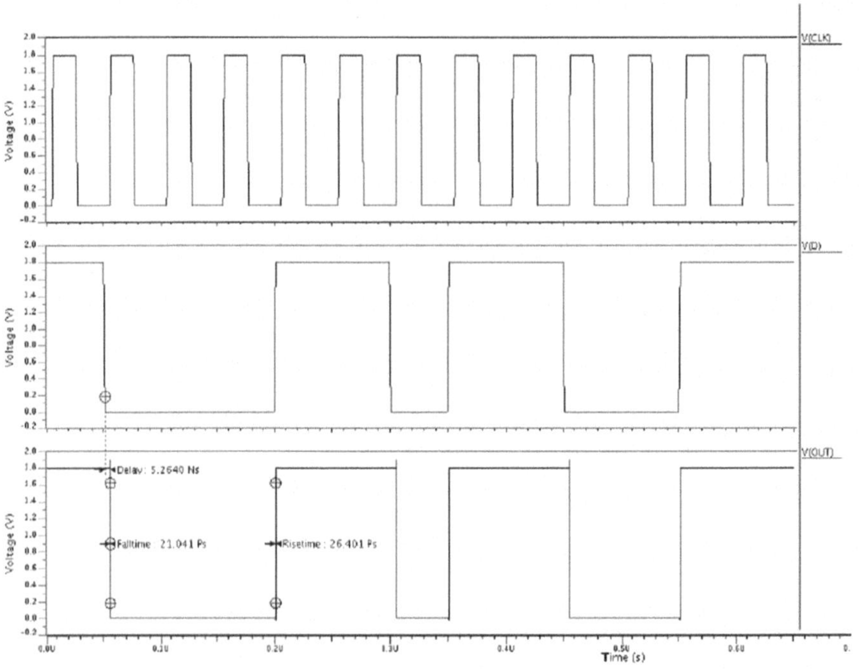


**Fig. 4** Output waveform of dynamic flip-flop

also there is an inverter in which NMOS and inverter have higher threshold voltage compared to the circuit. So the proposed circuit is modified DETFF with retention latch. The flip-flop is having lesser delay propagation and also consumes lesser power compared to the previous flip-flop. The proposed flip-flop has three stages for rising and falling edges where the first stage drives the input when the clock is high/low and the clock input changes the second stage passes the output to the third stage and then the last stage drives the output when the clock state is changed (Figs. 7 and 8).

The output waveform of the proposed circuit is shown in Fig. 5. The figure describes the waveform as when the sleep signal is low the data will be transferred same as input which means the flip-flop is in active state and when the sleep signal is high the flip-flop is in sleep mode and then it reduces the power dissipation or leakage power dissipation of the circuit. Also, when it is in sleep mode, the information is saved with the help of retention technique as during power-down there is possibility that information might get lost and when the circuit comes to its active state the information is restored back.

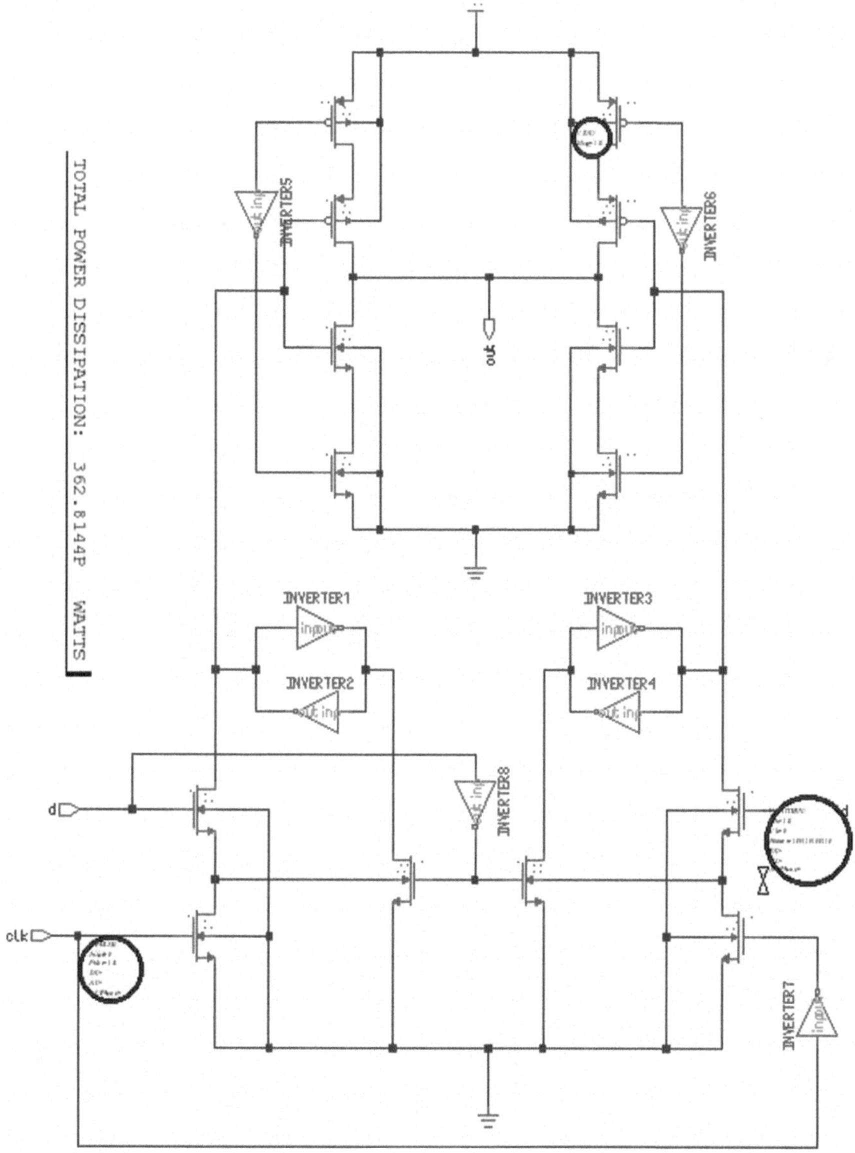


**Fig. 5** Low-voltage DETFF

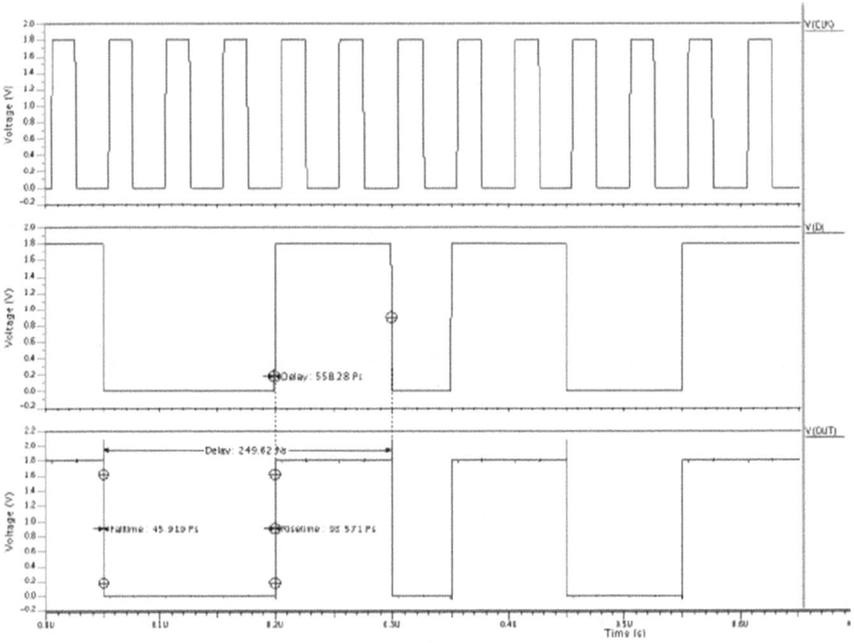

Fig. 6 Output waveform of DETFF using low voltage

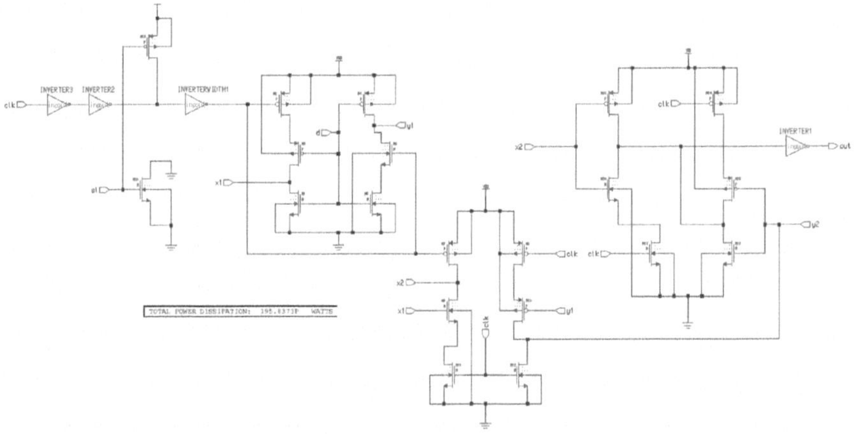

Fig. 7 Proposed DETFF with retention latch circuit

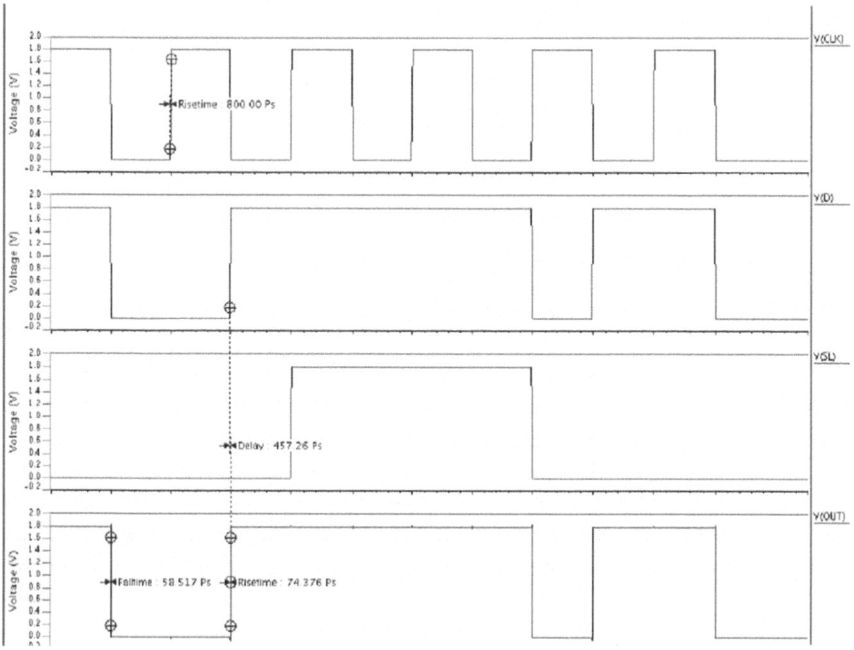


**Fig. 8** Output waveform of the proposed retention DETFF

# 3 Results and Discussion

For the purpose of simulation of various designs which are portrayed in this paper, several parameters such as supply voltage, clock pulse delay, width and period of clock pulse, rise time/fall time of clock pulse, length and width of transistor are mentioned below in the table. Besides this, various conditions for simulation such as temperature, input pattern and clock buffer are maintained. A constant supply voltage is provided to the circuit, and common input pattern (100110100110) is followed for all the circuits. The fixed length of transistor is provided while the width of each transistor is varying as per the threshold voltage. The propagation delay and power dissipation [12] are measured for all the circuit as these two factors are highly responsible for the degradation of speed of circuit and contribute to large area consumption. In this paper, the retention scheme is applied to the circuit as given in figure which is the combination of two DETFF and this thereby leads to less delay and less power dissipation out of other four dual edge-triggered flip-flop designs [13]. For the idea of the completion of design of dual edge-triggered flip-flop using retention technique is achieved by taking into account various parameters as mentioned in Table 1. Simulation of the circuit was performed in 180 nm CMOS technology with fixed input voltage of 1.8 V with fixed input values and input clock. The "Pyxis schematic tool by mentor Graphics" was used for design and simulation of all the circuit techniques.

Table 1 Parameters for simulation set-up

| Regular | Bold |
|---|---|
| Supply voltage | 1.8 V |
| Clock delay | 5n |
| Rise time/fall time | 1n |
| Clock width of the pulse | 20n |
| Period of pulse | 50n |
| Length of transistors | 0.18μ |
| Width of transistors | Variable |

The comparison results in terms of transistor count; rise and fall time; and delay and power dissipation are tabulated in Table 2 given. On analysing Table 2, it has been observed that the power dissipation of the proposed retention DETFF has dissipated power of 147.4642 pW which is much lower than other designs except single edge-triggered DETFF. But single edge-triggered DETFF design has some distorted output which is undesirable for any circuit. Therefore, the proposed design is best suited for low-power retention DETFF, and also, it has the lowest delay and best performance speed.

Table 2 Simulation results of the simulated circuits

| 180 nm Pyxis schematic IC platform | | | | |
|---|---|---|---|---|
| Flip-flop | Transistor count | Rise time/fall time (ps) | Delay (ns) | Power dissipation (pW) |
| Single edge-triggered flip-flop | 22 | RT = 90.546<br>FT = 60.222 | 5.2826 | 144.4161 |
| Dynamic design-based DETFF | 22 | RT = 26.400<br>FT = 21.039 | 5.2640 | 175.3268 |
| Low voltage-based DETFF | 26 | RT = 45.931<br>FT = 111.90 | 0.6240 | 265.2565 |
| Proposed DETFF w/o retention circuit | 20 | RT = 45.745<br>FT = 23.838 | 5.2513 | 118.5127 |
| Proposed DETFF with retention circuit | 28 | RT = 74.376<br>FT = 58.517 | 0.4572 | 147.4642 |

## 4 Conclusion

This paper uses dual edge-triggered flip-flop for low-power applications and is highly recommended for data retention techniques. This paper examines various designs of DETFF, and their results are evaluated and studied in order to provide the best technology meant for power-down applications [14]. The authors come up with the most desirable device based on the results of various DETFF designs. The proposed device incorporates DETFF along with retention circuit. In power-down mode, the various states of this device get locked in it and can be retrieved once the power is ON. Hence, this will lead to reduction in leakage power in standby mode. This circuit provides the total power dissipation of 145.352 pW. Furthermore, the speed of the circuit increases due to reduction in propagation delay. This circuit provides the breakthrough to the world of storage devices while using the effective data retention technique [15] and can be used widely in transiently powered systems.

Basically, this technique is useful for IoT sensor end node where conservation of energy is required for long-life of devices.

**Acknowledgements** The authors would like to thank CDAC, Noida, IP University for their support and constant motivation for the successful completion of this work and providing a platform to enhance our skills and practical implementation of this research work.

## References

1. Jayakumar, H., Raha, A., Kim, Y., Sutar, S., Lee, W.S., Raghunathan, V.: Energy-efficient system design for IoT. In: IEEE Design Automation Conference (ASP-DAC), 21st Asia and South Pacific. ISSN 2153-697X, pp. 25-28 (2016)
2. Kraijak, S., Tuwanut, P.: A survey on Internet of Things architecture, protocols, applications, security, privacy, real-world implementation and future trend. In: IEEE 16th International Conference on Communications Technology (ICCT 2015), pp. 1–6 (2015)
3. Wahbah, Maisam, Alhawari, M., Mohammad, B., Saleh, H., Ismail, M.: Characterization of human body-based thermal and vibrations energy harvesting for wearable devices. IEEE J. Emerg. Sel. Top. Circuits Syst. **4**(3), 354–363 (2014)
4. Dasari, M., Nikhil, R., Chavan, A.: Novel ultra-low power dual edge triggered retention flip-flop for transiently powered systems. In: IEEE International Advance Computing Conference, pp. 500–504 (2017)
5. Ma, K., Zheng, Y., Li, S., Swaminathan, K., Li, X., Liu, Y., Sampson, J., Xie, Y., Narayanan, V.: Architecture exploration for ambient energy harvesting nonvolatile processors. In: IEEE 21st International Symposium on High Performance Computer Architecture (HPCA), pp. 526–537 (2015)
6. Wu, J., Akingeneye, I., Yand, J.: Energy efficient optimum sensing with energy harvesting power sources. IEEE Access Big Data Green Commun. Comput. ISSN 2169-3536 (2015)
7. Chavan, A., MacDonald, E., Neff, J., Bozeman, E.: Radiation hardened flip-flop design for super and sub threshold voltage operation. In: Proceedings of IEEE aerospace conference, Big Sky, MT, pp. 1–6 (2011)
8. Chen, K.-H.: Power Management Techniques for Integrated Circuit Design, 1st edn. Wiley, Singapore (2016)

9. Mahmoodi-Meimand, H., Roy, K.: Data-Retention Flip-Flops for Power-Down Applications, pp. 677–679. Semiconductor Research Corporation (2004)
10. Chung, W., Lo, T., Sachdev, M.: A comparative analysis of low-power low-voltage dual-edge-triggered flip-flops. IEEE Trans. Very Large Scale Integr. (VLSI) Syst. **10**(6), 913–918 (2002)
11. Zhao, P., McNeely, J., Golconda, P., Bayoumi, M.A., Barcenas, R.A., Kuang, W.: Low-power clock branch sharing double edge-triggered flip-flop. IEEE Trans. Very Large Scale Integr. (VLSI) Syst. **15**(3), 338–345 (2007)
12. Zhao, P., Darwish, T.K., Bayoumi, M.A.: High performance and low-power conditional discharge flipflop. IEEE Trans. Very Large Scale Integr. (VLSI) Syst. **12**(5), 477–484 (2004)
13. Phyu, M.W., Goh, W.L., Yeo, K.S.: A low-power static dual edge triggered flip-flop using an output controlled discharge configuration. In: Proceeding of IEEE International Symposium on Circuits Systems (ISCAS), vol. 3, pp. 2429-2432 (2005)
14. Aliakbar, G., Hamid, M.: Dual-edge triggered static pulsed flip-flops. In: Proceeding of 18th International Conference on VLSI Design, pp. 846–849 (2005)
15. Strollo, A.G.M., De Caro, D., Napoli, E., Petra, N.: A novel high speed sense-amplifier-based flip-flop. IEEE Trans. Very Large Scale Integr. (VLSI) Syst. **13**(11), 1266–1274 (2005)

# Digital India

# Development of Slot Engine for Gaming Using Java

**Rahul Kumar Verma, Rajan Prasad Tripathi and Pavi Saraswat**

**Abstract** Slot games are games consisting of reels, appearing on the display randomly after the triggering of the spin symbol or the lever. It has various types of reel sets, most commonly used being 5 * 3. The game was played by inserting a coin, a ticket or a barcode in the slot provided in the machine. Thereafter, the reels displayed on the screen would start spinning for some time and then stop. As the reels will stop the symbols that will then be displayed on the screen, may–may not forms a winning pattern or a sequence. The player can either lose the bet that he played or win higher bonus depending on the sequence that gets displayed on the screen after the reels spin. Therefore, the basic principle behind all the casino games is a chance. This work throws light upon the development of the slot game which is coded in Java using various advanced Java techniques and Web server protocols.

**Keywords** Bonus · Putty · Hit rate · Cheat · Pay Lines · Total bet

## 1 Introduction

The development of the game begins after the development team is provided with the requirement gathered document [1]. The software called IntelliJ is used for writing the code and running the algorithm. Gradle is used as a build automation tool in the process. Vert.x is a polyglot event-driven application framework that runs on the Java Virtual Machine. WebSocket is a full-duplex communication protocol for computers which is used for communicating between the remote server and the client. This paper is organized as follows: The next section gives the information about the basic game development process [2, 3]. The paper explained the process used for the development of the game. The third section explains the features of the game and the payouts specified. The last section deals with the server testing of the test cases of the game.

R. K. Verma (✉) · R. P. Tripathi
Amity University Uttar Pradesh, Noida, India
e-mail: rkverma@amity.edu

P. Saraswat
Krishna Engineering College, Ghaziabad, India



P. K. Singh et al. (eds.), *Proceedings of ICRIC 2019*, Lecture Notes in Electrical Engineering 597, https://doi.org/10.1007/978-3-030-29407-6_61

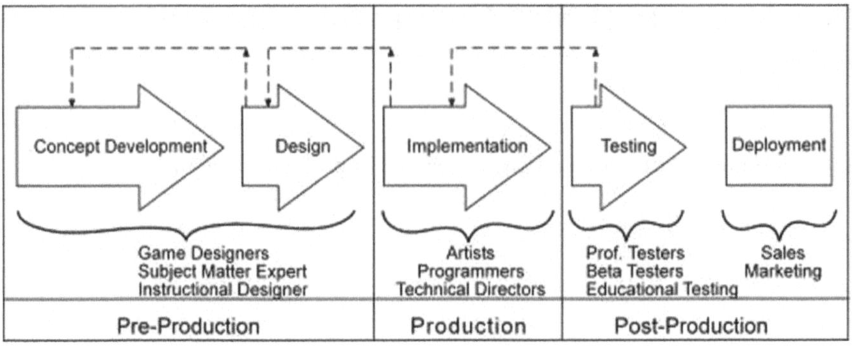


**Fig. 1** Flow diagram for game development

## 2 Flow Chart for Game Development

The process of development of the game can be divided into three major sections, namely pre-production, production and post-production. Pre-production consists of developing an idea about the game and designing the features which are included in it.

This phase can also be called as requirement gathering phase of development as all the requirements of the game are listed in a presentable format so that the developers can easily understand and develop the as per the requirements. Production phase is the actual implementation phase. It consists of algorithm building, coding and developing various softwares as per the requirement of the individual game. It converts the English language into the programming language. Post-production phase is the testing phase, and it checks if the execution of the code is correct or not. Here, all the test cases are checked with the help of cheats. After all the check reports pass the game is finally deployed and launched into the market or given to the customer (Fig. 1).

## 3 Flow Chart for Spirits of Zen

Algorithm building is the first task in the Production phase. Once this is done correctly, the coding begins. For writing Java codes, IntelliJ can be used and Java coding techniques like Gradle, Lambda are used. This consists of writing the main framework (game engine) of the game, where all the functions that are used in different files are developed. It requires a lot of accuracy and efficiency. Once the coding part is done, the testing of the game on the remote server is done to check if the game is executing the code correctly or not. For this software called PuTTY is used along with remote server protocol called WebSocket (Fig. 2).

**Fig. 2** Flow Chart

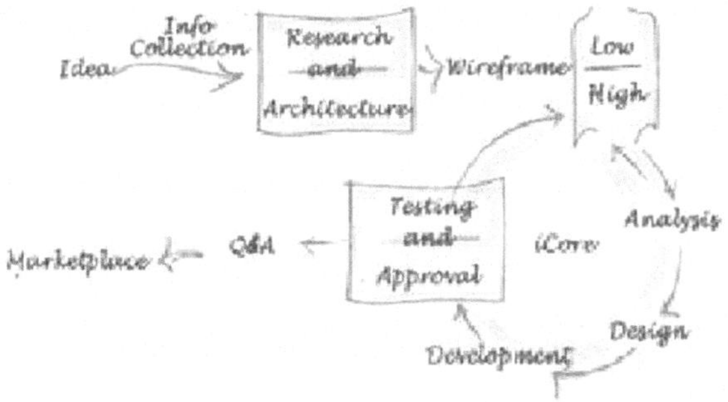


## 4 Features of the Game

The game will be available for both desktop and mobiles. Return to the player should be 94.52%. It will have five reels and three symbols per window. It is a ways game and it should have 243 ways to win. Win should be calculated from left to right except when free spin symbol is triggered. Stake can be selected by the player. It has cumulative reward system. The reel set for base game and the free game are same. It has total five reel sets. The game will have a spin button and a button for auto-play where the number of spins can be defined by the player.

### *4.1 Feature 1*

This feature gets triggered if three or more than three wilds occur on the screen. This symbol can substitute for any other symbol except free spins that is the scatter symbol. It is a wild card which will always help the player increase his win. In this game, these are hidden symbols and appear on the screen only if they are able to increase the win of the player or are able to create a win on the randomly chosen reel set and its symbols.

### *4.2 Feature 2*

This symbol randomly flies across the screen and causes the reels to spin again without triggering free spins. This reveals the winning symbols on the screen due to a re-spin. It has a fixed probability of flying and the direction of flying as well. It is triggered 25% of the time when the total win is three times or more than that.

**Table 1** Free spin bonus

| No. of scatter symbol | No. of free spins awarded |
|---|---|
| 3 | 10 |
| 4 | 15 |
| 5 | 20 |

### *4.3 Feature 3*

This feature provides the player another chance of winning as it spins the reels again to create a winning sequence. It efficiently hides the win and then the winning symbols are revealed after the shunt character has transformed the reel. This feature forms a part of the base slot [4] game RTP.

## 5 Free Spin

This feature will be triggered only if the scatter symbol appears on reel three, four and five. A re-spin will be triggered after every free spin feature gets triggered. Free spin should have a multiplier, that is, if a free spin is triggered for the second time then the total reward will get multiplied by two similarly, if free spin gets triggered for the fifth consecutive time, then the total reward money will get multiplied by five. This has only five consecutive multipliers after reaching this maximum the free spin end and the player returns to the base game.

## 6 Free Spin Bonus

This feature causes the cascading of free spin. Hereafter getting three or more scatter symbols on the grid, free spins get triggered (Table 1).

## 7 Reels and Win Information

This table explains the number of times the bonus gets multiplied when the multipliers get triggered in any of the features of the game. In Big Win, the total becomes ten times the actual value and so on and so forth (Table 2).

**Table 2** Multipliers

| Multipliers | Win |
|---|---|
| *10 | Big win |
| *20 | Huge win |
| *50 | Mega big win |

**Table 3** Payment table for symbols used

| Symbols | *3 | *4 | *5 |
|---|---|---|---|
| ! | 40 | 100 | 400 |
| @ | 25 | 60 | 200 |
| # | 20 | 50 | 150 |
| $ | 15 | 40 | 125 |
| % | 12 | 25 | 75 |
| ^ | 10 | 20 | 50 |
| & | 8 | 16 | 30 |
| * | 8 | 16 | 30 |
| + | 5 | 10 | 20 |
| / | 5 | 10 | 20 |

## 8 Payment Table

This table gives elaborates upon the weightage of each symbol. It is used for developing the maths file which eventually helps in giving the bonus points to the player whenever a feature gets triggered or a pattern gets displayed on the screen (Table 3).

## 9 Server Testing of the Game

Test cases represent the total number of features that the game has. Testing of test cases is done on a remote server [5]. The developer receives the request from the server and then checks if the correct response is getting generated. All test cases already have pre-defined responses. A session token gets generated from the platform, which is then copied on the server for processing the request. Session token ensures that the remote server is connected with game and the machine on which the game is being run. The session token therefore expires after some time and the user needs to authenticate itself again. This requires the username, operator Id, provider Id and game Id of the game that you want to test. The stake (the amount of money used in the game) is decided by the player, and the state (like spin, free spin, triggering of some feature) keeps on changing (Fig. 3).

After this, the pre-defined response appears which consist of the amount you have won or lost, the reel set that appeared, the symbols that appeared, the feature triggered

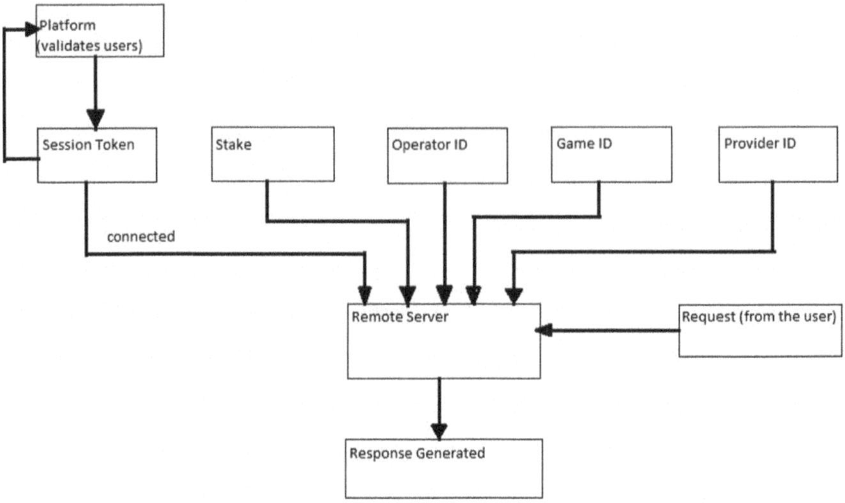


**Fig. 3** Block diagram for server testing of a game

and the matrix of reel set. This response helps us in checking the basic requirements of the game such as RTP and probability of all the features.

## 10 Conclusion

The development of the game was done successfully with the help of my industry guide. The development and understanding of the game development process were very interesting and enlightening. It helped me sharpen my programming skills. This internship gave me an idea about the actual industry targets, working environment and functioning. It was a learning experience for me and I enjoyed it a lot.

## References

1. Johnson, C., & Xiao, Z., Zhang, M., McGill, M., Bouchard, D., Bradshaw, M., Bucheli, V., Merkle, L., Scott, M., Sweedyk, Z., Ángel, J.: Game development for computer science education, pp. 23–44. https://doi.org/10.1145/3024906.3024908 (2016)
2. El Rhalibi, A et al.: 3D java web-based games development and deployment. In: International Conference on Multimedia Computing and Systems, Ouarzazate, pp. 553–559 (2009)
3. Saloni, J.: Developing games in java for beginners. In: International Journal for Research in Applied Science & Engineering Technology (IJRASET), vol. 4, issue III, March (2016)
4. Kilby, J., Fox, J., Lucas, A.F.: Casino operations management (2nd edn). Wiley (2005)
5. Cho, S.K., Sohn, K.M., Park, C.J., Kang, J.H.: Online game testing using scenario-based control of massive virtual users. In: The 12th International Conference on Advanced Communication Technology (ICACT), Phoenix Park, pp. 1676–1680 (2010)

# Hydroponics—An Alternative to Indian Agriculture System and Current Trends: A Review Study

**Ashish Aggarwal, Ratnakar Kumar, Sunil Kumar Chowdhary and Shailendra Kumar Jain**

**Abstract** India, the land of farmers, where agriculture has always been the primary occupation of the people, more than 50% of the population is still engaged in agriculture and its allied sectors. However, over the years, a significant rapid decline has been observed in the contribution by the agriculture sector toward India's GDP rate. In this paper, we aim toward identifying the gap between the ratio of high inputs and low yields by portraying the various loopholes in traditional Indian agriculture methods and how hydroponic agriculture is need of the hour for the growth of Indian agriculture. Also, the current trends in technology and research in the field of hydroponics around the world have been discussed to show how it can provide an ideal solution to the insufficiency of traditional farming, and how Indian farmers can adopt its implementation practices to boost their crop yield and income. Also, an IoT-based application has been proposed for monitoring and control of a hydroponic setup.

**Keywords** Hydroponic agriculture/hydroponics · Smart monitoring system · Internet of things (IoT) · Sensors

## 1 Introduction

India is a vast country in accordance with the geographical size, yet the amount of land suitable for agriculture is very limited. The movement of masses toward the cities and exponential increase in the population has resulted in the demand of land by many folds reducing the area for cultivation. Today, the contribution by this sector

A. Aggarwal · R. Kumar (✉) · S. K. Chowdhary · S. K. Jain
Amity University Uttar Pradesh, Noida, India
e-mail: ratnakar.kr0201@gmail.com

A. Aggarwal
e-mail: ashishagg888@gmail.com

S. K. Chowdhary
e-mail: skchowdhary@amity.edu

S. K. Jain
e-mail: skjain@amity.edu



P. K. Singh et al. (eds.), *Proceedings of ICRIC 2019*, Lecture Notes in Electrical Engineering 597, https://doi.org/10.1007/978-3-030-29407-6_62

to the GDP is as to near 17%, and its projection for the future is on the downside of it. The other major concern regarding the same is the imbalance in the ratio of people engaged in this sector to earnings yielded from it.

Hydroponics, or hydroponic agriculture [1–3], is the practice of growing crops in a solution medium. The solution generally consists of water-based solvent with minerals, nutrients, and other organic components dissolved in it to support plant growth. Unlike traditional farming culture, hydroponics does not involve the use of soil, sand, gravel, and other similar mediums. Since the nutrients and minerals are readily available for absorption by plants, the crop growth and yield are much higher compared to soil-based farming. Also, this method does not promote the use of commercial chemical fertilizers and pesticides. Rather, organic household and farm waste are used to maintain the nutrient balance of the solution. This organic approach gives crops higher nutrition values.

Though the hydroponic agriculture in India started as early as in the year 1946 in some parts of Bengal, yet it is a relatively unexplored technology to the farmers of India. This mapping of growth with the technology has suffered due to parameters which affect agriculture directly or indirectly.

This paper is divided into the following sections, hereafter:

- *Background*, which will cover the limitations of traditional farming practices in India. It will also talk about the benefits of hydroponics with reference to Indian perspective.
- *Literature Review*, which will talk about the innovations in the field of hydroponics by integrating it with the IT domain.
- *Proposed Solution*, which will discuss the application proposed for smart monitoring and control of hydroponic setup.
- *Discussion*, which will talk about the exigency of the proposed solution.
- *Conclusion*, which will summarize the paper and talk about the future scope of the topic.

## 2 Background

India is still chiefly an agrarian-based economy. However, the technological advancements that have taken place in this sector are fairly less and the number of farmers aware and using these technologies is even lesser. Despite huge demand, farmers in India are not able to generate enough income for themselves. The crops are regularly marred by climatic factors, inadequate or over-irrigation, pest infestations, overuse of chemical fertilizers, etc. Some of the factors that have restricted Indian farmers in meeting up with the market requirements and upgrading themselves with better technologies have been discussed below.

*Uneducated Farmers*. India's literacy rate is somewhat around 74%, which has seen a rise in recent years. Most of the farmers or laborers associated with agriculture are still uneducated and are dependent upon the techniques they acquired from their bloodline or adopted from neighbors, which in turn does not guarantee the appropriate or the most optimized agricultural techniques available. Lack of knowledge regarding the various techniques to replenish the fertility of soil, the types of fertilizers and pesticides, optimized amount of water that would be required to grow a particular crop, the plant which will be best suited for their soil are still some of the major challenges for them.

*Green Revolution*: Though it has been of great help to the agriculture industry yet a large number of problems came with it, a large quantity of chemical fertilizers, inorganic manures, and irrigation facilities is required for higher yield. This is believed to be the reason for the development of hazardous weeds and insects around the fields.

*Dependency on monsoon*: Since earlier times due to low irrigation supplies and facilities, a large population of farmers has always been dependent on monsoons which have never been steady. It varies season to season which has resulted in droughts in some areas while some areas saw a complete loss of cultivation for continuous years.

*Outdated Technology*: Though we are one of the tech-savvy giants, yet there are parts of India in the agriculture sector who have not yet witnessed a combine in their tenure. This is because the machinery is really costly and poor farmers with little land find it useless to invest in such machinery; furthermore, no machinery is provided on a rental basis by the government to save time.

*Limited Storage Facilities*: If there are years where monsoon is favorable and there is a surplus in that year, the amount of grains gets wasted due to the shortage of storage facilities. About 1.94 metric tons of grains were wasted from the year 2005–2013 as revealed by Food Corporation India against an RTI filed.

*Unhealthy Government reforms*: There are various reforms that the government introduces for the welfare of the farmers. On paper, they are definitely beneficial, but the ground reality is that farmers do not even get the promised minimum support price (MSP) which the government defines.

The rate at which the population and poverty in India are rising, soon we will be out of food to feed our population, and a large amount of population will be sleeping hungry or would not have adequate food to eat. Most of the crops and vegetables are grown in particular seasons only and then need to be stored in cold storage thus adding further to the costs.

Hydroponics, an approach of controlled environment agriculture (CEA), offers an effective solution as its various parameters can be manually controlled or/and can be automated to various levels with latest innovations to obtain a higher yield. It is beneficial in the following ways:

- The land requirement which is growing at an exponential pace along with the population leaves us with a limited yet depleting land area for agriculture. On the other hand, Hydroponic agriculture can be done in multistory buildings with each level as an individual and independent field, thus utilizing the area to the maximum.

**Table 1** Comparison of yields in hydroponic agriculture and soil-based agriculture, based on a case study in Bengal [15]

| Crop | Soil average per acre | Hydroponic equivalent per acre |
|---|---|---|
| Wheat | 5600 lb | 8000 lb |
| Potatoes | 8 tons | 70 tons |
| Maize | 1500 lb | 8000 lb |
| Rice | 750–900 lb | 12,000 lb |
| Cabbage | 13,000 lb | 18,000 lb |
| Cauliflower | 15,000–35,000 lb | 30,000 lb |
| Beetroot | 9500 lb | 20,000 lb |
| Lettuce | 9000 lb. | 21,000 lb |
| Okra | 5000–10,000 lb | 19, 000 lb |
| Oats | 1850 lb | 3000 lb |
| Soya beans | 600 lb | 1500 lb |

- Since only scheduled monitoring and maintenance need to be done once in a while after the plants have been set up, it, in turn, reduces the amount of labor required.
- There have been incidences where the wild animals or natural disasters destroy the crops in various areas and nothing substantial can be done to prevent it. Hydroponic agriculture, primarily being an indoor activity, reduces the risks considerably.
- The amount of water that is required by the hydroponic agriculture is much less than an equivalent field would require in normal soil field. Moreover, the water that is used can be recycled or reused for various other purposes or the same purpose by replenishing the depleted minerals.
- Since the minerals are available in the plenty within the water and days can be extended by using substitute lights to the sun, the growth of plants is at a higher rate than the traditional soil-based agriculture, as depicted in Table 1.
- Since there is no soil, there are no problems of soil-based diseases or growth of weed around the plant or breakout of pests and diseases which will affect the plant, and being free from all these things, there is no need spray pesticides on the plants.
- The crops or vegetables can be grown throughout the year, even when the weather is not favorable by controlling the indoor temperatures, thus providing greater growth and monetary benefits for the farmers.
- Since the vegetables grown through this technique are completely organic, it has a high amount of nutrients and essentials.

## 3 Literature Review

Hydroponics, as discussed earlier, is a soil-less farming culture. Being a controlled environment agriculture methodology, it is free from the adversities brought by nature, cattle, rodents, pests, weeds, etc. It also promises higher crop yields with higher nutrition values in less space and labor requirements. Thus, hydroponics offers a feasible solution to the limitations of traditional farming practices. However, it too has certain limitations associated with it. Some of the current challenges in hydroponic agriculture are: rapid change of nutrient concentrations in the setup; selective absorption or precipitation of nutrients; and requirement of routine monitoring and adjustment of the setup and its environment.

Researchers and innovators around the world are trying to technically enhance the hydroponics methodology to enhance the yields further. For this, smart and sophisticated "grow-rooms" have been proposed and designed by integrating IT domain to this agriculture sector. These grow-rooms are equipped with IT-enabled devices like IoT sensors, microcontrollers, wireless networks, and mobile applications that help to remotely monitor and control the environment parameters of the setup. Through this constant monitoring and control, favorable conditions are provided to the crops which result in their higher growth, better health, and higher nutrition values which are clearly evident from Table 2 given.

Further, this section will discuss the latest technologies and innovations that have been proposed or implemented to upgrade regular hydroponic farms to smart hydroponic farms.

Nishimura et al. [4] developed a sensor module based on simple oscillator circuits. The module was designed to be compatible with Arduino microcontroller and had the ability to measure the nutrient concentration and electrical conductivity (EC) of the water-based solution used in the hydroponic setup. Andaluz et al. [5] used Vernier sensors to monitor pH, EC, moisture, and temperature and developed a control mechanism for drip irrigation of Daniela tomato crop in a hydroponic environment.

Ruengittinun et al. [6] designed a hydroponic farming ecosystem (HFE) for farmers with relatively less knowledge. The HFE is basically an integration of IoT sensors coupled with Arduino to remotely monitor and control the parameters like water flow, temperature, and pH of the setup. Peuchpanngarm et al. [7] designed a mobile application integrated with sensors to achieve automation in control of hydroponics.

**Table 2** Comparison of produces of lettuce in smart and regular hydroponic agriculture [16]

| Feature | Regular farm | Smart farm | Change (%) |
|---|---|---|---|
| Plant height (cm) | 22.45 | 14.23 | −36.60 |
| Plant width (cm) | 15.40 | 14.23 | −7.54 |
| Root fresh weight (g) | 3.91 | 6.26 | 60.00 |
| Shoot fresh weight (g) | 46.17 | 62.09 | 34.46 |
| Stem diameter (mm) | 0.60 | 0.68 | 13.90 |
| Number of leaves | 23.00 | 26.96 | 17.20 |
| Leaf area ($cm^2$) | 67.89 | 65.11 | −4.10 |
| Nitrate content (mg/kg) | 3485.22 | 3198.18 | −8.24 |

The application included features to read data through sensors, manage them, and plan the harvest accordingly. Crisnapati et al. [8] developed Hommons, a web-based application to remotely monitor and manage an NFT hydroponic farm. The hardware components of Hommons were powered by solar energy. Umamaheshwari et al. [9] discussed a system aimed at integrating IoT sensors to achieve scheduled working and monitoring of hydroponic setup through automation. The system used the concepts of embedded systems. Helmy et al. [10] designed a wireless sensor network to remotely monitor a hydroponic setup implementing nutrient film technique (NFT) approach. The system used five sensors and was developed over the Raspberry Pi platform. Manju et al. [11] discussed a system with five sensors for measuring temperature, pH, ammonia, water level, and moisture in an aquaponic setup. The sensors were integrated with Lumisense IoT board to collect and save data over the internet for future analysis. Siregar et al. [12] designed a web-based application for remote monitoring of a hydroponic system using integration of various sensors, Open Garden modules, and Arduino Uno.

Melvix and Sridevi [13] identified technical faults to be the most critical hurdles in large-scale implementation of hydroponic agriculture. They proposed an algorithm based on fuzzy inference system for the purpose of grading of the solution in a hydroponic setup. The system parameters were optimized using genetic algorithms and the algorithm simulation was done through MATLAB and LabVIEW. The outcomes achieved were in form of higher resource utilization and better convergence efficiency. Yolanda et al. [14] proposed 12 inference rules of fuzzy logic to control the pH and EC of the solution in a hydroponic setup. The fuzzy control method controlled a number of tanks containing different nutrient solutions to be discharged into the system as required, to maintain the chemical composition of the solution flowing in the setup.

## 4 Proposed Solution

The objective of the proposed project is to develop a system comprising of an IoT-based application integrated with sensors that would be able to monitor the changes in a hydroponic agricultural setup like change in pH levels of the solution, change in nutrients' concentration in the solution, change in the temperature of setup's surrounding, etc. Further, these changes and the recommended actions to be taken will be notified to the farmer through the application.

In the solution we propose, as shown in Fig. 1, there will be sensors to measure the pH, temperature, and electro-conductivity (EC) of the solution flowing in the setup. These parameters are very significant for plant health and growth and hence these need to be monitored constantly. Additionally, there will be a water level sensor to identify the level of the solution flowing in setup and a camera which will give the live visual feed of the setup. All of the sensors and the camera will be integrated with a microcontroller which will coordinate with the Android application by sending the data to it through a Wi-Fi module. The application will allow users to access the

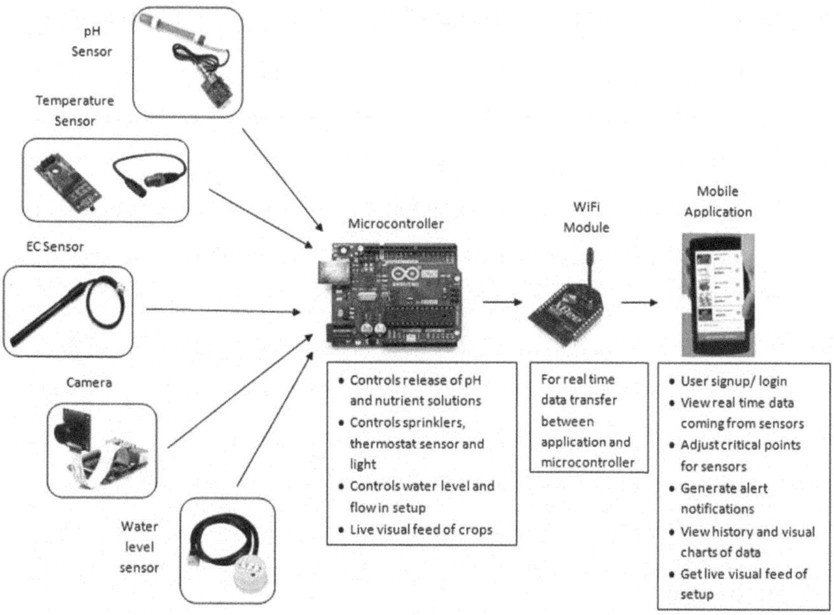


**Fig. 1** Schematic diagram of the proposed solution

data generated by sensors, visualize them in form of a graph over a period of time, view the data stored previously, and adjust settings of the sensors for generating alert notifications if a parameter rises above or falls below a specified alert limit.

Additionally, the microcontroller will be able to control the parameters in an automated manner. It will monitor the setup and perform the needful operations like balancing of pH level, nutrient concentration and EC of the system. Also, the water level will be kept in check and the water temperature if rises will be controlled through a sprinkler system controlled by the microcontroller.

## 5 Discussion

Some of the applications available for smart hydroponics on the Google Play Store are:

- *Hydroponic Nutrient Require* developed by JPT studio
- *Hydroponics Calendar* developed by GoMobileNowNet
- *BudLabs—Hydroponics Grow App* developed by advanced nutrients
- *Hydroponic Scale* developed by Harsh Technology.

However, the solution proposed by us would be better and efficient as compared to the existing ones, as clearly depicted in the following Table 3.

**Table 3** Comparison of existing applications (available on Google Play Store) and proposed solution

| | Existing applications | Proposed solution |
|---|---|---|
| Type | Mobile applications | Mobile application integrated with sensors to implement IoT |
| Features | • Calculation of pH, EC, and nutrient requirements for the plant growth based on crop<br>• Logging and planning of events during the crop growth cycle | • Real-time monitoring of parameter changes in the setup through sensors<br>• Live visual feed of the setup through a camera integrated with the microcontroller<br>• Generation of alert notifications in case of rapid/critical change in parameters<br>• Calculation of pH, EC, and nutrient requirements<br>• Logging and planning of events |
| Setup control | Completely manual | Partially automated |

## 6 Conclusion

In this paper, we have discussed the limitations associated with the current Indian agricultural methods and we have identified hydroponics as a viable and promising alternative. We have proposed a solution which aims at integrating ICT with hydroponics to increase productivity even further. The target end users of our proposed solution are Indian farmers, who have limited budgets for investing in advanced technical solutions. Future scope of this paper would include implementing the proposed solution which would provide an integrated platform to remotely monitor and control the setup to optimize the plant growth. This integrated solution will provide a platform for users to obtain an apt overview of the crops. The application would be developed in a manner such that it is easily affordable for the Indian farmers and they can easily interact with the application interface.

## References

1. Pal, S.: Growing Soil-Less with Hydroponics: An Introduction to Innovative Farming at Home. In: THE BETTER INDIA. (2016)
2. Hydroponic Vegetable Production—Guide. Department of Agriculture, Forestry and Fisheries. Republic of South Africa (2011)
3. Pandey, R., Jain, V., Pandey, K.S.R., Singh, K.P.: Hydroponics Agriculture: Its Status, Scope and Limitations. Division of Plant Physiology, Indian Agricultural Research Institute, New Delhi (2009)
4. Nishimura, T., Okuyama, Y., Matsushita, A., Ikeda, H., Satoh, A.: A compact hardware design of a sensor module for hydroponics. In 2017 IEEE 6th Global Conference on Consumer Electronics (GCCE). IEEE, pp. 1–4 (2017)

5. Andaluz, V. H., Tovar, A. Y., Bedón, K. D., Ortiz, J. S., Pruna, E. Automatic control of drip irrigation on hydroponic agriculture: Daniela tomato production. In: IEEE International Conference on Automatica (ICA-ACCA). IEEE, pp. 1–6 (2016)
6. Ruengittinun, S., Phongsamsuan, S., Sureeratanakorn, P.: Applied internet of thing for smart hydroponic farming ecosystem (HFE). In: 2017 10th International Conference on Ubi-media Computing and Workshops (Ubi-Media). IEEE, pp. 1–4 (2017)
7. Peuchpanngarm, C., Srinitiworawong, P., Samerjai, W., Sunetnanta, T.: DIY sensor-based automatic control mobile application for hydroponics. In: 2016 Fifth ICT International Student Project Conference (ICT-ISPC). IEEE, pp. 57–60 (2016)
8. Crisnapati, P.N., Wardana, I.N.K., Aryanto, I.K.A.A., Hermawan, A.: Hommons: hydroponic management and monitoring system for an IOT based NFT farm using web technology. In: 2017 5th International Conference on Cyber and IT Service Management (CITSM). IEEE, pp. 1–6 (2017)
9. Umamaheswari, S., Preethi, A., Pravin, E., Dhanusha, R.: Integrating scheduled hydroponic system. In: IEEE International Conference on Advances in Computer Applications (ICACA). IEEE, pp. 333–337 (2016)
10. Mahaidayu, M.G., Nursyahid, A., Setyawan, T.A., Hasan, A.: Nutrient Film Technique (NFT) hydroponic monitoring system based on wireless sensor network. In: 2017 IEEE International Conference on Communication, Networks and Satellite (Comnetsat). IEEE, pp. 81–84 (2017)
11. Manju, M., Karthik, V., Hariharan, S., & Sreekar, B.: Real time monitoring of the environmental parameters of an aquaponic system based on Internet of Things. In: 2017 Third International Conference on Science Technology Engineering & Management (ICONSTEM). IEEE, pp. 943–948 (2017)
12. Siregar, B., Efendi, S., Pranoto, H., Ginting, R., Andayani, U., Fahmi, F.: Remote monitoring system for hydroponic planting media. In: 2017 International Conference on ICT For Smart Society (ICISS). IEEE, pp. 1–6 (2017)
13. JSM, L.M., Sridevi, C.: Design of efficient hydroponic nutrient solution control system using soft computing based solution grading. In: 2014 International Conference on Computation of Power, Energy, Information and Communication (ICCPEIC). IEEE, pp. 148–154 (2014)
14. Yolanda, D., Hindersah, H., Hadiatna, F., Triawan, M.A.: Implementation of real-time fuzzy logic control for NFT-based hydroponic system on Internet of Things environment. In: 2016 6th International Conference on System Engineering and Technology (ICSET). IEEE, pp. 153–159 (2016)
15. Douglas, J.S.: Hydroponics: the Bengal system, 5th edn, pp. 80–190. Oxford University Press, USA (1975)
16. Changmai, T., Gertphol, S., Chulak, P.: Smart hydroponic lettuce farm using Internet of Things. In: 2018 10th International Conference on Knowledge and Smart Technology (KST). IEEE, pp. 231–236 (2018)

# Sports Policy Implementation by the IoT Platform

**Vishnu Priya Reddy Enugala and M. Abhinava Vinay Kumar**

**Abstract** In India, presently the meritorious sportspersons are not being benefited due to the non-implementation of the latest technologies; this can be achievable with the sports policy using IoT platform. By developing a special web address or a mobile application for each game to have the live telecast matches, an online score of matches with player details and along with online referee support method interlinking these through the internet and store all the information in the cloud accessible only to the sports policy authorities. Within this context, the contribution of this study is (i) Educating player in problem-solving for better performance. (ii) Player healthcare like nutrition, medical supports, fitness. (iii) High quality of digital and visualization technological skillful game training. (iv) Latest referee rules and regulations. (v) Identification of the meritorious sportsperson from the metropolitan area to agency area. This all is monitored to provide sports Excellency facilities and required sports benefits.

**Keywords** Internet of things · Sports policy · Sportsperson benefits · Monitoring the sportsperson records · Health care activities

## 1 Introduction

Sports is an essential activity in everyone life. But the Indian constitution does not show any priority by not mentioning sports policy in the constitution. Sports and games are not only the strength of every person but it is the power of the entire

V. P. R. Enugala (✉)
Department of Computer Science and Engineering, Carrier Point University, Rajasthan, India
e-mail: vishnupriya.enugala@gmail.com

M. Abhinava Vinay Kumar
Department of Sociology, Carrier Point University, Rajasthan, India
e-mail: vinayabhinava@gmail.com



P. K. Singh et al. (eds.), *Proceedings of ICRIC 2019*, Lecture Notes in Electrical Engineering 597, https://doi.org/10.1007/978-3-030-29407-6_63

nation. In India, so many sports organization, sports federations, and academies are running under the control of the sports ministry; out of these, some are running under Olympics associations and others are running as the individual federation/association. From abovementioned games, one of the game named cricket, which is private board of organization in India, which is the world's richest board called as Board of Control for Cricket in India (BCCI), and it is running under rules of Tamil Nadu State society registration act, even it does not comes under the Right to Information Act 2005. Due to the richest sports board, this BCCI is fighting against Supreme Court from last four years, as Lodha committee justices for the non-implementation of committee suggestions. As it is an economically sustained one, the richest board BCCI followed and implemented so many latest techniques and it is already updated with IoT technology in the cricket game. The day-wise cricket activities are monitored by the board officer barriers and succeeded in it by using of IoT. The player performance with irrespective of the age particulars at different stages, the score details, latest player track record, injuries, yoga fitness test, medical treatment information like doping test results, game facilities, and infrastructure of latest ground practice techniques these all are interlinked using IoT technology. By this reason, the BCCI is leading as first place in India, and it also covers as a business role with Indian premier league. By the implementation of IoT in the cricket, kabaddi, and many games, it affects the badminton premier league, the association/federations/board personal interest are only succeeded with introducing of the IoT system in those games. But by the implementation of IoT system in sports policy, its benefits reach to the real sportsperson like team's selection, awards, scholarships, academy administration, educations reservations, employment reservations, injury treatments, administration of association/federation/boards, monitoring. The sports authorities will have a glance at the expenditure, sports trainee, and in the utilization of sports cess of an event. Finally, sports insurance can also be provided by the effect of IoT.

## 2 Literature Review

Sports and games are one of the most rapidly growing areas for IoT technologies. Since the scope is very high, there are a lot of corporate companies which are targeting sports and trying to connect sports to the internet which has many benefits. Although IoT is yet to be incorporated on a large scale in sports, there are instances where it has been inculcated on a small scale.

## 2.1 Related Work Embedded IMU

Authors in [1–4] embedded IMUs in a cricket ball which is closest to our idea. In these, the authors gave a belief report on basic features such as angular velocity, time of flight, etc. These features are directly available from the sensors and no need to identify by the players/coaches. The author [5] mainly focused on the design of the ball with packaging for high impact which is also embedded IMUs. Author [6] explores spin-analytics in the context of a bowling ball. However, accelerometers and gyroscopes are usable due to low spin-rates and contact with the floor, which simplifies the problem in contrast to baseball and cricket.

## 2.2 Wearable, Camera, and Sports Analytics

Athletes are using wearable technology all the time in order to track their fitness. There are a few IoT-centered applications available in the markets which are beneficial to sportspeople such as Sensoria Smart Sock and Adidas MiCoach Smart Ball. Several startups like Zepp, MiCoach, and Ball are extracting motion patterns from wearable devices. The authors [7] have been proposed smart sensor shoes for analyzing soccer shots. The author in [8] Hawkeye gives an idea of camera-based tracker which is costly and most popular adopted in cricket, tennis, etc. The most popular IR technology hot spot [9] is used to determine contact points between ball and players. Video analytics efforts in [10–12] are processing video feeds to learn/predict game strategies and addressing a different set of problems.

## 2.3 Localization and Motion Tracking

Scope for indoor localization has rich literature [13–20] has mostly focused on human motion. The inertial sensor-based tracking has mostly been performed on humans, robots, and drones [21–26]. However, unlike iBall, can only address the space of freely falling objects. While work in [27] tracks ballistic missiles, the granularity of tracking is different both in time and space. iBall entails much finer granularities of tracking and appropriately formulates a global optimization problem for better accuracy unlike filtering techniques in [27].

In Swedish, ICT has initiated the Internet of Sports to fulfill the gap between sports and technology as it is huge in the sports industry. Mobile sensor-based products such as the Fitbit, Nike+, sports watches, and apps like Run keeper have been introduced pretty early over there [28] (Fig. 1).

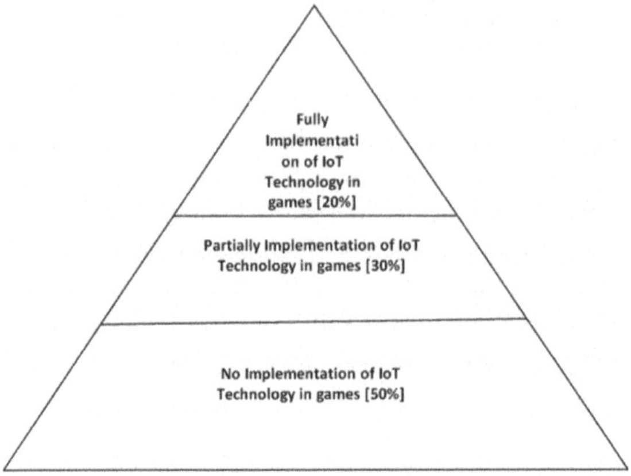


**Fig. 1** Pictorial representation of using IoT technology in sports and games

## 3 Purpose of the Study

Globally, India is a developing country leading as a top-level country in the education but within the sports activity (except private organization cricket game), it is in the down level. India is a mini world with different types of people with their physical activities, living in a different type of climatic weather conditions. But when it is compared with various small-sized countries with the same climatic weather conditions, the participation of sports and its achievements in India is least in positions. The main reason for the successful achievements of small-sized countries is only by the implementations of IoT-related sports activity methods. But in India till today, so many different types of sports organization are working on their interest bases only. The government aim in sports policy implementation is only for introducing and conducting the educations institutions, federation, associations, religions, zone level these all types of sports and games activities, but the original player's performance is struck by the lack of knowledge in the non-implementation of IoT. In the meantime, the small-sized countries with less in the population are utilizing the technique of IoT and getting success in gaining medals and in all aspects too.

Nowadays, sports are being developed in one type as in region with an only little specific type of games those are getting good results. By these others, game players are not been encouraged; this leads to the dissatisfaction of sports policy implementation. Particular of an area, the sports players are achieving good results like play court events Sindhu, Saina, Kashyap, Jwala, Sania Mirza, Bhupathi, Leander Paes; these are from south region players and those are getting good achievements in international. Similarly, shooting events Gagan, Abhinav Bindra, Rajyavardhan singh getting good achievements from the south Indian sports starts and at the hills station area, Mary Kom and like other boxers are fighting as natural players all from one side area are dominated. From the most forward, Maharashtra in cricket gave cricket god of Sachin and Sunil Gavaskar likewise, many players are been introduced.

The main aim is to introduce IoT technology in sports policy so that the meritorious sports players are easily identified and for whom the facilities are required from the agency area to metropolitan area as the player performance is monitored continuously and effectively. By this method local sports management, politics does not work to break the meritorious sportsperson performance. In some area from each person, as a sports development cess is collected to promote meritorious sportsperson, this can be overcome by the implementation of IoT in sports policy in sports and games.

## 4 Research Methodology Used in This Study

For the first time in India, the national sports policy was introduced in the year 1984. For every five years, the national sports policy was monitored and several goals were considered to encourage sports and promote to the national level and international level of sports achievements. From the year 2001, the modernized national sports policy is in use, i.e., with the combined effects of state government, national sports federation, and Olympics association. It also reconstructed the national sports policy, since from the starting of national sports policy year 1984 till today. The results of the research method used for the development considered under the different situations, facilities, weather atmosphere of India, and worldwide countries which is successfully methods in the implementation of sports policy by using IoT platform.

For this study, IoT-based relative information is chosen to develop a research methodology for district level, state level, national level, and international levels of all players' participation in sports and games. The players use internet, mobile applications, sensor technology, latest applications, like data analysis from a cloud and other related servers. These latest technologies can be identified using different types of IoT facilities, required for different types of players like indoor players, outdoor players, field track players, and etc. players. The following 20 major most required IoT-based technologies are considered for the research development for this study that are

(1) Online score, (2) Online draws, (3) Online referees facility, (4) Online scrutiny player details, (5) Online doping test and drug details, (6) Internet/Wi-Fi facility of grounds and participation, (7) Live telecast/recorded match facilities, (8) Recorded shouted match facilities, (9) E-fitness device, (10) E-health alert devices, (11) Online participation certificate record information, (12) Player E-commerce/banking details, (13) Online sports guidance/coach reviews, (14) RFID used for players information, (15) Speed-gun technology for calculation, (16) E-practice sessions, (17) E-medicine/treatment facility, (18) E-sports Id card (Aadhar/student id card/online age proof), (19) E-communication address, (20) E-photo identify/fingerprint/eye retina identity (like Aadhar) (Fig. 2).

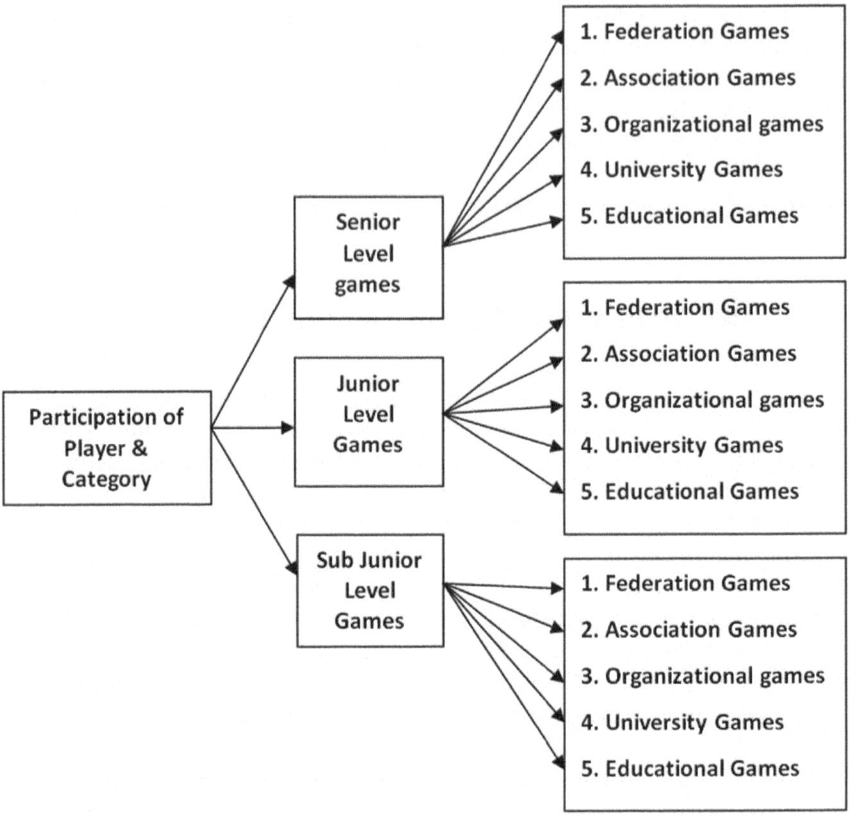


**Fig. 2** Participation of players levels in sports policy

Here from these, IoT-based 20 major elements make the player to achieve the goals of different games under different types of condition; to develop the research methodology, we consider the following point.

(i) Player better performance.
(ii) Player healthcare.
(iii) High-quality game training.
(iv) Latest rules and regulations.
(v) Identification player from a different area.

# 5 System Architecture for Smart Sports Policy

## *5.1 Implementation of Sports Policy with IoT Architecture for Play Field Games*

The sports policy implementation using IoT is called as "smart sports policy implementation system." This smart system consists of three major components they are Sensors, Connectivity, Storage, and Data Analysis. These components are the basic architecture of an IoT system. Each component is described as following.

**Sensors** As the sports policy is mainly for the players development so involves the player's data. The data or information is been taken using the sensors itself. For the data collection of a football player, whether he made the goal or not is been recorded using the goal-line technology which tells the ball has crossed the line or not and sensor are linked to the smart watch worn by the match referee. Similarly, for the water sports, data of simmer is taken from the sensor of with accelerometers and gyroscopes to measure the time, speed, acceleration, and velocity of swimmers combined with the video footage captured by cameras both above and below the waterline. For athletes, the smart watch senses the heartbeat, distance, speed, and the performance of the players. For volleyball player, wearable monitors can measure the athlete's jump height, average jump height, and amount of jumps while court monitoring systems can track both the player's and the ball's position too. In some games, even the shoes are represented with a sensor so that the speed and distance of the performer are recorded. The player details can be captured using RFID technology. The reward and certification of the player received at the events are also been captured and transmitted. The data collected from the sensors, it is being transmitted using the connectivity or through the gateway.

**Connectivity** Here, the connectivity is the gateway which allows the collection to send to the destination, i.e., for the storage and data analysis. The connectivity is also called a gateway; it can be wired or wireless connectivity. This connectivity is automatically accessed without any manual interruption. In sensing the data, different type of technologies and different sensor are used to transmit the data without the loss of data is done through wireless by Bluetooth technology, Wi-Fi technology, GSM technology using internet facilities. The gateway also provides secure transmitted as it involves the network security algorithms for the transmission of data. By using the internet, transmitted data from the sensor is sent for storage.

**Storage and Data Analysis** The data or information transmitted through the gateway is stored for future use. The data is stored in the cloud. As the sports policy has different levels, so the storage cloud for all level is considered. The cloud access is only under the sports authorities like for the state-level sport has a cloud which can be accessed by national wide sports authorizes so that player performance and activities can be monitored similarly, the national-level sports cloud has access for the worldwide sports authorities. The complete details of the player with the previous record of performance are been stored in the cloud. The data can also be monitored in the mobile application connected to the sensor. The data collected during the actual

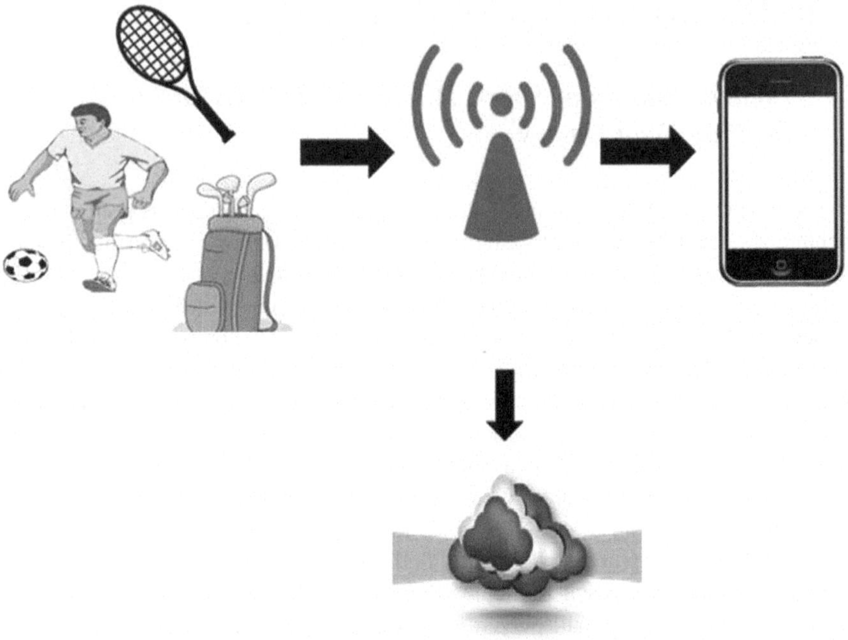

**Fig. 3** Generalized architecture for implementation of smart sports policy

event is also used to optimize training and performance for the trainee and sports authorities for future performance to rectify mistakes in the data analysis. During the data analysis, bogus certificates and fake sportsperson are identified. Misuse of sports policy can not happen by this (Fig. 3).

The work for the implementation of smart sports policy involves two types of classification of players

(1) District/State-Level Players
(2) National/International-Level Players.

(1) The district-level, state- players are a primary level source of sports policy players while a player comes from their part of the curriculum of educations so, here the players are not in a professional way in sports. But it is difficult to identify the player for the next level of national and international level of participation. This is the best level of encouragement in sports but due to lack of facilities, knowledge, the players are tampered at the level of state level only but here by the implementation of IoT it is possible. Even a single player does not have an option for learning techniques in games from the required educated trained coach but it will be provided through online training using IoT. The state-level player participation performance is monitored so that the merit one is identified for the national level through interlinking

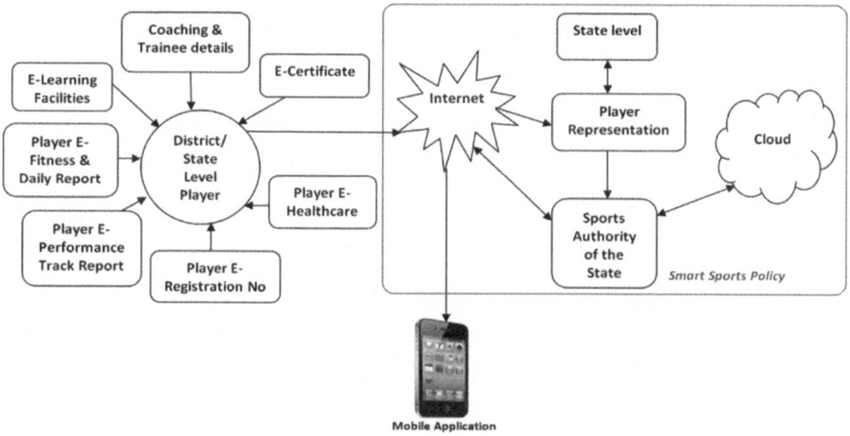


**Fig. 4** IoT architecture for district/state-level player in smart sports policy

of the internet, mobile application system, and the players details. Primary caching, E-fitness, E-performance, E-fitness track reports, E-learning facilities all are benefited to the players in the orientation by implementing IoT.

From the below architecture, the district/state-level player has following facilities in smart sports policy system. Using RFID, the player is given the registration number to record his sports details and to store in the cloud. The player has got a fitness tracker device which gives the health care details and daily fitness report. These are stored to the cloud using Bluetooth technology and can be monitored in the mobile application. The matches are under the surveillance camera; by this, player performance track report is identified. Coaching provided to the player is also recorded and the player has any access to learn from online using mobile application. The certificate received from level is been scanned using Zigbee transmission it uploads to a cloud so that record can be maintained of a particular player. The collected data is stored and in under access to sports authority of state (Fig. 4).

(2) The national level and international level of players play a major role in the IoT-based sports policy; here, the worldwide latest technologies and facilities can be adopted in the IoT system. Here, the player benefits like employment educational reservation, misused sports policy benefits, and Excellency of sports faculties are developed in the favor of sportsperson. This level of reaching players had faced major problem that as medical supports like injury E-treatment sports and nutrition of health care and monitoring of fitness coaches by E-track fitness devices, and mistakes can be identified and learned from the previous matches videos, etc., and all IoT-based technologies are to be benefited by this implementation system.

The architecture of national/international-level player basic details is already present in the cloud of sports authority of state. Using E-fitness tracker, the player fitness and health are recorded with the help of pure pulse technology, and this technology uses green light inserted into the blood which calculates the pulse rate, calories burned resting time of heart, heart stroke conditions. This information is

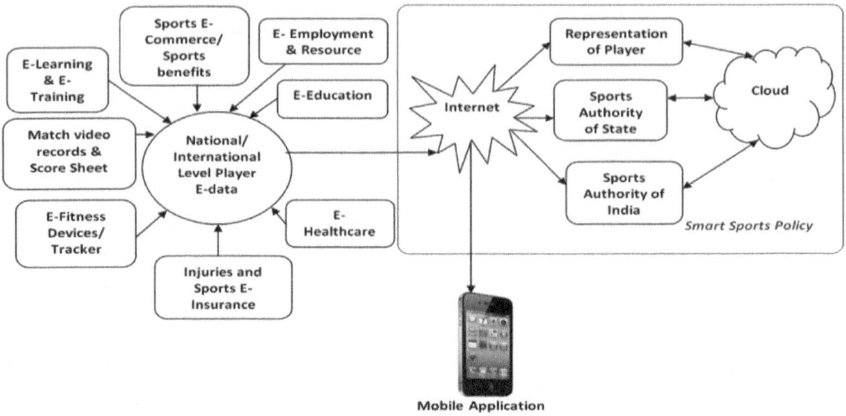


**Fig. 5** IoT architecture for national/international-level players in smart sports policy

automatically sent to the mobile application using Bluetooth technology and with the internet to the cloud storage. The complete record of the player is present so that if the player is injured, he/she is ready to claim the insurance directly by online records. The educational details of player are registered with their player Id so if he/she is interested in higher education, the reservation in the education is provided only for the meritorious sportsperson as all the certificates are updated. Similarly, in the employment purpose, the meritorious sportsperson is only benefitted. The player matches are captured using a high-speed camera with motion video acquisition technology, this is stored, and as per the performance, the required training is given through the mobile application at the players. The total data collected is in the access of sports authority of India (Fig. 5).

## 6 Findings and Discussion

The table consists of data collected by the standard questionnaires from the 300 players who participated at the state level, national level, and international level in India. Those who are elected from the five different zones come under the area of states that are the Telangana, Andhra Pradesh, Gujarat, Assam, Chandigarh, and Madhya Pradesh participated as meritorious players. For each state, only 50 players (25 male and 25 female) are considered and again 50 players are divided as 24 games/game players only as per G.O MS NO.84, Telangana state and Andhra Pradesh state issued sports policy in the year 2000, and also taken other one local-related state sport/game (Table 1).

**Table 1** Data analysis of sports at three levels of participation based on IoT system

| S. No. | Sports events | State-level sports | | National-level sports | | World-level sports | |
|---|---|---|---|---|---|---|---|
| | | IoT based | Non-IoT based | IoT based | Non-IoT based | IoT based | Non-IoT based |
| 1 | Olympics games | No | Yes | Yes | No | Yes | No |
| 2 | Non-Olympics games | No | Yes | No | Yes | Yes | No |
| 3 | World cup games | No | Yes | Yes | No | Yes | No |
| 4 | Asian games | No | No | Yes | No | Yes | No |
| 5 | Educational games | Yes | No | Yes | No | Yes | No |
| 6 | Invitational events | No | Yes | Yes | No | Yes | No |
| 7 | National federation cup | No | Yes | Yes | No | Yes | No |
| 8 | Prize money events | No | Yes | No | Yes | Yes | No |
| 9 | Rural sports | No | Yes | No | Yes | No | Yes |
| 10 | National games | No | Yes | Yes | No | – | – |
| 11 | Political influence sports | No | Yes | No | Yes | No | Yes |
| 12 | Sports employment events | No | Yes | Yes | No | – | – |
| 13 | Age-wise games | No | Yes | Yes | No | Yes | No |
| 14 | Rehabilitation sports | No | Yes | No | Yes | No | Yes |

### 6.1 Findings from Research Work

- The ground level (district level and state level) of sports policy is not even monitored by the sports authorities.
- 95% of rural area sportsperson does not have minimum facilities/sports promoters, they are been flourishing by the media publicity that to on their personal interest only. Due to this reason use, IoT platform makes the system automate but there is still a little bit hesitation in the people to use IoT systems that is mainly due to the lack of knowledge. 45% of people have the capability to participate in international standard but they are not promoted due to unawareness.
- Out of 100% using IoT-based systems who participated at state and national level of meritorious sportsperson, more than 90% players are only interested in E-fitness watch and more than 20% players using E-shoes, more than 65% of players are using mobile applications for specific participation in games, to calculate their performance, only below 10% of players are using his/her sports certificate enrolled on the IoT-based system, more than 30% players using players E-Id card, and more than 70% sports events of scored sheets and participation players roster list and proofs are not even saved.
- Each sports organization is different but every meritorious player participating from any organization of those respective game, participate in the events every time with many difficulties thus the player is not recognized as a merit sportsperson ultimately the player waste his/her valuable time and educational life so players parents are not interested in promotion in them into sports.
- 90% national-level player in his/her sports carrier has dropped his/her examination and same dates are clashed by conducting other same level events.

## 7 Conclusion

The quality of sports activities show the unity of nationality, as one nation with one sports policy is introduced in the entire nation but the powers of the national level of organization, authorities of varies organizations in educational, federations, associations are interfering and making sports like politics. Finally, the meritorious sportsperson is not identified by the sports policy misuse. Every sportsperson wants to calculate his/her regular performance but he/she would not have the previous performance details even with the authorities also, as the player concentrates only on the present performance; by this, meritorious sportsperson was not recognized yet. Due to this problem, the player will be unprompted and well-known coach players will be promoted automatically. India is huge in area and with heavy population, averagely so many talented players participated here, but the well-known player is only identified instead of meritorious sports player by the implementation of IoT in sports policy these all problems can be easily identified.

The old methodologies of physical educators and sports trainee course like the national institution of sports course, sports diet, sports medicine, sports management, etc. all sports-related course are re-modified by using of IoT. Sports facilities like RFID facilities for sports kits, E-fitness tags, E-gym, E-medicare, E-treatment, internet linking facilities for playgrounds, sports networking, online sports telecast like web porting system all these to be provided to identify the preliminary player, using fingerprint technology and eye retina with irrespective of age, along with the participation details of the events, and all the metadata is stored using IoT. This will help for the meritorious sportsperson development.

## References

1. Fuss, F.K., Lythgo, N., Smith, R.M., Benson, A.C., Gordon, B.: Identification of key performance parameters during off-spin bowling with a smart cricket ball. Sports Technol. **4**(3–4), 159–163 (2011)
2. Fuss, F.K., Smith, R.M.: Accuracy performance parameters of seam bowling, measured with a smart cricket ball. Procedia Eng. **72**, 435–440 (2014)
3. Fuss, F.K., Smith, R.M., Subic, A.: Determination of spin rate and axes with an instrumented cricket ball. Procedia Eng. **34**, 128–133 (2012)
4. Mcginnis, R.S., Perkins, N.C.: A highly miniaturized, wireless inertial measurement unit for characterizing the dynamics of pitched baseballs and softballs. Sensors **12**(9), 11933–11945 (2012)
5. Fuss, F.K., Ferdinands, R., Doljin, B., Beach, A.: Development of a smart cricket ball and advanced performance analysis of spin bowling. In: Advanced Technologies in Modern Day Sports (2014), Institute for Sports Research (ISR). ICSST, pp. 588–595 (2014)
6. King, K., Perkins, N.C., Churchill, H., Mcginnis, R., Doss, R., Hickland, R.: Bowling ball dynamics revealed by miniature wireless MEMs inertial measurement unit. Sports Eng. **13**(2), 95–104 (2011)
7. Zhou, B., Koerger, H., Wirth, M., Zwick, C., Martindale, C., Cruz, H., Eskofier, B., Lukowicz, P.: Smart soccer shoe monitoring football interaction with shoe integrated textile pressure sensor matrix. In: Proceedings of the 2016 ACM International Symposium on Wearable Computers ACM, pp. 64–71 (2016)
8. Hawk-eye. https://en.wikipedia.org/wiki/Hawk-Eye/
9. Hot spot. https://en.wikipedia.org/wiki/Hot_Spot_ (cricket)/
10. Halvorsen, P., Saegrov, S., Mortensen, A., Kristensen, D.K., Eichhorn, A., Stenhaug, M., Dahl, S., Stensland, H.K., Gaddam, V.R., Griwodz, C., et al.: Bagadus an integrated system for arena sports analytics a soccer case study. In: Proceedings of the 4th ACM Multimedia Systems Conference. ACM, pp. 48–59 (2013)
11. Seo, Y., Choi, S., Kim, H., Hong, K.S.: Where are the ball and players? Soccer game analysis with color-based tracking and image mosaic. In: International Conference on Image Analysis and Processing, Springer, pp. 196–203 (1997)
12. Yu, X., Xu, C., Leong, H.W., Tian, Q., Tang, Q., Wan, K.W.: Trajectory-based ball detection and tracking with applications to semantic analysis of broadcast soccer video. In: Proceedings of the Eleventh ACM International Conference on Multimedia. ACM, pp. 11–20 (2003)
13. Bahil, P., Padmanabhan, V.N.: Radar: an in-building RF-based user location and tracking system. In: Nineteenth Annual Joint Conference of the IEEE Computer and Communications Societies, INFOCOM Proceedings IEEE. IEEE, vol. 2, pp. 775–784 (2000)
14. Chintalapudi, K., Padmanabha Iyer, A., Padmanabhan, V. N.: Indoor localization without the pain. In: Proceedings of the Sixteenth Annual International Conference on Mobile Computing and Networking (2010), ACM, pp. 173–184

15. Kumar, S., Gil, S., Katabi, D., Rus, D.: Accurate indoor localization with zero start-up cost. In: Proceedings of the 20th Annual International Conference on Mobile Computing and Networking (2014), ACM, pp. 483–494
16. Niculescu, D., Nath, B.: Ad hoc positioning system (APS) using AOA. In: Twenty-Second Annual Joint Conference of the IEEE Computer and Communications, INFOCOM, IEEE Societies. IEEE, vol. 3, pp. 1734–1743 (2003)
17. Rai, A., Chintalapudi, K.K., Padmanabhan, V.N., Sen, R.: Zero-effort crowd sourcing for indoor localization. In: Proceedings of the 18th Annual International Conference on Mobile Computing and Networking, ACM, pp. 293–304 (2012)
18. Wang, H., Sen, S., Elgohary, A., Farid, M., Youssef, M., Choudhury, R.R.: No need to war-drive, unsupervised indoor localization. In: Proceedings of the 10th International Conference on Mobile Systems, Applications, and Services. ACM, pp. 197–210 (2012)
19. Xiong, J., Jamieson, K.: Array track a fine-grained indoor location system. In: Presented as Part of the 10th USENIX Symposium on Networked Systems Design and Implementation (NSDI 13), pp. 71–84 (2013)
20. Youssef, M., Agrawala, A.: The horus WLAN location determination system. In: Proceedings of the 3rd International Conference on Mobile Systems, Applications and Services ACM, pp. 205–218 (2005)
21. Lefferts, E.J., Markley, F.L., Shuster, M.D.: Kalman filtering for spacecraft attitude estimation. J. Guid. Control Dyn. **5**(5), 417–429 (1982)
22. Liang W.Y., Miao, W.T., Hong L.J., Lei, X.C., Chen, Z.: Attitude estimation for small helicopter using extended kalman filter. In: 2008 IEEE Conference on Robotics, Automation and Mechatronics. IEEE, pp. 577–581 (2008)
23. Madgwick, S.: An efficient orientation filter for inertial and inertial/magnetic sensor arrays. In: Report x-io and University of Bristol, UK (2010)
24. Mathony, R., Hamel, T., Pflimlin, J.M.: Nonlinear complementary filters on the special orthogonal group. IEEE Trans. Autom. Control **53**(5), 1203–1218 (2008)
25. Pflimlin, J.M., Hamel, T., Soueres, P.: Nonlinear attitude and gyroscope's bias estimation for a VTOL UAV. Int. J. Syst. Sci. **38**(3), 197–210 (2007)
26. Zhou, P., Li, M., Shen, G.: Use it free instantly knowing your phone attitude. In: Proceedings of the 20th Annual International Conference on Mobile Computing and Networking ACM, pp. 605–616 (2014)
27. Siouris, G.M., Chen, G., Wang, J.: Tracking an incoming ballistic missile using an extended interval kalman filter. IEEE Trans. Aerosp. Electron. Syst. **33**(1), 232–240 (1997)
28. https://www.sics.se/projects/internet-of-sports

# Bayesian Prediction on PM Modi's Future in 2019

**Aniruddh Sanga, Ashirwad Samuel, Nidhi Rathaur, Pelumi Abimbola and Sakshi Babbar**

**Abstract** Electing a Prime Minister (PM) is a process that occurs every 5 years in India, through conducting an election. Who will win the election is one of the most asked questions on everyone's tongue, as the election is just few months away. The objective of this research is to predict the likelihood of PM Narendra Modi's chances to continue as the Prime Minister of India using Bayesian network approach. The aim of this research is not to develop a new predictive algorithm, but to use the existing approach for making predictions on a real-life scenario. Our Bayesian modeling is based on public responses available on social media, India statistics, and news articles on the key policies undertaken by PM Narendra Modi during his current tenure. We explore using causal and diagnostic reasoning to find new insights on the factors shaping his win or no-win, verdict on his government strength and weakness. Our Bayesian model reveals that the current Prime Minister Modi has 61.4% chances of winning the upcoming 2019 elections.

**Keywords** Bayesian · PM Modi · Prediction · Election · Key policies

A. Sanga (✉) · A. Samuel · N. Rathaur · P. Abimbola · S. Babbar
Department of Big Data Analytics, School of Engineering, GD Goenka University, New Delhi, India
e-mail: Sangaaniruddh2@gmail.com

A. Samuel
e-mail: ashirwad.samuel@gmail.com

N. Rathaur
e-mail: nidhi030592@gmail.com

P. Abimbola
e-mail: Pelumi.abimbola@gmail.com

S. Babbar
e-mail: sakshi.babbar@gdgoenka.ac.in



P. K. Singh et al. (eds.), *Proceedings of ICRIC 2019*, Lecture Notes in Electrical Engineering 597, https://doi.org/10.1007/978-3-030-29407-6_64

## 1 Introduction

Leadership plays a vital role in the acceleration of any country's development [1]. India is a democratic country, in which people are allowed to voice their opinion and vote for what they believe in by the process of an election and where majority, subsequently, overrides the minority. The 17th Lok Sabha Prime Minister election which takes place every 5 years will be conducted in May 2019. Who will take the seat next, is one of the biggest questions on everyone's tongue? How is the current government performing, will it remain on sit or say good-bye?

The outcome of the election would appear to be very difficult to predict. Unlike, the first time PM Modi was elected, which was a clear win (his previous victory was a clear win). However, some predictions have been made on whether PM Narendra Modi will retain his seat or not by different individuals, while media and experts are busy making wild estimates based on what they think.

A Bayesian analytics network approach is proposed, to forecast the outcome of the 2019 election by making a probabilistic estimate from PM Modi's current 5-year accomplishments and (some form of) available data. In this research, causal and diagnostic reasoning is applied to the Bayesian network model to gain insight between the cause (happenings) and effect (impact). The reason for choosing the Bayesian network as a base model for the intended task is because it accurately captures our beliefs, and it also uses the graphical form to represent findings in a probabilistic manner, therefore, making the model easy to understand [2, 3].

No doubt that Modi's accomplishments during his current 5-year tenure will determine his next chances of winning. We will be seeing the outcome of that in the result. What is the people's primary concern or thought about this upcoming election? The election issues include an increase in the unemployment rate by 3.5% [4], Modi's policies like implementation of GST, demonetization, surgical strikes are the matter of concern to the voters, and will these factors play the key character in the election by determining Narendra Modi's victory? We will be combining data from social media, published public opinion polls, and India statistics from a reliable data source to develop a machine learning model for prediction tasks.

The rest of this paper is organized in section as follows: firstly, contribution to existing work, secondly, research methodology covering the factors and techniques to be used, thirdly, introduction of the features that are relevant and correlated with the election outcomes and propose our prediction model (experiment), fourthly, prediction results will be presented, and lastly discussion of the result and our beliefs, and conclusions from our experiment.

### 1.1 Contributions

We summarize our contributions below:

1. We propose the use of Bayesian network approach to predict the probability of PM Modi being reelected in Lok Sabha 2019 election. The aim of this work is not to propose or develop a new predictive algorithm, but to use the existing approach for making predictions on a real-life scenario.
2. Our proposed model can discover the key patterns or trends that may lead to PM Modi's win in 2019, i.e., analyze and draw meaningful inferences from the performance of Modi's current tenure, by using causal reasoning.
3. Using our predictive model, we can relate outcomes with valid statistical reasoning than making subjective predictions based on instincts/gut feelings.
4. Through our proposed model and analysis, PM Modi's government can learn in what areas or policies on which they need to improve on in other to strengthen their governance.
5. Our approach provides detail analyses, using diagnostic reasoning on why PM Modi may be victorious in the upcoming 2019 elections.

## 2 Research Methodology

In this age of data-driven revolution, many machine learning models have been developed with the capacity to analyze and forecast the outcome of an event based on statistical data, deductions, and people's emotions as expressed in the social media [5]. This ML models explore the data, make assumptions, and find interesting patterns and trends; therefore, predicting an outcome for any occurrence based on the probability of trend found in the different factors of the data [3].

India being the world's 41st largest democratic country in the world [6], all ears are interested in the outcome of the upcoming Prime Ministerial elections. We have proposed to integrate the Bayesian network machine learning model to predict PM Modi's chances to win.

For any machine learning model, relevant datasets are required [2], so we proceed to identify determinant key factors that will show the trend in our model, thereafter, acquiring the available datasets from a reliable source.

For our research, we have segregated the dataset into three clusters, namely external variable (which brings trends), influential variable (PM Modi's key policies), and decision variable. The causal interactions among variables are identified by available information from reliable data sources.

We have sorted the total voting population (i.e., citizens of India, who are eligible to vote: 18 years of age and above) across different cross sections of the society named population in external variables. The following groupings have been made under the assumption that a particular group has similar traits, opinions, and response (positive or negative impact) with respect to a particular situation or an event:

## 2.1 External Variables

**Population**: Human settlements are classified as rural or urban depending on the density of human-created structures and resident people in a particular area. Urban areas can include town and cities while rural areas include villages and hamlets.

*Reasoning*: The development of the respective regions and the understanding of a person in urban against a person in rural in choosing a leader to vote for.

**Occupation**: A person's usual or principal work or business, a means of earning for living are classified into this three groups: primary (agricultural and other allied activities such as forestry, farming, grazing, hunting, and gathering), secondary (industrial, government, and big business), and tertiary (private, small business, and services).

*Reasoning*: The deciding factors depend on the extent to which the work opportunities have increased or decreased, expanded or shrunk; the investments or policies affecting their job and income.

**Age**: Considered only the age of the citizens who are eligible to vote, classified into youth: between 18 and 35 years, middle-aged; between 35 and 55 years, and old; 55 years and above.

*Reasoning*: The decisive factor weighs on the points like the older people are relatively prejudiced while the youth are more open to change. On the other hand, the experience, knowledge of the past political scenarios, and maturity of the middle-aged and older people will fall less for deceptive political publicity.

**Religion**: The particular system of faith and worship practiced in India to be considered are Hindu, Islam, Sikh, Christian, Buddhist, Jain, and Others.

*Reasoning*: The propaganda of a specific religion to show itself or its demands superior over other, sometimes at the cost of others' religious beliefs can greatly determine the direction of a community or when a particular religion is targeted or attacked, verbally or physically.

**Employment**: The employment rate is classified into employed and unemployed.

*Reasoning*: The satisfaction of the employed and distraught of the unemployed.

## 2.2 Influential Variables

Policies or activities that were undertaken by the Modi government which are expected to have a substantial impact on the response of the public toward reelection of Modi government. Thus, all of these variables are taken as positive and negative (Table 1).

- Budget: Positive and Negative
- GST: Positive and Negative
- Demonetisation: Positive and Negative
- Leader: Positive and Negative
- Surgical Strikes: Positive and Negative

**Table 1** Description of the influential variables

| Influential variables | Description |
|---|---|
| Budget | This refers to the blueprint of the government's revenue and expenditure for the fiscal year, focusing mainly on the interim budget presented on February 1, 2019 [7] |
| GST | Launched on July 1, 2017, Goods and Services Tax is an indirect tax (or consumption tax) levied in India on the supply of goods and services [8] |
| Demonetisation | On November 8, 2016, the government of India announced the demonetization of all ₹500 and ₹1000 banknotes which was followed by prolonged cash shortages in the weeks that followed, which created significant disruption throughout the economy [9] |
| Leader | The leadership quality of PM Modi being the head political figure of the nation as perceived by the people |
| Surgical strikes | On September 29, 2016, India announced that it conducted "surgical strikes" against militant launch pads across the line of control in Pakistani-administered Kashmir [10] |
| Health awareness | Hospital facilities extended by the government, health programs to the poor, and yoga have been considered here |
| Research and development | Work directed toward the innovation, introduction, and improvement of products and processes |
| Foreign policies | Government's strategy in dealing with other nations or international relations |
| Digitalization | Launched on July 1, 2015, to ensure the government's services are made available to citizens electronically by improved online infrastructure and by increasing Internet connectivity or by making the country digitally empowered in the field of technology [11] |

- Health Awareness: Positive and Negative
- Research and Development: Positive and Negative
- Foreign Policies: Positive and Negative
- Digitalization: Positive and Negative.

## 2.3 *Decision Variable*

**Decision**: Win and No-win.

Next step was to choose a suitable machine learning model, and for this work, we have chosen Bayesian network approach.

**Bayesian networks** are a type of acyclic probabilistic graphical model that infers from probability computations to fix complicated relationships among variables in a direct manner by using Bayes' theorem, where $P(A/B)$ is the *posterior* that we are trying to estimate, $P(B/A)$ is the *likelihood* that an event will occur, $P(A)$ is the *prior,* *and* $P(B)$ is the marginal likelihood probability [12].

$$P(A/B) = (P(A) * P(B/A))\big/ P(B) \tag{1}$$

Bayesian networks aim to model conditional dependence, and therefore causation, by representing conditional dependence by edges in a directed graph. Through these relationships, one can efficiently conduct inference on the random variables in the graph through the use of factors [2, 3, 12].

The advantage of the Bayesian network here is given a limited amount of data from which we can "calibrate" a model [12–14]. Another advantage is that they are probabilistic and allow uncertain things to be "marginalized out." We do not need to have a clear idea of the exact forms and parameters of the distribution of some variables, but the framework is flexible enough to account for that uncertainty in the variables [3].

### Causal Interactions

The causal interaction model of causes and effect was developed graphically based on our understanding of the responses gotten from the different factors; we considered from the society news, articles, and twitter trends. As per our interpretation, the causal interactions of the influential variables (child nodes) are determined by the external variables (parent nodes) as shown in Table 2.

Probabilities taken are our interpretation of the combined response of each cross section for the causal interaction between the external variables to each influential variable. An example of our estimation of probabilities to one of the influential variable—demonetization—is given in Table 3.

Accordingly, the probabilities for each of the influential variables have been added to discover the relevant factor of interest, which will decide or determine the chances of PM Modi's win or no-win.

**Table 2** Descriptions of causal interactions

| Influential variables | External variables | | |
|---|---|---|---|
| Leader | Population | Occupation | Age |
| Budget | Occupation | Age | |
| GST | Occupation | Age | |
| Demonetisation | Occupation | Age | Employment |
| Surgical strikes | Age | Religion | |
| Health awareness | Age | Religion | |
| Research and development | Occupation | Age | |
| Foreign policies | Occupation | Age | Employment |
| Digitalization | Population | Occupation | Age |

**Table 3** Probabilistic estimations of the variables

| Demonetisation | | | Probability | |
|---|---|---|---|---|
| Occupation | Age | Employment | Positive | Negative |
| Primary | Middle-aged | Unemployed | 20 | 80 |
| Primary | Middle-aged | Employed | 30 | 70 |
| Primary | Youth | Unemployed | 40 | 60 |
| Primary | Youth | Employed | 70 | 30 |
| Primary | Old | Unemployed | 80 | 20 |
| Primary | Old | Employed | 40 | 60 |
| Secondary | Middle-aged | Unemployed | 20 | 80 |
| Secondary | Middle-aged | Employed | 30 | 70 |
| Secondary | Youth | Unemployed | 40 | 60 |
| Secondary | Youth | Employed | 50 | 50 |
| Secondary | Old | Unemployed | 40 | 60 |
| Secondary | Old | Employed | 20 | 80 |
| Tertiary | Middle-aged | Unemployed | 30 | 70 |
| Tertiary | Middle-aged | Employed | 35 | 65 |
| Tertiary | Youth | Unemployed | 50 | 50 |
| Tertiary | Youth | Employed | 90 | 10 |
| Tertiary | Old | Unemployed | 10 | 90 |
| Tertiary | Old | Employed | 60 | 40 |

# 3 Experiment

We now come to the interesting part, where we are going to speculate the different outcomes of the model, with respect to the changes in the response of the people to the (influential) key policies, for 2019 general election with the hot seat beckoning.

The detail of our predictions for 2019 General elections using Bayesian approach is discussed below. Firstly, we present a Bayesian model structured for the key policies in the PM Modi's era determining the outcome of the election as shown in Fig. 1. This is the model used to predict the outcome of the election as determined by the concerns of the voters to the key policies and their effect; we have further analyzed the dominant external factors, and where they are playing critical functions.

We have established a graphical structure to show the relationship between the parent nodes, which are the independent features, and the child nodes which are the dependent features based on the relevant factors for the election prediction. For each key policies, we have extracted corresponding information related to features in Bayesian network as shown in Fig. 1, and where we see that if all the factors are more or less the same, the chances of PM Modi repeating or getting another term as PM is 61.4%, since most of the influential variables (key policies) are positively affecting his chances of winning the election.

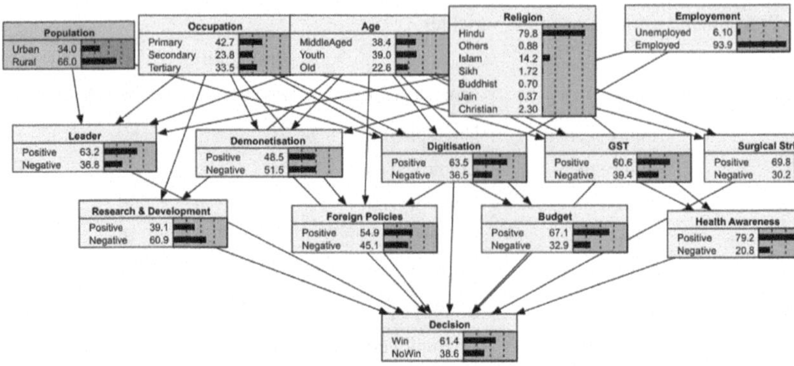


**Fig. 1** Normal Scenario where we see a clear indication that Modi is winning the 2019 general election

In the normal scenario, we observe that except demonetization, and other variables are considered as a positive factor toward PM Modi key polices, showing winning chance of 61.4% of PM Modi to retain his seat as Prime Minister (Fig. 2).

In this scenario, we observe that when voters are considered of Islamic population, which has the second highest population after Hindu, it is expected to result in a no-win scenario of 56.2%, with other factors as normal is giving a clear picture that they are not happy with the PM Modi as the Prime Minister (Fig. 3).

The above scenario shows that when every key policy is considered as negative factor except for surgical strike, then there is a 100% no-win case. It can be observed that it occurs when Hindu voters are lesser but bit higher for other minority communities, and who are of age 55 years and above (Fig. 4).

From the above scenario, we observe that after taking demonetization and GST as a negative factor and other factor as normal, the winning chances are at 49.2%.

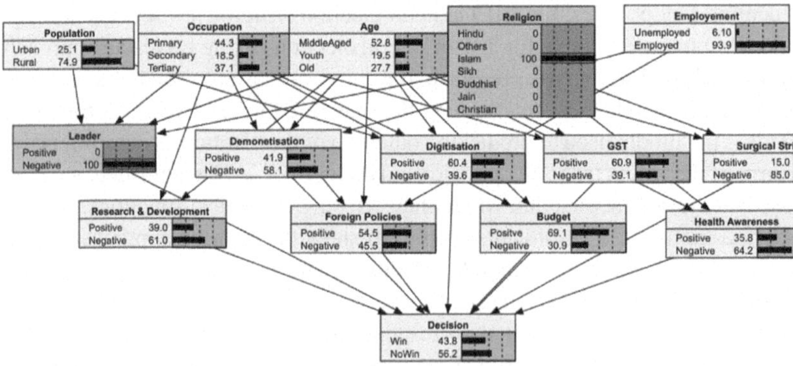


**Fig. 2** Scenario of Bayesian network when PM Modi is considered as a negative leader and the voters, Islam

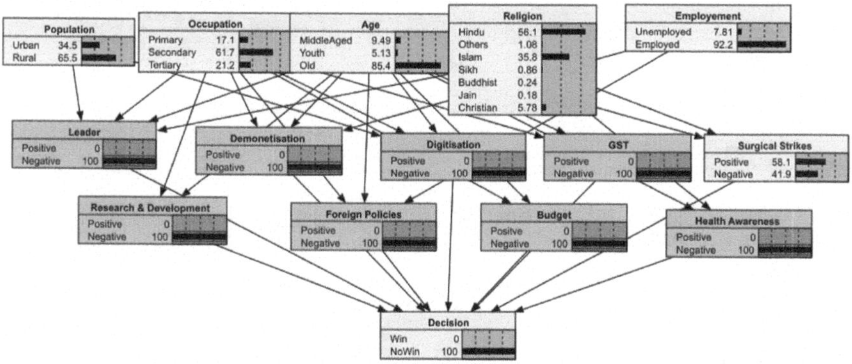


**Fig. 3** A no-win scenario when the key factors like leader, demonetization, digitization, health awareness, research, and development, foreign policies and budget are as a negative factor

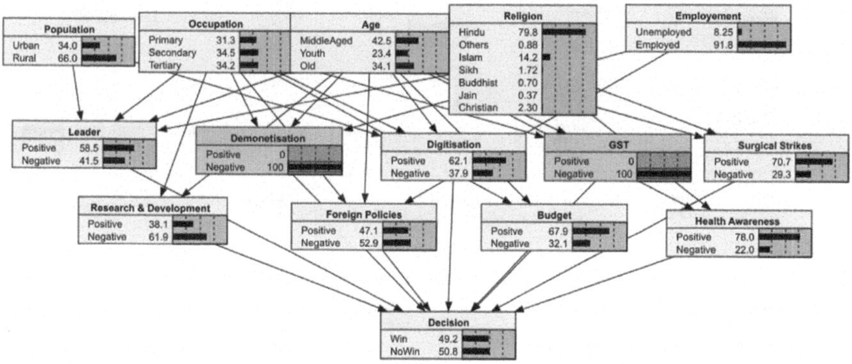


**Fig. 4** Scenario showing 50–50 chances when demonetization and GST has a negative factor

We observe that in this scenario, higher proportion of middle-aged businessmen is unhappy with the GST and demonetization policies.

However, the final outcome is about 50–50 chances, providing an insight that the demonetisation and GST are two vital variables, and where the response of the voters can give an edge on the final result (Fig. 5).

From the above scenario, we observe in the case of a Hindu voter of 55 years and above, the surgical strike is too dominating while GST is 50–50 and demonetization is slightly negative, and the final decision variable is 56% winning chances for PM Modi (Fig. 6).

From the above scenario, we clearly see that when Hindu voters of age between 18 and 35 years are employed in primary sector from urban area, the voters are giving a clear win with 90% positive outcome for PM as leader, 70% positive factor for demonetization, 80% positive factor for digitization, 60% positive factor for GST, 80% positive factor for surgical strike, 80% positive factor for foreign policies, 70%

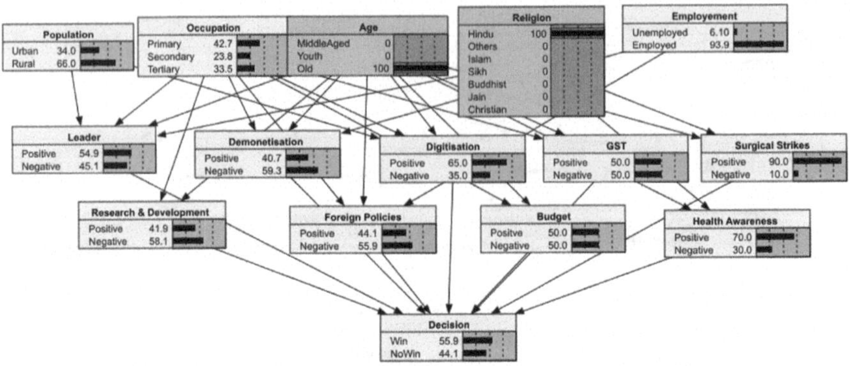


**Fig. 5** Scenario where we see the decision of old (55 years and above) people who are Hindu

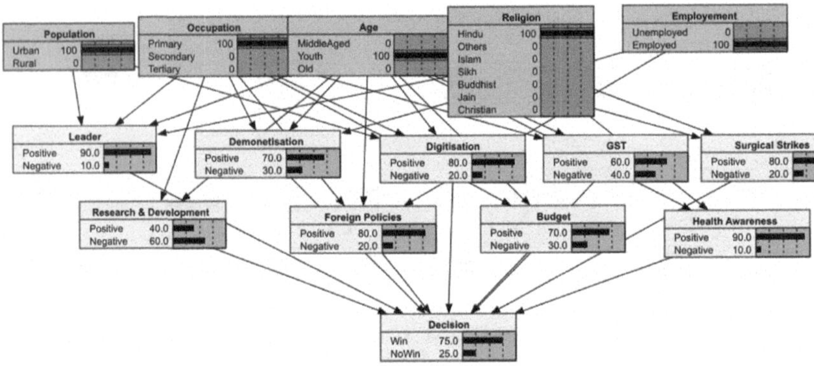


**Fig. 6** Scenario showing that 75% of Hindu youth working in the primary sector and living in an urban area is going to vote for Narendra Modi in this general election

positive factor for budget and 90% positive factor for health awareness, whereas 60% feel that research and development need more investment (Fig. 7).

From the above scenario, we see that when demonetization and GST are a negative factor, and other key policies are a positive factor, and then the winning chances are quite strong, where 96.9% are Hindu voters and the old voters seem to be a bit critical with the key policies (Fig. 8).

## 4 Discussion

The Bayesian analysis revealed that if PM Modi has to continue in the next term, he needs to focus on the rural population, showing them the necessary benefits of his key policies on a long run, policies like demonetization and GST implementation. It is also required that PM Modi targets the minority communities of India by offering

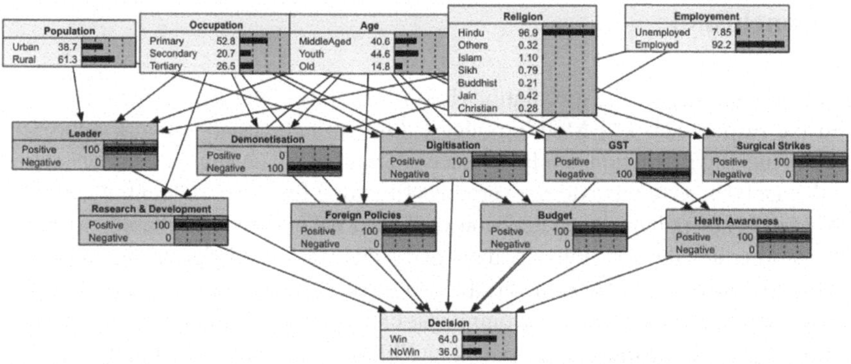


**Fig. 7** Scenario showing that when the leader, digitization, surgical strike, health awareness, research and development, foreign policies, budget are positive factors; GST and demonetization are negative factors, PM Modi still has considerable chance to win

**Fig. 8** Final outcome of the experiment with 61.4% of a win to 38.9% of no-win

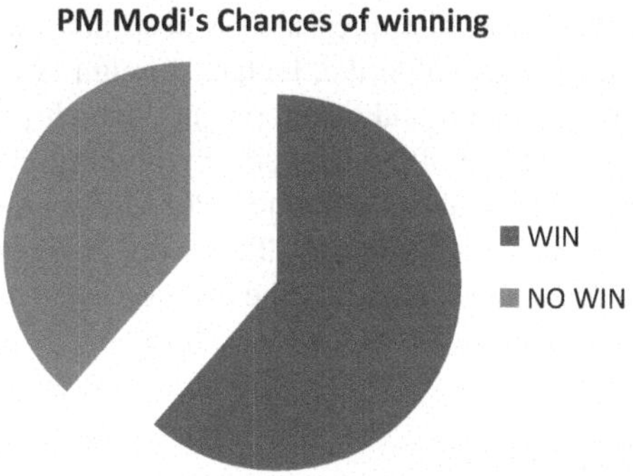


them several new and beneficial initiatives while maintaining harmony in the country and gaining their trust. Moreover, the majorities which consist of the youths that are exposed to social media seem to already have a good response to the key policies and prefer PM Modi to stay in power. PM Modi's activities both internationally and nationally have also positively contributed to the youths wanting him to remain as the Prime Minister.

If the opposition party wishes to win the election, then they really need to focus and attract the Hindu religion youths, which is the backbone factor of PM Modi's Government. They especially need to make their manifesto beneficial for youths and the urban population, and they need to explore and capitalize on some of the disadvantages of demonetization and GST among the workers. Another strategy that the opposition party should explore is to reach out to the youths through social media.

## 5 Conclusion

We have experimented the speculations for the outcome of the 2019 general elections with the response to PM Modi's policies through a Bayesian network approach, thereby making predictions on his chances of winning. The key factors considered for the analysis were the external variables (trending features), influential variables (PM Modi's key policies), and decision variables. We initiated a probabilistic model that incorporates the key policies, data, and statistical knowledge of the impact of the key policies on the decision variables to forecast the predictions. Based on this model, causal, and diagnostic reasoning was exploited in order to discover the cause and effect weight age of each key policy on PM Modi's chances of winning the 2019 elections.

The Bayesian network approach model predicted the outcome of the election to be in favor of the current Prime Minister Modi, with the chances of winning at 61.4%, which implies that PM Narendra Modi is going to repeat his term as PM for the next 5 years, according to our belief.

We have also presented a deep analysis on parameters like—GST, demonetization, surgical strike, budget, foreign policies, leader, health awareness, digitization, research and development—which proved to be the key factors in deciding the outcome of the network, by observing the current opinion of the public. Using Bayesian analysis technique, PM Modi can work on the factors negatively impacting his chances of winning. This work is a real-world application of the data mining approach to forecasting the election event.

## References


1. Zogjani, A., Llaci, S., Elmazi, E.: The role of power in effective leadership and followership: the Albanian Case. Rom. Econ. Bus. Rev. **9**(1) (2014)
2. Tang, Q., Wang, X., Huang, Z., Zheng, S.: Research of student model based on Bayesian Network. In: 2007 First IEEE International Symposium on Information Technologies and Applications in Education, 26 Dec 2007
3. Awasthi, S., Babbar, S., Predicting ICC Cricket World Cup 2015. Department of Computer Science & Engineering, Jaypee University of Information Technology, Waknaghat, Himachal Pradesh, India, vol. 2, Issue 5 (2015)
4. Anand. N.: Two years after demonetization, India is still struggling to create jobs. https://qz.com/india/1455596/indias-job-market-is-yet-to-recover-from-modis-demonetisation/, 9 Nov 2018
5. Liu, S., Wang, X., Liu, M., Zhu, J.: Towards a better analysis of machine learning models: a visual analytics perspective. Visual Inf. **1**(1), 48–56 (2017)
6. Democracy Index 2018, http://www.eiu.com/topic/democracy-index. Last accessed on 18 March 2019
7. Budget, https://en.wikipedia.org/wiki/2019_interim-Unnion_budget_of_India. Last accessed on 19 Feb 2019
8. GST, https://en.wikipedia.org/wiki/Goods_and_Services_Tax_(India). Last accessed on 19 Feb 2019

9. Demonetization, https://en.wikipedia.org/wiki/2016_Indian_banknote_demonetization. Last accessed on 19 Feb 2019
10. Surgical strikes, https://en.wikipedia.org/wiki/2016_Indian_Line_of_control_strike. Last accessed on 19 Feb 2019
11. Digitization, https://en.wikipedia.org/wiki/Digital_India. Last accessed on 19 Feb 2019
12. Subba, Patel, P., Girdhar, R., et al.: Bayes' Theorem, and operations research. Imperial J. Interdisc. Res. (IJIR) **3**(10) (2017). ISSN: 2454-1362. http://www.onlinejournal.in
13. Singh, A.K., Gupta, D.K., Singh, R.M.: Sentiment analysis of Twitter user data on Punjab Legislative Assembly Election, 2017. Int. J. Mod. Educ. Comput. Sci. **9**, 60–68 (2017)
14. Margaritis, D.: Learning Bayesian network model structure from data. May 2003. https://www.cs.cmu.edu/~dmarg/Papers/PhD-Thesis-Margaritis.pdf
15. Mega times group poll: 71.9% of Indians say they will vote for Narendra Modi as PM again in 2019. https://timesofindia.indiatimes.com/india/mega-times-group-poll-71-9-of-indians-say-they-will-vote-for-narendra-modi-as-pm-again-in-2019/articleshow. 26 May 2018
16. Age Structure And Marital Status, http://censusindia.gov.in/Census_And_You/age_structure_and_marital_status.aspx
17. India in Figures 2018, http://mospi.nic.in/sites/default/files/publication_reports/India_in_figures-2018_rev.pdf
18. www.twitter.com. Accessed on 9 Feb 2019
19. Making sense of the Aam Aadmi Party win in the Delhi Elections. https://www.rediff.com/news/report/making-sense-of-the-aam-aadmi-party-win-in-the-delhi-elections/20131223.htm, 24 Dec 2013

# Design and Analysis of Thermoelectric Energy Harvesting Module for Recovery of Household Waste Heat

**Shruti Jain, Vibhor Kashyap and Meenakshi Sood**

**Abstract** In recent years, there has been a lot of active research on energy recovery from waste heat obtained from various sources. A remarkable potential for harvesting the energy lies in the waste heat obtained during households daily chores. In this paper, waste heat obtained from chulhas has been exploited to generate energy with the help of thermoelectric energy system. Thermoelectric generators convert the waste heat to voltage are coupled with suitable heat sink, interfaced with DC–DC converter and storage circuit. The thermoelectric systems for waste heat recovery have low system efficiency, which prevents using them feasibly for the direct fuel energy conversion. This paper focuses on design of various key parameters, components and factors that determine the performance of DC–DC converter/energy harvester system. The theoretical results obtained showed good agreement compared with the simulation results using LTSPICE. This is a step toward greener source of energy as it does not hinder the normal working of the chulhas and the waste heat which is usually emitted out in the atmosphere and is of no use.

**Keywords** DC–DC converter · Energy harvesting · Thermoelectric generator · Waste heat recovery

## 1 Introduction

In recent years, energy harvesting has become a popular term in academic, household and industrial world [1]. The resources for energy harvesters are motion from human movement [2] waste heat from household [3], temperature gradient from the combustion engine [4–6], etc. Currently, there are different types of energy harvesting circuits which consist of pyroelectric energy harvesting, piezoelectric energy harvesting [7] waste heat recovery [3, 5] ambient-radiation energy harvesting, etc.

S. Jain (✉) · V. Kashyap · M. Sood
Department of Electronics and Communication Engineering, Jaypee University of Information Technology, Solan, Himachal Pradesh, India
e-mail: jain.shruti15@gmail.com

M. Sood
e-mail: meenusood9@gmail.com



P. K. Singh et al. (eds.), *Proceedings of ICRIC 2019*, Lecture Notes in Electrical Engineering 597, https://doi.org/10.1007/978-3-030-29407-6_65

However, current technologies of energy harvesting are capable of producing only limited power to drive relatively low-power electronics.

In general, the system is divided into four modules (shown in Fig. 1). The first module is system design in which a point was searched from where maximum waste heat can be obtained. The second module is harvesting generators. This module converts the heat energy into electrical energy. This energy can be in mV or few volts which is not enough for charging the battery. The third module is booster circuit. In this module, different voltages are boosted up. Booster circuits can be of different types. In this paper, we have used the LTC 3108 IC which is DC–DC converter/energy harvester circuit. Finally, the last stage is the storage.

From the third stage, enough voltage is obtained from where battery/supercapacitor can be charged. These batteries can be used anywhere in daily life. This paper mainly stresses on the designing of third module (booster circuits and harvesting ICs).

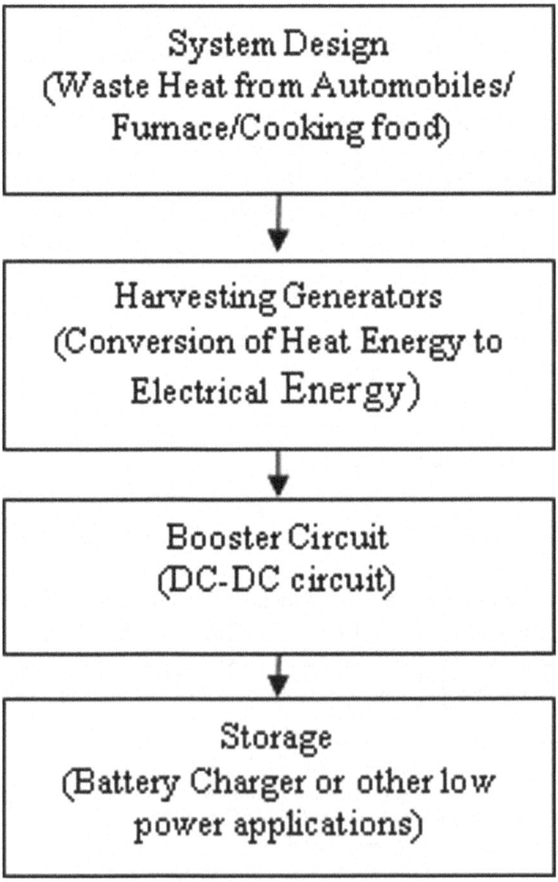


**Fig. 1** Flow diagram showing the different modules for the proposed algorithm

## 2 Materials and Methods

There are different types of booster and harvesting circuits. The proposed model operation and maintenance expenditures are very moderate. For the maintenance, we have to just take care of ICs and their replacement. In this paper LTC3108 IC, as DC–DC converter/energy harvester [8, 9] (the conventional diagram is shown in Fig. 2). Different output voltages ($V_{OUT}$) can be obtained which is the main advantage of this IC.

$V_{AUX}$: The circuits within the LTC3108 [7] are powered from $V_{AUX}$ and bypassed with a 1 μF capacitor. When using turns ratios of 1:50 or 1:20, larger capacitor values are required. The $V_{OUT}$ starts charging if the $V_{AUX}$ value becomes more than 2.5 V. As the $V_{OUT}$ reaches the regulated value, then there will be no charging current.

**Low Dropout Linear Regulator (LDO)**: The LDO requires a ceramic capacitor for stability. For stability, the 2.2 μF capacitor is used at the LDO. We can increase the values of capacitor but it takes maximum time to charge the output. 4 mA is the minimum LDO output current.

$V_{OUT}$: When the $V_{AUX}$ is greater than 2.5 V, then current starts flowing and output voltage drops below the regulated voltage. The different $V_{OUT}$ values which can achieve from the IC are 2.35, 3.3, 4.1 or 5 V. The $V_{OUT}$ is chosen with the help of $V_{S1}$ and $V_{S2}$. If we want 3.3 V as output voltage, then $V_{S1}$ should be connected to $V_{AUX}$ and $V_{S2}$ pin be connected to ground. Likewise, Table 1 shows the different combinations of various voltages.

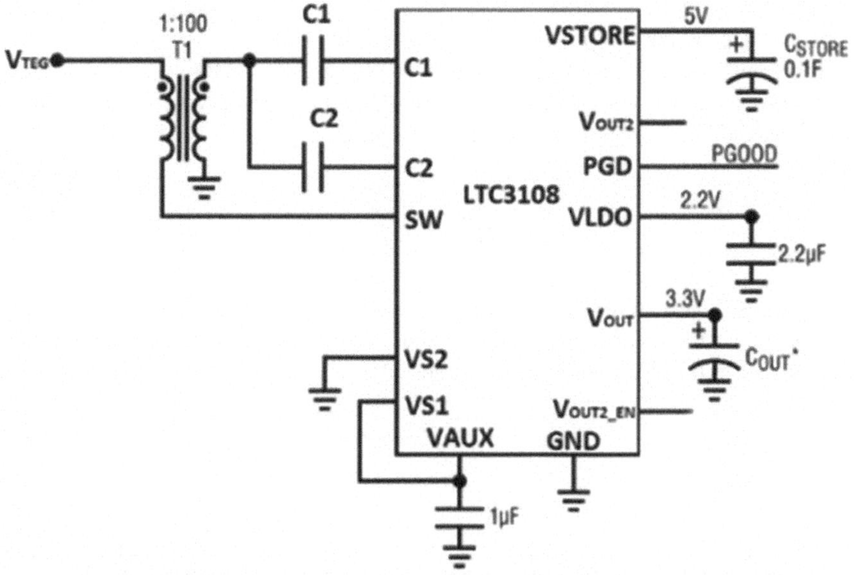


**Fig. 2** Conventional diagram of the DC–DC converter LTC3108 for 3.3 V

**Table 1** Different combinations to get output voltage

| S. No. | $V_{OUT}$ (V) | $V_{S1}$ | $V_{S2}$ |
|---|---|---|---|
| 1 | 2.35 | GND | GND |
| 2 | 3.3 | $V_{AUX}$ | GND |
| 3 | 4.1 | GND | $V_{AUX}$ |
| 4 | 5 | $V_{AUX}$ | $V_{AUX}$ |

**Power Good Comparator (PGOOD)**: This monitors the output voltage. This pin has a weak pull-up (1 MΩ) to the LDO voltage and an open-drain output voltage. The PGD output will be high if the $V_{OUT}$ is charged to 7.5% of regulated voltages but PGD is low if $V_{OUT}$ drops more than 9% of its regulated voltage.

$\boldsymbol{V_{STORE}}$: After $V_{OUT}$ has reached regulation, the $V_{STORE}$ output is used to charge a rechargeable battery or large storage capacitor. $V_{STORE}$ can be used to power the system if input source is unable to provide the current used by the $V_{OUT}$, $V_{OUT2}$ and LDO outputs.

## *2.1 Designing of Various Components*

Various steps for designing the different components are explained in this section. In this work, the voltage of interest is 3.3 V as $V_{OUT}$, so value of all components are calculated on this basis. (Maximum allowed voltage drop during the transmit burst is 10% of $V_{OUT}$.)

***Step 1***: Designing of $C_1$ and $C_2$ capacitor of LTC3108.

The transformer IC used in this paper is Wurths 74488540070 (1:100 Ratios) which gives start-up voltages as low as 20 mV and the primary winding has input capacitance of 220 μF. For a transformer, we have

$$\frac{N_S}{N_P} = \sqrt{\frac{Z_S}{Z_P}} \tag{1}$$

where $N_S$ is the no. of turns in secondary winding, $N_p$ is the no. of turns in primary winding, $Z_S$ is the secondary impedance and, $Z_P$ is the primary impedance. Further, $Z_S$ and $Z_P$ can be expressed by Eqs. (2) and (3).

$$Z_S = R_S + j(X_{LS} - X_{CS}) \tag{2}$$

$$Z_P = R_P + j(X_{LP} - X_{CP}) \tag{3}$$

Taking into consideration the values from datasheet; primary series resistance ($R_P$) as 0.085 Ω and secondary series resistance ($R_S$) as 205 Ω, primary inductance

($L_p$) having value 7.5 μH, secondary inductance ($L_s$) having value 7000 μH, and primary capacitance ($C_p$) having value 220 μF, we get

$X_{LS}$ ($2\pi f L_S$) as 43,982 Ω, $X_{LP}$ ($2\pi f L_p$) as 4.398 Ω,
$X_{CP}\left(\frac{1}{2\pi f C_P}\right)$ as 7.23 mΩ, and $X_{CS}\left(\frac{1}{2\pi f C_S}\right)$ as 1196.6 Ω.

Substituting these values further, we have calculated the value of $C_S$, the secondary capacitance as 1.33 nF, where $C_S$ is the parallel combination of $C_1$ and $C_2$ [10]. Simulations with different values of capacitor $C_2$ were simulated and the best results were obtained for $C_2$ as 330 pF (as per Fig. 3). To capture the effect of $C_1$, performance was evaluated by varying the value of $C_1$ but no significant change was observed, so $C_1$ is chosen as 1 nF.

***Step 2***: Calculation of storage capacitor ($C_{OUT}$): Storage capacitor is calculated by Eq. (4)

$$C_{OUT}(\mu F) \geq \frac{I_{LOAD}(mA) \times t_{PULSE}(ms)}{V_{OUT}} \tag{4}$$

Duration of transmit burst ($t_{PULSE}$) is presumed as 1 ms and total average current requirement during the burst ($I_{LOAD}$) as 40 mA. Putting all values in Eq. (4), we get

$$C_{OUT}(\mu F) \geq \frac{40\,\text{mA} \times 1\,\text{ms}}{0.33\,\text{V}} \tag{5}$$

Setting the value of $C_{OUT} = 150\ \mu F$.

***Step 3***: Calculation of maximum transmit rate ($t$): Maximum transmit rate is expressed by Eq. (6)

$$t_{trans} = \frac{C_{OUT} \times 0.33\,\text{V}}{\text{Charge current }(\mu A) - \text{Sleep current }(\mu A)} \tag{6}$$

(a) Calculation of average charge current $I_{CHG}$ where magnitude, duration and frequency of load current are fixed. $I_{CHG}$ is given by Eq. (7)

$$I_{CHG} \geq I_Q + \frac{I_{BURST} \times t_{PULSE}}{T} \tag{7}$$

where $I_Q$ is sleep current in between bursts, $I_{BURST}$ is total load current during the burst, $t_{PULSE}$ is the duration of transmit burst and $T$ is the period of transmit burst. For our design, we have considered $I_Q = 5\ \mu A$, $I_{BURST} = 100$ mA, $t_{PULSE} = 1$ ms and $T = 1$ s. Substituting all values in Eq. (7), average charge current comes out to be 105 μA.

(b) The average power is $V_{OUT} \times I_{CHG}$ which comes out to be 346.5 μW.

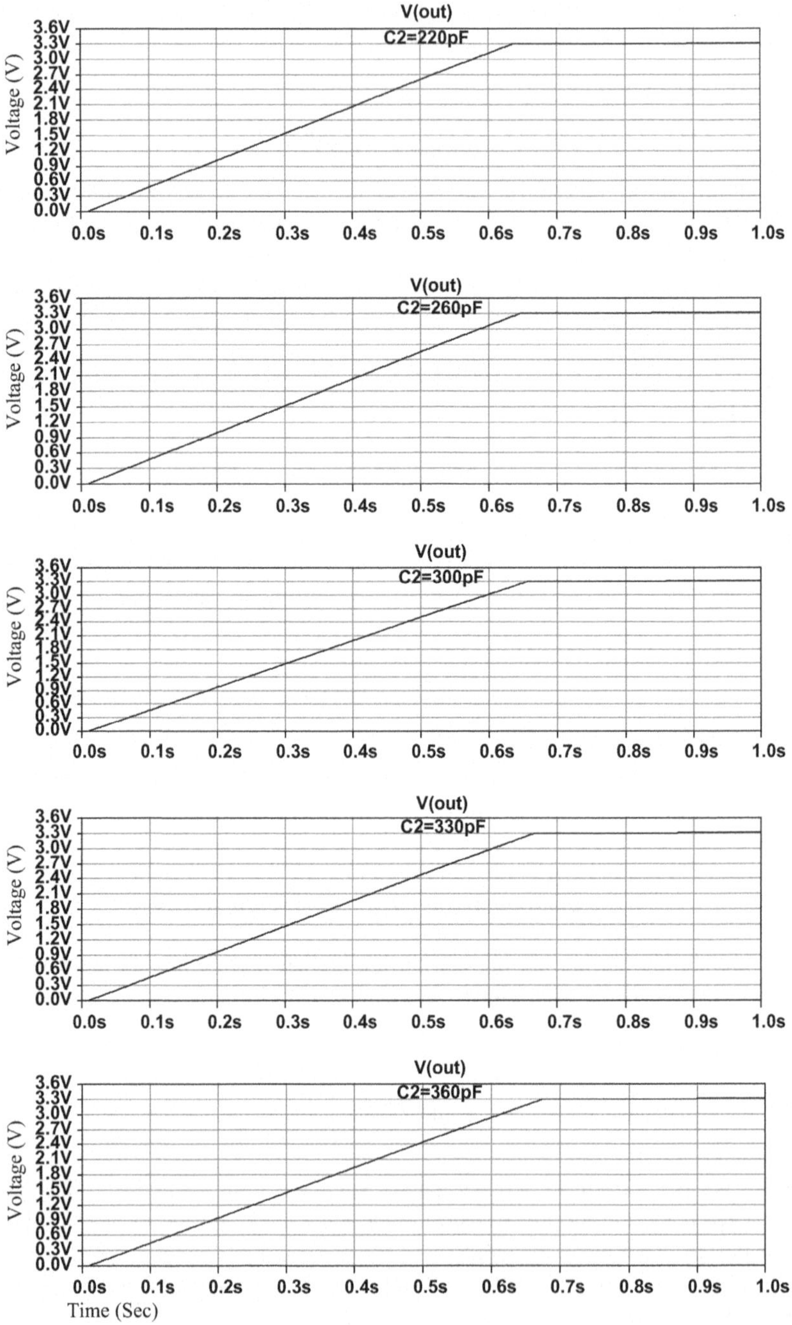


**Fig. 3** Effect of secondary capacitance on $V_{OUT}$

After calculations, $I_{\text{CHG}}$ was calculated as 105 μA and current drawn on $V_{\text{OUT}}$ in sleep state was 5 μA. Placing values of current in Eq. (6) we get:

$$t_{\text{trans}} = \frac{C_{\text{OUT}} \times 0.33\ \text{V}}{105\ \mu\text{A} - 5\ \mu\text{A}} = 0.495\ \text{s}$$

It signifies that circuit can support 1 ms transmit burst in every 0.495 s with $f_{\text{MAX}}$ = 2.02 Hz.

***Step 4***: Calculation of low drop output time ($t_{\text{LDO}}$): $t_{\text{LDO}}$ is the time that will charge the LDO capacitor and $V_{\text{OUT}}$ capacitor where $t_{\text{LDO}}$ is expressed by Eq. (8). With the help of datasheet, we get $C_{\text{LDO}}$ as 2.2 μF, $I_{\text{LDO}}$ as 5 μA and $V_{\text{LDO}}$ as 2.2 V.

$$t_{\text{LDO}} = \frac{V_{\text{LDO}}(V) \times \text{Capacitance at } V_{\text{LDO}}(\mu\text{F})}{I_{\text{CHG}} - I_{\text{LDO}}} \tag{8}$$

$I_{\text{CHG}}$ value is already calculated using Eq. (7) as 105 μA. Putting all values in Eq. (8), we have obtained $t_{\text{LDO}}$ as 48.4 ms.

***Step 5***: Calculation of time for $V_{\text{OUT}}$ to reach regulation ($t_{\text{VOUT}}$): $V_{\text{OUT}}$ is programmed for 3.3 V, the time for $V_{\text{OUT}}$ to reach regulation is given by Eq. (9)

$$t_{\text{VOUT}} = \frac{V_{\text{OUT}}\,(V) \times C_{\text{OUT}}}{I_{\text{CHG}} - I_{\text{VOUT}} - I_{\text{LDO}}} + t_{\text{LDO}} \tag{9}$$

All the values used in Eq. (9) are calculated above ($I_{\text{CHG}}$ as 105 μA, $I_{\text{LDO}}$ as 5 μA, $t_{\text{LDO}}$ as 48.4 ms, $C_{\text{OUT}}$ as 150 μF and $I_{\text{VOUT}}$ as 5 μA). Placing all values we get $t_{\text{VOUT}}$ as 5.25 s.

## 3 Results and Discussions

After calculating the values of all discrete components as in steps depicted above, the circuit is finally formulated and shown in Fig. 4. For performance analysis, different voltages ($V_{\text{AUX}}$, $V_{\text{LDO}}$, $V_{\text{OUT}}$, PGD and $V_{\text{STORE}}$ for 3.3 V shown in Fig. 5) and different current across various capacitors and inductors (shown in Fig. 6) were simulated using LTSPICE software.

Figure 6 shows the different current values at inductors and capacitors. $I(C_7)$ is the current at input capacitance, $I(L_1)$ is the current at primary inductance coil, $I(L_2)$ is the current at secondary inductance, $I(C_2)$ is the current across capacitance 1nF, $I(C_3)$ is the current across capacitance 330 pF, $I(C_1)$ is the current across $V_{\text{AUX}}$ capacitor, $I(C_4)$ is the current across $V_{\text{STORE}}$ capacitor, $I(C_5)$ is the current $V_{\text{OUT}}$ capacitor, $I(C_6)$ is the current across $V_{\text{LDO}}$ capacitor.

We have used the energy harvesting IC LTC3108 as a DC/DC convertor/energy harvester. Different simulation results were obtained using LTSPICE software. The output of the IC (3.3 V) as per designed requirement is obtained. The effect of

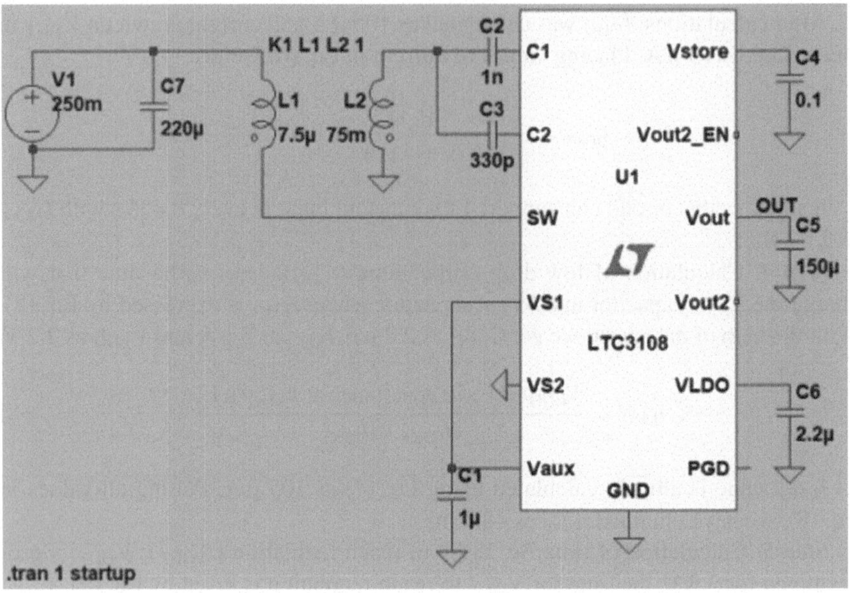


Fig. 4 Schematic diagram of energy harvesting circuit for 3.3 V output voltage

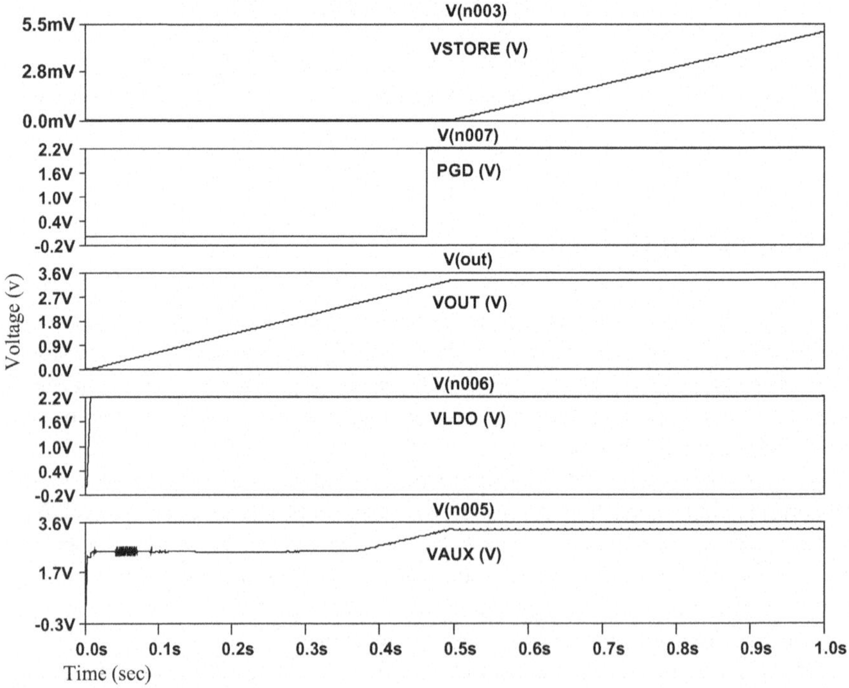


Fig. 5 Different output voltages of $V_{AUX}$, $V_{LDO}$, $V_{OUT}$, PGD and $V_{STORE}$ of 3.3 V

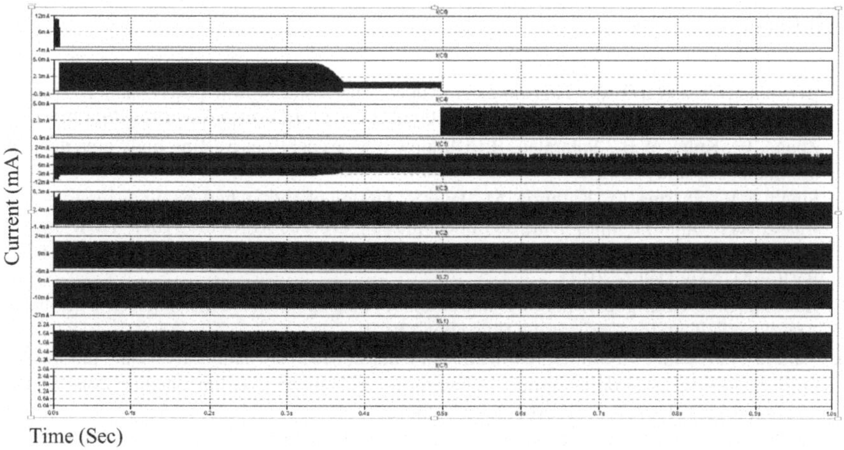


**Fig. 6** Different current values across various inductors and capacitor

secondary capacitance of transformer on $V_{OUT}$ is analyzed. The transient value of output voltage is increasing beyond the power good output signal, once the $V_{OUT}$ reaches the proper regulation the charging current is turned off. By making this design system, we can generate 3.3 V and can charge a supercapacitor/mobile. This is a promising technology for solving power crisis in remote area without polluting the environment.

## 4 Conclusion

The waste heat obtained has low efficiency in the original form, so the energy harvesting and booster circuit are to be designed for utilizing the recovered energy. In this paper, the design parameters of energy harvester module are calculated and validated. By designing the system, we could generate 3.3 V and can charge a supercapacitor/mobile. In this paper, we have calculated the values of all discrete components of the design and finally validated the results by simulation using LTSPICE. This is a promising technology for solving power crisis in remote area without polluting the environment.

**Acknowledgements** This work was supported and funded by Himachal Pradesh Council for Science, Technology & Environment (HIMCOSTE), H.P. vide project no. SCSTE/F(8)-1/2017 5085-5086.

## References


1. Beisteiner, C., Zagar, B.G.: Thermo-electric energy harvester for low-power sanitary applications. In: AMA Conferences SENSOR 2013, OPTO 2013, IRS 2013, pp. 471–476 (2013)
2. Zheng, C., Kuhn, W.B., Natarajan, B.: Ultralow power energy harvesting body area network design: a case study (2015)
3. Singh, K.R., Sharma, O., Kathuria, M., Jain, S.: Design of energy harvesting generators from waste heat of home chimney using thermocouples. In: Proceedings of the 11th INDIACom: 4th International Conference on Computing for Sustainable Global Development, pp. 6278–6281, BVICAM, New Delhi (2017)
4. Halloran, S.O., Rodrigues, M.: Power and efficiency measurement in a thermoelectric generator. American Society for Engineering Education (2012)
5. Kataria, A., Kaistha, P., Jain, S.: Utilization of exhaust heat from automotive using thermopile. In: 10th INDIACom: 3rd International Conference on Computing for Sustainable Global Development, pp. 2484–2487 BVICAM, New Delhi (2016)
6. Salerno, D.: Ultralow voltage energy harvester uses thermoelectric generator for battery free wireless sensors. J. Analog Innov. **20**(3), 1–11 (2010)
7. Mishra, R., Jain, S., Durgaprasad, C.: A review on piezoelectric material as a source of generating electricity and its possibility to fabricate devices for daily uses of army personnel. Int. J. Syst. Control Commun. **6**(3), 212–221 (2015)
8. Macrelli, E., Romani, A., Paganelli, R.P., Camarda, A., Tartagni, M.: Design of low-voltage integrated step-up oscillators with microtransformers for energy harvesting applications. IEEE Trans. Circuits Syst. I **62**(7), 1747–1756 (2015)
9. Reddy, B.R.: Body heat powered flashlight using LTC3108. Int. J. Eng. Res. Appl. **4**(8), 94–97 (2014)
10. Pang, F., Samaddar, T.: Charge pump circuit design. McGraw-Hill (2006)

# A Model of Information System Interventions for e-Learning: An Empirical Analysis of Information System Interventions in e-Learner Perceived Satisfaction

**Asif Ali and Jaya Bhasin**

**Abstract** Innovations in pedagogy and technology have lead to a new paradigm in teaching particularly in higher education. At the nexus of this new paradigm is blended learning and e-learning. Blended learning refers to combination of synchronous and asynchronous learning activities. While e-learning refers to learning system that utilize electronic means or information communication technology (ICT) to deliver information for education or training purposes. Though in infancy e-learning in India has garnered pace in last decade, introduction of SWAYAM portal for offering massive open online courses (MOOCs) has been an important initiative by Government of India in this direction. The present study conceptualizes various factors of e-Learner Satisfaction into one model. The data for study was collected from students of north Indian universities. A total of 266 responses were recorded out of which 25 responses were eliminated due to incomplete information. Final data analysis was conducted using 241 responses using structural equation modeling (SEM) (Amos 20). From the data, it was observed the constructs (predictors) namely Instructors Attitude, Learners Attitude, Course Quality, Technology Quality, Assessment Quality and Perceived ease of use explained 66.2% variance (adjusted $R^2 = 66.2\%$ and $p < 0.05$) in Perceived e-Learner Satisfaction.

**Keywords** Information systems · e-Learning · e-Learner Satisfaction

## 1 Introduction

Innovations in pedagogy and technology have lead to a new paradigm in teaching particularly in higher education. At the nexus of this new paradigm is blended learning and e-learning. Blended learning refers to the combination of synchronous and asynchronous learning activities. Blending learning is an approach to learning in which traditional face-to-face learning is complemented with online learning. The two most

A. Ali (✉) · J. Bhasin
Department of HRM & OB, Central University of Jammu, Jammu, India
e-mail: easif.101@gmail.com

J. Bhasin
e-mail: jayabhasincuj@gmail.com



P. K. Singh et al. (eds.), *Proceedings of ICRIC 2019*, Lecture Notes in Electrical Engineering 597, https://doi.org/10.1007/978-3-030-29407-6_66

common definitions of blended learning prevalent in literature are (i) combination of online and face-to-face instructions [11 and 23] (ii) combining instructional methods [9, 25].

While e-learning refers to learning system that utilizes electronic means or Internet communication technology (ICT) to deliver information for education or training purposes. e-learning is defined as learning assisted by utilization of digital tools and content that involves interactivity that includes online interaction between teacher and learners or peers [17].

The institutions of eminence (Harvard, MIT) of higher education in world have started to offer most of the courses in online mode with India being no exception [14, 22]. Though in infancy e-learning in India has garnered pace in last decade, introduction of SWAYAM portal for offering massive open online courses (MOOCs) has been an important initiative by Government of India in this direction. e-learning has not remained confined to the higher education industry only; however of late many global organizations have started to incorporate online training modules to employees as part of their overall training programmes [21].

## 2 Theoretical Framework

e-learning is delivery of education through various electronic media (activities pertinent to instructing, learning, and teaching) [13]. "e-learning is an approach to teaching and learning, representing all or part of the educational model applied, that is based on the use of electronic media and devices as tools for improving access to training, communication and interaction" [24].

e-learning being ubiquitous offers array of benefits to make learning possible any time anywhere at learners convenience as it cuts across the barriers of space and time. Being cost-efficient, it has received a wide acceptance across the world in higher education dissemination [1]. On account of high utility and other benefits, the adoption of e-learning has been on increase with 7% CAGR in last decade. e-learning industry was $107 Billion in 2015, and it has been forecasted to grow to $ 325 Billion in 2025 (Forbes, Jul 2018). Despite this high growth, adoption, and success of e-learning, the failures in e-learning do exist [5, 19]. The reasons for such failures can be varied sometimes learner centric other time instructor or environment centric. From the literature, numerous models have been proposed to access the success and adoption of information systems [8, 28]. Among various proposed models, user satisfaction emerges as one of the most important factors for adoption and success of information systems [8]. In e-learning context, various factors have been found associated with user satisfaction. With regard to present research per se, e-learning and blended learning have been used interchangeably. And the factors associated with user satisfaction have been clubbed into six dimensions: Learner, Instructor, Course, Technology, Course Design, and Environment as identified by Sun et al. [26].

## 3 Research Model and Hypotheses

From simple application software acceptance like spreadsheet applications [16] to complex e-learning [22], various factors in literature have been identified that determine user satisfaction. However, not much attention has been paid to integrate these factors into one framework/model. An attempt in this regard was made by Sun et al. [26]. Drawing inferences from previous studies and using model proposed by Sun et al. [26], the present study uses six dimensions, namely learner, instructor, course, technology, course design, and environment dimensions to access user satisfaction of e-learning. However, a little modification has been done to the model as each dimension is reduced to a factor. The learner dimension takes into account Learners Attitude/impression toward the e-learning activities by usage of computers. A positive attitude toward ICT will result in effective and efficient learning thus more satisfied users/learners [20]. Thus, we hypothesize

***H1: Learners Attitude will positively influence Perceived e-Learner Satisfaction.***

The Instructor dimension takes into account Instructors Attitude/impression toward the e-learning, and we hypothesize

***H2: Instructors Attitude toward e-learning positively influences Perceived e-Learner Satisfaction*** (Fig. 1).

The course dimension takes into account the e-learning course quality, technology dimension takes into account the technology quality, design dimension takes into account perceived ease of use of e-learning platform, and environment dimension takes into account diversity in assessment (quality of assessment) of e-learning. Thus, we hypothesize

***H3: Course quality positively influences Perceived e-Learner Satisfaction.***
***H4: Technology quality positively influences Perceived e-Learner Satisfaction.***
***H5: Perceived ease of use of e-learning positively influences Perceived e-Learner Satisfaction.***
***H6: Diversity in assessment of e-learning positively influences Perceived e-Learner Satisfaction.***

The actual simplified model used in the present study is reproduced in Fig. 2.

## 4 Research Methods

### 4.1 Instrument Development

Based on the research in the field of information systems and technology adoption, a questionnaire was adopted and modified to measure the various factors of the hypothesized model. The instrument with 30 items under 7 factors was adopted

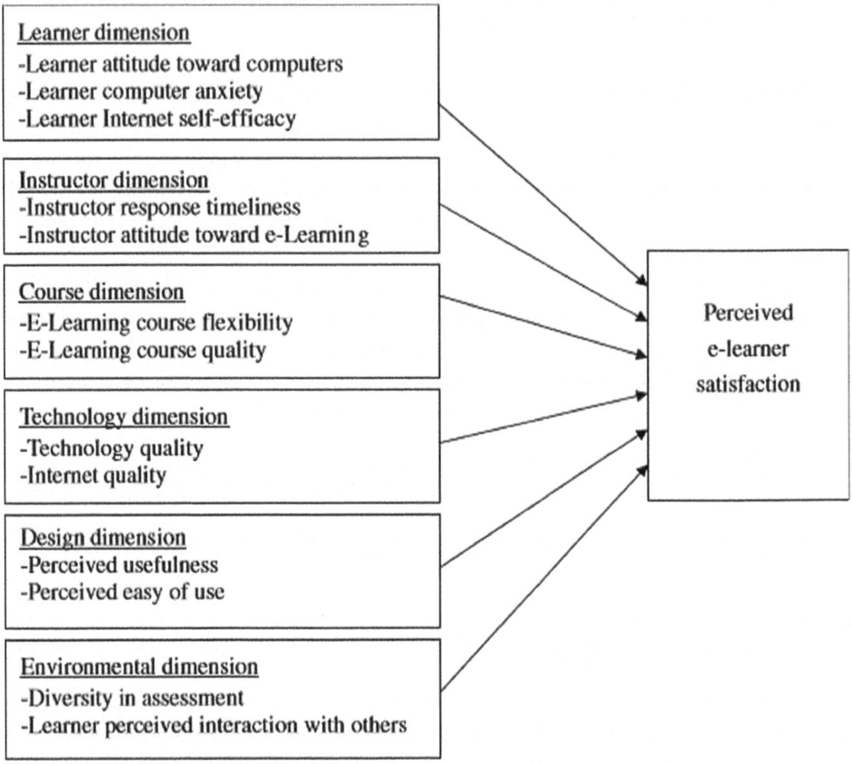


**Fig. 1** Dimensions and antecedents of e-learner satisfaction (adapted [26]

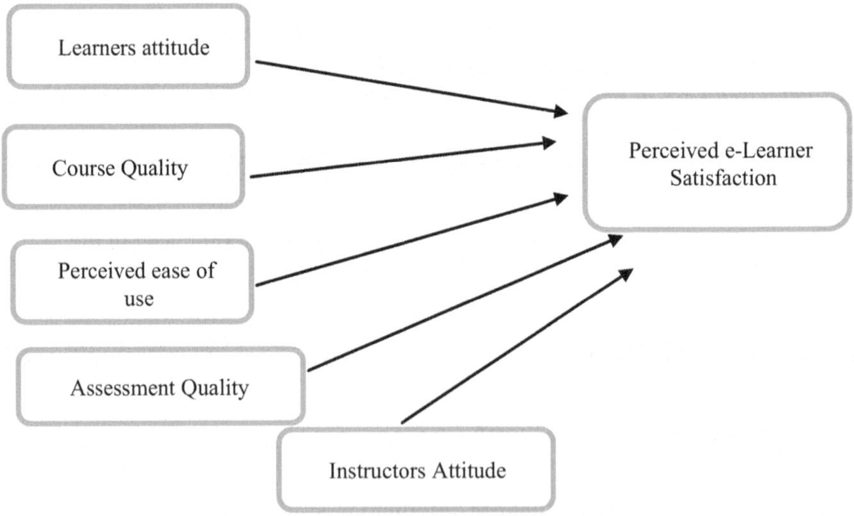


**Fig. 2** Simplified proposed research model

from Sun et al. [26] with modification by putting reverse order questions straightforward and by adding two additional items. The questionnaire consisting of 32 items measured under 7 factors is presented in Table 1 along with the references. Each measurement item on the scale was measured using seven-point Likert scale with (1) representing "very strongly disagree", (2) representing "strongly disagree", (3) representing "disagree", (4) representing "neutral", (5) representing "agree", (6) representing "strongly agree" and (7) representing "very strongly agree". The questionnaire was presented to two experts in information technology and marketing for expert comments and feedback. A pilot study was conducted with 30 postgraduate students with prior experience with e-learning systems.

**Table 1** Measurement instrument

| Construct | Item | Wording | Reference |
|---|---|---|---|
| Learners Attitude (LA) | LA1 | I believe that working with computers…is very easy | Gattiker and Hlavka [10] |
| | LA2 | is not complicated | |
| | LA3 | requires high technical skills | |
| | LA4 | does not give me psychological stress | |
| | LA5 | requires knowledge of programming language such as basic | |
| | LA6 | does not require patience | |
| | LA7 | makes a person more productive at his/her job | |
| | LA8 | is for people of any age | |
| Instructors Attitude (IA) | IA1 | Compared to traditional classrooms, how useful do you think your instructor considers web-based learning | Webster and Hackley [29] |
| | IA2 | Do you feel your instructor is comfortable with use of technology | (Self) Arbaugh [3] |
| Course Quality (CQ) | CQ1 | Conducting the course via the Internet improved the quality of the course compared to other courses | |
| | CQ2 | The quality of the course compared favorably to my other courses | |
| | CQ3 | I feel the quality of the course I took was largely not effected by conducting it via the Internet | |
| Technology Quality (TQ) | | I feel the information technologies used in e-Learning | Amoroso and Cheney [2] |
| | TQ1 | are very easy to use | |
| | TQ2 | have many useful functions | |
| | TQ3 | have good flexibility | |
| | TQ4 | are easy to obtain | |

(continued)

**Table 1** (continued)

| Construct | Item | Wording | Reference |
|---|---|---|---|
| Perceived ease of use (PE) | PE1 | Learning to operate an e-Learning system is easy for me | Davis [7] |
| | PE2 | It's easy for me to gain skill to operate e-Learning system | |
| | PE3 | I found e-Learning easy to use | |
| | PE4 | Overall e-Learning is easy to use | |
| Assessment Quality (AQ) | AQ1 | e-Learning course offered a variety of ways of assessing my learning (quizzes, written work, oral presentation, etc.) | Thurmond et al. [27] |
| | AQ2 | I am satisfied with diverse assessment options in e-Learning | |
| Perceived e-Learner Satisfaction (PS) | PS1 | I am satisfied with my decision to take this course via the Internet | (Self) Arbaugh [3] |
| | PS2 | If I had an opportunity to take another course via the Internet, I would gladly do so | |
| | PS3 | My choice to take this course via the Internet was a wise one | |
| | PS4 | I was very satisfied with the course | |
| | PS5 | I feel that this course served my needs well | |
| | PS6 | I will take as many courses via the Internet as I can | |
| | PS7 | I was disappointed with the way this course worked out | |
| | PS8 | If I had it to do over, I would not take this course via the Internet | |

## 4.2 Data Collection

For collecting data, an online survey was developed using online survey tools. The link to survey was emailed to postgraduate students of north Indian universities. The link to survey was also shared in three WhatsApp groups. A total of 275 links to survey were e-mailed. After initial and follow-up round of email, a total of 223 responses were generated. And 43 responses were also obtained from WhatsApp group. Out of the total responses, 25 responses were omitted due to incomplete information. A total of 241 responses (response rate = 87.63%) were put to analysis.

The demographic profile of the respondents is presented in Table 2. Out of total 241 respondents, 100 were females and 141 males with 23 years as an average age, while all the students were attending postgraduate course with most postgraduating students from management background followed by engineering and science background.

**Table 2** Demographic profile

| Category | | Frequency | Percentage |
|---|---|---|---|
| Gender | Male | 141 | 58.51 |
| | Female | 100 | 41.49 |
| | Total | 241 | 100 |
| Age | 21–25 | 210 | 87.13 |
| | 26–30 | 21 | 8.71 |
| | >30 | 10 | 4.14 |
| | Total | 241 | 100 |
| Course | Management | 102 | 42.32 |
| | Science | 41 | 17.01 |
| | Engineering | 75 | 31.12 |
| | Others | 23 | 9.54 |
| | Total | 241 | 100 |

Statistical Package for the Social Sciences version 23 (SPSS v.23.0) and Amos (V 20) was used for the statistical analysis in this research. Stepwise regression analysis was used wherein 6 constructs, namely Instructors Attitude, Learners Attitude, Course Quality, Technology Quality, Assessment Quality and Perceived ease of use were used as repressors, and Perceived e-Learner Satisfaction were used as regress.

Kaiser-Meyer-Olkin (KMO) and Bartletts Tests were used to verify the factorability of the data. KMO and Bartlett's sphericity test results were both significant ($p < 0.001$), and KMO value was found more than threshold (0.92).

# 5 Data Analysis and Results

## 5.1 *Factor Analysis (Exploratory and Confirmatory Factor Analysis)*

For scale development and validation, data was analyzed using two step process. In first step, an exploratory factor analysis was done on data. In exploratory factor analysis, data was subjected to principal component analysis with varimax rotation. All factor loadings above ±0.50 and with eigenvalues greater than 1 were considered. Item LA5, LA6, PS7, and PS8 had factor loadings less than 0.5 thus were dropped from further analysis. Seven (7) stable factors emerged with least or no cross loadings. Together, these 7 factors explained 65.6% variance (Table 3) in the dependent variable, namely Perceived e-Learner Satisfaction.

**Table 3** Exploratory Factor Analysis

| | Component | | | | | | |
|---|---|---|---|---|---|---|---|
| | 1 | 2 | 3 | 4 | 5 | 6 | 7 |
| LA1 | 0.702 | | | | | | |
| LA2 | 0.712 | | | | | | |
| LA3 | 0.780 | | | | | | |
| LA4 | 0.856 | | | | | | |
| LA7 | 0.821 | | | | | | |
| LA8 | 0.788 | | | | | | |
| IA1 | | | | | 0.788 | | |
| IA2 | | | | | 0.876 | | |
| CQ1 | | | | | | | 0.703 |
| CQ2 | | | | | | | 0.653 |
| CQ3 | | | | | | | 0.683 |
| TQ1 | | | | | | 0.724 | |
| TQ2 | | | | | | 0.642 | |
| TQ3 | | | | | | 0.688 | |
| TQ4 | | | | | | 0.768 | |
| PE1 | | 0.752 | | | | | |
| PE2 | | 0.732 | | | | | |
| PE3 | | 0.658 | | | | | |
| PE4 | | 0.742 | | | | | |
| AQ1 | | | | 0.862 | | | |
| AQ2 | | | | 0.881 | | | |
| PS1 | | | 0.742 | | | | |
| PS2 | | | 0.821 | | | | |
| PS3 | | | 0.862 | | | | |
| PS4 | | | 0.642 | | | | |
| PS5 | | | 0.668 | | | | |
| PS6 | | | 0.924 | | | | |
| Eigen value | >1 | >1 | >1 | >1 | >1 | | |
| Cumulative variance (%) | | | 65.6% | | | | |

## 5.2 Reliability and Validity

The instrument developed was shared with experts to improve the face and content validity of instrument. Also, reliability/internal consistency of data was measured using Cronbach's Alpha values for each construct. The Cronbach alpha coefficient of 0.7 or higher suggests high data reliability but a coefficient lower than 0.35 indicates poor reliability [6, 18]. All the values of Cronbach's alpha were found to be more than 0.8 which is higher than the minimum threshold thus indicating high reliability.

The construct validity refers to the extent to which two constructs are separate. Since seven stable constructs emerged in factor analysis using PCA with varimax rotation with all factor loads on the respective construct more than 0.6, the constructs thus exhibit convergent validity [12]. The discriminatory validity has been verified by checking the correlation between different constructs. In none of the cases, correlation greater than 0.85 between constructs was reported, indicating discriminating validity among the constructs.

## 5.3 Structural Model and Hypotheses Testing

To test the hypothesized research model among various constructs, regression analysis was performed to determine casual relations. Among the demographic variables namely gender, age, and course, no significant differences or correlations with dependent variable namely Perceived e-Learner Satisfaction were found; thus, demographic variables were not considered for further analysis. For regression analysis, six variables namely Instructors Attitude, Learners Attitude, Course Quality, Technology Quality, Assessment Quality and Perceived ease of use were used as independent variables (repressors), and Perceived e-Learner Satisfaction was used as dependent variables (regress).

For testing proposed hypothesized model, structural equation modeling (SEM) has been used as our sample size was greater than 200 respondents [15]. SEM Amos 20 with maximum likelihood method was used for model estimation. The various structural model indices results are shown in Fig. 3. The various model fit indices found are NFI = 0.954, GFI = 0.966, CFI = 0.921, AGFI = 0.942, RMSEA = 0.052, RMR = 0.031. With all good fitness indices above 0.9 in model [4], thus these path coefficients (standardized) can be used to test hypotheses.

Among the proposed hypotheses after testing, following hypotheses were found significant among various constructs. Learners Attitude was found significantly effecting Perceived e-Learner Satisfaction (H1). Instructors Attitude was found significantly effecting Perceived e-Learner Satisfaction (H2). Course quality was found significantly effecting Perceived e-Learner Satisfaction (H3). Technology quality was found significantly effecting Perceived e-Learner Satisfaction (H4). Perceived ease of use of e-learning was found significantly effecting Perceived e-Learner Satisfaction (H5). Diversity in assessment of e-learning was found significantly effecting

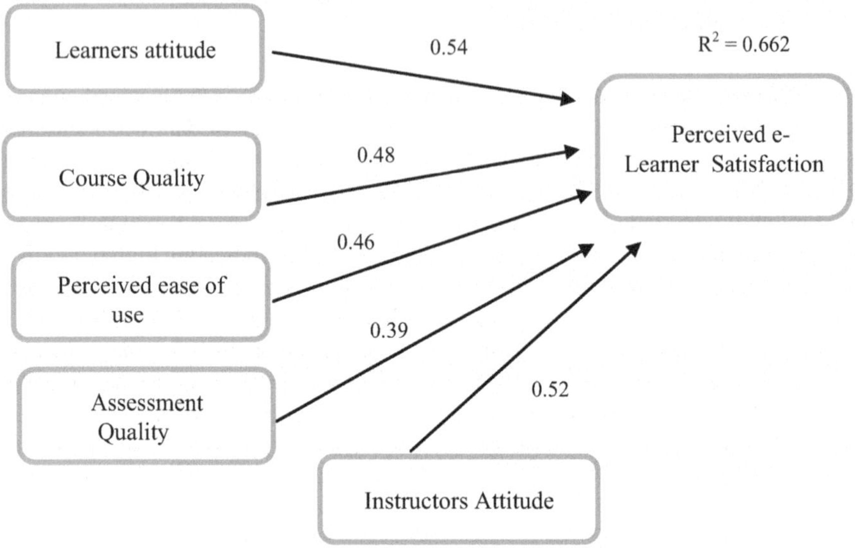


**Fig. 3** Structural model testing results ($p < 0.001$ in all relationships)

Perceived e-Learner Satisfaction (H6). The constructs (predictors) namely Instructors Attitude, Learners Attitude, Course Quality, Technology Quality, Assessment Quality, and Perceived ease of use explained 66.2% variance (Adjusted $R^2 = 66.2\%$ and $p < 0.05$) in Perceived e-Learner Satisfaction. The hypotheses results are enumerated in Table 4.

**Table 4** Hypotheses results

| Relationship | Hypotheses | Results |
|---|---|---|
| LA–PS | Technical Support is positively associated with Perceived usefulness | **H1**: Supported |
| IA–PS | Technical Support is positively associated with Perceived ease of use | **H2**: Supported |
| CQ–PS | Perceived ease of use is positively associated with Perceived usefulness | **H3**: Supported |
| TQ–PS | Perceived ease of use is positively associated with Attitude toward use | **H4**: Supported |
| PE–PS | Perceived usefulness is positively associated with Attitude toward use | **H5**: Supported |
| AQ–PS | Perceived usefulness is positively associated with Intention to use | **H6**: Supported |

## 6 Discussion and Implications

In present-day times, it has become important for educationists to impart or complement traditional learning with electronic media (e-learning) as it cuts across the time and space barriers. This present study tries to examine various factors that lead to students satisfaction by use of e-learning systems for knowledge dissemination or acquisition. All the hypotheses proposed in study were verified to be true. From the model, it can be observed that all the six proposed predictors namely Instructors Attitude, Learners Attitude, Course Quality, Technology Quality, Assessment Quality and Perceived ease of use have a significant positive influence on Perceived e-Learner Satisfaction. The results indicate good model fit indices with predictors explaining 66.2% variance in Perceived e-Learner Satisfaction. The prediction formula for the proposed model is presented under

$$\mathrm{PS} = \mathrm{LA}(w1) + \mathrm{IA}(w2) + \mathrm{CQ}(w3) + \mathrm{TQ}(w4) + \mathrm{PE}(w5) + \mathrm{AQ}(w6).$$

In the above formula, PS stands for Perceived e-Learner Satisfaction, IA stands for Instructors Attitude, LA stands for Learners Attitude, CQ stands for Course Quality, TQ stands Technology Quality, AQ stands for Assessment Quality, and PE stands Perceived ease of use where as $w1$, $w2$, $w3$, $w4$, $w5$, and $w6$ are empirical weights.

### *6.1 Limitations and Implications for Future Research*

This research makes a careful attempt to integrate various factors that influence Perceived e-Learner Satisfaction reported in literature into one model. Still, the proposed model may not be the comprehensive model to explain perceived satisfaction due to time, budget, and other resource constraint. The study identifies only single factor as dependent variable and measures it subjectively only, some objective measures like student score may be used in future research. The present study uses SEM for analysis in future study other tools like (stepwise regression, LISERL, PLS) and variance in results if any can be examined. Further, the other limitation of the study includes data which was collected from north Indian universities only which makes generalization of study little difficult. A cross-cultural study across continents may also be conducted that can highlight impact of culture in e-Learner satisfaction.

## References

1. Adomi, E.E., Kpangban, E.: Application of ICTs in Nigerian secondary schools (2010)
2. Amoroso, D.L., Cheney, P.H.: Testing a causal model of end-user application effectiveness. J. Manag. Inf. Syst. **8**(1), 63–89 (1991)
3. Arbaugh, J.B.: Virtual classroom characteristics and student satisfaction with internet-based MBA courses. J. Manag. Educ. **24**(1), 32–54 (2000)

4. Bentler, P.M., Bonett, D.G.: Significance tests and goodness of fit in the analysis of covariance structures. Psychol. Bull. **88**(3), 588 (1980)
5. Cook, D.A.: The failure of e-learning research to inform educational practice, and what we can do about it. Med. Teach. **31**(2), 158–162 (2009)
6. Cuieford, J.P.: Fundamental Statistics in Psychology and Education, 4th edn. McGraw Hill Publication, New York, N.Y. (1965)
7. Davis, F.D.: User acceptance of information technology: system characteristics, user perceptions and behavioral impacts. Int. J. Man Mach. Stud. **38**(3), 475–487 (1993)
8. DeLone, W.H., McLean, E.R.: Information systems success: the quest for the dependent variable. Inf. Syst. Res. **3**(1), 60–95 (1992)
9. Garrison, D.R., Kanuka, H.: Blended learning: uncovering its transformative potential in higher education. Internet High. Educ. **7**(2), 95–105 (2004)
10. Gattiker, U.E., Hlavka, A.: Computer attitudes and learning performance: issues for management education and training. J. Organ. Behav. **13**(1), 89–101 (1992)
11. Graham, C.R.: Blended learning systems. The Handbook of Blended Learning, 3–21(2006)
12. Hair, J.F., Black, W.C., Babin, B.J., Anderson, R.E., Tatham, R.L.: Multivariate Data Analysis, vol. 5, no. 3, pp. 207–219. Prentice Hall, Upper Saddle River, NJ (1998)
13. Koohang, A., Harman, K.: Open source: a metaphor for e-learning. Infor. Sci. **8** (2005)
14. Lee, M.C.: Explaining and predicting users' continuance intention toward e-learning: an extension of the expectation–confirmation model. Comput. Educ. **54**(2), 506–516 (2010)
15. Loehlin, J.C.: Latent Variable Models: An Introduction to Factor, Path, and Structural Analysis. Lawrence Erlbaum Associates Publishers (1998)
16. Mathieson, K.: Predicting user intentions: comparing the technology acceptance model with the theory of planned behavior. Inf. Syst. Res. **2**(3), 173–191 (1991)
17. Ministry of Communication and Technology of New Zealand: Digital strategy of the ministry of communication and technology (2008)
18. Nunnally, J.C., Bernstein, I.H.: Psychometric theory, vol. 226. McGraw-Hill, New York (1967)
19. Penna, M., Stara, V.: The failure of e-learning: why should we use a learner centred design. J. E-learning Knowl. Soc. **3**(2), 127–135 (2007)
20. Piccoli, G., Ahmad, R., Ives, B.: Web-based virtual learning environments: a research framework and a preliminary assessment of effectiveness in basic IT skills training. MIS Q., 401–426 (2001)
21. Roca, J.C., Gagné, M.: Understanding e-learning continuance intention in the workplace: a self-determination theory perspective. Comput. Hum. Behav. **24**(4), 1585–1604 (2008)
22. Roca, J.C., Chiu, C.M., Martínez, F.J.: Understanding e-learning continuance intention: an extension of the technology acceptance model. Int. J. Hum. Comput. Stud. **64**(8), 683–696 (2006)
23. Rooney, J.E.: Knowledge infusion. Assoc. Manag. **55**(5), 26 (2003)
24. Sangrà, A., Vlachopoulos, D., Cabrera, N.: Building an inclusive definition of e-learning: an approach to the conceptual framework. Int. Rev. Res. Open Distrib. Learn. **13**(2), 145–159 (2012)
25. Singh, H., Reed, C.: A white paper: achieving success with blended learning. Centra Software **1**, 1–11 (2001)
26. Sun, P.C., Tsai, R.J., Finger, G., Chen, Y.Y., Yeh, D.: What drives a successful e-Learning? An empirical investigation of the critical factors influencing learner satisfaction. Comput. Educ. **50**(4), 1183–1202 (2008)
27. Thurmond, V.A., Wambach, K., Connors, H.R., Frey, B.B.: Evaluation of student satisfaction: determining the impact of a web-based environment by controlling for student characteristics. Am. J. Distance Educ. **16**(3), 169–190 (2002)
28. Venkatesh, V., Morris, M.G., Davis, G.B., Davis, F.D.: User acceptance of information technology: toward a unified view. MIS Q., 425–478 (2003)
29. Webster, J., Hackley, P.: Teaching effectiveness in technology-mediated distance learning. Acad. Manag. J. **40**(6), 1282–1309 (1997)

The manufacturer's authorised representative in the EU is Springer Nature Customer Service Centre GmbH, Europaplatz 3, 69115 Heidelberg, Germany. If you have any concerns regarding our products, please contact ProductSafety@springernature.com

Printed and bound by CPI Group (UK) Ltd, Croydon, CR0 4YY
05/10/2026
02228131-0013